TRANSFUSION MEDICINE, APHERESIS, AND HEMOSTASIS

TRANSFUSION MEDICINE, APHERESIS, AND HEMOSTASIS

REVIEW QUESTIONS AND CASE STUDIES

Edited by

HUY P. PHAM
University of Alabama at Birmingham
Birmingham, AL, United States

LANCE A. WILLIAMS, III
University of Alabama at Birmingham
Birmingham, AL, United States

Academic Press is an imprint of Elsevier
125 London Wall, London EC2Y 5AS, United Kingdom
525 B Street, Suite 1800, San Diego, CA 92101-4495, United States
50 Hampshire Street, 5th Floor, Cambridge, MA 02139, United States
The Boulevard, Langford Lane, Kidlington, Oxford OX5 1GB, United Kingdom

Library of Congress Cataloging-in-Publication Data
A catalog record for this book is available from the Library of Congress

British Library Cataloguing-in-Publication Data
A catalogue record for this book is available from the British Library

ISBN: 978-0-12-803999-1

For information on all Academic Press publications visit our website at
https://www.elsevier.com/books-and-journals

Publisher: Mica Haley
Acquisition Editor: Tari Broderick
Editorial Project Manager: Lisa Eppich
Production Project Manager: Poulouse Joseph
Designer: Matthew Limbert

Typeset by Thomson Digital

To

Our parents

Hay P. Pham and DanVan T. Hoang

and

Lawrence and Mary Williams

and

Our teachers and mentors throughout our lives

for their love and support in the undertaking of this work,
for which we are deeply grateful.

Contents

Contents

Contributors

Jill Adamski Mayo Clinic Arizona, Phoenix, AZ, United States

Maksim Agaronov Kings County Hospital, Brooklyn, NY, United States

Beth M. Alden University of Iowa, Iowa City, IA, United States

Suzanne Arinsburg Icahn School of Medicine at Mount Sinai, New York, NY, United States

Evan M. Bloch Johns Hopkins University, School of Medicine, Baltimore, MD, United States

Michelle R. Brown University of Alabama at Birmingham, Birmingham, AL, United States

R. Pat Bucy University of Alabama at Birmingham, Birmingham, AL, United States

D. Joe Chaffin Blood Bank Guy Web Site (BBGuy.org), LifeStream Blood Bank, San Bernardino, CA, United States

Vishesh Chhibber Hofstra Northwell School of Medicine and Northwell Health, Manhasset, NY, United States

Jason E. Crane LifeSource Blood Center, Rosemont, IL, United States

Karen Dallas St. Paul's Hospital, Vancouver, BC, Canada

Helene DePalma City University of New York, Jamaica, NY, United States

Emmanuel A. Fadeyi Wake Forest University School of Medicine, Winston Salem, NC, United States

Richard O. Francis New York-Presbyterian Hospital—Columbia University, New York, NY, United States

George A. Fritsma The Fritsma Factor, Your Interactive Hemostasis Resource, Birmingham, AL, United States

Michael D. Gautreaux Wake Forest University School of Medicine, Winston-Salem, NC, United States

Eric A. Gehrie The Johns Hopkins Hospital, Baltimore, MD, United States

Javi L. Hartenstine University of California, Irvine, CA, United States

Chelsea Hayes Cedars-Sinai Medical Center, Los Angeles, CA, United States

Jeanne E. Hendrickson Yale University, New Haven, CT, United States

Yen-Michael S. Hsu New York-Presbyterian Hospital, New York, NY, United States

Tina S. Ipe Houston Methodist Hospital, Houston, TX, United States

Cyril Jacquot Children's National Health System and George Washington University, Washington, DC, United States

Jeffrey S. Jhang Icahn School of Medicine at Mount Sinai, New York, NY, United States

Susan T. Johnson Blood Center of Wisconsin, Marquette University, University of Wisconsin-Milwaukee, Milwaukee, WI, United States

Alesia Kaplan Institute for Transfusion Medicine, Pittsburgh, PA, United States

Theresa Kinard Mayo Clinic Arizona, Phoenix, AZ, United States

Robin G. Lorenz University of Alabama at Birmingham, Birmingham, AL, United States

Marisa B. Marques University of Alabama at Birmingham, Birmingham, AL, United States

Holli M. Mason Harbor-UCLA Medical Center, Torrance, CA, United States

Shanna Morgan American Red Cross, Saint Paul; University of Minnesota, Minneapolis, MN, United States

Theresa A. Nester University of Washington Medical Center and Bloodworks Northwest, Seattle, WA, United States

Monica B. Pagano University of Washington Medical Center, Seattle, WA, United States

Mona Papari LifeSource Blood Center, Rosemont, IL, United States

Seung Park Indiana University Health, Indianapolis, IN, United States

Huy P. Pham University of Alabama at Birmingham, Birmingham, AL, United States

Patricia M. Raciti New York-Presbyterian Hospital—Columbia University, New York, NY, United States

Swati Ratkal Hofstra Northwell School of Medicine and Northwell Health, Manhasset, NY, United States

Ronit Reich-Slotky New York-Presbyterian Hospital, New York, NY, United States

Annette J. Schlueter University of Iowa, Iowa City, IA, United States

John Schmitz University of North Carolina School of Medicine, Chapel Hill, NC, United States

Joseph Schwartz New York-Presbyterian Hospital—Columbia University, New York, NY, United States

Salima Shaikh Blood Centers of the Pacific, San Francisco, CA, United States

Beth H. Shaz New York Blood Center, New York, NY, United States

Rance C. Siniard The University of North Carolina at Chapel Hill, Chapel Hill, NC, United States

Jayanna Kay Slayten Indiana Blood Center SBB Program, Indianapolis, IN, United States

Christopher A. Tormey Yale University, New Haven, CT, United States

Mrigender Virk MedStar Georgetown University Hospital, Washington, DC, United States

Lance A. Williams, III University of Alabama at Birmingham, Birmingham, AL, United States

Edward C.C. Wong Children's National Medical Center, Center for Cancer and Blood Disorders; George Washington University School of Medicine and Health Sciences, Departments of Pediatrics and Pathology, Washington, DC, United States

YanYun Wu Bloodworks Northwest, Seattle, WA, United States

X. Long Zheng University of Alabama at Birmingham, Birmingham, AL, United States

About the Editors

Huy P. Pham, MD, MPH

Dr. Pham is currently an Assistant Professor in the Department of Pathology, Division of Laboratory Medicine, and serves as the Medical Director of Apheresis at the University of Alabama at Birmingham. He also regularly attends clinical service and provides resident/fellow teaching in the Transfusion Medicine, Apheresis, and Hemostasis. With research interests in statistics, mathematical modeling, and health economics in addition to the clinical aspects of the field, he has been the lead author or senior author for many original research and review articles as well as book chapters on different topics in Transfusion Medicine, Hemostasis, Apheresis, and Cellular Therapy. Nationally, he serves on multiple professional organization committees to provide guidance for clinical practice and research direction for advancing the field. Dr. Pham is board certified in both Clinical Pathology and Transfusion Medicine. Dr. Pham received his BS with high honors in bioengineering from the University of California, Berkeley, MD from the Chicago Medical School, and MPH in Biostatistics from Columbia University. He completed his Clinical Pathology residency at the New York-Presbyterian Hospital—Columbia University Medical Center and Transfusion Medicine fellowship training at the joint program between the New York Blood Center and Columbia University Medical Center.

Lance A. Williams, III, MD

Dr. Williams is currently an Assistant Professor in the Department of Pathology, Division of Laboratory Medicine, and serves as the Medical Director of Hemostasis Laboratory at the University of Alabama at Birmingham. He is also the Director of Transfusion Medicine for the Community Pathology Practice Program, which encompasses five hospitals throughout Alabama. Dr. Williams received his BS with high honors in Biology from Longwood University and his MD with honors from Ross University. He completed residency training at East Carolina University and two fellowships—Transfusion Medicine at Yale University and Hematopathology at Virginia Commonwealth University. Dr. Williams is an award-winning teacher, known for his innovative teaching methods and his dedication to elevating students of all levels. His publications include original research articles, review articles, and books, including *Quick Guide to Transfusion Medicine*, second edition. Nationally, he is known as an engaging and knowledgeable speaker and he serves on many committees committed to advancing the science of Transfusion Medicine, Apheresis, and Hemostasis.

Preface

With great excitement, we present this inaugural edition of *Transfusion Medicine, Apheresis, and Hemostasis: Review Questions and Case Studies*. This project is the collaborative effort that spanned a time period of 2 years and included more than 50 experts, many of whom are national leaders in their respected fields. It also represents the passion and privilege we feel to teach the next generation of physicians in Transfusion Medicine and Apheresis.

The main goal for this book is to help the readers build a solid foundation of both basic and advanced conceptual knowledge to prepare for the American Board of Pathology (ABP) certification exam in Transfusion Medicine. This book is not intended to be a substitute for textbooks, original research or review articles, and/or clinical training. Further, since the field of medicine, both from a scientific and regulatory perspective, rapidly changes, the readers are advised to continuously update their knowledge by attending national meetings and reading clinical journals.

To equip the readers with the basic knowledge in critical reading and data analysis, which is an essential skill in daily medical practice, a novel chapter titled "Data Interpretation in Laboratory Medicine" was included in this book. In this chapter, the readers are asked to make logical conclusions based on the given data and/or statistical results. Moreover, there is also a chapter on "Practical Calculations in Transfusion Medicine, Apheresis, and Hemostasis" to help consolidate all the necessary formulas commonly used in daily practice for easy reference. These chapters are unique to our book and will not be found in any other currently on the market.

All of the questions in this book were originally created by the authors of each chapter. Each question can either be standalone or part of a case scenario representing challenge cases in Transfusion Medicine, Apheresis, and Hemostasis. These questions often represent both rare and common clinical scenarios that the authors have seen during their clinical practice. Each question is then followed by five possible answers, with only one being correct (or the best answer). After the question, there is a conceptual explanation followed by a more factual explanation of the right and wrong answers. We gave the individual authors the freedom to choose how they explained the wrong answer choices. Some authors chose to be more direct (e.g., Answer A is incorrect because…), while other authors chose a more conversational style [e.g., Human resources (Answer A) includes staffing, selection, orientation, training, and competency assessment of employees]. This format is designed to help the student linking the conceptual and factual knowledge together to form a solid foundation for use in clinical practice. At the end of each chapter, there is a list of articles and textbooks that will prove useful to the motivated student who wishes to become an expert in the field. Another special feature to our textbook is the presence of a pretest and posttest, which are provided to help the readers with self-assessment.

As stated above, the main focus of this book is to help the readers preparing for the ABP certification exam in Transfusion Medicine. However, due to the interdisciplinary nature of the field of Transfusion Medicine, Apheresis, and Hemostasis, we believe that this book is also beneficial to and can be used by all clinicians involved in the management of complex transfusion, apheresis, and hemostasis issues, such as hematologists, anesthesiologists, surgeons, and critical care physicians. We further believe that it is a helpful guide for these specialists to prepare for their own specialty certification exam, when the topics are related to Transfusion Medicine, Apheresis, and Hemostasis.

Although the authors and editors have aimed for perfection in content, grammar, and syntax, we are realistic enough to know that there will be errors in our book. Therefore, we created an email for the readers to alert us such mistakes or criticisms and to discuss different viewpoints, as well as to provide suggestions for the next edition of the book. Please email us at: TMQuestionBook@gmail.com.

In conclusion, we are very grateful for the opportunity given to us by Elsevier. We are also deeply dedicated to the field of Transfusion Medicine, Apheresis, and Hemostasis and the education of trainee and future leaders in the field. We hope that this book will be used to help the readers to successfully pass the board certification exams, as a teaching tool for trainees, and to enhance knowledge for daily clinical practice.

Huy P. Pham, MD, MPH

Lance A. Williams, III, MD

Acknowledgment

We, the editors, would like to acknowledge the excellent technical and professional support of Lisa Eppich, Jeffrey Rossetti, Joseph Poulouse, and many other team members at Elsevier. Each of these individuals played critical role in the creation of the first edition of this question book, and we sincerely thank them.

We would like to thank our expert contributors and the department leadership at our institution for their collaboration and support of this project. We are indebted to all our teachers and mentors for their encouragement and mentorships provided in our professional careers. We are grateful to all our students, residents, fellows, and colleagues for their questions, scientific curiosity and discussions, and intellectual stimuli. We would like to acknowledge with gratitude, the support, patient, and love of our family members, especially Dr. Pham's parents (Hay P. Pham and Dan Van T. Hoang) and wife (Ning Jiang); and Dr. Williams' parents (Lawrence and Mary Williams), sisters (Christy Gill and Dana McGuire), and girlfriend (Beth Lett); and both editors' mentors (in alphabetical order: Jill Adamski, Jeffrey S. Jhang, Marisa B. Marques, Joseph (Yossi) Schwartz, Beth H. Shaz, John Smith, Edward Snyder, Steven L. Spitalnik, Christopher Tormey, YanYun Wu, and X. Long Zheng). Without their unconditional love, support, and mentorship, this project could not have come to fruition. There are also many others to thank and recognize, therefore, we offer those unnamed persons our most sincere "Thank You"!

Laboratory Reference Ranges

Hematology	Reference range
Hemoglobin (male)	14–17.5 g/dL
Hemoglobin (female)	12.3–15.3 g/dL
Hematocrit (male)	41%–53%
Hematocrit (female)	36%–46%
Red cell count (male)	4.3–5.9 million/μL
Red cell count (female)	3.5–5.5 million/μL
Mean corpuscular volume (MCV)	80–96 fL
Mean corpuscular hemoglobin (MCH)	27.5–33.2 pg/cell
Mean corpuscular hemoglobin concentration (MCHC)	33–36 g/dL
White blood cell count	4,500–11,000/μL
Platelets	150,000–400,000/μL

White blood cell differential (adult)	Mean %	Range of absolute counts
Segmented neutrophils	56	1800–7800/μL
Bands	3	0–700/μL
Eosinophils	2.7	0–450/μL
Basophils	0.3	0–200/μL
Lymphocytes	34	1000–4800/μL
Monocytes	4	0–800/μL
Hemoglobin A2	1.5%–3.5% of total hemoglobin	
Hemoglobin F	<2%	

Coagulation	Reference range
Prothrombin time	11–15 s
Activated partial thromboplastin time (aPTT)	25–35 s
International normalized ratio (INR)	0.9–1.2
Thrombin time	Depends on the concentration of thrombin reagent used, typically 17–25 s
Antithrombin activity	80%–120%
Coagulation factors	Typically 50%–150%
Fibrinogen	200–400 mg/dL
Plasma D-dimers	< 200 ng/mL [D-dimer units (DDUs)]
von Willebrand factor	50–150 U/dL
Ristocetin cofactor activity	50–150 U/dL
Thereapeutic anti-Xa (unfractionated heparin)	0.3–0.7 U/mL
Therapeutic anti-Xa (low–molecular weight heparin)	0.5–1.0 U/mL

Chemistry	Reference range
Chloride	95–105 mEq/L
Sodium	135–145 mEq/L
Potassium	3.5–5.0 mEq/L
Glucose	70–110 mEq/L
Serum calcium	8.4–10.2 mg/dL

Chemistry	Reference range
Ionized calcium	1.1–1.35 mmol/L
Creatinine	0.9–1.2 mg/dL
Blood urea nitrogen	7–18 mg/dL
Troponin-I	0.0–0.4 ng/mL
Iron studies	**Reference range**
Total serum iron (men)	50–180 μg/dL
Total serum iron (women)	40–190 μg/dL
Transferrin	188–341 mg/dL
Ferritin (men)	20–380 ng/mL
Ferritin (women)	20–288 ng/mL
Total iron-binding capacity (TIBC)	250–425 μg/dL
Liver function and hemolysis markers	**Reference range**
Aspartate aminotransferase (AST)	10–35 U/L
Alanine aminotransferase (ALT)	9–46 U/L
Alkaline phosphatase (ALKP)	40–115 U/L
Gamma glutamyl transferase (GGT)	3–10 U/L
Lactate dehydrogenase (LDH)	120–250 U/L
Total bilirubin	0.1–1.9 mg/dL
Direct bilirugin	0.0–0.3 mg/dL
Indirect bilirubin	0.2–0.8 mg/dL

CHAPTER

1

Pretest

Huy P. Pham, Helene DePalma**, Lance A. Williams, III**

***University of Alabama at Birmingham, Birmingham, AL, United States; **City University of New York, Jamaica, NY, United States**

This chapter provides a tool for you to evaluate your knowledge before reading the book. This will allow you to compare the knowledge you gain after completion. For the most accurate representation of your current knowledge, you should aim to complete this test within 80 min. All topics covered in this pretest have been covered throughout the book. At the end of the book, there is a posttest to be used as a comparison of the pretest results.

There are many ways to estimate the blood volume of a patient. For the purpose of this test, unless otherwise stated, please use 70 mL/kg (for adults) and 80 mL/kg (for neonates) when calculating the blood volume of a person.

Place your answers in the spaces provided. Check your answers at the end of the chapter and give yourself a final score.

1.____	11.____	21.____	31.____	41.____
2.____	12.____	22.____	32.____	42.____
3.____	13.____	23.____	33.____	43.____
4.____	14.____	24.____	34.____	44.____
5.____	15.____	25.____	35.____	45.____
6.____	16.____	26.____	36.____	46.____
7.____	17.____	27.____	37.____	47.____
8.____	18.____	28.____	38.____	48.____
9.____	19.____	29.____	39.____	49.____
10.____	20.____	30.____	40.____	50.____

Number Correct _____/Number Incorrect _____ × 100% = Final Score __________%

Transfusion Medicine, Apheresis, and Hemostasis. http://dx.doi.org/10.1016/B978-0-12-803999-1.00001-8

Please answer Questions 1–3 based on the following clinical scenario.

A 34-year-old female at 6 weeks-postpartum and without any other medical history is admitted to the emergency department (ED) with 2 days of fatigue, petechia, and dark urine. The following are laboratories drawn upon admissions: white blood count (WBC) 11,200/μL, hemoglobin (Hgb) 7.2 g/dL, platelet count 38,000/μL, unremarkable electrolytes except for Cr 2.6 mg/dL, and urinalysis showed the presence of Hgb without blood. Her LDH is 4 times the upper limit of the reference range. The blood film is shown in Fig. 1.1.

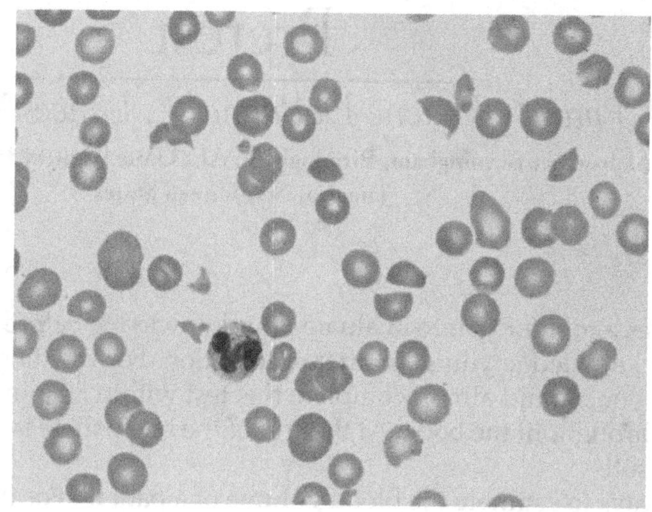

FIGURE 1.1 **Blood film for patient in Question 1.** *Source: Image courtesy of Dr. Vishnu Reddy, University of Alabama at Birmingham, Birmingham, AL.*

1. The ED physician would like to initiate emergent plasma exchange for the presumed diagnosis of thrombotic thrombocytopenic purpura (TTP). What laboratory test should be sent to confirm the diagnosis of TTP?
   A. ADAMTS13 level drawn prior to plasma exchange
   B. ADAMTS13 level drawn immediately after plasma exchange
   C. ADAMTS13 level drawn 24 h after the first procedure but immediately prior to the second procedure
   D. Complement levels drawn prior to plasma exchange
   E. Complement levels drawn immediately after plasma exchange

2. Which of the following replacement fluid has the highest risk of citrate toxicity during a plasma exchange procedure?
   A. 0.9% normal saline
   B. 5% albumin
   C. Lactated Ringer's solution
   D. Red blood cells (RBCs)
   E. Plasma

3. After five plasma exchanges, the patient's platelet count increased to 128,000/μL and her LDH is now within the reference range. However, her Cr continues to increase to 4.1 mg/dL. What is the best next step of management?
   A. Continue with plasma exchange without any further modification
   B. Continue with plasma exchange except exchanging 1.5 plasma volume instead of 1
   C. Continue with plasma exchange except using cryo-depleted plasma instead of plasma
   D. Stop plasma exchange and start eculizumab
   E. Stop plasma exchange and start rituximab and vincristine

End of Case

Please answer Questions 4 and 5 based on the following clinical scenario.

A 23-year-old male with history of severe hemophilia A is admitted to the emergency department (ED) with joint bleeding. He weighs 90 kg. His complete blood count (CBC) shows WBC 7,200/μL, Hgb 12 g/dL, hematocrit (Hct) 37%, and platelet count 338,000/μL.

4. The ED physician requested recombinant factor VIII to treat the joint bleeding. Assuming that his initial factor VIII level is <1%, then what is the appropriate dosage to be administered?
   A. 8,000 IU to achieve a goal of 100% factor VIII level
   B. 8,000 IU to achieve a goal of 80% factor VIII level
   C. 4,000 IU to achieve a goal of 100% factor VIII level
   D. 4,000 IU to achieve a goal of 80% factor VIII level
   E. 2,000 IU to achieve a goal of 80% factor VIII level

5. The patient received the appropriate dose to have his expected factor VIII level increase to ~100%. However, the peak factor VIII level drawn approximately 30 min after administration is still <1%. What is the appropriate next step of laboratory management and treatment?
   A. Perform a Bethesda assay to quantify the inhibitor and if detected, 180 μg/kg recombinant activated factor VII should be given
   B. Perform a Bethesda assay to quantify the inhibitor and if detected, 90 μg/kg recombinant activated factor VII should be given
   C. Perform a Bethesda assay to quantify the inhibitor and if detected, 45 μg/kg recombinant activated factor VII should be given
   D. Perform a dilute Russel viper venom test to quantify the inhibitor and if detected, 90 μg/kg recombinant activated factor VII should be given
   E. Perform a dilute Russel viper venom test to quantify the inhibitor and if detected, 45 μg/kg recombinant activated factor VII should be given

End of Case

Please answer Questions 6–8 based on the following clinical scenario.

A 69-year-old female admitted to the oncology unit for the treatment of acute myeloid leukemia. The patient received the standard induction chemotherapy. She is hemodynamically stable and is not currently bleeding. She also does not have any history of acute or chronic cardiac issues.

6. If the patient is medically stable, then what is the threshold for platelet transfusion?
   A. Not yet determined
   B. 5,000/μL
   C. 10,000/μL
   D. 20,000/μL
   E. 50,000/μL

7. This patient's blood type is O Rh negative. She is scheduled to receive an unrelated hematopoietic progenitor cell (HPC) transplant from an unrelated donor with a blood type of B Rh positive. Please classify the type of this HPC transplant as well as the ABO type of the blood products this patient needs to receive before the full RBC engraftment.
   A. Major mismatched; O RBCs and B or AB plasma-contained products are required
   B. Major mismatched; B RBCs and O plasma-contained products are required
   C. Minor mismatched; O RBCs and B or AB plasma-contained products are required
   D. Minor mismatched; B RBCs and O plasma-contained products are required
   E. Bidirectional mismatched; O or B RBCs and B or AB plasma-contained products are required

8. For transfusion, what modification to the cellular product is required?
   A. CMV negative
   B. Irradiation

C. Washed
D. Negative for hemoglobin S
E. Less than 7 days old

End of Case

9. Which of the following is an advantage of using ISBT 128 labeling?
A. It allows each facility to develop its own labeling according to its preference
B. It can track donations for up to 500 years
C. It has a safety mechanism that no addition or deletion of information for autologous donor is allowed
D. It has a safety mechanism by having a built-in self-checking character
E. It has a safety mechanism that it allows only one code to be read at one time

10. What are Current Procedural Terminology (CPT) codes?
A. Used to code medical diagnoses
B. Used to code adverse events
C. Used to code the amount of time a physician spent in diagnosis and treatment
D. Used to code if a physician trainee is involved in a medical procedure
E. Used to code medical procedures

11. A 39-year-old female comes to the clinic for a preoperative assessment prior to her scheduled hysterectomy. Her CBC shows Hgb 7.2 g/dL, a mean corpuscular volume of 70 fL, and a platelet count of 167,000/µL. Her ferritin level is 6 ng/mL (reference range: 11–307 ng/mL). Which of the following statements is true regarding correcting this patient's anemia?
A. A unit of RBC should be transfused prior to surgery to prevent intraoperative anemia
B. Blood salvage should be set up intraoperatively to prevent intraoperative anemia
C. Iron supplementation should be given to assist with RBC production
D. Vitamin B12 and folic acid should be given to assist with RBC production
E. No further management is necessary since the anemia is due to her underlying disease

12. Currently in the United States, which of the following antibodies is the most common cause of hemolytic disease of the fetus and newborn (HDFN)?
A. Anti-D
B. Anti-K
C. Anti-Jk^b
D. Anti-P1
E. Anti-A,B

13. In pediatric patients with beta thalassemia major, in order to suppress ineffective erythropoiesis, what should be the Hgb goal for transfusion?
A. 15–16 g/dL
B. 13–14 g/dL
C. 11–12 g/dL
D. 9–10 g/dL
E. 7–8 g/dL

Please answer Questions 14 and 15 based on the following clinical scenario.

A 3.5-kg full term newborn with HDFN (due to anti-c) is required 2-volume whole blood exchange transfusion. The patient's current Hct is 30%. The critical care team would like to have a reconstituted whole blood with an Hct of 50% using the freshest RBC unit possible that is lacking the c-antigen and extended crossmatched compatible with the mother's plasma.

14. The transfusion service identifies two RBC units that may be used for whole blood reconstitution and they are CPDA-1 units (average Hct ~70%). How much plasma should be used for this

procedure of whole blood reconstitution? Please use 85 mL/kg as an estimate for a newborn's blood volume.
A. 170 mL
B. 200 mL
C. 245 mL
D. 425 mL
E. 595 mL

15. Which of the following is a potential complication of whole blood exchange?
A. Vascular insufficiency of lower limbs
B. Seizure
C. Thrombocytosis
D. Bilirubin removal
E. Pancytopenia

End of Case

Please answer Questions 16–18 based on the following clinical scenario.

A 65-year-old woman G5P5005 admitted to the oncology unit for chemotherapy treatment of her relapsed acute myeloid leukemia (AML). For a platelet count of 7,500/μL, she was transfused with 1 unit of apheresis platelets. However, her 1-h posttransfusion platelet count was only 7,800/μL. Another unit of apheresis platelets was ordered. This time, an ABO-identical day 4 apheresis platelet unit was selected for the transfusion and 30-min posttransfusion, her platelet count went up to 8,600/μL.

16. The patient is currently stable without active signs of bleeding. In order to provide the platelets that may help her to achieve a reasonable increment as soon as possible, what is the next step of management?
A. ABO identical apheresis platelets
B. Day 1 apheresis platelets
C. Crossmatched compatible platelets
D. Apheresis platelets that are matched for the patient's HLA-A, B, and C antigens
E. Irradiated day 7 apheresis platelets after tested negative with Pan Genera Detection (PGD) assay

17. For a platelet count of 9,700/μL, this patient is transfused with a unit of crossmatched compatible platelets. She was stable throughout the procedure without any signs or symptoms of transfusion reaction. However, about 2 h after the transfusion, she develops severe respiratory distress and was intubated. An echocardiogram was performed and did not show any left ventricular dysfunction. Her chest X-ray is shown in Fig. 1.2.

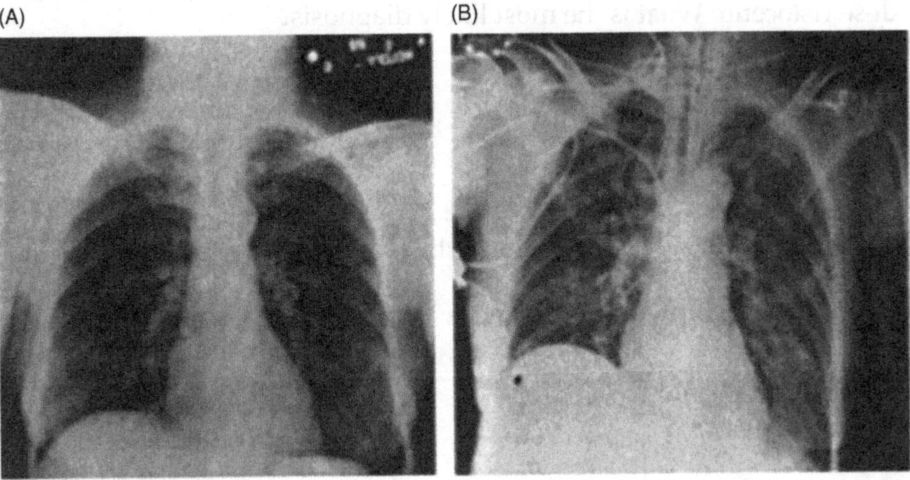

FIGURE 1.2 **Chest X-ray.** (A) Before transfusion; (B) 2 h after transfusion. *Source: Adapted from P.M. Kopko, P.V. Holland, Br. J. Haematol. 105 (1999) 322—329.*

What is the most likely cause of this patient's respiratory distress?
A. Allergic reaction
B. Anaphylactic reaction
C. Septic reaction
D. Transfusion associated circulatory overload (TACO)
E. Transfusion associated acute lung injury (TRALI)

18. Which of the following antibodies has been implicated as part of the pathogenesis of the reaction described in Question 17?
A. Anti-IgA
B. Anti-Jk[a]
C. Anti-ADAMTS13
D. Anti-HNA (human neutrophil antigen)
E. Anti-C5

End of Case

Please answer Questions 19–21 based on the following clinical scenario.

A 29-year-old female is admitted to the hospital for an elective cholecystectomy. However, upon interviewing, she revealed that she has a history of "excessive bleeding." Specifically, she usually has heavy menstruation. One time, she had prolonged bleeding after a dental surgery. She has no history of hemarthroses. She also stated that her mother and sister also tend to have heavy menstruation with easy bruising. Her current Hgb is 13.9 g/dL and her platelet count is 127,000/μL. There is no abnormality detected on the blood film. Her basic coagulation test results [prothrombin time (PT), and activated partial thromboplastin time (aPTT)] are within the reference range.

19. Based on her bleeding history and basic laboratory values, what is the most likely diagnosis?
A. Hemophilia A
B. Hemophilia C
C. Antithrombin III deficiency
D. Bernard-Soulier syndrome
E. von Willebrand disease

20. The physician ordered a von Willebrand panel. The results are as following: von Willebrand factor (vWF): 120%, Ristocetin Cofactor: 40%, and the von Willebrand multimer study shows a loss of high molecular multimer. A ristocetin-induced platelet aggregation demonstrates aggregation with both high and low dose ristocetin. What is the most likely diagnosis?
A. Type 1 von Willebrand disease
B. Type 2A von Willebrand disease
C. Type 2B von Willebrand disease
D. Type 2N von Willebrand disease
E. Type 3 von Willebrand disease

21. If the patient has bleeding during the operation, which of the following options is the best treatment modality?
A. Desmopressin
B. Recombinant vWF
C. Recombinant activated factor VII
D. Recombinant factor VIII
E. Cryoprecipitated AHF

End of Case

22. A company just developed a new medication for the treatment of TTP. They tested this medication in a clinical trial. The following are information and results from this clinical trial.
- Study question: Is the new medication better than TPE in the treatment of TTP?
- Study design: Clinical trial with subjects with TTP divided into two groups: Group 1: New medication and TPE; Group 2: TPE only
- Results: 40 subjects enrolled total, 20 in each group. The baseline characteristics were similar between the two groups. Please see Table. 1.1 for more detailed results.

TABLE 1.1 Detailed Results

	Group 1	Group 2	*P*-value
Mortality (%)	20%	10%	0.66
Average TPE days (± SD)	5 (3)	8 (2)	0.81
Exacerbation and relapse (%)	60%	25%	0.054
Significant adverse event (%)	70%	35%	0.056

Which of the following statements is correct regarding the clinical trial and its results?
A. The new medication should be used instead of TPE in the treatment of TTP
B. The new medication is safer than TPE
C. The new medication and TPE is safer than TPE alone
D. The new medication and TPE resulted in more exacerbation and relapse than TPE alone
E. If TPE is not available immediately, then the new medication should be used as a bridging therapy instead of plasma infusion

23. Which of the following storage temperature and length for the corresponding type of tissue used for transplant is correct?
A. Sclera at room temperature for 10 years
B. Frozen skin at room temperature for 5 years
C. Freeze dried bone at 4°C for 10 years
D. Corneas for keratoplasty at 2–8°C for 14 days
E. Cornea for procedures other than keratoplasty at room temperature for 5 years

24. A 36-year-old male received multiple fluid and blood products during resuscitation. Which of the following conditions make him suitable for a cell and tissue donation assuming that there is no preinfusion sample for infectious testing?
A. Infusion of 2 RBC units in the last 24 h
B. Infusion of 6 RBC units, 6 plasma units, and 1 apheresis platelet unit in the last 24 h
C. Infusion of 6 RBC units, 6 plasma units, and 1 apheresis platelet unit in the last 36 h
D. Infusion of 2 L normal saline in the last 30 min
E. Infusion of 1 L normal saline, 6 RBCs unit, and 4 plasma units in the last 8 h

25. Which of the following choices represent correctly the type of infectious test and its associated window period and residual risk of transfusion?

	Test	Window period (days)	Residual risk of transfusion
A.	HIV minipool nucleic acid test	21	1:1,467,000
B.	HIV minipool nucleic acid test	9	1:1,467,000
C.	HCV minipool nucleic acid test	21	1:1,149,000
D.	HCV minipool nucleic acid test	9	1: 750,000
E.	HTLV	21	1: 2,993,000

HCV, Hepatitis C virus; HIV, human immunodeficiency virus; HTLV, human T-lymphotropic virus.

26. You are providing medical support for a blood drive at a large urban college campus. There are many first-time donors from the faculty, staff, and students on campus. Which of the following blood donors would be acceptable for whole blood donation today?
   A. A 17-year-old female, weight 108 lbs, Hgb 12.0 g/dL, received an HPV (human papilloma virus) vaccine 4 weeks ago
   B. A 17-year-old male, weight 140 lbs, Hgb 13.5 g/dL, temperature 37.2°C, discontinued Accutane 3 months ago
   C. An 18-year-old male, weight 135 lbs, Hgb 12.5 g/dL, temperature 37.7°C, received the flu vaccine 2 weeks ago, tattoo 6 months ago in an unregulated facility
   D. A 48-year-old male, weight 160 lbs, Hgb 14.5 g/dL, temperature 37.1°C currently taking Finasteride
   E. A 64-year-old female, weight 145 lbs, Hgb 12.5 g/dL, received shingles vaccine 6 weeks ago, spent 6 months on sabbatical in London in 1995

27. What are the most common side effects of plerixafor and G-CSF when using for HPC mobilization?
   A. Diarrhea for plerixafor and bone pain for G-CSF
   B. Bone pain for plerixafor and diarrhea for G-CSF
   C. Nausea for plerixafor and anorexia for G-CSF
   D. Anorexia for plerixafor and nausea for G-CSF
   E. Fever for both plerixafor and G-CSF

28. Assuming everything else except for the conditions described in the following statements is the same, which one is correct?
   A. HPC derived from bone marrow has higher risk for graft versus host disease than HPC derived from peripheral blood
   B. HPC derived from peripheral blood usually has less CD34+ cells than HPC derived from umbilical cord
   C. The engraftment time for HPC derived from peripheral blood is longer than from bone marrow
   D. HPC derived from bone marrow tends to have the largest volume when comparing to HPC derived from peripheral blood or from umbilical cord
   E. HPC derived from peripheral blood tends to have higher hematocrit when comparing to HPC derived from bone marrow

29. A 47-year-old female blood donor presents at the donor center to donate apheresis platelets. Her last apheresis platelet donation was 7 days ago. Her predonation platelet count is 177,000/μL. She reports taking aspirin yesterday for joint pain. She received a hepatitis A vaccine (HAV) 3 days ago to prepare for an upcoming trip to Thailand. Which statement is true regarding her eligibility today?
   A. She is eligible to donate apheresis platelets today
   B. She is eligible to donate platelets again 14 days after her last donation
   C. She is temporarily deferred from platelet donation until 48 h after the aspirin dose
   D. She is temporarily deferred from platelet donation until 2 weeks after the HAV vaccine
   E. She is deferred from apheresis platelet donation due to her platelet count

30. In the absence of prior transfusion or pregnancy, the plasma of an individual with the Bombay phenotype will have which antibody?
   A. Anti-Rh 29
   B. Anti-Ku
   C. Anti-Jk3
   D. Anti-H
   E. None of the above, no antibody would be expected

Please answer Questions 31 and 32 based on the following clinical scenario.

31. The following results are obtained with a patient's plasma test against an antibody identification panel. Evaluate the panel results and choose the correct association with the antibody you identify and the next steps in your investigation.

	Rh						MNS				Lu		P1	Lewis		Kell		Duffy		Kidd		Result		
	D	C	E	c	e	f	M	N	S	s	Lu^a^	Lu^b^	P1	Le^a^	Le^b^	K	k	Fy^a^	Fy^b^	Jk^a^	Jk^b^	IS	PEG AHG	CC
1	+	+	0	0	+	0	+	+	+	+	0	+	+	+	0	0	+	0	+	0	+	0	2+	NT
2	+	+	0	0	+	0	0	+	0	+	0	+	0	0	+	0	+	0	+	+	+	0	2+	NT
3	+	0	+	+	0	0	+	+	+	+	0	+	+	0	+	+	+	+	+	+	0	0	2+	NT
4	+	0	0	+	+	+	0	+	0	+	0	+	+	0	0	0	+	0	0	+	0	0	0	2+
5	0	+	0	+	+	+	0	+	0	+	0	+	0	0	0	0	+	0	0	+	+	0	0	2+
6	0	0	+	+	+	+	+	0	+	+	0	+	+	+	0	0	+	+	+	0	+	0	2+	NT
7	0	0	0	+	+	+	0	+	+	+	0	+	+	0	+	+	+	+	0	+	+	0	0	2+
8	0	0	0	+	+	+	0	+	+	0	0	+	+	+	0	+	+	0	+	0	+	0	2+	NT
9	0	0	0	+	+	+	+	+	+	0	0	+	0	0	+	0	+	+	+	+	+	0	2+	NT
10	0	0	0	+	+	+	+	0	+	0	0	+	+	0	+	0	+	+	+	0	+	0	2+	NT
11	+	0	+	+	0	0	+	+	+	+	0	+	+	+	0	0	+	+	0	+	+	0	0	2+
Patient Red Blood Cells (Autocontrol)																						0	0	2+

CC, check cells; NT, not tested.

- **A.** IgM alloantibody directed against an enzyme sensitive antigen/Test against an enzyme pretreated panel
- **B.** Multiple alloantibodies/obtain an RBC phenotype and test additional selected red cells
- **C.** IgG alloantibody associated with an antigen sensitive to DTT treatment/test against a DTT-treated panel
- **D.** IgG alloantibody associated with a ficin sensitive antigen/test against a ficin pretreated panel
- **E.** IgM alloantibody/proceed with neutralization using blood group substance

32. What is the purpose of adding check cells to all negative AHG tubes?

- **A.** To ensure proper scoring of agglutination reactions
- **B.** To ensure adequate cell washing and addition of AHG reagent
- **C.** To check for hemolysis or reaction of complement
- **D.** To check for attachment of an IgM antibody
- **E.** To ensure there was no interruption during the washing steps

End of Case

33. Which of the following combinations correctly represents the antibody specificity with the expected reactivity?

	Antibody	Reactivity
A.	Anti-P1	Strongest reactions by indirect antiglobulin technique
B.	Anti-Jk^a^	Neutralized by hydatid cyst fluid
C.	Anti-E	Approximately 8 of 10 crossmatches would be incompatible
D.	Anti-M	Enhanced reactivity when tested against enzyme pretreated red cells
E.	Anti-K	Nonreactive when tested against DTT-treated reagent RBCs

34. A 67-year-old female with multiple myeloma received a unit of RBC 2 weeks ago. At that time, her antibody screen was negative. She is now presenting in the oncology clinic for follow-up. She is clinically stable and there is no evidence of hemolysis. However, her direct antiglobulin test (DAT) is positive (with anti-IgG but not with anti-C3d). Her autocontrol is also positive. Elution was performed and the eluate shows anti-E. Which of the following choices is correct?

	Diagnosis	Management	Selection of RBC if transfusion is needed
A.	Acute hemolytic transfusion reaction	Eculizumab	Extended crossmatched E-negative RBC units
B.	Delayed hemolytic transfusion reaction	Eculizumab	Extended crossmatched E-negative RBC units that are also phenotypically matched for K, Duffy, Kidd, and S/s
C.	Delayed hemolytic transfusion reaction	Red blood cell exchange	Extended crossmatched E-negative RBC units
D.	Delayed serologic transfusion reaction	Intravenous immunoglobulin	Extended crossmatched E-negative RBC units
E.	Delayed serologic transfusion reaction	Close observation	Extended crossmatched E-negative RBC units

35. Which of the following combinations represents correctly the interference of daratumumab with immunohematology tests and possible solution?

	Interference	Possible solution
A.	Causes panreactivity in antibody panel	Prewarm technique for antibody tests and crossmatches
B.	Causes panreactivity in antibody panel	Use DTT-treated reagent RBCs for antibody tests
C.	Mimics anti-K pattern in antibody panel	Alloadsorption with K-positive RBCs
D.	Mimics anti-Jk^a pattern in antibody panel	Alloadsorption with Jk^a-negative RBCs
E.	Mimics anti-M pattern in antibody panel	Prewarm technique for antibody tests and crossmatches

36. An adult patient's red cells phenotype as Le(a–b+). Based on these results, which statement is correct?
A. The patient's possible genotype is *Lele Hh sese*
B. The patient's possible genotype is *lele Hh Sese*
C. The patient is a nonsecretor
D. The patient is not at risk to make a Lewis antibody
E. The patient is at risk to make anti-Le^a, a clinically significant antibody

37. A type and screen specimen on a 37-year-old female is submitted to the blood bank. The patient's red cells type O Rh negative. Due to suspected anemia, the clinician orders a DAT and the red cells are positive with both polyspecific and anti-IgG antiglobulin reagents. Which of the following results should the technologist always expect?
A. The ABO front and back type will not agree
B. All crossmatches will be incompatible
C. The antibody screen will always be positive by PEG IAT
D. The antibody screen will always be positive by column agglutination method
E. The weak D test and Rh control will be positive

38. What is the general recommended dose for plasma administration?
A. 1–5 mL/kg
B. 5–10 mL/kg

C. 10–20 mL/kg
D. 20–40 mL/kg
E. 40–50 mL/kg

39. What is the minimum interval between double RBC donation?
A. 2 weeks
B. 4 weeks
C. 8 weeks
D. 12 weeks
E. 16 weeks

40. You have been asked to justify the expense associated with upgrading the methodology used for type and screen testing from semiautomated gel technology to an automated platform. Which of the following is considered an indirect expense?
A. Technologist salary
B. Service contract for the automated analyzer
C. Internet service for remote troubleshooting by the vendor
D. Reagents and disposables for the automated analyzer
E. Barcode scanner for specimen input

41. A Kleihauer-Betke stain shows a result of 2% for a 28-year-old female who recently delivered at 39 weeks of gestation. Her weight is 60 kg and her recent Hct is 33%. The patient is Rh negative and the baby is also Rh negative. She received one vial of RhIG at 28 weeks of pregnancy and another vial of RhIG at 35 weeks when she was involved in a minor car accident. Which of the following combinations is correct regarding RhIG dosage and time frame for administration?

	RhIG dosage	RhIG time frame for administration
A.	3 vials	Within 72 h from delivery
B.	3 vials	Within 24 h from delivery
C.	4 vials	Within 72 h from delivery
D.	4 vials	Within 24 h from delivery
E.	Not clinically indicated	Not clinically indicated

42. Transfusion-transmitted infection is an important concern in transfusion practice. Which of the following combinations is correct regarding the type of blood products associated with the highest rate of bacterial contamination and the associated fatality?

	Type of blood product	Type of bacteria most commonly contaminated with	If contaminated, type of bacteria that is more fatal
A.	Red blood cells	Gram positive bacteria	Gram positive bacteria
B.	Red blood cells	Gram negative bacteria	Gram negative bacteria
C.	Platelets	Gram positive bacteria	Gram negative bacteria
D.	Platelets	Gram negative bacteria	Gram positive bacteria
E.	Plasma	Gram positive bacteria	Gram negative bacteria

43. Which of the following situations correctly describes when a blood product can be transfused if the infectious disease markers are positive?
A. Anti-HBc repeat reactive; HBV minipool NAT negative
B. Anti-HBsAg repeat reactive; HBV minipool NAT negative

C. Anti-HBc initial reactive, duplicate testing demonstrates one nonreactive and one reactive, HBV minipool NAT negative
D. Anti-HBc initial reactive, repeat testing nonreactive on two samples; HBV minipool NAT negative
E. Anti-HCV initial reactive, duplicate testing demonstrates one nonreactive and one reactive, HCV minipool NAT negative

44. Additive solutions (AS) for blood products offer many advantages for product storage and cell viability. Which of the following statements regarding these solutions is correct?
A. AS red cells have a final hematocrit (Hct) of 55%–65%
B. Red cell additive solutions consist mainly of electrolytes and glucose
C. Platelet additive solution (PAS) is approved for both whole blood derived and apheresis platelets
D. PAS platelets have higher isohemagglutinin titers (anti-A and anti-B)
E. Mannitol is a key component of PAS

45. Blood product quality control has been submitted for your review. Based on the results provided, which product is considered a quality control failure?
A. Whole blood derived platelet with pH 6.8 and platelet count 5.7×10^{10}
B. Cryoprecipitated AHF with 250 mg fibrinogen and 85 IU Factor VIII
C. Apheresis platelets with pH 6.4 and platelet count 3.3×10^{11}
D. Apheresis red blood cells with 70 g of hemoglobin
E. Red blood cells, leukoreduced with 5.5×10^{6} residual leukocytes

46. The following are HLA typing results for a family:
Father: HLA A1, –; B57, 62; DR4, 7
Mother: HLA A7, 8; B3, 7; DR15, 17
Child 1: HLA A1, 7; B3, 57; DR 7, 15
Child 2: HLA A1, 8; B7, 62; DR 4, 17

Assuming there is no crossover and child 3 is the true child of this couple, then which of the following is a potential HLA typing for child 3?
A. HLA A1, 7; B3, 62; DR 7, 15
B. HLA A1, 8; B7, 57; DR 7, 17
C. HLA A1, 8; B3, 57; DR 7, 17
D. HLA A1, 7; B3, 57; DR 4, 15
E. HLA A1, 7; B7, 57; DR 7, 15

47. Which of the following interpretations of lymphocyte crossmatch is correct?

	T cell	B cell	Interpretation
A.	Positive	Positive	Class I ± Class II antibody
B.	Positive	Positive	Class II antibody only
C.	Positive	Negative	Class II antibody only
D.	Negative	Positive	Class I antibody only
E.	Negative	Negative	Technical error, needs to repeat test

48. The evening shift supervisor in the transfusion service of a busy tertiary care facility is retiring after 20 years of service. This a bench supervisor position with responsibility for equipment quality control, workflow coordination and review of the testing/products required for the following day's surgical and outpatient schedules. The job description requires a minimum of 6 years' experience with SBB certification preferred. The human resources department has identified a candidate for you to interview. Which of the following questions are you allowed to ask the candidate?

A. What childcare arrangements do you have in place to allow working an evening shift schedule?
B. What year did you graduate from high school?
C. Are you willing to provide weekend coverage?
D. What religious holidays do you celebrate?
E. Did you request any family and medical leave at your last job?

49. For which of the following issues should a blood product deviation report (BPDR) be submitted to the FDA?

A. A unit of RBCs is irradiated prior to shipment to a children's hospital. The irradiation indicator on the product shows a successful irradiation cycle; however, the red cell expiration date was not changed prior to shipping
B. An antigen negative red cell unit is ordered by St John's Hospital and is erroneously shipped to St Joseph's Hospital, resulting in a service delay
C. A hospital received its shipment from their blood supplier. The packing slip states that 8 units of O Rh negative red cells were shipped; however, there are only 7 units in the box
D. A reagent quality control failure was overlooked by the novice night shift technologist. The supervisor noted the failure on the next morning and the test run is repeated before the blood products were shipped
E. Quality control is due for the refrigerated centrifuge by December 31. Due to an oversight, the QC is not performed until January 2. The blood center initiates a deviation report. No whole blood units were processed between December 31 and January 2

50. A physician from the adult hematology/oncology service contacts the transfusion service director because one of his patients was transfused a unit of RBCs that was not irradiated. He insists that the request for irradiated products was submitted when the patient was admitted; however, the transfusion service had no record of the request. The best way to investigate the nonconformance is the use of which tool?

A. Ishikawa (fishbone) diagram
B. Develop a control chart for the process
C. Prepare a run chart
D. Design a Pareto chart
E. Prepare a histogram

ANSWERS AND BRIEF EXPLANATIONS

1. **Answer**: *A*—ADAMTS13 level should be drawn prior to plasma exchange. Refer to Chapter 14 Question 18 for more information.
2. **Answer**: *E*—Plasma, as a replacement fluid, has the highest risk of citrate toxicity in an apheresis procedure. Refer to Chapter 14 Question 4 for more information.
3. **Answer**: *D*—The patient may have atypical hemolytic-uremic syndrome (aHUS) and thus, eculizumab should be attempted. Refer to Chapter 14 Question 20 for more information.
4. **Answer**: *C*—Approximately, 4,000 IU of recombinant factor VIII will bring the patient's factor VIII level to ~100%. Refer to Chapter 20 Question 3 for more information.
5. **Answer**: *B*—The inappropriate response to a factor VIII administration should raise the suspicion of the presence of an inhibitor. Thus, a Bethesda assay should be performed to quantify the inhibitor level. If the inhibitor level is >5 BU, then recombinant activated factor VII should be considered (90 μg/kg). Additionally, Factor Eight Inhibitor Bypassing Agent (FEIBA) can be an alternative. Refer to Chapter 13 Questions 5 and 23 for more information.
6. **Answer**: *C*—In medically stable patients without active bleeding and acute or chronic cardiac issues, the threshold for platelet transfusion is 10,000/μL based on the AABB recommendations. Refer to Chapter 8 Question 25 and 26 for more information.
7. **Answer**: *A*—This case is an example of major mismatched HPC transplant. The patient should receive O RBCs and B or AB plasma-contained products before the full RBC engraftment. Refer to Chapter 17 Questions 9 and 11 for more information.
8. **Answer**: *B*—Irradiation should be performed for all cellular blood products (except HPC products) used for transfusions in order to prevent transfusion associated graft versus host disease (TA-GVHD). Refer to Chapter 8 Question 4 for more information.
9. **Answer**: *D*—ISBT 128 labeling has a built-in mechanism for self-checking errors through a check character intended to confirm the accurate entry of the DIN (donor identification number) when a manual keyboard entry is performed. Refer to Chapter 19 Question 9 for more information.
10. **Answer**: *E*—CPT codes are used to code medical procedures. Refer to Chapter 19 Question 19 for more information.
11. **Answer**: *C*—This patient has laboratory evidence of iron deficiency. Iron supplementation should be given to avoid the need of transfusion if the patient develops perioperative anemia. Refer to Chapter 9 Question 3 for more information.
12. **Answer**: *E*—With the routine use of Rh immunoglobulin (RhIG), HDFN due to ABO incompatibility is now the most common cause of HDFN in the United States. Refer to Chapter 10 Question 2 for more information.
13. **Answer**: *B*—Only transfusion of RBC to an Hgb goal of 13–14 g/dL can suppress ineffective erythropoiesis in pediatric patients with beta thalassemia major. Refer to Chapter 10 Question 16 for more information.
14. **Answer**: *A*—The total blood volume of this newborn is 85 (mL/kg) × 3.5 = 297.5 mL; thus, the total volume of the reconstituted unit is 297.5 × 2 = 595 mL. Since this unit must have an Hct of 50%, the volume of CPDA-1 unit must be used is 595 mL × 50%/70% = 425 mL. Therefore, the plasma volume is 595−425 mL = 170 mL. Refer to Chapter 20 Question 20 for more information.
15. **Answer**: *A*—Vascular insufficiency of the lower limbs can be a complication of whole blood exchange. Other complications are akin to the ones observed with massive transfusion. Refer to Chapter 10 Question 6 for more information.
16. **Answer**: *C*—This patient has evidence of immune-mediated platelet refractoriness. Therefore, providing crossmatched compatible platelets may be the quickest method to help her achieve a reasonable platelet count increment with transfusions. Refer to Chapter 15 Question 3 for more information.
17. **Answer**: *E*—The signs and symptoms that this patient experienced are consistent with the diagnosis of TRALI. Refer to Chapter 12 Question 19 for more information.

18. **Answer**: *D*—Anti-HNA and anti-HLA antibodies are implicated in the pathogenesis of TRALI. Refer to Chapter 12 Questions 19 and 21 for more information.
19. **Answer**: *E*—This patient's clinical history and laboratory results suggest that she may have von Willebrand disease (vWD). Refer to Chapter 13 Questions 22 and 24 for more information.
20. **Answer**: *C*—This patient has low ristocetin cofactor comparing to the vWF antigen level. Furthermore, an abnormality in the von Willebrand multimer as well as platelet aggregation to both high and low dose of ristocetin in the ristocetin-induced platelet aggregation test are consistent with the diagnosis of a type 2B vWD. Of note, this patient also has mild thrombocytopenia, which is another feature of this diagnosis. Refer to Chapter 13 Question 24 for more information.
21. **Answer**: *B*—This patient has vWD, and thus, should be treated with recombinant vWF. Humate P, which is a factor VIII/vWF complex derived from human plasma, can also be used. Refer to Chapter 13 Questions 22–24 for more information.
22. **Answer**: *D*—The new medication and TPE (group 1) resulted in more exacerbation and relapse than TPE alone (group 2; 60% vs. 25%, respectively). This is trending toward statistical significant (*P*-value = 0.054). Refer to Chapter 21 for more information.
23. **Answer**: *D*—Corneas for keratoplasty should be stored at 2–8°C for 14 days. Refer to Chapter 18 Question 6 for more information.
24. **Answer**: *A*—Transfusions of blood products or infusions of colloidal or crystalloid solutions may cause a plasma dilution effect significant enough to alter the results of communicable disease testing. Postinfusion sample can be used for infectious testing if the patient (>12 year-old) receives <2,000 mL of whole blood, red blood cells, and/or colloids within 48 h or <2,000 mL of crystalloid within 1 h, or any combination. Refer to Chapter 18 Question 36 for more information.
25. **Answer**: *B*—Choice B represents the correct window period and residual risk of transfusion-transmitted infection for HIV minipool nucleic acid test (NAT). Refer to Chapter 11 Questions 32–35 for more information.
26. **Answer**: *B*—The donor meets weight and hemoglobin requirements. Accutane deferral is 1 month, therefore, sufficient time has elapsed. Refer to Chapter 4 Questions 1, 4, 13, and 14 for more information.
27. **Answer**: *A*—The most common side effects for plerixafor and G-CSF when using for HPC mobilization are diarrhea and bone pain, respectively. Refer to Chapter 17 Question 19 for more information.
28. **Answer**: *D*—HPC derived from bone marrow tends to have the largest volume when comparing to HPC derived from peripheral blood or from umbilical cord. Other choices are in reverse. Refer to Chapter 17 Question 12 for more information.
29. **Answer**: *C*—The donor is not a candidate for apheresis platelet donation today due to aspirin (defer for 48 h). Refer to Chapter 4 Questions 18 and 35 for more information.
30. **Answer**: *D*—The red cells of an individual with the Bombay phenotype (O_h) lacks the H antigen and anti-H will be detected in their plasma. Refer to Chapter 7 Question 6 for more information.
31. **Answer**: *D*—The pattern of reactivity fits anti-Fy^b, an IgG antibody directed against a ficin sensitive antigen. Refer to Chapter 7 Questions 21 and 24 for more information.
32. **Answer**: *B*—Check cells ensure sufficient washing and that the AHG reagent is still active and has not been neutralized by unbound globulins. Refer to Chapter 7 Question 10 for more information.
33. **Answer**: *E*—Kell system antigens are denatured by sulfhydryl compounds, such as dithiothreitol (DTT), therefore, anti-K would not react with DTT-treated RBCs. Refer to Chapter 6 Questions 9 and 34 for more information.
34. **Answer**: *E*—This clinical scenario represents a delayed serologic transfusion reaction. A close observation (without any treatment) is all that is necessary at this time. If transfusion is clinically indicated, then the patient should receive extended crossmatched E-negative RBC units. Refer to Chapter 12 Question 10 for more information.
35. **Answer**: *B*—Daratumumab can cause panreactivity in antibody testing due to the presence of CD38 on RBC surface. DTT denatures CD38, and thus, removes the interference. Refer to Chapter 15 Question 20 for more information.

36. **Answer**: *D*—Individuals whose red cells are Le(a–b+) are not expected to make Lewis antibodies. Refer to Chapter 6 Questions 9 and 19 for more information.
37. **Answer**: *E*—Many serological complications can result when RBCs have a positive DAT; however, the only finding that will always be encountered is the weak D and Rh control, since the RBCs were coated with IgG molecules prior to the testing. Refer to Chapter 7 Question 25 for more information.
38. **Answer**: *C*—Based on the transfusion guidelines, most experts recommend the dose between 10 and 20 mL/kg for plasma transfusion. This dose of plasma would be expected to increase coagulation factors by ~10%–20% immediately after infusion. Refer to Chapter 8 Question 15 for more information.
39. **Answer**: *E*—Sixteen weeks is the minimum interval between double RBC donation. An allogeneic donor is deferred from all types of blood donation for 8 weeks after whole blood donation. See Chapter 4 Question 6 for more information.
40. **Answer**: *C*—Internet service, utilized for multiple purposes in the laboratory, is considered overhead or an indirect expense. Refer to Chapter 2 Question 6 for more information.
41. **Answer**: *E*—Both the baby and the mother are Rh negative; thus, RhIG is not indicated in this clinical scenario. Refer to Chapter 20 Question 18 for more information.
42. **Answer**: *C*—Platelets are the blood products that have the highest rate of bacterial contamination. Most often, Gram positive bacteria are the cause. However, the fatality rate is higher when the platelet unit is contaminated with Gram negative bacteria. Refer to Chapter 11 Question 1 and Chapter 12 Question 40 for more information.
43. **Answer**: *D*—The unit collected from a donor with anti-HBc initial reactive, repeat testing nonreactive on two samples, and HBV minipool NAT negative may be used for transfusion. Refer to Chapter 11 Question 24 for more information.
44. **Answer**: *A*—AS red cells have a lower final Hct, approximately around 55%–65%. Refer to Chapter 5 Questions 6, 13, and 14 for more information.
45. **Answer**: *E*—Leukoreduced RBC units must have $<5 \times 10^6$ residual leukocytes. Refer to Chapter 5 Table 5.4 for more information.
46. **Answer**: *B*—From the family's HLA typing, the father's genotype is HLA A1, B57, DR7; HLA A1, B62, DR4. The mother's genotype is HLA A7, B3, DR15; HLA A8, B7, DR17. HLA is inherited as haplotype; thus, HLA A1, 8; B7, 57; DR 7, 17 is a potential HLA for the next child. Refer to Chapter 16 Question 15 for more information.
47. **Answer**: *A*—Since B cells express both class I and class II HLA antigens while T cells only express Class I HLA antigen, a positive leukocyte crossmatch with both T and B cells means that there is a Class I ± Class II antibody presence in the serum. Refer to Chapter 16 Question 2 for more information.
48. **Answer**: *C*—Interview questions that could potentially discriminate against age, sex, religion, or other factors are not allowed. Refer to Chapter 2 Questions 7 for more information.
49. **Answer**: *A*—Incorrect labeling, such as an expiration date, on a product that is shipped out of the facility is an example of an error that would require filing a BPDR with the FDA. Refer to Chapter 3 Question 24 for more information.
50. **Answer**: *A*—An Ishikawa diagram, also known as a fishbone or cause and effect diagram would be most appropriate to investigate the various causes contributing to the error, such as methods, people, equipment, and environment. Refer to Chapter 3 Question 28 for more information.

CHAPTER

2

Statistics and General Principles of Laboratory Management

Huy P. Pham, Emmanuel A. Fadeyi**, Lance A. Williams, III**

*University of Alabama at Birmingham, Birmingham, AL, United States; **Wake Forest University School of Medicine, Winston Salem, NC, United States

Laboratory medical directors must understand and adhere to many basic laboratory principles. Basic principles include regulatory requirements and economical evaluation of new assays. They should also be able to interpret statistical tests/analyses, as an integral part of method comparison and assay validation. This chapter provides a review of laboratory principles and statistical methods, as well as a brief discussion on different study designs. Of note, this may be an advanced chapter for some students and you may wish to cover this chapter later on in your test preparation.

Attention: Some diseases and scoring systems are used as examples in this chapter; however, intimate knowledge of these diseases and systems is not required to answer the questions. They are used to provide a reference for the statistical or laboratory management concepts. Further, the data used for the questions in this chapter are only for illustration—they are not actual data, and thus, the conclusions (i.e., right answers) are only for the purpose of concept/calculation demonstration, and should not be used to make clinical decisions.

Please answer Questions 1–9 based on the following scenario:

You were recently appointed as the Medical Director for the Hemostasis Laboratory at your hospital. You have noticed that there are many patients with thrombotic thrombocytopenic purpura (TTP) admitted and being treated at your hospital. Currently, the assay to measure ADAMTS13 (A Disintegrin and Metalloproteinase with a Thrombospondin type 1 motif, member 13) a key enzyme in the pathogenesis of TTP, is a send-out test and it has a 1–3 day turnaround time. You would like to evaluate the possibility of performing the assay in your laboratory. You have several options for this assay, including several commercially available kits, as well as an in-house developed assay.

1. You find that there are currently no Food and Drug Administration (FDA)-approved ADAMTS13 assays available. According to the Clinical Laboratory Improvement Amendments (CLIA), this test is classified as follows:
   A. Waived test
   B. CLIA-exempt
   C. Moderate-complexity
   D. Moderately high-complexity
   E. High complexity

Transfusion Medicine, Apheresis, and Hemostasis. http://dx.doi.org/10.1016/B978-0-12-803999-1.00002-X

Concept: Congress passed CLIA in 1998 to ensure accuracy, reliability, and timeliness of patient test results regardless where the tests are performed. The FDA determines the classification for all CLIA-regulated tests. There are several test categories: waived, moderate, and high complexity. Essentially, tests are classified by the potential risk for an incorrect result.

For commercially available FDA-cleared or approved tests, the test complexity is determined during the premarket approval process. Of special note, the phrase "FDA cleared" does not mean that a test is waived. Tests developed by the laboratory or that have been modified from the manufacturer's approved instructions are defaulted to high complexity per CLIA regulations.

Answer: *E*—Since this is not an FDA-approved assay, it is high complexity by default. Waived tests (Answer A) are defined as "simple laboratory examinations and procedures that have an insignificant risk of an erroneous result." Examples of waived tests include urine pregnancy tests and hemoglobin measurement by the copper sulfate method (which by the way is not an acceptable method to qualify blood donors). As above with "FDA-cleared," "CLIA-exempt" does not mean that a test is waived. CLIA-exempt (Answer B) is an older term used to describe a laboratory where the state laws/laboratory requirements are equal to or more stringent than that of the Center for Medicare and Medicaid Services (CMS). Thus, the test was considered CLIA-exempt. There is no category called "low-complexity" (Answer C). Provider-performed microscopy (PPM) is an example of a moderately complex test (Answer D). Examples of PPM include vaginal wet mounts and pinworm examinations.

2. Which of the following agencies oversees CLIA?
A. Center for Medicare and Medicaid Services (CMS)
B. The Joint Commission for Hospital Accreditation (JCAHO)
C. Food and Drug Administration (FDA)
D. Center for Disease Control and Prevention (CDC)
E. US Department of Health and Human Services (HHS)

Concept: CMS oversees the CLIA program. However, CMS does not necessarily inspect laboratories. Rather, CMS recognizes several accrediting agencies as being capable of inspecting laboratories to ensure quality and safety.

Answer: *A*—CMS oversees the CLIA program. CMS officially recognizes several accrediting agencies and will issue the certificate of accreditation to laboratories accredited by organizations, such as the AABB, American Osteopathic Association (AOA), American Society for Histocompatibility and Immunogenetics (ASHI), Commission on Office Laboratory Accreditation (COLA), College of American Pathologists (CAP), and the Joint Commission on Accreditation of Health Care Organization (JCAHO) (Answer B). The FDA (Answer C) is responsible for protecting the public health by assuring the safety and efficacy of drugs, biological products (such as blood products), medical devices (such as apheresis machines), food supply, cosmetics, and products that emit radiation. The CDC (Answer D) is a federal agency that detects and responds to new and emerging health threats. It is also responsible to promote healthy and safe behaviors and to train the public health workforce, including disease investigators. The mission of HHS (Answer E) is to enhance and protect the health of Americans by providing effective health and human services and fostering advances in medicine, public health, and social services.

3. You performed a study to compare the performance between a commercially available ADAMTS13 assay and the assay you developed in your research laboratory. Receiver operating characteristic (ROC) curves for both the methods are shown in Fig. 2.1.

Based on the information, at what point (A,B,C,D or E) will the commercial assay give you the most diagnostic accuracy (i.e., both high sensitivity and specificity) and at what point will your in-house developed assay more sensitive than the commercial one?

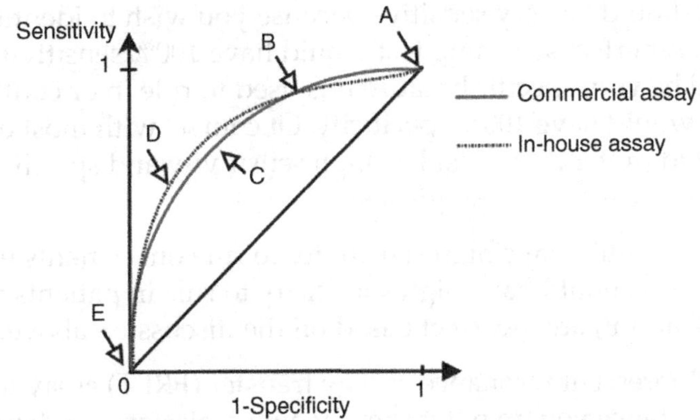

FIGURE 2.1 ROC curves for commercial assay and in-house assay.

	Cutoff point with the most diagnostic accuracy for commercial assay	Cutoff point that gives the in-house developed assay more sensitivity than commercial one
A.	A	B
B.	B	D
C.	B	E
D.	C	D
E.	A	E

Concept: An ROC curve represents the sensitivity and specificity of the assay at different cutoff points. The area under the curve (AUC) represents the overall diagnostic ability to discriminate between disease and non-disease. For example, if the AUC is 95%, it means that 95% of the time the test will be able to discriminate between disease and non-disease in two random subjects. An area of 1 represents the perfect test, while an area of 0.5 (i.e., a 45 degree diagonal line or a line of no discrimination) represents a "useless" test because it cannot distinguish between disease and no disease (i.e., it is not better than a random guess). The point on the curve closest to the upper left corner [i.e., at coordinate (0,1) on the graph] represents the cutoff with the greatest diagnostic value (i.e., highest sensitivity while still maintaining a low false positive rate).

Answer: *D*—Using the commercial assay's ROC curve, we see that point C is closest in distance to the coordinate (0,1); thus, it is the cutoff point with the most diagnostic value for this particular assay. At cutoff point D, your in-house assay is more sensitive than the commercial one. Point A represents the most sensitive (but yet, least specific) cutoff. On the contrary, point E represents the most specific but least sensitive cutoff. Point B is just the cutoff that the sensitivity and specificity of your in-house assay is the same as the commercial one. Based on the figure and the above explanation, all the other choices (Answers A, B, C, and E) are incorrect.

4. One of your brilliant residents suggests doing a serial test for the patients who have a clinical suspicion of TTP. Which of the following statements is true?
A. Screening test should be very sensitive to rule out patients with disease
B. Screening test should be very specific to rule out patients with disease
C. Screening test should be very specific to rule in patients with disease
D. Confirmation test should be very specific to rule out patients with disease
E. Confirmation test should be very sensitive to rule in patients with disease

Concept: Screening tests should be very sensitive because you wish to identify all potential individuals with disease. A perfect screening test would have 100% sensitivity. On the other hand, confirmation tests should be very specific because it is used to rule in or confirm the disease. A perfect confirmation test would have 100% specificity. Of course, with most of the testing we perform in the laboratory, we would prefer that a test has high sensitivity and specificity, so we are able to both detect and confirm a disease with a single test.

Answer: *A*—screening test should have high sensitivity to rule out patients with disease. On the other hand, confirmation test should have high specificity to rule in patients with disease. All other choices (Answers B, C, D, and E) are incorrect based on the discussion above.

5. You are considering the fluorescent resonance energy transfer (FRET) assay to measure ADAMTS13 activity in your laboratory. Since you do not currently own a platform to detect fluorescent signals, you plan to lease it for $5,000 monthly. The reagent cost for each test is $50, and the reimbursement is $100 per test. Assuming that there are no additional costs to consider, how many tests do you have to perform each month to break-even?
A. 50
B. 75
C. 100
D. 125
E. 150

Concept: Before bringing a test into a laboratory, the laboratory director should perform a break-even analysis. This analysis takes into account all costs associated with performing the test and the expected profit. In the question above, we only take into account the reagent cost and the reimbursement. However, in real life, you may need to account for technologist time/salary, maintenance contracts, the cost of test tubes, etc. If the expected profit does not exceed the cost, then you may need to reconsider bringing the test in-house and consider the cost-benefit of sending the sample to a reference laboratory. Of course, sometimes the clinical need may override a break-even analysis. In those cases, you may take a financial loss on this test and make up the difference on another, more commonly ordered test.

Answer: *C*—You must perform 100 tests every month to break even.
Break-even calculation: Fixed costs/(Price – Variable Cost) = Breakeven Point. In this case, this calculates as follows: 5,000/(100–50) = 100. Thus, # of tests to break-even = 100. Other choices (Answers A, B, D, and E) are wrong based on the calculation.

6. If you decide to bring the FRET assay for ADAMTS13 activity into your laboratory, which of the following is considered a variable direct cost?
A. Internet service
B. Reagent cost
C. Technologist salary
D. Contract maintenance
E. Electricity

Concept: Direct cost is the cost that can be attributed to a specific test (FRET assay for ADAMTS13 in this case). It can be fixed or variable. Fixed direct cost is a constant cost regardless of the test volume. Variable cost varies with the test volume. Examples of variable costs are reagent and supply costs. Indirect costs (i.e., overhead costs) are the costs that benefit more than one test, which is often difficult to quantify.

Answer: *B*—Reagent cost is an example of a variable direct cost. Based on the definition, technologist salary (Answer C) or maintenance of the contract (Answer D) is example of fixed direct cost. Internet service and electricity are examples of indirect costs since they are required to run more than one test in a laboratory.

7. Your break-even analysis for the FRET assay for ADAMTS13 activity shows that you will not perform enough tests to profit from the test itself. However, further evaluation reveals that performing the result in-house will result in a quicker turn-around-time. Considering the cost of therapeutic plasma exchange (TPE) procedures, you determine that quicker ADAMTS13 results will allow the hospital to save money by avoiding unnecessary TPE in patients that do not have TTP. This will also enhance patient safety and quality of care since they would not be exposed to unnecessary procedures or transfusion. Therefore, you decide to offer the test in-house and begin hiring staff. You are interviewing a technologist whose partial responsibility will be running this test Monday through Friday, 8:00 a.m. to 5:00 p.m. Which of the following questions can be asked during the interview?

A. Are you married?
B. What is your religion?
C. How old are you?
D. Where are you originally from?
E. Please tell me the principle behind the FRET-based ADAMTS13 measurement

Concept: It is illegal to discriminate future employees based on their age, gender, religion, marital status, disability, language, country of origin, and sexual orientation. However, it is perfectly acceptable to ask questions about the capacity of the potential employees to perform the job.

Answer: *E*—This question assesses the ability of this applicant to perform the job and is thus, a legally acceptable question. Questions that involve an applicant's marital status, sexual orientation, religion, current or future plans to have children, and age are illegal (Answers A, B, and C) Asking about the applicant's origin (Answer D) is also illegal; however, the employer can ask if the applicant is authorized to work in the United States. If the employer concerns about the possibility of an applicant to work during holidays and weekends, then instead of asking about marriage and religion, it is acceptable to ask if the applicant is willing to work during holidays and weekends, if necessary. For example, "what hours can you work?" or "do you have any other responsibilities that may interfere with job requirements?" are acceptable. Please see http://www.businessinsider.com/11-illegal-interview-questions-2013-7?op=1 for further education on this topic.

8. Your father is a clinical laboratory physician scientist and an entrepreneur. He recently opened a private reference laboratory that specializes in hemostasis testing, including ADAMTS13 activity. He approaches you with an idea to perform FRET-based ADAMTS13 activity for you and your hospital at a cost of $35 per test, with a turnaround time of 12 h or less. From your analysis (Question #5 above), running the test in-house is costing you $50 just to buy the reagents, without considering other direct and indirect costs. How should you answer your father's proposal?

A. Agree to send all samples to him because he offers a good price and turnaround time
B. Only send up to 50% of the nonurgent samples to him
C. Only send the samples to him if his laboratory information system can interface with your hospital system
D. Negotiate the price and only send the samples if the price is less than $25 per test
E. Do not send any sample to him

Concept: Stark's law governs the physician self-referral to Medicare and Medicaid patients. This law prohibits physicians from making patient or laboratory referral for services payable by Medicare and Medicaid to an organization that they or their immediate family has a financial relationship. Though definitions vary, immediate family usually includes spouses, children, parents, siblings, and first cousins.

Answer: *E*—Since your father owns this private laboratory, you cannot send the samples to him based on Stark's law. Answers A, B, C, and D are incorrect because they would involve sending samples to your father's laboratory. Although having the reference laboratory system interfaces with your hospital system is important, it will not prevent you from violating the Stark's law if you send the samples there.

9. You decide to do a method comparison study between your in-house developed ADAMTS13 assay (assay 2) and ADAMTS13 from a reference laboratory (not your father's laboratory) (assay 1). You run 40 samples, and below is the plot of ADAMTS13 activity measurements, using the two assays. Deming regression was performed, and it shows [assay 2 measurement] = 0.81 + 1.49 × [assay 1 measurement]. The 95% confidence interval for the slope is [1.48, 1.51] and the intercept is [0.28, 1.47].

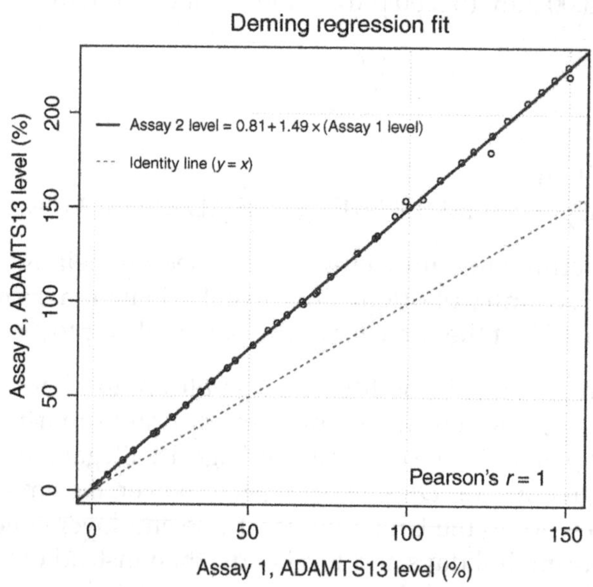

Which of the following provides the best interpretation of both the plot and Deming regression?

A. Assay 1 and assay 2 have excellent correlation and no bias
B. Assay 2 has significant proportional bias compared to assay 1
C. Although assay 1 and assay 2 have excellent correlation, assay 2 has a significant constant bias compared to assay 1
D. Although assay 1 and assay 2 have excellent correlation, assay 2 has both significant proportional and constant bias comparing to assay 1
E. There is no relationship between assay 1 and assay 2

Concept: Correlation studies and Deming regression are two common methods used for assay comparison. Correlation determines how well the two methods correlate linearly with each other. Mathematically, correlation is measured by r, which ranges from -1 to 1. If r is between -1 and 0, then it is called negative association. If r is between 0 and 1, then the relationship is positive association. The closer r to 1 (or -1), the stronger the linear association is. Nonetheless, if $r \sim 0$, then it only means that the two variables does not have a good linear relationship. It does not necessary mean that there is no relationship since the relationship can be polynomial or log-log, which is not indicated by r. In this example, the two measurements are perfectly correlated ($r = 1$). However, just because the two assays correlate well, it does not mean you can accept them as equivalents without evaluating the Deming regression.

Deming regression should be used in this case instead of simple linear regression because Deming regression allows for both assay 1 (X variable) and assay 2 (Y variable) to be subjected to measurement errors. Simple regression only allows Y variable to be measured with error. Since the analyte is measured by both assays and neither is perfect, both measurements are subject to errors. Thus, Deming regression should be used.

In this case, Deming regression demonstrated [assay 2 measurement] = 0.81 + 1.49 × [assay 1 measurement] with a 95% confidence interval for the slope [1.48, 1.51] and the intercept [0.28, 1.47]. The slope in Deming regression represents proportional bias while the intercept represents constant bias. If the 95% confidence interval does not contain 1 for the slope, then there is a proportional bias between the assays. Likewise, if 0 is not in the 95% confidence interval, then there is evidence of constant bias between the assays. In this example, there is statistically significant evidence that there is both proportional difference (because the 95% confidence interval for the slope did not contain 1) and constant bias (because the 95% confidence interval for the intercept did not contain 0). This is important information because if assay 1 is the "gold standard," then modifications, such as recalibration, must be made to assay 2 measurements before it can be used in the clinical laboratory.

Answer: *D*—Although assay 1 and assay 2 have excellent correlation, assay 2 has both significant proportional and constant bias compared to assay 1 based on the results of the Deming regression analysis. The remaining choices (Answers A, B, C, and E) are incorrect interpretations based on the explanation above.

End of Case

Please answer Questions 10–12 based on the following scenario:

You are the Medical Director for the Peripheral Blood and Bone Marrow Hematopoietic Progenitor Cell Processing Laboratory at your hospital. You currently have one controlled-rate freezer for slow freezing of your stem cells prior to cryopreservation in a liquid nitrogen (LN_2) (≤ −160°C) reservoir for storage. Per FACT (Foundation for the Accreditation of Cellular Therapy) standards, you are required to have a backup freezer, such as a mechanical freezer with −80°C storage capacity, in case your primary freezer fails. You would like to evaluate your mechanical freezer to see if it meets FACT standards.

10. To comply with FACT standards, which of the following is required before the mechanical freezer before it can be used?
A. Validation studies on the freezer
B. Reliability studies on the freezer
C. Heat tolerance studies
D. Linearity studies
E. Durability studies on the freezer

Concept: Before an instrument is placed into use in a clinical laboratory, a validation study must be performed to demonstrate that the instrument will meet specifications and fulfill the intended purpose. Validation will test the mechanical freezer to ensure that the freezing process will work similar to the controlled-rate freezer in the actual live environment, as part of the required contingency plan. If the controlled-rate freezer fails, the stem cells can be placed at −80°C in the mechanical freezer before transfer to the LN_2 reservoir for cryopreservation and storage.

Answer: *A*—As explained earlier, the mechanical/backup freezer must be validated before being accepted for use. The purpose of the validation study in this case is to have a contingency plan in place per FACT standards. Reliability studies (Answer B) are performed on clinical laboratory tests, when assessing accuracy and precision while temperature-controlled studies are not conducted on a mechanical freezer. Linearity studies are most commonly done for analytes, to determine if the instrument measurements are consistent with expected values (Answer D). Heat tolerance studies (Answer C) would not test the desired function of the freezer, and durability studies (Answer E) are carried out by the manufacturer.

11. Who is responsible for overseeing the validation study and its final approval?
A. Director, biomedical engineering department
B. Chairman, pathology department
C. Medical technologist, stem cell processing laboratory

D. Manager, stem cell processing laboratory
E. Medical director, stem cell processing laboratory

Concept: Validation of laboratory equipment is important not only for meeting regulatory requirements, but also for producing high-quality results and patient care. Each member of the laboratory participates in this process with varying degrees of responsibility.

Answer: *E*—The medical director, the manager, and the technologist(s) in the stem cell laboratory should all be involved in writing a protocol for the validation study. The protocol should include the purpose of validation, the process description, responsibilities, the materials required, test samples required, testing conditions, data collection, acceptance criteria, and conclusions. The study results are reviewed by the laboratory manager and the medical director; however, the ultimate responsibility and approval rests with the medical director. All of the other choices (Answers A, B, C, and D) are incorrect, even though some of them contain personnel that might be involved in a validation (e.g., the manager of the stem cell laboratory).

12. Your laboratory tests for stem cell viability using the trypan blue method. To meet regulatory standards, the laboratory participates in proficiency testing (PT) for this test. When the PT specimen arrives, the stem cell laboratory technologist performs the test and repeats the test to make sure they are correct before reporting the results. Which of the following is true?
A. PT may not be performed more than once
B. PT is required only for CLIA-regulated tests
C. The stem cell laboratory may compare its results with other laboratories before reporting the results
D. If a laboratory fails a PT test, it must discontinue performing the test
E. PT is required for moderate complexity tests only

Concept: As mentioned previously, CLIA is the agency that oversees laboratory accreditation. Participation in proficiency testing (PT) is a condition of CLIA accreditation for any CLIA regulated testing that is performed. For example, CAP sends a sample with an unknown stem cell viability to your laboratory. Your technologist should test/treat the sample as they would treat a normal sample. The results are recorded and sent back to the CAP for scoring/grading. If your results are similar to the CAP's results, your laboratory is deemed competent to continue performing the testing. If your results are not in the CAP's acceptable range, then you have to troubleshoot the problem and may need to halt testing depending on how many failures you have reported for the particular test.

Answer: *B*—PT testing is required for CLIA-regulated tests [moderate (Answer E) and high-complexity]. Other "home-brew" tests do not require PT testing, but may have other quality measures to meet. In the absence of an approved program, laboratories must have a system of determining accuracy and reliability of test results. If required, PT testing samples must be handled and tested in the same manner as regular patient samples. Repeat testing is permitted provided that the patient samples are tested in similar manner (Answer A). Laboratories may not discuss a proficiency test results with other laboratories (i.e., laboratories with a different CLIA number) during the testing period (Answer C).

Failure to achieve a satisfactory score requires corrective action or suspension of testing (Answer D). The corrective action must be approved by the accrediting program. According to the College of American Pathologists CAP-PT (published August 2015), PT failures of 2 out of 3 testing events on a CLIA regulated analytes is reported to Center for Medicare and Medicaid services (CMS) and CAP accreditation programs and requires immediate corrective action. The corrective action will be prepared to include retraining of the testing personnel to achieve competency and the action plan must be approved by the accrediting agency. Failure to attain a satisfactory score on 3 out of 4 testing events is considered critical proficiency testing performance and requires immediate suspension of testing. Before the laboratory can resume testing, it must determine the reason for the PT failures, determine whether patient results were impacted by the PT failures, develop, and implement a corrective action plan, perform two testing events of successful reinstatement of PT and submit to CAP appropriate documentation. A nonregulated analyte requires the same corrective action plan for PT failures as a regulated analyte, however; the report is only made to CAP accreditation programs.

End of Case

13. Which of the following parameters represents the precision of an assay?
A. Mean
B. Mode
C. Coefficient of variation
D. Analytical sensitivity
E. Type I error

Concept: An assay should be both accurate and precise. Accuracy describes how close the measurement is to the "true" result measured by the "gold standard" method. Precision, on the other hand, is the reproducibility of the measurement.

Answer: C—Since precision is the repeatability of the measurement, it can be conveniently expressed with the coefficient of variation (CV). CV is the standard deviation (SD) normalized by the magnitude of the signal being measured. Mathematically, CV is expressed as

$$CV = \frac{\textit{Standard deviation}}{\textit{Mean measurement of the analyte using the assay}}$$

Mean (Answer A) and mode (Answer B) are part of the descriptive statistics of a continuous variable. Mean commonly refers to the arithmetic mean of the measurements, which can be expressed as $mean = \left(\frac{1}{n}\right)\Sigma_1^n x_i$ (where n is the number of measurements, and the ith measurement is represented as x_i). Mode is the measurement that appears the most in a set of data. For example, given the following five measurements: 1, 2, 2, 2.4, and 2.9, the mode is 2, and the mean is $\left(\frac{1}{5}\right)(1+2+2+2.4+2.9)=2.06$.

The sample SD can be calculated as $\sqrt{\left(\frac{1}{n-1}\right)\Sigma_1^n (x_i - \bar{x})^2}$ where $\bar{x}$ is the sample mean. Thus, SD for the sequence is $\sqrt{\frac{1}{5-1}\left[(1-2.06)^2+(2-2.06)^2+(2-2.06)^2+(2.4-2.06)^2+(2.9-2.06)^2\right]}=0.699$. Thus,

$$CV = \frac{0.699}{2.06} = 0.34.$$

Analytical sensitivity (Answer D) is the lowest value that an assay can reliably detect. A common method to making this decision is to measure the blank standard (i.e., containing 0 analyte) multiple times (usually 10 blanks and 3 nonblanks) and then calculate the SD of the measurements. The lowest detection threshold can be set at the level of 2 or 3 times the SD. Then, this level is compared with the manufacturer's claim (if available), and if it is less than the manufacturer's claim, then it passes the "limit of detection" test.

Type I error (Answer E) occurs when a researcher incorrectly rejects the null hypothesis. In the majority of statistical tests in medicine, type I error is set at 5%. Type II error (equivalent to 1-power) occurs when the researcher incorrectly fails to reject the null hypothesis. In many randomized trials in medicine, the power used in the sample size calculation is usually >80% (or type II error <20%).

14. The following information represents your laboratory's PT/INR assay:
- International Sensitivity Index (ISI) of the reagent is 1.5
- Geometric mean of the reference plasma PT is 12 s

Your patient is currently taking warfarin, and his PT is 30 s. Which of the following represents the correct calculated INR?

A. 2.5
B. 3.75
C. 3.95
D. 6.25
E. Cannot be determined from the given information

Concept: The director of the Hemostasis Laboratory is responsible for ensuring the reported PT/INR is correct. Since patients on warfarin need to be monitored for coagulation status and PT may be differed among laboratories since they use different reagents, INR was developed to standardized results for patients on warfarin among laboratories. Many laboratories switch reagent lots every 6 months or so; thus, there must be a process to ensure that the correct ISI is used when lots are switched. The formula to calculate INR is:

$$INR = \left(\frac{PT_{patient}}{PT_{geometric\,mean}}\right)^{ISI}$$

Answer: C—Applying the above formula, we have:

$$INR = \left(\frac{30}{12}\right)^{1.5} = 3.95$$

Of note, geometric mean is different from arithmetic mean (discussed in Question 13). Given n number of measurements, and the ith measurement is represented as x_i, geometric mean is calculated as follows:

$$Geometric\ mean = \rho = \sqrt[n]{x_1 x_2 \ldots x_n} = (\Pi_1^n x_i)^{\frac{1}{n}}$$

Mathematically, $\ln(\rho) = \frac{1}{n}[\ln(\Pi_1^n x_i)] = \frac{1}{n}\Sigma_1^n \ln(x_i) = \frac{1}{n}\left[\ln(x_1) + \ln(x_2) + \ldots + \ln(x_n)\right]$

Thus, $Geometric\ mean = \rho = e^{\frac{1}{n}[\ln(x1)+\ln(x2)+\ldots+\ln(xn)]}$. In other words, the geometric mean is the log average of a set of data. It calculates the arithmetic mean of the log transformation of data x_i [i.e., the mean of $\ln(x_i)$] and then uses exponentiation to compute the original (as demonstrated earlier).

For example, the geometric mean of four measurements of 1, 3, 2.5, and 1.6 is

$$\sqrt[4]{(1)(3)(2.5)(1.6)} = [(1)(3)(2.5)(1.6)]^{\frac{1}{4}} = e^{\frac{1}{4}[\ln(1)+\ln(3)+\ln(2.5)+\ln(1.6)]} = 1.86$$

Please answer Questions 15–18 based on the following scenario:

You and your colleague are working on identifying new TTP biomarkers. You have developed a new ADAMTS13 assay in your laboratory. Your colleague also independently developed assays for two new novel TTP biomarkers, namely biomarkers X and Y, in her laboratory. She has proven that both markers X and Y play important roles in the interaction between ADAMTS13, von Willebrand factor (vWF), and platelet adhesion. She hypothesizes that doing sequential tests for both markers X and Y will improve the diagnosis of TTP.

You decide to investigate the performance characteristics of your new ADAMTS13 assay and have collected the following data:

	TTP present	TTP not present
Test positive for TTP	24	5
Test negative for TTP	1	99

15. What is the sensitivity and specificity of your new assay in detecting TTP?
A. Sensitivity = 96%; Specificity = 99%
B. Sensitivity = 96%; Specificity = 95%
C. Sensitivity = 83%; Specificity = 99%
D. Sensitivity = 83%; Specificity = 95%
E. Sensitivity = 83%; Specificity = 96%

Concept: For 2 × 2 tables:

	Disease present	**Disease not present**
Test positive for disease	True positive (TP)	False positive (FP)
Test negative for disease	False negative (FN)	True negative (TN)

Sensitivity and specificity are inherent properties of the assay. It does not change based on the prevalence of the disease. Sensitivity is the probability of the test is positive given the patient has the disease. On the other hand, specificity is the probability of the test is negative given the patient does not have the disease. They are calculated as follows:

$$Sensitivity = \frac{True\ positive}{True\ positive + False\ negative}$$

$$Specificity = \frac{True\ negative}{True\ negative + False\ positive}$$

Another way to think about these formulas is as follows: for sensitivity, you are taking into account the true positive over all the results that either are positive (TP) or should be positive (FN); for specificity, you are taking into account the true negative over all the results that either are negative (TN) or should be negative (FP).

Answer: *B*—Based on the above formula:

$$Sensitivity = \frac{24}{24+1} = 96\%$$

$$Specificity = \frac{99}{99+5} = 95\%$$

All the other choices (Answers A, C, D, and E) are incorrect based on the formula.

16. What is the positive predictive value (PPV) and negative predictive value (NPV) of this new assay?
A. PPV = 96%; NPV = 99%
B. PPV = 96%; NPV = 95%
C. PPV = 83%; NPV = 99%
D. PPV = 83%; NPV = 95%
E. PPV = 83%; NPV = 96%

Concept: Unlike sensitivity and specificity, PPV and NPV calculations depend on the prevalence of the disease in the population. PPV is the probability that the patients having the disease given that they have positive test. NPV is the probability that the patients do not have the disease given that they have negative tests. They are calculated as follows:

$$PPV = \frac{True\ positive}{True\ positive + False\ positive}$$

$$NPV = \frac{True\ negative}{True\ negative + False\ negative}$$

Answer: C—Based on the above formula:

$$PPV = \frac{24}{24+5} = 83\%$$

$$NPV = \frac{99}{99+1} = 99\%$$

To recall these formulas, remember that for PPV, you are taking into account the true positive over all positives, whether they are TP or FP. Similarly for NPV, you are taking into account the true negative over all negatives, whether they are TN or FN. All the other choices (Answers A, B, D, and E) are incorrect based on the formulas.

17. You now offer your assay described in Question 15 for clinical use, after a rigorous validation process, in accordance with applicable rules and regulations. You also have the disclaimer that your assay is not FDA-approved when reporting the results. One of your colleagues in obstetrics just learned from your TTP lecture that pregnancy can be a risk factor for TTP. He asked you to measure ADAMTS13 level for a 24-year-old patient with no previous medical history who is 8 weeks pregnant. The patient has a hemoglobin level of 10.1 g/dL and a platelet count of 140,000/μL (which dropped from 350,000/μL 2 weeks ago). You advised him against the measurement since ADAMTS13 is not a screening test for TTP in the general population, but he insisted on performing the test in your lab. The assay came back <5%, which is consistent with a laboratory diagnosis of TTP. You repeat the test and the result is the same. Given a 0.4% prevalence of TTP in this population, what is the probability that this patient has TTP given this test result?

A. 0.004
B. 0.08
C. 0.83
D. 0.96
E. 0.99

Concept: As discussed earlier, sensitivity is the probability of the test is positive given the patient has the disease (Pr(positive test | disease)). However, this question asks the reverse—given a positive test, what is the probability of the patient having the disease (Pr(disease | positive test))? We use Bayes' theorem to solve this problem as following:

$$\Pr(disease \mid positive\ test) = \frac{\Pr(positive\ test \mid disease) * \Pr(disease)}{\Pr(positive\ test \mid disease) * \Pr(disease) + \Pr(positive\ test \mid no\ disease) * \Pr(no\ disease)}$$

$$= \frac{Sensitivity * Prevalence}{Sensitivity * Prevalence + (1 - Specificity) * (1 - Prevalence)}$$

Answer: *B*—Using the sensitivity and specificity of the assay calculated in Question 15, we have:

$$\Pr(disease \mid positive\ test) = \frac{Sensitivity * Prevalence}{Sensitivity * Prevalence + (1 - Specificity) * (1 - Prevalence)}$$

$$= \frac{0.96 * (0.004)}{0.96 * 0.004 + (1 - 0.95) * (1 - 0.004)} = 0.08$$

Given the prevalence is very low in this general population, it intuitively makes sense that this positive result is most likely false positive. However, let say that you find out this patient actually had an episode of TTP in the past and assuming the prevalence of TTP relapse is 40%, then the positive result would be more concerning given the calculation below:

$$\Pr(\text{disease} \mid \text{positive test}) = \frac{\text{Sensitivity} * \text{Prevalence}}{\text{Sensitivity} * \text{Prevalence} + (1 - \text{Specificity}) * (1 - \text{Prevalence})}$$

$$= \frac{0.96 * (0.4)}{0.96 * 0.4 + (1 - 0.95) * (1 - 0.4)} = 0.93$$

All the other choices (Answers A, C, D, and E) are incorrect based on the formula.

18. Your colleague hypothesizes that sequential testing for TTP using her novel biomarkers (X and Y) may improve patient care. Her results for biomarker X are presented in the following table:

	TTP present	TTP not present
Biomarker X positive for TTP	25	10
Biomarker X negative for TTP	1	50

If the test for biomarker X is positive in biomarker X, then biomarker Y will be measured (in the above example, this would equal 35 people). The results for biomarker Y are presented in the following table:

	TTP present	TTP not present
Biomarker Y positive for TTP	23	1
Biomarker Y negative for TTP	2	9

What is the overall sensitivity and specificity for serial testing of *both* biomarkers X and Y in the diagnosis of TTP?

A. Sensitivity = 96%; Specificity = 83%
B. Sensitivity = 92%; Specificity = 90%
C. Sensitivity = 88%; Specificity = 98%
D. Sensitivity = 88%; Specificity = 90%
E. Sensitivity = 96%; Specificity = 98%

Concept: In real-life, tests may be done sequentially. For example, in the laboratory diagnosis of heparin induced thrombocytopenia (HIT), the screening test is the ELISA assay, which is used to measure the presence of the heparin antibodies bound to the platelet factor 4 (PF4)-heparin complex. If the HIT ELISA is positive, a serotonin release assay (SRA) will be performed as a confirmatory assay. Usually the screening test has a high sensitivity but low specificity, and the reverse is true for the confirmatory assay.

Using the formula in Question 15, the assay for biomarker X has a 96% sensitivity and 83% specificity, and the assay for biomarker Y has 92% sensitivity and 90% specificity.

$$\text{Net sensitivity} = \frac{\text{True positive}}{\#\text{TTP patients}} = \frac{23}{26} = 88\%$$

$$\text{Net specificity} = \frac{\text{True negative}}{\#\text{patients without TTP}} = \frac{50+9}{60} = 98\%$$

Note: There is no addition in the numerator when calculating net sensitivity because the test result is not considered to be positive until the second test (assay for biomarker Y) confirms the positivity. Thus, only 23 results are considered as true positives.

Answer: C—Based on the calculation above, with sequential testing, the net sensitivity is 88% and the net specificity is 98%. For sequential testing, the net sensitivity decreases compared to the sensitivity of the screening test, but the net specificity increases. All the other choices (Answers A, B, D, and E) are incorrect based on the formula.

End of Case

Please answer Questions 19–23 based on the following scenario:

Your hemostasis laboratory offers measurement for both the vWF antigen and vWF activity by the ristocetin cofactor assay.

19. For quality control (QC) and quality assurance (QA) purposes, you are reviewing the Levy-Jennings plot below for your vWF analyte "normal control" for the past 20 days. Review the "normal control" plot below and answer the question that follows. The plot of the "low control" is perfectly acceptable and thus, is not shown here.

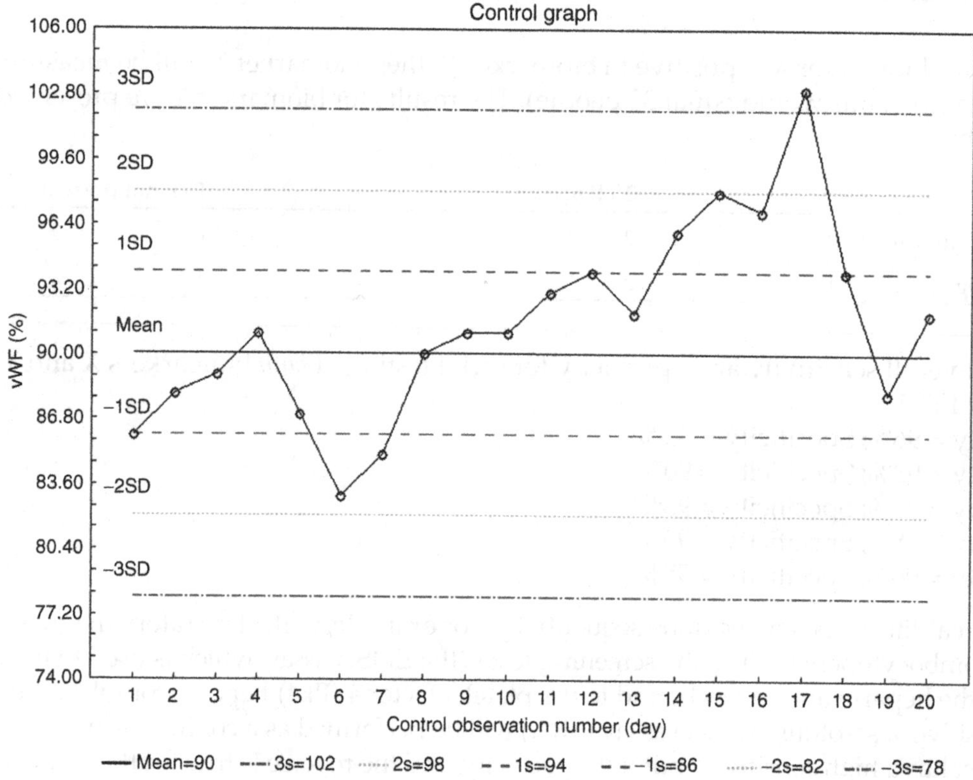

Applying the Westgard rules, at what day the QC analyte first failed and the technologist should have rejected the run?

A. Day 15
B. Day 16
C. Day 17
D. Day 18
E. Day 19

Concept: Westgard rules are statistically based and designed to assess if a test system is within the realm of random/normal/day to day variation. It should only be used when two levels of control materials are analyzed per run. In this example, a normal and a low control are performed, but only the plot for the normal control is provided since the plot of the "low control" is perfectly acceptable. Table 2.1 explains how to use these rules to determine if the QC is in control or not.

TABLE 2.1 Westgard Rules

Westgard rule	Explanation
1_{2s}	Warning rule—One of the two control results falls outside ±2SD (standard deviation)
1_{3s}	Reject run—One of the two control falls outside ±3SD
2_{2s}	Reject run—two consecutive QC levels fall outside of ±2SD or both controls on the same run exceed ±2SD
R_{4s}	Reject run—The range between two results exceed 4SD or one control exceed the mean by +2SD and the other exceeds the mean by -2SD
4_{1s}	Reject run—Four consecutive QC results are outside of ±1SD or both levels of control have consecutive results that are outside of ±1SD
10_x	Reject run—10 consecutive QC results are on one side of the mean or both levels of control have five consecutive results that are on the same side of the mean

Answer: C—Using Table 2.1, day 17 is the first time the run violated the Westgard rules (1_{3s} rule was violated). All the other choices are incorrect based on the rules in Table 2.1.

20. You are also reviewing the QC data for your factor VIII assay. The following is one of the QC level's Levy-Jennings curves.

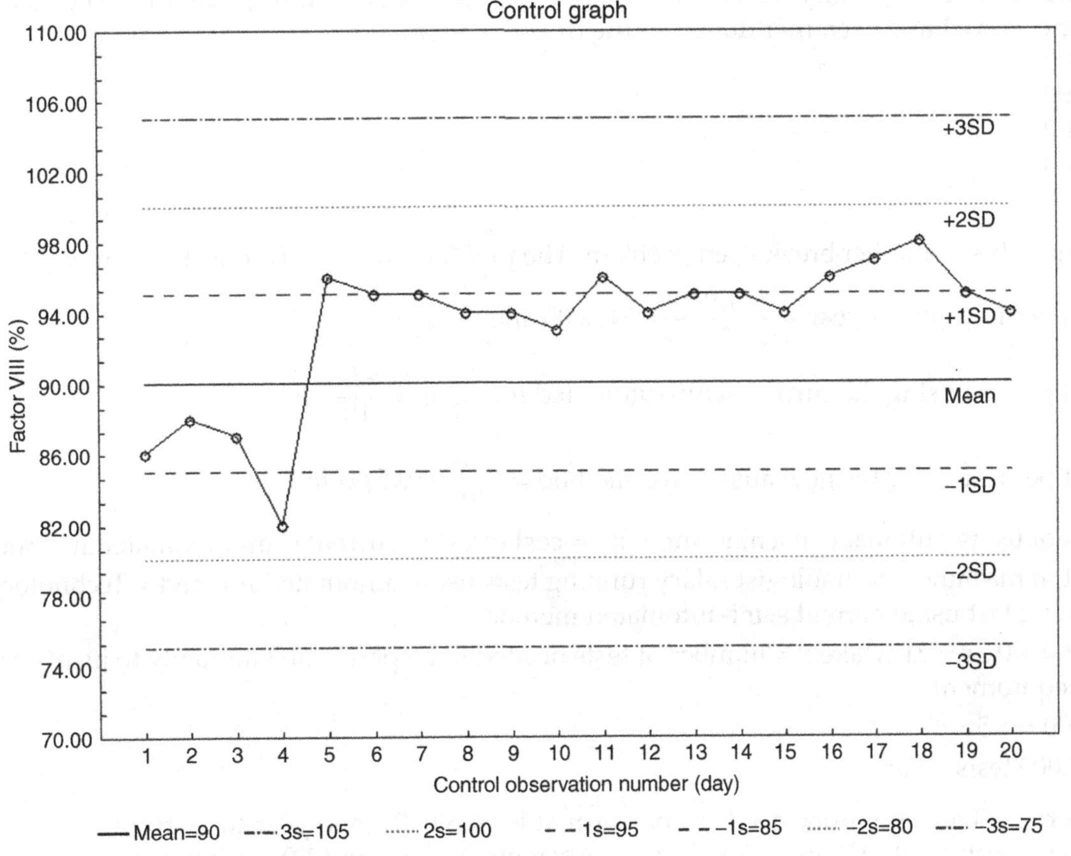

Please describe what might had happened between day 4 and 5?

A. This indicates a shift—possibly due to change of reagent

B. This indicates a shift—possibly due to progressive change in instrument function

C. This indicates a trend—possibly due to change of reagent

D. This indicates a trend—possibly due to progressive change in instrument function
E. This indicates a trend—possibly due to instrument maintenance

Concept: Levy-Jennings plots can also be used to detect trends and shifts in the system. Trends are usually due to a progressive change in instrument function (such as from day 8 to 17 in the Levy-Jennings curve in Question 19), and usually requires attention to prevent erroneous patient results. Shifts are sudden changes in result patterns, usually due to a change of reagents or equipment maintenance. The laboratory needs to evaluate the clinical significance of the new results. If the change is clinically significant, recalibration of the equipment should be considered.

Answer: *A*—This pattern indicates a shift—possibly due to a reagent change. The new level is around 95%, compared to 90% for the previous reagent. This is probably not clinically significant for a factor VIII level, but the decision is ultimately up to the medical director of the laboratory. Answers C, D, and E are incorrect based on the definition of shifts and trends. Answer B is wrong because a progressive change in instrument function usually results in a trend, and not a shift.

21. Your hemostasis laboratory also offers a lupus anticoagulant panel. There is new equipment that can do this panel automatically. This new machine has a 7-year lifetime and costs $84,000. It can perform 40 tests/hour. Currently, with the semiautomated method performed in your laboratory, a technologist can do 10 tests/hour. Your technologist earns $20/hour. How many tests need to be done annually to justify the cost of this new equipment, assuming that all the costs associated with the assay have been mentioned in the question stem?
A. 4,000
B. 5,000
C. 6,000
D. 7,000
E. 8,000

Concept: This is another break-even problem. The problem can be solved as following:

$$\text{Cost of machine per year} = \frac{\$84{,}000}{7} = \$12{,}000 \text{ annually}$$

$$\text{Cost per test using the current semi-automated methods} = \frac{\$20}{10} = \$2 \, \textit{per test}$$

$$\text{Cost per test using the new automated method} = \frac{\$20}{40} = \$0.5 \, \textit{per test}$$

Cost of using automatic machine annually = cost of using current semi-automated method
Cost of machine + technologist salary running tests using automated method = Technologist salary running test using current semi-automated method
$12{,}000 + 0.5x = 2x$ where x = number of tests needed to be performed annually to justify the cost of this equipment
$12{,}000 = 1.5x$
$x = 8{,}000$ tests/year

Answer: *E*—The laboratory needs to perform at least 8,000 tests/year to justify the cost of this new equipment. All the other choices (Answers A, B, C, and D) are incorrect based on the calculations.

22. Your laboratory is in the process of switching to a new coagulation analyzer and you would like to verify the reference range provided by the manufacturer for your prothrombin time (PT) test. Twenty samples from healthy individuals were collected and the data are shown below.

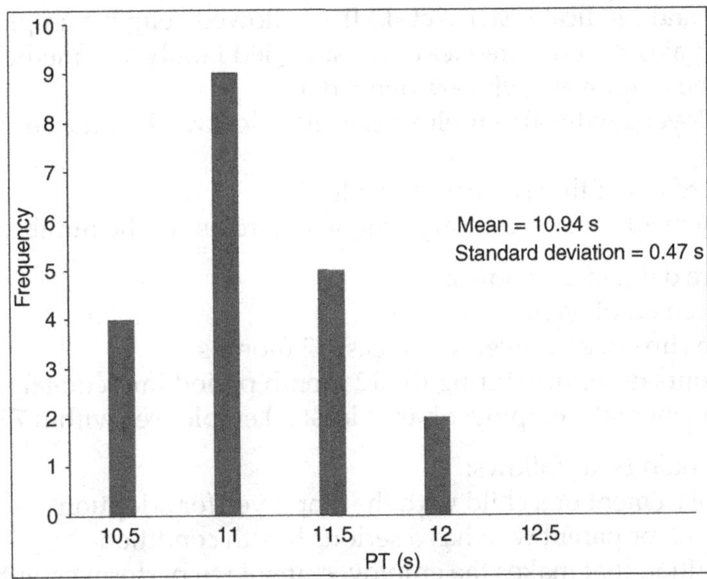

If you want your PT reference range to capture 95% of the normal population and your sample represents the healthy individuals, what is your reference range?

A. 10.5–12 s
B. 10.47–11.41 s
C. 10.5–11.5 s
D. 10–11.88 s
E. 9.53–12.35 s

Concept: Reference usually refers to the range of results that define a condition of health (or lack thereof). Reference range does not only mean normal range since reference range can also be used to refer to therapeutic ranges for certain medications, such as immunosuppressants. Normal range is a subset of reference range, and refers to the range of the assay measurements that the majority of healthy individuals will fall into. The laboratory can establish this normal range by collecting samples from at least 120 healthy individuals. More commonly however, the manufacturer of the analyzer already has a recommended range, and thus, the laboratory usually only needs to validate this normal range by collecting samples from at least 20 individuals. This process is done in this example.

Answer: *D*—In this case, the distribution seems to follow Gaussian distribution. Thus, the mean ± 2SD will capture the central 95% of the normal population. Hence, if you want your PT reference range to capture 95% of the normal population and your sample represents the healthy individuals, then the range will be 10.94 ± 2(0.47) or 10–11.88 s. All the other choices (Answers A, B, C, and E) are incorrect based on the explanation above.

Note: If the distribution is non-Gaussian, then the range can be established by calculating the 2.5th and 97.5th percentile of the measurements.

23. One of your experienced technologists is now pregnant and expects to deliver her first baby soon. She has worked at your institution for 5 years and now would like to request some time off to take care of her baby. Based on regulatory rules, what is the maximum number of weeks that she can take leave for her stated purpose in a 12-month period?

A. 6 weeks
B. 12 weeks
C. 18 weeks
D. 24 weeks
E. 26 weeks

Concept: The Family and Medical Leave Act (FMLA) allowed "eligible employees of covered employers to take unpaid, job-protected leave for specified family and medical reasons." This act only applies to covered employers, who are defined as:

- Private sector employers, with 50+ employees in 20+ workweek in the current or preceding calendar year
- Public agency, regardless of the number of employees
- Public or private elementary or secondary school, regardless of the number of employees

Eligible employees are defined as follows:

- Works for the covered employers
- Has worked for the current employer for at least 12 months
- Has at least 1250 hours of service during the 12 month period immediately preceding the leave
- Works at a location where the employer has at least 50 employees within 75 miles

The reason for leave could be as follows:

- Birth of a child or placement of a child with the employee for adoption
- Care for spouse, child, or parent who has a serious health condition
- Serious health condition that makes the employee unable to perform essential job duty
- Any qualifying exigency arising out of the fact that a spouse, child, or parent is a military member on covered active duty or call to covered active duty status

Of note, an eligible employee may also take up to 26 weeks of leave during a single 12-month period to care for a covered service member with serious injury or illness, when the employee is the spouse, child, parent, or next of kin of the service member. In other words, FMLA guarantees that an eligible employee is allowed to take the specific time off and come back to the same job afterwards. The employee cannot be let go due to a qualified extended leave.

Answer: *B*—Based on the above FMLA guidelines, she is an eligible employee, and thus, qualifies to take 12-weeks of unpaid leave for a medical reason. All of the other choices (Answers A, C, D, and E) do not represent the correct time frame.

End of Case

Please answer Questions 24–27 based on the following scenario:

Though there are available laboratory tests, HIT remains a clinical diagnosis. Although the incidence is low, true HIT can result in severe morbidity and mortality, including organ and limb ischemia. Failure to diagnose and treat HIT in a timely manner has been the basis for many medical malpractice lawsuits. The local hematologist is using the HIT expert probability (HEP) score, with a score of 2 or more being considered clinically suspicious of HIT. The sensitivity is 100% and the specificity is 60% for the HEP score.

24. What is the likelihood ratio (LR) of positive and negative test for this HEP score?

	LR of a positive test	LR of a negative test
A.	1	0.6
B.	1.67	0.6
C.	0.6	0.6
D.	2.5	0
E.	0	2.5

Concept: LR is used to assess the value of performing a test. LR of a positive test (or positive LR) is the ratio of probability of a positive test positive in a patients with and without the disease. Essentially, it is the odds of a positive result in a patient with disease as opposed to a patient without the disease. The higher the positive LR, the more useful information the physician can get from a positive test. Similarly, LR of a negative test (or negative LR) is the ratio of probability of a negative test in patients with and without the disease. Essentially, it is the odds of a negative result in a patient without the disease, as opposed to a patient with the disease. They are calculated as follows:

$$LR\ of\ a\ positive\ test = \frac{Sensitivity}{1 - Specificity}$$

$$LR\ of\ a\ negative\ test = \frac{1 - Sensitivity}{Specificity}$$

LR is a characteristic of the assay or measurement device (e.g., HEP score), and it is not affected by the prevalence of the disease. A $LR > 1$ indicates the test result is associated with the disease. A $LR < 1$ indicates that the result is associated with absence of the disease. Tests where LR ~ 1 have little practical significance.

Answer: D—Using the above formula:

$$LR\ of\ a\ positive\ test = \frac{Sensitivity}{1 - Specificity} = \frac{1}{1 - 0.6} = 2.5$$

$$LR\ of\ a\ negative\ test = \frac{1 - Sensitivity}{Specificity} = \frac{1 - 1}{0.6} = 0$$

All the other choices (Answers A, B, C, and E) are incorrect based on the formula.

25. The prevalence of HIT in your specific population is 5%. A patient was selected at random from the population. What are the odds of this patient having HIT?

A. 20
B. 19
C. 0.053
D. 0.1
E. It cannot be calculated from the information given

Concept: Odds is the ratio of the probability of having a disease over the probability of not having the disease.

$$Odds = \frac{Probability\ of\ an\ event}{1 - Probability\ of\ an\ event}$$

Answer: C—In this question, the event is the prevalence of having a disease

$$Odds = \frac{Probability\ of\ an\ event}{1 - Probability\ of\ an\ event} = \frac{0.05}{1 - 0.05} = 0.053$$

All the other choices (Answers A, B, D, and E) are incorrect based on the formula.

26. The patient from Questions 24 and 25 has a HEP score of 3. What is the probability of this patient having HIT?

A. 0.05
B. 0.12
C. 0.13
D. 0.15
E. 0.30

Concept: LR can be combined with prevalence to calculate the posttest probability as follows:

Posttest odds of disease = Pretest odds of disease × LR of a positive result

$$\text{Posttest probability} = \frac{\textit{Posttest odds of disease}}{1+\textit{Posttest odds of disease}}$$

A high LR increases the probability of a positive test predicting disease. A clinically useful test will have a high LR of a positive test and a low LR of a negative test.

Answer: *B*—Using the above formula and the answers from Questions 24 and 25, we have:

Posttest odd of disease = Pretest odds of disease × LR of a positive result = 0.053 × 2.5 = 0.1325

$$\text{Posttest probability} = \frac{\textit{Posttest odd of disease}}{1+\textit{Posttest odds of disease}} = \frac{0.1325}{1+0.1325} = 0.12$$

Thus, the clinician should consider discontinuing heparin and switching to a direct thrombin inhibitor, as well as sending the patient's sample for a HIT ELISA assay to screen for the presence of the heparin antibodies bound to the PF4-heparin complex since the probability of having HIT increased from 5% to 12% after we know that this patient has a HEP score of 3. All the other choices (Answers A, C, D, and E) are incorrect based on the formula.

27. Let us now say that the patient from Question 26 has a HEP score of 1. An orthopedic surgeon asks you if she should stop heparin and send the patient's sample for a HIT ELISA assay. What will you recommend?

A. Do not stop heparin and do not send for ELISA assay
B. Do not stop heparin but send for ELISA assay
C. Stop heparin and switch to bivalirudin but do not send for ELISA assay
D. Stop heparin, switch to bivalirudin, and send for ELISA assay
E. Stop heparin, switch to warfarin, and send for both ELISA and SRA assay

Concept: Since the HEP score of 1 portrays a low pretest probability, posttest probability calculation may be helpful in this scenario. Using results from Questions 24 and 25, we have:

Posttest odds of disease = Pretest odds of disease × LR of a negative result = 0.053 × 0 = 0

$$\text{Posttest probability} = \frac{\textit{Posttest odds of disease}}{1+\textit{Posttest odds of disease}} = \frac{0}{1+0} = 0$$

Answer: *A*—The posttest probability of having HIT given a low HEP score is 0, which essentially rules out HIT. Thus, the surgeon does not need to stop heparin, and ELISA assay should not be performed (Answers B, D, and E). In this scenario, a HEP score is a good screening test because it rules out HIT if the score is <2. A HEP score ≥2 should prompt discontinuation of heparin and a laboratory workup for HIT. Nonetheless, HIT remains a clinical diagnosis, and thus, the decision to whether send for ELISA and/or stop heparin remains a clinical decision. Laboratory assays (e.g., HIT ELISA) and HEP score may help with the decision, but at the end, clinical judgement should prevail. Since the HEP score is low in this case, neither discontinuation of heparin nor switching to an alternative anticoagulant (Answer C) in warranted.

End of Case

28. A 58-year-old male presented to the ED with reports of hematemesis and weakness. His initial blood pressure is 80/60, so the ED resident orders a 500 mL bolus of saline, which raises his blood pressure to 110/70. His initial hemoglobin (Hgb) was 10.8 g/dL; however, after two witnessed episodes of hematemesis in the ED, his Hgb has dropped to 6.9 g/dL and he is complaining of lightheadedness. The ED resident decided to admit the patient to the floor after transfusing one unit of blood. Once the patient arrived on the floor, he went into cardiac arrest and eventually died.

Review of the case reveals that though the blood was ordered and released, it was not transfused into the patient before he left the ED, as intended by the resident. Since this event potentially resulted in the death of the patient, you are asked to participate in a root cause analysis. Which of the following is true about a root cause analysis?

A. It is focused on identifying the individual at fault
B. Brainstorming is a crucial component of the process
C. The analysis is complete once the fishbone diagram is done
D. The root cause is almost always five levels deep
E. The final data analysis is the responsibility of the ED director

Concept: Root cause analysis (RCA) is a structured method used to analyze serious adverse events. Initially developed to analyze industrial accidents, RCA is now widely deployed as an error analysis tool in health care. A central tenet of RCA is to identify underlying/process problems that increase the likelihood of errors, while avoiding the trap of focusing on mistakes by individuals. RCA uses a systematic approach to identify both active errors (i.e., errors occurring at the point of interface between humans and a complex system) and latent errors (i.e., the hidden problems within health care systems that contribute to adverse events).

Answer: *B*—To minimize the risk of a faulty analysis, a group brainstorming session is required. Ideally, all stakeholders are represented. Although, a good facilitator will challenge the group by asking "Why?" 5 times, the "root" cause is typically less than five levels deep (Answer D). A fishbone diagram (Answer C) is a useful tool to document the output of the brainstorming session; however, further analysis is required to rule in or rule out each potential root cause with data.

RCAs should generally follow a prespecified protocol that begins with data collection and reconstruction of the event in question through record review and participant interviews. A multidisciplinary team, not one individual (Answer E), should then analyze the sequence of events leading to the error, with the goal of identifying how the event occurred through identification of active errors and why the event occurred through systematic identification and analysis of latent errors. The ultimate goal of RCA, of course, is to prevent future harm by eliminating the latent errors that so often underlie adverse events and is not focused on finding fault in any one individual (Answer A).

29. A nurse's aide performed a glucose point-of-care test on a 43 year patient in the ED with confusion and lethargy. The result is normal. The patient is discharged, and later returns to the ED, eventually being diagnosed with diabetic ketoacidosis. The patient dies in the ICU the next day. Autopsy reveals that the patient had undiagnosed type 2 diabetes mellitus, which lead to cerebral edema, raised intracranial pressure, and death. Investigation of the event by laboratory staff reveals that the nurse's aide performed the test incorrectly, and that there is no evidence of a defect in any of the components of the point-of-care glucose test system. Which of the following is true?

A. The event must be reported only to the FDA
B. There is no legal requirement to report this event, since the cause has been identified
C. The event must be reported only to the device manufacturer
D. The event must be reported to FDA directly and to the device manufacturer
E. There is no legal requirement to report this event, since the device was not defective in any way

Concept: When deaths occur in the hospital that could be the result of human or device error, regulatory agencies must be notified as soon as possible. A root cause investigation should take place, and once the cause is identified, further notification may become necessary (e.g., notifying the device manufacturer).

Answer: *D*—Since the investigation concluded the error and subsequent patient death was likely related to a medical device, it becomes necessary to notify the FDA and the device manufacturer of the findings. Essentially, the facility is obligated to report device-related deaths both to FDA and the device manufacturer. This is true even in the absence of a malfunction and even if the event was cause by user error per Medical Device Reporting (MDR) regulation (21 CFR 803). The choices (Answers A and C) are incorrect because they specify that only one of the parties be notified. The choices (Answers B and E) are incorrect because the event must be reported regardless of the individual cause and even if they device was not defective.

Please answer Questions 30–33 based on the following clinical scenario:

Your research group is interested in investigating the risk of cardiac disease associated with levels of ADAMTS13. You have several study design options.

30. In order to gather preliminary data, you decid to do a retrospective case-control study. You select 100 patients admitted to the hospital with chest pain and myocardial infarction (MI). You also select 500 control patients with similar demographics, who were also admitted with chest pain, but did not have any evidence of a MI. ADAMTS13 level was measured on all 600 patients. The results are as follows:

	Cases	Controls
ADAMTS13 level < 25th percentile	85	20
ADAMTS13 level ≥ 25th percentile	15	480

Based on this data, what are the odds of having MI given a patient's ADAMTS13 levels < 25th percentile compared to patients with ADAMTS13 levels ≥ 25th percentile?

A. 7
B. 24
C. 85
D. 136
E. 512

Concept: A case control study is an observational study. Classically, it is a study that starts with the identification of people with the disease and a suitable control group without the disease. The relationship of an exposure to a disease with regard to how frequently the exposure is present or, if quantitative, the level of the exposure, in each of the groups, will be quantified.

The appropriate measure of association for a case-control study is the odd ratio. In a 2 × 2 table:

	Disease positive	Disease negative
Exposure positive	A	B
Exposure negative	C	D

The odd ratio (OR) is calculated as follows:

$$OR = \frac{A * D}{B * C}$$

Answer: *D*—Based on the above data:

$$\text{OR} = \frac{A * D}{B * C} = \frac{85 * 480}{20 * 15} = 136$$

Thus, the odds of having MI is 136 times higher if a patient has an ADAMTS13 level < 25th percentile compared to those with ADAMTS13 levels ≥ 25th percentile. All the other choices (Answers A, B, C, and E) are incorrect based on the formula.

31. Given the data in Question 30, what is the risk ratio for having an MI with an ADAMTS13 level < 25th percentile compared to those with ADAMTS13 level ≥ 25th percentile?
A. 21.25
B. 26.7
C. 85
D. 136
E. Risk ratio is not an appropriate measure for association in this study

Concept: Risk ratio (RR) is defined as the ratio of the probability of having the disease given the exposure over the probability of having the disease given a negative exposure. It is used as measure of association for both randomized controlled trials and cohort studies. On the other hand, the OR can be used as measure of association for randomized controlled trials, cohort studies, and case-control studies. However, in epidemiological studies, the risk is the main variable of interested (and not the odds).

Answer: *E*—In a case-control study, since the numbers of cases and controls are preselected, it does not make sense to compute RR. If the disease is rare (prevalence <10%), the OR will approximate RR. However, if the disease is not rare, the OR will overestimate RR.

If this was a cohort study (and not a case-control study as in this case), then RR can be computed as follows:

$$\mathrm{RR} = \frac{\Pr(\mathit{Disease+} \mid \mathit{Exposure\ positive})}{\Pr(\mathit{Disease+} \mid \mathit{Exposure\ negative})} = \frac{A/(A+B)}{C/(C+D)} = \frac{85/(85+20)}{15/(15+480)} = 26.7$$

All the other choices (Answers A, B, C, and D) are incorrect based on the formula.

32. Which of the following is an advantage of a case-control study?
A. Rate of disease in exposed and nonexposed group can be determined
B. Selection of the control group is relatively easy
C. Complete control over confounder variables
D. Well suited to study rare disease or disease with long latency
E. Sometimes it has to rely on recall information

Concept: For a classical case-control study, you should select cases based on a precise definition or diagnosis of the disease. There are two types of cases: incident cases and prevalent cases. Incident cases are newly diagnosed cases, which can take an extended amount of time to collect, if you are studying a rare disease. Prevalent cases are patients who have already been diagnosed with the disease and are thus already available to study. Selecting controls is the most difficult and the most important part of the case-control design. Controls are selected from the same population that gives rise to the actual cases. The purpose of having a control group is to provide the exposure distribution of the source population. The methods used to select the control will affect the validity and the generalizability of the study. In addition, selecting the control will also affect the cost of the study.

Answer: *D*—Case control studies have several advantages and disadvantages.

The advantages include the following:
1. Well-suited to study rare disease or diseases with long latency periods
2. Relatively inexpensive and can be conducted quickly based on available existing records
3. Requires relatively fewer subjects with minimal risks to the subjects
4. Allows study of multiple disease etiologies

Disadvantages include the following:
1. Sometimes it has to rely on recall of information (Answer E)
2. Control of other confounders may be incomplete (Answer C)

3. Selection of control group can be difficult (Answer B)
4. Rate of disease in exposed and non-exposed group cannot be determined (Answer A)

33. Which of the following is an advantage of a randomized controlled trial (RCT)?
   A. It has mechanisms to protect against biases
   B. Relatively inexpensive to conduct
   C. It allows multiple clinical questions to be answered using one RCT
   D. Relatively easy to conduct
   E. It is always possible to conduct

Concept: An RCT is the gold standard method in both epidemiological and biomedical research to answer a clinical question. However, it does not mean that an RCT is easy to conduct, inexpensive, always possible to conduct, or is the only way to answer a clinical question. However, it should be conducted if there is a genuine clinical equipoise exists, and its result will change clinical practice. A well-designed RCT has the following seven features to protect against bias:

1. It is a prospective study; thus, it prevents recall and selection bias.
2. It has at least an experimental group.
3. It has a concurrent control group. Without a control, it is impossible to know if the treatment works. Furthermore, having a concurrent control and experiment group avoids differences in quality of data collection, disease definition, and standard of treatment.
4. It uses randomization to allocate subjects to being in a control or experimental group; thus, it controls both measured and unmeasured confounders/biases at baseline—this is the major strength of the RCT design.
5. It has blinding mechanisms to minimize both conscious and nonconscious bias.
6. It uses intention-to-treat analysis to prevent statistical bias because intention-to-treat is an exact, nonparametric test that allows a valid inference about the *P*-value, without any assumption regarding the distribution.
7. It has mechanisms to minimize loss of follow-up to reduce selection bias.

In real life, many RCTs do not have all seven features listed earlier. For example, it may not be possible to design a trial that is blinded to the treatment group if it is a procedure (i.e., apheresis). An RCT can only answer one clinical question, given its tight control on both type I and type II errors. RCTs usually report both primary and secondary outcomes for efficacy and adverse events. However, only the primary outcomes should be used to change clinical practice because all the allowable type I error has been used to treat the primary hypothesis for the primary outcome.

Answer: *A*—RCT has multiple mechanisms to protect against biases per the discussion above. The other answer choices (Answers B, C, D, and E) actually represent weaknesses or limitations of conducting an RCT.

End of Case

34. What is the fundamental assumption underlying the design of a cohort study?
   A. Controls are selected from the same population that gives rise to the cases
   B. Subjects must be free of the disease of interest at the beginning of the study and also must be at risk for developing the disease
   C. Known and unknown biases are balanced between the cases and the controls
   D. It is the best design to study rare diseases
   E. There is no selection bias

Concept: A cohort study is an observational study. It is usually a prospective study, but it can be retrospective. Classically, it is a study in which the investigator selects a group of exposed individuals and a group of nonexposed individuals and then follows them into future and monitors them for development of the disease under study.

In many aspects, the principles of experimental studies are similar to the principles of a cohort study. For example, both make comparisons across exposure groups. Both follow participants longitudinally and monitor rates of one or more than one outcome. Both select groups to achieve comparability, although the relative proportion of subjects in compared groups do not reflect that of the general population. Nonetheless, there are key differences. For instance, experimental study investigators allocate the exposure themselves, and cohort study subjects choose their exposures. Experimental studies use randomization to achieve comparability while cohort studies must carefully select groups and measure potential confounders to achieve comparability.

Answer: *B*—The fundamental assumption underlying the design of all cohort study is that the subjects in a cohort study are disease-free at the beginning of the study and they also must be at risk for developing the disease.

There are several advantages of the cohort study. Because it is usually a prospective study, it ensures temporality. This is important because it ensures the postulated cause is assessed before the occurrence of the disease. It also reduces selection and recall biases. A cohort-design study also allows us to study multiple outcomes at the same time, as well as estimate the incidence of the disease in a population. Nonetheless, it also has several limitations. In order to have high internal validity, there must be a mechanism to ensure high follow-up rate. In a cohort, loss to follow-up is a type of selection bias (Answer E). Therefore, it is essential to choose the populations that are well-defined, so it is possible to gain access to data through centralized sources and minimize loss to follow-up.

Answer A describes the fundamental assumption underlying the case control study. Since both groups in the cohort study do not have the disease at the beginning, and if the disease is rare, it would take a long time to acquire enough cases; thus, the cohort study is not the best design to study rare diseases (Answer D). Case control design may be a better design for rare diseases or disease with long period of latency. Answer C is one of the unique features of a RCT to protect from biases (see Question 33).

Please answer questions 35–38 based on the 30th edition of the AABB Standards:

35. How long records must be kept for a suspected transfusion reaction evaluation/investigation?
A. 3 years
B. 5 years
C. 10 years
D. 20 years
E. Indefinitely

36. How long records must be kept for physician statement releasing emergency blood?
A. 3 years
B. 5 years
C. 10 years
D. 20 years
E. Indefinitely

37. How long records must be kept for clinically significant antibodies?
A. 3 years
B. 5 years
C. 10 years
D. 20 years
E. Indefinitely

38. How long records must be kept for annual review of policies, processes, and procedures?
A. 3 years
B. 5 years

C. 10 years
D. 20 years
E. Indefinitely

Concept: *(Questions 35–38)* According to the 30th edition of the AABB Standards, "blood banks or transfusion service shall have policies, processes, and procedures to ensure that documents are identified, reviewed, approved, and retained and that records are created, stored, and archived in accordance with record retention policies."

Answers:
Question 35 *B* (5 years for a suspected transfusion reaction evaluation/investigation)
Question 36 *C* (10 years for physician statement releasing emergency blood)
Question 37 *E* (Indefinitely for clinically significant antibodies)
Question 38 *B* (5 years for annual review of policies, processes, and procedures)

End of Case

Suggested Reading

[1] B.H. Shaz, C.D. Hillyer, M. Roshal, C.S. Abrams (Eds.), Transfusion Medicine and Hemostasis: Clinical and Laboratory Aspects, second ed., Elsevier, San Diego, CA, 2013.
[2] R.A. McPherson, M.R. Pincus (Eds.), Henry's Clinical Diagnosis and Management by Laboratory Methods, twenty second ed., Elsevier, Philadelphia, PA, 2011.
[3] S.T. Bennett, C.M. Lehman, G.M. Rodgers (Eds.), Laboratory Hemostasis: A Practical Guide for Pathologist, Springer, New York, NY, 2007.
[4] L. Gordis, Epidemiology, fifth ed., Elsevier, Philadelphia, PA, (2014).
[5] B.B. Gerstman, Basic Biostatistics: Statistics for Public Health Practice, second ed., Jone & Bartlett Learning, Burlington, MA, (2014).
[6] E. Vittinghoff, D.V. Glidden, S.C. Shiboski, C.E. McCulloch, Regression Methods in Biostatistics: Linear, Logistic, Survival, and Repeated Measures Models, second ed, Springer, New York, NY, (2012).
[7] A. Agresti, An Introduction to Categorical Data Analysis, second ed, John Wiley & Sons, Inc, Hoboken, NJ, (2007).
[8] L.M. Friedman, C.D. Furberg, D.L. DeMets, Fundamentals of Clinical Trials, fourth ed., Springer, New York, NY, (2010).
[9] Westgard Q.C. Tools, technologies, and training for healthcare laboratories. Available from: https://www.westgard.com/.
[10] Centers for Disease Control and Prevention. Test complexities. Available from: https://wwwn.cdc.gov/clia/Resources/TestComplexities.aspx.
[11] United States Department of Labor. Family and Medical Leave Act. Available from: http://www.dol.gov/whd/fmla.
[12] Stark Law. Available from: http://starklaw.org.
[13] Cuker, et al. The HIT Expert Probability (HEP) Score: a novel pre-test probability model for heparin-induced thrombocytopenia based on broad expert opinion, J. Thromb Haemost. 8 (12) (2010) 2642–2650.
[14] C.H. Kim, S.C. Simmons, L.A. Williams, E.M. Staley, X.L. Zheng, H.P. Pham, ADAMTS13 test and/or PLASMIC clinical score in management of acquired thrombotic thrombocytopenia purpura—a cost-effectiveness analysis, Transfusion (2017), doi: 10.1111/trf.14230 [Epub ahead of print].

CHAPTER

3

Quality Assurance and Regulatory Issues

Maksim Agaronov, Swati Ratkal**,*
*Vishesh Chhibber***

***Kings County Hospital, Brooklyn, NY, United States;**
****Hofstra Northwell School of Medicine and Northwell Health, Manhasset, NY, United States**

Facilities that collect and/or transfuse blood products must have quality management plans. A focus on quality is critical in order to meet the needs of customers and ensure the safety, purity, and potency of the blood products transfused to patients. Regulatory agencies and accrediting organizations provide oversight for these processes by scheduling inspections and setting up proficiency testing. This chapter reviews quality and regulatory concepts in blood banking and introduces some principles related to disaster management.

1. Which of the following is not a basic element of a quality management system (QMS)?
A. Human resources
B. Customer focus
C. Equipment and process management
D. Minimizing cost of testing
E. Process improvement

Concept: A quality management system encompasses all processes required for an organization to deliver the desired service and consistently meet the needs and expectations of the customers. The three most important activities of such a system are planning, control, and improvement.

Answer: *D*—Minimizing cost of testing is important, but is not a necessary element of a QMS. The following are basic elements of a QMS:

- Organization and leadership—where the roles and responsibilities for the provision of tests, products, and services are clearly defined
- Customer focus (Answer B)—includes customer requirements for their satisfaction and review of feedback for continuing improvement
- Facilities, work environment, and safety—includes safe and clean work environment for employees, emergency preparedness, and communication systems
- Human resources (Answer A)—includes staffing, selection, orientation, training, and competency assessment of employees
- Suppliers and material management—includes supplier qualifications, agreement reviews, inventory management, and acceptance or rejection of materials
- Equipment management (Answer C)—selection, installation, calibration, and maintenance
- Process management and improvement (Answers C and E)—process development and validation studies, quality control program, root cause evaluation, corrective and preventative action plans

Transfusion Medicine, Apheresis, and Hemostasis. http://dx.doi.org/10.1016/B978-0-12-803999-1.00003-1

- Documents and records/information management—standardized document and record creation, periodic review, storage, and retrieval
- Management of nonconforming events—detection, complaint filing, investigations, and immediate actions
- Monitor and assessment—data analyses, review quality indicators, internal and external assessments, and proficiency testing

2. The success of any organization lies in its QMS. The basis of QMS is the concept of the customer and supplier working together for their mutual benefit. Which of the following tools in QMS is effective in achieving customer satisfaction?
A. The complaints from customers should be acted upon twice every year
B. Flexibility in service and uniformity in production
C. Only positive customer feedback should be taken into consideration
D. Meet the changing needs of customers and compromise the quality, if necessary
E. Written or oral contracts and agreements are not necessary

Concept: Customers have a variety of needs and expectations and along with producing high quality products, customer satisfaction should be an ongoing and evolving priority in an organization.

Answer: *B*—Uniformity in services helps to meet customer requirements with ease and fewer errors. The service providers should have flexibility to meet the changing needs of customers and adapt to new requirements.

Customer satisfaction is a constantly moving target and minor modifications to the services may be necessary to meet the changing needs. However, there should not be any compromise in the quality of the products and services provided (Answer D).

Customer feedback through complaint analysis, opinion surveys, and regular contacts ensures that the provider is meeting the expectations (Answer A). Both positive and negative feedback from customers is important in improving the services (Answer C).

Records of events along with corrective and preventive actions should be maintained to serve as an opportunity to improve quality management. The requirements of clients should be adequately defined, documented, and understood. The service provider should have procedures in place to ensure that it has the capability of meeting these requirements before entering into a contract. Revised agreements should be documented and effective communication is the key to success (Answer E).

3. Which of the following concepts in quality is correctly matched?
A. Quality assurance is the measurable aspect of processes or outcomes that provide an indication of the condition or direction of performance over time
B. Quality indicators are measurable aspects of only outcomes which can be used objectively to evaluate a process
C. Quality management involves interrelated processes in the context of an organization and its relations with customers and suppliers
D. Quality indicators include operational techniques and activities that are used to monitor and eliminate causes of unsatisfactory performance at any stage of a process
E. Quality control is a dynamic model that includes retrospective reviews and analyses of operational performance data to detect shifts or trends that require attention

Concept: Quality includes various aspects, such as quality control, quality assurance, quality indicators, and a quality management system.

Answer: *C*—*Quality management* (QM) includes all activities of an organization to attain/maintain quality and ensure customer satisfaction. QM considers processes required for the organization to deliver the desired service and consistently meet the needs and expectations of the customer.

QM encompasses the relationships between the organization, its suppliers and customers, and the management of resources.

Quality assurance (Answer A) is a dynamic model that includes retrospective reviews and analyses of operational performance data to detect shifts or trends that require attention. It includes systematic measurement, comparison with standards, monitoring of processes, and an associated feedback loop that confers error prevention. This can be contrasted with quality control (QC) (Answer E), which is focused on process output. It is a system designed and implemented to ensure that testing is consistently performed in such a way as to yield a product of consistent quality. *Quality control* includes operational techniques and activities used to monitor and eliminate causes of unsatisfactory performance at any stage of a process. The purpose of the QC system is to prevent, detect, and remedy errors in the analytic testing process. Laboratory control procedures must include adequate provisions for monitoring the reliability, accuracy, precision, and performance of laboratory test procedures and instruments.

Quality indicators (Answers B and D) are used to monitor the progress toward stated goals and may be process-based or outcome-based (measurable aspects of processes and outcomes). They provide an indication of direction of performance over time.

4. A formal quality assurance program is a regulatory requirement by which of the following?

A. AABB
B. Commission on Office Laboratory Accreditation (COLA)
C. College of American Pathologists (CAP)
D. Center for Medicare and Medicaid Services (CMS) under Clinical Laboratory Improvement Amendments (CLIA)
E. The Joint Commission (TJC)

Concept: The CLIA regulations address specific QA requirements. Laboratories must establish and follow written policies and procedures for a comprehensive quality assurance program that is designed to monitor and evaluate the ongoing and overall quality of the total testing process. The QA program must:

- Assess the effectiveness of the laboratory's policies and procedures.
- Identify and correct problems.
- Assure the accurate, reliable, and prompt reporting of test results.
- Assure the adequacy and competency of the staff.

Answer: *D*—A formal QA program is required by CMS under CLIA, as well as by the FDA under Code of Federal Regulations (CFR) Title 21.

CLIA laws regulate laboratory testing and require clinical laboratories to be certified by the state and CMS.

AABB (Answer A) is an international, not-for-profit organization, that represents institutions and individuals involved in transfusion medicine and cellular therapies. Its mission is to develop and deliver standards, accreditation, and educational programs that focus on patient and donor care and safety.

COLA (Answer B) is an independent, not-for-profit, physician-driven organization, which accredits more than 7,000 medical laboratories in the United States. It promotes excellence in laboratory medicine and patient care through voluntary education, consultation, and accreditation.

CAP (Answer C) is an organization composed of board-certified pathologists and it serves patients, pathologists, and the public by advocating excellence in the practice of pathology and laboratory medicine through its leadership, inspections, and proficiency testing initiatives.

TJC (Answer E) is an independent, not-for-profit organization, that accredits and certifies more than 20,000 health care organizations and programs in the United States.

5. A batch of 15 patient samples for type and screen are loaded on the automated analyzer. At the end of the run, you note that the positive control on forward typing with anti-A reagent gives erroneous results. This test run was labeled as "control failure." Which of the following steps does NOT represent an appropriate action during such an event?
   A. Check the records for documentation that the correct steps of testing were followed
   B. Report and release the test results, but with a footnote stating failure of the positive control
   C. Check the expiration date of the anti-A reagent
   D. Check the quality control log to see if there was a similar problem in previous runs
   E. Temporarily suspend testing on the automated analyzer and switch to performing type and screens using tube testing until the issue has been resolved

Concept: QC results are used to validate whether the instrument is operating within predefined specifications, inferring that patient test results are reliable. Internal QC is a set of procedures undertaken by the staff of a laboratory for continuously and concurrently assessing the test system and the test results, to decide whether they are reliable enough to be released. It primarily evaluates the accuracy of the method.

Answer: *B*—In event of a failure of internal control, the corrective actions should be taken without delay and the results should not be reported until the error is fixed.

QC records should be created concurrently with the performance of each significant step. This allows retrospective root cause analysis in case of QC failure. This QC log can be electronic or paper-based. The log should identify the name of the test, the instrument, units, the date on which the test is performed, the initials of the person performing the test, and the results for each level of control assayed. Other choices (Answers A, C, D, and E) represented appropriate actions to handle "control failure".

6. Quality control is the backbone of any quality assurance program. Which of the following is true regarding the performance and analysis of quality control?
   A. Quality control records and logs are reported using a Levey-Jennings chart in the paper or electronic-based system
   B. Obtaining quantitative data is the primary objective of quality control analysis
   C. Random error is inherent in every test system and does not have to be investigated
   D. QC data revealing an abrupt change in its mean value is defined as a trend in statistical analysis
   E. External quality control is performed only when internal quality control is suboptimal

Concept: Quality control records provide evidence that critical steps in a procedure have been appropriately performed and that products and services conform to special requirements. QC analysis is an objective assessment of daily operations and detects any deviations, such as trends or shifts, very early. It can be a quantitative or qualitative analysis. Quantitative data, for example, sample dilution, and qualitative data, for example, operator performance, are complementary in a quality control analysis and together provide the best opportunity for detecting errors. Certain errors may be detected by analyzing qualitative data while other errors may be detected by analyzing quantitative data or both.

The director must define and oversee the overall quality control program for the laboratory. This includes goals, policies, procedures, delegation of functions, and regular review by appropriate levels of personnel. Unacceptable QC results must be investigated and corrective actions must be implemented.

Answer: *A*—The quality control results are inspected to assure the quality of the analytical run and this documentation is maintained in a QC log using a Levey-Jennings Chart. The dates of analyses are plotted along the X-axis and control values are plotted on the Y-axis. The mean and standard deviation limits are also marked on the Y-axis. Inspecting the pattern of plotted points provides a simple way to detect random error and shifts or trends in calibration.

Errors can be systematic or random. Systematic error is evidenced by a change in the mean of the control values. Random error (Answer C) is any deviation from an expected result. A trend (Answer D) indicates a gradual loss of reliability in the test system. Trends are usually subtle. Abrupt changes in the control mean are defined as shifts. All errors are significant and have to be acted upon without delay. The results should not be reported until corrective actions have been taken.

External QC methods (Answer E) are performed periodically for accuracy and/or precision with the contribution of an outside institution (referral laboratory, scientific associations, diagnostic industry, etc.). It checks primarily the accuracy of the laboratory's analytical methods. Other terms for external QC are: interlaboratory comparisons and proficiency testing (PT).

7. During routine antibody identification using tube testing, three blood bank technologists obtained different results for the same sample with only one technologist being correct. The medical director ordered competency assessment of the technologists performing the testing. Which of the following about assessing competency is correct?
   A. Competency assessments are performed only by direct observation of activities
   B. Competency assessments include reviewing work records and information system records
   C. Competency assessments are targeted to each individual test performed by the employee
   D. Competency assessments are done semiannually during the first year and every 5 years thereafter
   E. Performance of competency assessments is optional

   Concept: Core competencies represent a set of skills desirable for the broad practice of public health that professionals need to possess as they work to protect and improve the health of their patients. Employees receive an initial orientation to the position and the organization's policies and procedures and are also trained in their new duties. The ultimate result of orientation and training is to make employees competent to independently perform the duties in their job description. To maintain these skills, it is a requirement to conduct regular competency evaluations (Answer E). All the staff members whose activities affect the quality of the laboratory testing and the manufacturing of blood products should be assessed. Management personnel should also undergo competency assessment.

   Answer: *B*—Although training should be provided for each procedure for which an employee is responsible, an assessment need not be targeted to each individual test or procedure performed (Answer C). They can be grouped together to assess similar techniques or methods. The Joint Commission requires analyses of aggregate competence assessment data. For all testing personnel, any/all of the following methods can be used when required, as per CMS:
   - Direct observation of activities (Answer A)
   - Written tests to evaluate problem solving skills
   - Review of work records and information system records (Answer B)
   - The assessments must be performed semiannually during the first year and annually thereafter (Answer D)

8. A validation study demonstrates the capability of testing to achieve consistent and reliable results. Which of the following about validation is correct?
   A. Validation should be done only in a prospective manner
   B. Revalidation is not required after changes are made to a validated process
   C. The only minimum CLIA requirement for test system performance specification is accuracy
   D. Validation documents should include an implementation timeline, results, and interpretations along with any necessary corrective actions
   E. Validation results should be reviewed only by the staff performing the validation study

   Concept: Validation includes system validation, equipment validation, and computer system validation. Each facility should have its own validation plan. Quality oversight personnel should review and approve validation plans before the activities are carried out. A well-documented plan is important for the success of the validation. They should also review the results and corrective actions.

Answer: *D*—The activities performed as part of a validation should be documented and basic elements of such a document are:
- Results and interpretations
- Conclusions and limitations
- Corrective action for unexpected results
- Explanation for any deviations from the validation plan
- Implementation timeline

Prior to carrying out the actual validation testing, there needs to be documentation of how the validation results will be interpreted.

Process validation can be done in a prospective, concurrent, or retrospective manner (Answer A). When no data can be available without performing a live process, a concurrent validation may be carried out. When modifications are necessary, a revalidation of a validated process may be needed based on the nature and extent of the modifications (Answer B).

For a test method validation, the minimum CLIA requirement for such system performance includes accuracy, precision, analytical sensitivity, analytical specificity, and reference interval (Answer C). For new equipment validation, a process should include installation, operational, and performance qualifications. For computer system validation, the interfaces between systems should be validated in the environment in which they will be used and an end-user validation should be performed (Answer E).

Please refer to Question 9 for an explanation of installation, operational, and performance qualifications.

9. A new instrument using gel technology for antibody testing and identification is being validated at a community hospital blood bank. Upon delivery and installation of the instrument, the manager initiates the validation process by reading the operation manual to confirm that the new instrument has been installed according to the manufacturer's specifications. The manager subsequently selects several patient samples that were flagged for inconsistent antibody identification results by different testing methods and tests them using gel technology. The test results meet the criteria provided by the manufacturer. Finally, the manager introduces the new instrument during the day shift and performs the first 15 patient samples requiring antibody testing and identification. What is the correct designation for the equipment validation process described earlier?
A. Operational qualification; performance qualification; installation qualification
B. Installation qualification; operational qualification; performance qualification
C. Performance qualification; operational qualification; installation qualification
D. Performance qualification; installation qualification
E. Installation qualification; Operational qualification

Concept: Validation of an instrument for testing ensures that the results obtained are correct since the necessary steps prior to perform testing on the instrument have been taken. When validating new instruments these qualification steps are installation qualification, operational qualification, and performance qualification.

Answer: *B*—Installation qualification; operational qualification; performance qualification. Installation qualification is met when one can demonstrate that the equipment has been installed according to objective specifications such as those set by the manufacturer.

Operational qualification is met when one can demonstrate that the equipment operates effectively within the established control limits for testing set by the manufacturer.

Performance qualification is met when one can demonstrate that the equipment is able to function during routine daily testing of patient samples. This testing is carried out by the hospital blood bank's staff using their own supplies and following the standard operating procedures established by the blood bank. The other choices (Answers A, C, D, and E) are incorrect based on the definitions earlier and the order of the validation steps.

10. Documents provide how processes are intended to work, where they must be controlled, what their requirements are, and how to implement them. Which of these is true about documents and document management systems?

A. Documents should be developed in a format that primarily conveys information to the management personnel
B. Documents and records must always be on paper
C. Documents should be legible and available in the locations in which they will be used
D. Documents should be verified for adequacy and once created should never be edited
E. Document management systems should retain the most updated version and discard the previous versions

Concept: A few examples of documents include: policies (goals, objectives, and intent); processes (sequence of actions and control points); procedures (work instructions and job aids); forms (templates and records); and labels. In essence, documents state what "should" happen, while records state what "did" happen.

Answer: C—Access to the use of documents should be made available to all staff in the area in which the corresponding job tasks are performed (Answer A). The oversight personnel should monitor the adequacy of documents and verify them. Document management systems provide assurance that documents are legible, comprehensive, and current. The use of standardized formats for documents helps staff know where to find specific elements and facilitates implementation. Documents and records can be paper-based or electronic (Answer B). Documents should be periodically updated and all the previous versions should be retained and retrieved, whenever required (Answers D and E). It is important to protect documents and records from alteration, damage, or unintended destruction.

Records provide evidence that critical steps in a procedure were performed appropriately and that the final results conform to specific requirements. Records are created during each step of performance of a test by the individual performing it and should be accurate and complete.

11. Records are important to maintain as they provide proof that a process occurred and contain information necessary to assess process and service quality. Records should be stored for a designated amount of time. According to AABB Standards, what is the correct minimum retention time for the corresponding record?

A. Donor consents—5 years
B. Physician order to collect blood for autologous use—8 years
C. Inspection of incoming blood and blood components—10 years
D. Parental permission for donation—2 years
E. Donor acknowledgment that educational materials have been read—5 years

Concept: Retention period for records is one of several items, such as creation and identification of records, protection from unauthorized modification, or storage and retrieval of records which need to be addressed in the process of managing them.

Answer: C—The minimum record retention time for the incoming blood and blood component inspection forms is 10 years.

The minimum record retention time for all of the other answer listed is 10 years. Please refer to the current version of the AABB Standards for the minimum retention times required for other documents.

12. The continued success of an organization lies in its well-planned QMS. These activities should be carried out by a person designated to provide quality oversight who reports directly to the executive management. Which of the following is NOT a responsibility of this individual?

A. Establish policies and processes that are applicable to the departments overseen by this individual
B. Provide quality oversight for the operational work performed, except for work that they themselves performed
C. Create SOPs, develop and execute validation plans, and initiate corrective actions
D. Develop a QMS plan that addresses all matters related to the compliance with federal, state, and local regulations and accreditation standards
E. Must provide QMS plans to other departments not under their oversight

Concept: Executive management is ultimately responsible for a facility's QM, but the quality activities are often carried out by an individual designated to provide oversight. The activities should be directed toward continuous improvement in quality, including identifying actions that can anticipate and prevent problems. These activities should be coordinated between the operational and quality teams.

Answer: *E*—The individual designated to provide quality oversight has several responsibilities, such as establishing policies and processes that are applicable to the departments they oversee (Answer A). However, this person does not need to provide the blueprint of the QMS plan to departments that they do not directly oversee. They should develop a QMS plan that addresses all matters related to the compliance with federal, state, and local regulations and accreditation standards (Answer D). Oversight personnel should also develop standard operating procedures (SOPs), carry on validation plans, review suppliers, and monitor/initiate corrective action plans when appropriate (Answer C). However, individuals with dual operational and quality responsibilities cannot provide quality oversight for the operational work they have performed (Answer B). The oversight functions can be shared among existing staff, departments, and facilities, or performed by an outside firm under a contract.

13. A facility is interested in switching from a semiautomated to a fully automated non-FDA approved hematology analyzer. Which of the following correctly describes a test system's performance specification and validation?
A. Accuracy is a level of measurement that yields consistent results when repeated
B. Reportable range is a range of values for physiologic measurement in healthy persons
C. Precision is a level of measurement that yields true and consistent results
D. Analytical sensitivity is the smallest amount of substance in a sample that can be accurately measured by an assay
E. Reference interval is the range of values that a method can directly measure on the specimen without any pretreatment that is not part of the usual assay process

Concept: CLIA requires a minimum performance specification for nonwaived tests that are FDA approved. These are established by the manufacturer and verified before reporting on patients.

For tests that are *not* FDA approved, for laboratory developed tests, or for modified-FDA approved tests, the laboratory must establish test system performance which includes: accuracy, precision, reportable range, reference range, analytical sensitivity, and analytical specificity.

Answer: *D*—Analytical sensitivity is the smallest amount of substance in a sample that can be accurately measured by an assay. Analytical specificity, on the other hand, refers to the ability of an assay to measure only the substance of interest.

Accuracy (Answer A) is a level of measurement that yields true (no systemic errors) results.

Precision (Answer C) is a description of a level of measurement that yields consistent results when repeated. Though accurate/true results are also desired, precise results are not always accurate.

A reference range or reference interval (Answer E) is the range of analyte values physiologically seen in healthy persons.

Reportable range (Answer B) is the range of analyte values that a method can directly measure on the specimen without any dilution, concentration, or other pretreatment that is not part of the usual assay process, and corresponds to the CLIA reportable range.

14. A newly appointed medical director of a 300-bed community hospital is preparing his blood bank for an inspection by the FDA. He realizes that his facility is approved for reimbursement by CMS but is not registered. Which of the following activities performed by the blood bank would require the facility to register with the FDA?

A. Occasionally diverting whole blood derived platelets into a 20 mL syringe for neonatal transfusions
B. Having collected 100 units of whole blood over the past 5 years at times of severe shortages
C. Occasionally irradiating units of red blood cells for pediatric patients
D. Having to rarely pool 3–5 units of whole blood derived platelets into one bag
E. Occasionally converting plasma to recovered plasma to extend the shelf-life

Concept: Whole blood, blood components, and derivatives are biological products which come from living sources and are regulated by the FDA. Registration with the FDA by hospital blood banks or blood centers is dependent upon the scope and extent of manipulation of the blood products performed by these facilities.

Answer: *C*—In order to be able to irradiate blood products, a facility needs to be registered with the FDA and be subject to their inspections. If the hospital receives irradiated blood units from an outside institution and does not have the capability of irradiating units on their own, then they are not required to register with the FDA. Hospital blood banks that routinely collect blood or perform post collection procedures, such as washing, irradiation, leukocyte reduction, deglycerolization, freezing, or rejuvenation of components must be registered with the FDA. The registration must be renewed annually to maintain these privileges. If a facility performs their own infectious disease testing on donated blood samples, then it must register with the FDA as well, unless they are approved by CMS to perform this testing.

The other choices (Answers A, B, D, and E) are regarded as basic preparation activities performed by any transfusion service facility. The facility performing these activities is exempt from the FDA registration requirement, as long as, they are approved for reimbursement by CMS. In contrast to FDA registration, obtaining a FDA product license is required for facilities participating in interstate shipment of blood products. Moreover, a FDA license signifies that the blood is safe, pure, potent, and that the facility where the product was produced meets cGMP standards and undergoes FDA inspections.

15. Your hospital blood bank is anticipating an inspection by the state and federal government. Which agency regulates all the medical laboratories in the United States through CLIA?

A. AABB
B. CAP
C. COLA
D. CMS
E. TJC

Concept: Quality laboratory testing is essential to produce accurate and precise results. CLIA regulations establish quality standards for laboratory testing performed on specimens from humans for the purpose of diagnosis, prevention, or treatment of disease, or for assessment of health. A single government agency in the United States has the authority to regulate medical laboratories through CLIA.

Answer: *D*—CMS regulates all US medical laboratories through CLIA. All laboratories must register with CMS, submit to inspection by CMS, and obtain recertification every 2 years. The law and regulations establish the requirements and procedures for laboratories to be certified under CLIA as both a general requirement and prerequisite for receiving Medicare and Medicaid payments. To be certified, laboratories must have adequate facilities and equipment, supervisory and technical personnel with training and experience appropriate to the complexity of testing, a quality management system, and successful ongoing proficiency testing.

Blood banks and transfusion services can obtain a CLIA certificate through one of the following pathways; (1) certificate of compliance or (2) certificate of accreditation or (3) CMS-exempt status. The other choices (Answers A, B, C, and E) represent accrediting, not regulatory agencies. See the next question for a full explanation of the difference between these terms.

16. On the day of the unannounced CMS inspection, the blood bank staff was surprised to see the affiliation of the group of inspectors. Which of the following is an example of an accrediting organization qualified to perform the inspection for CMS?
A. Food and Drug Administration (FDA)
B. Occupation Safety and Health Administration (OSHA)
C. CMS
D. The Joint Commission
E. Center for Biologics Evaluation and Research (CBER)

Concept: A governing agency, either a part of the state or federal government, controls a laboratory's operations through regulations or laws. The regulations are often enforced via an accrediting agency. The governing agency, such as CMS or one of its designated representatives may conduct an unannounced inspection of the laboratory when there are complaints alleging noncompliance with the regulations. Initial and biennial recertification inspections are made known to the laboratory prior to the date of inspection.

Answer: *D*—The Joint Commission is an accrediting organization that has been approved for having requirements that meet CMS regulations (Answer C) and having deemed status. This implies that a visit and potential findings by these agencies are accepted in place of an inspection by the CMS itself. Accreditation is the process of verifying the laboratory by a recognized accrediting agency and granting permission to perform tests with a defined scope. It is regarded as a reliable indicator of technical competence and also provides formal recognition and an effective marketing tool. These accrediting entities include the AABB, CAP, TJC, American Society for Histocompatibility and Immunogenetics (ASHI), American Osteopathic Association, and COLA.

The accreditation process is voluntary and the entity that the laboratory chooses is based on their personal preference. Laboratories can have either all or part of their testing and calibration activities accredited. Failure to comply with accreditation will not result in sanctions imposed on the laboratory. However, significant actions of noncompliance with regulations of a regulatory agency, such as CMS (Answer C), may result in strict sanctions, such as license revocations or criminal prosecution.

OSHA (Answer B) is a federal agency whose goal is to provide safe working conditions for employees by setting and enforcing standards and providing educational training programs. The Occupational Safety and Health Act of 1970 states that employers must provide their employees with a workplace that is free from recognized hazards likely to cause death or serious physical harm. OSHA has several approaches to reduce occupational fatalities in America's workplaces, which include fair enforcement of standards, proactive education, and compliance assistance to small and medium-sized businesses. Noncompliance with OSHA standards leads to issuance of appropriate citations and fines to the employers.

The FDA (Answer A) is a regulatory agency within the Department of Health and Human Services. The FDA works to ensure the safety, efficacy, and security of human and veterinary drugs, biological products, and medical devices; and also ensures the safety of the nation's food supply, cosmetics, and products that emit radiation. CBER (Answer E) is a regulatory division within the FDA. CBER focus on the regulation of biological products for human use.

17. At the local hospital's quarterly transfusion committee meeting, the usage of blood products during cardiac surgeries was reviewed among the interdisciplinary team. This is an example of what type of assessment and is required by what regulatory agencies or accrediting organizations?
A. External assessment; AABB and TJC
B. External assessment; AABB and CMS

C. Internal assessment; AABB and CAP
D. Internal assessment; AABB, TJC, CMS
E. Internal assessment; TJC, AABB, CAP, FDA

Concept: Assessments are systematic examinations of an organization's processes. The assessment investigates how well the organization is able to comply with the actions necessary to complete the process efficiently and achieve a set goal. The assessments can be carried out by either members within the same organization or by an outside agency.

Answer: *D*—Whenever the blood utilization review committee meets, their evaluations are considered an internal assessment, since it is performed by members within the same organization. Assessments can include quality assessments, peer reviews, self-assessments, or proficiency testing. The transfusion committee is a type of peer review committee and is required by the AABB and TJC for hospital accreditation. It is also required by the CMS for hospitals to qualify for Medicare reimbursement. An external assessment is carried out by accrediting organizations, such as AABB, CAP, COLA, TJC, FACT or regulatory agencies, such as CMS and FDA. These external assessments include inspections, surveys, and audits. All the other choices (Answers A, B, C, and E) are incorrect based on the explanation earlier.

18. All laboratory tests are rated for complexity by the FDA as waived or nonwaived. Which of the following is an example of a waived test?
A. Antibody identification using an 11-cell donor panel
B. Finger stick test for measurement of a donor's hemoglobin level
C. Type and screen
D. Type and crossmatch
E. Red blood cell antigen phenotyping

Concept: Waived tests are simple and need a limited amount of training to perform. Nonwaived tests are classified as being moderate or highly complex.

Answer: *B*—A finger stick hemoglobin level test does not require extensive training, preparation, or interpretive skills to perform and is therefore, considered a waived test. Waived tests are simple tests with an insignificant risk if an incorrect result is reported. Nonwaived tests require personnel with a higher level of training since these tests are moderately or highly complex. All the other choices (Answers A, C, D, and E) include nonwaived tests performed by immunohematology laboratories that are either of moderate or high complexity based on a scoring system, which is centered on the need for training required in order to perform the test and interpret the results.

19. Which of the following is true about waived tests?
A. Extensive training is mandatory to perform these tests
B. Laboratories with a CLIA Certificate of Waiver are always subject to a routine inspection under the CLIA program
C. Laboratories performing only waived tests must have a valid CLIA Certificate of Waiver
D. Annual proficiency testing is a requirement for waived tests
E. Laboratories with a valid CLIA Certificate of Waiver can perform both waived and nonwaived tests

Concept: Laboratories performing only waived tests are subject to less stringent regulations as opposed to laboratories that perform both waived and nonwaived or moderate and high complexity tests.

Answer: *C*—Laboratories that perform only waived tests must register with CMS for a CLIA Certificate of Waiver and they cannot perform any tests other than those listed as waived tests. Laboratories with a CLIA Certificate of Waiver are not subject to a routine inspection under CLIA regulations, but may be surveyed in response to a complaint or if they are performing nonwaived testing without approval (Answers B and E).

Proficiency testing (PT) is not required for any test that is waived (Answer D). PT is required for ongoing laboratory certification of nonwaived testing. However, enrolling in a PT program and performing PT on waived tests will provide an excellent indication of the accuracy of the tests and thus improve the quality of testing provided to the patients. Waived tests are generally less complex than nonwaived tests and do not generally require extensive training to perform (Answer A).

20. Proficiency testing (PT) is a tool used by laboratories to assess the accuracy and reliability of their testing. Which of the following is true about PT?

A. CMS requires laboratories to participate in an approved PT program for CLIA-regulated testing
B. PT is not required for nonwaived tests
C. PT samples require special handling and testing that is different from routine patient or donor samples
D. AABB is the only CMS approved PT program for blood banking certification
E. PT samples are typically sent to a participating laboratory once per year

Concept: Under CLIA and Section 353 of the Public Health Service Act, CMS regulates all US medical laboratories. Therefore, laboratories that perform CLIA-regulated testing have to comply with regulations set forth by CMS including proficiency testing.

Answer: *A*—CMS requires laboratories to enroll in CLIA-approved PT programs which include CAP, Medical Laboratory Evaluation Program (MLE), or Wisconsin State Laboratory of Hygiene (WSLH) among several others that are available. PT is required for nonwaived tests (Answer B) and involves testing unknown samples using routine work processes and conditions. However, unlike patient samples, PT samples cannot undergo repeat testing. PT samples cannot be sent to a reference laboratory or another laboratory with a different CLIA number, even if the two laboratories are within the same organization (Answer C).

In blood banking, examples of nonwaived tests are ABO and Rh grouping, antibody detection and identification, and compatibility testing. PT samples are typically sent to participating laboratories 3 times per year (Answer E). The AABB (Answer D) is not the only CMS approved PT provider. For example, many laboratories utilize the College of American Pathologists (CAP) for PT.

21. A medical director from another CLIA-certified laboratory has contacted you for assistance regarding a currently distributed PT sample. What information can be shared?

A. You can test another laboratory's (which has a different CLIA number) PT sample in your laboratory, but cannot compare the results
B. You can only test your own PT samples and then compare the results with another laboratory (which has a different CLIA number)
C. Laboratories may discuss PT sample test results, but cannot do testing for each other
D. No information can be shared between these two laboratories
E. Laboratories may discuss only the analyzers that they used to obtain the results for a particular PT sample, but not discuss the test results

Concept: The CMS considers proficiency testing a high priority in being able to assess a laboratory's ability to provide accurate/reliable test results. Therefore, the laboratory performing the testing should rely only on their own facility and expertise to complete the testing on PT samples. Thus, only laboratories that share the same CLIA number can communicate about PT samples, or perform testing of PT samples.

Answer: *D*—A CLIA-certified laboratory is restricted from sharing information about a current PT sample or sending the sample to a laboratory with a *different CLIA number* for testing, even if that laboratory is within the same organization. Sending a PT sample to another laboratory is considered PT referral and this action is strictly prohibited by CLIA regulations (Answers A, B, C, and E). Violations can result in serious penalties, such as loss of the laboratory's CLIA certificate for 1 year or suspension of the laboratory's director from running a laboratory for 2 years.

22. You are the blood bank director of an academic medical center that performs platelet crossmatch studies. Today you find out that a testing reagent used in this test is being temporarily taken out of production by the manufacturer. Your standard operating procedure requires this reagent for testing. Without this reagent available, you subsequently send your samples for platelet crossmatch studies to a reference laboratory which uses a different reagent. After a few months of testing, a proficiency test (PT) sample arrives for platelet crossmatch studies. Which of the following is the correct action taken with regard to the PT sample?
   A. Send the PT sample to the same reference laboratory with a different CLIA number that has been performing platelet crossmatch studies for your laboratory in the past few months
   B. Perform all of the testing on the PT sample except for the part that requires the missing reagent which can be done by another laboratory
   C. Acquire the reagent used by the reference laboratory and perform the test on the PT sample at your laboratory
   D. Send the PT sample to a laboratory with a different CLIA number that is still performing the platelet crossmatch studies according to your laboratory's standard operating procedure with the same testing reagent that your laboratory ran out of
   E. Acquire the same testing reagent you previously used from another laboratory to perform the PT sample testing entirely in your laboratory

Concept: Proficiency testing is an important measure of a laboratory's performance, since correct results obtained on the PT samples verify that the laboratory is producing accurate and reliable results for their patient samples as well. Therefore, it is critical for a laboratory to treat PT samples as they would treat a routine patient sample and must never send a PT sample to another laboratory for testing under any circumstances.

Answer: *E*—Acquiring the same testing reagent used by your laboratory from another laboratory, with the same or different CLIA number, is allowed as long as testing on the PT sample is entirely performed by your laboratory and there is no sharing of the results prior to the PT program's event cut-off date.

PT referrals are represented by the choices in Answers A, B, and D, and are thus not allowed. Regulations governing PT under CLIA state that a laboratory is prohibited from sending PT samples or any portion of the PT samples to another laboratory with a different CLIA number for testing, even if it would normally send a patient's specimen to that same laboratory for testing.

Since your laboratory's standard operating procedures require a different testing reagent from the one that the reference laboratory is using to test the PT sample, you should not acquire the reagent used by the reference laboratory (Answer C).

For PT, reflex testing and confirmatory testing are prohibited unless they are performed by the same laboratory that performed the initial test, are included in that laboratory's standard operating procedure, and the results are reported as part of the proficiency testing program.

23. Which of the following scenarios best describes a market withdrawal?
   A. An apheresis platelet donor calls the day after her donation to report that she had taken aspirin in the morning prior to donation
   B. A hospital transfusion service is made aware by the manufacturer of the antibody screening cells that it is an impure product and cannot be used
   C. A manufacturer of a blood infusion set notifies its end-users that the filter contains inappropriately placed filter material which may be infused into the patients
   D. A regional blood center retrieves a donor's donation history to investigate a possible case of transfusion-transmitted hepatitis
   E. A regional blood center reports lack of assurance that plasma collected over the past 3 days was done in an aseptic manner and the product has already been distributed

Concept: Hospitals and other organizations receiving blood and blood components from blood centers rely on the blood centers to promptly report any products that deviate from collection, processing, and/or production rules, standards, specifications, or cGMP regulations. The type of method used to regain control of the product that has already been distributed is either through a product recall *or* a market withdrawal. Assessing the severity of the violation and the impact it may have on the products' recipients will help to categorize the event.

Answer: *A*—An apheresis platelet donor informing the donor center the following day after donation about an antiplatelet drug, an event that the donor center could not have known prior to donation unless the donor disclosed it to them. This platelet product is therefore considered to be in minor violation of the law and market withdrawals are required of such products. Products that undergo market withdrawal typically pose a low risk to the recipient and the event is not subject for reporting to the FDA. Postdonation information from donors are often beyond the control of the blood center and are the most common reasons for market withdrawals of blood components.

Three of the other choices (Answers B, C, and E) are scenarios depicting products that are in violation of the law in the CFR, and are therefore subject to legal action. The method of removal of these products from the market is by product recall which applies to distributed products only.

An efficient quality management system for timely recall of the products known or suspected to be nonconforming is of utmost importance. There should be written protocols, with all activities described in detail and responsibility of those involved in the process precisely defined. They should be regularly reviewed, revised, and approved by the manager or medical director.

Product recalls involve about 1 in 5800 blood components in the United States and are much less common than market withdrawals. Product recalls are classified based on severity with Class I recalls involving products that may cause serious or fatal outcomes, Class III recalls are products that are not likely to cause adverse health consequences, and Class II recalls are for products that may cause temporary adverse effects or rare serious adverse consequences. Product recalls are reportable to the FDA by the blood supplier. Some of the product recalls identified by the FDA include improper collection sterility and arm preparation, incorrect product storage temperature, failure to produce the product according to cGMP, or release of products before completion of all testing, and testing error that results in release of an incorrect unit.

An example of a lookback is represented by Answer D. Lookbacks are a review of a donor's donation history to ascertain whether there are products from previous donation that would require quarantine, further testing and/or notification of transfusion facilities and recipients.

24. A blood center has been cited for errors in labeling their products incorrectly. Which of the following events associated with labeling is not a biological product deviation (BPD) and therefore, does NOT need to be reported to the FDA?

A. A red blood cell unit with no volume on the label but the weight is present on the label
B. A platelet unit is labeled as CMV negative although the product was never tested for CMV
C. A granulocyte unit is labeled with a 3-day expiration date from time of collection
D. A frozen red blood cell unit that was not labeled as deglycerolized red blood cells post thawing
E. A fresh frozen plasma unit that listed the correct expiration date but did not have a collection date on the label

Concept: The FDA's key regulations on blood are found in Title 21 of the CFR. Under 21 CFR 606.171, blood and plasma establishments are required to report adverse events associated with the manufacturing, which includes testing, processing, packing, labeling, storing, or distribution of blood or a blood component. If an event which affects the safety, purity, or potency of a product that has been distributed has occurred, then this event is a BPD and needs to be reported to the FDA.

Answer: *E*—A fresh frozen plasma unit that is labeled with the right expiration date but does not show the collection date is not a reportable event because the safety, purity, or potency of the product is not affected. Events associated with labeling that are not required to be reported to the FDA include a product labeled with a shortened expiration date, a product labeled lacking the name of the collection facility or the collection date, as long as the expiration date is printed correctly.

The other choices (Answers A, B, C, and D) are labeling events which require reporting because either critical information is missing, product label indicated testing was performed although the product was never tested, the label revealed a wrong expiration date, the label displayed an incorrect product name, or the product name is missing. A BPD needs to be reported as soon as possible and within 45 calendar days from the time it is identified.

25. A red blood cell unit that expired the previous day was issued by the hospital transfusion services and the unit was transfused without any events. What is the best classification for this event?

A. Disaster
B. Nonconformance
C. A missed event
D. Adverse event
E. Accident

Concept: The services provided by a hospital blood bank must fulfill specific requirements, which may be defined by federal law or by practice standards established by the facility or accrediting agencies (e.g., AABB). A blood bank that does not follow standard operating procedures and deviates from written procedures is not in compliance with these requirements.

Answer: *B*—The red blood cell unit that expired should have been discarded and never issued according to regulations/policies for expired blood products. This event is considered a nonconformance, as it is a failure to meet established requirements. Though it may have been an "accident" (Answer E), the hospital must have policies and procedures in place to prevent such occurrences. The severity of the incident does not meet the criteria for a disaster (Answer A).

A missed event (Answer C) would have occurred if the unit was detected prior to being transfused to the patient. In contrast, the case mentioned earlier represents a near-missed event, since it did not adversely affect the outcome, but could have resulted in a potentially serious adverse event.

An adverse event (Answer D) would have occurred if the patient who received the expired unit had a complication directly related to the transfusion. Adverse events may occur in relation to donation, a transfusion, or a diagnostic or therapeutic procedure.

26. The hospital's blood bank transfusion safety officer is reviewing the adverse events reported to the FDA at her institution over the past 5 years. Which of the following reportable events is also considered a sentinel event?

A. The donor room staff failed to use aseptic technique prior to collection of a whole blood unit resulting in having to discard the unit
B. The blood center accidentally distributed a collected red cell unit with repeatedly reactive viral markers but the mistake was identified by the hospital blood bank before release
C. Transfusion of a platelet unit that was expired with no adverse patient outcome
D. A batch of red cell units had to be discarded as the storage refrigerator temperature reached above the 1–6°C range for 4 h over the weekend
E. Transfusion of group A blood to a group B patient resulting in severe hemolysis and a prolonged stay in the intensive care unit but eventually the patient recovered

Concept: The FDA expects notification from all blood establishments when deviations from cGMP or established specifications have occurred which may affect the safety, purity, or potency of a biological product. The determination of whether this deviation from accepted policies, processes, and procedures is a sentinel event or a nonconforming event depends predominantly on whether or not it is associated with patient injury.

Answer: *E*—The Joint Commission defines a sentinel event as a patient safety event that reaches the patient and results in either death or permanent harm or temporary harm. A hemolytic transfusion reaction resulting from a major incompatible blood group transfusion, regardless of whether an injury to a patient occurs is an exception to the earlier definition and is considered a sentinel event by TJC. All sentinel events need to be reviewed by the hospital or the facility where it occurred and a root cause analysis needs to be performed followed by an action plan to prevent this event from occurring again. This comprehensive systematic analysis and action plan describing the hospital's risk reduction strategies postsentinel event need to be submitted within 45 business days of the event or becoming aware of the event. Failure to submit this report within an additional 45 days (90 days total) of its due date may result in loss of TJC accreditation.

The other choices (Answers A, B, C, and D) are nonconforming events that could have become sentinel events if the affected blood products were transfused and led to adverse patient outcomes. Conformance is defined as fulfillment of requirements as indicated by standards. These standards can be policies, processes, or procedures established by the facility, AABB, or other applicable regulations. Nonconforming events are deviations from these accepted standards.

27. Shortly after receiving a third unit of red blood cells, a 95-year-old male with congestive heart failure had severe respiratory distress and succumbed to respiratory failure. What is the correct FDA-mandated protocol, initial and follow-up, for reporting fatalities related to collection or, transfusion of blood, cellular or tissue-based products?

A. As soon as possible followed by a written report within 3 days, including measures taken to avoid recurrence
B. As soon as possible followed by a written report within 7 days
C. As soon as possible followed by a written report within 7 days including measures taken to avoid recurrence
D. Report within 48 h followed by a phone call or written report within 7 days
E. Report within 48 h followed by a phone call or written report within 7 days, including measures taken to avoid recurrence

Concept: The current good manufacturing practice (cGMP) regulations for blood and blood components require that fatalities related to blood transfusion or collection are reported to the FDA's CBER. CBER wants to be aware of the fatality so that an early and quick assessment can be made regarding the potential risk to the public using the blood products. CBER can then make recommendations or intervene in the distribution of the blood supply, if necessary, to avoid a health risk to the public. CBER investigates transfusion related fatalities to exclude any other plausible causes other than transfusion. CBER's goal is to continuously decrease the risk associated with transfusion, as well as blood collection.

Answer: *B*—FDA's CBER division should be notified of a transfusion related fatality as soon as possible, preferably within the first 24 h, either by email, which is the preferred method, fax, telephone, or express mail. CBER has a specific email address and telephone number dedicated to fatality reporting (http://www.fda.gov/BiologicsBloodVaccines/SafetyAvailability/ReportaProblem/TransfusionDonationFatalities/). A 7-day follow-up written report sent either by email, fax, or express mail is required. There is no required form or format to use but specific details, such as your name, institution name, date of transfusion, blood component implicated, name and address of facility providing the blood, and brief description of the events surrounding the fatality are some of the details that are helpful for the ensuing investigation. The 7-day report should provide additional findings or conclusions, if available, from the institution's own investigation. The other choices (Answers A, C, D, and E) do not represent the correct timeframe and/or sequence of events.

28. A junior blood bank technologist issued 1 unit of red blood cells without a crossmatch even though the patient had a positive antibody screen. The problem was discovered after a few days when a suspected delayed hemolytic reaction was being investigated. The technologist along with the rest of the team created a diagram looking at potential problems with materials, methods, environment, or human factors that may have contributed to this error. Which of the following best depicts the resource used by the blood bank team?

A. Run chart
B. Pareto chart

C. Control chart
D. Cause and effect diagram
E. Repetitive why

Concept: Root cause analysis helps assesses a process and attempts to identify faults in the process to make it a better one. Finding and understanding the reasons why a problem occurred will potentially prevent it from happening again. This will reduce the potential for error committed by individuals who are carrying out the process. Identification of problems and their causes requires a team to analyze the problem by using various resources to ideally depict the scope, relative frequency, and potential effects of the problem on the organization's operations.

Answer: *D*—A cause and effect diagram, also known as Ishikawa diagram, illustrates all the major component parts of a process, such as equipment, materials, methods, and human factors. It shows what problems can arise within each component and how they can influence the outcome. In this case, the technologist was working in the night shift and there is inadequate staffing during the night, resulting in him feeling rushed and committing an error.

A Pareto chart (Answer B) is a bar graph which can effectively display the relative frequencies of a problem's causes from highest to lowest. The longest bars are on the left side of the chart and the shortest are on the right. Pareto charts along with process flowcharts, repetitive why, and cause and effect diagrams are used in root cause analysis to identify the cause of the problem. The repetitive why (Answer E) is when one repeatedly asks, "Why did this happen?", until all the known information related to the cause is obtained. A control chart (Answer C) is a tool used in quality control to analyze the variables in a process and monitor the effect of these variables on performance of a process. A run chart (Answer A) is a graph of data over time and helps to assess the effectiveness of a change in the process. A run chart is similar to a control chart but it does not show the control limits of a process.

29. During a major power outage, a blood bank technologist issued a unit of blood without a requisition form. The error was discovered on the following day during the morning administrative rounds in the blood bank. The blood bank team performs a root cause analysis and implements corrective action. Which of the following is an example of corrective action?
A. The technologist is suspended for 2 weeks and told to review the standard operating procedures
B. The technologist is given a quiz on various case scenarios to assess his knowledge of standard operating procedures
C. The technologist is told to shadow a senior technologist for 2 weeks
D. The team's root cause analysis revealed that the error partly occurred due to the lack of staff during the night, so the hospital hires two additional technologists for the night shift
E. The technologist was trained on where to find a paper blood requisition form and how to fill it out

Concept: The QMS of the hospital transfusion services should include a process for detecting, investigating, and responding to the events that result in deviations from accepted policies, processes, and procedures. The technologist deviated from the established policy and created this nonconforming event. For each of these events, the facility should define how to implement preventive and corrective actions and how to evaluate their effectiveness.

Answer: *E*—Training the technologist on how to proceed during a power outage and where to find the requisition forms is a corrective action. Corrective action is defined as an activity performed to eliminate the cause of an existing nonconformance or other undesirable situation in order to prevent recurrence. Since the corrective action plan is a significant component of root cause analysis, it needs to be documented and reviewed.

Hiring additional technologists (Answer D) would be considered a preventative action, since it would allow for sharing of work, additional expertise, and more time for handling any single blood order. Preventive action is defined as an action taken to reduce the potential for nonconformances or other undesirable situations.

Remedial actions (Answers A, B, and C) which are defined as alleviating the symptoms of existing nonconformances. By having the technologists take a quiz, review the standard operating procedure, or shadow a senior technologist, the hope is that the technologists can acquire the skills necessary to perform their daily work tasks, which includes having a requisition form filled out when issuing blood products. However, the actual cause of the problem, such as the technologist doesn't know where the requisition form is or how to fill it out, is not addressed.

30. A new Rh reagent is purchased by your blood bank which is reported to be more accurate in identifying patients with D variants. A validation study at your blood bank reveals the old Rh reagent to be superior in identifying D variants. You continue using the old reagent for patient samples and temporarily remove the new reagent from testing. After contacting the manufacturer and reviewing the test procedure, you realize that your technologists are using only one drop instead of the recommended two drops of this new reagent. You immediately revise the standard operating procedure and over the next few weeks conduct daily short meetings to inform all the different work shifts of this new procedure. Which of the following answer correctly pairs the action taken by the supervisor?

A. Preventive; immediately revising the standard operating procedure to reflect the recommended two drops when using the new Rh reagent
B. Remedial; continue using the old Rh reagent for patient samples until the issue with the new Rh reagent is resolved
C. Corrective; switching from the old Rh reagent to the new Rh reagent to identify more weak D patients
D. Remedial; immediately revising the standard operating procedure (SOP) to reflect the recommended two drops when using the new Rh reagent
E. Corrective; continue using the old Rh reagent for patient samples until the issue with the new Rh reagent is resolved

Concept: Evaluation and analysis of nonconformances resulting in near-miss events or actual adverse events are necessary steps for identifying the problem that led to the event. Once the issue is known, the next step is to correct it and put measures in place to prevent it from occurring again.

Answer: *B*—Remedial action is taken with the goal of minimizing the damage of any existing nonconformance. It is a reactive approach to an existing problem in order to improve the current outcome. By temporarily using only the old reagent until the issue with the new reagent was resolved, the supervisor created more time to figure out the problem and avoided missing additional patients with D variants. Once the problem is fully elucidated, the SOP will certainly need to be changed to reflect the proper procedure for the new agent (Answer D), but this would not solve the immediate issue.

Preventive action (Answer A) reduces or eliminates the potential for an error or an undesirable situation from happening when the problem is still nonexistent. By introducing the new reagent to the blood bank, the supervisor was hoping to identify more D variant patients that may have been missed by the old reagent.

Corrective action (Answers C and E) aims to eliminate the direct cause of a nonconformance with the goal of preventing it from occurring again.

31. The hospital transfusion service is required to have a process in place to handle, store, and transport blood, blood components, and critical materials in a way that prevents damage, limits deterioration, and meets AABB Standards requirements for storage, transportation, and expiration. Which of the following products is not part of this standard?

A. Anti-A, Anti-B, Anti-AB, Anti-D
B. Antibody screening cells
C. Blood bank refrigerator

D. Automated cell washer
E. AHG testing reagents

Concept: Critical equipment, supplies, and services are defined as components which have the potential to impact the quality, safety, efficacy, or purity of products or services. Incoming reagents and critical supplies must be received, inspected, and tested prior to using them, given their potential impact on the outcome of services provided or final product.

Answer: *D*—Cell washing is a necessary component in carrying out routine and complex blood bank testing; however, the washing step can be performed manually and an automated cell washer does not have to be used; therefore it is considered a noncritical component.

The other choices (Answers A, B, C, and E) represent critical reagents and equipment that are essential for being able to provide blood bank services and must be functioning properly, readily available, and used as intended by the manufacturer unless another method has been validated and approved by the medical director.

32. You are a medical director of a small regional blood center which just experienced major flooding and subsequent damage to your donor center after a hurricane. You are in charge of leading a team to carry out your blood center's disaster management program at this critical time. What is an example of a mitigation strategy within your disaster management program?
A. Given your facility's critical role in relief efforts, you request an emergency electrical generator from the city
B. You increase the number of blood drives in anticipation of increased demand for blood
C. You organize an emergency team that meets every 4 h to update the needs of your customers, status of your employees, and conduct field operations
D. You draft a proposal to have quarterly drills, which simulate actions to be taken during a potential disaster
E. You move your file cabinets with donor records from the basement to the first floor and install more drains in the basement

Concept: For an organization, a disaster may be defined as a serious event that disrupts the routine ways that the organization functions, in such a way that is beyond what the organization can handle on their own. The way an organization responds to a disaster depends in part on how prepared they were before it occurred. To help in preparation, a disaster management plan should be developed and distributed within the organization. This plan includes strategies or efforts, such as mitigation, preparedness, response, and recovery which allow for more efficient operations during a disaster.

Answer: *E*—Mitigation efforts include making permanent changes, such as with a building's structure, physical property within the facility, or environment directly outside the building. These strategies aim to provide more safety to the organization's employees as well as protect the physical property within the facility. For example, building more drains, performing quality and safety checks on equipment, creating a new shelter for employees, or protecting vital records by moving file cabinets to a relatively flood-safe level are examples of mitigation strategies.

Requesting an emergency generator (Answer A) refers to recovery or restoring a critical component of a facility's operations, such as electrical power which is certainly necessary to run a blood center. Recovery is the last step in the disaster management cycle and during this step, more ideas for mitigation strategies are identified. Organizing an emergency team (Answer C) refers to a response effort in which you maintain communication with your employees and customers to continue critical operations and regain some form of stability. Drafting a proposal for quarterly drills (Answer D) refers to a preparedness plan whereby the organization routinely simulates an unexpected event through drills to reinforce what needs to be done as well as to identify inconsistencies or lack of critical measures that need to be corrected. Increasing the number of blood drives (Answer B) is another example of preparedness.

33. During a natural disaster, large amounts of blood products may be required. Which of the following is true regarding testing blood products in case of catastrophic events whereby blood is needed immediately to save lives?

A. Full panel of routine donor tests for infectious diseases is always required
B. Infectious disease testing can be done retrospectively in certain circumstances
C. Testing blood for infectious diseases is required from first-time donors but not from your repeat donors
D. Testing should be only done to determine the donor's ABO and Rh type
E. Testing of plasma and platelet products for infectious diseases but not red cell products is always required

Concept: A blood center is fully committed for making the blood supply as safe as possible by continuously adding sensitive and specific infectious disease tests as part of donor testing. However, an exception to full testing is acceptable in certain circumstances that make holding back on releasing blood products life-threatening.

Answer: *B*—When blood is in critical need for keeping patients alive, blood supplies are exhausted and resupply is not possible, full testing for infectious disease agents is not required (Answers A and C). However, samples of donor blood should be saved for testing as soon as the circumstances of the disaster are under control. The blood center should attempt to do as much donor testing as time allows including donor ABO and Rh type (Answer D). If the blood can be tested just for a limited number of infectious disease agents then it should be done. The physicians who ordered the blood products must be notified of what tests have and have not been completed on the distributed units of blood products. Once testing is performed, any positive results of infectious disease markers must be communicated to the ordering physicians. The testing is not limited to plasma and platelets since all products have some infectious risk (Answer E).

34. The blood bank director along with his team of managers and supervisors has created an emergency communications plan (ECP). Which of the following individuals or organizations does not need to be aware of this ECP?

A. Medical Director
B. Overnight lead technologist
C. Blood center distribution division
D. Immunohematology reference laboratory
E. Hospital's Department of Radiology

Concept: Communication is a key element of managing an emergency and must allow for critical information on supplies, personnel, or even building structural stability to be clearly and quickly shared among all of those involved. An emergency communications plan (ECP) should be developed by organizations to make sure that their staff can communicate effectively both internally and externally. If there are more staff members who know and understand the ECP, a critical event can be managed more efficiently.

Answer: *E*—The hospital's Department of Radiology is not part of the external audience, which routinely uses blood products so it does not need to be necessarily made aware of the blood bank's ECP. All the other choices (Answers A, B, C, and D) represent key personnel/agencies that need to be aware of the ECP.

All the members of a blood bank staff need to be included in the ECP and all employees need to receive information on how the organization is responding to the disaster, as well as plan for the future. The employees need to know the operational status of the blood bank and when to report back to work. The blood bank staff members are the internal audiences for the ECP and external audiences include vendors and blood centers who have an ECP of their own (for interacting with the blood bank which relies on them).

35. You are a Medical Director of a major trauma center's hospital blood bank and your region has experienced a severe winter snowstorm creating a state of emergency. You anticipate your use of blood products to increase. According to the AABB, how should requests for blood from your blood supplier be made and carried out in a disaster?

A. Call your blood center and ask that it is mandatory for your order to be filled since you are a major trauma center

B. Reach out to major blood centers in states which are not affected and make your blood order requests directly with them

C. The hospital will make the initial contact with the blood center to relay their current inventory and potential need for blood products

D. Start an immediate blood drive at your hospital and create a temporary donor room

E. The blood center will make the initial contact with the hospital to assess their need for blood

Concept: There needs to be a central organization that assesses the need for blood products based on the nature of the event, injuries expected, and current inventory at the blood center. The AABB's Interorganizational Task Force (ITF) is an organization that was formed in January 2002 to make sure that blood collection efforts after a disaster are managed properly and to make sure that safe and adequate blood product inventories are in place at all times.

Answer: *E*—The hospital needs to determine their immediate and short-term need for blood after a disaster. The blood center whose hospital or customer is affected will contact the hospital blood bank to determine the impact of the event, such as types of injuries expected, current group O RBC's in inventory, and the number of current and expected hospital admissions. If there is an immediate need for blood products, the blood center will distribute the blood from its inventories to the hospital.

The blood center needs to contact the ITF to report on current and expected hospital admissions and the blood center's group O RBC inventory levels. The ITF will then assess the situation and determine the need for blood in excess of what is available at the affected blood center. If there is a need for blood, then the ITF will coordinate the immediate transport of blood between blood centers. The hospital will maintain communication with the blood center on their inventory levels and need for blood throughout the disaster.

The other choices (Answers A, B, C, and D) are inappropriate since the situation represents a widespread emergency and the actions represented in those choices would cause more stress on the system as a whole.

Suggested Reading

[1] M.K. Fung, et al. Technical Manual, eighteenth ed., AABB Press, Bethesda, MD, (2014).

[2] Standards for Blood Banks and Transfusion Services, thirtieth ed., AABB Press, Bethesda, MD, 2016.

[3] B. Shaz, et al. Transfusion Medicine and Hemostasis: Clinical and Laboratory Aspects, second ed., Elsevier, London, UK, (2013).

[4] C.D. Hillyer, et al. Blood Banking and Transfusion Medicine: Basic Principles and Practice, second ed., Elsevier, Philadelphia, PA, (2007).

[5] Disaster Operations Handbook, vol. 2.0, AABB Press, Bethesda, MD, 2008.

[6] Guidance for Industry: Biological Product Deviation Reporting for Blood and Plasma Establishments. Food and Drug Administration. Center for Biologics Evaluation and Research. Available from: http://www.fda.gov/BiologicsBloodVaccines/GuidanceComplianceRegulatoryInformation/Guidances/Blood/ucm073455.htm.

[7] Sentinel Events. Comprehensive Accreditation Manual for Hospitals. The Joint Commission. Available from: http://www.jointcommission.org/assets/1/6/CAMH_2012_Update2_24_SE.pdf.

[8] Clinical Laboratory Improvement Amendments (CLIA). Available from: https://www.cms.gov/Regulations-and-Guidance/Legislation/CLIA/index.html.

[9] COLA's Insights into Proficiency Testing and Transfusion Services. Available from: http://newsletter.cola.org/2013/November/insights.pdf.

[10] Part 493 Laboratory Requirements. Electronic Code of Federal Regulations. Available from: http://www.ecfr.gov/cgi-bin/text-idx?rgn=div5&node=42:5.0.1.1.9.

38. Noura is Medical Director of a major trauma center's hospital-based bank and your region has experienced a severe winter snowstorm creating a state of emergency. Your hospital anticipates use of blood products to increase. According to the AABB how should requests for blood from your blood supplier be made and carried out in a disaster?

A. Call your blood center and ask that it is mandatory for your order to be filled since you are a major trauma center.
B. Reach out to a few blood centers in states which are not affected and make your blood order requests directly with them.
C. The hospital will make the initial contact with the blood center to relay their current inventory and potential need for blood products.
D. Start an immediate blood drive at your hospital and create a temporary donor room.
E. The blood center will make the initial contact with the hospital to assess their need for blood.

Concept: There needs to be a central organization that assesses the need for blood products based on the nature of the event, injuries expected, and current inventory at the blood center. The AABB's Interorganizational Task Force (ITF) is an organization that was formed in January 2002 to make sure that blood collected before or after a disaster are managed properly and to make sure that safe and adequate blood product inventories are in place at all times.

Answer: E—The hospital needs to determine their immediate and short-term need for blood after a disaster. The blood center whose hospital or customer is affected will contact the hospital's blood bank to determine the impact of the event, such as types of injuries expected, current group O RBC inventory, and the number of current and expected hospital admissions. If there is an immediate need for blood products, the blood center will distribute the blood from its inventories to the hospital. The blood center needs to contact the ITF to report on current and expected hospital admissions and the blood center's group O RBC inventory levels. The ITF will then assess the situation and determine the need for blood in excess of what is available at the affected blood center. If there is a need for blood, then the ITF will coordinate the immediate transport of blood between blood centers. The hospital will maintain communication with the blood center on their inventory levels and need for blood throughout the disaster.

The other choices (Answer A, B, C, and D) are inappropriate since the situation represents a widespread emergency and the actions represented by those choices would cause an overwhelming of the system as a whole.

Suggested Reading

[1] M.K. Fung, et al. Technical Manual, eighteenth ed., AABB Press, Bethesda, MD, 2014.
[2] Standards for Blood Banks and Transfusion Services, [illegible] ed., AABB Press, Bethesda, MD, [illegible].
[3] B. Shaz, [illegible] Transfusion Medicine and Hemostasis: Clinical and Laboratory Aspects, [illegible] (2013).
[4] [illegible] Transfusion Medicine [illegible]
[illegible]

CHAPTER

4

Blood Donation and Collection

Shanna Morgan*,**, Salima Shaikh†

*American Red Cross, Saint Paul, MN, United States; **University of Minnesota, Minneapolis, MN, United States; †Blood Centers of the Pacific, San Francisco, CA, United States

There are many steps involved in protecting the safety, purity, and potency of the blood supply, while at the same time ensuring both donor and recipient safety. The donor health questionnaire is designed to help evaluate the donors' health history and capture broad health, lifestyle, and disease risk questions that may impact currently recognized safety risk factors. Other aspects of the blood donation process include a brief examination of vital signs and hemoglobin. Policies are subject to revision by the FDA as new risk factors emerge and others are reassessed. At any given time, recommendations are based off the best available scientific evidence that would maintain or improve the safety of blood. An important note is that the criteria for allogeneic, autologous, and directed donation may vary. This chapter focuses on blood donation and collection and will review important concepts of eligibility. Tables are included throughout the chapter to help consolidate donor criteria into helpful categories.

1. Which of the following donors is acceptable for allogeneic donation today?
   A. A 16-year-old male athlete, height 5′6″, weight 130 lbs, hemoglobin 11.0 g/dL by earlobe puncture, BP 140/88, Pulse 47 beats per minute (bpm)
   B. A 17-year-old male, height 5′7″, weight 140 lbs, hemoglobin 11.0 g/dL by venous sample, BP 135/85, Pulse 101 bpm
   C. A 18-year-old male, height 5′8″, weight 170 lbs, hemoglobin 11.0 g/dL by fingerstick puncture, BP 120/83, Pulse 85 bpm
   D. A 39-year-old female, height 5′4″, weight 120 lbs, hemoglobin 12.5 g/dL by earlobe puncture, BP 115/75, Pulse 75 bpm
   E. A 79-year-old female, height 5′8″, weight 135 lbs, hemoglobin 12.5 g/dL by fingerstick puncture, BP 95/55, Pulse 55 bpm

   Concept: In addition to the donor history questionnaire, a patient's age, vital signs, and hemoglobin level are evaluated to determine donation eligibility.

   Answer: *E*—There is no upper age limit for blood donation, as long as the donor is in good general health, feeling well, meets all donation criteria, and has no restrictions or limitations to activities. Therefore, the donor in Answer E meets all the requirements for donation.

   In general, one must be at least 17 years old to donate; however, some states allow a 16-year-old to donate if they have parental/guardian consent (Answer A). Donors must weigh at least 110 lbs, according to most blood center regulations. The hemoglobin threshold for acceptable blood donation varies from nation to nation. In the United States, the minimal hemoglobin level of 13.0 g/dL for males and 12.5 g/dL for females was established in for the purpose of ensuring donor safety and also to ensure collection of a potent product (Answers A, B, and C). Predonation qualification of

Transfusion Medicine, Apheresis, and Hemostasis. http://dx.doi.org/10.1016/B978-0-12-803999-1.00004-3

hemoglobin levels should be obtained by either fingerstick or venous sampling. The determination of hemoglobin levels by earlobe puncture (Answers A and D) method is unreliable, as samples give higher values compared to fingerstick or venous sampling, and it is therefore no longer considered an acceptable method. There are additional requirements for hemoglobin levels, height, and weight that apply for double red cell donations and for donors 18 years old and younger. Blood pressure requirements state that the donor is acceptable as long as the systolic is below 180 mmHg and the diastolic is below 100 mmHg at the time of donation. Medications for high blood pressure do not disqualify a donor. Low blood pressure is acceptable if it is at least 80/50 mmHg, though some blood centers require a minimum of 90/50, and the donor must feel well. The acceptable range for pulse is 50–100 bpm (Answers A and B); though a pulse of less than 50 bpm is acceptable if the donor is an athlete who participates in regular exercise.

2. Which of the following donors is eligible for autologous donation?
A. Male donor with a hemoglobin 10.8 g/dL
B. Female donor with fever and chills who is taking antibiotics for urinary tract infection
C. Female donor who would like one autologous red blood cell unit to be available for her surgery tomorrow
D. Male donor with a hemoglobin 11.5 g/dL
E. Female donor who does not complete the donor history questionnaire

Concept: A patient that does not wish to receive blood from another person may opt to donate for themselves before a surgical procedure. This is termed autologous donation. Though autologous blood avoids the risk of foreign antigen exposure, certain precautions are still necessary to provide a safe product. Additionally, logistical coordination is required to ensure that the patient actually receives their donation, as opposed to an allogeneic unit from a volunteer donor. Another type of donation is directed donation, where the donation is considered allogeneic, but they wish the unit of blood to be used for a specific person, typically a family member (Table 4.1 describes the criteria for autologous donation).

TABLE 4.1 Eligibility, Testing, and Labeling Requirements for Autologous Donation

Physician order	Required
Hemoglobin or hematocrit	Minimum 11 g/dL and 33%
Pregnancy	Not a contraindication
Frequency	For preoperative, all donations must be completed > 72 h before anticipated surgery
Condition with risk of bacteremia	Defer
ABO and Rh typing	Performed by the collecting facility and confirmed by the transfusing facility, if different from the collecting facility
Infectious disease testing	HBsAg, anti-HBc, NAT for HBV DNA, anti-HCV, NAT for HCV RNA, anti-HIV-1/2, NAT for HIV-1 RNA, anti-HTLV-I/II, NAT for ZIKA RNA and syphilis must be performed on at least the first unit collected during a 30-day period, if unit is to be transfused outside the collecting facility
Labeling	"Autologous Donor," on tie-tag or label "For Autologous Use Only," "Biohazard," for confirmed positive or repeatedly reactive infectious disease tests "Donor Untested," if infectious disease testing not performed "Donor Tested within the Last 30 Days," when the unit not tested but donor had been tested within 30 days of collection

Reprinted with permission from L.A. Williams III, M.G. Fritsma, M.B. Marques, Quick Guide to Transfusion Medicine, second ed., AACC Press, Washington, DC, 2014, pp. 13–26.

Answer: *D*—The male donor with a hemoglobin of 11.5 g/dL meets one of the requirements for autologous donation. The minimum hemoglobin and hematocrit for autologous donors is 11 g/dL and 33% (Answer A), which is less than the 12.5 g/dL (female) and 13.0 g/dL (male) hemoglobin that is required for allogeneic donors. Autologous donors must complete the donor history questionnaire and physical examination prior to blood donation (Answer E), but do not have to meet all the same requirements as allogeneic donors.

Autologous donors are usually patients who need blood for an upcoming procedure or surgery. They are not as healthy as allogeneic donors and are not expected to adhere to the higher hemoglobin and hematocrit requirements. Autologous donors are not eligible to donate if they appear to have signs or symptoms of infection (same as allogeneic donors), since blood that could possibly be carrying microorganisms and should not be transfused to anyone, including an autologous donor (Answer B).

An autologous collection may not be collected less than 72 h before the donor's date of procedure, surgery, or anticipated need of transfusion (Answer C). This time interval allows the donor to begin recovery before the anticipated need of transfusion and furthermore allows appropriate time for infectious disease testing, processing, and shipment of the unit to the hospital by the blood center.

3. A female patient donates two autologous red blood cell units for an upcoming hip surgery. The units are collected at the blood center and shipped to the hospital where they are placed in a special area in the blood bank apart from other units. The surgery proceeds without incident and neither of the red blood cell units are transfused. The postoperative hemoglobin is 10.2 g/dL. She has no other significant medical conditions and is generally healthy. What decision should be made regarding the disposition of the autologous units?

A. Both units should be transfused to the patient, even though her postoperative hemoglobin is above the suggested transfusion level guidelines
B. Both units should be discarded after the patient is discharged
C. One unit should be saved for the patient for potential future use, and the other unit should be entered into the community blood supply
D. Both units should be entered into the community blood supply
E. Both units should be frozen for future use by the patient

Concept: Autologous units are not held to the same test result regulations as volunteer allogeneic donor units and consequently they cannot enter the community blood supply.

Answer: *B*—Autologous donors are not held to the same donor eligibility requirements as allogeneic donors. The hemoglobin and hematocrit requirements are lower for autologous versus allogeneic donors. Also, autologous donors are allowed to have certain blood-borne diseases by way of the Americans with Disabilities Act, which would deem allogeneic donors ineligible. Therefore, autologous blood can only be used for the donor who has donated the units and it cannot be crossed over to the community blood supply.

Each hospital blood bank decides if they are willing to accept or reject autologous units that are positive for infectious disease markers. The decision is often taken as a precautionary measure as some hospitals have concern that an infectious autologous unit may inadvertently be issued to another patient. Another precaution that the blood bank laboratory takes to prevent this mistake is to quarantine the autologous units in a special area until the time of issue to the intended recipient.

Unless the autologous donor is a rare blood donor or has a medical condition that necessitates saving donated units for future use by freezing them, both unused autologous units should be discarded after the patient is discharged, thereby eliminating the remaining answer choices (Answers A, C, D, and E).

4. Which of the following list of criteria meets the requirements for donation today as an allogeneic versus an autologous donor?

A. Temperature: less than or equal to 99.5°F (37.5°C) for allogeneic; less than or equal to 100.5°F (38.5°C) for autologous

B. Blood pressure: systolic BP 90–180 mmHg for allogeneic; systolic BP 100–180 mmHg for autologous

C. Pulse: 50–100 bpm for allogeneic; 50–110 bpm for autologous

D. Weight: greater than or equal to 110 pounds for allogeneic; any weight acceptable for autologous

E. Hemoglobin: greater than or equal to 13.0 g/dL for allogeneic male (12.5 g/dL for allogeneic female); greater than or equal to 11 g/dL for autologous

Concept: Certain donation requirements, such as the hemoglobin level, are less restrictive for autologous donors.

Answer: *E*—According to the current AABB Technical Manual (18th edition, 2014), the only criteria on the physical exam that is specified for allogeneic versus autologous donors is the hemoglobin. For allogeneic male donors the lowest acceptable hemoglobin is 13.0 g/dL and for allogeneic females it is 12.5 g/dL. For autologous donors, both male and female, the lowest acceptable hemoglobin is 11 g/dL. The temperature for allogeneic donors must be less than or equal to 37.5°C, but no range is given for autologous donors. Blood pressure, pulse, and weight can be determined by the collecting blood center, for both allogeneic and autologous donors. All the other choices (Answers A, B, C, and D) are incorrect based on the discussion above.

5. A local television station promotes a community appeal for blood and a 57-year-old female presents for donation. She mentions that she had dental work yesterday. She also mentions that she was in a motor vehicle accident in the United States 10 months ago and received one unit of red blood cells. Additionally, she stopped hormone replacement therapy 11 months ago. Which of the following statements represents the time when she will be eligible to donate?

A. Today; she is not deferred

B. Three days after her dental procedure (2 days from now)

C. Twelve months after cessation of hormone therapy (1 month from now)

D. Twelve months after receiving a blood transfusion in the United States (2 months from now)

E. Three years after receiving a blood transfusion in the United States (2 years, 2 months from now)

Concept: Each donor may present a unique set of variables that may either make them acceptable or unacceptable for blood donation. Often, the donor may be temporarily ineligible, but is able to return at a later date.

Answer: *D*—Due to receiving an allogeneic blood transfusion, she is ineligible to donate for 12 months after the transfusion (Answers A and E). A permanent deferral is in place for people who have received a blood transfusion since 1980 in the United Kingdom or France (The United Kingdom consists of the following countries: England, Wales, Scotland, Northern Ireland, Channel Islands, Isle of Man, Gibraltar or Falkland Islands) due to concerns about variant Creutzfeldt-Jakob (vCJD).

The donor is considered acceptable after dental procedures as long as there is no infection present (Answer B). If an infection is present, the donor must wait until finishing antibiotics. Although the donor does not have to wait for simple dental procedures, the donor must wait 3 days after having oral surgery. There is no deferral period for women who are on hormone replacement therapy for menopausal symptoms and prevention of osteoporosis (Answer C).

6. A 20-year-old male would like to donate allogeneic blood. Today is June 23rd and he most recently donated one unit of whole blood on May 21st of this year. He has never donated platelets. Which blood product(s) is he eligible to donate today?
   A. Platelets and RBCs
   B. Double RBCs
   C. No blood products
   D. Whole blood
   E. Granulocytes

   Concept: Donation intervals differ depending on the type of blood product donated or the amount of product donated, in addition to minimum laboratory values.

   Answer: *C*—Since the donor donated whole blood less than 8 weeks (or 56 days) ago, he is not eligible to donate any blood products today. An allogeneic donor is deferred from all types of blood donation for 8 weeks after whole blood donation, 16 weeks after double RBC donation, 4 weeks after infrequent plasma donation, and over 2 days after plasma, platelet, or granulocyte donation. All the other choices (Answers A, B, D, and E) are thus, incorrect.

7. A 45-year-old potential first time donor reports that when she was younger she was told that she has a hole in her heart that did not close. She is asymptomatic, has no activity restrictions, and is not taking any medications. Her current pulse is 67 bpm and her other vital signs are stable. When is she eligible to donate?
   A. Today, as long as she has been medically evaluated and treated (if need be), is symptom free, and meets other criteria
   B. Today, if she possesses a written copy of a consult note from a cardiologist giving her clearance
   C. She would require a 1 month waiting period from day of presentation at the donor site to ensure that she is stable
   D. She would require a 6 month waiting period from day of presentation at the donor site to ensure that she is stable
   E. The risk of stroke is too high and therefore the donor is indefinitely deferred

   Concept: Past medical conditions could potentially exclude a person from donating. Conditions that have been resolved and are currently not causing any symptoms or restrictions on activity are generally not a reason for deferral.

   Answer: *A*—Most likely the donor has a patent foramen ovale (PFO), which is a hole between the left and right atria (upper chambers) of the heart. This hole is supposed to close shortly after birth; however, sometimes it fails to close. The majority of people with PFOs are asymptomatic and do not have restrictions on activity. Assuming that the donor has no other underlying cardiac defects and no symptoms, she would be allowed to donate as long as she had been medically evaluated and if necessary, be treated. This would not require a written note from a cardiologist but rather an approval by the blood center medical director (Answers B and C). If she had a procedure or surgery on her heart, was symptomatic, or had any change in medications, she would be required to wait 6 months (Answer D), according to criteria used by most blood centers. There is no known risk of stroke in this patient (Answer E).

8. A 60-year-old male is visiting his son at college and upon his arrival there is a blood drive in progress. He used to donate blood back when his kids were young and would like to donate again. Upon taking the blood donor questionnaire, one of his responses is that about 2 years ago he began to feel a fluttering sensation in his chest along with dizziness and a feeling as though his heart was racing. He confirms that during a period of extreme discomfort, he had a one-time visit to the emergency department and from there he was referred to a cardiologist.

The cardiologist diagnosed him with an arrhythmia and he later had an ablation (4 months ago). The donor states that since the ablation, he has had no further symptoms nor does he have limits or restrictions on his activities. His current pulse is 80 bpm. When is this man eligible to donate?

A. Today, as long as he has been medically evaluated and treated, is asymptomatic, and meets other criteria.
B. Today, as long as he has been medically evaluated and treated and has been asymptomatic for greater than 3 months.
C. He is deferred for 2 more months as the donor must wait 6 months after being medically evaluated and treated for a heart condition. If he is asymptomatic and still meets other criteria in 2 months, then he is allowed to donate.
D. He is deferred for 8 more months as the donor must wait a total of 12 months after being medically evaluated and treated for a heart condition. If he is asymptomatic and still meets other criteria in 8 months, then he is allowed to donate.
E. He is deferred for a total of 3 years after an invasive medical procedure.

Concept: Most past medical conditions that have been treated will only result in temporary deferrals.

Answer: C—In general, donors with a history of a heart condition who are currently asymptomatic with no restrictions on daily living, are able to donate 6 months after they have been medically cleared. It is necessary to wait at least 6 months following an episode of angina, a heart attack, after bypass surgery, angioplasty, and after a change in the heart condition that resulted in a change to any medications. If donors have a pacemaker, they are eligible to donate as long as their pulse is between 50 and 100 bpm with no more than a small number of irregular beats, and they have met the other heart disease criteria. The other choices (Answers A, B, D, and E) are incorrect based on the timeframe given.

9. A professional football player is a potential blood donor for a single unit of red blood cells. He states that he woke up with a stuffy nose but he otherwise feels well. He denies a fever, and does not have a productive cough. He is not on antibiotics. He has the following vital signs: pulse 47 bpm, BP 186/90 mmHg, respirations 12. His hemoglobin is 14.0 g/dL and he weighs 285 pounds. According to current standards, which of the following is correct?

A. He should be deferred due to stuffy nose
B. He should be deferred due to pulse rate
C. He should be deferred due to his blood pressure
D. He should be deferred due to his weight
E. He is an acceptable donor

Concept: Though some of a donor's laboratory results and vital signs may be abnormal per normal reference ranges, they are only excluded from donation when the abnormality reaches levels that would be unsafe during donation.

Answer: C—The donor has hypertension and at a level of 186/90, he will not be able to donate today (Answer E). Acceptable levels for blood pressure are as follows: systolic below 180 mmHg and diastolic below 100 mmHg. Low blood pressure is acceptable as long as it is at least 80/50 mmHg (some blood centers require 90/50 mmHg) and the donor feels well. The donor would be deferred if he has a productive cough, fever, or does not feel well on the day of donation (Answer A). If he was on antibiotics for sinus, throat, or lung infection, he would be deferred until he finished the course of treatment. The donor pulse should be between 50 bpm and 100 bpm, however well-conditioned athletes, such as this donor, may be accepted with a lower heart rate (Answer B). The hemoglobin is at an acceptable level and he easily meets the minimal acceptable weigh criteria for most blood centers (at least 110 pounds). There is no maximum weight that would disqualify a donation so long as the donor center chairs can accommodate the donor's weight and they can fit comfortably in the chair (Answer D).

10. A male donor would like to donate one unit of red blood cells for his brother, who has an upcoming spinal surgery. He successfully answers the questions on the donor history questionnaire, and his physical exam results are as follows: temperature 98.8°F, blood pressure 138/78 mmHg, pulse 79 bpm, hemoglobin 12.2 g/dL. Is this donor eligible to make a directed donation for his brother today?
A. Yes, because he passed the donor history questionnaire
B. Yes, because his vital signs fulfill criteria for a directed donation
C. No, because his vital signs do not fulfill criteria for a directed donation
D. No, because his hemoglobin is below the threshold required for a directed donation
E. No, because his hemoglobin is higher than the threshold required for a directed donation

Concept: Directed donors must fulfill the same donation criteria as allogeneic donors, since the unit will become part of the community blood supply if not transfused to the intended recipient.

Answer: *D*—This donor's vital signs meet allogeneic donor criteria, but his hemoglobin value is 12.2 g/dL, which is below the current threshold of 13.0 g/dL for male allogeneic donors (Answer E). Therefore, this donor is not eligible to make a directed donation today. A directed donor is a donor who donates blood products for a specific individual, such as a friend or family member. This is in contrast to an allogeneic donor, whose blood becomes part of the community blood supply and may go to any patient in need of a transfusion.

Directed donations are tagged by the blood center for the intended recipient, but if the blood is not transfused to that recipient, it is returned to the blood center and becomes part of the community blood supply. For this reason, a directed donor must meet the same donation criteria (questionnaire (Answer A), vital signs (Answers B and C), hemoglobin, and negative infectious disease test results) as other allogeneic donors. Patients may desire a directed donation from a known person because of the perception that a known person's blood is "safer" than community blood. However, directed donors are often under pressure to donate for the patient, and consequently might intentionally lie on the donor history questionnaire to avoid revealing high risk behavior or other information that might cause them to be disqualified from donation. In addition, directed donors are often first-time donors and have a higher rate of positive infectious disease markers upon testing in comparison to community blood donors who are usually repeat blood donors. Of note, directed donations should be irradiated, especially if coming from close relatives, to prevent transfusion-associated graft-versus host disease (TA-GVHD).

Table 4.2 summarizes the allogeneic donor criteria discussed in Questions 1–10.

11. A 49-year-old male states that his job requires travel. He has lived around the world, including the United Kingdom from 1993 to 1994, in Rome from 1998 to 1999, in India from 2006 to 2009, Puerto Rico from 2010 to 2012, and in Minnesota from 2013 to the present.

Which of the following would prevent this donor from donating today?
A. Living in the United Kingdom from 1993 to 1994
B. Living in Rome from 1998 to 1999
C. Living in India from 2006 to 2009
D. Living in Minnesota from 2013 to the present
E. Living in Puerto Rico from 2010 to 2012

Concept: Donors are often deferred after travel to foreign countries because certain areas of the world are endemic for diseases that may have a long window period. This is true even if they do not currently have any signs of the disease.

TABLE 4.2 Summary of Criteria for Allogeneic Blood Donation

Criteria	Deferral period
Age	At least 17 years of age; 16 years of age acceptable if allowed by state law
Blood pressure	Systolic: ≤180 mm Hg; diastolic: ≤100 mm Hg
Pulse	50–100 bpm; no irregularity; < 50 bpm in healthy athlete
Temperature	≤37.5°C (99.5°F)
HGB/HCT	≥12.5 g/dL/ ≥38% for females (without FDA variance) ≥12.0 g/dL/ ≥36% for females (with FDA variance) ≥13 g/dL/ ≥39% for males
Major organ disease (e.g., heart, liver, lungs); cancer; abnormal bleeding tendency	Deferral, unless determined suitable by Medical Director of blood center (see heart and lung disease deferral criteria below)
Donation interval	• 8 weeks (≥56 days) after whole blood donation • 16 weeks (≥112 days) after double red cell collection • 4 weeks (≥28 days) after infrequent plasmapheresis • ≥2 days after plasmapheresis, plateletpheresis, or leukapheresis
Heart/lung disease	• Deferral of 6 months from last symptom/medical evaluation/restrictions on activities for most heart diseases such as murmur, angina, valvular insufficiency, coronary artery bypass surgery, arrhythmia, aneurysm, cardiac catheterization, congestive heart failure, heart attack, stroke, transient ischemic attack, pacemaker • Temporary deferral if symptomatic or with restrictions on activities for most lung diseases
Infectious diseases	Tables 4.6 and 4.7
Immunizations	Table 4.4
Medications	Table 4.5
Pregnancy	• Deferral for 6 weeks following delivery • Deferral of 12 months if delivery required blood transfusion
Sickle cell disease	• Permanent deferral • Sickle cell trait acceptable
Skin disease and rash	• Donors individually evaluated • Phlebotomy site must be free of rash/skin disease
Surgery	Acceptable if: • Underlying illness does not disqualify donor • Stitches/staples dissolved or removed • Donor has resumed normal activity and is feeling well • Donor did not receive a blood transfusion or transplant
Tattoo/body piercing	Tattoo acceptable if: • Applied in a state that regulates tattoo facilities, donor states tattoo was applied by a regulated facility, and tattoo was applied with sterile, single-use needles and ink, or • More than 12 months since procedure Ear/body piercing (including acupuncture and electrolysis) acceptable if: • Sterile, single-use needles were used, or • More than 12 months since procedure
Transfusion/transplant/needle-stick	• Deferral of 12 months • Receipt of human dura mater is indefinite deferral
Travel	Table 4.3

weight/height	• Minimum 110 lbs. for whole blood donation
	• Maximum of 10.5 mL of blood collected per kilogram of donor weight
	• Minimum for double red cell donation:
	  ◦ Males: at least 130 lbs. and 5′1″ in height
	  ◦ Females: at least 150 lbs. and 5′5″ in height
	• Maximum acceptable weight on most donation chairs is 350 lbs. Few locations have chairs suitable to accommodate donors who weigh > 350 lbs.

Reprinted and modified with permission from L.A. Williams III, M.G. Fritsma, M.B. Marques, Quick Guide to Transfusion Medicine, second ed., AACC Press, Washington, DC, 2014, pp. 13–26.

Answer: *A*—The potential donor is ineligible to donate because he has spent more than 3 months in the United Kingdom (UK) from 1993 to 1994. Any donors living in the UK from 1980 to 1996 are permanently deferred due to the risk of transmitting vCJD. He currently lives in Minnesota (Answer D) where *Ixodes scapularis* (the blacklegged tick or deer tick) exists and possibly could transmit *Babesia microti*, a protozoan parasite which can also be transmitted by blood transfusion. However, living in *Babesia microti* endemic area in and of itself is not a current reason for deferral. The remaining choices (Answers B, C, and E) are incorrect and will be expanded upon after Question #12.

12. A 35-year-old exdiplomat presents for donation at his workplace in Minneapolis, Minnesota. His diplomatic duties included serving in India from 2002 to 2014. He moved to Minneapolis in 2014 and visited India again for 6 months in 2016. When is the donor eligible to donate?

A. Never
B. Twelve months after leaving India
C. Six months after leaving India
D. Two years after leaving India
E. Today

Concept: Donation deferrals for malaria depend on how long the potential donor lived in the endemic region and if they were infected.

Answer: *B*—The donor lived in a malaria risk area for over 5 years, and is eligible to donate 3 years or more after leaving the area. However, an additional deferral time of 1 year is required if the donor returns to the malaria risk area within 3 years of leaving. The other choices (Answers A, C, D, and E) are incorrect based on the 12 month deferral for this history and region of travel.

13. A donor mentions to the collection staff that he suffers from attention deficit hyperactivity disorder (ADHD). He requests that the donor center staff read the donor history questionnaire to him. During the questionnaire he demonstrates understanding of the questions being asked and responds to a question on travel outside the United States by revealing that he lived in Iraq from 2004 to 2007, visited his daughter in London from June to September of 2010, and that he sometimes travels to Kenya to visit his wife's family. His last visit to Kenya was from 2011 to 2012. Is this donor eligible to donate today?

A. No, he was stationed in Iraq and should be deferred for leishmania risk
B. No, he had to have the DHQ read to him by donor center staff, due to ADHD
C. No, he visited Kenya from 2011 to 2012 and should be deferred for malaria risk
D. Yes, his lifestyle and travels pose no reasons for deferral
E. No, he visited his daughter in London in 2010 and should be deferred for vCJD risk

Concept: Donors may have personal or travel histories that present multiple potential reasons for deferral. Each of these must be evaluated individually to determine donation eligibility.

Answer: *D*—The donor is allowed to ask staff to read the DHQ to him/her, and as long as the donor demonstrates understanding of the verbal DHQ and qualifies to donate, there is no reason for deferral and he is eligible to donate blood today. ADHD is not a reason for deferral (Answer B). Being stationed in Iraq carries a 12 month deferral from the date of departure from the region due to leishmania risk (Answer A). Visiting a malaria endemic country like Kenya carries a 12 month deferral from the date of departure (Answer C). Visiting London in 2010 does not carry a deferral since it is outside the vCJD risk period for the United Kingdom (1980–96) (Answer E). Table 4.3 covers the current travel deferrals.

TABLE 4.3 Deferral Criteria for Travel Outside of the United States

Endemic for malaria	Defer for 12 months
Travel or residency in Iraq with possible Leishmaniasis exposure	
Previous residency in country endemic for malaria	Defer for 3 years, if no symptoms of malaria
Cumulative time spent (visited or lived) of 3 months or more in the United Kingdom (UK) from January 1, 1980 through December 31, 1996 (CJD risk)	Defer indefinitely
Former or current members of the US Military, civilian military employee, or dependent of a member of the US Military who spent a total time of 6 months or more on or associated with a military base in • Germany, Belgium, and the Netherlands (Holland) from 1980 through 1990 • Greece, Turkey, Spain, Portugal, and Italy from 1980 through 1996 • (CJD risk)	
Cumulative time spent (visited or lived) of 5 years or more in Europe (UK from 1980 through 1996) from January 1, 1980 to the present (CJD risk)	
Born in, traveled to, or lived in the following African countries after 1977: Cameroon, Central African Republic, Chad, Congo, Equatorial Guinea, Gabon, Niger, or Nigeria (HIV risk)	
Received blood transfusion in above African countries (HIV risk)	
Received blood in the UK or France from 1980 to the present (CJD risk)	

Reprinted and modified with permission from L.A. Williams III, M.G. Fritsma, M.B. Marques, Quick Guide to Transfusion Medicine, second ed., AACC Press, Washington, DC, 2014, pp. 13–26.

14. A prospective college student received vaccinations for measles, mumps, and rubella (MMR) and human papilloma virus (HPV) 2 weeks ago. What is the recommended deferral period for this donor?
A. No deferral for either
B. Two weeks after the date of immunization of both MMR and HPV
C. Twenty-one days after the date of immunization of MMR; 2 weeks after the date of immunization of HPV
D. Four weeks after the date of immunization of MMR; no deferral period for HPV
E. Wait a total of 8 weeks (56 days) after the dates of immunization for the combination of MMR and HPV

Concept: Vaccines carry different deferral periods due to known or unknown risk of infectivity to others through blood donation after the injection.

Answer: *D*—The donor must wait 4 weeks after the date of immunization for German measles (rubella), MMR, chicken pox, and shingles vaccinations (Answer A). There is no deferral period for the HPV vaccination (Answers B, C, and E), as donors infected with HPV have no deferral despite that HPV is a sexually transmitted disease. An expanded explanation of vaccine deferrals is present in the answer to Question 15.

15. A female graduate student states that 2 weeks ago she was hanging a picture in her apartment when she dropped the nail and stepped on it. She went to the emergency department and received a tetanus booster. She stated that she had a slight fever after the vaccination and that it lasted approximately 1 day. She is now asymptomatic. What is the recommended deferral period for this donor?
A. No deferral
B. Two weeks after the date of immunization
C. Twenty-one days after the date of immunization

D. Four weeks after the date of immunization
E. Wait 8 weeks (56 days)

Concept: Vaccines carry different deferral periods due to known or unknown risk of infectivity to others through blood donation after the injection.

Answer: *A*—The donor is considered acceptable if they were vaccinated for influenza, tetanus (includes Tdap vaccine), meningitis, and the human papilloma virus (HPV) vaccine, providing they are asymptomatic and afebrile. This donor is currently asymptomatic and afebrile after receiving a tetanus booster; therefore, she is eligible to donate today. The donor must wait 2 weeks (Answer B) after immunizations for red measles (rubeola), mumps, polio (oral form), and yellow fever vaccine. The donor must wait 21 days (Answer C) after immunization for hepatitis B, as long as they were not given the immunization for exposure to hepatitis B. The donor must wait 4 weeks (Answer D) after immunizations for German measles (rubella), MMR (measles, mumps and rubella), chicken pox, and shingles. There is a wait period of 8 weeks (56 days) (Answer E) from the date of having a smallpox vaccination, as long as the donor did not have any complications. If the donor experienced complications (these may include skin reactions beyond the vaccination site or general illness related to the vaccination), the donor must wait 14 days after all vaccine complications have resolved or 8 weeks (56 days) from the date of having had the smallpox vaccination, whichever is the longer period of time. Table 4.4 expands upon the deferral period for vaccinated donors.

TABLE 4.4 Deferral Criteria Based on History of Immunizations

Toxoids, synthetic or killed vaccines: Anthrax, cholera, diphtheria, HAV, influenza, Lyme disease, paratyphoid, pertussis, plague, pneumococcal, polio, Rocky Mountain Spotted Fever, rabies, tetanus, typhoid (injection)	No deferral, if symptom-free and afebrile
Red measles (rubeola)	Defer for 2 weeks
Mumps	
Oral polio	
Typhoid (oral)	
Yellow fever	
HBV	Defer for 21 days if no previous exposure to HBV
German measles (rubella)	Defer for 4 weeks
MMR (measles, mumps, and rubella)	
Varicella (chicken pox)	
Smallpox	Refer to FDA guidance
Hepatitis B immune globulin	Defer 12 months
Other vaccines, including unlicensed	Defer 1 year or refer to medical director of blood collection facility

Reprinted and modified with permission from L.A. Williams III, M.G. Fritsma, M.B. Marques, Quick Guide to Transfusion Medicine, second ed., AACC Press, Washington, DC, 2014, pp. 13–26.

16. A potential male donor had taken finasteride for several years for male pattern hair loss prior to undergoing surgical hair implants. His surgery was successful and he no longer is taking the medication. Though he does not recall the date of the last dose, his surgery was over 1 year ago and he states his last dose was around that time. The donation site calls you (the medical director of the blood center) asking for a statement on what to tell this man with regards to his medication. You tell the donation site that he is eligible to donate at what time?

A. Fourteen days after his last dose
B. One month (30 days) after his last dose
C. Six months after his last dose
D. Three years after his last dose
E. He is unable to give blood at any time

Concept: Certain medications will require donor deferrals. Some medications may have teratogenic potential (e.g., finasteride), while others, such as antibiotics pose no immediate harm, but signify a potential infectious risk.

Answer: *B*—Due to the risk of teratogenicity if his blood is transfused to a pregnant female, this donor must wait 1 month after the last dose of finasteride, thus, he would technically be eligible to donate today. Please see Questions 17 and 18 for an expanded explanation of medication deferrals and for an explanation of why the other choices (Answers A, C, D, and E) are incorrect.

17. A male college student discloses on his donor history questionnaire that he took isotretinoin for severe acne on and off for 3 years while in high school. His last dose was over 1 year ago. When is this man eligible to donate?

A. Fourteen days after his last dose
B. One month after his last dose
C. One year after his last dose
D. Three years after his last dose
E. He is unable to give blood at any time

Concept: The deferral period for medications is often based on the risk of adverse effects to a potential recipient based on worst case scenario being a pregnant female or developing fetus and the elimination of the pharmaceutical product from the body.

Answer: *B*—Certain medications interfere with the function of the blood components, while other medications could result in adverse effects to certain recipients such as a developing fetus of a pregnant recipient. Isotretinoin is a potential teratogen and there is not a separate blood supply for pregnant females or developing fetuses. The drug resembles retinoic acid, a vitamin A derivative known to control embryonic development. For these reasons, transfusion of products manufactured from donors receiving certain medications requires waiting periods after their last dose. The waiting period for isoretinoin, finasteride, or any of the brand derivatives is 1 month from the last dose. The waiting period for dutasteride is 6 months from the last dose. The donor must wait 3 years (Answer D) after taking the last dose of acitretin. A person who has taken etretinate at any time is not eligible to donate blood because it is stored and released from adipose tissue and can be released long after the last dosage. Prasugrel, clopidogrel, and ticlopidine require a wait period of 14 days (Answer A) after taking this medication before donating platelets by apheresis. The remainder of the choices from Questions 15 and 16 are incorrect based on the Table 4.5, seen further in the explanation to Question 18.

18. A regular blood donor was recruited to donate an apheresis platelet. His past medical history includes coronary artery disease but he has not had any symptoms in over 1 year and has no restrictions on activities of daily living. He currently takes 81 mg aspirin once daily for preventative measures and he states that he has had no further complications. In addition, he also takes several nutritional supplements in order to "keep in shape." Which of the following is true regarding the eligibility of this man?

A. He is eligible to donate platelets today
B. He is eligible to donate platelets 2 days after his last dose of nutritional supplements and he can keep taking the aspirin

TABLE 4.5 Deferral Criteria Due to Medications

Aspirin, piroxicam (Feldene) Clopidogrel (Plavix), prasugrel (Effient) Other antiplatelet medications	Platelet apheresis donors must wait 48 h after taking aspirin or aspirin products; 14 days for clopidogrel and prasugrel. Other antiplatelet medication deferral as defined by medical director. No deferral for whole blood donations.
Antibiotics (oral or intramuscular)	No deferral if course of medication is completed and donor is asymptomatic
Prophylactic antibiotics	No deferral
Antifungals	No deferral if course of medication is completed and donor is asymptomatic
Antivirals	No deferral if course of medication is completed and donor is asymptomatic
Anticoagulants	Defer for 7 days since last dose
Accutane, Absorica, Amnesteen, Claravis, Myoris, Sotret, Zenatane (isotretinoin)	Defer for 1 month
Proscar, Propecia (finasteride)	Defer for 1 month
Avodart, Jalyn (dutasteride)	Defer for 6 months from last dose
Soriatane (acitretin)	Defer for 3 years from last dose
Pituitary-derived human growth hormone; pituitary-derived thyrotropin	Defer indefinitely
Tegison (etretinate)	Defer indefinitely
Bovine insulin	Defer indefinitely if, since 1980, received an injection of bovine (beef) insulin made from cattle from Europe

Reprinted and modified with permission from L.A. Williams III, M.G. Fritsma, M.B. Marques, Quick Guide to Transfusion Medicine, second ed., AACC Press, Washington, DC, 2014, pp. 13–26.

C. He is eligible to donate platelets 2 days after his last dose of aspirin and he can keep taking the nutritional supplement
D. He is eligible to donate platelets 7 days after his last dose of aspirin and he can keep taking the nutritional supplement
E. He is eligible to donate platelets 7 days after his last dose of aspirin and 2 days after the last dose of nutritional supplements

Concept: Selected medication deferrals include medications that may affect the quality of the product. This is most commonly encountered with antiplatelet medications.

Answer: C—There is no waiting period for aspirin with whole blood donation; however, there is a 48-h waiting period after taking aspirin or after taking a medication that contains aspirin when donating apheresis platelets (Answers A, B, D, and E). If the donor is on warfarin, heparin, dabigatran, rivaoxaban, enoxaparin, or other prescription blood thinners, they should not donate blood unless a doctor discontinues the treatment and then they should wait 7 days before donation. If a donor is on clopidogrel, they must wait 14 days after the last dose before donating platelets by apheresis. Over-the-counter oral homeopathic medications, herbal remedies, and nutritional supplements are acceptable to take and require no waiting period (Answers B, D, and E).

19. An adult male goes with his mother (a regular donor) to donate blood for the first time. He discloses that he was diagnosed with genital herpes and was recently detained for 48 h, released home for

48 h, and subsequently jailed for 14 additional consecutive days. When is this man eligible to donate blood?

A. Today
B. Thirty days from the day of release from the second jail term
C. Sixty days from the day of release from the second jail term
D. Twelve months from the day of release from the second jail term
E. Twelve months from the genital herpes diagnosis

Concept: Lifestyle risk factors for obtaining a transfusion transmittable disease are criteria that are evaluated during donor screening. Such factors may include intravenous drug abuse or extended stays in a correction facility. The latter is because as the stay extends, there is a greater likelihood of sexual activity. However, some infectious diseases are not transmitted by blood transfusion (e.g., herpes) and are not a reason for exclusion.

Answer: *D*—A person who has been detained or incarcerated in a facility (juvenile detention, lockup, jail, or prison) for greater than 72 consecutive hours (3 days) is considered to be at higher risk for exposure to infectious diseases and therefore they are deferred for 12 months from the date of last occurrence (Answers A, B, and C). A diagnosis of chlamydia, venereal warts (human papilloma virus), or genital herpes (Answer E) are not a cause for deferral, as long as the donor is feeling well and meets other eligibility requirements.

20. A potential first time blood donor joined her coworkers at their workplace blood drive. On the donor health questionnaire, she reports that today she is healthy, but reveals that about 15 years ago she suffered headaches, blurry vision, and a seizure. She was diagnosed with a meningioma for which she underwent resection and received a dura matter graft in the United States. She is symptom free and has no restrictions on daily activities. She also says that she is on doxycycline for dry eyes for the past 2 months and remembers getting a belly button piercing on a trip to Las Vegas several years ago and doesn't recall whether the facility was licensed. When is she eligible to donate blood?

A. Accept her today; there are no eligibility issues
B. Accept her today knowing that it has been over 12 months from the date of tissue transplant and assume the piercing facility was licensed
C. Accept her today knowing that it has been over 12 months from the date of tissue transplant and also since it has also been over twelve months from the date of piercing (to be safe since she cannot verify whether the facility was licensed)
D. Defer until she is off doxycycline for 2 weeks
E. Defer her indefinitely

Concept: Donors may have multiple reasons for deferral even if they are currently "healthy." An indefinite deferral may be overturned at a later date, if the Food and Drug Administration (FDA) decides to shorten the deferral time based on recently reviewed scientific or medical evidence. A permanent deferral, by contrast, is not expected to be overturned by the FDA.

Answer: *E*—Since this donor received a dura mater transplant or graft, she is indefinitely deferred and cannot donate today (Answer A). This requirement is due to the association with dura mater transplants and Creutzfeld-Jacob Disease (CJD). Donors that receive an allogeneic organ transplant must wait 12 months before donating blood. Body piercings (Answers B and C) are acceptable if they are performed with the use of sterile or single-use equipment. If there is any question on the use of sterile or single-use equipment, the donor must wait 12 months. Antibiotic use (Answer D) must be evaluated to determine if the donor could transmit an infection and a donor with an infection should not donate. Prophylactic antibiotic use (e.g., acne, antiinflammatory for dry eyes) is acceptable.

21. Which of the following donors is acceptable for donation today?

A. A female who received human derived pituitary-derived growth hormone as a child, no sexual history as donor is practicing abstinence

B. A male who reports having sex with another man (MSM) one time with a latex condom (not sheepskin) 6 months ago and has had a negative HIV test 1 month ago, no other infectious risks disclosed

C. A female who had sexual contact with her husband who has hemophilia and he received clotting factor concentrates a few months ago when he had bleeding into his joints

D. A male who reports having MSM several times during 1990 but not since then, discloses that he was treated for both *Treponema pallidum* (syphilis) and *Neisseria gonorrhea* (gonorrhea) 14 months ago, and currently has genital warts (Human Papilloma Virus or HPV)

E. A male who has paid "in the thousands of dollars" for sexual entertainment with the last encounter being 3 years ago, just finished pharmacologic treatment for *Neisseria gonorrhea* (gonorrhea) yesterday with the infection cleared (no longer experiencing pus-like discharge)

Concept: One of the main goals of the donor history questionnaire is to detect risk factors in potential donors that might identify them as a higher risk for donating blood that is potentially infectious. There are two reasons for this. First, a donor may be in the "window period" for a disease. This is where an infection is present, but not yet detectable by screening tests. Second, there may be no reliable screening test available for a disease and for that reason screening questions are the best way to identify high-risk behavior.

Answer: *D*—Though the potential donor in answer choice D has some high-risk behavior, none of it disqualifies him from donating at this time. In the past, men who have had sex with other men (MSM) at any time since 1977, which marks the beginning of the acquired immunodeficiency syndrome (AIDS) epidemic in the United States, were placed on an indefinite deferral as blood donors; however, per recent FDA rule changes, this is now a 12 month deferral from the date of the most recent MSM encounter (Answer B). The FDA's primary responsibility is to assure blood safety and protect both the blood donors and blood recipients. FDA modifications are possible for indefinite deferrals when it is supported by new scientific data showing that a change in policy would not present a significant and preventable risk to the patients that receive blood products.

A donor who has had either *Treponema pallidum* (syphilis) or *Neisseria gonorrhea* (gonorrhea) (Answer E), or who has been treated for either syphilis or gonorrhea, is deferred for 12 months from completion of treatment or from day of presentation if treatment has not yet begun. Though HPV is a sexually transmitted disease, it is not known to be transmissible by blood transfusion, and therefore a donor with HPV is acceptable.

If a donor has received a dura mater transplant or pituitary growth hormone (Answer A), they are indefinitely deferred. This is because as a group, these donors have an increased risk for Creutzfeld-Jacob Disease (CJD). CJD appears to be an infectious disease as it has been transmitted from infected humans to patients through the dura mater transplantation, the use of contaminated brain electrodes, and injection of human pituitary-derived growth hormones. A donor is also considered to be at a higher risk of CJD if they have had a biologic relative who has been diagnosed with CJD. Currently, there is nothing that suggests CJD transmission through blood transfusions; however, there is no test for CJD used to screen blood donors and so special precautions must be in place to keep CJD out of the blood supply.

Body piercings are acceptable if they are performed with the use of sterile or single-use equipment. If there is any question on the use of sterile or single-use equipment, the donor must wait 12 months. A person who has had sexual contact with someone who has hemophilia and has used clotting factor concentrates is deferred for 12 months from the date of last sexual occurrence due to a theoretical risk of infectious disease transmission. A person who has had sexual contact with someone who has hemophilia and has not used clotting factor concentrates is acceptable (Answer C). A donor who has had sexual contact with someone at high risk of having HIV, such as a prostitute, is deferred for 12 months from the date of the last occurrence. However, the prostitute, as a person that has been given money, drugs, or other forms of payment is indefinitely deferred from donating blood (Tables 4.6 and 4.7).

TABLE 4.6 Deferral Criteria to Prevent HIV Transmission

Criteria	Deferral period
From date of release from lock-up, jail, prison, or juvenile detention center (including work release) if held more than 72 consecutive hours Mucous membrane or skin penetration exposure to another person's blood	Defer 12 months
Male who has ever had sex, even once, with another male since 1977 Any person who has had sexual contact with someone with HIV or at high-risk of HIV	Defer 12 months from date of last sexual contact
Any person who has ever had a positive test for HIV Any person who has AIDS or one of its symptoms, which include: • Unexplained weight loss (10 lbs or more in < 2 months) or night sweats • Blue or purple spots on or under the skin or in the mouth • Long-lasting white spots or unusual mouth sores • Lumps in the neck, axilla, or groin for more than a month • Fever higher than 100.5 F (38.5°C) for more than 10 days • Diarrhea lasting over a month • Persistent cough and shortness of breath Any person who has used needles, even once, to take drugs, steroids, or anything not prescribed by a doctor Any person who has been given money, drugs, or other payment for sex since 1977 Any person who was born in or lived in Cameroon, Central African Republic, Chad, Congo, Equatorial Guinea, Gabon, Niger, or Nigeria since 1977 for 5 years or more or: • Any person who received blood transfusions in any of these countries since 1977 • Any person who had sex with anyone who was born in or lived in any of these countries Any person who received clotting factor concentrates for therapy of a bleeding disorder such as hemophilia	Defer indefinitely

Reprinted and modified with permission from L.A. Williams III, M.G. Fritsma, M.B. Marques, Quick Guide to Transfusion Medicine, second ed., AACC Press, Washington, DC, 2014, pp. 13–26.

22. A donor has polycythemia rubra vera, a neoplastic disease of the bone marrow resulting in excess red blood cell production. Her physician would like her to undergo periodic therapeutic phlebotomy to prevent her hemoglobin from rising to life-threatening levels. The physician's order indicates that the donor should present for phlebotomy once a month and should have one unit of blood drawn when her hemoglobin is greater than 15 g/dL. Can this donor's units be used for allogeneic transfusion?

A. No, because she is not a volunteer donor (i.e., she has a physician's order requiring therapeutic phlebotomy)
B. Yes, as long as she passes the donor history questionnaire, physical exam, and infectious disease testing
C. Yes, because her hemoglobin is above the 12.5 g/dL threshold for female allogeneic transfusion
D. No, because she is being phlebotomized more frequently than once every 56 days (the normal interval for whole blood donation)
E. No, because she has a neoplastic blood disease

Concept: Donors with a neoplastic blood disease are not acceptable as allogeneic donors. Theoretically, the safety, quality, potency, and purity of the blood products collected from these donors might be affected by their disease and there is concern over the possibility of transmission to the recipients, although no cases of transfusion transmitted blood neoplasia have ever been reported.

Answer: *E*—Allogeneic donation is not allowed from donors with neoplastic blood diseases as there is concern for potential transmission of the blood neoplasia to recipients and there is no such variance from the FDA allowed for these diseases. Therapeutic collections of polycythemia rubra vera patients should be discarded at the collecting facility. The donor history questionnaire specifically asks if a donor has a blood disorder or blood disease. Certain disorders, such as hereditary hemochromatosis, might allow for allogeneic donation if the collecting blood center has a variance from the FDA that allows for collections from these donors.

TABLE 4.7 Deferral Criteria to Prevent Transmission of Infections Other than HIV

Cold, flu, or sore throat	Defer temporarily for active cold or flu symptoms, such as fever, sore throat, productive cough, or generalized fatigue on day of donation
Creutzfeldt-Jakob disease (CJD)	Defer indefinitely if at increased risk for, a history of, a diagnosis of, or any blood relatives diagnosed with CJD Table 4.6 (Travel outside the US)
Hepatitis/Jaundice	• Defer indefinitely for Hepatitis B, Hepatitis C, and any other known viral hepatitis except Hepatitis A
Exposure to hepatitis/ HIV/AIDS	Defer 12 months: • From last contact with another person's blood into open wound and/or nonintact skin and/or mucous membrane (this includes sharing razors or toothbrushes, and occupational exposure) • Following human or primate (ape family) bite that broke the skin • After accidental exposure to needle/instrument contaminated with blood from another person/primate • From last sexual contact with anyone diagnosed with HIV/AIDS, hepatitis, or who is a hepatitis B carrier • From last contact if lived with (resided in same house, apartment, dormitory, etc.) person with hepatitis or is chronic hepatitis B carrier • From date of release from lock-up, jail, prison, or juvenile detention center (including work release) if held more than 72 consecutive hours
Malaria	Defer 12 months after travel to endemic area Defer 3 years after living 5 consecutive years in country endemic for malaria Defer 3 years after symptom/treatment of malaria
Syphilis	Defer 12 months after positive test and treatment completion
West Nile Virus (WNV)	Donor diagnosed with or with suspected acute WNV infection: defer for 120 days following diagnosis or onset of illness, whichever is later Donor with may have transmitted WNV infection: Defer for 120 days following date of donation
Babesiosis	Defer indefinitely
Chagas disease	Defer indefinitely
Leishmaniasis	Defer 12 months after travel to Iraq
Zika	Defer for at least 120 days from the date of the reactive test or resolution of symptoms, whichever time frame is longer
Ebola	Defer indefinitely

Reprinted and modified with permission from L.A. Williams III, M.G. Fritsma, M.B. Marques, Quick Guide to Transfusion Medicine, second ed., AACC Press, Washington, DC, 2014, pp. 13–26.

If the donor does not qualify for allogeneic donation, a physician's order is needed. The order should specify the indication for therapeutic phlebotomy, how often therapeutic phlebotomy should be performed, the quantity of blood phlebotomized at each visit, and the minimum hemoglobin threshold below which the collecting facility should not phlebotomize (Answer A). The other choices (Answers B, C, and D) are incorrect because due to her condition, she is not allowed to donate, regardless of meeting other criteria.

23. A blood center receives a request from a hospital to provide granulocyte products for transfusion to a patient. The patient is very ill with an invasive Aspergillosis infection that is not responding to antifungal medications and the absolute neutrophil count (ANC) is 388/uL. The blood center has a database of qualified donors to select from for granulocyte donation. What type of donors should be listed in this database?

A. First-time donors
B. Infrequent plasmapheresis donors

C. Therapeutic phlebotomy donors
D. Female donors with HLA antibodies
E. Regular plateletpheresis donors

Concept: Patients with widespread bacterial or fungal infections not responding pharmacotherapy and who have an ANC less than 500/uL, might benefit from granulocyte transfusions. Granulocyte products expire 24 h after collection. Thus, the product will be collected before donor infectious disease test results are available. For this reason, granulocyte donors must be chosen from a pool of donors who have regularly tested negative for infectious diseases by the blood center, such as regular plateletpheresis donors. Of note, because they are being transfused to immunocompromised patients, granulocyte products are irradiated to prevent TA-GVHD.

Answer: *E*—To collect granulocytes, a donor typically receives medication, such as dexamethasone, prednisone, or granulocyte colony stimulating factor. These medications stimulate the circulation of granulocytes into the peripheral blood to allow for apheresis collection. Nonstimulated donors produce a product containing roughly 1×10^{10} granulocytes; donors stimulated with steroids produce a product containing roughly 2×10^{10} granulocytes; and donors stimulated with both steroids and G-CSF produce a product containing roughly 4×10^{10} granulocytes. Granulocyte collections must be transfused within 24 h after collection. This is in contrast to other blood products, such as platelets and red blood cells, which have longer shelf-lives of 5 days and 35–42 days, respectively. Infectious disease test results are not available by the time the granulocytes expire, so the donor must have repeatedly tested negative for infectious disease markers in recent past donations. Recent (within 30 days) regular plateletpheresis donors are the best choice as they can donate platelets twice in a 7 day period and are tested for infectious disease markers each time they donate.

A first time donor (Answer A) would not be a good choice as there is no record of past infectious disease testing. An infrequent plasmapheresis donor (Answer B) would also not be a good choice, as these donors do not fulfill an established record of frequent repeated negative donations. Therapeutic phlebotomy donors (Answer C) are often patients with either hereditary hemochromatosis or polycythemia vera and should not be chosen. Females with HLA antibodies (Answer D) should never be chosen, since HLA antibodies may attack the HLA on the patient's granulocytes or lungs.

24. A physician is performing donor record review at a large commercial plasmapheresis collection center, where source plasma is collected for manufacturing into albumin, intravenous immunoglobulin, and other blood-derived products. The physician is currently reviewing the record of a donor who has donated source plasma once every week for the past 2 months, for a total of eight plasma collections. The volume of RBCs lost by the donor (including samples taken for testing) during each of these donations is as follows: 25 mL, 23 mL, 25 mL, 24 mL, 22 mL, 22 mL, 24 mL, 44 mL. During the most recent donation, which occurred today, the apheresis line was accidently pulled out of the donor's arm at the end of the procedure, leading to the 44 mL total RBC loss. When is the donor eligible to donate *Whole Blood* next, and why?

A. After 48 h, because source plasma donors can donate twice per week with 2 days between collections
B. After 2 weeks (14 days), because over 25 mL of RBCs was lost during the most recent source plasma collection
C. After 4 weeks (28 days), because a total of over 200 mL of RBCs was lost during source plasma collections
D. After 8 weeks (56 days), because a total of over 200 mL of RBCs was lost during source plasma collections
E. After 16 weeks (112 days), because a total of over 200 mL of RBCs was lost during source plasma collections

Concept: The total RBC loss for source plasma donors must be less than 200 mL over 8 weeks. If the total RBC loss exceeds this amount, the loss is treated like a whole blood donation and the donor is deferred from donating whole blood for the corresponding time period.

Plasma collected for manufacturing is called source plasma. Source plasma collection and product regulations are mandated by the FDA and the Plasma Protein Therapeutics Association (PPTA). Source plasma donors are paid by the plasma collection center and are sometimes referred to as "frequent plasma donors" as they can donate more frequently than every 4 weeks. Transfusable plasma donors are volunteer unpaid donors and cannot donate plasma more than once every 4 weeks.

Answer: *D*—Source plasma donors can donate twice per week with at least 48 h between donations (Answer A), and have limits on the annual volume of plasma collected based on weight. The annual volume of plasma collected cannot exceed 12,000 mL for donors weighing ≤175 pounds and 14,400 mL for donors weighing > 175 pounds. Red blood cell volume loss is also measured and total RBC loss during the collection should not exceed 25 mL per week or no more than 200 mL of RBCs over 8 weeks (Answers B, C, and E). If the RBC loss is greater than 200 mL over 8 weeks with source plasma donation, the cumulative loss is treated as whole blood donation, and the donor will be deferred from whole blood donation for 8 weeks or 56 days. In this example, the source plasma donor lost a total of 209 mL RBCs over 8 weeks, so the donor will be deferred from whole blood donation for 8 weeks. Source plasma donors must also be tested for total protein and serum protein electrophoresis (SPEP) or quantitative immunoglobulins, at the initial plasmapheresis and at 4 month intervals and the results must be within normal limits to remain eligible.

25. The Medical Director at a blood donor center is notified that a positive BacT alert has occurred after 5.9 h of incubation. The unit was held and subsequently sent out for reculture and microbial speciation which identified *Streptococcus bovis*. Which of the following is true regarding *Streptococcus bovis*?

A. It is a common skin commensal or environmental contaminant
B. It is associated with incomplete treatment of osteomyelitis
C. It denotes an important public health significance for an agent of bioterrorism and is considered a national notifiable disease
D. This is likely a false positive
E. It represents a bacteremia associated with colon cancer

Concept: Initial BacT results can include both true and false positives. A false positive may result from a quality failure or a false signal from the instrument. A True positive will confirm with reculture and microbial speciation. This may occur with skin flora contamination or possibly be the result of true donor bacteremia.

Answer: *E*—*Streptococcus bovis* is known to be associated with colon cancer, since the cancer allows the bacteria to enter the blood stream. It is not associated with osteomyelitis (Answer B). Endogenous bacteremia in the donor (e.g., *Staphylococcus aureus* with incompletely treated osteomyelitis, *Escherichia coli* with a number of conditions, or *Streptococcus bovis* with colon cancer) results in true positive results (Answers A and D). True positive results most often occur due to incomplete phlebotomy scrubs, skin plugs, or environmental contaminants (e.g., *Staphylococcus epidermidis*), but can also be due to actual infections in the donor.

When bacteria are detected by culture and speciation, the Medical Director of the blood collection facility must notify the donor of the abnormal test result. This is because some organisms are of public health significance (*Bacillus anthracis, Yersinia pestis, Francisella tularensis, Clostridium botulinum*) (Answer C), some are of national surveillance importance (*Listeria monocytogenes, Salmonella* spp., *Shigella*, Group A *streptococcus, Streptococcus pneumonia, Neisseria meningitides, Neiserria gonorrheae*, among others), and some have donor health implications, such as recommendation of a colonoscopy upon detection of *Streptococcus bovis. Streptococcus bovis* has also been associated with endocarditis, urinary tract infections, meningitis, biliary tract infections, and noncolorectal cancer, but its association with colon cancer has been more widely studied in blood donors.

26. A donor underwent venipuncture of the right antecubital fossa for donation of one unit of whole blood. The donation was successful after three attempts at venous access. The donation took approximately 5 min, and after the needle was removed, the donor stated that he had pain radiating down his right arm worse with flexion, along with paresthesia described as pins and needles. The most common reason for this scenario in the context of postdonation adverse reactions is which of the following?

A. Hematoma/bruise
B. Peripheral nerve injury/irritation
C. Superficial thrombophlebitis
D. Citrate reaction
E. Arterial puncture

Concept: The majority of donors tolerate the blood donation process without incident; however, complications of variable severity may occur during or after the donation process.

Answer: *B*—This case is consistent with peripheral nerve injury/irritation. Peripheral nerves are positioned in close proximity to the veins in the antecubital fossa. Nerve injuries as a result of venipuncture have been described as burning, shooting, or electric-type pain that arises when the needle is either inserted or withdrawn, but can occur several hours thereafter. Paresthesias, such as numbness, tingling, or burning, can occur alone or in combination with the pain. Symptoms may be worse with flexion.

Hematomas (Answer A) present as bruising, swelling, and local pain due to the accumulation of blood in the tissues surrounding the vessels. Superficial thrombophlebitis (Answer C) is inflammation of the vein characterized by redness, swelling, and tenderness that follows the course of the vein just under the skin. Citrate reactions (Answer D) are more characteristic of apheresis donations or procedures and are most typically caused by low ionized calcium. Presentation encompasses a wide range of symptoms including paresthesias, metallic taste, shivering, muscle twitching, nausea, vomiting, irregular pulse and cardiac arrest depending on the severity of hypocalcemia. Arterial puncture (Answer E) results in a lighter color than usual or "bright red" blood in the collection bag. The needle and tubing may pulsate and the bag could fill in less than 3 min. Other reactions include prolonged recovery, prefaint, loss of consciousness, allergic, and infiltration of the vein.

27. A male first-time donor presents at his workplace for a blood drive. Per the donor history questionnaire and his physical examination, he is eligible to donate and proceeds to donate one unit of whole blood. He returns to his office and approximately 4 hours later, he collapses on the floor in the middle of a meeting. Paramedics are called, but are not able to revive the donor. The donor's colleague also donated at the blood drive that day. The colleague calls the blood center to inform them of the fatality and says that the family has requested an autopsy to be performed on the donor. What should the blood center do next?

A. Nothing, as the donor is deceased and no further actions are needed
B. Notify the Food and Drug Administration (FDA) Center for Biologics Evaluation and Research (CBER) of the fatality as soon as possible
C. Gather as much information about the fatality as possible over the next few days and then notify FDA CBER of the fatality
D. Wait for the donor's autopsy report findings and then notify FDA CBER of the fatality
E. Notify the blood center's corporate headquarters medical office only—FDA CBER notification is not needed

Concept: If a blood center is notified of a donor fatality that is temporally related to blood donation, the FDA CBER must be notified immediately. Most blood centers define "temporally" as within 24 h of blood donation.

Answer: *B*—CBER should be initially notified of the fatality as soon as possible, without waiting for additional information such as an investigation, autopsy, etc. (Answers A, C, D, and E). The CBER division of the FDA investigates all reported fatalities that are possibly related to blood donation or blood transfusion in the United States. Current good manufacturing practice (CGMP) regulations for blood and blood components require that fatalities related to blood collection or transfusion must be reported to CBER, according to the FDA Code of Federal Regulations (21 CFR 606.170(b)). Section 606.170(b) states: "When a complication of blood collection or transfusion is confirmed to be fatal, the Director, Office of Compliance and Biologics Quality, Center for Biologics Evaluation and Research, shall be notified by telephone, facsimile, express mail, or electronically transmitted mail as soon as possible.

A written report of the investigation shall be submitted to the Director, Office of Compliance and Biologics Quality, Center for Biologics Evaluation and Research, within seven days after the fatality by the collecting facility." This written report should consist of any new findings or information relevant to the fatality, including results of the follow up investigation and conclusions. If an autopsy is to be performed on the donor, the autopsy results should be reported to FDA CBER when available and the case file will be amended to include this additional information. Details of fatality reporting to FDA CBER can be found in the FDA Guidance for Industry: Notifying FDA of Fatalities Related to Blood Collection or Transfusion (September 2003).

Please answer Questions 28–31 in response to the following case scenario:

An adult female presents for whole blood donation 3 weeks after giving birth to her first child. She is feeling healthy and well and denies any major organ disease. She takes metoprolol and hydrochlorothiazide for hypertension and is on day five of seven of a course of oral cephalosporin to treat an ear infection. She received an influenza vaccination 4 months ago. She reports a family history of melanoma and colon cancer; however, she denies any history of cancer in herself. She is a healthcare worker and reports an accidental needlestick injury 6 months ago during phlebotomy of a pediatric patient. She did not report this to her employer since the patient is "just a toddler and looked healthy." She also confirms a sexual encounter 11 months ago with a man who has symptomatic hepatitis C. She states it was a onetime event and she denies any symptoms. She denies travel outside of the United States.

28. Is she eligible to donate whole blood today?

A. Yes, she is eligible

B. No, she is not eligible due to the following: she is taking metoprolol and hydrochlorothiazide for blood pressure control; she had a needlestick injury 6 months ago; she has been pregnant in the past 6 weeks; she received an influenza vaccination 4 months ago; she is taking an antibiotic for treatment of an ear infection; she has had sexual relations less than 1 year ago with a man who has symptomatic hepatitis

C. No, she is not eligible due to the following: she has been pregnant in the past 6 weeks; she is taking an antibiotic for treatment of an ear infection; she has a family history of melanoma and colon cancer; she has never traveled outside the USA; she has had a needlestick injury 6 months ago; she has had sexual relations with a man who has symptomatic hepatitis C less than 1 year ago

D. No, she is not eligible due to the following: she has been pregnant in the past 6 weeks; she is taking an antibiotic for treatment of an ear infection; she has had a needlestick injury 6 months ago; she had sexual relations with a man who has symptomatic hepatitis C less than 1 year ago

E. No, she is not eligible due to the following: she is 35 years old; she received an influenza vaccination 4 months ago; she is taking metoprolol and lisinopril for blood pressure control; she has had a needlestick injury 6 months ago; she has been pregnant in the past 6 weeks; she is taking an antibiotic for treatment of an ear infection

Concept: Potential donors may have multiple reasons to exclude them from donation for various lengths of time. Along with improved screening tests, deferral policies have been very effective at reducing the risk of transfusion transmitted diseases in the blood supply.

Answer: D—According to the current edition of AABB Standards for Blood Banks and Transfusion Services (30th edition, 2016), a prospective donor may not donate until 6 weeks after the conclusion of a pregnancy, or 1 year if the delivery required a blood transfusion. This is a donor safety issue and the 6 weeks allows the new mother's body to heal properly, stay hydrated, and replenish their hemoglobin. Many new mothers may not meet the appropriate hemoglobin criteria nor will they likely be in good physical condition to donate within 6 weeks of delivery.

A donor may donate 24 h after the last dose of an antibiotic course so long as they have no signs of infection (many prophylactic antibiotics are allowed, e.g., antibiotics for acne). Other deferrals include the following: 12 months for a needlestick injury contaminated with untested blood, and 12 months for sexual relations with someone with symptomatic hepatitis C. The donor's age, use of blood pressure medications, receipt of an influenza vaccine, family history of cancer, and lack of travel outside the USA are not reasons for deferral. The other choices (Answers A, B, C, and E) are incorrect based on donor criteria.

29. Which deferral period is correct?

A. For sexual contact with individual with HIV: defer for 9 months after date of last sexual contact
B. For oral cephalosporin used for treatment of bacterial infection: defer 1 month after last dose and donor must be asymptomatic
C. For needlestick injury: defer for 12 months after date of occurrence
D. For sexual contact with individual with symptomatic hepatitis C: defer for 6 months after date of last sexual contact
E. For influenza vaccine: defer for 1 week from the date of vaccination

Concept: Deferral periods are designed to provide the maximum risk reduction for recipients of blood transfusion. Some are based on knowledge of a specific disease's life cycle, while others are based on a "best guess."

Answer: C—According to the current edition of AABB Standards for Blood Banks and Transfusion Services (30th edition, 2016), if a donor has experienced a needlestick injury contaminated with untested blood, the donor must be deferred for 12 months from the date of occurrence. This deferral period allows for seroconversion and rise of disease markers to detectable levels in the event that the donor has been exposed to bloodborne diseases that are routinely tested for in blood donors (e.g., HIV, hepatitis B, hepatitis C, HTLV I/II, and others).

Pregnant women are ineligible to donate blood and must wait a minimum of 6 weeks after the conclusion of the pregnancy. Donors on antibiotics (Answer B) for treatment of an infection should be deferred for 1 day after the last dose and the donor must not have any signs or symptoms of an acute or chronic infection. A donor who has had sexual contact with an individual with symptomatic hepatitis C (which increases the risk of hepatitis C transmission, compared to individuals with asymptomatic disease) or HIV (Answers A and D) are deferred for 1 year after the date of last sexual contact. This allows an appropriate time for seroconversion and allows the disease markers to be detected, similar to the rationale for the needlestick injury deferral. No deferral is required after receiving the influenza vaccine (Answer E) as long as the donor is symptom-free. Neither the influenza shot nor the intranasal form is a reason for blood donation deferral since there is no risk of transmitting the influenza virus from the vaccines.

30. Based on the information provided by the donor, when will she be eligible to donate whole blood again?

A. Six months from now; 12 months from the date of needlestick injury occurrence
B. Two years from now
C. One month from now; 12 months after sexual contact with a man with symptomatic hepatitis C
D. Three days from now; 1 day after completion of antibiotic treatment
E. Four weeks after receiving the influenza vaccine

Concept: When deciding the time frame of deferral, the longest disqualifier is the deciding factor and the deferrals are not additive.

Answer: *A*—The donor will be eligible to donate again after the longest applicable deferral period has ended, which in this case will be 12 months after the needlestick injury (6 months from now). The other deferral periods that apply to this donor are 6 weeks after conclusion of her pregnancy, one day after completion of antibiotic treatment (Answer D), and 12 months after her sexual contact with a man with symptomatic hepatitis C (Answer C). She does not have a deferral after receiving the influenza vaccine (Answer E) as long as the she is symptom-free. She has no reasons for a 2 year deferral (Answer B).

31. The collections staff calls you and reports that the donor is willing to undergo infectious disease testing at her primary care clinic if that would allow her to donate blood products. You offer to speak to the donor. How would you counsel her?

A. Explain to the donor that she is encouraged to undergo infectious disease testing as soon as possible and is welcome to come back and donate, providing her test results are negative.
B. Explain to the donor that she is encouraged to undergo infectious disease testing at her primary care clinic; however, the results may not be accurate until after the deferral period is over and she remains ineligible to donate until the deferral period ends.
C. Explain to the donor that she is encouraged to undergo infectious disease testing in 3 months, and she can come back to donate right after that, providing that her test results are negative.
D. Explain to the donor that she is encouraged to undergo infectious disease testing in 6 weeks, and she can come back to donate in 2 months, providing that her test results are negative.
E. Tell the donor to postpone infectious disease testing and to come back and donate after her deferral period ends. The blood center will test for infectious diseases at that time point and will notify her of any issues.

Concept: The reason for many deferrals within the donor history questionnaire is due to extended window periods or because no good screening test exists for the particular disease.

Answer: *B*—Certain deferral periods are in place with the intent to avoid window period false negatives for infectious bloodborne diseases. The window period is the time from potential exposure to the time of disease detection. This is the timeframe that donor may not have seroconverted or have serologic markers at detectable levels. For example, the window period for hepatitis B varies with different serologic markers and can be several months for some markers, while the window period for HIV with nucleic acid testing (NAT) is approximately 9–12 days after exposure.

A deferral period of 12 months is often chosen when a donor may have been exposed to infectious bloodborne diseases to safely ensure all window periods are met. If this donor undergoes infectious disease testing after the deferral period is over (12 months after the needlestick injury and six months from now), the test results should be accurate. Donors are not encouraged to donate blood products for verification of infectious disease status. If the donor has infectious disease risk factors, they should be encouraged to undergo earlier testing so that in the event of any positive results, they may receive earlier treatment. All the other choices (Answers A, C, D, and E) are incorrect based on this donor's deferral criteria and for the reasons explained above.

End of Case

Please answer Questions 32–35 in response to the following case scenario:

An allogeneic male donor presents for apheresis platelet collection. He donated platelets 3 months ago. During his interview, he acknowledges taking one aspirin 9 days ago. The donor's hemoglobin is 14.3 g/dL, weight is 210 pounds, blood pressure (BP) is 140/75 mmHg, pulse is 88 bpm with regular rate and rhythm, and temperature is 98.9°F. Inspection of his antecubital areas reveals no evidence of needle tracks, rashes, or other pathology.

The phlebotomist identifies an appropriate venipuncture site and scrubs the area thoroughly with disinfectant. A blood sample for platelet count is obtained and the donor is placed on the apheresis instrument. The sample diversion pouch fills with blood and the donor's blood volume circulates throughout the instrument. Midway through the procedure, the donor experiences perioral tingling, nausea, and muscle cramping. The collection staff performs therapeutic interventions according to the standard operating procedures (SOPs) and the procedure continues. The remainder of the collection proceeds without incident. The final product is approximately 300 mL of platelets suspended in plasma.

32. What is the purpose of the sample diversion pouch mentioned in this case?

A. The antecubital skin plug is drawn into the sample diversion pouch, to prevent bacterial contamination of the collected product
B. Blood samples for hemoglobin testing are drawn from the sample diversion pouch
C. The blood in the sample diversion pouch is added to the collected product, for transfusion
D. The blood in the sample diversion pouch is stored only for use in adverse event investigation
E. The blood in the sample diversion pouch is not used for any purpose and is simply discarded

Concept: During blood donation, the first 40–50 mL of blood goes into what is called a diversion pouch. This has proven to be highly effective in reducing bacterial contamination of blood products.

Answer: *A*—The phlebotomist thoroughly disinfects the donor's antecubital areas to prevent skin contaminant microorganisms (usually gram positive cocci such as *Staphylococcus* or *Streptococcus*) from entering the collected blood product after venipuncture. However, the needle also comes into contact with a skin plug underneath the donor's skin during venipuncture. This skin plug is not accessible to the arm scrub and may carry skin flora, which could contaminate the collected blood product unless it is diverted into another bag with the first 40–50 mL of the initial blood draw. The sample diversion pouch is used for this purpose. Integration of this method has significantly reduced the risk of bacterial contamination. The greatest reduction in rate of bacterial contamination is seen with platelet products, which are prone to contamination as they are stored at room temperature in which Gram positive cocci grow easily. Samples for infectious disease testing are also drawn from the sample diversion pouch, but it is not used in adverse event testing or for testing the donor's hemoglobin level (Answers B and D), nor is it simply discarded without any further use (Answer E). Blood in the pouch should never be used for transfusion (Answer C).

33. This donor's pre-donation platelet count result is 170,000/μL. Which statement about platelet counts for apheresis platelet donors is correct?

A. A blood sample should be collected 2 weeks prior to each plateletpheresis procedure to determine the donor's platelet count
B. The predonation platelet count result is not necessary to qualify the donor for a same day platelet apheresis donation
C. Platelet count results performed prior to or after the procedure may not be used to qualify the donor for the next procedure
D. Based on this donor's predonation platelet count, he is not eligible to donate platelets
E. A triple platelet apheresis product can be drawn from a first time platelet donor, if certain requirements are met

Concept: To ensure a quality product and donor safety, donors of apheresis platelets have additional criteria to meet before being allowed to donate.

Answer: *B*—According to the current edition of AABB Standards for Blood Banks and Transfusion Services (30th edition, 2016), a donor is deferred if the platelet count is less than 150,000 u/L, until the platelet count is measured to be greater than or equal to 150,000 u/L. Therefore, this donor is eligible to donate platelets today (Answer D). The platelet count can be performed prior to or after the collection procedure as the guidelines do not specify the timeframe for when the blood sample has to be drawn (Answer A). Since this person is a repeat donor, the results can be used to qualify the donor for the next platelet donation (Answer C). First time platelet donors should be tested prior to the first donation whenever possible. A triple platelet apheresis product should not be collected from a first time platelet donor, but can be drawn from repeat platelet donors if the donor's platelet count meets certain thresholds, depending on the apheresis instrument used (Answer E).

34. What types of interventions did the collections staff person most likely perform, once the donor started experiencing symptoms (perioral tingling, nausea, and muscle cramping) during the collection procedure?

A. Slow down the calcium infusion rate and administer citrate to the donor
B. Increase the citrate infusion rate
C. Continue the procedure without citrate infusion
D. Slow down the citrate infusion rate and administer oral or intravenous calcium to the donor
E. Call a neurologist to assess the donor's symptoms

Concept: Citrate is an effective anticoagulant for apheresis procedures. It works by binding calcium in the blood, which is necessary for a critical step in the coagulation cascade. If too much citrate reenters the body, especially in patients with poor liver or renal function, the citrate will cause hypocalcemia in the donor.

Answer: *D*—The donor's symptoms of perioral tingling, nausea, and muscle cramping is most consistent with a hypocalcemia reaction to the citrate anticoagulant in the apheresis device. Citrate chelates calcium, and to a lesser extent magnesium, in order to prevent clotting of blood within the apheresis instrument and stored blood products. Because citrate is used in the instrument, the donor becomes exposed when the donor's blood is returned from the device. This occurs throughout the procedure.

Symptoms of citrate toxicity can include nausea, tingling, cramping, hypotension, decreased pulse pressure, arrhythmias, mental status changes, coagulopathy, twitching, and other symptoms associated with hypocalcemia or hypomagnesemia. Sequela of citrate exposure is dependent upon rate and duration of citrate administration, along with donor size and rate of metabolism. Interventions may include pausing the procedure, slowing down the citrate infusion rate and replacement of calcium, either orally or intravenously. If the donor's symptoms continue or increase in severity, call the blood donor center medical director, who will decide whether to stop the procedure. None of the other interventions are appropriate for treating citrate toxicity (Answers A, B, C, and E).

35. The donor wants to return for another platelet donation tomorrow. Which of the following represents the correct response to his request?

A. No, he cannot return tomorrow for another platelet donation. Plateletpheresis can be performed once per week, not to exceed 24 times in a rolling 12 month period. The interval between procedures should be at least 3 days.
B. No, he cannot return tomorrow for another platelet donation. Plateletpheresis can be performed once a month, not to exceed 12 times in a rolling 12 month period. The interval between procedures should be at least 2 days.
C. No, he cannot return tomorrow for another platelet donation. Plateletpheresis can be performed twice in a 7 day period, not to exceed 24 times in a rolling 12 month period. The interval between procedures should be at least 2 days.
D. Yes, he can return tomorrow for another platelet donation. Unlike red blood cells, platelets can be donated at any time as long as the donor's platelet count is greater than 150,000 u/L.
E. Yes, he can return tomorrow for another platelet donation. He is no longer symptomatic from his reaction so he is eligible to donate platelets again.

Concept: To promote donor safety and product quality, platelet donors can only present for donation as specified intervals.

Answer: C—According to the current edition of AABB Standards for Blood Banks and Transfusion Services (29th edition, 2014), a platelet apheresis donor can be collected a maximum of 2 times in a 7 day period, not to exceed 24 times in a rolling 12 month period. The interval between the two collections should be at least 2 days. These guidelines are mandated by Food and Drug Administration (FDA) guidance's for industry to prevent excessive donor platelet and plasma loss. Blood centers will defer a donor once they have reached these limits, until they are eligible again to donate. In unusual circumstances, the blood center medical director can make an exception to allow a donor to donate platelets outside of these guidelines. Based on the criteria discussed above, all the other choices (Answers A, B, D, and E) are incorrect.

Suggested Reading

[1] AABB, Medication deferral list, 2013. Available from: http://www.aabb.org/tm/questionnaires/Documents/dhq/v1-3/MedicationDeferralListv1.3.pdf

[2] FDA Guidance to Industry: Recommendations for deferral of donors and quarantine and retrieval of blood and blood products in recent recipients of smallpox vaccine (vaccinia virus) and certain contacts of smallpox vaccine recipients, 2002. Available from: http://www.fda.gov/downloads/BiologicsBloodVaccines/GuidanceComplianceRegulatoryInformation/Guidances/Blood/ucm080371.pdf

[3] FDA Guidance for Industry: Eligibility determination for donors of human cells, tissues, and cellular and tissue-based products/HCT/Ps, 2007. Available from: http://www.fda.gov/BiologicsBloodVaccines/GuidanceComplianceRegulatoryInformation/Guidances/Tissue/ucm073366.htm

[4] FDA Final Rule: Requirements for blood and blood components intended for transfusion or for further manufacturing use, 2015. Available from: https://www.federalregister.gov/articles/2015/05/22/2015-12228/requirements-for-blood-and-blood-components-intended-for-transfusion-or-for-further-manufacturing

[5] Food and Drug Administration (FDA). Code of Federal Regulations, Title 21, Food and Drugs, Pt. 600–799. College Park, MD: National Archives and Records Administration, Office of the Federal Register, 2015.

[6] M.K. Fung, B.J. Grossman, C.D. Hillyer, C.M. Westoff (Eds.), Technical Manual, eighteenth ed., AABB Press, Bethesda, MD, 2014.

[7] D.M. Harmening, Modern Blood Banking and Transfusion Practices, sixth ed., F.A. Davis Company, Philadelphia, PA, (2012).

[8] P.W. Ooley (Ed.), Standards for Blood Banks and Transfusion Services, thirtieth ed., AABB, Bethesda, MD, 2016.

[9] B.H. Shaz, C.D. Hillyer, M. Roshal, C.S. Abrams (Eds.), Transfusion Medicine and Hemostasis: Clinical and Laboratory Aspects, Elsevier, London, UK, 2013.

[10] T.L. Simon, E.L. Snyder, B.J. Solheim, C.P. Stowell, R.G. Strauss, M. Petrides, Rossi's Principles of Transfusion Medicine, fourth ed., AAAB Press, Bethesda, MD, (2009).

CHAPTER

5

Blood Component Preparation and Storage

Jason E. Crane, Mona Papari*, Alesia Kaplan***

*LifeSource Blood Center, Rosemont, IL, United States; **Institute for Transfusion Medicine, Pittsburgh, PA, United States

The preparation of blood components involves several steps of manufacturing, which begin as soon as the donation of the blood product is complete. Donations may be either whole blood or apheresis. This chapter reviews the processes for manufacturing, storage conditions, and the required quality control (QC) for red blood cells, platelets, plasma, and cryoprecipitated AHF. Modifications to the products, such as irradiation, leukoreduction, and pathogen inactivation are also discussed. Additionally, granulocyte collections are reviewed, as they are an infrequent but important blood product for severely neutropenic patients with infections. Throughout the chapter, several scenarios which mimic situations one may see in either a blood center or in the hospital, are included.

1. There are two major approaches in preparation of blood components—preparation of blood components from whole blood and by apheresis technique. Which of the following basic principles is similar for these two techniques?
   A. Intermittent flow centrifugation
   B. Sedimentation
   C. Continuous flow centrifugation
   D. Separation by centrifugation
   E. Centrifugal elutriation

   Concept: Blood components can be prepared by two main methods: from whole blood (WB) by centrifugation and by apheresis. Even though these are two different methods, they rely on the same basic principal: a separation of blood components occurs during centrifugation based on gravity. The most dense blood elements migrate to the bottom and the least dense ones remain at the top [i.e., from bottom to top: RBCs (most dense) → WBCs → Platelets → Plasma (least dense)].

   To prepare the components from WB, large, high-speed centrifuges accommodating 4–6 units of WB are usually used. The recovery of cells from whole blood depends on the rotor size, centrifuge speed, duration of centrifugation, and acceleration/deceleration protocol. The WB can be manipulated differently and it depends on what components are desired—RBC, platelets, and plasma or RBC and plasma. Platelet rich plasma (PRP) is a method for platelet preparation from WB in the United States that uses low g-force centrifugation as a first step (e.g., separates WB into RBC and PRP) and high g-force centrifugation as a second step (separates PRP into platelets and plasma). If whole blood-derived platelets are not desired, then WB is separated into RBC and plasma.

   Apheresis is another method of blood component preparation. During apheresis procedure, whole blood is spun in the chamber of the apheresis machine and is separated into red cells, leukocytes,

Transfusion Medicine, Apheresis, and Hemostasis. http://dx.doi.org/10.1016/B978-0-12-803999-1.00005-5

platelets, and plasma. Any of these components or a combination of several components can then be selected for collection, while the remaining blood components are returned to a donor. Examples of apheresis equipment for component collection are represented in Table 5.1.

TABLE 5.1 Apheresis Equipment for Different Component Collection

		Component produced						
Apheresis equipment	**Centrifugation methods**	**RBCs**	**Platelets**	**Plasma**	**Granulocytes**	**MNCs**	**PBSCs**	**Component combinations**
TerumoBCT (COBE) Spectra	CFC		x	x	x	x	x	x
TerumoBCT Trima Accel	CFC	x	x	x				x
Fenwal Alyx	CFC	x						RBCs and plasma only
Fresenius Com.Tec	CFC		x	x	x	x	x	x
Fenwal Amicus	CFC	x	x	x		x	x	x
Haemonetics Cymbal	IFC	x						
Haemonetics PCS2	IFC			x				
Haemonetics MCS+ LN 9000	IFC	x	x	x	x			x

CFC, Continuous flow centrifugation; IFC, Intermittent flow centrifugation.

Answer: *D*—Separation by centrifugation is the main principle of WB separation into components, as well as separation and component collection through an apheresis machine.

Sedimentation (Answer B) is another method to prepare blood components from whole blood. However, it's not currently used in developed countries because it is relatively inefficient. If a unit of whole blood is left untouched for several hours, red blood cells sediment and RBCs and plasma can be collected without centrifugation. However, it is not a very efficient method since RBCs are not well packed and have a low hematocrit. In addition, plasma recovery is diminished.

Discontinuous or intermittent flow centrifugation (IFC) (Answer A) and continuous flow centrifugation (CFC) (Answer C) refer to two centrifugation methods used by different apheresis machines. IFC works in *cycles—draw and return*—drawing whole blood in a bowel, spinning/separating into different layers, collecting a desired component, and returning the rest of it back to a donor. It requires only one venipuncture (i.e., single access) since it is done through a single line. In comparison, CFC uses two venipunctures (one for draw, one for return line, i.e., double access) and whole blood is collected, spun, and returned continuously (see examples of IFC and CFC apheresis machines used in donor setting, Table 5.1).

Centrifugal elutriation or counter-flow centrifugal elutriation (Answer E) is a cell separating technique used in some apheresis machines. Cell separation is based on cell size and density. Cells are separated according to their rate of sedimentation, using gravity during centrifugation, where the liquid containing the cells is made to flow against the gravitational force. Therefore, cells are subjected to two opposing forces within the separation chamber: the centrifugal and the counter flow of the fluid in the opposite direction. Historically, this method was used for T-cell depletion of hematopoietic progenitor cells. Elutriation is still used for leukoreduction of blood components in Trima Accel (TerumoBCT, Lakewood, CO). It is used in extracorporeal photopheresis equipment Therakos Cellex (Mallinckrodt Pharmaceuticals, West Chester, PA) for elutriating WBC prior to photo treatment with ultraviolet lights.

2. If platelets are to be prepared from whole blood, at what temperature should the blood be maintained after collection, prior to processing?

A. 1–6°C
B. 1–10°C
C. 20–24°C
D. 35–37°C
E. The temperature depends on the anticoagulant-preservative solution

Concept: As described earlier, blood components can be prepared from whole blood by centrifugation methods. However, during this process, each product's storage temperature must be kept in mind. For example, exposure to cold temperatures may change the hemostatic properties of platelets, leads to platelet activation and shape change, and may cause a rapid platelet clearance in vivo.

Answer: *C*—If platelets are to be prepared from WB, the blood must be kept at room temperature (20–24°C), until the different components are separated. If only RBCs (i.e., packed RBCs) and plasma are to be prepared from WB, the blood is kept between 1 and 6°C. Answers A, B, D, and E are incorrect.

3. Which one of the following statements describes a part of the ideal storage conditions for whole blood-derived platelets?

A. Platelets are stored between 20–24°C to prevent bacterial contamination
B. Platelets-storage containers are designed to permit gas exchange
C. Platelets need agitation during storage to prevent ATP production
D. Interruption of agitation for maximum of 24 h is detrimental to platelet recovery
E. Maximum pH of platelets at the end of storage should be 6.2

Concept: Storage conditions and the type of storage container are critical for the shelf-life and clinical efficacy and safety of all blood products. Due to their relatively short in vivo half-life of 66–73 h, as well as high risk for bacterial contamination during storage, stored platelets require specific conditions for maximum function and safety once transfused.

Answer: *B*—Platelet storage containers are designed to allow circulation of gases through permeable plastics: oxygen can enter the bag, in order to support oxidative phosphorylation, and carbon dioxide can exit the bag. Agitation of the platelets during storage contributes to the effectiveness of this gas exchange and minimizes the accumulation of lactic acid, which is a product of glycolysis, while maintaining pH. Interruption of agitation for long periods of time results in increased lactic acid production and a subsequent drop in pH.

In addition to the storage container, platelets have the following storage requirements: must be kept at temperatures between 20 and 24°C (not between 1 and 6°C like RBCs), with continuous agitation. This storage at room temperature makes platelets more susceptible to bacterial contamination (Answer A). Agitation facilitates the gas exchange to maintain the pH. Agitation has no effect on ATP production (Answer C). Agitation can be discontinued for up to 24 h, for example, during transportation, and in vitro studies showed that platelets are not damaged by such action (Answer D). AABB Standards require a pH of 6.2 or higher for whole blood-derived and apheresis platelets at the end of the storage period (Answer E).

4. According to AABB Standards, quality control (QC) of leukocyte-reduced apheresis platelets has which of the following requirements?

A. At least 90% of sampled units should contain a minimum of 3.0×10^{11} platelets
B. At least 75% of sampled units should have a pH of at least 6 at the end of storage period
C. 100% of units should have a residual leukocyte count of $< 5 \times 10^6$
D. At least 75% recovery of the original platelet yield after leukoreduction
E. At least 90% of units should have $> 3.0 \times 10^{11}$ platelets and a residual leukocyte count of $< 5 \times 10^6$

Concept: QC of blood products is essential for Good Manufacturing Practices (GMP) and for assuring compliance with regulatory standards. AABB Standards require compliance with the FDA and the Code of Federal Regulations.

Answer: *A*—QC of leuko-reduced apheresis platelets must demonstrate that 90% of the units sampled contain $\geq 3.0 \times 10^{11}$ platelets.

Additional requirements include 90% of the units should have a pH ≥ 6.2 at end of storage (Answer B), and that at least 95% of units sampled should have a residual leukocyte count of $< 5 \times 10^6$ (Answers C and E). The platelet recovery should be at least 85% of the content prior to filtration (Answer D).

5. After receiving approximately 100 mL of a pooled, whole blood platelet product, a 45-year-old male patient experienced severe chills and rigors, in addition to the following changes in his vital signs:

	Pretransfusion	During transfusion
Temperature	37.5°C	38.9°C
Heart rate	74 beats per min	82 beats per min
Blood pressure	135/90	80/50
Respiratory rate	12 per min	14 per min
O_2 saturation	99%	97%

Which of the following answers represents a change made to reduce the risk of such reactions?

A. Disinfection of the phlebotomy site and collection of first aliquot of blood into a diversion pouch
B. Use of leukoreduction filters
C. Anaerobic culture-based primary testing performed on a sample of the platelet product
D. Testing for bacterial contamination should be done within 8 h after collection
E. Primary testing of platelets with a rapid bacterial detection device

Concept: This patient has a transfusion reaction associated with fever, chills, rigors, and hypotension, which may indicate a transfusion-transmitted infection (e.g., a septic transfusion reaction). Bacterial contamination of platelets is of higher concern when compared to other blood products because the higher storage temperature with agitation facilitates proliferation of bacteria. Different methods are in place in order to prevent the initial contamination during the collection process, as well as detection of bacterial contamination prior to transfusion.

Answer: *A*—The venipuncture site should be properly disinfected with approved antibacterial agents, such as povidone iodine and/or chlorhexidine; however, bacteria residing deeper in the skin are not destroyed by disinfectants. AABB Standards require for collection containers to have a diversion pouch if platelets are being prepared or collected from the unit being collected. The pouch is intended to capture the skin plug with the first few millimeters of blood, decreasing the rate of bacterial contamination. The blood from the diversion pouch may be used for infectious disease testing.

Leukoreduction filters have a pore size of about 4 μm and would allow majority of bacteria to pass through (Answer B). After collection, blood collection facilities perform initial/primary testing of platelets using FDA-approved culture-based devices that can detect both aerobic and anaerobic bacteria (Answer C), such as BacT/ALERT system (bioMerieux, Durham, NC), eBDS system (Haemonetics, Braintree, MA), and Acrodose systems (Haemonetics, Braintree, MA). Platelets should be tested at least 24 h after collection, and products can be released to transfusion services after at least 24 h from inoculation, or according to instructions from the bacterial detection device used (Answer D). Studies have shown that the rate of transfusion-related septic reactions and fatalities associated with transfusion of platelets increases with duration of storage. Thus, platelets that have not undergone pathogen reduction have an expiration date of 5 days from the initial collection.

Secondary testing of platelet products may be performed by transfusion services by rapid bacterial detection tests performed within 24 h prior to transfusion. An example is the Verax Platelet PGD test (Verax Biomedical, Marlborough, MA) (Answer E). The FDA recently cleared the use of the Verax Platelet PGD test to extend the shelf-life of platelets to 7 days as long as the platelets are stored in plasma in approved platelet collection bags and the test is performed within 24 h of transfusion. Further, these platelets with extended shelf-life can be shipped intrastate but not interstate.

6. Which one of the following statements is accurate for apheresis platelets with added platelet additive solution (PAS)?

A. Platelet count of each unit should be a minimum of 3.5×10^{11} platelets, as there is a poor in vivo recovery
B. Platelets are stored in equal amounts of platelet additive solution and plasma
C. The platelet additive solution contains electrolytes and glucose
D. PAS platelets contain a minimum of 5×10^{6} leukocytes
E. PAS platelets cannot be divided into smaller aliquots

Concept: Platelets with PAS added are collected by apheresis and suspended in a storage solution consisting of plasma and PAS in variable proportions; they are approved by the FDA for use in clinical settings. It has been reported that the rate of allergic transfusion reactions is lower with PAS platelets, due to the reduction of plasma. Different studies report that the rate of allergic reactions associated with PAS platelet transfusions is about one third of that seen with platelets stored in plasma. Another benefit of PAS platelets is the reduction of anti-A and anti-B titers, which should decrease the risk of hemolysis if a patient receives ABO-incompatible plasma. Reduction in plasma volume may also decrease the risk of TRALI due to a decrease in the level of anti-HLA and anti-HNA antibodies and storing platelets in additive solutions is considered a "potential strategy" for TRALI mitigation by AABB. A potential downside to PAS platelets is that they cost more than standard apheresis platelets.

Answer: *C*—FDA-approved platelet additive solutions are electrolyte-based formulas which contain sodium chloride, acetate, phosphate, citrate, magnesium, potassium, gluconate, and glucose to provide nutrients for the platelets while in storage.

PAS platelets are stored in a mixture of approximately 65% additive solution and 35% plasma (Answer B) and have a shelf-life of 5 days when stored at 20–24°C. The platelet count of a PAS platelet should be at least 3×10^{11}, same as for apheresis platelets (Answer A). PAS platelets are leuko-reduced and one unit should contain less than 5.0×10^{6} leukocytes (Answer D). PAS platelets can be further processed, including dividing the unit into smaller aliquots (Answer E). At the time of writing, unlike platelets stored in plasma, PAS platelets' shelf-life cannot be extended to 7 days with Verax Platelet PGD test.

7. Compared to fresh frozen plasma (FFP), plasma frozen within 24 h after phlebotomy (PF24) has which of the following characteristics?

A. Longer shelf-life in frozen state
B. Storage temperature is −65°C or colder
C. Can be stored as thawed plasma for 8 days
D. Markedly decreased levels of factor V
E. Decreased levels of factor VIII

Concept: Storage conditions for plasma products are essential for preservation of coagulation factors activity after thawing. PF 24 is frozen within 24 h of collection, where FFP is frozen within 8 h of collection.

Answer: *E*—Levels of factor VIII are approximately 20% lower in PF 24 comparing to FFP, but factor VIII replacement in patients with factor VIII deficiency is usually accomplished by factor VIII concentrates rather than plasma transfusion.

Plasma frozen within 24 h of phlebotomy has the same frozen shelf-life as FFP, which is 1 year (Answer A) when stored at temperatures of −18°C or lower (Answer B). Similar to FFP, PF 24 is thawed at 30–37°C and either can be transfused immediately or stored for 24 h between 1 and 6°C. Additionally, if the product is relabeled "thawed plasma," the FDA allows for storage up to 5 days at 1–6°C (Answer C). PF 24 is considered to be a therapeutically appropriate substitute to FFP in regards to levels of most coagulation factors and can be used interchangeably with the FFP. Factor V is at normal levels in PF 24 (Answer D).

8. Liquid plasma is currently used for the treatment of coagulopathies in patients undergoing massive transfusions. Which of the following characteristics best describes liquid plasma?
A. Contains viable lymphocytes that can cause graft versus host disease in patients at risk
B. Can be separated up to 7 days after expiration of the whole blood
C. Levels of coagulation factors are stable throughout the storage period
D. Can be stored at room temperature, without agitation
E. Can be further processed for manufacturing of cryoprecipitate

Concept: Liquid plasma is the plasma derived from whole blood but is never frozen. It is used for the treatment of patients with massive hemorrhages and associated coagulopathies. Advantages of liquid plasma are that it does not necessitate thawing and is therefore, immediately available in emergency situations, and it has a longer shelf-life compared to thawed plasma. Liquid plasma is not yet widely available in the United States.

Answer: *A*—One disadvantage of liquid plasma is that contains viable lymphocytes, and it puts susceptible patients at risk for transfusion-associated graft versus host disease.

Liquid plasma is separated from whole blood within 5 days of expiration of the whole blood (Answer B) and it is stored at temperatures between 1 and 6°C (Answer D). Although coagulation profile is similar to thawed plasma after manufacturing, levels of coagulation factors decline with increased storage time, after 2 weeks coagulation factors being at about 50% of the initial levels (Answer C). Liquid plasma cannot be used for cryoprecipitated AHF manufacturing, as this process requires initial freezing of plasma (Answer E).

9. Cryoprecipitate reduced plasma (i.e., "cryopoor" plasma) has which of the following characteristics?
A. Is prepared by refreezing thawed plasma
B. May be refrozen within 5 days of thawing the plasma
C. It is relatively deficient in Factor VIII and fibrinogen
D. Does not contain ADAMTS13
E. Has normal amounts of von Willebrand factor (vWF)

Concept: Cryoprecipitate reduced plasma (i.e., cryopoor plasma) is manufactured by removing of the cryoprecipitate, and it may be used for plasma exchange procedures in some refractory TTP cases.

Answer: *C*—Cryopoor plasma is relatively deficient in those factors that are mainly contained in cryoprecipitate, namely factor VIII, vWF (Answer E), fibrinogen, factor XIII, and fibronectin.

Other plasma proteins and coagulation factors, including ADAMTS13, are found at levels similar to other plasma components (Answer D). Cryopoor plasma is prepared by thawing FFP at 1–6°C, followed by centrifugation and removal of cryoprecipitate (Answer A). It must be refrozen within 24 h at less than –18°C (Answer B).

10. Which of the following statements is true about volume-reduced platelets?
A. Achieves an in vitro recovery rate of more than 90% of initial platelet count
B. Maximum of 4 h storage time if an open system was used
C. Maximum of 4 h storage time if a closed system was used
D. Platelet morphology is affected by the volume-reduction process
E. Decreased platelet aggregation in vitro

Concept: Volume–reduction of platelets may be indicated for patients at risk for transfusion-associated circulatory overload, for intrauterine transfusions, or for limiting the amount of isoagglutinins in platelet products. Volume-reduction of whole-blood derived platelets is done by centrifugation.

Answer: *B*—Volume-reduced platelets may be stored for a maximum of 4 h when manufactured by an open system.

There is no determination of the maximum allowable storage time of a closed manufacturing system (Answer C). The in vitro recovery rate for volume-reduced platelets is about 85% (Answer A). Platelet morphology and mean platelet volume are generally not affected by the volume-reduction process (Answer D). Platelet aggregation is also maintained, and it may be increased after centrifugation (Answer E) because lowering the pH of the platelet product by addition of anticoagulant prior to centrifugation prevents aggregation.

11. How many milliliters of anticoagulant-preservative solution does one 500 mL whole blood unit contain?
A. 50
B. 63
C. 70
D. 100
E. 250

Concept: Whole blood is collected either in a 450 or 500 mL plastic bags containing anticoagulant-preservative solution. Anticoagulants are agents that prevent blood from clotting. Anticoagulants used for blood collections contain citrate that prevents coagulation by binding to calcium ions and blocking the coagulation cascade. Reduced amount of anticoagulant could lead to clotting of the unit and its wastage. Excessive amount of anticoagulant would potentially cause unnecessary citrate toxicity in a patient.

Answer: *C*—To insure the appropriate anticoagulant-to-WB ratio, WB and anticoagulant-preservative solution volumes are specified by the manufacturer. One 500 mL whole blood unit contains 70 mL of anticoagulant-preservative solution. The other choices (Answers A, B, D, and E) represent the incorrect amount of anticoagulant-preservative solution.

12. RBCs preserved in CPD and CPDA-1 can be stored at 1–6°C for how many days, respectively?
A. 21 and 35
B. Both for 21
C. 28 and 35
D. 28 and 42
E. 35 and 42

Concept: Various anticoagulant-preservative combinations result in different storage duration for RBCs. Membrane, hemoglobin, and energetics of RBC metabolism are targeted during storage with anticoagulant-preservative solutions. Storage lesions that occur during RBC preservation include depletion of ATP, 2,3-DPG, decrease in RBC deformability, membrane lipids, and intracellular potassium, increase in internal RBC viscosity and hemolysis, microvesiculation and macroaggregate formation. Each component of an anticoagulant-preservative solution has a specific function and targets different elements of RBC metabolism during biopreservation. Citrate is added to prevent blood clotting and it also acts as a membrane stabilizer. RBCs metabolize glucose to make ATP. Glucose is an essential nutrient for the retention of cellular metabolism. Phosphate in CPD, CP2D, CPDA-1, and additive solutions acts as buffer and as substrate and leads to a higher pH during storage precluding early loss of 2,3-DPG compare to ACD solution. CPDA-1 and additive solutions contain adenine, which increases the adenylate pool and shifts the equilibrium conditions toward

ATP production. Therefore, energy for the first half of the glycolysis pathway and membrane integrity is maintained.

Answer: *A*—ACD, CPD, and CP2D are approved anticoagulant-preservative solutions for storage of RBCs at 1–6°C for 21 days, while CPDA-1 is approved for 35 days. Additive solutions are preserving solutions that are added to the original RBCs that are either collected in CPD or CP2D after removal of plasma. Major components of AS are saline, dextrose, adenine, and variable mannitol (for membrane preservation) content. Additive solutions (Adsol, Nutricel, Optisol, SOLX) are approved in the United States and extend RBC storage up to 42 days. The other choices (Answers B, C, D, and E) represent the incorrect range. Refer to Table 5.2 for a summary of these solutions.

TABLE 5.2 RBC Anticoagulant-Preservative and Additive Solutions Currently Used in the United States

Anticoagulant-preservative solutions	Abbreviation	Storage time (days)
Acid-citrate-dextrose	ACD	21
Citrate-phosphate-dextrose	CPD	21
Citrate-phosphate-double dextrose	CP2D	21
Citrate-phosphate-dextrose-adenine	CPDA-1	35
Additive solutions	AS-1[a], AS-3[a], AS-5[a], AS-7[a]	42

[a]*AS-1 (Adsol, Fresensius-Kabi, Fenwal), AS-3 (Nutricel, Haemonetics), AS-5 (Optisol, TerumoBCT), AS-7 (SOLX, Haemonetics).*

13. There are different types of red blood cell additive solutions (AS) currently used. In addition to extending RBC shelf-life up to 42 days, which of the following is another advantage?
A. Increase of extracellular potassium level
B. Complete prevention of TRALI
C. Improved 24-h posttransfusion RBC recovery at the end of storage
D. Extends storage time of frozen RBCs
E. Limit donor exposure in neonates

Concept: Additive solution RBC units are prepared by collecting whole blood in CPD or CP2D bags. After centrifugation and expression of plasma, 10–50 mL of plasma remains. The final step is addition of AS: 100 mL of AS per 450 mL bag and 110 mL of AS per 500 mL bag. The components of AS aid the RBCs in surviving longer during storage, which improves inventory management and adds financial and clinical benefits.

Answer: *E*—RBCs stored in AS may limit donor exposure in neonates, since many institutions assign 1 unit to 1 patient and aliquot over an extended period of time. Use of 1 unit, reduces donor exposures and leads to decreased risk of potential transfusion-transmitted infections and decreased risk of alloimmunization.

Multiple in vitro and in vivo studies report that "storage lesions" (Answer A) formed during storage of RBCs in AS equal to "storage lesions" of RBCs preserved in other preservatives. Examples of storage lesions include decreased deformability, decrease of ATP and 2,3-DPG levels, nitric oxide release, cytokine accumulation, and potassium leakage from RBCs. In addition, RBCs in AS similar to other preservatives follow the same FDA criteria for ≤ 1% hemolysis and at least 75% 24-h posttransfusion RBC recovery at the end of storage (Answer C).

Since RBCs in AS have a decreased residual volume of plasma compare to other preservatives, it reduces the rate of allergic and febrile nonhemolytic transfusion reactions. For the same reason, it has lower isohemagglutinin titers and may be beneficial for posttransplant transfusion in minor and bidirectional ABO incompatible hematopoietic progenitor cell or bone marrow transplant recipients.

Unfortunately, TRALI (Answer B) (the number one cause of transfusion related mortality) is not prevented by use of AS because even a minimal amount of plasma in a blood product is enough to cause TRALI. Other advantages of AS include increased plasma yield during manufacturing, and improved blood supply and inventory management due to the longer shelf-life. Additive solutions do not prolong storage time of frozen RBCs (Answer D).

14. Which of the following represents the approximate range of hematocrit in an AS unit of RBCs?

A. 33%–44%
B. 45%–50%
C. 55%–65%
D. 65%–80%
E. 85%–90%

Concept: Because more fluid is added to a unit with additive solution, the hematocrit is lower than that of units preserved with other anticoagulant-preservative solutions.

Answer: C—Additive solution RBC units are prepared by adding the AS to whole blood collected in CPD or CP2D bags. The final hematocrit of an AS unit is 55%–65%. In contrast, CPD, CP2D, or CPDA-1 units have approximately 50–80 mL of residual plasma and hematocrit of 65%–80% (Answer D). Hematocrit of a unit of whole blood is 33%–44% (Answer A). Answers B and E are incorrect.

15. You are notified that your regular shipment of red blood cells (RBCs) will be late 2 h due to a major accident along the driver's route. Upon arrival, the temperature of the shipment is checked. What is the highest temperature the RBCs can reach to be acceptable for use?

A. 6°C
B. 7°C
C. 8°C
D. 10°C
E. 15°C

Concept: Knowledge of the dating/storage/transport conditions of the various components, as outlined in the current AABB Standards, is important. Also, one should recognize that the transport conditions may differ from the storage conditions. Table 5.3 illustrates dating/storage/transport for the most common components.

TABLE 5.3 Storage and Transport Conditions of Selected Blood Components

Component	Storage (°C)	Transport (°C)	Expiration
Red Blood Cells	1–6	1–10	Depending on storage solution Table 5.2
Red Blood Cells, Frozen	≤65	Frozen	10 years
Platelets	20–24 with continuous gentle agitation	20–24, no more than 24 h without agitation	5 days
Plasma	≤ −18 or ≤ −65	Frozen	1 year (at ≤ −18°C) or 7 years (at ≤ −65°C)
Cryoprecipitate d AHF	≤ −18	Frozen	1 year

Answer: *D*—Standards require red cells to be stored at 1–6°C; however, during shipping the range is 1–10°C. Answers A, B, C, and E are incorrect.

16. The high glycerol (40%) concentration method of freezing of RBCs is generally preferred by blood centers over low glycerol (20%) concentration method for which of the following reasons?
A. Need to control freezing rate
B. It doesn't require a trained technician
C. It doesn't require liquid nitrogen
D. It results in lower rate of hemolysis
E. It doesn't require deglycerolization

Concept: Glycerol (high and low concentration) is an FDA approved cryopreservative for RBCs that prevents freezing injury. Glycerol is added to the bag of RBCs and it penetrates the cell membrane slowly and prevents movement of the intracellular fluid to the extracellular space. If glycerol is not added, crystal formation in the extracellular space will cause movement of the intracellular fluid out of the cell by osmotic force and result in cell shrinkage. High concentration glycerol method is somewhat simpler compared to a low concentration technique. After RBCs are mixed with 40% glycerol, the bag is placed in the mechanical freezer at −80°C and it can be stored at −65°C or below for 10 years according to FDA. It can be shipped on dry ice. Low concentration technique requires liquid nitrogen for freezing, storing, and shipping, and also the freezing rate needs to be controlled during freezing. Both methods require trained personnel, and deglycerolization after thawing with extensive washing since glycerol can cause in vivo and in vitro hemolysis. According to AABB standards, a quality control step should be in place to check for free hemoglobin and residual glycerol in the final product. A mean recovery of ≥80% of the preglycerolization red cells should be achieved.

Answer: *C*—Advantage of a high glycerol concentration method includes need only for a mechanical freezer for freezing and storing of RBCs and no requirement for liquid nitrogen. Answers A, B, D, E are incorrect.

17. Your blood bank receives an order for 2 units of red blood cells for an outpatient transfusion for a sickle cell patient with anti-U and anti-k antibodies. The appointment is scheduled for 8 days from now in the afternoon. After an exhaustive search of the Rare Donor Inventory, 2 antigen-negative frozen units are identified in a donor center across the country. Your blood bank has the capability to thaw/deglycerolize on-site with an open system. What is the best course of action to address the needs of the patient?
A. Notify the patient's physician that the blood bank will not be able to obtain the units
B. Have the blood center thaw/deglycerolize the units and ship so they are immediately available
C. Have the units shipped frozen, then thaw/deglycerolize immediately
D. Have the units shipped frozen, confirm appointment with clinician, thaw 2 days prior to the appointment to ensure availability for the appointment
E. Have the units shipped frozen, confirm appointment with the clinician, thaw/deglycerolize the units at the morning of the appointment

Concept: The use of glycerol, as discussed in Question 17, allows for freezing red cells at −65°C for up to 10 years. This allows for longer term storage of rare units, such as the ones noted in this question. As these are very rare units (making up < 0.1% of the donor population), care should be taken to ensure they are used appropriately. As the method used to thaw/deglycerolize involves an open system, the units will expire 24 h after thaw (there is an approved closed method, both when adding glycerol and removing it, allowing for up to 2 weeks expiration postthaw but that is not specified here).

Answer: *E*—This option provides the approach most likely to avoid wasting the units. Answer A is not correct, as the hospital's blood supplier likely would arrange for shipment of the units. The other choices (Answers B, C, and D) are incorrect because the units would expire prior to the appointment based on the use of an open system.

Please answer Questions 18 and 19 using the following process map:

18. Which process (X) is described in Fig. 5.1?

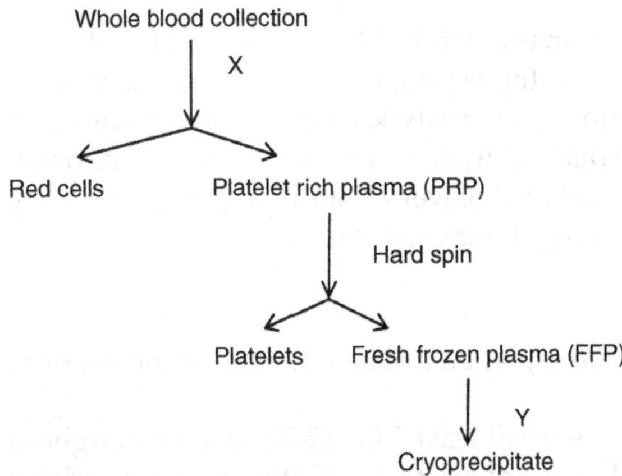

FIGURE 5.1 **Blood collection process map.**

A. Pasteurization
B. Centrifugation at low g-force (soft spin)
C. Centrifugation at high g-force (hard spin)
D. Irradiation
E. Leukoreduction

Concept: This question highlights the processing of a unit of whole blood into component parts. Through a stepwise series of manufacturing processes as outlined earlier, whole blood is separated into RBCs, plasma, platelets, and cryoprecipitated AHF (or commonly known as cryoprecipitate).

Answer: *B*—In the United States, the initial spin is a soft spin to make RBCs and platelet rich plasma. This is followed by a hard spin to separate platelets and FFP. European countries use a method known as the buffy coat method, which begins with a hard spin (Answer C). The buffy coat and RBCs are then separated, and the buffy coats are pooled in batches of 4–5 units with a unit of plasma, then centrifuged with a soft spin to separate the platelet concentrates. Pasteurization (Answer A) is a process used to manufacture albumin, which is pasteurized at 60°C for 10 h to inactivate bacteria and viruses. Irradiation (Answer D) is performed on cellular product to prevent transfusion-associated graft versus host disease (TA-GVHD). Leukoreduction (Answer E) is the removal of leukocytes (white blood cells) from the product to a concentration of less than 5×10^6 leukocytes per unit. Prestorage leukoreduction has been demonstrated to reduce red blood cell alloimmunization, HLA alloimmunization, febrile reactions, and CMV transmission. Given these factors, many hospitals have 100% leukoreduced inventories (known as universal leukoreduction).

19. Which process (seen as Y in Fig. 5.1) is the final step in cryoprecipitate production?
A. Pasteurization
B. Thaw at 1–6°C followed by centrifugation
C. Irradiation
D. Thaw at room temp followed by filtration
E. Thaw at 1–6°C followed by filtration

Concept: This question highlights the final manufacturing processes in cryoprecipitate production. A summary of the process is as follows: whole blood is "soft spun" into red blood cells and PRP; then the PRP is "hard spun" to isolate the plasma and platelets; finally the plasma is processed to isolate cryoprecipitate (explained in more detail later). It is important to note cryoprecipitate can only be manufactured from FFP, and not from any other type of plasma (e.g., PF 24, etc.).

Answer: *B*—Cryoprecipitate is manufactured by thawing FFP at 1–6°C, which causes formation of the cryoprecipitate. The unit is then spun, causing the supernatant plasma to be transferred to another bag (supernatant is now commonly known as "cryo-poor plasma"), and the cryoprecipitate resuspended in residual plasma and frozen. Answers A and C are used in different processes, as explained in the question earlier. Answers D and E are incorrect for either incorrect thawing conditions or process for isolating the cryoprecipitate.

End of Case

20. According to the AABB Standards, when indicated, what is the proper dose of irradiation for blood components?
A. 10 Gy (1,000 cGy) centrally with at least 2 Gy (2,000 cGy) throughout
B. 15 Gy (1,500 cGy) centrally with at least 5 Gy (5,000 cGy) throughout
C. 20 Gy (2,000 cGy) centrally with at least 10 Gy (1,000 cGy) throughout
D. 25 Gy (2,500 cGy) centrally with at least 15 Gy (1,500 cGy) throughout
E. 30 Gy (3,000 cGy) centrally with at least 20 Gy (20,000 cGy) throughout

Concept: Irradiation of cellular components is required to prevent TA-GVHD by inactivating the donor T-lymphocytes. TA-GVHD is a serious consequence of transfusing nonirradiated lymphocytes into immunosuppressed recipients, with a mortality rate > 90%. Sources of radiation include gamma rays, supplied by cesium-137 and cobalt-60, and X-rays, which may be from either standalone units or linear accelerators used in radiation therapy. Many centers use X-ray units, since gamma irradiators require increased surveillance/monitoring by the US Nuclear Regulatory Commission.

Answer: *D*—As per the AABB Standards, 25 Gy (2500 cGy) centrally with a minimum dose of 15 Gy (1500 cGy) throughout the product. These requirements must be verified annually for cesium-137, semiannually for cobalt-60, and upon installation, major repair, or relocation of any machine. Alternate radiation sources must be checked periodically, as recommended by the manufacturer. The other choices (Answers A, B, C, and E) are all incorrect because they do not represent the correct Gy level or target. Additionally, the shelf-life of an irradiated RBC unit is a maximum of 28 days from the time of irradiation. For example, if the unit had a shelf-life of 32 days before irradiation, the new expiration date is 28 days. However, if the unit had a shelf-life of 23 days before irradiation, the postirradiation shelf-life is still 23 days.

21. You are reviewing recent QC results for cryoprecipitate. Which unit would NOT meet the FDA's minimum requirements?
A. 250 mg fibrinogen, 90 IU Factor VIII
B. 250 mg fibrinogen, 85 IU Factor VIII
C. 150 mg fibrinogen, 85 IU Factor VIII
D. 150 mg fibrinogen, 80 IU Factor VIII
E. 150 mg fibrinogen, 70 IU Factor VIII

Concept: QC requirements ensure products are safe, pure, and potent prior to distribution. Table 5.4 summarizes several of the requirements.

Answer: *E*—The unit fails QC due to low Factor VIII. Answers A, B, C, and D pass QC since the values are greater than or equal to the FDA requirements.

TABLE 5.4 Blood Product QC Requirements

Product	Quality control requirements
Red blood cells	Final hematocrit ≤ 80%
Red blood cells, LR	85% of original red blood cells retained $<5 \times 10^6$ residual leukocytes
Apheresis red blood cells	≥60 g hemoglobin (180 mL red cell volume) per unit ≥95% of units sampled must have 50 g hemoglobin (150 mL red cell volume)
Apheresis red blood cells, LR	Mean hemoglobin ≥ 51 g (153 mL red cell volume) $< 5 \times 10^6$ residual leukocytes ≥ 95% of units must have > 42.5 g hemoglobin (128 mL red cell volume)
Random donor platelets	≥ 90% of units sampled contain ≥ 5.5×10^{10} platelets ≥ 90% of units sampled have pH ≥ 6.2 at end of storage
Random donor platelets, LR	≥ 75% units sampled contain ≥ 5.5×10^{10} platelets ≥ 90% units sampled must have pH ≥ 6.2 at end of storage ≥ 95% of units sampled must have < 8.3×10^5 leukocytes
Pooled platelets, LR	Same QC as for Random Donor Platelets plus residual leukocytes < 5×10^6; assume potency = # units in pool × 5.5×10^{10}
Apheresis platelets	90% of units sampled have 3×10^{11} platelets and pH ≥ 6.2 at end of storage
Apheresis platelets, LR	Apheresis platelets requirement plus residual leukocytes < 5×10^6
Apheresis platelets platelet additive Solution added LR	90% of units sampled contain 3×10^{11} platelets 95% of units sampled contain < 5×10^6 leukocytes
Cryoprecipitate	≥ 150 mg fibrinogen in each unit ≥ 80 IU Factor VIII per unit If testing prestorage pooled, must be multiple of the above values times the bags in the pool
Apheresis granulocytes	≥ 75% of units tested must have ≥ 1×10^{10} Granulocytes

LR, Leukoreduced.

22. A unit of apheresis leukoreduced red blood cells fails QC with because of too many residual leukocytes. Which test would most likely quickly identify the most common cause of filter failures?

A. Sickle cell testing
B. Type and Screen
C. Hemoglobin content
D. Direct antibody test (DAT)
E. Bacterial culture

Concept: Leukoreduction of cellular products is recommended to reduce red cell alloimmunization, CMV transmission, febrile nonhemolytic transfusion reactions, and HLA alloimmunization. Filters require assessment if products do not pass the QC and one should be familiar with possible factors leading to poor leukoreduction.

Answer: *A*—Sickle cell trait is responsible for at least a third of leukoreduction filter failures, for which electron microscopy has shown increased adherence to the filter by the sickled cells, which may impact the ability to remove the white cells. Additional parameters thought to impact filtration efficacy include temperature, cell–cell interactions, number of leukocytes prior to filtration, protein content, velocity of the product through the filter, and the storage age of the product. Blood type and presence of antibody (Answer B) are not known to have an impact on filtration. A very high hemoglobin content, or extensive bacterial contamination theoretically could increase the protein content; however, on screening, a very high hemoglobin would likely be identified beforehand with questions of suitability or by hemoglobin measurement prior to donation, and bacteremia identified via culture may take several hours (Answers C and E). Lastly, DAT result has no impact on filtration failure (Answer D).

23. Which of the following is the correct description of thawed plasma?

A. Plasma that has never been frozen
B. Any thawed plasma can be labeled as "thawed plasma"
C. Plasma thawed at 30–37°C and issued within 24 h postthaw
D. Apheresis-derived plasma only
E. Plasma thawed and issued >24 h after thawing but ≤ 5 days from date it was thawed

Concept: Thawed plasma is prepared from FFP, PF24, or PF24RT24 using aseptic techniques (functionally closed system). It is thawed at 30–37°C, and maintained at 1–6°C for up to 5 days from date it was thawed. Thawed plasma contains stable coagulation factors, such as Factor II and fibrinogen in concentrations similar to those of FFP, but variably reduced amounts of other factors. The smaller percentage of these other factors does not appear to be clinically relevant.

Answer: *E*—Thawed plasma is prepared from FFP, PF24, or PF24RT24 by thawing these products out in a water bath at 30–37°C. After it is thawed out and kept in refrigerator for more than 24 h at 1–6°C, it can be relabeled as *thawed plasma* and can be used for additional 4 days. Of note, thawed plasma is not a licensed product unlike other plasma products, and is not described in the Code of Federal Regulations (CFR). Practically speaking, most blood banks using thawed plasma, label the product as thawed plasma as soon as it is thawed, so they do not have to relabel the product at 24 h. Thus, the product is good for 5 days from the time of thawing. The other choices (Answers A, B, C, and D) are incorrect based on the earlier description.

24. What is the function of hydroxyethyl starch (HES) in granulocyte collections?

A. Enhance RBC sedimentation; thus, facilitating collection of granulocytes
B. Used as an adjunct with corticosteroids in donor stimulation
C. Used as an adjunct with G-CSF in donor stimulation
D. Used to prolong granulocyte activity in the collected product
E. Prevention of hypocalcemia in the donor

Concept: Granulocytes are collected using a leukapheresis procedure. A suggested blood volume of 8–10 L is processed during the collection, and the AABB Standards require 75% of the collections to contain at least 1×10^{10} granulocytes in the final product. This question highlights the use of HES in the collection of granulocytes. HES causes rouleaux formation of red cells to enhance the separation of RBCs and buffy coat; thus, it may increase the efficiency of granulocyte collection. HES is a colloid solution which acts as a volume expander. Possible side effects, such as headache and peripheral edema, may occur. Its use should be avoided in critically ill patients, patients with severe liver disease, patients with kidney disease, and in patients with bleeding disorders.

Answer: *A*—HES is a sedimentation agent. It enhances RBC sedimentation and thus, increases the efficiency of granulocyte collection. HES is used during the collection process of granulocytes and not the stimulation process (Answers B and C). Typically, only corticosteroids and/or G-CSF (also known as filgrastim) are used for stimulation. There is nothing in use to prolong granulocyte activity (Answer D), hence the 24 h expiration date. Hypocalcemia (Answer E) in the donor during the leukapheresis procedure is prevented/treated with oral and/or IV calcium replacement. The collection rate of the procedure may also be reduced.

25. Which of the following donors would be the best candidate for a granulocyte collection where G-CSF and dexamethasone are used for donor stimulation?

A. A donor who admits to type 2 diabetes mellitus and is not compliant with his metformin
B. A donor who donated platelets 5 days ago with the same blood type as the recipient

C. A donor who donated whole blood 1 week ago with the same blood type as the patient
D. A first-time donor who wants to donate for his friend
E. A frequent platelet donor who is 2 weeks postoperative from a coronary artery bypass graft (CABG)

Concept: Granulocyte collections require knowledge of a donor's medical and donation history, since medications with known side-effects are prescribed as part of the procedure. Additionally, the product must be released prior to the completion of infectious disease testing, due to the short shelf-life of the product (24 h). Consideration must also be given to recent donation history, since red cell loss is associated with this donation. After donating granulocytes, the donor is deferred for 56 days. Furthermore, because granulocytes do contain red cells, they must be ABO-compatible and crossmatch-compatible with the recipient. Irradiation of the granulocyte unit is also recommended to prevent TA-GVHD.

Answer: *B*—Granulocyte donors are typically recruited from recent (and repeated) platelet donors (who have recently been tested negative for infectious disease markers and have a history of multiple negative tests), though a new test is required at each donation. The rationale is that a recent platelet donor has a recent negative ID test panel, has an established blood type, and has passed the physical screening, and therefore, is low risk in terms of infecting the patient or not tolerating the procedure.

A donor with possibly uncontrolled diabetes (Answer A), would be made worse by dexamethasone. A recent whole blood donor (Answer C) cannot donate for 56 days after the donation. A first time donor (Answer D) has no infectious disease testing history and it is unknown how he will tolerate the procedure. His blood type is also unknown. In a case of medical urgency, this donor could be considered but if a recently tested donor is available that should be the first choice. A donor with recent major surgery may not be able to tolerate the fluid shifts which occur during the procedure and HES administration (Answer E). Most blood centers would defer this individual for any type of donation.

26. You receive a shipment of granulocytes 8 h after collection. They are packed in a cooler with RBCs. The cooler temperature was measured and noted to be 5°C at the time of delivery. What is the appropriate disposition for the granulocytes?
A. Place them in the 1–6°C cooler
B. Place them in the 1–6°C cooler and notify the floor as they expire within 24 h from receiving
C. Place at –20°C and notify the floor as they expire within 24 h from receiving
D. Place at room temperature and notify the floor as they expire within 48 h from receiving
E. Notify the clinical team that the product will need to be disposed due to improper storage

Concept: As previously noted, knowledge of the various storage and transport conditions of the blood components is required. Granulocytes, typically intended for patients with neutropenia (and with hope of marrow recovery) and a documented bacterial or fungal infection unresponsive to maximal dose of antibiotics, should be stored at 20–24°C, *without agitation*, with transport as close as possible to those storage conditions. In these patients, the bone marrow failure causing the neutropenia may improve. Granulocytes may also be used in patients with hereditary neutrophil function defects.

Answer: *E*—This granulocyte unit violated the storage requirements. Granulocytes are known for rapid degradation; therefore, use as soon as possible after collection is advised, with expiration of 24 h after collection. Because of this short shelf-life, it is not possible to complete infectious disease testing prior to distribution, and the clinical team must acknowledge that fact either before or after issue of the product. Any abnormal test result must be reported to the hospital as soon as possible and the medical director of the transfusion service should communicate promptly with the clinical team. Stimulation with dexamethasone and/or G-CSF is started on the day prior to collection. Dexamethasone consists of dosing the day prior and the day of the collection, G-CSF is only dosed the day prior to the collection. The other choices (Answers A, B, C, and D) are incorrect because you cannot keep or infuse an improperly stored product.

27. What severe reaction has been associated with the use of G-CSF?
A. Cataract formation
B. Neutropenia
C. Splenic rupture
D. Tendon rupture
E. Pancreatitis

Concept: G-CSF is used to stimulate granulocyte donors and HPC donors by apheresis, and may be used in conjunction with dexamethasone. Dexamethasone alone may also be used to stimulate donors, though the granulocyte yield is typically higher with G-CSF stimulation. The recently published RING study, while not definitive in recommending for or against granulocytes overall, did find in a subset analysis that patients who received a higher dose of granulocytes tended to have better outcomes. These findings support the use of G-CSF in combination with dexamethasone in donor stimulation.

Answer: *C*—Splenic rupture, though rare, is a serious complication of G-CSF use, marked by most commonly abdominal pain, and may include left shoulder pain and hypotension. Fatal ruptures have occurred in *patients* taking G-CSF and nonfatal ruptures have occurred in healthy donors being mobilized for a HPC collections. Donors taking G-CSF, which has a half-life of 3.5 h, should not participate in contact sports until the drug is cleared. The other choices (Answers A, B, D, and E) are all potential adverse reactions from different drugs but have not been seen with G-CSF.

28. Which of the following side-effects are associated with the long-term use of dexamethasone stimulation in the collection of granulocytes?
A. Hypertension
B. Fluid retention
C. Hyperglycemia
D. Cataracts
E. Weight loss

Concept: As noted earlier, dexamethasone is also used in stimulation of granulocyte donors (though not for HPC donors by apheresis technique). While its use as the sole drug for donor stimulation is falling out of favor, its use continues and one should be aware of the side effects.

Answer: *D*—Cataracts are a side effect of long term usage of dexamethasone. A single granulocyte donation will not lead to cataracts; however, in the setting of repeated donations, this side effect should be given consideration. The choices (Answers A, B, and C) are side effects associated with short term use. Dexamethasone is associated with weight *gain* rather than loss (Answer E). In assessing granulocyte donors, it is important to consider the history of the donor before prescribing this medication. Donors with uncontrolled diabetes, or hypertension, should not be allowed to donate given these side effects.

29. Solvent/detergent treated plasma (S/D plasma) was recently approved for coagulopathy treatment in the United States and can be manufactured by which of the following processes?
A. Pooling of liquid plasma collected by apheresis that underwent detergent treatment during collection
B. Plasma units of different ABO types can be pooled together
C. Pooling of plasma from multiple donors of the same ABO type
D. Can be refrozen after thawing
E. It is not necessary to remove the solvent/detergents after treatment

Concept: S/D plasma may be used as replacement of different coagulation factors in coagulopathies or as a replacement fluid for plasma exchange in patients with TTP. Knowledge of manufacturing processes and storage conditions are important for understanding the properties of S/D plasma.

Answer: *C*—S/D plasma is manufactured from pooling plasma of a single ABO blood group (A, B, AB, or O).

Pooling occurs after thawing of individual units of frozen plasma, followed by a filtration step and treatment with S/D reagents (Answer A). The S/D reagents are subsequently removed from the pooled plasma product (Answer E). The storage temperature is below −18°C. S/D plasma needs to be thawed prior to administration either by water bath method or by a dry tempering method (such as heated air technologies or radio wave-based technologies), and cannot be refrozen after thawing (Answer D); the thawed S/D plasma may be used within 12 h after thawing if stored at 2–4°C or within 3 h after thawing if stored at room temperature (20–25°C); by contrast, thawed plasma may be used for up to 5 days after thawing if stored between 1 and 6°C. S/D plasma is administered according to ABO compatibility, and pooling of plasma units of different ABO types is not acceptable (Answer B).

30. Solvent/detergent treated plasma (S/D plasma) has a higher risk of thrombosis due to deficiency of which of the following?
A. α2-antiplasmin
B. Fibrinogen
C. Von Willebrand factor
D. Protein S
E. Protein C

Concept: It is important to be aware of adverse effects associated with administration of S/D plasma, as some of the coagulation factor levels are affected by the manufacturing processes.

Answer: *D*—There is a higher risk of thrombosis due to lower levels of Protein S, which is sensitive to the S/D treatment.

The other protein that is decreased by the S/D treatment is α2-antiplasmin which may result in excessive bleeding due to hyperfibrinolysis (Answer A). The levels of all other coagulation factors are similar to levels found in other plasma products, including fibrinogen (Answer B), von Willebrand factor (Answer C), and protein C (Answer E).

31. Octaplas plasma (Octapharma USA, Hoboken, NJ) has been approved by FDA for the management of the following patient populations: preoperative or bleeding patients that who require replacement of multiple plasma coagulation factors; patients with coagulation deficiencies due to hepatic disease or those undergoing cardiac surgery or liver transplantation; and patients with thrombotic thrombocytopenic purpura (TTP). What is the storage time of Octaplas at ≤−18°C?
A. 6 months
B. 1 year
C. 3 years
D. 5 years
E. 10 years

Concept: Octaplas is manufactured from human plasma collected in US licensed donation centers. Plasma is pooled and treated with solvent/detergent (S/D) reagents [1% tri(*n*-butyl) phosphate (TNBP) and 1% octoxynol for 1–1.5 h at +30°C (86°F)] to inactivate *enveloped* viruses. This method is not adequate for inactivating of nonenveloped viruses, such as Hepatitis A, Hepatitis E, and Parvovirus B19. The S/D reagents are removed by sequential oil and solid phase extraction procedures. Glycine is added to adjust the osmolality. Plasma with glycine is applied to a column filled with affinity ligand resin intended for selective binding of prion protein (PrP). However, the effectiveness of this step in removal of prion infectivity from the product has not been established.

The finished product is tested for coagulation factors II, V, VII, VIII, X, and XI, Protein C, Protein S, α2-antiplasmin (also known as Plasmin Inhibitor), fibrinogen, and ADAMTS13, and all the levels are comparable to FFP except lower levels of α2-antiplasmin and Protein S. Octaplas can be stored for up to 3 years at ≤−18°C.

Answer: C—Octaplas treated plasma can be stored for up to 3 years at ≤−18°C. The other choices (Answers A, B, D, and E) are incorrect based on the earlier discussion.

32. Pathogen inactivation methods currently approved in the United States are best described by which of the following?
A. Denaturation of peptidoglycans in microbial wall
B. Crosslinking of nucleic acids
C. No effect on T-cell lymphocytes
D. Accelerated cytokine synthesis
E. Effectiveness only against enveloped viruses

Concept: Pathogen inactivation methods are currently for use on certain platelet and plasma products, and knowledge of the mechanism by which different pathogens are inactivated is important for understanding the benefits and limitations offered by these methods.

Answer: *B*—The currently approved methodology for platelet inactivation, INTERCEPT (Cerus Corporation, Concord, CA) uses treatment with amotosalen followed by an illumination step in order to induce intercalation of the amotosalen molecule between the DNA or RNA strands of different pathogens, and subsequent crosslinking, which prevents replication of the nucleic acids.

T-cell lymphocytes are undergoing a log reduction of more than 5.4, which is more than the reduction induced by gamma irradiation, meaning that this technology may replace gamma irradiation for prevention of transfusion-associated graft versus host disease (TA-GVHD) (Answer C). The synthesis of cytokines is completely inhibited by the Intercept system during platelet storage, a fact that may translate into reducing the number of febrile nonhemolytic transfusion reactions (Answer D). Pathogen inactivation for blood products does not involve a possible mechanism by which peptidoglycans that are forming the bacterial wall are denatured (Answer A). Pathogen inactivation is effective against a broad variety of microorganisms, including enveloped and nonenveloped viruses, Gram-positive and Gram-negative bacteria, spirochetes, and protozoa (Answer E).

33. Different methods of pathogen inactivation exist. Recently, amotosalen based pathogen inactivation technology (INTERCEPT Blood System for Platelets, Cerus Corporation, Concord, CA) for platelets and plasma was approved in the US trials with amotosalen based technology for RBC and whole blood are ongoing. What is the major obstacle that had to be overcome in developing amotosalen UVA-light dependent pathogen inactivation for RBC?
A. Anticoagulant-preservative solutions
B. Red blood cell membranes
C. Hemoglobin
D. DNA and RNA cross-linking
E. Alteration of RBC antigen expression

Concept: Emergent pathogens constantly threaten our blood supply. One of the earliest pathogen inactivation technologies included solvent-detergent (SD) for large plasma pools or by the addition of methylene blue (MB) for single plasma products in small-sized blood establishments. However, these technologies were not acceptable for use for platelets or RBCs. New technologies were developed, such as INTERCEPT (Cerus Corporation, Concord, CA, United States of America), Mirasol (Terumo BCT, Lakewood, CO, United States of America), and THERAFLEX UV (Macopharma, Mouvaux, France). These pathogen inactivation technologies (PITs) can be used not only for plasma, but also for platelets. PITs for RBC and whole blood are developed but still undergoing clinical trials and not in routine use.

INTERCEPT technology is based on supplementation with synthetic psoralen (amotosalen), which penetrates cellular and nuclear membranes and reversibly binds to nucleic acids, especially to pyrimidines in DNA and RNA. Upon illumination with UVA light, an interstrand crosslink is formed that blocks transcription, translation, and growth of the pathogen. The hemoglobin present in erythrocytes prohibits the use of this technology for pathogen inactivation in RBC concentrates or whole blood due to the efficient absorption of UVA light by hemoglobin. Therefore, a new system with S-303 compound independent of UVA light was developed. In short, S-303 is a modular compound that enables nucleic acid targeting and crosslinking, thereby preventing nucleic acid replication. The nonreactive byproduct formed after reaction of the S-303 compound with nucleic acids or by decomposition is S-300. S-303 has the potential to react with macromolecules, such as proteins. To minimize these nonspecific reactions with proteins, glutathione (GSH), a naturally occurring antioxidant, is included in the process. Since GSH is distributed extracellular only, GSH binds to extracellular S-303 and neutralizes it.

Riboflavin (vitamin B2) based pathogen inactivation technology (Mirasol, Terumo BCT, Lakewood, CO, United States of America) uses vitamin B2 and UVB light. RBCs are combined with Riboflavin that crosslinks nucleic acids preventing pathogen from growth. RBCs are exposed to UVB light, a wavelength with minimal absorbance by hemoglobin. Riboflavin pathogen inactivated RBCs and whole blood are not currently in routine use and clinical trials are still ongoing.

Answer: C—The main obstacle that had to be overcome with amotosalen UVA light based PIT was absorbance of UVA light by hemoglobin. The new technology using UVA light independent compound S-303 was developed. The other choices (Answers A, B, D, and E) are incorrect based on the earlier discussion.

Please answer Questions 34–36 based on the following clinical scenario:

34. During a local mobile high-school blood drive, a 17-year-old female who is a first-time donor and eligible for donation of whole blood felt light-headed, weak, and nauseous during the donation. Her blood pressure dropped from 120/80 to 90/65 mmHg, and pulse decreased from 75 to 60. What type of reaction does this describe?

A. Citrate toxicity
B. Allergic reaction to antiseptic
C. Presyncope (vasovagal) reaction
D. Bleeding from arterial puncture
E. Seizure

Concept: Between 2% and 5% of all donors experience adverse reactions. They can occur pre-, intra-, and postdonation. Although the majority of reactions are mild, moderate to severe reactions that lead to loss of consciousness, seizure-like activity, and severe injury do occur. The most common systemic donor reaction is presyncope or a vasovagal reaction characterized by dizziness, nausea, vomiting, pallor, weakness, sweating, and hyperventilation. Vital sign changes include hypotension and bradycardia. It can progress to loss of consciousness and seizures. Risk factors include young age, first time donation, low weight, and an inattentive phlebotomist. Interventions include cold compresses around the neck and shoulder area, recumbent position, loosening of tight clothes, and reassurance from collection staff. Vasovagal reactions should be differentiated from hypotension due to arterial puncture and rapid blood loss. Arterial puncture occurs when the needle penetrates through the vein and enters the underlying artery. Hypotension and tachycardia, not bradycardia, accompany this type of reaction. In addition, severe pain and rapid (pulsatile) filling of the bag with bright red blood occur during an arterial puncture. Treatment consists of discontinuing the donation and applying pressure to the site for a long period of time.

Answer: C—The signs and symptoms described in the clinical vignette are consistent with a vasovagal reaction. This is a whole blood donation and thus, the donor is not exposed to citrate (Answer A). There was no description of active bleeding (Answer D). The symptoms are not consistent with allergic reaction or seizure (Answers B and E).

35. Due to donor's signs and symptoms, the whole-blood collection was stopped and only 375 mL of whole blood was collected in a 450 mL collection bag. What is the most appropriate next step?
A. Discard this undercollected unit
B. Nothing special needs to be done, treat it as a regular unit
C. Call medical office and ask for an approval to use it
D. Send a notification to FDA stating that unit was undercollected because of a donor reaction
E. This unit can be used for RBCs only manufacturing and should be labeled as "RBCs Low Volume"

Concept: When 300–404 mL of whole blood is collected into a 450 mL collection bag or when 333–449 mL of whole blood is collected into a 500 mL collection bag, red cells prepared from the resulting unit shall be labeled "Red Blood Cells Low Volume." No other components should be made from a low-volume collection.

Answer: *E*—The unit collected meets a standard for RBCs Low Volume unit and only RBC should be manufactured from this unit. Plasma and platelets should be discarded. Answers A, B, C, and D are incorrect.

36. Unfortunately, an appropriate label "RBCs Low Volume" was not affixed to the final product and RBCs were released into general inventory. One of the hospital blood bank laboratory assistants noticed the low volume of RBCs and called up the blood establishment that distributed the unit. After investigation, it was clear that the final RBC product was missing "RBCs Low Volume" label. The product was not transfused to a patient. What is the most appropriate next step for the hospital blood bank and/or the blood establishment to take?
A. The blood establishment that collected RBC unit needs to file an FDA report regarding this labeling event
B. The hospital needs to file an FDA report since the missing label was discovered there
C. Return RBC to the blood establishment, nothing needs to be done since RBCs weren't transfused
D. A hospital blood bank can add a missing label and issue the unit to a patient as "RBCs Low Volume"
E. Call medical office at the blood establishment and receive an approval for transfusion of this unit as is

Concept: Under 21 CFR 606.171, the collection facility is required to report certain events associated with the manufacturing, to include testing, processing, packing, labeling, or storage, or with the holding or distribution of blood or a blood component, which may affect the safety, purity, or potency of a distributed product. Under 21 CFR 606.171(c), the facility should submit reports as soon as possible (but not to exceed 45 calendar days from the date of discovery of information). Under 21 CFR 606.171(a), the manufacturer that had control over the product when the deviation from current good manufacturing practice, applicable regulations, applicable standards, or established specifications or an unexpected or unforeseeable event that my affect the safety, purity, or potency occurred must submit a report. According to 21 CFR 606.3 (1) "control" is defined as having responsibility for maintaining the continued safety, purity, and potency of the product and for compliance with applicable product and establishment standards, and for compliance with current good manufacturing practice (CGMP). Labeling events include those that occur during the labeling process, which you didn't discover until after you have distributed the product. Labeling events include incorrect, missing, or misleading information on any labeling pertaining to the product. Under 21 CFR 606.171(b), the blood center must submit a report when there is an event (a deviation or unexpected or unforeseeable event) during labeling that may affect the safety, purity, or potency of a product it distributed.

Answer: *A*—The blood establishment that had control of the unit and distributed it without an appropriate label needs to report to FDA about this labeling event as soon as possible and not to exceed 45 calendar days from the date of discovery of this event. The report can be submitted electronically or in paper form by mail. The other choices (Answers B, C, D, and E) are incorrect based on the earlier discussion.

End of Case

Suggested Reading

[1] AABB, American Red Cross, America's Blood Centers, Armed Services Blood Program. Circular of information for the use of human blood and blood components. AABB, Bethesda, MD, 2013.

[2] G. Akyol, et al. A rare but severe complication of filgrastim in a healthy donor: splenic rupture, Transfus. Apher. Sci. 50 (2013) 53–55.

[3] A. Bartlett, R. Williams, M. Hilton. Splenic rupture in infectious mononucleosis: a systematic review of published case reports. Injury 47 (3) (2016) 531–538.

[4] Code of Federal Regulations. Title 21, CFR Part 606.3 and 606.171. Government Printing Office, Washington, DC, 2013 (revised annually).

[5] M.K. Fung, B.J. Grossman, C.D. Hillyer, C.M. Westoff (Eds.), Technical Manual, eighteenth ed., AABB Press, Bethesda, MD, 2014.

[6] J.R. Hess, An update on solutions for red cell storage, Vox Sang. 91 (1) (2006) 13–19.

[7] J.R. Hess, T.G. Greenwalt, Storage of red blood cells: new approaches, Transfus. Med. Rev. 16 (4) (2002) 283–295.

[8] Hetastarch [package insert]. B. Braun Medical Inc., Bethlehem, PA, 2013.

[9] http://www.businesswire.com/news/home/20150309005120/en/Octapharma-USA-FDA-Approves-Octaplas%E2%84%A2-License-Supplement.

[10] INTERCEPT blood system for platelets and INTERCEPT blood system for plasma [package insert]. Cerus Corporation, Concord, CA, USA.

[11] InterSol [package insert]. Fenwal, Inc., Lake Zurich, IL, 2014.

[12] P.W. Ooley (Ed.), Standards for Blood Banks and Transfusion Services, thirtieth ed., AABB, Bethesda, MD, 2016.

[13] S. Marschner, R. Goodrich, Pathogen reduction technology treatment of platelets, plasma, and whole blood using Riboflavin and UV light, Transfus. Med. Hemother. 38 (1) (2011) 8–18.

[14] Neupogen [package insert]. Amgen Inc., Thousand Oaks, CA, 2015.

[15] Octaplas [package insert]. Octapharma, Inc., Vienna, Austria.

[16] T.H. Price, et al. Efficacy of transfusion with granulocytes from G-CSF/dexamethasone-treated donors in neutropenic patients with infection, Blood 126 (18) (2015) 2153–2161.

[17] M.E. Reid, C. Lomas-Francis, M.L. Olsson (Eds.), The Blood Group Antigen FactsBook, third ed., Academic Press, London, UK, 2012.

[18] A.N. Schuetz, et al. Leukoreduction filtration of blood with sickle cell trait, Transfus. Med. Rev. 18 (3) (2004) 168–176.

[19] K.L. Scott, et al. Biopreservation of red blood cells: past, present, and future, Transfus. Med. Rev. 19 (2) (2005) 127–142.

[20] T.L. Simon, E.L. Snyder, B.G. Solheim, C.P. Stowell, R.G. Strauss, M. Petrides (Eds.), Rossi's principles of transfusion medicine, fourth ed., Wiley-Blackwell, West Sussex UK, 2009.

[21] A.B. Zimrin, J.R. Hess, Current issues related to the transfusion of stored red blood cells, Vox Sang. 96 (2) (2009) 93–103.

C H A P T E R

6

Blood Group Antigens and Antibodies

Susan T. Johnson, Michelle R. Brown**, Jayanna Kay Slayten*[†]

*Blood Center of Wisconsin, Marquette University, University of Wisconsin-Milwaukee, Milwaukee, WI, United States; **University of Alabama at Birmingham, Birmingham, AL, United States; [†]Indiana Blood Center SBB Program, Indianapolis, IN, United States

Blood groups can be divided into those consisting of carbohydrate antigens and those that are protein in nature. Carbohydrate blood group systems include ABO, H, Lewis, P1Pk, and Globoside. Protein blood group systems may be single-pass proteins (e.g., MNS and Kell), multipass proteins (e.g., Rh, Duffy, Kidd, and Diego), and glycosylphosphatidylinositol (GPI)-linked proteins (e.g., Cromer, Cartwright, and Dombrock), integral proteins (e.g., Lutheran), or water channel proteins/aquaporins (e.g., Colton and GIL).

The heart of immunohematology involves the detection of antibodies formed in response to foreign antigen exposure. As a general rule, those who lack carbohydrate antigens may produce "naturally" occurring or non-red cell stimulated IgM antibodies without exposure to foreign RBC antigens. In contrast, those individuals who lack protein antigens may produce IgG alloantibodies when exposed to foreign RBC antigens, typically via transfusion or pregnancy. Of course, the immune system "does not read the books" and there are examples of IgG antibodies to carbohydrate antigens and IgM antibodies to protein antigens, albeit unusual. This chapter provides an overview of blood groups, their unique characteristics, and the problems that arise during testing.

1. Review the antibody detection test (screen) summary below. What class of antibody is most likely causing the reactivity in the serum?

	IS	37°C	IAT
I	0	0	1+
II	0	0	2+
III	0	0	0√

A. IgM
B. IgG
C. Both IgG and IgM
D. There are no antibodies present in the patient's serum
E. This antibody screen is invalid because check cells were not run on cells I and II

Concept: Antibody classes (e.g., IgM and IgG) can be suspected based on the phase in which they demonstrate reactivity. During testing, samples are spun and checked for hemolysis and/or agglutination immediately after mixing (i.e., "Immediate Spin" or "IS" phase); after a 37°C incubation ("37°C" phase); and after mixing the sample with antihuman globulin (AHG or IAT

Transfusion Medicine, Apheresis, and Hemostasis. http://dx.doi.org/10.1016/B978-0-12-803999-1.00006-7

phase) to aid IgG antibody binding. If a cell is negative in all phases, check cells (containing IgG) are added and agglutination should be observed; thus, ensuring that AHG was indeed added to the sample. Positive cells do not require this step. Check cell completion is denoted by a check in the last box of the screen, as seen earlier.

Answer: *B*—The pattern in this question reflects the characteristics of IgG and the positivity reflects that antibodies are indeed present (Answer D). IgG antibodies react most often in the indirect antiglobulin test (antihuman globulin phase) after incubation at 37°C and the addition of enhancement media such as low ionic strength solution (LISS) or polyethylene glycol (PEG). IgG is a monomer and requires the addition of AHG to bring multiple RBCs together in order for agglutination to be observed. In contrast, IgM (Answers A and C), which is a pentamer, does not usually require the addition of anti-IgG for agglutination to be observed. Therefore, IgM typically reacts in the immediate spin (IS) phase. However, it can also react at 37°C phase and IAT/AHG phase, if the antibody has reactivity at body temperature. As mentioned earlier, positive results do not require check cells, so this screen is valid (Answer E).

2. Which immunoglobulin class is most likely to cause *intravascular* hemolysis?
   A. IgA
   B. IgG
   C. IgM
   D. IgD
   E. IgE

Concept: Intravascular hemolysis (IgM related) takes place when complement is activated and causes direct lysis of the cells within the vascular system. This process is brisk and can be life-threatening if enough hemolysis takes place in a short period of time. In contrast, extravascular hemolysis (IgG related) takes place when an antibody binds to the cell and "marks" it for destruction by the macrophages in the spleen. This is a longer process and toxic substances, such as free hemoglobin, are unlikely to be massively released as in intravascular hemolysis, and thus, in general, it is considered to be less severe as compared to intravascular hemolysis.

Answer: *C*—Complement activation, leading to intravascular hemolysis, requires two Fc regions of an immunoglobulin to be in close proximity to each other. A single IgM molecule provides adjacent Fc regions since they are pentameric. IgA, IgD, or IgE (Answers A, D, and E) are not known to cause hemolysis. Patients severely deficient in IgA, who have also formed anti-IgA, can have severe anaphylactic reactions, if exposed to IgA via blood products (see Chapter 12 for an in-depth discussion of this reaction). IgG (Answer B) is less efficient at binding complement because at least two IgG molecules are required to be in close proximity to activate the pathway. In many cases, hundreds of IgG molecules are needed to bind to a RBC in order to have two in close proximity to activate the complement cascade. IgG antibodies are more likely to lead to extravascular hemolysis, as discussed in the concept above.

3. A pregnant mother has a mixture of both IgG and IgM antibodies in her serum. Which of the following treatment pairs would be useful to distinguish between the two antibodies?
   A. Dithiothreitol and Choloroquine
   B. β-2-mercaptoethanol and Papain
   C. Chloroquine and β-2-mercaptoethanol
   D. Dithiothreitol and β-2-mercaptoethanol
   E. β-2-mercaptoethanol and EDTA-Glycine

Concept: During pregnancy, it is very important to identify the class and specificity of any alloantibodies present. For example, IgG antibodies can cross the placenta and cause hemolytic disease of the fetus and newborn, if the fetuses' RBCs have the corresponding antigen present (e.g., anti-D from mother can bind and hemolyze fetal Rh positive RBCs). Whereas IgM antibodies are typically considered too large to cross the placenta and therefore, are usually not clinically significant. Specialized techniques and reagents are available to help distinguish the classes of antibodies.

Answer: *D*—Sulfhydryl reducing reagents can dissociate pentameric (i.e., IgM) immunoglobulins by cleaving the bonds that connect the monomeric subunits and the J chain. Therefore, IgM antibodies are destroyed through this process. Dithiothreitol and β-2-mercaptoethanol are both sulfhydryl reducing reagents that destroy IgM antibodies, leaving IgG antibodies intact and reactive. Papain (Answer B) is proteolytic enzyme useful in denaturing red cell antigens, such as Fy^a, Fy^b, M, N, and variably S and s. Chloroquine and EDTA glycine-acid (Answers A, C, and E) remove IgG antibodies from red cells which have a positive DAT so that phenotyping can be performed.

4. Paroxysmal cold hemoglobinuria (PCH) is an uncommon type of hemolytic anemia that most often occurs in children following a viral infection. Which of the following is associated with this condition?
A. Usually caused by autoanti-P
B. Hemolysis only occurs at 30°C
C. Is a monophasic hemolysin
D. Is caused by a defect in phosphatidylinositol glycan A (PIGA)
E. Is associated with the absence of CD55 and CD59 on RBCs

Concept: Hemolysis in PCH occurs when an autoantibody, most often autoanti-P, binds to red blood cells at lower temperatures, fixes complement, and lyses the red blood cells at 37°C. This two-phase mechanism gives this antibody the name "biphasic hemolysin". It requires exposure to both a lower temperature, during which the antibody binds, followed by a higher temperature to cause hemolysis. The Donath-Landsteiner test is used to confirm the presence of this antibody. This test requires a sample collected and maintained at 37°C, followed by a biphasic incubation of patient serum and reagent red cells with the P antigen—first in an ice bath around 0–4°C, then at 37°C. A 0–4°C control and a 37°C control must be performed in parallel with the biphasic incubation. This test is the reason the antibody is often called the Donath-Landsteiner (D-L) antibody.

Answer: *A*—Autoanti-P, Donath-Landsteiner antibody, and biphasic hemolysin are all names for the antibody associated with PCH. As mentioned above, the antibody binds at 30°C (Answer B), but hemolysis occurs when the temperature is 37°C, thus is it not a monophasic hemolysin (Answer C). PIGA defects (Answer D) and absence of CD55 /59 (Answer E) are associated with another hemolytic condition, termed paroxysmal nocturnal hemoglobinuria (PNH). Essentially, in PNH, the lack of cell surface makers that normally protect the cell from destruction by complement are absent, leading to complement binding and eventual hemolysis. PNH patients usually have a negative direct antiglobulin test (DAT) while patients with PCH typically have a positive DAT with a negative elution.

5. Which of the following is *not* an example of a passively acquired antibody?
A. Antibody in a fetus that crossed the placenta from the mother
B. Anti-E in a person who received two units of RBCs three months ago
C. Antibody from an intravenous immunoglobulin infusion
D. Low titer anti-B in a patient with blood type B who has received several units of type O platelets
E. Anti-D present in a women who received an antenatal injection of Rh immune globulin (RhIG)

Concept: Passive immunity is when antibodies are transferred to a person from another person, animal, and/or medications. Active immunity is when antibodies are formed in response to a foreign antigen, such as in response to a transfusion.

Answer: *B*—Anti-E is an alloantibody that can be actively produced when a person that is negative for the E antigen receives E antigen positive blood via transfusion or pregnancy. Another example of active immunity is the presence of anti-A in a person with type B blood. This anti-A is considered a "naturally occurring" antibody because it is produced without exposure to red cells and is likely an immune response to bacteria and other carbohydrate containing substances in the environment during early development.

In contrast to active immunity, examples of passive transfer of antibodies includes maternal IgG antibody that crosses the placenta into fetal circulation (Answer A), infusion of intravenous immunoglobulin (Answer C), administration of RhIG (Answer E), and transfusion of nontype specific blood components are transfused (e.g., type O plasma has the ability to transfer both anti-A and anti-B to the recipient (Answer D).

6. An "antiglobulin crossmatch" performed in a test tube includes both immediate spin and 37°C/IAT(AHG) phases. An order for three units of blood for a surgery patient was sent. The antibody screen (detection test) is negative, but the patient has a history of anti-C. Three C-antigen negative units from the inventory were chosen for crossmatching. However, the immediate spin phase demonstrates agglutination of all 3 units. What is the most likely cause of this result?
   A. Donor units are not ABO compatible
   B. Patient has an unexpected IgG antibody
   C. Patient has an antibody to a low frequency IgM antibody
   D. Donor units have a positive direct antiglobulin test (DAT)
   E. Antihuman globulin was neutralized

   Concept: As mentioned in Question 1, each phase of tube testing has a specific purpose. Immediate spin positivity generally signifies IgM antibodies, while positivity at 37°C and IAT/AHG phases signify IgG antibodies.

   Answer: *A*—The donor units are likely incompatible due to selecting the wrong ABO blood type from the refrigerator. For example, group A red blood cells could have mistakenly been crossmatched to a group O recipient. The immediate spin portion of an antiglobulin crossmatch is used to determine ABO compatibility. Since most ABO antibodies are IgM, any reactivity at IS phase (done at room temperature) signifies that IgM is reacting and warrants further investigation. IgG (Answer B) antibodies will most likely be detected at 37°C/IAT. Donor units with a positive DAT can cause a positive reaction, but this will likely occur at 37°C/IAT and only in few units (Answer D). The remaining choices (Answers E and C) are incorrect because AHG has not been added to the test yet and panreactivity is not consistent with an antibody to a low frequency antigen.

7. A primary immune response is characterized by initial production of which class of immunoglobulin?
   A. IgG
   B. IgM
   C. IgA
   D. IgD
   E. IgE

   Concept: A primary immune response is characterized by initial production of IgM by the B cell. The B cell then switches heavy chain production from μ (IgM) to γ (IgG). When this switch occurs; the IgM titer decreases as the IgG titer increases.

   Answer: *B*—IgM is the initial class of immunoglobulin produced in a primary immune response. The other choices (Answers A, C, D, and E) are incorrect based on the description of IgM property above.

8. Antibody titers can decrease over time. This can result in antibody levels that are undetectable during antibody detection (screen). However, if red blood cells with the corresponding antigen are

transfused, the antibody titer can increase very quickly. What is a term used for this type of antibody response?

A. Anamnestic response
B. Primary immune response
C. Tertiary immune response
D. Rapid induction response
E. Hyperacute response

Concept: When an antigen stimulates a B cell, the B cell either matures into an antibody-producing plasma cell that lives for a few weeks or a memory B cell that survives for years. Though antibody levels may diminish or disappear (i.e., evanesce), the memory cells are prepared to rapidly produce antibody upon subsequent exposure to the antigen. This is called a secondary immune response. During this secondary immune response, memory cells produce antibody and titers rise quickly. Another name for this response is anamnestic since it occurs without forgetting or without amnesia.

Answer: *A*—The terms secondary immune response and anamnestic response are synonymous in this case. The other choices (Answers B, C, D, and E) are incorrect based on the description above.

9. Which of the following techniques is useful if a Lewis antibody (e.g., anti-Le^a or anti-Le^b) is suspected of complicating an antibody identification?

A. Ficin treatment to destroy the Lewis antigen
B. Elution to remove antibody from the patient cells
C. Autoadsoprtion because Lewis antibodies can mimic autoantibodies
D. Direct antiglobulin test to determine if patient cells are coated with antibody
E. Neutralization since Lewis antigens are soluble

Concept: Sometimes during immunohematologic workups, it is necessary to work around nuisance reactivity, in order to detect clinically significant activity. Anti-Le^a and anti-Le^b are examples of this necessity, since they do not typically cause hemolysis, but rather can mask the reactivity of a more clinically significant antibody. Lewis antigens are not intrinsic to the red blood cell membrane, rather, they are glycosphingolipids found in plasma that adsorb onto red blood cells. Lewis substance (soluble antigen) can be added to the serum to neutralize Lewis antibodies. When performing neutralization, it is important to determine whether loss of antibody reactivity is due to neutralization rather than dilution. Therefore, a control, consisting of patient serum with saline, must be run in parallel with the neutralization. Table 6.1 provides additional soluble antigens that can be used to neutralize antibodies.

TABLE 6.1 Source of Blood Group Antigen Substance

Antigen	Source of substance
Le^a/Le^b	Human saliva Also commercially prepared
P1	Pigeon eggs or hydatid cyst fluid Also commercially prepared
Sd^a	Pooled human urine
Chido and Rogers	Human plasma from Ch+, Rg+ individuals

Answer: *E*—If Lewis antibodies are reacting at 37°C and complicating the identification of clinically significant antibodies, the Lewis antibodies can be neutralized with Lewis substance.

Ficin (Answer A) does not destroy Le^a or Le^b antigens, but rather enhances antibody reactivity along with Rh, Kidd, I, ABO, and P antigens. Ficin has no effect on the Kell group and destroys Fy^a/Fy^b and MNS. Direct antiglobulin test (DAT) (Answer D) determines if antibody is coating the cells in vivo and would not aid this antibody identification.

Elution, typically performed when the DAT is positive (Answer B) removes and identifies the specificity (if there is one) of the antibody attached to red cells. Autoadsorption (Answer C) removes autoantibodies from serum, but Lewis antibodies do not mimic autoantibodies.

10. Which of the following antigens is most likely to stimulate antibody production?
A. D
B. K
C. Fy^a
D. S
E. M

Concept: The ability of an antigen to stimulate an immune response is called immunogenicity and blood group antigens vary in their tendency to stimulate antibody production.

Answer: *A*—The most immunogenic antigen is D. Alloimmunization rates are highest for D, E, K, and c. Alloimmunization rates are affected by both immunogenicity and antigen exposure. The percentage of people who lack the D antigen who are likely to form an antibody after being transfused with one unit of D positive red blood cells is approximately 50%. The next most immunogenic antigen is K (Answer B), in which 5% of K-negative people who are transfused with one unit of K-positive red blood cells will form the antibody. The percentages for c, E, and k are 2.05%, 1.69%, and 1.50%, respectively. The other choices (Answers C, D, and E) represent antigens with extremely low immunogenicity.

11. Per regulatory and accreditation standards, methods to detect clinically significant antibodies must be used prior to transfusion of red blood cells. Acceptable methods to detect these antibodies include which of the following?
A. Test tube method reading at immediate spin only with a follow-up DAT
B. Electronic crossmatch only if no transfusion in the past month
C. Test tube method with low ionic strength saline (LISS)
D. Electronic crossmatch only, if last positive screen was 3 years ago
E. DAT only

Concept: Upon exposure to foreign antigens, a person's immune system may respond by producing antibodies. On subsequent transfusion, these antibodies in the patient's serum can bind with the transfused red blood cells and cause hemolysis. Therefore, it is imperative to detect these antibodies prior to transfusion.

Answer: *C*—The test tube method using a LISS additive, incubation at 37°C, followed by an indirect antiglobulin test will detect IgG antibodies. Regulatory agencies require testing prior to transfusion that includes methods to detect clinically significant antibodies. Various technology platforms have been introduced that comply with these regulations and, in some cases, have increased sensitivity and specificity over tube testing. Examples of the latter include column agglutination (also called gel testing) and solid phase red cell adherence or protein A coated wells. Tube test at IS only (Answer A) is insufficient to meet regulatory requirements. Electronic crossmatches cannot be performed without a test to detect unexpected antibody. (Answers B and C). The DAT (Answer E) will detect antibody coating the patients RBCs, but will not detect antibody present in an individual's plasma.

12. Which of the following ABO blood types possess the most amount of H antigen?
A. A_1
B. B
C. A_1B
D. O
E. A_2B

Concept: The most important blood group system in transfusion is ABO. It was discovered in 1901 by Karl Landsteiner. There are four major blood types within the ABO blood group system, A, B, AB and O. A and/or B antigens are produced as result of inheriting an *A* or *B* gene that codes for the presence of a transferase (enzyme) with the ability to attach a specific carbohydrate to another carbohydrate.

H is the precursor to A and B antigen. The H antigen results from inheritance of the *H* gene which codes for an α-2-L-fucosyltransferase. This enzyme attaches a fucose to a type II chain. Type II chains are found on red blood cells.

Type II Chain **Gal ß1-4GlcNAcß1-3Galß1-4GlcNAcß1-3 – R**
R=additional carbohydrate sequence

The H Type II chain must be present in order for A and/or B transferase to attach the appropriate immuodominant sugar, α-3-*N*-acetyl-galactosamine (GALNac) for A or α-3-D-galactose for B to the Type II H chain. As group O individuals inherit a nonfunctional gene, their red cells possess only H.

Answer: *D*—Blood type O possesses the greatest amount of H antigen as none is converted to A and/or B antigen. The B transferase is not quite as effective at converting H to B as A1 is in converting H to A. In A_1B heterozygotes, there is competition for H between the A1 and B transferase. Nearly all H is converted to A1 and B. H-deficient individuals, also known as Bombay type, inherit *hh* resulting in no fucosyl-transferase (H transferase) to convert type II chains to H antigen.

Essentially, the amount of H antigen on the different blood types can be remembered as follows:

GREATEST amount of H **O>A2>B>A2B>A1>A1B** LEAST amount of H

The other choices (Answers A, B, C, and E) are incorrect based on the order above.

13. Evaluate the following ABO/Rh typing results.

Forward:			Reverse:	
Anti-A	Anti-B	Anti-D	A1 Cells	B Cells
4+mf	0	4+	0	4+

mf, Mixed field

What should be your first step in evaluating the mixed field (mf) result noted in the forward typing?

A. Check patient's clinical history
B. Have the medical laboratory scientist repeat the ABO/Rh type
C. Do nothing, the forward and reverse type match
D. Investigate this typing as a subgroup of A
E. Repeat the forward type with a new lot of reagent anti-A

Concept: Mixed field agglutination indicates the presence of two cell populations (Fig. 6.1). Anti-A, Anti-B, and Anti-D will agglutinate RBCs that possess corresponding antigen and not agglutinate RBCs that do not have antigen. This characteristic agglutination pattern may be observed by the medical laboratory scientist performing the testing.

Answer: *A*—Since the forward and reverse types do not perfectly match (Answer C), the patient's history should be checked for possible explanations including recent transfusion and/or hematopoietic progenitor cell (HPC) transplant. These types of occurrences should be recorded in the patient's record, especially if they are being treated at the same institution. The most common cause of two cell populations is transfusion of group O RBCs to an A, B, or AB individual. A_3 individuals may show mixed field reactivity due to weaker than normal expression of A antigen present on the RBCs.

Mixed field may also indicate the patient has had a HPC transplant and has not yet fully engrafted RBCs. In the scenario mentioned, the patient may have been group O pretransplant and is converting to group A. These types of treatments often occur in large, tertiary care, or academic medical centers. If this patient is now being seen in a community hospital this information may not be in the

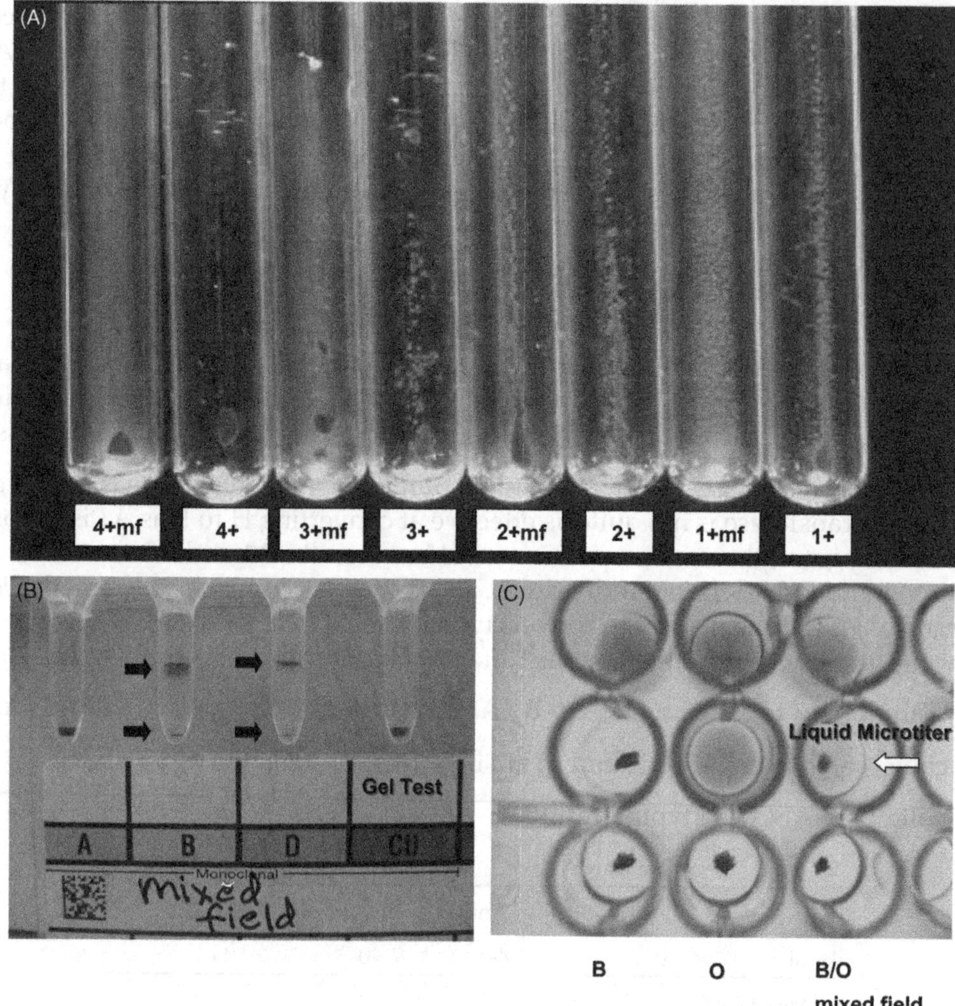

FIGURE 6.1 (A) Mixed field agglutination in test tube methods. (B) Mixed field agglutination in gel testing—Anti-B and Anti-D—agglutination is observed at top of the column while unagglutinated RBCs are in the bottom of the column. (C) Mixed field agglutination in liquid microplate testing—results in third column show agglutinated RBCs in a background of unagglutinated RBCs with anti-B. *mf*, Mixed field. *Source: Images courtesy of MR Combs.*

electronic medical records at the transfusing facility and a phone call may be required to determine the true cause of the mixed field reaction. This may lead to uncovering additional special transfusion needs, such as irradiated or CMV "safe" blood.

Finally, mixed field may indicate the individual is a true chimera. True chimerism can be due to either twin chimerism or tetragametic (dispermic) chimerism. Fraternal twins may have an exchange of hematopoietic stem cells by the formation of a vascular bridge. If these cells engraft in the bone marrow two different cell populations will be produced. Twin chimerism is limited to blood cells. Tetragametic chimerism is caused when two eggs are fertilized by sperm and a fusion of the two occurs causing a single individual with two cell lineages. This occurs in all cell types as opposed to only the blood cells of twin chimerism. A true chimera is very rare and difficult to prove.

When unusual findings are observed, repeating the ABO/Rh type (Answer B) and/or requesting a repeat sample is always good practice to ensure accurate results. *AABB Standards for Blood Banks and Transfusion Services, 30th ed., 2016,* 5.16.2.2 requires *two* determinations of the recipient's ABO group. This can include testing a second current sample, or comparing with previous records, or retesting the same sample if patient identification was verified using an electronic identification system or another process validated to reduce the risk of misidentification. In this case, there is no need to perform a third type, a check of the patient's history is most important.

If there is nothing unusual in the patient's history, mixed field in a group A individual would most likely be due to a subgroup, but checking the patient's history would be the first step (Answer D). Repeating the ABO type (Answer E) with a new lot number of anti-A would provide little additional useful information, as mixed field would still likely be observed.

14. Evaluate the following ABO/Rh type obtained by tube testing.

Forward:			**Reverse:**	
Anti-A	Anti-B	Anti-D	A1 Cells	B Cells
4+	0	4+	0	4+

Your transfusion service records indicate the patient's blood type is B Rh positive.

Which of the following might be most useful for solving this discrepancy with the historic type?
A. Perform an autocontrol
B. Perform a DAT
C. Repeat ABO, Rh type on a new sample drawn from patient
D. Have three different technologists test the sample
E. Repeat the ABO/Rh type using gel

Concept: Many laboratories are instituting a second sample draw when there is no ABO, Rh type on record in a facility. This aids in detection of the wrong person being drawn and detecting wrong blood in tube (WBIT).

Answer: *C*—A new sample was drawn and the ABO, Rh type was repeated. The ABO type, group A was confirmed. No mixed field was observed. A check of the patient's clinical history revealed she had received a HPC transplant 1 year ago at the university medical center. She was a B Rh positive pre-transplant and her donor was A Rh positive. She has completely converted to her new type. Performing a DAT or autocontrol (Answers A and B) would not address the discrepancy in this case, nor would having multiple technologist repeat the testing (Answer D) or using a different methodology (Answer E).

15. Evaluate the following ABO/Rh typing results.

Forward:			**Reverse:**	
Anti-A	Anti-B	Anti-D	A1 Cells	B Cells
4+	0	4+	1+	4+

The patient has a negative antibody detection test (screen). What testing should be performed next?
A. Direct antiglobulin test
B. Type patient red cells with anti-A1 lectin
C. Repeat testing with a new lot number of anti-A
D. Investigate as a possible acquired B phenotype
E. No further testing is necessary, report the type as A positive

Concept: The forward and reverse typing needs to match prior to transfusion of type-specific red blood cells. In this case, the patient forward type is interpreted as A and reverse type is O. This is a discrepant ABO typing result and further investigation is indicated. When evaluating ABO discrepancies, often, but not always, weaker reactions should be investigated first.

Answer: *B*—A weak reaction on the reverse with A_1 cells paired with a 4+ reaction with anti-A indicate that this could be an A subgroup. A good first step in the serological investigation of A subgroups is to use anti-A1 lectin. Anti-A1 lectin is an extract of the *Dolichos biflorus* plant. Anti-A1 lectin is only positive with individuals possessing the *A1* gene resulting in A1 antigen on their RBCs.

If a patient's cells do not react with anti-A1 lectin, they are a subgroup other than A_1. Approximately 80% of type A people are A_1 and 20% are A_2. The remaining subgroups are rarely seen. Subgroup A_3 shows characteristic mixed field agglutination which is not present in our patient. Investigating possible subgroups of A always require additional investigation (Answer E). Table 6.2 lists typical serological reactions of some A subgroups.

TABLE 6.2 Characteristic Reactions of Some A Subgroups

	Percent of subgroups	Anti-A	Anti-B	Anti-A,B	Anti-A1 lectin	Anti-H
A_1	80	4+	0	4+	3+	0
A_2	20	4+	0	4+	0	2+
A_3	<1	2+ mf	0	2+ mf	0	3+
A_x	rare	0	0	1+	0	3+

A DAT (Answer A) would not address the discrepancy in this case, nor would repeating testing with a different lot number of antisera (Answer C).

Acquired B phenotype (Answer D) is identified by a type A patient having an additional reaction with anti-B antisera. In the acquired B phenotype, the terminal sugar, *N*-acetylgalactosamine, is deacetylated, converting the sugar to galactosamine. This is similar enough to the terminal sugar of the B antigen, galactose, to react with some anti-B. The patient in this example does not have any reactivity with anti-B and does not fit the serological profile for a patient with an acquired B phenotype.

16. What cold-reactive autoantibody is often seen secondary to *Mycoplasma pneumonia*?

A. Autoanti-IH
B. Autoanti-I
C. Autoanti-i
D. Autoanti-P
E. Autoanti-M

Concept: Cold-reactive autoantibodies are IgM and fairly common. In fact, many healthy individuals have cold autoantibodies reactive at room temperature (22–25°C) or below. These antibodies are usually not clinically significant, since they do not bind to RBCs at body temperature, (generally considered >30°C). Occasionally, a cold-reactive autoantibody will be produced secondary to infection, bacterial or viral, that is capable of binding to RBCs in vivo. When this occurs the patient also develops a positive direct antiglobulin test (DAT) with complement (C3) binding.

Answer: *B*—Autoanti-I is most often found secondary to *Mycoplasma pneumonia*. Autoanti-i (Answer C) is associated with infectious mononucleosis. Autoanti-P (Answer D) is most often associated with paroxysmal cold hemoglobinuria (PCH). Autoanti-IH (Answer A) is seen in healthy individuals or in hospitalized patients when it becomes reactive at room temperature and causing a positive antibody detection test (screen). It is most often found in group A individuals and occasionally those who are group B and or AB. Autoanti-IH does not cause autoimmune hemolytic anemia. Finally, while autoanti-M (Answer E) has been reported, it is very rare and not known to cause immune hemolytic anemia.

17. Which of the following selected red cells is most often used to differentiate an autoanti-I from an autoanti-i?

A. Screening cells
B. Cord cells
C. PP1Pk-negative cells
D. Patient cells
E. I-negative cells

Concept: I and i antigens reside on the same type II chains that serve as the basis for H, A, and/or B antigens. i antigen consists of linear (straight) type II chains with at least one repeating Gal β 1-4GlcNAc. Shortly after birth most individuals inherit the *I* gene (*GCNT2*), which results in production of β 1-6 *N*-acetyl-glucosaminyltransferase that converts i antigen (straight type II chains) to I antigen (branched type II chains). I-negative adults, also known as adult i, are very rare with a frequency reported to be 0.01%–0.003%.

Answer: *B*—Cord cells have i antigen and are for the most part considered I-negative because the branching enzyme (β 1-6 *N*-acetyl-glucosaminyltransferase) is just becoming active at birth. Cord cells are the key to differentiating autoanti-I from autoanti-i. Autoani-I will be non-reactive when tested with cord cells because they lack I antigen. Autoanti-i would be strongly positive with cord cells. While I-negative (Answer E) cells would be helpful, they are very rare and would not be available for routine use. Patient cells and screening cells (Answers A and D) will be I positive and unhelpful in this case (Table 6.3). PP1Pk-negative cells (Answer C) would only be used if a rare antibody was suspected to an antigen(s) in the P1Pk or Globoside blood group system. They would not be useful to differentiate Autoanti-I from Autoanti-i (Table 6.3).

TABLE 6.3 Serologic Reactivity of Autoanti-I, −i, and −IH

	A1	A2	B	O	O cord cells
Autoanti-I	4+	4+	4+	4+	0−2+
Autoanti-i	0−1+	0−1+	0−1+	0−1+	3+
Autoanti-IH					
Found in A1, B, and AB individuals	0	1−2+	1−2+	3+	0-2+

18. Of the following antibodies, which is most commonly identified?

A. Anti-p
B. Anti-P
C. Anti-Pk
D. Anti-PP1Pk
E. Anti-P1

Concept: Over the years there has been much confusion over the P1Pk and Globoside blood group systems. P1 and Pk antigens are in the P1Pk system and P is in the Globoside system. The most commonly seen antibody is anti-P1, since approximately 20% of Caucasians lack the P1 antigen. The other antibodies to antigens in these two systems are to antigens of high prevalence (Pk, P, and PP1Pk) and are therefore rarely seen.

Briefly, *P1PK* gene codes for an α-1,4-galactosyltransferase (A4GALT) which adds a galactose to the same type II chain i, I, H, A, and B are built on to create the P1 antigen and to add a second carbohydrate chain (lactosylceramide) to create the Pk antigen. A majority of individuals also inherit a ß-1,3-*N*-acetylgalactosamyl transferase 1 (B3GALNACT1) that adds a galactose to the Pk antigen to make P antigen.

Answer: *E*—Anti-P1 is most often detected in routine pretransfusion testing given the frequency of being P1 negative is the most common of this group.

Anti-P1 is an IgM antibody and similar to antibodies to other carbohydrate antigens, it can be formed with no prior exposure to red blood cells through transfusion or pregnancy. It is clinically insignificant, since it is an IgM antibody reactive at colder temperatures, but nonreactive at body temperature. Anti-P1 can be challenging to identify because antigen expression varies between individuals. Antigen expression depends on whether the panel donor has a one or two *P1PK* genes.

Manufacturers do not indicate on the antigen profile whether there is strong or weak expression of the antigen so results of testing can be confusing on first review.

The antibody panel below (Table 6.4) demonstrates the classic pattern of reactivity seen with anti-P1. Panel cell 1 is P1 positive, but is non-reactive. Panel cell 3 is P1 positive and strongly reactive (3+) at immediate spin and cells 4 and 5 are only weakly reactive (1+). In order to enhance the reactivity of the antibody and to increase confidence in identification, the panel can be incubated at room temperature and 18°C.

After incubation, all P1+ red cells are positive and P1− red cells are negative, consistent with identification of anti-P1 (Table 6.5).

Occasionally the manufacturer will indicate on the antigen profile unusual P1 antigen expression, as indicated by w for weak and s for strong (Table 6.6).

TABLE 6.4 Antibody Identification Panel Showing Classic Anti-P1 Reactivity

	Rh						MNS				Lu		P	Lewis		Kell		Duffy		Kidd		Saline		
	D	C	E	c	E	f	M	N	S	S	Lu^a	Lu^b	P1	Le^a	Le^b	K	K	Fy^a	Fy^b	Jk^a	Jk^b	IS	37	IAT
1	+	+	0	0	+	0	+	+	+	+	0	+	+	+	0	0	+	0	+	0	+	0	0	0
2	0	0	0	+	+	+	+	0	+	0	0	+	0	0	+	+	0	+	+	+	0	0	0	0
3	0	0	+	+	0	0	0	+	0	+	0	+	+	0	+	+	+	+	0	0	+	3+	3+	2+
4	+	0	+	+	0	0	+	+	+	+	0	+	+	0	+	0	+	0	+	+	+	1+	0	0
5	+	0	0	+	+	+	+	+	0	+	0	+	0	0	0	0	+	0	0	+	0	0	0	0
6	0	0	0	+	+	+	+	0	+	+	0	+	+	+	0	0	+	+	+	0	+	1+	0	0

TABLE 6.5 Antibody Identification Panel Incubated at Room Temperature and 18°C, Demonstrating Anti-P1 Pattern

	Rh						MNS				Lu		P	Lewis		Kell		Duffy		Kidd		Saline		
	D	C	E	c	E	f	M	N	S	s	Lu^a	Lu^b	P1	Le^a	Le^b	K	K	Fy^a	Fy^b	Jk^a	Jk^b	IS	RT	18C
1	+	+	0	0	+	0	+	+	+	+	0	+	+	+	0	0	+	0	+	0	+	0	1+	2+
2	0	0	0	+	+	+	+	0	+	0	0	+	0	0	+	+	0	+	+	+	0	0	0	0
3	0	0	+	+	0	0	0	+	0	+	0	+	+	0	+	+	+	+	0	0	+	3+	4+	4+
4	+	0	+	+	0	0	+	+	+	+	0	+	+	0	+	0	+	0	+	+	+	1+	2+	3+
5	+	0	0	+	+	+	+	+	0	+	0	+	0	0	0	0	+	0	0	+	0	0	0	0
6	0	0	0	+	+	+	+	0	+	+	0	+	+	+	0	0	+	+	+	0	+	1+	2+	3+

TABLE 6.6 Antibody Identification Panel Denoting Weak and Strong Reactions for P Antigen

	Rh						MNS				Lu		P	Lewis		Kell		Duffy		Kidd	
	D	C	E	c	E	f	M	N	S	s	Lu^a	Lu^b	P1	Le^a	Le^b	K	K	Fy^a	Fy^b	Jk^a	Jk^b
1	+	+	0	0	+	0	+	+	+	+	0	+	+w	+	0	0	+	0	+	0	+
2	0	0	0	+	+	+	+	0	+	0	0	+	0	0	+	+	0	+	+	+	0
3	0	0	+	+	0	0	0	+	0	+	0	+	+s	0	+	+	+	+	0	0	+
4	+	0	+	+	0	0	+	+	+	+	0	+	+	0	+	0	+	0	+	+	+
5	+	0	0	+	+	+	+	+	0	+	0	+	0	0	0	0	+	0	0	+	0
6	0	0	0	+	+	+	+	0	+	+	0	+	+	+	0	0	+	+	+	0	+

Anti-P (Answer B) and anti-PP1Pk are antibodies made by rare P-negative or PP1Pk-negative individuals. Anti-P and anti-PP1Pk would be strongly positive with all panel cells tested except the autocontrol (patient's own cells). These antibodies are IgG, reactive at 37°C and associated with acute hemolytic transfusion reactions and spontaneous abortions. Anti-Pk does not occur as a separate antibody. Anti-Pk (Answer C) is found only in combination with anti-PP1Pk (Answer D) and anti-p (Answer A) has never been reported.

19. If an individual's phenotype is Le(a+b−), which of the following genes do they possess?

A. *Le, Sese*
B. *le, Se*
C. *le, sese*
D. *Le, sese*
E. *Le, SeSe*

Concept: Lewis blood group antigens are unique because they are produced by epithelial cells released into plasma and are adsorbed onto the red blood cell membrane versus most other antigens that are formed as part of the red cell membrane. Lewis blood group antigens are also unique because antigen expression is dependent on the presence of *LE* and presence or absence of the Secretor (*SE*) genes. The *LE (FUT3)* gene codes for the presence of *fucosyltransferase* 3, which attaches a fucose to a Type I chain found in the secretions. The difference between a Type I and Type II chain is the linkage of the galactose (Gal) to the *N*-acetyl-glucosamine (GlcNac). A Type I chain is a ß 1-3 linkage while a Type II chain is a ß 1-4 linkage.

Type I Chain **Gal ß1-3GlcNAcß 1-3Galß1-4GlcNAcß1-3 – R**

R = additional carbohydrate sequence

When an individual possesses the *LE (FUT3)* gene, but is a non-secretor (*sese*), the fucose is added to the sub-terminal sugar, the *N*-acetyl-glucosamine (GlcNac) in a α1-4 linkage and the resulting structure is the Le^a antigen (Fig. 6.2).

The *SE (FUT2)* gene codes for the presence of fucosyltransferase 2 (α-1,2-fucosyltransferase), which adds a fucose to the terminal galactose of a type I chain in an α1-2 linkage.

If the individual is *lele*, that is they do not have a *LE (FUT3)* gene, they will lack Le^a and Le^b (Fig. 6.3).

LE (FUT3) ➡ fucosyltransferase 3

Le^a Type I Chain   Gal ß1-3GlcNAc ß1-3Gal ß1-4GlcNAcß1-3 – R
4
1
Fuc

Phenotype: Le(a+b−)

FIGURE 6.2 **Le(a+b−) Phenotype.**

SE (FUT2) ➡ fucosyltransferase 2

H Type I Chain   Gal ß1-3GlcNAc ß1-3Galß1-4GlcNAc ß1-3 – R
2
1
Fuc

Phenotype: Le(a−b−) Or,

sese

Type I Chain Galß1-3GlcNAc ß1-3Galß1-4GlcNAc ß1-3 – R

Phenotype: Le(a−b−)

FIGURE 6.3 **Le(a b) Phenotype.**

When both *LE (FUT3)* and *SE (FUT2)* genes are inherited both enzymes are present to add fucose to both the sub-terminal (GlcNAc) and terminal (Gal) sugars of the Type I chain. This results in a new antigenic structure, Le^b.

While the Le^a immunodominant sugar is present on the carbohydrate chain it is not recognized by anti-Le^a in routine laboratory testing, therefore their red blood cells will type Le(a−b+) (Fig. 6.4).

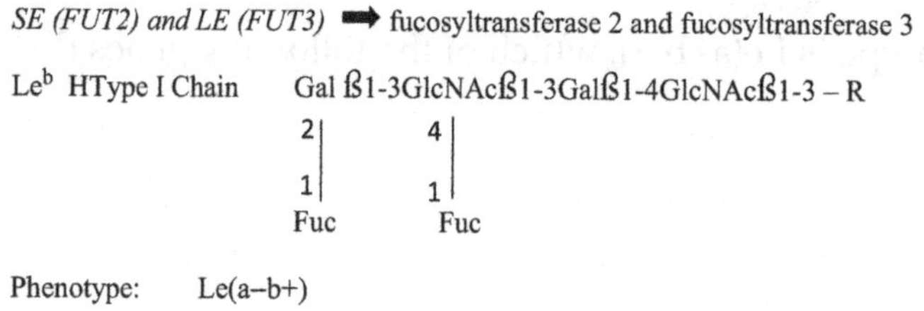

SE (FUT2) and LE (FUT3) ➡ fucosyltransferase 2 and fucosyltransferase 3

Le^b HType I Chain Gal ß1-3GlcNAcß1-3Galß1-4GlcNAcß1-3 – R

2 | 4 |

1 | 1 |

Fuc Fuc

Phenotype: Le(a–b+)

FIGURE 6.4 Le(a−b+) Phenotype.

Answer: *D*—If an individual types Le(a+b−) they only have the *LE (FUT3)* gene which adds the fucose to the subterminal sugar. When Le^a is found in serological typing the individual is a non secretor, *sese* indicating no terminal fucose was added to the Type I chain changing the antigenic structure to Le^b.

20. Rh phenotypes are inherited based on which of the following genes?

A. One gene that codes for all Rh antigens (e.g., Rh_o)
B. Three genes D/d, C/c, E/e
C. Two genes, *RHD* and *RHCE* with little genetic variation
D. Two genes, *RHD* and *RHCE* with many genetic variants
E. Many *RH* genes

Concept: The Rh nomenclature used today is based on theories of inheritance proposed in the 1940s. The R/r terminology arose from the theory proposed by Weiner that one gene was responsible for producing an agglutinogen containing a series of at least three blood factors.

Fisher and Race postulated that the antigens of the system were produced by three closely linked sets of alleles, D/d, C/c, and E/e.

It is now known the Rh system is encoded by genes *RHD* (encodes D) and *RHCE* (encodes CEce) on chromosome 1. These two genes are 97% homologous, which lend to the extremely polymorphic nature of the system. There are at least four genetic mechanisms leading to Rh variant phenotypes: (1) nonsense mutations, (2) specific point mutations in *RHD* or *RHCE* genes leading to amino acid changes, (3) rearrangements in *RHD and RHCE* exons 1–10, and (4) nucleotide deletion which causes a frame shift and premature stop codon.

Answer: *D*—Two genes, *RHD* and *RHCE* with many genetic variants are responsible for expression of over 50 Rh antigens (Answer E). Weiner's theory of inheriting one gene (Answer A) and Fisher-Race's theory of inheriting three separate but closely linked genes (Answer B) were both incorrect. Only two genes (Answer C) with little genetic variation would not explain all the antigens in the Rh blood group system.

21. If a D positive patient has anti-c in his serum, which of the following units of red cells will be compatible with this patient?

A. R_1R_1
B. R_1R_2
C. R_1r
D. R_2r'
E. rr

Concept: When a patient has a clinically significant antibody, it is vital to transfuse red blood cells that lack the corresponding antigen. In this case, the patient has anti-c, therefore, he must receive c negative red blood cells. The Weiner nomenclature must be converted to Fisher-Race to determine which predicted genotype will result in red blood cells that lack the c antigen. If a person is D positive, using Weiner nomenclature, his predicted genotype is designated with a capital R. If he is D negative, a lower case r is used.

An individual who has the C antigen is indicated by a 1 or a single prime ('). Lowercase c is implied when there is *no* 1 or ' indicated. (It is assumed that the third antigen is e) The presence of E antigen is indicated by the number 2 or double prime ("). Lowercase e is implied when there is *no* 2 or " indicated

(Table 6.7) can be used for the conversion:

TABLE 6.7 Conversion Between Weiner and Fisher-Race

Weiner terminology	Fisher-Race terminology	Antigens
R_0	Dce	D, c, e
R_1	DCe	D, C, e
R_2	DcE	D, c, E
R_z	DCE	D, C, E
r	dce	c, e
r'	dCe	C, e
r"	dcE	c, E
r^y	dCE	C, E

Answer: *A*—The R_1R_1 phenotype, using Fisher-Race nomenclature is DCe/DCe. This phenotype does not have the c antigen and can therefore, be transfused to a patient with anti-c. R_1R_2 (DCe/DcE) (Answer B), has c antigen and should not be transfused. R_1r (DCe/dce) (Answer C), has c antigen and should not be transfused. R_2r' (DcE/dCe) (Answer D) also has the c antigen and should not be transfused. Finally, rr (dce/dce) (Answer E) is homozygous for c and should not be transfused (Table 6.7).

22. Which of the following is the most common predicted Rh genotype in African American patients?

A. R_1R_0
B. R_1R_1
C. R_1R_2
D. R_0r
E. R_1r

Concept: The incidence of *RH* genes varies with geography and ethnicity. Below is a table (Table 6.8) with the incidence of the more common Rh phenotypes in D positive people.

TABLE 6.8 Rh Genotype Prevalence in Rh Positive Individuals

Predicted genotype	Prevalence in European ancestry (%)	Prevalence in African ancestory (%)
R_1r	31.1	8.8
R_1R_0	3.4	15.0
R_1R_1	17.6	2.9
R_1R_2	11.8	3.7
R^0r	3.0	22.9
R_0R_0	0.2	19.4
R_2r	10.4	5.7

Answer: *D*—The table above demonstrates the prevalence of the more common predicted Rh genotypes in those of African ancestry. Antigen typing and determining possible and most probable predicted genotypes can allow transfusion medicine specialists to make assumptions about a patient's true genotype. However, molecular techniques should be used to accurately determine this information. The other choices (Answers A, B, C, and E) are incorrect based on the table above.

23. Review the panel of selected cells. What is the antibody pattern detected?

	D	C	E	c	e	IAT
1	+	+	0	0	+	2+
2	+	+	0	0	+	2+
3	0	0	+	+	+	0√
4	0	+	0	+	+	2+
5	0	+	0	+	+	2+
6	0	0	0	+	+	0√
7	+	0	0	+	+	2+
8	+	0	0	+	+	2+

A. Anti-C
B. Anti-E and anti-D
C. Anti-E
D. Anti-D
E. Anti-G

Concept: The G antigen is present on D or C antigen positive cells. Anti-G should be suspected whenever a pattern of anti-D and anti-C is detected in red cell antibody identification panel testing. Rh antibodies are clinically significant and anti-D and anti-C/suspected anti-G should be further investigated, especially in the prenatal setting. Detection of a pattern of anti-D and anti-C should reflex to further serologic testing to verify if anti-G is present, due to the risk of HDFN and the potential benefit of Rh immunoglobulins (if the patient does not have anti-D). Anti-G can be verified when tested with a rare D-C-G+ cell or by double adsorption/elution techniques. Verification of the presence of anti-D/anti-C or anti-D/anti-C and anti-G is not generally pursued outside of the obstetric population since the provision of D-C- antigen negative units would be appropriate for a safe transfusion.

Answer: *E*—Anti-G pattern is suspected due to all D+ or C+ cells being reactive. One may not distinguish anti-D, anti-C and anti-G without further testing (i.e., serial adsorption and elution). If anti-E (Answer C) was present panel cell no. 3 should be positive. Anti-D (Answer D) alone would not explain the positive reaction with panel cells no. 4 and 5. The negativity of panel cell six rules out a combination of E and D (Answer B) and the negativity of panel cells seven and eight rules out Anti-C alone (Answer A).

24. Of the following antigens, it is most difficult to locate antigen-negative RBCs for a patient with antibody to:
A. M
B. N
C. S
D. s
E. M^g

Concept: The MNS blood group antigens are carried on glycophorin A (GPA) and glycophorin B (GPB). *GYPA* codes for the M and N antigens. If an individual lacks the M and N antigen, then the individual lacks *GYPA* resulting in the rare En(a-) phenotype. This individual may produce the corresponding antibody to the high incidence En[a] antigen, anti-En[a]. En(a-) red cells have reduced levels of sialic acid (40% of normal).

GYPB codes for the S, s and U antigens. The U antigen is present when the S or s antigen is inherited. If an individual lacks both S and s antigens, then the individual has an alteration or absence of *GYB* resulting in S-s-U- phenotype. The U-negative phenotype is predominantly found in individuals of African ancestry. These individuals may produce the corresponding antibody to a high incidence antigen; anti-U. Changes in single amino acids of *GYPA* and *GYPB* genes lead to hybrids caused by unequal crossing over and gene conversion. Next to the Rh system, the MNS is the second-most polymorphic blood group system. The frequency of the common antigens in this system, M, N, S, s, U, is useful to know for provision of a safe transfusion.

Anti-M and anti-N are generally not considered clinically significant and a compatible AHG-crossmatch may be used as an indicator of transfusion safety. Alternately, anti-En[a] has been documented in the literature to be a clinically significant antibody. Anti-U, anti-S and anti-s are generally IgG in nature and clinically significant. Provision of En(a−) or U− red cells is challenging, since 99%–100% of the population is En(a+) and U+. To match a patient with anti-En[a] or anti-U would require the assistance of an AABB accredited Immunohematology Reference Lab (IRL). The IRL would determine if rare donor units, En(a−) or U−, are available for transfusion locally/regionally or the IRL may contact the American Rare Donor Program (ARDP) to find a compatible donor nationwide. Only AABB accredited or American Red Cross Advanced IRLs may request units through the ARDP. Requests for antigen-negative blood from the ARDP must be related to the patient's corresponding antibody. ARDP is most often contacted to locate blood which is less than 1%–5% of the general population, such as En(a−) or U−.

Answer: *D*—The s antigen is present in 89% of the population; therefore, only 11% of the population would be compatible with the patient who has anti-s. Approximately 1 in 10 individuals are s antigen negative. Matching for the s antigen may be relatively difficult (10% seronegative in the general population). The frequency of antigen matched or antigen negative units in the population for the other MNSs antigens are as follows: M (22%) (Answer A), N (28%) (Answer B), S (45%) (Answer C), and s (11%). M[g] (Answer E) is a low prevalence antigen in the MNS system.

25. Anti-Kp[a] is an antibody to a low prevalence antigen. If rare antisera is unavailable to confirm red cell units as Kp(a−), what is the best choice for provision of blood for the patient with anti-Kp[a]?

A. AHG Crossmatch
B. Immediate spin crossmatch
C. Suggest to not transfuse the patient
D. Ask for family members to donate to find a compatible unit
E. Call the rare donor program

Concept: Anti-Kp[a] is an antibody to a low prevalence antigen. The Kp[a] antigen is present in less than 2.3% of the Caucasian population and less than 1% of the black population. Therefore, 97.7%–100% of the population will be compatible with a patient who has anti-Kp[a]. Since the antibody is IgG in nature (detected by IAT method) then the same method may be used for compatibly testing in lieu of antigen matched units. An immediate spin crossmatch verifies the donor unit is ABO compatible. An IAT method used for antibody detection may then be used for the crossmatch testing to detect if the donor unit is Kp[a] positive.

Answer: *A*—The patient may be safely transfused with donor units when antigen confirmation is not available since a high percentage of units will be compatible with an IAT method. An immediate spin crossmatch (Answer B) would not detect this antibody, as it is normally IgG reacting best in an IAT. There is no reason to suggest this patient should not be transfused based on these test results

(Answer C). Finding compatible blood with an IAT method would be simple to do and the family members do not need to donate (Answer D), nor would the rare donor registry need to be contacted (Answer E).

26. A 20-year-old woman is 21 weeks pregnant and initial antibody detection (screen) results are below. What antibody do you suspect based off of the screen results?

Screen cell	D	C	E	c	e	K	k	Fy^a	Fy^b	IAT
I	+	+	0	0	+	0	+	0	+	0√
II	+	0	+	+	0	0	+	0	+	0√
III	0	0	0	+	+	+	+	+	+	3+

A. Anti-D
B. Anti-C
C. Anti-E
D. Anti-K
E. Anti-Fy^b

Concept: Anti-K is an IgG reactive antibody, which is known to be clinically significant. An antibody is considered clinically significant if it may destroy antigen positive cells leading to hemolytic transfusion reaction and HDFN. Anti-K does not bind complement, but the antibody is able to cause extravascular hemolysis. Anti-K is the third most common cause of HDFN after ABO and Rh antibodies. Anti-K can cause fetal anemia due to suppression of fetal RBC synthesis.

Answer: *D*—The pattern in the antibody screen is most consistent with anti-K. Anti-K is a clinically significant risk in pregnancy, especially since Anti-K inhibit the RBC precursors. Therefore, anti-K is typically considered clinically significant at a titer of 8, in contrast to other antibody groups that are considered significant at a titer of ~16. Clinical and laboratory monitoring is highly recommended for the patient. None of the other antibodies are consistent with the screen results (Answers A, B, C, and E).

27. Lutheran (LU) antibodies are NOT usually implicated in HDFN primarily because of which of the following?
A. LU antigens are not antigenic
B. LU antigens are present on other tissues and are weak on cord cells
C. LU antibodies do not cross the placenta barrier
D. Most mothers are Lu(a+b+)
E. LU antibodies are IgM and not clinically significant

Concept: The Lutheran blood group system was the fifth blood group system defined after ABO, MN, P and Rh. The blood group is made up of antithetical pairs of high and low prevalence antigens coded by the *LU* gene located on chromosome 19. Antibodies in the Lutheran system are generally not considered clinically significant. If a Lutheran antibody is associated with a hemolytic transfusion reaction, the reaction is generally mild. Lutheran antibodies have not been reported in the literature to be related to severe HDFN but mildly affected infants requiring phototherapy have been reported.

Lutheran antigens demonstrate variable strength observed in serologic testing, which accounts for the characteristic mixed-field agglutination patterns observed. Lutheran antigens are present on blood vessels and tissues including brain, heart, kidney, liver, lung, placenta, pancreas, atrial wall, tongue, trachea, skin, skeletal muscle, cervix, ileum, colon, stomach, and gall bladder, but are not present on lymphocytes, granulocytes, monocytes and platelets. The antigens are detected on fetal cells at 10–12 weeks gestation and are poorly developed at birth. Since Lutheran antigens are present on placental tissue, this may result in absorption of maternal antibody reducing the risk of HDFN.

Answer: *B*—Lutheran antibodies are adsorbed by the Lutheran antigens present on the placenta and Lutheran antigens are only weakly expressed on cord cells. The most common Lutheran phenotype is Lu(a−b+), 92.4%, while only 7.4% of the population is Lu(a+b+) (Answer D). Lutheran antibodies can be formed (Answer A) are primarily IgG (Answer E) in nature; therefore, these antibodies have the ability to cross the placenta (Answer C) and potentially cause HDFN. Lutheran antigens are able to stimulate IgG clinically significant antibodies, evidenced in a mild to moderate transfusion reactions which have been documented in the literature.

28. The most common phenotype in the Duffy blood group system in the Caucasian population is which of the following?

A. Fy(a−b+)
B. Fy(a+b+)
C. Fy(a+b−)
D. Fy(a−b−)
E. Fy^x positive

Concept: The Duffy system is encoded on the *FY* or *DARC* gene. The system is made up of two common antigens which result in the common phenotypes of Fy(a+b−)/ in 20% of the Caucasian population, Fy(a+b+) in 48% of the Caucasian population and Fy(a−b+) in 32% of the Caucasian population. A null expression Fy(a−b−) is associated with black phenotypes with up to 67%–70% of African Americans lacking the Duffy antigens. Fy3 and Fy5 are high incidence antigens in the Duffy blood group system.

Answer: *B*—Although Fy(a−b−) is the most common phenotype in the African/Black population (Answer D), the Fy(a+b+) phenotype is most common in the European/white populations (83%). Fy(a+b−) (Answer C) is the second most common phenotype in the Caucasian population. Fy(a−b−) phenotypes (Answer D) are most common in individuals of African descent. Fy^x is associated with Caucasians. The Fy^x (Answer E) allele correlates with an unusual weak expression of Fy^b antigen in a Caucasian.

29. Of the following phenotypes, the most common among individuals of African ancestry is which of the following?

A. Jk(a+b−)
B. Jk(a+b+)
C. Jk(a−b+)
D. Jk(a−b−)
E. JK null

Concept: It is important to understand the differences in the expression of the blood antigen among different ethnic groups. The Kidd blood group system is made up of three antigens Jk^a, Jk^b, and Jk3. There are four phenotype combinations Jk(a+b−), Jk(a+b+), Jk(a−b+) and Jk(a−b−). These phenotypes have notable racial expression differences. Jk(a+) phenotypes are found in 91% of African/Black populations while only 77% of the European/Caucasian population. Another example is Jk(b−) phenotype is found in 57% of African/Black population while only 28% of the European/Caucasian Anti-Jk3 made to a high prevalence antigen is produced by individuals who are Jk(a−b−) phenotype. This expression is more prevalent in Philippino, Polynesian, Chinese, Japanese, and Indonesian populations.

Answer: *A*—Ethnic differences are demonstrated in the Kidd blood group system. Most Caucasians are Jk(a−) while the majority of black individuals are Jk(b−). The rare Jk(a−b−) (Answer D) phenotype is associated with Pacific islander, Japanese and Chinese populations. None of the other choices (Answers B, C, and E) are correct for this ethnic group.

30. In what percentage does Co^a occur in most populations?

A. >99
B. 90
C. 66
D. <0.5
E. 52

Concept: Colton blood group antigens are present on aquaporin-1, a protein responsible for transporting water. Studies by Peter Agre on aquaporin-1 lead to the discovery of the first water transporter in man and only the second Nobel Prize in Transfusion Medicine. The first was awarded to Karl Landsteiner for discovery of ABO.The Colton blood group system is made up of four antigens (Co^a, Co^b, Co3, and Co4). These antigens are one antithetical pair (Co^a and Co^b) and two high incidence antigens (Co3 and Co4). The most common antigen is Co^a (99.5%), while Co^b is rarely expressed (0.5%). Co^a and Co^b are an example of a high/low prevalence antithetical pair. In application, the majority of the population would be incompatible with a patient who has anti-Co^a (99.5%). On the other hand if a patient has an antibody to a low prevalence antigen, such as Co^b, most of the population will be compatible.

Colton antibodies are known to cause acute and delayed transfusion reactions. In addition, they are able to cause severe hemolytic disease of the fetus and newborn.

Answer: *A*—Co^a is a high incidence antigen present in >99.5% of the population. None of the other choices (Answers B, C, D, and E) are correct based on the information above.

31. Anti-Yt^a is identified in a 72-year-old male and the attending physician wants to transfuse urgently. What option may be considered?

A. Transfuse random group compatible RBCs
B. Wait for results of Cr^{51} testing
C. Send the sample for IgG subclass assay
D. Locate Yt(a−) RBCs from the Rare Donor Registry
E. Call in family members to identify potential donors

Concept: The YT blood group system, also referred to as Cartwright blood group system, was first defined in 1964 and is made up of the high/low antithetical pair Yt^a (high) and Yt^b (low). Antibodies in this system are IgG in nature and most often IgG4. YT antibodies have not been associated with HDFN, but transfusion reactions reported in the literature have ranged from none to severe hemolysis. Thus, the best *immediate* action is to provide IAT crossmatch compatible blood. Since anti-Yt^a is an antibody to a high prevalence antigen, the majority of the population will be incompatible with the patient's sample. However, rare Yt(a−) red blood cell units may be difficult to find locally and may require a search with the national rare donor registry.

Answer: *A*—The AHG-crossmatch/full crossmatch should be used as an indicator of transfusion safety for immediate/urgent transfusion of a patient with anti-Yt^a. However, the clinical significance of the antibody is unknown without further advanced testing, such as antibody subclass testing, Monocyte Monolayer Assay, or chromium labeled studies. It may be helpful to know the IgG subclass (Answer C), but this will not provide information to determine clinical significance. The ^{51}Cr (Answer B) red cell survival studies are rarely performed to assess antibody significance and given the urgency would not be completed for this type of antibody of variable clinical significance.

One would only need Yt(a−) donor units through a national search (Answer D) if the patient's antibody has demonstrated decreased red cell survival in previous analysis. The family (Answer E) is a possibility for compatible donors, but in the scenario provided this is an urgent provision of blood. There is not time for screening, collection, and provision of dedicated donors for this patient.

32. What type of antigenic determinant and antibody class is most likely present in the given antibody detection (screen) results below?

Screen cell	D	C	E	c	E	K	k	Fy^a	Fy^b	IS	37C	IAT
I	+	+	0	0	+	0	+	0	+	0	0	4+
II	+	0	+	+	0	0	+	0	+	0	0	4+
III	0	0	0	+	+	+	+	+	+	0	0	0√

A. Carbohydrate, IgM
B. Protein, IgM
C. Lipid, IgM
D. Protein, IgG
E. Carbohydrate, IgM and Protein, IgG

Concept: Carbohydrate type blood group antibodies, such as ABO, Lewis, MN or P1, are usually IgM in nature. These are "naturally occurring" antibodies, not stimulated by red cell exposure. IgM antibodies are often reactive in immediate spin/direct testing. These antibodies can cause incompatible immediate spin crossmatches and ABO discrepancies.

Protein type blood group antibodies, such as Rh, Kell, Duffy, or Kidd are typically IgG in nature. These antibodies are not generally naturally occurring, but stimulated by red cell exposure from transfusion, transplantation, and/or pregnancy. These antibodies will cause incompatibilities detected in the antibody detection test (screen) and identified with antibody panel testing at IAT. The Transfusion Service chooses red cell donor units which lack the specificity that is detected to prevent acute and/or delayed hemolysis in the presence of the IgG antibody.

Answer: *D*—The antibody detection test (screen) is non-reactive at immediate spin; therefore, there is not an IgM directed to carbohydrate blood group (Answers A and E) specificity detected. The antibody screen is positive at 37°C (body temperature) and at IAT, indicating the antibody is IgG in nature. Protein blood group system antibodies are more likely to be IgG, whereas carbohydrate blood group antibodies are likely to be IgM (Answer B). No blood groups exist that are lipid based (Answer C).

33. Pretransfusion testing on a 68-year-old African American women showed her to be A Rh positive with a positive antibody detection test (screen) reactive with all screening cells in IAT/AHG. An antibody identification panel showed similar results (weak + to 2+ positive) with all panel cells tested. The autocontrol was negative. A sample was sent to the Immunohematology Reference Laboratory and an anti-McC^a was identified. All other antibodies were ruled-out. What type of blood would be appropriate for transfusion?
A. AHG crossmatch compatible RBCs
B. McC(a−) crossmatch compatible RBCs
C. IS crossmatch compatible RBCs
D. Call the rare donor program
E. Serologically incompatible RBCs

Concept: Antibodies to high prevalence antigens in the Knops (Kn^a, McC^a, Sl1, Yk^a), Chido/Rogers (CH/RG), and John Milton Hagen (JMH) blood group systems typically fit into a group of antibodies traditionally called "high titer, low avidity" or HTLA. This term is used to describe their typical serologic reactivity. These IgG antibodies can be of high titer, >16 and are of low avidity, showing weak, variable reactivity (Table 6.9) with all RBCs tested except those rare individuals who lack the antigen.

Antibodies to antigens in these systems are typically considered clinically insignificant. They do not cause hemolytic transfusion reactions or hemolytic disease of the fetus or newborn. When transfusion is required, serologically incompatible RBCs are selected.

TABLE 6.9 Doubling Dilutions of Patient Serum With Saline

Undiluted	1:2	1:4	1:8	1:16	1:32	1:64	1:128
Donor 1	1+	1+	Weak+	Weak+	Weak+	0	0
Donor 2	2+	1+	1+	Weak+	Weak+	0	0

Answer: *E*—Serologically incompatible RBCs are selected for transfusion. AHG crossmatch compatible RBCs (Answer A) would be difficult to find as the antigen frequency for most of the antigens in these blood groups systems is >92%. Given the high frequency of the antigens virtually all crossmatches will be incompatible. As these antibodies do not cause transfusion reactions, McC(a−) RBCs (Answer B) or calling the rare donor program (Answer D) to locate a donor/units are not necessary. All crossmatches will be compatible at IS (Answer C) as these antibodies are IgG. An IS crossmatch would avoid reactivity but standard practice is to perform a full crossmatch when a "potentially" clinically significant antibody is present.

34. Two units of blood are requested for a 55-year-old male with multiple myeloma. He has received several cycles of anti-CD38 (daratumumab). He types O Rh positive and the antibody detection test is positive. The antibody identification panel shows similar reactivity.

	Rh						MNS				Lu		P	Lewis		Kell		Duffy		Kidd		PEG		
	D	C	E	c	e	f	M	N	S	S	Lu^a	Lu^b	P1	Le^a	Le^b	K	K	Fy^a	Fy^b	Jk^a	Jk^b	IS	37	IAT
1	+	+	0	0	+	0	+	+	+	+	0	+	+	+	0	0	+	0	+	0	+	0	NH	2+
2	0	0	0	+	+	+	+	0	+	0	0	+	0	0	+	+	0	+	+	+	0	0	NH	2+
3	0	0	+	+	0	0	0	+	0	+	0	+	+	0	+	+	+	+	0	0	+	0	NH	2+
4	+	0	+	+	0	0	+	+	+	+	0	+	+	0	+	0	+	0	+	+	+	0	NH	2+
5	+	0	0	+	+	+	+	+	0	+	0	+	0	0	0	0	+	0	0	+	0	0	NH	2+
6	0	0	0	+	+	+	+	0	+	+	0	+	+	+	0	0	+	+	+	0	+	0	NH	2+

Which of the following approaches to deal with the serologic reactivity of anti-CD38 would carry the risk of missing an antibody to a high prevalence antigen?

A. Use DTT/AET treated RBCs
B. Use RBCs from a patient on anti-CD38
C. Use cord cells
D. Genotype the patient and provide phenotypically similar RBCs
E. Phenotype the patient and provide phenotypically similar RBCs

Concept: Anti-CD38, such as daratumumab, is used to treat refractory multiple myeloma. CD38 is present on myeloma cells but is also on RBCs causing a positive screen and panel making it impossible to rule-out clinically significant antibodies without additional testing.

There are several strategies to work around the anti-CD38 reactivity. DTT (Dithiothreitol) or AET (2-aminoethylisothiouronium bromide), sulfhydryl reagents, denature CD38 on RBCs. The downside is that these chemicals also denature other blood group antigens. (See Question 39 in this chapter for a table detailing these reagents). If an antibody is against one of these antigens, it will not be detected.

An antibody screening cell set can be prepared from a set of cord cells as CD38 is not yet developed on these cells. Another option, prepare a screening cell set from patients on anti-CD38. Anti-CD38 treatment causes decreased expression of CD38 on patient's RBCs.

The patient may be genotyped or phenotyped if they have not been transfused and all subsequent transfusions provided are phenotype-matched.

Answer: *A*—DTT denatures several blood group antigens, notably all Kell system antigens as well as Do[a]/Do[b] and Yt[a]. Testing DTT treated RBCs runs the risk of missing an antibody to an antigen denatured. Anti-K is the most common so laboratories choosing to test DTT/AET-treated RBCs provide K-negative blood. Testing a screening cell set prepared from patients taking anti-CD38 (Answer B) or cord cells (Answer C) will allow ruling-out all antibodies to antigens present on these RBCs. Genotyping (Answer D) or phenotyping (Answer E) the patient pretransfusion and providing antigen matched RBCs avoids the risk of missing clinically significant antibodies.

Please answer questions 35–37 based on the following clinical scenario:

A 40-year-old African-American woman, G3P1001, presents at a local clinic with generalized weakness. Her hemoglobin is 6.5 g/dL and her provider has admitted her to a local hospital for observation and additional testing. Per her medical record, the patient was previously transfused for anemia of unknown origin. The patient received blood products 2 years earlier. Due to the current anemia, the provider ordered 2 units of leukocyte-reduced red blood cells to be transfused.

Routine pretransfusion testing showed she was B Positive and the antibody detection test (screen) was positive with all three screening cells tested using an automated Gel test. An antibody identification panel was tested by PEG-IAT/AHG. Results of the PEG panel are shown below.

	Rh						MNS				Lu		P1	Lewis		Kell		Duffy		Kidd		PEG	
	D	C	E	C	e	f	M	N	S	s	Lu[a]	Lu[b]	P1	Le[a]	Le[b]	K	k	Fy[a]	Fy[b]	Jk[a]	Jk[b]	IS	IAT
1	+	+	0	0	+	0	+	+	+	+	0	+	+	+	0	0	+	0	+	0	+	0	3+
2	+	+	0	0	+	0	0	+	0	+	0	+	0	0	+	0	+	0	+	+	+	0	3+
3	+	0	+	+	0	0	+	+	+	+	0	+	+	0	+	+	+	+	+	+	0	0	3+
4	+	0	0	+	+	+	0	+	0	+	0	+	+	0	0	0	+	0	0	+	0	0	4+
5	0	+	0	+	+	+	0	+	0	+	0	+	0	0	0	0	+	0	0	+	+	0	0√
6	0	0	+	+	+	+	+	0	+	+	0	+	+	+	0	0	+	+	+	0	+	0	2+
7	0	0	0	+	+	+	0	+	+	+	0	+	+	0	+	+	+	+	0	+	+	0	2+
8	0	0	0	+	+	+	0	+	+	0	0	+	+	+	0	+	+	0	+	0	+	0	0√
9	0	0	0	+	+	+	+	+	+	0	0	+	0	0	+	0	+	+	+	+	+	0	2+
10	0	0	0	+	+	+	+	0	+	0	0	+	+	0	+	0	+	+	+	0	+	0	2+
11	+	0	+	+	0	0	+	+	+	+	0	+	+	0	+	0	+	+	0	+	+	0	3+
	Patient red blood cells (autocontrol)																					0	0√

35. Which of the following is the most likely explanation for the antibody detection and panel results?

A. Warm autoantibody
B. Cold autoantibody
C. Multiple alloantibodies
D. Alloantibody to high prevalence antigen
E. Rouleaux

Concept: Based on the two different agglutination reaction patterns, 4+ cells compared to 2+, it is likely that antibodies to multiple antigens are present. Literature suggests alloimmunization to red cell antigens may be influenced by genetic factors, clinical/medical effects of treatment course or infection and volume of units transfused. Antibody detection and identification are important to verify specificity.

Answer: C—Variable strength and negative reactions observed in the panel with a negative autologous control suggests multiple alloantibodies. Additional selected panel testing (referred to as selected cell panel testing) should be pursued to verify the specificities detected.

Since the patient's autocontrol was non-reactive, the panel testing does not suggest a warm or cold autoantibody (Answers A and B). Rouleaux (Answer E) may occur when the patient has abnormally high protein level in their plasma. It causes reactivity when testing at immediate spin and/or 37°C. In this case there was no reactivity at IS. With two negative cells on the panel an antibody to a high prevalence (incidence) antigen (Answer D) may be excluded. Alloantibody to high prevalence antigen presents with consistent panel reactions with all cells, with a negative autocontrol test.

Evaluate the patient's sample is tested in a ficin panel, as shown below to continue this case.

	Rh						MNS				Lu		P1	Lewis		Kell		Duffy		Kidd		PEG		Ficin
	D	C	E	C	e	f	M	N	S	s	Lu^a	Lu^b	P1	Le^a	Le^b	K	k	Fy^a	Fy^b	Jk^a	Jk^b	IS	IAT	IAT
1	+	+	0	0	+	0	+	+	+	+	0	+	+	+	0	0	+	0	+	0	+	0	3+	4+
2	+	0	0	+	+	+	0	+	0	+	0	+	+	0	0	0	+	0	0	+	0	0	3+	4+
3	0	+	0	+	+	+	0	+	0	+	0	+	0	0	+	0	+	0	0	+	0	0	0√	0√
4	0	0	+	+	+	+	+	0	+	0	0	+	+	+	0	+	+	0	+	0	+	0	0√	0√
5	0	0	0	+	+	+	+	+	+	0	0	+	0	0	+	0	+	+	+	+	+	0	2+	0√
6	0	0	0	+	+	+	+	0	+	0	0	+	+	0	+	0	+	+	+	0	+	0	2+	0√

Patient red cell phenotype: D+C+E+c+e+; K-k+, Fy(a−b+), Jk(a+b−), S-s+.

36. Given the above results, which of the following antibodies is most likely present?

A. Anti-D, anti-Fy^a
B. Anti-K, Anti-Kp^a
C. Anti-D, Anti-Js^a
D. Anti-C, anti-E, anti-K
E. Anti-Js^b

Concept: Enzyme panel testing is a secondary investigation method in the blood bank. The enzyme testing results may be used alongside the initial panel testing. Enzyme testing may allow for separation of antibody specificities based on their enhancement or elimination of reactivity. The elimination of the anti-Fy^a pattern in the selected cell panel testing, verifies anti-Fy^a in the sample. See Table 6.10 below of common antigen specificity and their reactions with enzymes.

TABLE 6.10 Expression of Common Blood Group Antigens When Treated With Enzymes

Rh						MNS				Lu		P1	Lewis		Kell		Duffy		Kidd	
D	C	E	c	e	f	M	N	S	s	Lu^a	Lu^b	P1	Le^a	Le^b	K	k	Fy^a	Fy^b	Jk^a	Jk^b
↑	↑	↑	↑	↑	↑	↓	↓	(↓)	(↓)	NC	NC	↑	↑	↑	NC	NC	↓	↓	↑	↑

Destroyed, ↓, enhanced, ↑, no change, NC, variable, (↓).

Answer: *A*—Ficin testing may enhance anti-D (cells 1 and 2), while elimination of reactivity in panel cells 5 and 6 is consistent with anti-Fy^a. Anti-C and anti-E (Answer A) were suspected, but excluded on non-reactive cells (cell 3 and cell 4). The Kell blood group antigens (Answers B, C, D, and E) are not affected by ficin treatment.

37. The patient's ABO/Rh results are below. Which of the following answer choices explains how an Rh positive individual can form anti-D?

Forward:		**Reverse:**		**Rh typing:**				
Anti-A	Anti-B	A1 cells	B cells	Anti-D	Rh control	Weak D IAT	Rh control IAT	Interpretation
0	4+	4+	0	1+	0	3+	0√	B Positive

A. Autoantibody with a preference for D+ cells
B. Partial D Phenotype
C. Weak D types 1, 2, and 3
D. DAT positive autologous cells
E. D^{el} phenotype

Concept: There are three D phenotypes which may produce anti-D: (1) D^{el} (rarely), (2) some Weak D other than type 1, 2, and 3 and Partial D phenotypes. After the administration of passive anti-D/Rh immune globulin (no documented evidence of receiving the product) and exclusion of the presence of an autoantibody (autocontrol and DAT negative), then additional serologic or molecular testing may be used to verify if the patient has altered D expression.

Table 6.11 demonstrates a typical serologic investigation with multiple sources of D antisera as part of the verification of partial D phenotype. Since the advent of molecular testing for D variants, such investigations at hospital transfusion service laboratories are not typically performed today. However, the comparison analysis can support the reason for the atypical D expression, but only red cell molecular testing can verify the difference between a Weak D typing and Partial D or D Variant expression.

TABLE 6.11 Serologic Investigation of Weak D Versus Partial D/D Variant

Anti-D reagent	Anti-D	Rh control	Weak D IAT	Rh control IAT
Vendor 1	0	0	3+	0√
Vendor 2	1+	0	4+	0√
Vendor 3	W	0	2+	0√
Vendor 4	3+	0	3+	0√

Answer: *B*—The best answer is Partial D phenotype. These red cells lack a portion of the D antigen, leading to an altered expression of the D phenotype. There are four primary mechanisms which lead to the Rh variant expression (1) nonsense mutation, (2) point mutations in the *RHD* or *RHCE* genes, (3) rearrangements of the paired genes *RHD* and *RHCE*, and (4) nucleotide deletion which changes the expression of the antigen.

Since the patient's autocontrol is nonreactive, this pattern of reactivity is not related to an autoantibody with a preference for D+ cells (Answer A). The patient autocontrol testing usually correlates with coating of IgG on the patient's red cells detected when performing a direct antiglobulin test (DAT). Since the autocontrol is negative, this is not related to a positive DAT (Answer D).

D^{el} (Answer E), Weak D, and Partial D phenotypes have been documented to produce anti-D. D^{el} phenotype pattern of reactivity is not detected by standard Rh testing. The D antigen is only detected by elution methods, thus the designation D^{el}. Weak D types 1, 2, and 3 (Answer C) have not been documented to produce anti-D, but rarely other Weak D types have been documented to produce anti-D.

38. The *RHD* and *RHCE* genotype testing was completed for the patient's sample. The patient was verified as DAU genotype. What type red blood cells should this patient receive?
A. B, Rh Positive, Fy(a−)
B. O, Rh Positive, Fy(a−)
C. B, Rh Negative, Fy(a−)
D. O, Rh Negative, Fy(a+)
E. This patient should not be transfused

Concept: As a partial D phenotype, this patient has produced alloantibody to the portion of the Rh(D) antigen that they lack. Therefore, the Rh(D) antigen should be honored for provision of blood.

Answer: C—Since the patient has produced anti-D and anti-Fy[a], the patient must receive ABO compatible, extended IAT crossmatch compatible, D and Fy[a] antigen negative cells.

Although the patient's molecular testing indicates that there is an altered expression of Rh(D) antigen, molecularly-matched units of the *DAU* genotype are not necessary. The most practical approach is to provide Rh negative blood. Neither Rh positive blood (Answers A and B) nor Fy[a] positive blood (Answer D) should be given to this patient. There is no reason to avoid transfusion in this patient (Answer E).

Of special note, if a pregnant patient is proven to have a weak D type 1, 2, or 3 by genotyping, RhIg administration is not indicated. However, patients with Partial D or other type of D variants are at risk of forming anti-D and RhIg administration is indicated.

Please answer questions 39–43 based on the following clinical scenario:

A 78-year-old woman with myelodysplastic syndrome is in clinic for evaluation. Her hemoglobin is 6.8 g/dL and her physician has ordered 2 units of leukocyte-reduced red blood cells to be transfused in outpatient clinic tomorrow. She is of Eastern European ancestry and has four children.

Routine pretransfusion testing showed she is A Rh positive and the antibody detection test (screen) was positive with all three screening cells tested, using an automated solid phase test. An antibody identification panel was tested by solid phase and all panel cells were positive (3+) with the positive and negative controls reacting as expected. A second antibody identification panel was tested using a polyethylene glycol (PEG) tube test. Results of the initial PEG panel are as follows:

	Rh						MNS				Lu		P1	Lewis		Kell		Duffy		Kidd		PEG		
	D	C	E	c	E	f	M	N	S	s	Lu[a]	Lu[b]	P1	Le[a]	Le[b]	K	k	Fy[a]	Fy[b]	Jk[a]	Jk[b]	IS	37C	IAT
1	+	+	0	0	+	0	+	+	+	+	0	+	+	+	0	0	+	0	+	0	+	0	NH	3+
2	+	+	0	0	+	0	0	+	0	+	0	+	0	0	+	0	+	0	+	+	+	0	NH	3+
3	+	0	+	+	0	0	+	+	+	+	0	+	+	0	+	+	+	+	+	+	0	0	NH	3+
4	+	0	+	+	0	0	0	+	+	0	0	+	+	0	+	0	+	+	0	+	0	0	NH	3+
5	0	+	0	+	+	+	0	+	0	+	0	+	0	0	0	0	+	0	0	+	+	0	NH	3+
6	0	0	+	+	+	+	+	0	+	+	0	+	+	+	0	0	+	+	+	0	+	0	NH	3+
7	0	0	0	+	+	+	0	+	+	+	0	+	+	0	+	+	+	0	+	+	+	0	NH	3+
8	0	0	0	+	+	+	0	+	+	+	0	+	+	+	0	0	+	0	+	0	+	0	NH	3+
9	0	0	0	+	+	+	+	+	+	0	0	+	0	0	+	0	+	+	+	+	+	0	NH	3+
10	0	0	0	+	+	+	+	0	+	0	0	+	+	0	+	0	+	+	+	0	+	0	NH	3+
11	+	0	+	+	0	0	+	+	+	+	0	+	+	0	+	0	+	+	0	+	+	0	NH	3+
	Patient red blood cells (autocontrol)																					0	NH	0√

NH, No hemolysis.

39. Which of the following is the most likely explanation for the antibody detection and panel results?

A. Warm autoantibody
B. Alloantibody to high prevalence antigen
C. Multiple alloantibodies
D. Method-dependent antibody
E. Cold autoantibody

Concept: There are antigens of high frequency in nearly all human blood groups. When rare, genetic mutations occur the resulting product is an individual who lacks the antigen and makes the corresponding antibody. Examples include, in the ABO system, anti-H in the H-deficient (Bombay) phenotype, anti-Js^b, or anti-Kp^b in the Kell blood group or anti-U in the MNS blood group system. These antibodies to high prevalence antigens are challenging to identify in the routine Transfusion Service and normally require sending to an Immunohematology Reference Laboratory who can test rare RBCs, antisera and chemically treated RBCs to assist in identifying these antibodies.

Answer: *B*—Equal reactivity with all screening cells and panel cells along with a negative autocontrol are consistent with reactivity commonly seen with an alloantibody to a high prevalence, also known as high frequency antigen.

Multiple antibodies (Answer C) typically present as showing variable reactivity and often have at least one negative panel cell.

Positive reactions were seen using solid phase and PEG test tube methods ruling out the possibility of a method-dependent antibody.

If this were a warm or cold-reactive autoantibody (Answers A and E) the autocontrol should be positive. In addition, a cold-reactive antibody would be showing reactivity at immediate spin (IS). There could be multiple antibodies if there is an antibody to a high prevalence antigen along with an underlying "common" antibody. Though the different detection methods do have different analytic sensitivities, the antibody in this case is equally present in all methods; thus, this is not a method dependent antibody (Answer D)

Evaluate the following panel to continue this case...

Antibody Identification Panel—Patient Serum versus Untreated (UNT), Ficin-Treated, and DTT-Treated Panel Cells

	Rh						MNS				Lu		P1	Lewis		Kell		Duffy		Kidd		Saline		
	D	C	E	c	e	f	M	N	S	s	Lu^a	Lu^b	P1	Le^a	Le^b	K	k	Fy^a	Fy^b	Jk^a	Jk^b	UNT IAT	Ficin IAT	DTT IAT
1	+	+	0	0	+	0	+	+	+	+	0	+	+	+	0	0	+	0	+	0	+	3+	3+	0√
2	+	+	0	0	+	0	0	+	0	+	0	+	0	0	+	0	+	0	+	+	0	3+	3+	0√
3	+	0	+	+	0	0	+	+	+	+	0	+	+	0	+	+	+	+	+	+	0	3+	3+	2+
4	+	0	+	+	0	0	0	+	+	0	0	+	+	0	+	0	+	+	0	+	0	3+	3+	2+
5	0	+	0	+	+	+	0	+	0	+	0	+	0	0	0	0	+	0	0	+	+	3+	3+	0√
6	0	0	+	+	+	+	+	0	+	+	0	+	+	+	0	0	+	+	+	0	+	3+	3+	2+
7	0	0	0	+	+	+	0	+	+	+	0	+	+	0	+	+	0	0	+	+	+	3+	3+	0√
8	0	0	0	+	+	+	0	+	+	+	0	+	+	+	0	0	+	0	+	0	+	3+	3+	0√
9	0	0	0	+	+	+	+	+	+	0	0	+	0	0	0	0	+	+	+	+	+	3+	3+	0√
10	0	0	0	+	+	+	+	0	+	0	0	+	+	0	+	0	+	+	+	0	+	3+	3+	0√
11	+	+	0	0	+	0	+	+	+	+	0	+	+	0	+	0	+	+	0	+	+	3+	3+	0√
	Patient red blood cells (autocontrol)																					0√	0√	0√

40. Given the results of testing the patient's serum against ficin and DTT-treated RBCs which of the following antibodies choices is most likely?
A. Anti-k
B. Anti-Kp^b
C. Anti-Fy^a
D. Anti-M
E. Anti-s

Concept: A common approach to evaluating a patient with a possible antibody to a high prevalence antigen is to test the patient's serum/plasma with chemically modified RBCs. Proteolytic enzymes such as ficin or papain are often used. Dithiothreitol (DTT) or 2-aminoethylamine bromide (AET), sulfhydryl chemicals, is also used to denature certain blood group antigens. Results obtained after testing with chemically treated RBCs provides clues as to the specificity of the antibody (Table 6.12).

TABLE 6.12 Effect of Enzymes and Chemicals on Reactivity of Antibodies to High Prevalence Antigens

	Enzyme/Chemical			Enzyme/Chemical	
Anti-	Ficin	DTT/AET	Anti-	Ficin	DTT/AET
Co^a	+	+	Kp^b	+	0
Cs^a	+	+/0	Kn^a	W	0
Di^b	+	+	McC^a	W	0
Ge2	0	+	Sl^a	W	0
Ge3	+	+	Lu^b	+	0
Gy^a	+	W	LW^a	+	0
Hy	+	W	U	+	+
In^b	0	0	Vel	+	+
Js^b	+	0	Yk^a	+	0
K	+	0	Yt^a	0	0

W, Reactions are weaker or a weak antibody may be nonreactive.

Answer: *B*—Kell blood group system antigens are destroyed by sulfydryl compounds, such as DTT. Negative reactions with panel cells 1, 2, 5, 7, 8, 9, 10, 11 indicate the antigen is destroyed consistent with a high prevalence Kell system antigen (e.g., Kp^b and Js^b). Anti-Kp^b, -Js^b and –k would be negative with the DTT-treated panel, however, anti-k is ruled-out since panel cell no. 7 is k-negative and was 3+ (Answer A). The Duffy and MNSs groups are destroyed by ficin, which would not fit with the results in this case (Answers C, D, and E).

41. What antibody is also likely to be present?
A. Anti-S
B. Anti-Js^b
C. Anti-E
D. Anti-c
E. Anti-Jk^a

Concept: Given that blood group antigens are inherited characteristics an individual's ethnicity influences their phenotype. Determining an individual's ethnicity aids in identifying the specificity of antibodies.

Answer: C—Anti-Kp^b and anti-E are the most likely antibodies present, based on the testing results and the patient's ethnicity. Some blood group antibodies are associated with specific ethnic groups because being antigen negative is predominantly or only found in individuals of that ethnicity. This information aids in confirmation of the antibody. For example, the Js(b-) phenotype (Answer B) has been found exclusively in individuals of African background.

This patient is of Eastern European ancestry. The history was provided in this case; however, getting such information often requires further investigation with a call to the patient's nurse or a visit with the patient. Anti-E is probable as panel cells no. 3, 4, and 6 are E-positive and reactive with DTT treated RBCs. Anti-Jk^a (Answer E) can be eliminated based on negative reaction with panel cell no. 2 which is Jk(a+b−). If anti-c (Answer D) were present, panel cells no. 5, 7, 8, 9, and 10 would have been positive with the DTT treated RBCs. Anti-S (Answer A) would be destroyed by ficin and is therefore, incorrect.

42. Which of the following antibodies CANNOT be ruled-out using the DTT-treated panel?
A. Anti-S
B. Anti-K
C. Anti-e
D. Anti-Fy^a
E. Anti-Jk^b

Concept: One of the advantages of using chemically-treated RBCs is that you may use them for ruling-in and ruling-out other antibodies. However, when performing the rule-out, one must keep in mind which common antigens are also destroyed by various chemicals.

Answer: *B*—The K antigen is destroyed by DTT, so panel cell no. 7 is essentially K−k− therefore anti-K cannot be ruled-out. Anti-S (Answer A) is ruled-out on panel cells 1, 7, 8, 9, and 10, anti-e on panel cells 1, 2, 5, and 7–10. Anti-Fy^a (Answer D) is eliminated based on negative reactions with panel cell 11 and anti-Jk^b with panel cells 1, 8, and 10. Anti-e is ruled out based on negative reactions in panel cells 1, 2, 5, 7, 8, 9, 10, and 11 (Answer C). Anti-Jk^b is ruled out based on negative reactions in panel cells 1, 8, and 10 each having a double-dose (homozygous) expression of Jk^b antigen. (Answer E)

43. In this case, what is the phenotype of the RBCs that should be used to adsorb the anti-Kp^b?
A. K−, Kp(b−)
B. K+, Kp(b−)
C. K−, Kp(b+)
D. K+, Kp(b+)
E. It is impossible to adsorb anti-Kp^b

Concept: It is nearly impossible to rule-out anti-K in the presence of anti-Kp^b because of the lack of K+, Kp(a+b−) red cells. The common phenotype in the Kell system is positive for the high prevalence antigens, K−k+Kp(a−b+). It would be a highly unusual genetic event to have an individual who has inherited two low frequency alleles, K and Kp^a.

Another method used to rule-out underlying antibodies in the presence of an antibody to a high prevalence antigen is allogeneic adsorption. If the anti-Kp^b in this case is adsorbed from the patient's serum/plasma, anti-K can be eliminated when testing selected K+ red cells.

Performing allogeneic adsorption is time consuming and generally performed in immunohematology reference laboratories. Allogeneic adsorption would allow for known antibodies to be adsorbed/removed from the plasma to verify if any other antibody specificities are present. Allogeneic adsorption may be used to differentiate reaction patterns by choosing adsorption cells to split-out the reaction patterns. Table 6.13 demonstrates an allogeneic adsorption results when anti-Kp^b and anti-K are suspected.

TABLE 6.13 Allogeneic Adsorption Results

Antibody present	Adsorbing cell K+Kp(b−)	Adsorbing cell K−Kp(b+)	Adsorbing cell K−Kp(b−)
Anti-Kp^b	Leave anti-Kp^b	Adsorb out anti-Kp^b	Leave the anti-Kp^b
Anti-K	Adsorb out anti-K	Leave the anti-K	Leave the anti-K

An alternative to performing adsorptions, especially to rule-out anti-K, is to provide K-negative RBCs for transfusion. If one is unable to rule-out other antibodies or if the phenotype/genotype of the patient is known an alternative is to provide phenotype-matched blood. This can be challenging if the patient is negative for many antigens, for example, c−, E−, Fy(a−), Jk(b−), s− but relatively easy if the patient is negative for C, E, and K only.

Answer: *C*—The goal of an allogeneic adsorption is to remove the anti-Kp^b and leave anti-K. The RBCs must be Kp(b+) to adsorb the anti-Kp^b and K-negative to leave behind anti-K. Thus, the other choices (Answers A, B, D, and E) are incorrect due to the lack of the necessary antigen combinations.

Suggested Reading

[1] K. Rittenhouse-Olson, E. DeNardin, Contemporary Clinical Immunology and Serology, Pearson, Boston, (2012).
[2] H. Schonewille, Á. Honohan, L.M.G. van der Watering, F. Hudig, P.A. te Boekhorst, A.W.M.M. Koopman-van Gemert, A. Brand, Incidence of alloantibody formation after ABO-D or extended matched red blood cell transfusions: a randomized trial (MATCH study), Transfusion 56 (2016) 311–320.
[3] S. Zalpuri, R.A. Middelburg, H. Schonewille, K.M.K. de Vooght, S. le Cessie, J.G. van der Bom, J.J. Zwaginga, Intensive red blood cell transfusions and risk of alloimmunization, Transfusion 54 (2014) 278–284.
[4] M.K. Fung, B.J. Grossman, C.D. Hillyer, et al. (Eds.), Technical Manual, eighteenth ed., AABB, Bethesda (MD), 2014.
[5] D. Harmening (Ed.), Modern Blood Banking and Transfusion Practices, sixth ed., F.A. Davis, Philadelphia (PA), 2012.
[6] M.E. Reid, C. Lomas-Francis, M.L. Olsson, The Blood Group Antigen Facts Book, third ed., Elsevier, Oxford (UK), (2012).
[7] G. Daniels, Human Blood Groups, third ed., Wiley-Blackwell, West Sussex (UK), (2013).
[8] Petz, Garratty, Immune hemolytic anemias, second ed., Churchill-Livingstone, Philadelphia, (2004).
[9] M.E. Reid, MNS blood group system: a review, Immunohematology 25 (3) (2009) 95–101.
[10] D.G. Lutheran, Immunohematology 25 (4) (2009) 152–159.
[11] L. Dean, The Kell blood group, in: Blood Groups and Red Cell Antigens, NCBI, Bethesda, MD.
[12] N.D. Avent, M.E. Reid, The Rh blood group system: a review, Blood 95 (2000) 375–387.
[13] R.N. Makroo, et al. Anti-G antibody in alloimmunized pregnant women: report of two cases, Asian J. Transfus. Sci. 9 (2) (2015) 210–212.
[14] C.M. Westoff, The Rh blood group in review: a new face for the next decade, Transfusion 44 (2004) 1663–1673.
[15] W.J. Judd, S.T. Johnson, J. Storry, Judd's Methods in Immunohematology, third ed., AABB, Bethesda (MD), (2008).

CHAPTER

7

Pretransfusion Testing

Lance A. Williams, III, Patricia M. Raciti**, Helene DePalma[†], Huy P. Pham**

*University of Alabama at Birmingham, Birmingham, AL, United States; **New York-Presbyterian Hospital—Columbia University, New York, NY, United States; [†]City University of New York, Jamaica, NY, United States

Many tests are completed to ensure the safety of blood transfusions. The basic tests include ABO/Rh typing, antibody screening, and crossmatching. For more complex patients, such as those with allo- and/or autoantibodies, specialized techniques, such as antibody panels, elutions, enzymatic treatment, and adsorptions are often necessary to correctly identify the antibody and allow for the transfusion service to provide blood products suitable for transfusion. This chapter will provide a review of pretransfusing testing in the blood bank.

1. Which of the following describes the minimum information required on a blood sample for pretransfusion testing?
   A. Patient medical record number and date of birth
   B. Two unique patient identifiers, time of sample, and name of physician
   C. Patient name, blood type, and current location
   D. Two unique patient identifiers, phlebotomist name, and date/time of the sample
   E. Patient name, phlebotomist name, and date/time of the sample

   Concept: Accrediting agencies mandate that blood samples for pretransfusion testing have sufficient information to uniquely identify the patient and also a way to verify when the sample was collected and by whom. In most hospitals, the patient is identified by their name and their medical record number. The blood sample is also labeled with the phlebotomist's legible name or signature and the date/time of collection.

   Answer: *D*—This answer choice fulfills all the requirements of a properly labeled specimen, according to the AABB Standards. All the other choices (Answers A, B, C, and E) represent the incorrect combination of requirements or insufficient requirements.

2. Today is *June 8, 2016*. Which of the following samples can still be used for pretransfusion testing in a patient with a history of red blood cell (RBC) transfusion 82 days ago?
   A. A sample from June 1, 2016 collected at 11:00 pm
   B. A sample from June 2, 2016 collected at 11:00 pm
   C. A sample from June 3, 2016 collected at 11:00 pm
   D. A sample from June 4, 2016 collected at 11:00 pm
   E. A sample from June 5, 2016 collected at 11:00 pm

   Concept: If a patient was transfused, transplanted, or pregnant within the past 3 months, they should be tested every 3 days with a new type and screen, if transfusion is anticipated. This is to guard

Transfusion Medicine, Apheresis, and Hemostasis. http://dx.doi.org/10.1016/B978-0-12-803999-1.00007-9

against missing newly formed alloantibodies that could subsequently cause a hemolytic transfusion reaction.

Answer: *E*—The "3-day rule" applies to patients with a history of antibodies, those with a history of transfusion, transplantation, or pregnancy in the past 3 months, and/or those with an unclear history of transfusion, transplantation, or pregnancy. These factors imply an increased risk of new alloantibody formation, thus the need for more frequent testing, if transfusion is anticipated. The "3-day rule" can be confusing because the sample collection day is considered "day 0". So, in the question above, June 5th is "day 0'. Regardless of the exact collection time, the sample will expire at midnight on "day 3", which is June 8th. All the other choices (Answers A, B, C, and D) do not adhere to the 3 day rule.

3. What is the minimum amount of time that both recipient and donor samples must be retained following transfusion?
A. 3 days
B. 5 days
C. 7 days
D. 14 days
E. 30 days

Concept: Both recipient and donor samples must be retained a specific amount of time according to AABB Standards in case an adverse event occurs that requires investigation and possible testing of the sample used for pretransfusion testing.

Answer: *C*—Both recipient and donor samples must be retained for a *minimum* of 7 days following transfusion of blood products. Therefore, most blood banks retain samples for 10–14 days from the original collection date (for patient samples) or transfusion date (for donors' segments). For example, a patient sample collected on June 2nd will typically be refrigerated until June 12th. Therefore, if the patient is transfused on June 5th (the last day the sample is eligible for use in pretransfusion testing per the "3-day rule") and develops evidence of delayed hemolytic transfusion reaction on June 11th, the sample is still available for use in the transfusion reaction workup. All the other choices (Answers A, B, D, and E) do not meet the time requirements.

4. Evaluate the following ABO/Rh typing results and choose the answer that best matches the patient's blood type *AND* the standard color of the anti-A and anti-B reagents used during testing.

Forward			**Reverse**	
Anti-A	Anti-B	Anti-D	A1 Cells	B Cells
0	0	4+	4+	4+

A. Type AB Rh negative; anti-A = yellow, anti-B = purple
B. Type O Rh positive; anti-A = blue, anti-B = yellow
C. Type ABO Rh positive; anti-A = blue, anti-B = green
D. Type AB Rh negative; anti-A = blue, anti-B = yellow
E. Type O positive; anti-A = yellow, anti-B = blue

Concept: An ABO type is performed by using the patient's RBCs and plasma/serum to identify the presence of A and/or B antigen on the RBCs and the anti-A and/or anti-B antibodies in the plasma/serum. This is accomplished by using "forward and reverse" types. Table 7.1 demonstrates what results are like for a patient with blood type A positive:

TABLE 7.1 Pattern of Reactivity for an A Rh Positive Patient

Forward			Reverse	
Anti-A	Anti-B	Anti-D	A_1 Cells	B Cells
4+	0	4+	0	4+

In the "forward" type, patient RBCs are mixed with anti-A or anti-B reagent in separate test tubes. This allows for the detection of A or B antigen on the surface of the RBC. Agglutination of the RBCs constitutes a positive reaction. If neither tube demonstrates agglutination, then neither A nor B antigen is present, thus the patient is front-typed as an "O". In the example, there is a 4+ reaction when anti-A and anti-D is mixed with the patient's RBCs. Thus, the forward reaction is read as A Rh positive. ABO/Rh typing can also be performed by microplate or column agglutination technology.

The "reverse" type is a confirmation of the forward type. The reverse type is performed by mixing reagent RBCs containing either A_1 or B antigen with the patient's plasma/serum in separate test tubes. This will test for the presence of "naturally occurring" anti-A and/or anti-B. In the example aforementioned, the patient's forward type is A Rh positive, thus the expected reaction is for the patient to express anti-B in the reverse type.

Naturally occurring anti-A and/or anti-B are formed during childhood in response to gastrointestinal bacteria that have antigens mimicking the A and/or B antigens, thus the term "naturally occurring" is a bit of a misnomer. For example, a patient with blood type A would be expected to express A antigen on the RBCs and conversely express anti-B in their plasma/serum. They are not expected to have any naturally occurring anti-A. Remember, humans form antibodies to antigens that they lack and/or to foreign antigens to which they are exposed.

The Rh type only involves the forward type with the patient's RBCs being tested with anti-D reagent. Agglutination is interpreted as the patient being Rh positive. No reverse type is necessary because anti-D is not "naturally occurring," and should not be present without prior exposure to the D-antigen.

Anti-A and Anti-B typing reagents are purposefully colored differently to avoid ABO typing errors. The reagents are labeled as anti-A and anti-B, but the bright reagent colors give another visual clue to the technologist performing the testing.

Answer: *B*—There is no reaction/agglutination when the patient's RBCs are mixed with anti-A and anti-B reagent, indicating that the patient is likely blood type O. Patients with blood type O have no A or B antigens on their surface. There is a reaction when the patient's RBCs are mixed with anti-D, indicating the presence of the D antigen. The reverse type concurs with the forward type as the patient's plasma/serum contains both anti-A and anti-B that reacts against the reagent RBCs with A_1 and B antigen on their surface.

For the second part of the answer: anti-A is colored blue, while anti-B is colored yellow. One way to remember this is to associate "*B*ananas" with the Anti-*B*, thus remembering the color of Anti-*B* is yellow. All of the other choices (Answers A, C, D, and E) are incorrect based on the interpretation of the result and the color of the reagents.

5. Evaluate the following ABO discrepancy. Which disease process is most commonly associated with this discrepancy?

Forward			Reverse		Antibody Screen	Autocontrol
Anti-A	Anti-B	Anti-D	A_1 Cells	B Cells		
4+	2+	0	0	4+	0	0

A. Immunodeficiency
B. Testicular cancer

C. Colon cancer
D. Gilbert syndrome
E. Hashimoto's thyroiditis

Concept: When the forward and reverse type reactions do not agree, this is termed an ABO discrepancy. Weaker than expected or unexpectedly negative/positive results during ABO typing should prompt further investigation by the technologist. Discrepancies often happen when there is something missing or added to the patient's RBCs or plasma. The possible causes are numerous, but focused problem solving, testing, and review of the patient's medical/medication history often reveal the cause for the discrepancy.

Answer: C—This patient has positive reactions for anti-A and anti-B in the forward type, and a negative reaction with anti-D, leading one to believe that they could be blood type AB Rh negative. However, in the reverse type, the patient's plasma/serum only reacts against B cells; thus, it is more consistent with blood type A. The 2+ reaction with anti-B is the "discrepancy" in this case.

This pattern is most commonly seen in what is termed "acquired B". It is often seen in patients with colon cancer or gastrointestinal infection. The colon cancer causes damage to the gastrointestinal tract, which releases Gram-negative bacteria into the patient's blood stream. A deacetylase enzyme from the bacteria converts the A antigen sugar (*N*-acetyl-D -galactosamine) into D-galactosamine. This sugar is similar enough to the dominant B antigen sugar (D -galactose) to react weakly with the anti-B reagent, thus causing the discrepant positive result. This discrepancy can be resolved by using a more specific (monoclonal) anti-B reagent or acidified (pH 6) human anti-B. It should also be confirmed by evaluating the patient's history for colon cancer. If there is no such history, then the blood bank physician should alert the primary physician to the possibility of an occult colon carcinoma.

Immunodeficiencies (answer A) can cause type discrepancies but these will most likely be negative or extremely weak reactions in the reverse type. This is because these patients (e.g., primary immunodeficient patients, newborns, elderly) are not producing sufficient immunoglobulins to produce a "normal" reaction.

The conditions in the other choices (Answers B, D, and E) do not typically cause type discrepancies.

6. The type and screen results below reveal a patient with apparent blood type O Rh negative. However, the antibody screen is 4+ with every cell, as is the subsequent antibody panel that was completed. The technologist decides to attempt to crossmatch random O Rh negative and O Rh positive units with the patient, all of which are also 4+ incompatible. This leads you to suspect that the patient may express the rare Bombay phenotype. Which of the following is the only blood type that will be compatible with this patient?

Forward			**Reverse**		**Antibody screen**	**Autocontrol**
Anti-A	Anti-B	Anti-D	A_1 cells	B cells		
0	0	0	4+	4+	4+	0

A. Type A_2
B. Type B
C. Type A
D. Type AB
E. Type O_h

Concept: People with Bombay phenotype (O_h) completely lack H substance on their RBCs. As a review, H substance is required for blood cells to synthesize A and B antigens. Those that do not covert H substance into A or B antigen are labeled type O. However, in the case of Bombay phenotype, there is no H substance present. Thus, they form abundant amounts of anti-H. This is a problem because people *without* Bombay phenotype always have a varying amount of H substance on their RBCs. The order of the amount of H substance present on RBC blood groups is as follows from most to least: O > A_2 > B > A_2B > A_1 > A_1B > O_h.

Therefore, people with Bombay phenotype can *only* receive blood products from other people with Bombay phenotype to prevent a severe hemolytic transfusion reaction. *Ulex europaeus* is a lectin that can be used in cases of suspected Bombay phenotype. This lectin binds to RBCs with H antigen present. In Bombay phenotype, there would be no agglutination present.

Answer: *E*—At first, the patient aforementioned appears to be blood type O. However, the antibody screen is 4+, indicating the need for an antibody panel. The antibody ID panel will also demonstrate 4+ results, but the autocontrol will be negative, indicating that this is not an autoantibody. All of the panel cells are reacting 4+ because they are all type O and therefore, have abundant H substance for the patient's anti-H to react against. If you do not recognize this as a possible Bombay phenotype and label the patient at blood type O, transfusion with type O blood would result in a severe, acute, hemolytic transfusion reaction and possibly death. This is because type O blood has the most H substance present on it. In essence, Bombay patients can only receive blood products from other individuals with the Bombay phenotype.

However, any blood type that you mix with this patient will be incompatible, unless the blood is from another patient with Bombay phenotype (O_h). The other choices (Answers A, B, C, and D) would not have the pattern of reactivity seen previously.

7. Evaluate the following ABO discrepancy and choose the answer that combines how to resolve it and what ABO type you would recommend, if transfusion with RBCs is needed.

Without prewarm technique:

Forward			**Reverse**		**Antibody screen**	**Autocontrol**
Anti-A	Anti-B	Anti-D	A_1 Cells	B Cells		
4+	0	0	1+	4+	0	0

With prewarm technique (at 37°C):

Forward			**Reverse**		**Antibody screen**	**Autocontrol**
Anti-A	Anti-B	Anti-D	A_1 Cells	B Cells		
4+	0	0	1+	4+	0	0

A. *Ulex europaeus*; type O or AB
B. Hydatid cyst fluid; type O or A_1
C. *Arachis hypgaea*; type O or A_2
D. *Dolichos bifloris*; type O or A_2
E. *Salvia horminum* type O or A_1

Concept: Type A people can have many subgroups. The most common subgroups are A_1 (~80%) and A_2 (~20%). The remaining subgroups are very rare. *Dolichos bifloris* lectin can distinguish between group A_1 and the remaining subgroups because it only binds to A_1 cells. If it does not bind, then it is assumed that an "A" subgroup is present. There is also an A_2B subgroup. Of note, 1%–8% of A_2 patients develop anti-A_1, whereas 22%–35% of A_2B patients develop anti-A_1. Anti-A_1 usually reacts only at room temperature, and thus is usually considered clinically insignificant. However, if the reactivity is still present 37°C, then it is clinically significant, as in the question above.

The most common way to safely transfuse an A_2 patient is to place a comment in the blood-bank computer system that the patient should receive type O RBCs. This blood will be safe for the patient because there is no "A" antigen on type O RBCs. Otherwise, a special request will have to be made to the blood collection center to locate type $A_{subgroup}$ blood for this patient because his anti-A_1 reacts at 37°C. An additional reason that this is a reasonable strategy (i.e., transfusing with O RBCs) is that *Dolichos biflorus* lectin does not definitively identify A_2, it simply identifies the antigen as something other than A_1. So, in rare cases, the patient could be A_3, A_{el}, and so forth. Additional monoclonal reagents or special techniques can be useful for resolving ABO discrepancies due to weak subgroups. An extremely weak subgroup may not be detectable by serologic techniques, but can be resolved by ABO genotyping.

Answer: *D* – *D. biflorus* binds to A_1 antigen, but not to other subgroups of the "A" antigen. Thus, it is useful to distinguish patients with "A" subgroups. The anti-A_1 reactivity remains after testing at 37°C, and thus, this patient should only be transfused with group O or A_2 RBCs. Patients with "A" subgroups are most commonly transfused with group O RBCs.

U. europaeus (answer A) is used to bind to H antigen, as explained in Question 6 aforementioned. Hydatid cyst fluid (answer B) is used to neutralize antibodies to P1 antigen. *A. hypogaea* (answer C) is a peanut lectin used to identify T, Th, and Tk antigens, which sometimes develop on RBCs after infection with organisms in the Pneumococci family. *S. horminum* (Answer E) is used to differentiate the Tn antigen from T, Th, and Tk. Tn antigen develops secondary to infections with *E. coli*, among other organisms.

8. As a safety measure, blood banks are required to compare a patient's previous and current records/results. Which of the following is a required component of the comparison process?

A. Number of units transfused
B. ABO/Rh typing
C. Pregnancy status
D. The date and type of any upcoming surgery
E. Current medication list

Concept: It is important for the blood bank technologist to compare the current ABO/Rh results with historical results. Errors in testing, patient identifications (ID), or both may be detected when a discrepancy is identified. Many times, this is the only way to detect sample mix-ups, also called "wrong blood in tube" or WBIT. WBITs, if not identified can lead to severe, acute hemolytic transfusion reactions due to transfusion of ABO incompatible blood.

The problem may also arise if this is the first time the blood bank is testing the patient because there is no historical ABO/Rh type on the patient. This potential error can be mitigated by requiring a second sample be drawn at a different time from the initial sample, for confirmation of the ABO/Rh type.

Answer: *B*—The AABB Standards require that results of ABO/Rh testing on a current sample must be compared with previous transfusion service records. The standards used to require that this be performed on samples tested within a 12 month time frame; however, recent editions have left out that time frame, likely due to electronic health records that allow for checks that extend well beyond a 12 month time frame. The Standards also require records be reviewed for the presence of clinically significant antibodies, for difficulties in testing, for the occurrence of adverse reactions, and for special transfusion requirements. The comparison must be documented and any discrepancies must be investigated and appropriate action taken before a unit is issued for transfusion. While they certainly represent a thorough patient history, the other choices (Answers A, C, D, and E) are not part of the requirements.

9. An obstetric physician consults you regarding the ABO/Rh type of one of his newly pregnant patients. The patient has been a blood donor for 5 years, and she was repeatedly told by her local-blood center that she is O Rh positive; however, using tube testing, your lab identified her as being O Rh negative. A repeat of the testing on a different sample from the patient is also read as O Rh negative. What is the most likely cause of the discrepancy in this case?

A. The patient has weak D antigen
B. The blood center's results are an erroneous
C. Your lab's results are erroneous
D. The sample at your lab is from another person
E. The sample at the blood center is from another person

Concept: Individuals with "weak D" have fewer D antigens on their RBCs than D-positive individuals whose red cells express a conventional RhD protein. Weak D types result from a single amino acid change that likely affects the insertion of the protein in the red cell membrane. Standard Rh testing methodologies, such as tube testing, often do not detect this small amount of D antigen on the surface of the RBC.

Whenever a donor types as Rh negative at a blood collection center, weak D testing must be performed by indirect antiglobulin technique. The addition of antihuman globulin (AHG) strengthens the reaction of the donor's RBCs to the "reagent anti-D". If the reaction becomes positive with the addition of AHG, the donor is labeled as Rh positive.

This is done mainly for two reasons. First, to prevent even a small amount D antigen exposure to females of childbearing age who are Rh negative, who if exposed to D antigen could then form anti-D and possibly develop hemolytic disease of the fetus and newborn (HDFN). Second, Rh negative patients who have already formed anti-D in response to previous transfusions or pregnancies should not receive any products with D antigen present to prevent a potential hemolytic transfusion reaction.

Historically, weak D testing is not performed in the patient setting because there is no anticipated consequence in labeling a patient as Rh negative and transfusing Rh negative blood products, even if they do have weak D. However, new guidelines from the College of American Pathologists' Transfusion Medicine Resource Committee recommend that RHD genotyping be performed on patients of all types (e.g., pregnant females, newborns, transfusion recipients) if the initial testing is suggestive of a weak or variant D. This policy is meant to protect pregnant women from receiving RhIG unnecessarily and to prevent weak/variant D individuals from using up the Rh negative blood supply when they could safely receive Rh-positive products.

Answer: *A*—Donor centers perform weak D testing to prevent anyone with even a small amount of D antigen on their RBCs from being labeled as Rh negative to prevent exposure of Rh negative females of childbearing age and to those patients who have already formed anti-D. However, in a patient setting, the only consequence of labeling a "weak D" patient as Rh negative is that the patient will require Rh negative product, thus AHG is not added to Rh testing that is initially read as Rh negative. Therefore, the patient in the question likely has weak D and was recorded as being Rh positive at the donor center due to the addition of AHG. This patient will be listed as Rh negative in the patient/hospital setting. Therefore, there was no error by your lab or the blood center (answers B and C), nor is the sample from the wrong person (Answers D and E).

10. What is the role of the antibody screen in preventing transfusion reactions?

A. To screen for the presence of ABO antibodies
B. To screen for the presence of alloantibodies and ABO antibodies
C. To screen for the presence of alloantibodies and autoantibodies
D. To screen for the presence of reagent related antibodies
E. To screen for the presence of high titer low avidity antibodies

Concept: An antibody screen is used to check for the presence of unexpected allo- or autoantibodies in the patient's plasma/serum. With tube methodology, it is performed by labeling three tubes as I, II, III. To each of the tubes, reagent RBCs labeled as I, II, III are added to the appropriate tube. Then, the 2–3 drops of the patient's plasma/serum is added to each test tube. The sample is mixed, centrifuged to form a RBC "button", and then evaluated for incompatibility via agglutination at room temperature or the immediate spin (IS) phase, at body temperature (37°C), and at the AHG phase for agglutination/positivity. Agglutination at any of the phases is graded w+ to 4+. Typically, cold/IgM antibodies will react at the IS phase because their pentameric structure allows them to easily bind to RBCs and to one another causing visible agglutination. Due to their configuration as monomers, IgG antibodies will typically agglutinate only after AHG is added. AHG binds to the IgG, allowing the IgG-RBC complexes to form visible agglutinates.

If an antibody screen is positive, a full antibody panel will be necessary to identify the antibody present. If there is no reaction observed, then check cells (cells that have IgG bound) are added to ensure that the technologist performed all the steps and added AHG. If the check cells do not react, then the test must be repeated.

In many instances, enhancers, such as low-ionic saline solution (LISS) or polyethylene glycol (PEG), also added to the mixture to increase the sensitivity of the method. In some hospitals, a 2-cell screen (cells I and II) are used rather than the 3-cell screen (cells I, II, and III).

It is very important to know that ALL reagent RBCs used in antibody screens and panels are ABO type O. This is done purposefully to avoid any false positive reactions due to ABO antibodies. For example, if type A RBCs were used as the reagent cells, then a type B patient would react with every panel cell, but this will be due to their naturally occurring anti-A, not an allo- or autoantibody.

Answer: C—There can be many causes for a positive antibody screen, but the clinically significant reasons that must be investigated by performing a full antibody ID panel. A person exposed to foreign RBCs, either through transfusion, transplant, or pregnancy is at risk of forming alloantibodies. In contrast to alloantibodies, any person can form an autoantibody, though patients with autoimmune diseases tend to be at a higher risk of this. The other choices (Answers A, B, D, and E) do not represent the main reason for performing an antibody screen.

11. Nontube test methods are commonly used as an alternative to tube testing, as they are more easily automated, offer barcode capability for positive patient ID and provide a less subjective endpoint than tube testing. Select the correct statement:

A. A strong positive reaction is indicated when all indicator cells are in a solid button at the bottom of the microplate well in a solid phase red cell adherence test
B. IgG coated indicator cells are added following the washing step in column agglutination technology
C. In a solid phase red cell adherence test, check cells are added to all negative wells
D. The smallest red cell agglutinates are trapped at the top of the dextran acrylamide gel in column agglutination technology
E. Column agglutination techniques require no washing steps to demonstrate an IgG antibody

Concept: Nontube test methods for antibody detection/ID include solid phase red cell adherence tests and gel/column agglutination technology. In a solid phase red cell adherence test (Fig. 7.1), microplate wells are coated with red cells or red cell membranes and patient plasma/serum is added.

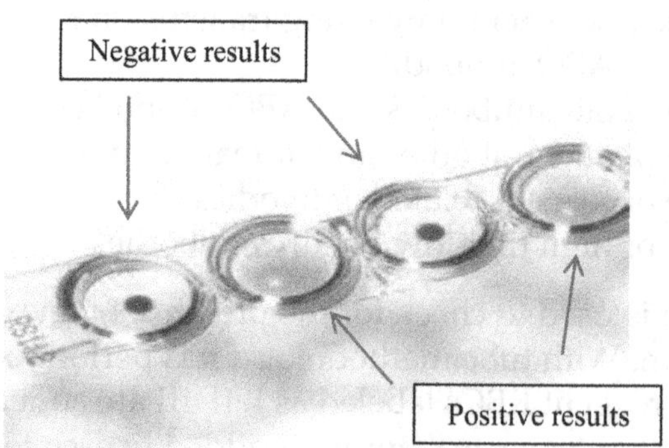

FIGURE 7.1 **Solid phase wells.** Capture Solid phase technology. *Source: Image courtesy of Immucor, Inc.*

Following 37°C incubation, the wells are washed to remove unbound antibodies and anti-IgG-coated indicator red cells are added. After centrifugation, a positive reaction is demonstrated by the diffuse adherence of the indicator cells around the microplate well. In a negative reaction, the indicator cells pellet to the bottom of the microplate well.

In gel/column agglutination technology (Fig. 7.2), a suspension of RBCs and the patient plasma/serum is added to the chamber at the top of the column. Following 37°C incubation, the columns undergo centrifugation to separate the agglutinated from nonagglutinated RBCs. Agglutinated RBCs are too large to migrate through the gel matrix while nonagglutinated RBCs can pass through the gel pores to the bottom of the column.

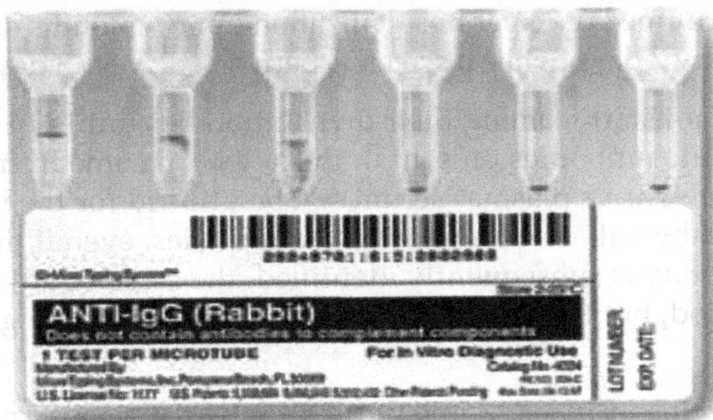

FIGURE 7.2 **Gel columns.** ID-MTS Gel technology: Reaction grades in the 6 microtubes (left to right): 4+, 3+, 2+, 1+, 0, 0. *Source: Image courtesy of Ortho Clinical Diagnostics.*

Answer: *E*—Column agglutination technology detects antibodies when the red cell agglutinates are trapped in a dextran acrylamide gel, following incubation at 37°C and centrifugation. The technique does not require a washing step.

Answer A is incorrect as this describes a negative test by solid phase red cell adherence; a strong positive would show a diffuse pattern of the indicator cells over the entire surface of the microwell. Answer B is incorrect because column agglutination technology requires neither, IgG coated indicator cells or washing steps; these are requirements of solid phase technology. Answer C is incorrect because check cells are not required for either solid phase red cell adherence tests or column agglutination technology. Answer D is incorrect because the largest red cell agglutinates are trapped at the top of the dextran acrylamide gel in column agglutination technology.

12. A patient previously tested in your blood bank comes in for a type and screen 12 days prior to undergoing a coronary artery bypass graft surgery. His blood type is confirmed as B Rh positive and his antibody screen is negative. He has no history of transfusions in the past three months and no history of alloantibodies. How soon will the patient require a new type and screen?

A. The time is at the discretion of the medical director
B. The time is at the discretion of the blood-bank supervisor
C. The time must be within 3 days of the expected surgery
D. The time must be within ten days of the expected discharge date
E. The time is at the discretion of the clinician

Concept: In patients with no previous exposure to foreign RBCs via transfusion or pregnancy, the risk of alloantibody formation is extremely low, thus the expiration date of the type and screen can be extended to more than 3 days. Current accreditation agencies do not give any strict guidance on this situation. Thus, the medical director can set the length of time that this presurgery specimen can be used for crossmatching. Typically, this ranges between 14 and 30 days.

Answer: *A*—The medical director decides the time a presurgery specimen is viable for pretransfusion testing in patients with no history of antibodies and no history of transfusion or pregnancy in the past 3 months. The other choices (Answers B, C, D, and E) do not represent the correct policy, according to regulatory and accrediting agencies.

13. Testing of neonates prior to transfusion consists of ABO/Rh typing using the neonate's RBCs and an antibody screen using either the neonate's or mother's plasma/serum. Which of the following describes the testing requirements for neonates prior to transfusion?
A. Type and screen is performed every 3 days
B. Type is performed every 3 days, while a screen is performed every 7 days
C. Type and screen is performed once per hospitalization up to 4 months of age
D. Type and screen is performed once per hospitalization up to 6 months of age
E. Type is performed every 7 days, while a screen is performed every 3 days

Concept: Newborns do not start forming their own antibodies until 4–6 months of age. Therefore, any allo- or autoantibodies present in the newborn come from the mother. Thus, either the mother's or newborn's plasma/serum can be utilized for the initial antibody screen. Additionally, since the baby will not form any new antibodies, even if the initial screen is positive and an alloantibody is subsequently identified, the screen does not need to be repeated during the neonatal period, but the newborn should receive antigen negative RBCs for the alloantibody identified.

Answer: *C*—Newborn testing includes an initial ABO/Rh type (using the newborn's RBCs) and an antibody screen (using either the newborn's or mother's plasma/serum). Of note, in contrast to ABO/Rh testing on adult patients, a reverse blood type is not necessary in newborns because they have not yet formed their "naturally occurring" ABO antibodies.

Some students may ask, "why aren't the mother's ABO antibodies present in the newborns serum?" This is because ABO antibodies are mostly IgM, which are too large to cross the placenta. The exception to this is mothers that are blood type O. These mothers possess both IgM and IgG antibodies (anti-A,B) that can cross the placenta. If the fetus is type A or B, this can lead to hemolytic disease of the fetus and newborn (HDFN) due to ABO incompatibility between the mother and fetus/newborn. In fact, in the United States, since Rh Immune Globulin is so widely used to prevent maternal formation of anti-D, HDFN is most commonly caused by ABO incompatibility between the mother and fetus rather than anti-D.

In contrast, alloantibodies, such as anti-D or anti-K, are typically IgG and can easily cross the placenta and into the fetus' circulation where they can then attack the fetus' RBCs if the corresponding antigen is present. The other choices (Answers A, B, D, and E) are incorrect based on the information aforementioned.

14. Which of the following describes the infectious disease testing requirements for autologous blood?
A. Infectious disease testing must be performed if the patient has a history of HIV
B. Infectious disease testing must be performed if the product will be transfused outside of the collecting facility
C. Infectious disease testing must be performed if the patient has recently returned from Africa
D. Infectious disease testing must be performed if the patient has no previous history of infectious disease
E. Infectious disease testing must be performed if the product will be transfused within the same facility at which it was collected

Concept: *Allo*geneic blood products are those that are collected from a donor to be used in another person. This includes volunteer donors who do not know the intended recipient and "directed donors" who donate for a specific patient (e.g., when a family member is going for a surgery and asks a relative to donate blood for them).

Autologous products are those that the patient donates for themselves. This typically occurs when the patient is expected to require blood during a surgery, but does not wish to receive allogeneic blood products. This blood can ONLY be used for the autologous donor and must have a special label to indicate "AUTOLOUS USE ONLY." If the unit is unused after the patient is discharged, it must be discarded and cannot be placed into the general inventory of blood products for use in another patient.

Answer: *B*—If the autologous unit will be transfused within the collection facility, no transfusion transmitted disease (TTD) testing is required, but the unit must be labeled "NO INFECTIOUS DISEASE TESTING COMPLETED". For example, if the unit will be collected and stored in the same hospital that the surgery will take place, then TTD testing is not required. Autologous blood products require TTD testing if the unit will be transfused outside of the collecting facility. TTD testing on such donors should take place every 30 days if repeat collections are necessary. The other choices (Answers A, C, D, and E) are incorrect based on standard policies for the handling of autologous units.

15. Which of the following patients meets the requirements for computer crossmatching?

A. A patient with no antibody history per the blood bank at a neighboring hospital
B. A patient with a nonspecific antibody identified four months ago, but the current antibody screen is negative
C. A patient who is a regular blood donor with a confirmed ABO type
D. A patient with no antibody history, a confirmed ABO type, and a negative antibody screen
E. A patient who has a confirmed ABO type and a weak positive antibody screen

Concept: A crossmatch is typically performed by mixing patient's plasma/serum with donor RBCs. This is a final safety check for patient to donor compatibility. If any agglutination takes place, the unit is considered "incompatible".

There are three basic types of crossmatching as follows:

Computer crossmatching is performed by a computer with validated software for this procedure after a technologist has confirmed that ABO/Rh type of the patient and confirmed that the antibody screen is negative. This is done by a computer program and no mixing of the patient's serum/plasma and donor RBCs takes place. The requirements of patient qualification are listed in the answer later mentioned. As you can imagine, this is the least rigorous of the crossmatching procedures, but is generally safe for patients with a confirmed ABO type, no antibody history, and a negative antibody screen.

IS crossmatching is designed to confirm the ABO compatibility between patient's plasma/serum and donor RBCs. It involves mixing two drops of the patient's plasma/serum in a tube with one drop of a 2%–5% suspension of donor RBCs. If agglutination occurs, the unit is incompatible. The qualification requirements for IS crossmatching are similar to those of computer crossmatching. The difference is that IS crossmatching involves a technologist performing the actual testing.

Lastly, AHG (i.e., "extended" or "full") crossmatching is required whenever a patient has a past history of antibodies or a positive antibody screen. Extended crossmatching is performed by mixing the patient's plasma/serum, donor RBCs, and AHG (similar to performing the antibody screen). AHG is added to facilitate the crosslinking of IgG antibodies together, so that agglutination can be detected visually by the technologist. Additionally, it is termed "extended" crossmatching because the mixture of donor RBCs and patient's plasma/serum is allowed to incubate for 30 min to 1 h at 37°C before AHG is added and sample is checked for agglutination.

Enhancement media such as LISS or PEG can be used to reduce the incubation time for extended crossmatches. Gel/column agglutination and solid phase methodologies can also be used for crossmatches. It is important to remember that no technique will detect all antibody specificities, thus each has advantages and disadvantages. For example, PEG is considered a more sensitive enhancement media compared to LISS, when used in tube testing. Both gel and solid phase are considered more sensitive than tube testing and have the advantage of well-defined endpoints.

Answer: *D*—To qualify for computer crossmatching, a patient must have no antibody history, a negative antibody screen, and an ABO type that has been confirmed according to AABB Standards. None of the other choices (Answers A, B, C, and E) meet these requirements. From the blood bank's perspective, the computer system must contain the following information on the component to be transfused: ABO group (from the label and manually confirmed by technologist); Rh type (no confirmation necessary); component name (e.g., thawed plasma); and the unit number. The following patient specific information must also be in the computer system: two unique identifiers; ABO group; Rh type; and the results of the antibody screen. The software functionality for the selection of only ABO-compatible red cells for transfusion must have been validated on site by the facility. If there is any mismatch in the information, there must be a way for the computer to alert the technologist and/or for the technologist to correct the data.

16. A 47 year old male is admitted to the emergency department (ED) due to GI bleeding and two units of RBCs are ordered. The patient was transfused 6 months ago and an anti-E antibody was identified at that time. The current type and screen specimen is tested and found to be O Rh positive and the antibody screen is negative. Which crossmatch procedure is indicated?
A. IS crossmatch of E- RBCs
B. LISS antiglobulin crossmatch of O Rh negative RBCs
C. Computer crossmatch, since the antibody screen is negative
D. PEG antiglobulin crossmatch of O Rh positive units that are E-
E. Ficin crossmatch of O Rh positive units that are E-

Concept: Once a clinically significant antibody has been identified, the patient is no longer eligible for an abbreviated crossmatch procedure, such as IS crossmatch or computer crossmatch. Selection of RBC units lacking the corresponding antigen is required, even in the absence of a currently demonstrable antibody.

Answer: *D*—A full or extended antiglobulin crossmatch is required. Selection of E-negative donor units is indicated to prevent stimulation of the anti-E by transfusion of E-positive donor units and the subsequent reduced survival of the transfused RBCs due to an anamnestic response. The patient is not eligible for the abbreviated crossmatch procedures (Answers A and C). Answer B is incorrect because the O Rh-negative units would need to be tested for E antigen prior to crossmatch. Answer E is incorrect because ficin crossmatches are not performed, since proteolytic enzymes like ficin denature some red cell antigens.

17. Which of the following blood products requires a crossmatch?
A. 5% albumin
B. Platelets with 3 mL of RBCs
C. Plasma with 1.5 mL of RBCs
D. Platelets with 1.5 mL of RBCs
E. Plasma derived factor VIII concentrate

Concept: The goal of the crossmatch is to detect antibodies in the patient that might hemolyze transfused RBCs. If a product does not contain a significant amount of RBCs, then any hemolysis that does occur should not be clinically significant.

Answer: *B*—Any product with $\geq$2 mL of RBCs requires crossmatching due to the risk of clinically significant hemolysis, if any incompatibility exists. 5% albumin and plasma derived factor VIII concentrates do not contain a significant amount of RBCs to warrant crossmatching (Answers A and E). Answers C and D contain $<$2 mL of RBCs, and thus, do not require a crossmatch.

18. A patient is admitted to the hospital with a diagnosis of multiple myeloma. He also has a history of anti-Jk^b at a previous hospital. However, in 3 years of testing at your blood bank, the patient has always had a negative antibody screen. With hemoglobin of 6.5 g/dL, a unit of RBCs was requested. A technologist accidentally releases Jk^b positive RBCs for this patient. Shortly after the initiation of transfusion, the patient complains of acute onset severe back pain, as well as pain at the infusion site. His vitals are stable except the temperature went up from 98.6 to 100.9°F. What is the most likely cause of these signs/symptoms?

A. The patient experienced an acute hemolytic transfusion reaction due to an anamnestic response
B. The patient experienced a hemolytic transfusion reaction due to the formation of a new alloantibody
C. The patient experienced a febrile nonhemolytic transfusion reaction
D. The patient did not have a transfusion reaction. His fever is suggestive of neutropenic fever. His pain is consistent with his underlying diagnosis of multiple myeloma
E. The patient does not experience a transfusion reaction. His fever, back pain, and pain at the infusion site are consistent with his history of malingering

Concept: Kidd antibodies (anti-Jk^a and Jk^b) are considered evanescent antibodies. This means that they are notorious for falling below the detection limit of blood bank antibody screening, but then return briskly via an amnestic response once reexposure to Kidd antigens occurs, either through a blood transfusion or pregnancy. Kidd antibodies are also unique because they can cause both acute and delayed hemolytic transfusion reactions. They can cause acute transfusion reactions due to their ability to activate complement.

Answer: *A*—Fever is the most common symptom in a hemolytic transfusion reaction. Additionally, the patient's back pain and pain at the site of infusion are also suggestive of hemolysis. Thus, this patient is likely experiencing a hemolytic transfusion reaction due to the amnestic response of anti-Jk^b. The timing of the reaction is too soon for new alloantibody formation (Answer B). Though a febrile reaction (Answer C) is a possibility, the patient's other symptoms point toward a more serious reaction. The patient has no reason for neutropenia (Answer D) or a history of malingering (Answer E) at this time and the other symptoms are more suggestive of acute hemolysis.

19. A 22 year-old female patient with sickle cell disease presents to the outpatient clinic for her monthly transfusion therapy. She had a red cell exchange transfusion with 8 units of RBCs 3 weeks ago at another facility. The current type and screen shows the patient is B Rh positive, the antibody screen is positive and DAT is negative. The antibody ID appears to indicate anti-e; however, the transfusing facility has Rh phenotype for the patient indicating her RBCs are e positive. What is the most appropriate course of action?

A. Repeat the e typing on the patient's RBCs using a monoclonal reagent
B. Repeat the e typing after the patient's RBCs have been treated with acid glycine/EDTA (EGA)
C. Perform an autoadsorption to remove the anti-e
D. Submit the patient's sample for molecular-based testing to evaluate the e antigen
E. No further testing is indicated—provide e positive blood for any future transfusions

Concept: Patients with sickle cell disease present serological challenges for the blood bank as they are commonly alloimmunized, often with multiple antibodies. As these patients are frequently transfusion dependent, obtaining an accurate red cell phenotype may require the use of special techniques. Availability of a full RBC phenotype can inform current and future transfusion decisions and aid in the management of these patients.

Answer: *D*—The patient's red cells may express a variant e antigen, explaining the findings of an anti-e in a patient reported as e positive. Since the patient has been recently transfused, evaluating the red cells by testing with different monoclonal antisera to detect a variant antigen would not provide accurate information regarding the patient's RBCs. Additionally, some variant antigens can't be detected or distinguished serologically. Since molecular-based testing is performed on nucleated cells from the patient, most commonly from a whole blood sample, the transfused donor red cells do not interfere. The phenotype of the patient's RBCs is predicted based on the genotype obtained.

Molecular-based testing can be used to determine a predicted phenotype when the RBCs have a positive DAT.

Answer A is incorrect as the transfused donor cells would interfere with the testing. Answer B is incorrect; the DAT is negative therefore EGA treatment is not indicated. Answer C is incorrect; autoadsorptions cannot be performed on a recently transfused patient and the negative DAT suggests the anti-e is not an autoantibody. Answer E is incorrect as further investigation of the e antigen would be needed before assuming the anti-e is an autoantibody.

20. Please provide the most likely interpretation for each of the following antibody panels. Each answer may be used once, more than once, or none.

Panel 1				Panel 2				Panel 3				Panel 4				Panel 5			
	IS	37°C	AHG		IS	37°C	AHG		IS	37C	AHG		IS	37°C	AHG		IS	37°C	AHG
1	0	0	3+	1	3+	0	0√	1	0	0	2+	1	0	0	w+	1	1+	1+	0√
2	0	0	3+	2	3+	0	0√	2	0	0	0√	2	0	0	w+	2	1+	1+	0√
3	0	0	3+	3	3+	0	0√	3	0	0	0√	3	0	0	w+	3	1+	1+	0√
4	0	0	3+	4	3+	0	0√	4	0	0	0√	4	0	0	w+	4	1+	1+	0√
5	0	0	3+	5	3+	0	0√	5	0	0	2+	5	0	0	w+	5	1+	1+	0√
6	0	0	3+	6	3+	0	0√	6	0	0	0√	6	0	0	w+	6	1+	1+	0√
7	0	0	3+	7	3+	0	0√	7	0	0	2+	7	0	0	w+	7	1+	1+	0√
8	0	0	3+	8	3+	0	0√	8	0	0	0√	8	0	0	w+	8	1+	1+	0√
9	0	0	3+	9	3+	0	0√	9	0	0	0√	9	0	0	w+	9	1+	1+	0√
10	0	0	3+	10	3+	0	0√	10	0	0	0√	10	0	0	w+	10	1+	1+	0√
AC	0	0	3+	AC	3+	0	0√	AC	0	0	0√	AC	0	0	0√	AC	1+	1+	0√

	Answer choices
Panel 1 ___	A. Single alloantibody
Panel 2 ___	B. Cold autoantibody
Panel 3 ___	C. Reagent related antibody
Panel 4 ___	D. Warm autoantibody
Panel 5 ___	E. HTLA-like antibody

Concept: Immediate recognition of certain patterns will speed up your performance in the real world and during exams. Some key features may help in quickly identifying the antibody ID. First, the autocontrol tells you if the patient is reacting against their own RBCs. This test is performed by mixing the patient's RBCs with their own plasma/serum. If agglutination occurs, this could signify an autoantibody or recent transfusion. Second, the phase of reactivity is important. If the antibody reacts only at IS phase, a cold/IgM antibody is likely present. However, if the antibody is only present at the AHG phase, a warm/IgG antibody is likely present. Third, the number of reactive cells present. If only a few cells are reactive, then an alloantibody could be present. If every cell is positive, this could either mean the presence of an autoantibody, multiple alloantibodies, antibody to a high frequency antigen, or reagent antibody. In the case of an autoantibody, the antibody is against an antigen common to most human RBCs, so the panel will be "panreactive" and the autocontrol will be positive.

Answer: *Panel 1* is a likely a warm autoantibody (*answer D*). The key features are the positive autocontrol, the panreactive panel, and the reactivity only at AHG phase. Additionally, the DAT should be positive for polyspecific and IgG. An adsorption procedure will need to be completed to detect any underlying alloantibodies (*see description later in Questions 21 and 26)*.

Panel 2 is likely a cold autoantibody (*answer B*). The key features are the positive autocontrol, the panreactive panel, and the reactivity only at IS phase. Additionally, the DAT should be positive for polyspecific and C3. Adsorption procedures are not usually necessary for cold autoantibodies because they do not interfere with the AHG phase of testing where most clinically significant alloantibodies are detected. Additionally, cold autoantibodies are present in about 30%–40% of the population and are usually only clinically significant if the patient is undergoing a surgery where the operating room or instrument is cooled to a very low temperature. If the cold autoantibody is interfering with IS crossmatching, the patient's plasma/serum can be warmed to 37°C prior to testing, thus eliminating the interference.

Panel 3 is likely a single alloantibody (*answer A*). The key features are the negative autocontrol, the few numbers of reactive cells, and the reactivity only at AHG phase. This panel will need to be solved to identify the antibody.

Panel 4 is either a high-titer low avidity antibody (now termed a HTLA-like antibody) or an antibody to a high-prevalence antigen (also known as high-frequency antigen—for the purpose of this book, these terms are used interchangeably) (*answer E*). The key features are the negative autocontrol, the weakly (w+) panreactive panel, and the reactivity only at AHG phase. HTLA-like antibodies will remain the same strength even after serial dilutions/titer levels. In contrast, an antibody to a high frequency antigen will weaken with serial dilutions. HTLA-like antibodies are usually clinically insignificant, whereas antibodies to high frequency antigens can be clinically significant and may require antigen negative blood for safe transfusion.

Panel 5 is likely a false positive reaction secondary to reagent related antibodies (*answer C*). The key features are the positive autocontrol, the panreactive panel, and the reactivity only at the IS and 37°C phases. Additionally, the DAT will be negative because a washing step occurs that removes the reagent before testing. This panel will need to be repeated without the reagent or with an alternative reagent.

21. Evaluate the following panel and choose the correct association with the antibody you identify.

	D	C	c	E	e	f	V	C^W	K	k	Kp^a	Kp^b	Js^a	Js^b	Fy^a	Fy^b	Jk^a	Jk^b	Le^a	Le^b	P_1	M	N	S	s	Lu^a	Lu^b	IS	37	AHG
1	+	+	0	0	+	0	0	+	0	+	0	+	0	+	+	0	+	+	0	+	+	+	+	+	+	0	+	0	0	3+
2	+	+	0	0	+	0	0	0	+	+	0	+	0	+	0	+	+	0	0	0	+	+	0	+	+	0	+	0	0	0√
3	+	0	+	+	0	0	0	0	0	+	0	+	0	+	+	+	0	+	0	+	+	+	0	+	+	0	+	0	0	3+
4	+	0	+	0	+	+	+	0	+	0	0	+	0	+	0	0	+	0	0	0	+	+	+	0	+	0	+	0	0	0√
5	0	+	+	0	+	+	0	0	0	+	0	+	0	+	0	+	0	+	+	0	0	+	+	+	+	0	+	0	0	3+
6	0	0	+	+	+	+	0	0	0	+	0	+	0	+	+	+	+	+	0	+	+	0	+	0	+	0	+	0	0	3+
7	0	0	+	0	+	+	0	0	+	+	0	+	0	+	0	+	+	+	+	0	0	+	0	+	0	0	+	0	0	3+
8	0	0	+	0	+	+	0	0	0	+	0	+	0	+	+	0	+	0	+	0	+	0	+	+	+	0	+	0	0	0√
9	0	0	+	0	+	+	0	0	0	+	0	+	0	+	0	+	0	+	0	+	+	+	+	+	+	0	+	0	0	3+
10	+	0	+	+	0	+	0	0	0	+	0	+	0	+	0	0	+	+	+	0	+	+	+	+	0	0	+	0	0	3+
11	+	+	0	0	+	0	0	0	0	+	0	+	0	+	0	+	+	+	+	0	+	+	0	+	0	0	+	0	0	3+
PC																												0	0	0√

A. This antibody is known for its evanescent properties and potential for intravascular hemolysis
B. This antibody is normally cold reacting but can also react at body temperature
C. This antibody is associated with an antigen that serves as a receptor for Malaria
D. This antibody was previously the most common cause for HDFN
E. This antibody is associated with an antigen that serves as a receptor for *H. pylori*

Concept: This is a fairly straight-forward antibody panel that reveals the presence of anti-Jk^b. The general process of solving an antibody panel is as follows:

1. Review the overall pattern of reactivity and the phases in which the reactivity occurs.
2. Review the results of the autocontrol.
3. Use negative cells to rule out antigens that are present on the panel cell.
4. After you have ruled out as many cells as possible, see if any of the remaining antibodies fit the pattern of positive reactions that remain.
5. If necessary, choose another panel to complete the rule out process and positively identify the antibody present.

Answer: *A*—Kidd antibodies (Jk^a and Jk^b) are notorious for disappearing from detection and then coming back briskly when reexposure to the antigen occurs via transfusion or pregnancy. This feature is termed evanescence. Often, these antibodies will not even be detected by an initial antibody screen. If the patient moves to a different area and their new hospital blood bank is not aware of the presence of a previous Kidd antibody, transfusion of red cells could lead to a hemolytic transfusion reaction.

Answer B describes typically cold reacting antibodies such as anti-M. These antibodies typically react at cold temperatures and are usually not clinically significant; however, sometimes they can become reactive at 37°C and require that antigen negative blood be provided. Answer C describes the Duffy blood group (Fy^a and Fy^b). These antigens serve as receptors for Malaria, which explains why many African Americans lack these antigens on their RBCs, as an evolutionary protection against Malaria.

Answer D describes anti-D. Before the use of Rh Immune Globulin, anti-D was the most common cause for HDFN. However, with the success of Rh Immune Globulin, ABO incompatibility between the mother and fetus, typically and O mother to an A or B fetus, is now the most common cause of HDFN in the United States. Answer E describes anti-Le^b. Le^b antigen acts as the receptor for *H. pylori*, along with the H antigen.

22. Evaluate the following antibody panel and choose the answer that best describes the antibody(s) that are present.

	D	C	c	E	e	f	V	C^W	K	k	Kp^a	Kp^b	Js^a	Js^b	Fy^a	Fy^b	Jk^a	Jk^b	Le^a	Le^b	P$_1$	M	N	S	s	Lu^a	Lu^b	IS	37	AHG
1.	+	+	0	0	+	0	0	+	0	+	0	+	0	+	+	0	+	+	0	+	+	+	+	+	+	0	+	0	0	3+
2.	+	+	0	0	+	0	0	0	+	+	0	+	0	+	0	+	+	0	0	0	+	+	0	+	+	0	+	0	0	0√
3.	+	0	+	+	0	0	0	0	0	+	0	+	0	+	+	+	0	+	0	+	+	+	0	+	+	0	+	0	0	1+
4.	+	0	+	0	+	+	+	0	+	0	0	+	0	+	0	0	+	0	0	0	+	+	+	0	+	0	+	0	0	0√
5.	0	+	+	0	+	+	0	0	0	+	0	+	0	+	0	+	0	+	+	0	0	+	+	+	+	0	+	0	0	0√
6.	0	0	+	+	+	+	0	0	0	+	0	+	0	+	+	+	+	+	0	+	+	0	+	0	+	0	+	0	0	1+
7.	0	0	+	0	+	+	0	0	+	+	0	+	0	+	0	+	+	+	+	0	0	+	0	+	0	0	+	0	0	0√
8.	0	0	+	0	+	+	0	0	0	+	0	+	0	+	+	0	+	0	+	0	+	0	+	+	+	0	+	0	0	3+
9.	0	0	+	0	+	+	0	0	0	+	0	+	0	+	0	+	0	+	0	+	+	+	+	+	+	0	+	0	0	0√
10.	+	0	+	+	0	+	0	0	0	+	0	+	0	+	0	0	+	+	+	0	+	+	+	+	0	0	+	0	0	0√
11.	+	+	0	0	+	0	0	0	0	+	0	+	0	+	0	+	+	+	+	0	+	+	0	+	0	0	+	0	0	0√
PC																												0	0	0√

A. Anti-E and anti-Fy^a are both present
B. Anti-Fy^a is present alone
C. Anti-Jk^b and anti-e are both present
D. Anti-Fy^a and anti-K are both present
E. Anti-E is present alone

Concept: Some blood group systems demonstrate "dosage" effect. Dosage effect is characterized by very strong reactions to "double-dose"/homozygous antigen expression, but weak reactions to "single-dose"/heterozygous expression. Note: some authorities recommended using homozygous and heterozygous only when describing genes or alleles. Essentially, the more antigens present, the stronger the agglutination reaction will be. Antibodies to the Rh, Kidd, Duffy, MNS, and Lewis blood group systems can demonstrate dosage. Thus, caution should be used in attempting to rule these antibodies out on a panel cell with heterozygous expression alone.

Additionally, since the strengths of the reactions vary, dosage effect can be confused with the presence of multiple alloantibodies, if careful attention is not paid to the homozygosity or heterozygosity of the panel cells.

Answer: *B*—Anti-Fy^a is demonstrating dosage effect in this panel with 3+ agglutination reactions for homozygous cells, Fy(a + b−) and 1+ reactions with heterozygous cells, Fy(a + b +). The other choices (answers A, C, D, and E) do not match the pattern of this panel.

23. The blood bank receives a preoperative type and screen specimen from a 73 year-old male patient scheduled for hip replacement surgery. The specimen tests as B Rh positive and the antibody screen by column agglutination technology is negative. The IS crossmatch with both B Rh positive RBC units is 1+ incompatible. What should the technologist do next?

A. No further serological testing is required since the antibody screen was negative. Perform a computer crossmatch with O Rh positive RBCs
B. Perform an antibody ID by column agglutination technology to identify the antibody directed against a low prevalence antigen
C. Perform an antibody ID by tube method, including an IS phase, to identify the likely IgM antibody
D. Perform an antibody ID by enzyme technique
E. Repeat the crossmatches using a prewarm technique and issue any units found compatible

Concept: Unexpected results can be obtained when a crossmatch is performed when the antibody present is not detected by the antibody screen method. Column agglutination technology generally detects IgG antibodies. The cause of the incompatible crossmatches needs to be investigated before RBCs are issued for transfusion.

Answer: *C*—Identify the antibody responsible for the incompatible IS crossmatches. As IgM antibodies react best at IS/room temperature, including this phase of reactivity will provide the optimal reaction conditions to demonstrate the antibody. Once identified, the appropriate crossmatch method can be determined.

Answer A is incorrect because the incompatible IS crossmatch can't be ignored, even in the absence of a negative antibody screen. Answer B is incorrect because it is unlikely both RBC units would be incompatible if an antibody directed against a low prevalence antigen was involved. Answer D is incorrect because some antibodies are not detected by enzyme technique because the antigens they are directed against are denatured. This would not be the sole technique to choose for antibody ID. Answer E is a possible choice after the antibody has been identified, however it should not be performed without identifying the antibody, as some clinically significant antibodies can be missed by using a prewarm technique.

24. You are attempting to solve an antibody panel and you think there are potentially two antibodies present, an anti-E and anti-Fy^a. However, the reactions between these two antibodies sometimes occur on the same reagent red cell. What special technique could you use to help clarify the antibody ID?

A. Heat treatment
B. Hydatid cyst fluid treatment
C. Acid treatment

D. *D. bifloris* treatment
E. Ficin treatment

Concept: Treatment of red cells with enzymes, such as ficin, can uncover antigen sites that can then react with antibodies present in the patient's plasma/serum. This leads to enhanced reactivity of antibodies to ABO, Rh, Kidd, I, P, and Lewis blood groups. Ficin can also destroy some antigens, leading to decrease to absent reactivity to antibodies against MNS and Duffy blood groups. Therefore, ficin is often utilized to clarify panels where two alloantibodies are present.

Answer: *E*—Both anti-E and anti-Fy^a are potentially present in this antibody panel. Treatment of the panel cells with ficin will enhance/strengthen the reactivity of the anti-E and the anti-Fy^a, will not be detected as the Fy^a antigens are destroyed by ficin, thus helping to clarify the antibody ID.

The other answer choices are used as follows:

Heat or acid treatment (Answers A and C) can both be used in elutions to release antibodies from the RBC surface. Hydatid cyst fluid (Answer B) neutralizes anti-P_1. *D. biflorus* lectin (Answer D) binds to A_1 antigen to help solve ABO discrepancies in patient that are blood group A subgroups.

25. Direct and indirect antiglobulin tests (DAT and IAT) are important techniques utilized in many aspects of pretransfusion testing. There are many considerations for the appropriate application and interpretation of the techniques. Choose the correct statement:
A. The DAT detects in vitro RBC sensitization and does not require washing steps
B. Antibody screen tests to detect antibodies in a patient's serum/plasma utilize the DAT
C. The IAT requires 37°C incubation and is used to detect in vivo RBC sensitization
D. The use of IgG sensitized control cells (Coombs control cells) is required for negative IATs but not when the DAT is negative
E. An IAT requires saline washing after 37°C incubation to remove unbound globulins

Concept: The DAT is a one-step test, requiring washing but no incubation, and is used to detect in vivo RBC sensitization—it can detect 100–500 molecules of IgG/RBCs and 400–1100 molecules of C3d/RBCs. Up to 15% of hospitalized patients have positive DATs. The predictive value of positive DAT is 83% in patients with hemolysis and only 1.4% without hemolytic anemia. Thus, it should only be ordered in patients with signs, symptoms, and/or laboratory evidence of hemolysis, as well as in the diagnostic process of in transfusion reaction investigations and detection of hemolytic disease of the fetus and newborn (HDFN).

The IAT requires 37°C incubation and is used to detect in vitro RBC sensitization and is used in antibody screen and antibody ID procedures to detect/identify antibodies in a patient's serum/plasma. The IAT is also used when a weak D test is performed or when required for antisera for RBC phenotyping. Whenever a DAT or IAT is negative, IgG-sensitized control cells (Coombs control cells) are subsequently added to the test to verify that the AHG reagent was added and/or is still active.

Answer: *E*—An IAT requires 37°C incubation, as this is the temperature where IgG antibodies will attach to RBCs. Prior to adding antihuman globulin reagent (AHG), sequential washing steps are performed with saline to avoid the neutralization of the AHG by unbound serum/plasma globulins.

Answer A is incorrect, as the DAT detects in vivo RBC sensitization and washing steps are required. Answer B is incorrect, as the antibody screen employs the IAT, not the DAT. Answer C describes the correct incubation temperature for the IAT, but is an incorrect choice because the test detects in vitro, not in vivo RBC sensitization. Answer D is incorrect as the use of IgG sensitized control cells is required for both DAT and IAT tests that are found negative.

26. Which of the following could explain a patient who has hemolysis and with a negative antibody screen, a positive DAT (Poly 4+ and IgG 3+), and a completely negative eluate? Note: The patient has no previous transfusions or pregnancies.
A. The positive DAT is due to contaminated reagents
B. The elution was performed using acid instead of heat

C. The elution was performed using heat instead of acid
D. Type AB panel cells were used to test the elution
E. The positive DAT is due to a medication the patient is taking

Concept: A positive DAT result means that an "antibody" has attached itself to the patient's RBCs. A DAT is performed by adding a few drops of AHG to a sample of the patient's washed RBCs. If agglutination occurs, this is considered a positive reaction. There are many causes of a positive DAT, therefore investigation of the cause most commonly includes a review of the medical/medication history and performance of an elution. An elution is a laboratory technique that uses either heat or acid to remove the bound antibody from the surface of the RBCs. The resulting solution is termed an "eluate". This eluate is then tested against antibody ID panel cells to see if a specific RBC antibody can be identified.

Answer: *E*—In the case above, a strong positive DAT (poly and IgG) with a negative eluate, should prompt an evaluation of the patient's current medications. There are hundreds of medications that can cause drug-induced hemolytic anemia. Antibodies are formed against the medications in vivo. If the drug then attaches to the RBC membrane, the antibody attaches to the RBC and destroys it as a bystander. The elution is negative in these cases because the panel RBCs do not have any medication associated with them and thus are not "attacked" by the drug specific antibody.

Answer A, while possibly correct, would most likely lead to panreactive eluate due to the contamination. Answer D is incorrect because all panel cells are blood type O to avoid reactions with ABO antibodies. Other differential diagnoses for a positive DAT with a negative eluate include paroxysmal cold hemoglobinuria and ABO incompatible transfusion. Answers B and C are incorrect because elutions can be performed using either heat or acid.

Please answer Questions 27–30 based on the following clinical scenario:

A trauma patient is brought to your ED with an estimated blood loss of 10%. The patient's initial vital signs are stable and fluid resuscitation is initiated. A sample is sent to the blood bank for a STAT type and screen. A call to the patient's local hospital reveals a history of anti-C. Before the type and screen can be completed, the ED calls for 3 units of emergency release RBCs because the patient vomited 1 L of blood and her vital signs are unstable with a blood pressure of 80/38, a heart rate of 137, an O_2 saturation of 77%, and a respiratory rate of 32.

27. In an emergent case, such as this when there is no time for a crossmatch or confirmation of the blood type, what is the safest ABO type of RBCs that can be provided based on the patient's antibody history?
A. O Rh negative
B. A Rh negative
C. AB Rh positive
D. B Rh negative
E. O Rh positive

Concept: Emergency situations often demand quick thinking on the part of the blood bank physician and technologist. In this case, the easy part is, knowing that blood type O is the universal blood type for RBC transfusions. However, one must remember that the majority of Rh negative units will be positive for the e and c antigens, and therefore negative for C and E antigens. Therefore, O Rh negative RBCs are the safest blood type for this patient with history of anti-C.

Answer: *A*—O Rh negative RBC products are the generally the 'safest' products available in emergency circumstances when there is no time for ABO/Rh type and screen, or even for crossmatching of the product. However, just because the unit is O Rh negative does not mean it is universally compatible with every antibody a patient could have. This is a common misconception of clinicians. They think that the "negative" means that it is negative for everything. The reason that

O Rh negative is the right choice for this patient is because O type blood is the universal donor and the Rh-negative units will likely lack the C antigen, thus potentially avoiding an acute hemolytic transfusion reaction. All the other choices (Answers B, C, D, and E) could put the patient at risk of a hemolytic reaction.

28. The patient is transfused with 2 units of emergency release blood. Before transfusion, samples for blood bank testing were delivered to the blood bank. Subsequent testing revealed the following results:
ABO type: A Rh negative
Antibody screen: positive
DAT: negative
Antibody panel: anti-C and anti-K identified
Knowing that C and K antigens are present in approximately 70% and 9% of the donor population respectively, what percent of the donor blood products should be compatible with your patient?
A. <1%
B. 15%
C. 22%
D. 27%
E. 41%

Concept: The likelihood of finding antigen negative units is calculated by multiplying the percentage of the population that lacks each individual antigen times one another. This percentage is obtained by 1- the percent prevalence in the population.

Answer: *D*—In the question you are given a 9% prevalence of K-antigen in the donor population. Therefore, the percentage of donors that will lack the K-antigen is $1 - 0.09 = 0.91$. This means that 91% of units should be negative for the K-antigen. Similarly, for the C antigen, the calculation $1 - 0.70 = 0.30$ means that 30% of units should lack the C antigen. Multiplying these two percentages together ($0.91 \times 0.30 \sim 0.27$) gives us a 27% likelihood of finding C-, K-units in our blood bank. All the other choices (Answers A, B, C, and E) are incorrect based on the above calculation.

29. If you were to do a "random" search of your blood bank inventory of RBC units, how many units would the technologist be required to screen before finding 4 compatible RBCs that are negative for C and K antigens?
A. 8
B. 10
C. 15
D. 22
E. 30

Concept: Not every unit in the blood bank will lack the antigens needed to safely transfuse this patient. The number of units that will need to be screened is calculated using the following formula: number needed to screen = (number of units needed/% of compatible units in the population)

Answer: *C*—In this example, we calculate the number of units we need to screen to dividing the number of units we need (4) by the likelihood of finding units negative for C and K in the donor population $4/0.27 = 14.81$. Therefore, in a random search, approximately 15 units of RBCs will need to be screened in order to find 4 compatible units for this patient. The other choices (Answers A, B, D, and E) are incorrect based on this calculation.

In real-life emergent situations, the technologist will probably not do a random search. Instead, he or she will search the Rh negative inventory for K-antigen negative units by phenotyping methods. This process is much more efficient because the Rh-negative units likely lack C-antigen. If the units are indeed K-antigen negative, the technologist will confirm that they are also C-antigen negative via phenotyping. Thus, in this scenario, the technologist only needs to screen ~4/0.91 or 5 units to find 4 units that are negative for both the C- and K-antigen.

However, if the patient had an anti-e, it would be much more challenging. The e antigen is prevalent in 98% of the population; therefore, screening for e antigen negative units in a blood bank would usually be a fruitless endeavor. Thus, most blood banks depend on blood centers to provide such hard to find units. Blood centers accomplish this by antigen typing selected blood donors and keeping a list of donors with rare combinations of blood group antigens.

30. During the patient's hospitalization, a historic snowstorm shuts down transportation in your state for 5 days, but now the roads are finally clear. The patient now requires blood transfusion again, due to a severe bleeding episode, but you have no more compatible units in your blood bank. A desperate call to the blood center reveals that they have compatible units that they can send, but the transfusion transmitted disease (TTD) testing has not yet been completed. Which of the following statements describes the regulations regarding this scenario?

A. The pathologist can approve this emergent measure, but the physician must assume responsibility for transfusing an untested product into the patient
B. The patient should receive notification within 30 days of any positive test results
C. The clinician does not need to be notified of positive test results after the transfusion is completed
D. The clinician can approve this emergent measure, but the pathologist must assume responsibility for transfusing an untested product in the patient
E. The patient and clinician should receive notification within 30 days of any positive test results

Concept: Severe weather can create severe blood shortages, especially in a large hospital that utilizes a lot of blood products on a daily basis. In very rare circumstances, urgent needs for blood can require the blood bank medical director to consider accepting blood products that have not been fully tested for TTDs. This is a consideration that must be discussed with both the treating clinician and the patient to determine if the risk to benefit ratio outweighs waiting for a fully tested product.

Answer: *A*—Ultimately, the clinician must accept the responsibility for transfusing an untested product into his patient. The pathologist and blood bank staff will ensure that the product is compatible with the patient, but cannot make any claims as to the potential infectivity of the product. If the TTD testing comes back positive, it is the responsibility of the blood-bank director to notify the responsible clinician as soon as possible, so treatment can be initiated immediately for any resulting infection. All the other choices (Answers B, C, D, and E) are incorrect based on the information above.

End of Case

Please answer Questions 31-35 based on the following clinical scenario:

31. A patient presents to your ED with a 2-week history of fatigue and shortness of breath. A CBC reveals hemoglobin and hematocrit of 4.5 g/dL and 16% respectively. The patient is currently having mild difficulty breathing and is somewhat pale. A blood sample is sent to the blood bank with a type and crossmatch order for 4 units of RBCs. The results of gel testing are as follows:

ABO/Rh type: B Rh positive
Antibody screen: 2+ in all screening cells, including the autocontrol
DAT: Poly 2+; IgG 2+; C3−
What blood bank technique should be utilized for this patient with a positive antibody screen and a positive DAT?

A. Elution technique to remove the antibody from the plasma/serum
B. Switch to a less sensitive method, such as tube testing
C. Ficin treatment to remove the interference
D. Elution technique to remove the antibody from the RBCs
E. Prewarm the plasma/serum to allow the antibodies to fully bind to the RBCs

Concept: A positive DAT means there is antibody currently attaching to the RBCs. The DAT is not specific for a certain type of antibody. To identify the antibody, it must first be removed from the

RBC. This is typically accomplished in an elution procedure where heat or acid is used to dissociate the antibody from the RBCs. The RBCs are then separated from the solution and an "eluate" remains that contains the previously attached antibodies. This eluate can then be tested against a RBC panel to see if any pattern of reactivity emerges.

Answer: *D*—An elution will remove the antibody from the RBCs and potentially allow for antibody ID.

Answer A (elution technique to remove the antibody from the plasma/serum) does not accurately describe an elution, but rather describes the purpose of an adsorption procedure. Answer B (tube testing) might lead to weaker reactivity or even no reactivity, but you might also miss a clinically significant antibody by switching to a less sensitive testing method. Answer C (ficin treatment) might remove the reactivity and prevent discovery of a clinically significant antibody. Answer E (prewarming) would have not been useful in this case. Prewarming is sometimes used when a cold antibody is interfering with antibody ID or IS crossmatches.

32. What technique should be utilized in a patient (not transfused or pregnant within the last 3 months) with a positive DAT (IgG), a panreactive eluate, and a positive autocontrol?

A. Neutralization
B. Solid phase testing
C. Autoadsorption
D. Ficin treatment
E. Alloadsorption

Concept: A panreactive eluate with a positive autocontrol and a positive DAT (IgG) is most consistent with an autoantibody. An autoantibody is an antibody with specificity for a common antigen on most human RBCs. These antigens are often those in the Rh blood group system. Autoantibodies will react similarly to all panel cells because the common antigen is so prevalent. The major concern with a panreactive autoantibody is not necessarily hemolysis, but rather that the reactivity may be concealing a clinically significant alloantibody, such as an anti-E. Therefore, we use a special technique called absorption to remove as much of the autoantibody as we can without also removing any alloantibodies.

This is accomplished in two ways. First, in a patient who has no interfering RBCs (i.e., no transfusions or pregnancies in the last 3 months) an autoadsorption can be performed by mixing the patient's RBC and plasma/serum in a test tube and letting them incubate for 30 min to 1 h. During this time, the autoantibodies should adsorb onto the patient's RBCs and move out of the plasma/serum leaving any alloantibodies behind for detection.

After each adsorption, the remaining adsorbed serum can be tested against a RBC panel. There are three possible outcomes for an adsorption procedure, as follows: (1) the reactivity can completely disappear indicating that the autoantibody has been successfully adsorbed and no alloantibodies were being concealed. (2) The panreactivity is no longer present, but scattered panel cells are still present. This indicates that the autoantibody has been successfully adsorbed and underlying alloantibodies are present. The panel will need to be solved to positively identify the underlying alloantibody(s). (3) The panreactivity remains. In this case, the adsorption procedure can usually be completed up to three times. If the adsorption remains panreactive after this, the patient's RBCs may be already have too many antibodies attached and they cannot absorb any more antibody. Thus, the sample may need to be sent to a reference laboratory for more specialized testing, such as an alloadsorption.

An alloadsorption may be necessary if the autoadsorption procedure is necessary or if there are interfering RBCs present from a recent transfusion or pregnancy. Alloadsorptions utilize specially selected, DAT negative, RBCs with a specific mix of antigens present on the RBCs to adsorb autoantibodies in the plasma/serum, leaving behind clinically significant alloantibodies, if any. While expensive and time-consuming, these techniques often clarify the antibody ID.

Answer: *C*—An autoadsorption is necessary to adsorb the autoantibody out of the patient's plasma/serum to allow for detection of any underlying alloantibodies.
Answer A (neutralization) is commonly performed with Lewis antibodies because these antibodies are not typically clinically significant, but can mask the reactivity of other clinically significant antibodies. Soluble forms of some blood group antigens, such as Lewis, can be found in bodily fluids or are available as commercially prepared substances. When these soluble antigens are mixed with the patient's plasma the antibody is neutralized, thereby confirming Lewis specificity. The neutralized plasma can also be used to detect the presence of any additional antibodies.

Answer B (solid phase testing) is a very sensitive antibody detection system, similar to gel technology, and is not likely to help in this case. Generally, autoantibodies react strongly in solid phase testing, therefore it would not be a useful technique to distinguish an underlying alloantibody. Answer D (ficin treatment) may enhance or destroy some reactivity and may confuse the technologist even more. Also, autoantibodies usually react strongly by enzyme techniques, therefore it would not be a useful method to distinguish an underlying alloantibody. Answer E (alloadsorption) may be necessary if autoadsorption is unsuccessful or if the patient was transfused or pregnant within the last three months.

33. After autoadsorption, the patient's antibody screen is negative, thus ruling out any underlying alloantibodies. What do these results imply about the potential for in vivo hemolysis if this patient is subsequently transfused?

A. The chances of the patient hemolyzing the transfused RBCs are exactly 50% less than the chance that they will hemolyze their own RBCs
B. The chances of the patient hemolyzing the transfused RBCs are generally the same as the chance that they will hemolyze their own RBCs
C. The chances of the patient hemolyzing the transfused RBCs are exactly 50% more than the chance that they will hemolyze their own RBCs
D. The patient will not hemolyze the transfused RBCs if they are phenotypically matched with the patient's Rh and K antigen-status
E. The patient will not hemolyze the transfused RBCs if full phenotypically matched RBC (i.e., matching with the patient's Rh, K, Duffy, Kidd, and S/s antigen-status) units are provided

Concept: The strength of a DAT or antibody screen is generally not predictive of in vivo hemolysis. This is especially true of plasma/serum where the autoantibody has been adsorbed out. Remember, only the plasma/serum in the test tube has been adsorbed, the plasma/serum in the patient's body has not. Thus, the reactivity is still present in vivo, as is the risk of hemolysis.

Answer: *B*—Autoantibody action in vivo cannot be predicted by the strength of DAT or antibody panel reactivity. The other choices (Answers A, C, D, and E) are wrong because they try to predict a hemolytic response in vivo. Phenotypically matched (either partial with Rh and K, or fully matched) units are not required or proven to be beneficial/safer in these patients. However, phenotypically matched units may protect the patient from developing alloantibodies that may be difficult to detect in the future if they are masked by the reactivity of the autoantibody. Generally, the matched unit has the same risk of hemolysis as an unmatched unit.

34. If transfusion is clinically indicated, who is responsible for signing and documenting acceptance of responsibility for the transfusion of 'least incompatible' RBCs for this patient?

A. The clinician and the patient
B. The pathologist
C. The clinician, the pathologist, and the patient
D. The clinician and the pathologist
E. The clinician

Concept: Incompatible crossmatched blood in this case implies that an autoantibody is present in the patient's plasma/serum that could lead to in vivo hemolysis. Since there are different numbers of antigen sites on different donors' RBCs, a blood bank technologist may find crossmatched units that range between w+ and 4+ incompatible. If that is the case, then at many institutions, the technologist will choose the units that are the "least incompatible" with the patient's plasma/serum. This is a controversial term that is not embraced by all blood bank specialists.

Answer: *E*—The pathologist (Answers A, B, C, and D,) will often call the clinician to educate them about the autoantibody presence; however, it is ultimately up to the clinician to decide if the benefit of transfusion outweighs the potential risk of hemolysis. The patient (Answers A and C) is not usually involved in this decision because of the many variables that are involved. For example, the clinician may decide that the benefit is worth the risk, but decided to treat the patient with steroids to further to dampen the immune system reaction to the transfused blood. Or the clinician may decide the risk is too great and may treat the patient with other therapy, such as iron, to correct the anemia. In this case, this patient has a severe and potentially life-threatening anemia. Thus, transfusion should not be withheld and the pathologist should strongly advise the clinician to transfuse the patient.

35. You release type specific (B Rh positive), 'least incompatible' RBCs for the patient. Soon thereafter, the clinician calls the blood bank saying he is returning the blood and will only accept O Rh negative, washed RBCs for the patient, in order to decrease the chance of hemolysis. What is an appropriate response to this situation?

A. Apologize for the mistake and release O Rh negative blood, but explain that washed RBCs are not necessary

B. Explain that O Rh negative blood does not decrease the degree of hemolysis due to an autoantibody

C. Apologize for the mistake and release O Rh negative, washed RBCs

D. Explain that O Rh negative blood only slightly decreases the degree of hemolysis due to an autoantibody

E. Explain that O Rh negative blood may cause harm to the patient since their ABO/Rh type is B positive

Concept: Some clinicians assume that O Rh negative blood is safe for every patient, regardless of how many or what type of antibodies they have. Therefore, some may assume that giving O Rh negative blood in the case of an autoantibody may add an additional level of protection against hemolysis. This is clearly not the case because blood type O and Rh-negative units are likely to still have the antigen that the autoantibody is binding to.

Answer: *B*—Releasing O Rh negative blood does not decrease the chance of hemolysis in this case. The clinician should be told to keep the B Rh positive RBCs and transfuse if clinically indicated, while monitoring the patient closely for any signs of hemolysis. Additionally, "washing" the RBCs will not decrease the chance of hemolysis because this process does not remove any RBC antigens. RBC washing is typically done to reduce plasma content for patients with severe IgA deficiency, and severe, repeated allergic/febrile transfusion reactions. All the other choices (Answers A, C, D, and E) are incorrect based on the information mentioned previously.

End of Case

Suggested Reading

[1] M.R. Brown, M.G. Fritsma, M.B. Marques, Transfusion safety: what has been done; what is still needed?, Med. Lab. Obs. 37 (2005) 20–26.

[2] K.D. Blaney, P.R. Howard, Basic and Applied Concepts of Immunohematology, second ed., Elsevier-Mosby, St. Louis, MO, (2008).

[3] K.D. Blaney, P.R. Howard, Basic and Applied Concepts of Blood Banking and Transfusion Practices, third ed., Elsevier-Mosby, St. Louis, MO, (2013).

[4] M.K. Fung, B.J. Grossman, C.D. Hillyer, C.M. Westhoff (Eds.), Technical Manual, eighteenth ed., AABB Press, Bethesda, MD, 2016.
[5] C.D. Hillyer, R.G. Strauss, N.L.C. Luban, Handbook of Pediatric Transfusion Medicine, Elsevier, San Diego, CA, (2004).
[6] P. W. Ooley (Ed.), AABB Standards, 30th ed., AABB, Bethesda, MD, 2016.
[7] J.E. Reardon, M.B. Marques, Laboratory evaluation and transfusion support of patients with autoimmune hemolytic anemia, Am. J. Clin. Pathol. 125 (Suppl. 1) (2006) S71–S77.
[8] M.E. Reid, C. Lomas-Francis, M.L. Olsson, The Blood Group Antigens Facts Book, third ed., Elsevier, London, UK, (2012).
[9] C.A. Wheeler, L. Calhoun, D.P. Blackall, Warm reactive autoantibodies: clinical and serologic correlations, Am. J. Clin. Pathol. 122 (2004) 680–685.
[10] Sandler SG1, W.A. Flegel, C.M. Westhoff, et al. It's time to phase in RHD genotyping for patients with a serologic weak D phenotype. College of American Pathologists Transfusion Medicine Resource Committee Work Group, Transfusion 55 (3) (2015) 680–689.

C H A P T E R

8

Adult Transfusion—Principles and Practice

Theresa A. Nester, Monica B. Pagano**, YanYun Wu*[†]

*University of Washington Medical Center and Bloodworks Northwest, Seattle, WA, United States;
**University of Washington Medical Center, Seattle, WA, United States; [†]Bloodworks Northwest, Seattle, WA, United States

Blood component therapy plays a critical role in the management of patients who are experiencing reduced hematopoiesis, increased peripheral destruction of cells, or generalized bleeding. Whole blood is collected from donors; however, the general practice is to process the whole blood into its components and then store these components under their optimal conditions. Apheresis instruments may also be used to collect blood components. This chapter highlights principles of transfusion support of the adult patient. This includes component therapy for each blood component, including granulocytes. Where appropriate, the following topics are covered: indication, dose, and ABO/Rh compatibility of the component. Selected clinical scenarios are used to illustrate transfusion support decisions in adult patients. Also covered in this chapter are explanations intended to help guide the decision to irradiate, wash, or leukocyte-reduce a cellular blood component.

1. A 65-year-old woman presents to the emergency department complaining of progressive fatigue and dyspnea for the last 4 days. The patient has a history of atrial fibrillation and is receiving oral anticoagulation with warfarin. Physical examination demonstrates pale skin and mucosa and vital signs remarkable for tachycardia with normal blood pressure. On examination the patient is not actively bleeding. Laboratory results show a hematocrit of 19%, platelet count of 32,000/μL, white blood count (WBC) of 8,000/μL, and INR of 4. Which is the most appropriate blood product to administer?

A. Red blood cells
B. Platelets
C. Granulocytes
D. Plasma
E. No blood product at this time

Concept: When a patient has signs and symptoms of anemia, red blood cell (RBC) transfusion should be considered. Clinical assessment of the patient, in addition to reviewing laboratory values is imperative. In this case, the patient has signs and symptoms of anemia (i.e., progressive fatigue and dyspnea with tachycardia) and thus, additional oxygen-delivery capacity in the form of RBC transfusion may be clinically beneficial. It is important to remember that there is not a single sign or symptom that is a perfect indicator for increased oxygen carrying capacity requirements.

Transfusion Medicine, Apheresis and Hemostasis. http://dx.doi.org/10.1016/B978-0-12-803999-1.00008-0

Answer: *A*—Transfusion of RBCs is indicated to increase the patient's oxygen delivery capacity and improve tissue oxygenation, which in turn may alleviate the signs and symptoms of anemia. The patient is maintaining normal blood pressure and does not have life-threatening bleeding; thus plasma in not necessary (Answer D). In a nonbleeding patient with a warfarin-induced INR elevation, the warfarin should be temporarily discontinued and monitored for a return to therapeutic range (such as typically between 2 to 3 in patients with atrial fibrillation).

Though the patient has a decreased platelet count, there is no evidence of bleeding and the patient has no obvious reason for her platelets to be dysfunctional (e.g., aspirin therapy) (Answer B). Studies in hematology-oncology patients have demonstrated that patients do not tend to have spontaneous bleeding until the platelet count decreases below 10,000/μL. Of course, if a patient has evidence of bleeding, then platelet transfusion triggers change based on the severity and/or location of bleeding. Other reasons for platelet transfusion include the following: preparation for surgical/invasive procedures; congenital, acquired, or iatrogenic platelet dysfunction; and lack of platelet production by the bone marrow. The patient has a normal white blood cell count, thus granulocyte transfusion is not needed (Answer C).

2. A 45-year-old man was admitted to the intensive care unit with acute respiratory distress syndrome secondary to viral pneumonia and is currently on mechanical ventilation. The patient has a previous history of diabetes and hypertension that are well controlled with medication. He has no history of cardiovascular disease. Laboratory results this morning demonstrate a Hemoglobin (Hb)/Hematocrit (Hct) of 7.2g/dL and 22%, respectively. Which of the following hemoglobin levels is considered a reasonable indication for transfusion of red blood cells in nonbleeding, yet critically ill patients without any acute or chronic cardiovascular disease, as described in this case?

A. Hb <5 g/dL
B. Hb <6 g/dL
C. Hb <7 g/dL
D. Hb <8 g/dL
E. Hb <9 g/dL

Concept: Hematocrit and hemoglobin levels are usually used as triggers for transfusion. Currently, there are no other proven/100% reliable indicators of tissue hypoxia used to determine the indication of transfusion. As evidenced by multiple randomized control trials, such as the TRICC trial (Transfusion Requirements in Critical Care), a "restrictive" strategy (transfuse when Hb <7–8 g/dL) is clinically noninferior to a "liberal" transfusion strategy (transfusion when Hb <9–10 g/dL) in critically ill patients with hemodynamically stable anemia, except possibly in patients with acute myocardial ischemia. The AABB recommends adhering to a restrictive transfusion strategy (7–8 g/dL) in hospitalized, stable patients. Nonetheless, whether it is safe to use a "restrictive" strategy is safe in patients with acute or chronic cardiovascular diseases remains unclear at this point.

Answer: *C*—Hb of 7 g/dL is an accepted trigger for transfusion in nonbleeding, critically ill patients without acute or chronic cardiovascular diseases. All other choices (Answers A, B, D, and E) represent the wrong hemoglobin triggers.

3. A 65-year-old male newly diagnosed with acute myeloid leukemia is currently undergoing chemotherapy. He is critically ill, with Gram negative bacteremia and the clinical team has been increasing his oxygen supplementation. Laboratory results demonstrate an Hct 22%, Hb 7.5 g/dL, white cell count (WBC) 200/μL, and platelets 6,000/μL. Which of the following options is an evidence-based Hb trigger for transfusion in patients with hematological malignancies or primary bone marrow failures?

A. Hb <6 g/dL
B. Hb <7 g/dL
C. Hb <8 g/dL

D. Hb <9 g/dL
E. Not yet determined

Concept: There is reasonable evidence to support a restrictive strategy in several clinical situations, such as in patients who are critically ill, neonatal patients, patients with gastrointestinal bleeding, postoperative patients, and patients undergoing bypass surgery. However, there are no randomized controlled trials that support one Hb level as a transfusion trigger for patients with bone marrow failure secondary to hematological malignancies.

Answer: *E*—There is no evidence-based Hb trigger for transfusion of patients with hematological malignancies. The most commonly used trigger is Hb of ~7–8 g/dL (Answers B and C), deriving from the trials in critically ill patient with stable anemia without acute or chronic cardiovascular diseases. Other choices (Answers A and D) are incorrect based on the current evidence.

4. A patient with newly diagnosed acute leukemia is admitted. Which of the following options is the best combination for cellular blood product modification(s) for this patient?
   A. Leukoreduced, irradiated
   B. CMV negative, irradiated
   C. Leukoreduced, CMV negative, washed
   D. Leukoreduced, irradiated, washed
   E. No special modification is indicated

Concept: Transfusion-associated graft versus host disease (TA-GVHD) is a rare and often fatal (~90%–99%) complication of blood transfusion, associated with the transfusion of viable lymphocytes that can engraft and develop an immune response against host tissue. It often results from an immune-incompetent recipient's inability to defend against transfused lymphocytes. Patients at risk for TA-GVHD are those with absent T cell immunity, as seen in Table 8.1.

TABLE 8.1 Typical Irradiation Indications

Irradiation indications	
Indicated	**Not indicated**
• Status post total body irradiation [such as with hematopoietic progenitor cell (HPC) transplantation] • DiGeorge's syndrome • Low birth weight, premature infants • Intrauterine transfusion • Medication (e.g., Fludarabine, Alemtuzumab) • Hodgkin's disease • Neuroblastoma • Hematologic malignancy • Directed donations from close relatives	• Diagnosis of HIV • Solid organ transplantation

Irradiation of cellular blood components is performed either by gamma or X-ray irradiation to reduce the risk of transfusion-associated graft versus host disease (TA-GVHD). Irradiation damages the DNA of lymphocytes, preventing activation and engraftment. Transfusion services may favor X-ray irradiators, as the Federal requirements for having an irradiator with gamma irradiation are quite extensive due to fear of terrorists being able to weaponize the Cobalt or Cesium source.

While leukoreduction (LR) does not prevent TA-GVHD, it has been shown to have three effects:

- Decreased rate of HLA-alloimmunization, via the removal of HLA class II antigen-presenting cells from the cellular blood product
- Reduction of febrile nonhemolytic transfusion reactions (FNHTR) by reducing the amount of cytokines being secreted by leukocytes

- Reduction of the risk of transfusion-transmitted CMV infection, since CMV survives in a latent state within white cells.

In many centers, a leukoreduced cellular product is considered equivalent to a CMV seronegative product (i.e., CMV safe), with both having a ~1–3% chance of CMV transmission. Currently, many hospitals provide 100% leukoreduced products, but implementation is not yet universal.

Answer: *A*—A patient with acute leukemia should receive leukoreduced products to reduce alloimmunization against HLA antibodies, reduce the risk of CMV transmission. The cellular products should also be irradiated to prevent TA-GVHD (Answer E). Irradiation is also usually placed on the patient's record at the time of diagnosis, in case the patient receives chemotherapy, such as fludarabine, and/or total body irradiation as part of an HPC transplant preparation regimen.

Although CMV seronegative product and leukoreduction are considered equivalent, leukoreduction has other benefits compared to CMV negative product, such as reduction in HLA alloimmunization, and thus, will confer benefit to this patient (Answer B).

There is no indication to provide washed products if the patient has not experienced severe allergic transfusion reactions or repeated, severe FNHTRs (Answers C and D). Washing is performed to remove the plasma content of cellular blood components. The most common indications for washed products include: history of anaphylactic reactions (removal of allergenic plasma proteins); prevention of hyperkalemia in newborns and small children (removal of potassium present in the plasma); and removal of antibodies (e.g., removal of maternal antibodies when using maternal platelets for a neonatal transfusion for neonatal alloimmune thrombocytopenia). Although questionable, it is generally accepted that patients with complete IgA deficiency plus the presence of anti-IgA should also receive washed products with the goal of removing as much IgA as possible. Current guidelines recommend washing the product 6 times for sufficient IgA removal. Of note, washing is typically only done for RBC products, since washing plasma is impossible. For platelets, washing will both reduce the number of platelets, as well as activate them, leading to degranulation, and thus, washing platelet products may lead to an inferior product with questionable benefit.

5. A 65-year-old female patient requires 2 units of RBCs for elective cardiac surgery. The antibody screen is positive and additional testing demonstrates the presence of anti-k. There are no compatible units available. The patient's sister is tested and is k-negative, ABO compatible, and free of transfusion transmissible diseases. She offers to donate blood for her sister. Which of the following is correct?
   A. This donor is not required to go through the standard donor screening and testing process
   B. The unit should be crossmatch compatible and irradiated
   C. The unit should be crossmatch compatible and leukoreduced
   D. The unit should be crossmatch compatible with no modifications
   E. Family members are not allowed to donate blood components due to the risk of alloimmunization

Concept: The patient has an antibody against a high incidence antigen. More than 99% of the donor population expresses k on their RBC surface, which makes it difficult to find compatible donors. Although directed donations are usually discouraged, family members are a good resource for compatible blood in these situations. The main risk associated with donation from family members is TA-GVHD. The pathophysiologic mechanism involves the proliferation of viable donor lymphocytes in the recipient, due to the recipient being unable to mount an immune response against donor lymphocytes. Patients sharing an HLA haplotype with an HLA homozygous donor (i.e., the recipient cannot recognize the donor lymphocytes as foreign, but the donor lymphocytes can recognize the recipient (foreign HLA), such as in blood-related family members, are at risk. See Table 8.1 for a list of other indications for irradiation of blood products.

Answer: *B*—Gamma (or more recently, X-ray) irradiation of cellular blood components from HLA matched or related donors is recommended to prevent TA-GVHD (Answer D). The minimum required dose of gamma radiation is 2500 cGy (25 Gy) to the center of the blood product bag, and 1500 cGy (15 Gy) to the remainder of the blood product bag (Fig. 8.1).

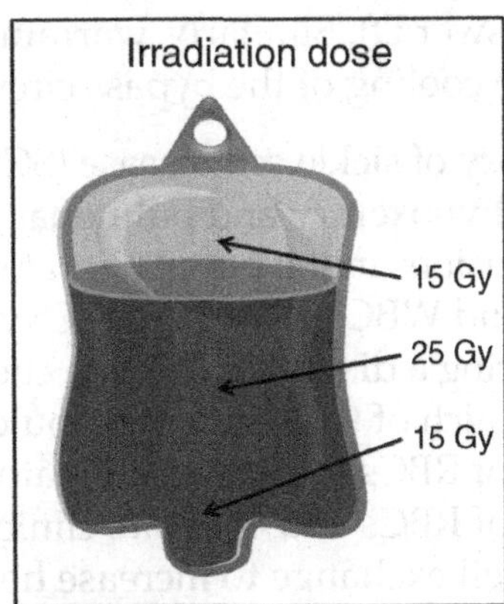

FIGURE 8.1 **Irradiation dose, as required by AABB/FDA.**

Leukoreduction alone (Answer C) cannot prevent TA-GVHD. In the case of rare blood, family members should be encouraged to donate for the patient (Answer E). Nonetheless, they have to go through the standard screening process and testing as with any other allogeneic blood donors (Answer A). Moreover, these units must be irradiated.

6. A 52-year-old man is scheduled to undergo open heart surgery in 2 days. The antibody screen is positive at the immediate spin phase; antibody identification shows that he has anti-M. Typically, anti-M is a naturally occurring alloantibody that is detected at room temperature. In the setting of open heart surgery, the reactivity of the antibody can become stronger as the patient is cooled down during the surgery. This can result in cells being agglutinated along tubing, which may be a concern. What is one of the common approaches used to address this?

A. Postpone the surgery until the antibody is no longer detectable
B. Heat up the operating room to 37°C
C. Provide M-negative red cells
D. Recommend that the patient travel to a warmer climate to have the surgery
E. Use a blood warmer for the case

Concept: Anti-M is usually not clinically significant. However, it may cause disruption to the tubing during cardiac surgery if the patient is cooled down and received M positive RBC. There are two common approaches to address this concern. One can perform a thermal amplitude test to determine the temperature range over which the antibody reacts. If, for example, the antibody does not react above 32°C, the anesthesiologist may choose to cool the patient down only to 32°C (rather than 30°C). Also, since about 20% of blood donor red cells do not express the M-antigen, it may be possible to provide M-negative red cell units for the surgery. Once the patient is off bypass and back to 37°C, the anti-M would not be expected to bind red cells; thus, M-antigen untested units (but crossmatch compatible) may be transfused. Note that if the antibody screen was positive at the AHG phase, one might want to do further testing (such as prewarm compatibility testing) to evaluate if an IgG component is present, or if this is just "carry over" from a cold-reacting anti-M.

A similar situation occurs when a patient has hemolysis due to a cold autoantibody. If transfusion is clinically indicated, then transfusion via a blood warmer should be given. Blood warmers are limited to a maximum of 42°C based on AABB recommendations. In patients with severe life-threatening hemolysis due to this condition, therapeutic plasma exchange may be considered. During plasma exchange, a blood warmer for both draw and return lines, as well as heating up the room to 37°C, should be considered.

Answer: C—Providing M-negative RBCs during the operation is a common choice to avoid tubing disruption during the operation. Open heart surgery will typically not be postponed due to an anti-M (Answer A). Having the patient move to a warmer climate will have no effect on the cooling of the bypass circuit for surgery (Answer D). Similarly, warming up the operating room or using a blood warmer will also not affect the cooling of the bypass circuit (Answers B and E).

7. A 25-year-old male patient with history of sickle cell disease (SCD, Hemoglobin SS, HbS), is being evaluated for new onset of dyspnea, hypoxemia, and pulmonary infiltrates on chest radiograph. The patient has not required many transfusions in the past. Laboratory data reveals a hemoglobin of 7.2 g/dL, hematocrit of 22%, HbS of 40%, and WBC of 22,000/μL. Oxygen saturation is 93% on 2 L nasal cannula. The clinical team is considering a diagnosis of acute chest syndrome, and consults transfusion medicine about treatment options. Which of the following would you recommend for this patient?
   A. Proceed with simple transfusion of RBCs to increase hematocrit to 40%
   B. Proceed with simple transfusion of RBCs and evaluate clinical response
   C. Proceed with an automated red cell exchange to increase hematocrit to 30% and reduce HbS to less than 30%
   D. Provide symptomatic treatment and avoid transfusion to reduce the risk of hyperhemolysis syndrome
   E. Provide symptomatic treatment and avoid transfusion to reduce the risk of alloimmunization

Concept: Causes of acute chest syndrome include infection, pulmonary infarction, and fat embolism. Nonetheless, the etiology of many acute chest syndromes remains unknown. Treatment includes hydration, respiratory support, antibiotics, and transfusion to reduce HbS content and prevent further sickling. It is common practice to start with simple transfusion for less severe cases, and to reserve red cell exchange for more severe cases, or cases that do not respond well to simple transfusion.

The NHLBI consensus guidelines are as follows:

- In patients with sickle cell disease (SCD), give simple blood transfusion (10 mL/kg red blood cells) to improve oxygen carrying capacity to people with symptomatic ACS whose hemoglobin concentration is >1 g/dL below baseline. If baseline hemoglobin is 9 g/dL or higher, simple blood transfusion may not be required (Weak Recommendation, Low-Quality Evidence).
- In all persons with SCD, perform urgent exchange transfusion with consultation from hematology, critical care, and/or apheresis specialists when there is rapid progression of ACS as manifested by oxygen saturation below 90% despite supplemental oxygen, increasing respiratory distress, progressive pulmonary infiltrates, and/or decline in hemoglobin concentration despite simple transfusion (Strong Recommendation, Low- Quality Evidence).

Answer: *B*—This patient has a clinical presentation suggestive of acute chest syndrome. The first treatment approach is to perform a simple transfusion to increase oxygen carrying capacity, and to decrease hemoglobin S, especially in patients with an initial Hb <9 g/dL. Simple transfusion is considered to be first line therapy to treat mild to moderate acute chest syndrome (Answers D and E). However, if the patient has a higher Hb or starts to deteriorate clinically, red blood cell exchange should be considered.

With simple transfusion, the goal to achieve a hematocrit of 40% (Hb ~13 g/dL) is typically considered to be too high in patients with SCD (Answer A). This is because sickle cells have altered adhesion molecules, which make them "stickier" than normal red cells. This can result in blood with a high hematocrit becoming too viscous, which predisposes the patient to more microvascular sickling/thrombosis. Thus, it is recommended that for patients with SCD who are not chronically transfused and have high percentage of HbS, the red cell transfusion goal should be targeted to avoid Hb levels above 10–11 g/dL (or ~30%–33% hematocrit).

In relatively stable patient, such as this one, the management algorithms typically call for this procedure after the patient fails to improve with simple transfusion (Answer C). Although intuitively one might believe red cell exchange to be superior to simple transfusion because sickle cells are being removed out of circulation (and at the same time, being replaced with non-sickle RBCs), there is no data to date indicating red cell exchange is superior to simple transfusion in mild to moderate acute chest syndromes.

Hyperhemolysis is an infrequent but severe, and sometimes fatal, complication associated with RBC transfusions. It is usually observed in patients with SCD, but cases in other disease states have been reported. Hyperhemolysis is suspected when the hematocrit decreases after transfusion and laboratory parameters are consistent with hemolysis. The mechanism is not well understood, but there is hemolysis of both the transfused RBCs and the patient's own RBCs. In these cases, the Hb can reach critical levels, 2 or 3 g/dL. The most common therapeutic strategy is to avoid transfusion and administer intravenous immunoglobulin (IVIG) and steroids. This complication can make treatment of the patient with sickle cell disease very difficult.

Alloimmunization against red cell antigens can occur with transfusions. Because African-American individuals usually lack blood group antigens in the Rh, Kell, and Duffy blood group systems more often than Caucasian individuals, SCD patients frequently make these antibodies when transfused with red cells from Caucasian donors. Because this is a known risk, most transfusion services that support sickle cell patients will prophylactically match the red cells for these patients for major antigens in the Rh, Kell and sometimes the Kidd and Duffy blood group systems. The addition of the latter groups is considered extended phenotype matching. There is evidence that suggests if cells matching Rh and K are provided, then the patient is less likely to make red cell antibodies to other blood groups as well. However, a 2013 paper by Chou et al. did not find such a benefit, even in racially matched units, leading to some suggestions that only genotypically-matched units will protect against alloantibody formation.

8. A 54-year-old woman is admitted through the emergency department after a high speed motor vehicle accident. Past medical history is not available. Upon arrival she is combative and confused. Her vital signs demonstrate a heart rate of 120 beats per minute and a blood pressure of 85/45. Laboratory results show a hematocrit of 17%, WBCs 14,000/μL, platelets 95,000/μL, and INR 3.2. Fluid resuscitation with crystalloids is initiated and two units of RBCs are requested emergently, while the patient is transferred to the operating room. Which of the following is the recommended blood component transfusion strategy to prevent the progression of trauma associated coagulopathy?

A. Transfusion of RBCs, plasma, platelets in a 1:1:1 ratio
B. Transfusion of RBCs to keep the hematocrit > 24%
C. Transfusion of plasma to correct the INR to <1.6
D. Infusion of tranexamic acid
E. Transfusion of platelets

Concept: Shock trauma can result in massive tissue destruction, hypovolemia, hypotension, hypothermia, and acidosis which can lead to the coagulopathy of trauma. In these cases, initiation of a massive transfusion protocol with a fixed ratio of blood products (RBC:Plasma: PLT 1:1:1) has been shown to reduce the occurrence of trauma-associated coagulopathy. The 1 for platelet is in reference to a whole blood derived unit of platelets, which is much smaller in volume and platelet number than a unit of single donor apheresis platelets. Of note, it usually takes a pool of approximately 5–6 units of whole blood derived platelets to be equivalent to 1 unit of single donor apheresis platelets.

Answer: *A*—This patient is in hemorrhagic shock; thus, receiving RBCs, plasma, and platelets in 1:1:1 ratio could reduce the risk of trauma associated coagulopathy versus transfusing RBCs alone.

Transfusion of single components will not address the anemia or coagulopathy (Answers B, C, and E). In a trauma setting, early initiation of tranexamic acid infusion has been shown to decrease the need of blood transfusions and reduce mortality, and this treatment strategy may be a part of the massive transfusion protocol, not a separate component (Answer D).

9. A 24-year-old Caucasian male is admitted to the trauma center after a motor vehicle crash. He has a femur fracture and requires uncrossmatched group O cells until his ABO type is known. Per policy, the transfusion service issues group O Rh positive cells. Which of the following is correct regarding the percentage of Rh positive in both patient and donor populations?

	Percent of Caucasian individuals who are Rh positive (%)	Percent of blood donors who are group O Rh negative (%)
A.	12	16–18
B.	24	13–15
C.	60	10–13
D.	85	7–9
E.	95	5–6

Concept: Group O Rh negative red cells are frequently in short supply, as only 7%–9% of blood donors will have this blood type. Approximately 15% of people are Rh negative, including those that are group O (Table 8.2). Special recruiting efforts may raise the available inventory of Rh negative cells to a higher level, but this is a short-term gain since the whole blood donation interval is 8 weeks. Due to the relatively low number of group O Rh negative individuals (and therefore, blood donors) in the population, it is impossible to support 100% of individuals needing uncrossmatched group O negative RBCs with these supply limitations.

TABLE 8.2 Frequency of ABO and Rh Type in Caucasian Population

		Rh positive (%)	Rh negative (%)
ABO type	O	36	9
	A	34	8
	B	8	2
	AB	2.5	0.5

From a resource management perspective, transfusion services will typically reserve group O Rh negative uncrossmatched RBCs for female patients with child bearing potential (typically 50 years of age or less or for those with a known history of anti-D). This is to prevent sensitization to the D antigen in a woman who is Rh negative, thus avoiding subsequent risk of hemolytic disease of the fetus and newborn due to anti-D. As indicated above, 85% of white male and female trauma patients are Rh positive. In Asian populations, this number is closer to 95%. This is one of several reasons why getting the patient's sample for ABO/D type early in the resuscitation process is important for ensuring optimal use of the community blood supply. Further, without a pre-transfusion specimen, after a few rounds of resuscitation, it may be difficult to interpret the ABO/Rh and antibody screen results due to the coexistence of both patient's and transfused RBCs.

Answer: *D*—In a typical blood donor population, about 15% are Rh negative. The ABO distributions are as follows: O (~45%), A (~42%), B (~10%), and AB (~3%). Since the inheritance of ABO and Rh are independent, the incidence of being O Rh negative is the product of the incidence of being O and the incidence of being Rh negative (or 45%×15% = ~7%). Thus, the answer is D and all the other choices (Answers A, B, C, and E) are incorrect.

Please answer Questions 10–12 based on the following clinical scenario.

A 35-year-old female with no previous medical history presents to an urgent care clinic with complaints of increasing fatigue for the last 3 weeks, which worsened over the last 48 h. The patient reports that she is now unable to perform her daily activities, and feels fatigued even at rest. The patient is admitted to the hospital for evaluation and treatment. Her Hct is 17%, with normal platelet and white blood cell counts. Further laboratory testing reveals an elevated lactate dehydrogenase, an elevated total bilirubin, and an undetectable level of haptoglobin.

10. Regarding further laboratory testing, the next step in the management is to order a/an:
A. Direct antiglobulin test (DAT)
B. Serum creatinine
C. ADAMTS13
D. Serum ferritin
E. Flow cytometry for CD55 and CD59

Concept: In clinical cases that are suspicious for hemolytic etiologies, understanding the possible mechanisms involved in red cell destruction can help guide testing and diagnosis. Hemolysis can be classified as immune mediated vs. non immune mediated, and by the site of hemolysis (i.e., intravascular vs. extravascular). The DAT demonstrates the presence or absence of IgG and/or complement attached to the RBCs surface. Although there are multiple causes of a positive DAT, in the right clinical context, the DAT can help determine whether the hemolysis is immune mediated or not.

Intravascular hemolysis results in increased indirect bilirubin and lactate dehydrogenase. The haptoglobin is typically decreased, since it binds free hemoglobin, which is then picked up by the liver, and the peripheral blood smear usually demonstrates schistocytes (Fig. 8.2). Extravascular hemolysis usually presents with less evident changes, since the RBCs destruction takes place in the reticuloendothelial system, resulting in the appearance of spherocytes (Fig. 8.3) on the peripheral blood smear.

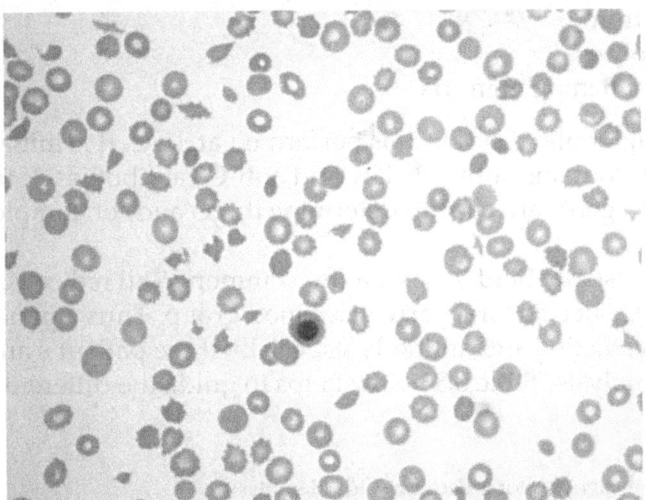

FIGURE 8.2 **Schistocytes.** *Source: Image courtesy of Dr. Sindhu Cherian, University of Washington Medical Center, Seattle, WA.*

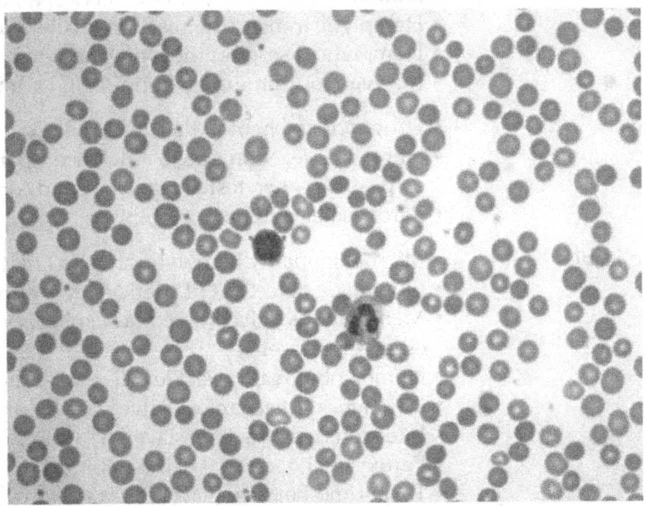

FIGURE 8.3 **Spherocytes.** *Source: Image courtesy of Dr. Sindhu Cherian, University of Washington Medical Center, Seattle, WA.*

Answer: *A*—To narrow the differential diagnosis, the DAT is the next logical test to order. If the DAT is positive, this suggests immune-mediated hemolysis and an elution would be the next likely step.

Liver and renal function tests, such as creatinine and AST are often useful, but will not help to classify this patient's anemia (Answer B). Serum ferritin will not help to elucidate the etiology of an anemia with this type of presentation (Answer D).

This patient has anemia with normal platelet count. Thus, ADAMTS13 is not useful because there is no suspicion of microangiopathic hemolytic anemia and thrombocytopenia.

CD55 and CD59 are markers of paroxysmal nocturnal hemoglobinuria (PNH, Answer E). While this is a form of hemolytic anemia, it is rare. Patients with PNH classically notice red urine, especially after waking up in the morning. They may also present with abdominal pain, pain with swallowing, and unusual thromboses. Nonetheless, a DAT should be ordered as the first test in any patient with hemolysis.

11. The patient's lab results thus far are suggestive of hemolysis and the DAT is positive for poly and IgG, but is negative for C3. Her antibody screen is positive and an eluate demonstrates a panagglutination at antihuman globulin (AHG) phase, with no reactivity at room temperature. At this time, what is the most likely diagnosis?

A. Paroxysmal cold hemoglobinuria
B. Warm autoimmune hemolytic anemia
C. Cold agglutinin syndrome
D. Drug mediated hemolysis
E. Paroxysmal nocturnal hemoglobinuria

Concept: The direct antiglobulin test (DAT) is performed at 37°C. It is initially performed with a polyspecific reagent that contains both anti-IgG and anti-C3 antibodies. If the polyspecific reagent is positive, monospecific reagents are used to determine if IgG and/or complement (C3d) is coating the RBCs.

Warm autoantibodies usually bind to something common to all red cells, such as the Band3 protein, rather than being specific for a particular blood group. Thus a panagglutinin at 37°C (AHG phase) suggests a warm reacting autoantibody, especially if the patient's autocontrol is also positive.

In the presence of hemolysis, Table 8.3 below helps to guide the differential diagnosis:

TABLE 8.3 Typical Differential Diagnoses for a Patient With Hemolysis

DAT	Eluate	Typical differential diagnoses
Positive	Positive	• Autoimmune hemolytic anemia (warm, cold, or mixed) • Hemolytic transfusion reaction (acute or delayed) • Hemolytic disease of the fetus and newborn • Intravenous immunoglobulin (IVIG) administration
Positive	Negative	• Drug induced hemolytic anemia • Hemolytic reaction due to ABO incompatible transfusion (eluate will be positive when tested against A and/or B cells) • Paroxysmal cold hemoglobinuria
Negative	Not applicable	• Microangiopathy hemolytic anemia (TTP, HUS, DIC) • Enzymatic defect (G6PD deficiency, pyruvate kinase deficiency, etc.) • Membranous defect (hereditary spherocytosis, hereditary elliptocytosis, etc.) • Hemoglobinopathy (sickle cell disease, hemoglobin C disease, etc.) • Mechanical (mechanical valve, ECMO, etc.) • Infection (malaria, babesia, etc.) • Paroxysmal nocturnal hemoglobinuria • Toxins and heavy metal (Wilson's disease, etc.)

Answer: *B*—The presence of hemolysis, a positive DAT for IgG, and a panagglutinin in the eluate are suggestive of a warm autoimmune hemolytic anemia (WAHA).

This patient has a positive DAT and a positive elution. Thus, based on Table 8.3, other choices (Answers A, C, D, and E) are wrong. Paroxysmal cold hemoglobinuria (PCH) and cold agglutinin syndrome are typically due to antibodies that cause complement to coat red cells, rather than IgG. Paroxysmal nocturnal hemoglobinuria might not have a positive DAT, as the cells lyse easily in the presence of complement. If the DAT were positive, it would be with complement (C3), not IgG. The classic pattern for drug induced hemolysis is a positive DAT with a negative elution. However, some medications give the pattern for a warm autoimmune hemolytic anemia. A good history and knowledge of medications that can cause hemolysis are the keys to determining whether the anemia is idiopathic or due to a medication.

12. The primary physician orders 1 unit of RBCs due to severe symptomatic anemia. Her Hct is 17% and her clinical status has deteriorated significantly since admission. The blood bank informs the clinician that it was possible to obtain the patient's extended phenotype (D+, C+, e+, E−, K−, Fy(a + b +), Jk(a + b−), S+ s+), but the presence of a clinically significant underlying alloantibody has not been entirely ruled out because autoadsorption was unable to remove the reactivity of the autoantibody. Thus, all red cell units will be incompatible because of the autoantibody. Which of the following options is the next best step in transfusing this patient?

A. Transfuse a red blood cell substitute (i.e., hemoglobin based oxygen carrier)
B. Transfuse phenotypically matched, as available
C. Transfusion genotypically matched, as available
D. Transfuse random units since all will be incompatible
E. Avoid transfusion until a compatible unit is available

Concept: Because warm-reacting autoantibodies bind some structure that is common to all red cells, every red cell unit tested will appear crossmatch incompatible. A priority when deciding which units to transfuse these patients is to rule out the presence of underlying clinically significant alloantibodies. This is essentially done by extracting the autoantibody out of the patient serum in vitro, and then testing the serum against a cell panel(s). The extraction is achieved by performing a series of auto- (if the patient has not been transfused or pregnant with the past 3 months) or allo- (if the patient has been transfused or pregnant with the past 3 months) adsorptions and elutions. This is a lengthy process, and if the autoantibody is bound too tightly to the cells, it does not always completely remove the activity/interference.

Answer: *B*—Ideally, when the presence of an alloantibody has not been ruled out, it is recommended to perform an extended phenotype of patient's red blood cells, and transfuse donor units that are phenotypically matched. Antigen matched units will still be crossmatch incompatible, and the patient will hemolyze transfused red blood cells at a similar rate to the rate at which their own red blood cells are being hemolyzed. However, if phenotypically matched cells to common clinically significant antigens (such as Rh, Kell, Kidd, Duffy, and S/s) can be provided, the risk of hemolysis due to an alloantibody will be reduced. In the case of the presence of a warm autoantibody, EDTA glycine acid (EGA) treatment of the RBCs can be helpful to obtain the RBC phenotype of the patient. Providing a phenotypically matched RBC may be safer than providing a completely random unit (Answer D).

The warm-reacting antibody may be demonstrable for days to weeks, depending on the success of treatments with immunosuppressive agents, such as steroids. If the patient is tolerating the anemia extremely well, transfusion can be withheld. If the patient has signs and symptoms of anemia (as in this case), transfusion should be attempted (Answer E). Slow transfusion with close monitoring in the setting of WAHA is appropriate, particularly if the transfusion is done with a unit that does not match the patient's full phenotype.

Hemoglobin-based oxygen carriers (Answer A) have been evaluated as an alternative to red blood cell transfusions for patients who cannot, or refuse to, receive red blood cell transfusions (i.e., multiple alloantibodies requiring rare red cells, personal or religious preferences). These cell-free hemoglobins are structurally modified to receive and release oxygen while in circulation. One main side effect is severe vasoconstriction (NO scavengers). At the time this chapter was written, there were no FDA approved hemoglobin-based oxygen carriers available in the United States. There are a handful of case reports of patients who have received these products under compassionate use with positive results.

It will take a while to obtain the genotype. Thus, if the patient needs red cells relatively quickly, as is often the case in WAIHA, the time required to obtain a genotype (predicted phenotype) is typically too long (Answer C).

End of Case

13. A 26-year-old female with diagnosis of SCD has been supported by your laboratory for the last 10 years. She has a history of alloimmunization including anti-K and anti-Jk^a. Her phenotype shows that her RBCs do not express the following antigens: D, K, Jk^a, S, or s. She was last transfused 4 months ago and at that time the antibody screen was negative. The patient now comes for a scheduled transfusion of RBCs. The antibody screen demonstrates 3+ reactivity against all tested cells. The autocontrol is negative. Which of the following statements most likely represent this new reactivity?

A. The patient has developed antibodies against the reagents
B. The patient has developed an autoantibody
C. The patient has developed an alloantibody against a high frequency antigen
D. The patient has developed an alloantibody against a low frequency antigen
E. The patient has developed an alloantibody against a high titer low avidity antibody (HTLA)

Concept: Reactivity against all tested cells is concerning for the presence of an antibody directed against a high incidence antigen, multiple alloantibodies, reagent antibodies, or autoantibodies. The auto control is negative, which indicates that the newly formed antibody does not bind patient's own RBCs, excluding the presence of an autoantibody or an antibody directed against reagents.

Answer: C—In this case, the unexpected antibody was directed against a high frequency antigen, anti-U from the MNS system blood group. The U antigen (from the almost universal distribution) has an occurrence of 99.9% on the Caucasian population and 99% on African Americans. The U negative phenotype is associated with an absence of glycophorin B (GPB) which carries the S, s and U antigens, or with an altered form of GBP. Patients with alloantibodies against U are at risk of transfusion reactions and hemolytic disease of the fetus and newborn. The autocontrol is negative; thus, there is no evidence of an autoantibody
(Answer B) or an antibody against reagent (Answer A). All the tested cell reacted; thus, the antibody is more likely to be against a high frequency antigen, and not a low frequency antigen (Answer D). Although a HTLA will react with all panel cells, the strength of reactivity is typically (very) weak compared to an alloantibody directed against a high incidence antigen (Answer E).

14. Which of the following clinical scenarios has the most compelling reason to administer plasma?

A. Patient with ascites and an INR of 1.6 needs liver biopsy
B. Patient with thrombotic thrombocytopenic purpura (TTP)
C. Patient with malnutrition and an INR of 4.0, not bleeding
D. Patient with slow lower gastrointestinal hemorrhage and an INR of 1.7, not on warfarin
E. Patient with a lower extremity fracture and congenital factor VII deficiency

Concept: Plasma contains all of the pro- and anticoagulation factors as well as the fibrinolytic proteins. The coagulation proteins are the basis for secondary hemostasis. The end result of the coagulation cascade is precipitation of a fibrin meshwork which seals the platelet plug in place. Currently, there are very few accepted indications for plasma infusion. The main indications for plasma include bleeding in the setting of massive transfusion and for bleeding patients who are taking warfarin when other agents (such as prothrombin complex concentrates) are unavailable. Other indications are bleeding in the setting of a factor deficiency that does not have specific factor concentrate available (e.g., factor XI deficiency). There are several studies suggesting that prophylactic transfusion of plasma is not indicated. This includes patients with liver disease requiring a liver biopsy.

Answer: *B*—Plasma is always indicated in TTP. This is because plasma contains ADAMTS13 (a disintegrin and metalloproteinase with a thrombospondin type 1 motif, member 13), a metalloprotease required to break von Willebrand multimers into small and medium fragments. A main part of the pathophysiology of TTP is that patients lacking ADAMTS13 (either a congenital deficiency or, more commonly, an acquired deficiency following formation of an autoantibody). This leads to the circulation of only large von Willebrand multimers, with subsequent platelet activation and microthrombi formation. Typically, plasma is used as replacement fluid during plasma exchange. If line placement for plasma exchange is delayed, plasma (30 mL/kg for the first 24 h followed by 15 mL/kg each day) may be given as bridging therapy before plasma exchange can be initiated. Diuretics may be considered if the patient cannot tolerate large volume of plasma infusion as bridging therapy.

There are many studies indicating that prophylactic infusion of plasma in patient with liver disease and mild elevation of INR is not warranted (Answer A). Additionally, the PT/INR is not a good indicator of bleeding risk in a patient with liver disease. Patients with end stage liver disease typically have normal thrombin generation, and thus, a normal ability to form a clot. This is because both coagulation proteins and fibrinolytic proteins are decreased in liver disease. A "rebalancing" of these proteins occurs, such that thrombin generation is close to normal. Such a patient has less bleeding than the INR would predict, compared to a person without liver disease.

In a nonbleeding patient with malnutrition, the physician should start with vitamin K administration (Answer C). IV vitamin K may be given if the underlying cause for malnourishment is malabsorption, since vitamin K absorption depends on the ability of the patient to absorb fat. This may be all that is needed to correct the mild coagulopathy. In patients with slow lower gastrointestinal hemorrhage and an INR of 1.7, who are not on warfarin, finding and controlling the source of bleeding is more important than correcting the mild elevation of INR (Answer D). In patients with congenital factor VII deficiency and a lower extremity fracture, the agent of choice is activated factor VII.

15. In conditions other than TTP, what is the recommended dosage for plasma transfusion?

A. 1–10 mL/kg
B. 10–20 mL/kg
C. 20–30 mL/kg
D. 30–40 mL/kg
E. 40–50 mL/kg

Concept: The laboratory response to a dose of plasma is not intuitive. This is because if one were to graph it, the relationship between the decrease in PT/INR and the percent increase in coagulation factors with plasma infusion is logarithmic, rather than linear (Fig. 8.4). Thus, for the same dose of plasma, the INR reduction will be larger for a higher starting INR (i.e., the impact on an INR of 9 will be much greater than for an INR of, say, 1.8). A mathematical describing the relationship between INR and plasma infusion by Pham H.P. et al. suggested that the final INR after infusion can be approximated as

$$INR_{predicted} = \left[\frac{(INR_{initial})^{-0.9196}}{1+\beta} + \frac{\beta}{(0.8039)(1+\beta)} \right]^{-1.0874} \quad \text{where } \beta = \frac{\text{Plasma volume transfused}}{\text{Plasma Volume}_{initial}}.$$

Answer: *B*—A typical dose of plasma is 10–20 mL/kg as recommended by many experts. This dose of plasma would be expected to increase coagulation factors by ~10%–20% immediately after infusion. Additionally, the effect of plasma on PT/INR reduction diminishes as more plasma is administered (Fig. 8.4). Thus, if given >20 mL/kg, the risk of transfusion associated cardiac overload (TACO) may outweigh the benefit of INR reduction (Answers C, D, E). Answer A represents plasma underdose.

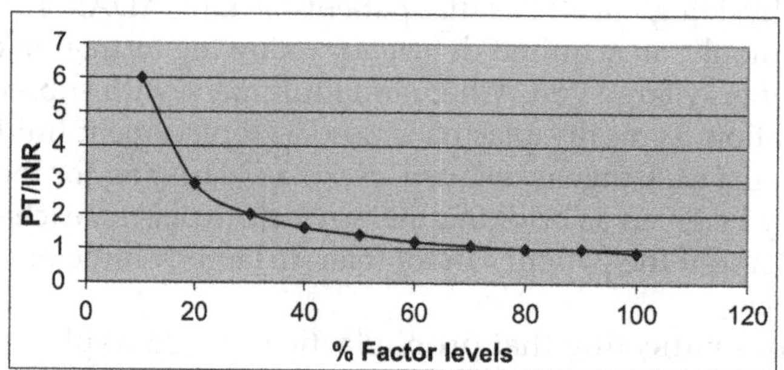

FIGURE 8.4 **PT/INR versus percent coagulation factor levels.** *Source: Used with permission from Dr. Wayne Chandler, Seattle Children's Hospital, Seattle, WA.*

16. When one replaces coagulation factors using plasma administration to help manage bleeding, the target level of coagulation factor goal to establish adequate repletion is which of the following?
A. 10%
B. 40%
C. 50%
D. 80%
E. 100%

Concept: Individuals in good health will routinely have 100% or greater levels of circulating coagulation factors. However, these levels are more than is actually needed to form a clot. There is a buffer zone (or reserve zone) for coagulation proteins, such that we only need approximately 40% total coagulation factors for hemostasis to occur. Note that many texts will cite a 30% level; however, these studies were for single coagulation factor deficiencies, rather than the multiple factor depletion that arises with most forms of bleeding. The practical application of this is that our recommended plasma doses are smaller than would be needed if the goal were replacement to 100% levels. In fact, due to the dilutional effect of plasma transfusion, it is impossible (mathematically and practically) to achieve 100% coagulation factor postplasma transfusion.

Answer: *B*—40% level of coagulation factor is typically enough for hemostasis. Other choices (Answers A, C, D, and E) represents the wrong target based on the above discussion.

17. The difference between FFP and PF24 is described by which of the following?
A. FFP must be frozen sooner after collection than PF24
B. Only FFP can be converted to thawed plasma
C. Only PF24 can be converted to thawed plasma
D. PF24 is never frozen, but remains in a liquid state following collection
E. PF24 is less effective than FFP for transfusion support of trauma patients

Concept: There are several types of plasma products. These include FFP, PF24, thawed plasma, and liquid plasma. Fresh frozen plasma (FFP) is plasma that has been frozen within 8 h of the collection time (typically frozen between 6 and 8 h), whereas PF24 has to be frozen within 24 h (typically they are frozen between 9 and 24 h after collection). There is an additional product, FP24RT24, which is plasma that has been held for up to 24 h at room temperature (rather than on wet ice or refrigerated) after collection, and then frozen. Any of these products may be thawed, relabeled as thawed plasma, and stored under refrigerated conditions for up to 5 days. If these products are NOT converted to thawed plasma, they expire 24 h after thawing.

The FDA directs that each of these products needs a unique label to reflect the fact that the coagulation factors are highest in FFP, slightly lower in PF24 and lower still in thawed plasma (factors degrade slightly during refrigeration). PF24 has slightly lower factor V (~10% less) and factor VIII (20%–50% less) compared to FFP. For thawed plasma, roughly 50% of the amount of coagulation factors remained after 5 days compared to the time of collection. Nonetheless, these levels are enough for critical steps in hemostasis to occur. Liquid plasma, the newest type of plasma available, is plasma that is never frozen following collection. It requires storage under refrigerated conditions and it expires 5 days after the expiration date of the whole blood, depending on the anticoagulant the whole blood is collected in.

Answer: *A*—FFP needs to be frozen within 8 h of collection (vs. 24 h for FP 24. Both products can be converted to thawed plasma (Answers B and C). Studies have demonstrated that most of the coagulation factors in PF24 remain at levels similar to FFP. Both products contain hemostatic levels of clotting factors (Answer E). Answer D is incorrect because it describes liquid plasma and not PF24 or FFP.

18. A male trauma patient is group B Rh positive. Given the answer choices below, what ABO type of RBCs and plasma are compatible with his blood type?

A. Group O RBCs and group O plasma
B. Group AB RBCs and group A plasma
C. Group B RBCs and group O plasma
D. Group B RBCs and group AB plasma
E. Group A RBCs and group O plasma

Concept: When transfusing RBCs, one must be very careful to transfuse cells that do not express the ABO antigen(s) to which the patient has "naturally occurring" antibodies. Anti-A and anti-B are very potent antibodies that, once bound to their respective antigen, activate complement. This may result in intravascular hemolysis, which can be lethal to the patient. When transfusing plasma, you are trying to avoid the infusion of passive isoagglutinins that could bind the patient's RBCs and cause "reverse" hemolysis (Table 8.4). For example, a donor that is blood type O has no A or B antigens on their RBCs, making them a universal RBC donor; however, their plasma will have both anti-A and anti-B. Now, if this person became a patient in the future and required RBC transfusion, they can only receive type O blood because of the anti-A and B they possess, but they are universal recipients as far as plasma transfusion because their RBCs lack any A or B antigens.

TABLE 8.4 ABO Compatibility for RBCs and Plasma

Blood type	Red cells	Plasma
A	A or O	A or AB
B	B or O	B or AB
O	O	Any
AB	Any	AB

Answer: *D*—This patient is type B; thus, he can receive B or O RBCs and B or AB plasma.

Although group O and B RBCs are compatible, group O plasma (Answers A and C) contains anti-B (in addition to anti-A), and is therefore, not compatible with the patient's type B RBCs.

Answer choice B has incompatible RBCs, but technically incompatible plasma. The caveat here is that due TRALI mitigation efforts, there is a nationwide shortage of AB plasma. Therefore, many trauma centers will use group A plasma for the initial part of the resuscitation (before the patient is ABO-typed). This is based on the fact that about 80% of patients (either type A or type o) will be compatible and group A individuals do not typically have high titer anti-B in circulation. Empirically, patients can tolerate small amounts of incompatible plasma as often seen in patients receiving ABO incompatible platelets. A potential contributing factor is that ABO antigens lining endothelium, and soluble ABO antigens in people with secretor status, may help to neutralize the incompatible passive antibody. Answer E is not correct based on Table 8.4.

19. Which of the following is true regarding cryoprecipitated AHF?
A. One unit has more fibrinogen than a unit of plasma
B. Its main use today is as a concentrated source of fibrinogen
C. It contains all of the same coagulation proteins as plasma, but in a higher concentration
D. It contains factors I, VII, VIII, and XIII
E. It contains factors I, VIII, X, and von Willebrand factor

Concept: One unit of cryoprecipitated AHF (commonly known as cryoprecipitate) is produced from 1 unit of FFP (and not from other types of plasma, such as PF24, FP24RT24, thawed plasma, or liquid plasma). The FFP is allowed to thaw at 1 to 6°C and the precipitate is collected (and this is cryoprecipitated AHF). The unit is then centrifuged to concentrate this proteinaceous portion. The cryo-depleted plasma (more commonly known as cryopoor plasma) is then removed, and both products may be refrozen. In order to use cryoprecipitated AHF, the frozen precipitate is resuspended in saline and thawed prior to transfusion. For transfusion in adults, 5–10 units are typically pooled together and transfused in order to achieve the desired increase in serum fibrinogen level. A newer process is to create "prepooled" cryoprecipitated AHF. This process involves resuspending several units of cryoprecipitated AHF in plasma (typically 5), pooling in a closed system, and then freezing the pool. Preparation when transfusion is needed is then faster, with the thaw step being the only step needed prior to issue of the pool. Units pooled at the time of use expire in 4 h, whereas prepooled units expire in 6 h.

Answer: *B*—In the United States, the main use of a cryoprecipitated AHF is as a concentrated source of fibrinogen. Nonetheless, fibrinogen concentrate is also available for fibrinogen supplementation; however, it is more expensive than cryoprecipitated AHF [24]. Cryoprecipitated AHF can also be used to treat factor XIII deficiency, which is a rare genetic disorder. At this time, recombinant factor XIII is also FDA-approved for use in the treatment of congenital factor XIII deficiency. Although cryoprecipitated AHF can be used to treat von Willebrand disease and hemophilia A (factor VIII deficiency), concentrates are available and are typically more effective than cryoprecipitated AHF. Likewise, factor VIII concentrates are typically used to treat hemophilia A (or factor VIII deficiency).

Overall, the original unit of plasma will have a higher amount of fibrinogen (Answer A). Nonetheless, every mL of cryoprecipitated AHF has more fibrinogen than plasma. In certain clinical states, such as DIC, cryoprecipitated AHF may be more effective than plasma in increasing the fibrinogen to hemostatic levels quickly with much lower risk of volume overload.

There are five proteins that compose cryoprecipitate [fibrinogen (factor I), factors VIII, von Willebrand factor, factor XIII, and fibronectin]; thus, Answers C, D, and E are incorrect.

20. A 30-year-old postpartum woman experiences excessive postpartum hemorrhage. Her PT, aPTT, and D-dimer are all markedly elevated, fibrinogen is 40 mg/dL, and her platelet count is 32,000/μL; which of the listed blood products or factor concentrates is likely to be part of her treatment at this time?
A. Recombinant activated factor VII
B. Factor VIII concentrate

C. Prothrombin complex concentrate
D. Cryoprecipitated AHF
E. Cryodepleted or cryopoor plasma

Concept: An obstetric patient with postpartum hemorrhage may experience marked disseminated intravascular coagulopathy (DIC), as the placenta is a significant source of tissue factor. DIC is typically a thrombin generating process, which disproportionately lowers the level of circulating fibrinogen compared to other coagulation factor levels. When circulating fibrinogen is very low, both the PT and PTT will be markedly prolonged, as in this patient.

Answer: *D*—Although a fibrinogen concentrate is FDA-approved to be used as fibrinogen supplementation in patients with congenital fibrinogen deficiency, it is not approved to be used for fibrinogen replacement in acquired bleeding patients. Furthermore, it is very expensive; thus, cryoprecipitated AHF is mainly used in the United States as a source of fibrinogen replacement currently. This is very effective in raising the fibrinogen level, and thus, may be beneficial in bleeding control, while the clinical team decides what further measures they will take to stop the hemorrhage. Fibrinogen is a critical protein in maintaining hemostasis by (1) acting as a precursor of fibrin, (2) mediating platelet aggregation via the GpIIa/IIIa receptor, and (3) providing a matrix and mesh network to support and maintain a fibrin clot.

Plasma will also be necessary—because all coagulation factors are being consumed, some plasma replacement will likely be indicated if the patient continues to bleed and/or DIC persists for a period of time. Platelets and RBCs may also be clinically indicated, depending on the severity and length of time of the bleeding diathesis.

Since this patient is in active DIC as the laboratory results suggest, she is at high risk for thrombosis and thus, recombinant activated factor VII and prothrombin concentrate complex may not be appropriate as first-line therapies, since they could increase thrombus formation, especially in higher doses (Answers A and C). This patient is not deficient in factor VIII; thus, factor VIII concentrate is not clinically indicated (Answer B).

Cryodepleted plasma is plasma without the cryoprecipitate, and therefore, much of the fibrinogen, removed (Answer E). It will be quite ineffective compared to cryoprecipitated AHF to rapidly replace the depleted fibrinogen. Currently, there are very few uses for this product with the exception of use during therapeutic plasma exchange (TPE) for refractory TTP, though the usefulness of this approach is debatable.

21. If the obstetric patient has an ABO/D type of AB Rh negative, which group of the listed cryoprecipitate units is acceptable for transfusion?
A. O Rh negative
B. O Rh positive
C. AB Rh negative
D. A Rh negative
E. Any ABO/Rh type

Concept: Because cryoprecipitated AHF is re-suspended in a small amount of plasma (10–15 mL), any ABO type of cryoprecipitated AHF can be issued for transfusion for adult patients. Since cryoprecipitated AHF has been frozen, it is acellular and the Rh type does not matter. This is also true for Rh compatibility and plasma that has been previously frozen.

Answer: *E*—Any type is acceptable for transfusion. Answer choices A, B, C, and D are incorrect.

22. For platelet transfusion, which of the following is correct?
A. Platelets serve a critical role in primary hemostasis
B. Platelets serve a critical role in secondary hemostasis
C. Optimal platelet storage temperature is 37°C
D. Platelet shelf-life is 8 days from the time of collection
E. Apheresis-derived platelets have greater efficacy compared to whole blood-derived platelets

Concept: Platelets play a very important role in primary hemostasis by binding to von Willebrand factor (VWF), which results in formation of a platelet plug.

Answer: *A*—Platelets are involved in primary hemostasis. Together with von Willebrand factor, they form the platelet plug. Platelets also provide phospholipid surface for coagulation factors during secondary hemostasis in order to generate a more stable fibrin clot. Secondary hemostasis is the end result of the coagulation cascade, where crosslinked fibrin seals down the platelet plug to form the more stable fibrin clot (Answer B). Coagulation factors are the main players in secondary hemostasis. The optimal platelet storage is at 20–24°C, with gentle agitation (Answer C).

They can be stored for 5 days from the time of collection (Answer D). A recent FDA draft guidance indicates that if a transfusion service has the appropriate storage bags and an ability to test for bacteria using an FDA-approved assay, the service may be able to extend the shelf-life to 7 days. This may prove to be an option for transfusion services. The caveat is that the assay must be performed within 24 h of transfusion. Apheresis-derived and whole blood-derived platelets (given in an equivalent dose) have the same efficacy (Answer E).

23. One unit of apheresis-derived platelets is ordered for a 70-kg man with a platelet count of 10,000/μL due to chemotherapy for multiple myeloma. If the patient is not bleeding and is not refractory to platelet transfusion, what is the estimated/expected increment posttransfusion?
A. 15,000/μL
B. 20,000/μL
C. 50,000/μL
D. 80,000/μL
E. 100,000/μL

Concept: One unit of apheresis-derived platelets must contain at least 3×10^{11} platelets per unit, per AABB Standards. This dose would be expected to raise the platelet count by 30,000–60,000/μL in an average size patient. If one were using whole blood-derived platelets, 6 units are typically considered to be an equivalent dose (to 1 unit apheresis-derived platelets). Although each whole blood platelet must contain $(3 \times 10^{11})/6 = 5.5 \times 10^{10}$ platelets/unit, processing techniques often produce a higher yield, so 4–6 whole blood-derived platelets in a pool may actually be equivalent to an apheresis-derived platelet dose. The disadvantages of using whole blood-derived platelets are multiple donor exposures and slightly higher risk of bacterial contamination without bacterial testing.

Answer: *C*—One unit of apheresis-derived platelets would be expected to raise the platelet count by 30,000–60,000/μL in an average size patient. The other choices (Answers A, B, D, and E) are not correct based on the expected increment discussed above.

24. A 52-year-old man is day +6 status post-hematopoietic progenitor cell (HPC) transplant from an allogeneic donor. The patient has unremarkable coagulation values (PT, aPTT), an Hb of 9 g/dL following RBC transfusion, and a platelet count of 8,000/μL. How many unit(s) of apheresis platelets should be recommended to prevent spontaneous hemorrhage?
A. 1
B. 2
C. 3
D. 4
E. 5

Concept: A high dose of platelets for prophylactic transfusion to prevent spontaneous hemorrhage is likely not needed. This idea comes largely from results of one randomized controlled trial, which showed that patients who received a low dose of platelets had the same incidence of WHO grade 2 or higher bleeding than the patients who received a medium or high platelet dose (low dose: 1.1×10^{11} platelets, medium dose: 2.2×10^{11} platelets, high dose: 4.4×10^{11} platelets per square meter of body surface area).

Answer: *A*—Several studies have indicated that a platelet count of ≥10,000/μL is sufficient to reduce the risk of spontaneous bleeding in the setting of normal PT, aPTT, and fibrinogen. Therefore, assuming that this patient is not bleeding and not refractory to platelet transfusion, 1 unit of apheresis platelet is enough to raise his platelet count to >10,000/μL. Based on the expected increment (see Question 23), other choices (Answers B, C, D, and E) are thus, wrong.

25. A 55-year-old woman with AML, status-post high dose chemotherapy. She has a platelet count of 14,000/μL. Per the report of the oncology floor nurse, she has no fever and no signs of bleeding. No invasive procedure is planned. What is the recommended platelet transfusion threshold for this setting per AABB guidance?
A. 10,000/μL or less
B. 20,000/μL or less
C. 50,000/μL or less
D. 100,000/μL or less
E. 200,000/μL or less

Concept: Similar to the situation with coagulation factor levels, nature typically produces platelets in excess of the number actually needed to form a clot. Typically, the lower limit of a normal platelet count is 150,000/μL. The number actually needed to prevent spontaneous hemorrhage (assuming normal PT, PTT, and fibrinogen) is 10,000/μL, and a least one study suggests that even a number between 5,000 and 10,000/μL may be adequate.

Answer: *A*—According to a recent AABB guideline on platelet transfusion therapy, platelets should be transfused prophylactically to reduce the risk for spontaneous bleeding in hospitalized adult patients with therapy-induced hypoproliferative thrombocytopenia. For hospitalized adult patients, transfusion should be given with a platelet count of 10,000/μL or less. In terms of other platelet transfusion "triggers," 50,000/μL is often used if a patient is going to need general surgery (Answer C). For neurosurgery or eye surgery, a trigger of 100,000/μL is allowed, as these are spaces where a small amount of bleeding could have devastating consequences (Answer D). Answers B and E are wrong based on the AABB guidelines (see Table 8.5).

26. A 55-year-old woman with AML status post high dose chemotherapy and with platelet count of 25,000/μL. She will have a central line placed today. What is the recommended platelet transfusion threshold for this setting per AABB guidance?
A. 10,000/μL or less
B. 15,000/μL or less
C. 20,000/μL or less
D. 30,000/μL or less
E. 50,000/μL or less

Concept: The platelet count needed for invasive line placement can require much negotiation with the clinical team.

Answer: *C*—Based on the AABB published guideline, prophylactic platelet transfusion is indicated for patients having elective central venous catheter placement with a platelet count less than 20,000/μL (weak recommendation, low quality evidence). Answers A, B, D, and E are wrong based on the AABB guidelines (Table 8.5).

27. You are working with the cardiac surgery team on formulating a hospital wide policy regarding platelet transfusion indications. Which of the following would be appropriate recommendation based on the 2014 AABB guidelines on platelet transfusion in cases where cardiopulmonary bypass is needed?
A. Transfuse 2 apheresis platelet units as prophylaxis prior to cardiac surgery
B. Transfuse 2 apheresis platelet units as prophylaxis postcardiac surgery
C. Transfuse 2 apheresis platelet units prior to cardiac surgery, only for patients who take an antiplatelet agent within 21 days

D. Transfuse 2 apheresis platelet units prior to cardiac surgery, only for patients who take an anticoagulant agent within 7 days
E. Transfuse platelets only to patients who exhibit perioperative bleeding with thrombocytopenia and/or evidence of platelet dysfunction

Concept: The cardiopulmonary bypass circuit used in open-heart surgery has a major effect on platelets, either causing them to activate as they come into contact with the plastic in the circuit, or rendering them dysfunctional. In general, the longer the patient is on the cardiopulmonary bypass circuit, the more dysfunctional the platelets become. Patients not on any presurgery antiplatelet or anticoagulant medications do not typically bleed while on the bypass circuit. The bleeding will manifest shortly after coming off of the circuit. The platelet count may appear normal, while the platelets themselves may not be functioning well. There are other reasons that the patient may be bleeding postbypass (heparin rebound due to the shorter half-life of protamine vs. half-life of heparin, low or dysfunctional fibrinogen, surgical bleeding, etc.), but dysfunctional platelets should be in the differential.

Answer: *E*—The AABB guideline on platelet transfusion (Table 8.5) suggests platelet transfusion for patients having cardiopulmonary bypass who exhibit perioperative bleeding with thrombocytopenia and/or evidence of platelet dysfunction. Typically, 1 unit of apheresis platelets is sufficient.

TABLE 8.5 AABB Clinical Practice Guidelines

Clinical situation	Threshold for platelet transfusion	Grade
1. Therapy-induced hyperproliferative thrombocytopenia	10×10^9/L (prophylactic)	Strong recommendation; moderate quality evidence
2. Elective central venous catheter placement	$< 20 \times 10^9$/L	Weak recommendation; low quality evidence
3. Elective diagnostic lumbar puncture	$< 50 \times 10^9$/L (prophylactic)	Weak recommendation; very low quality evidence
4. Major elective nonneuraxial surgery	$< 50 \times 10^9$/L (prophylactic)	Weak recommendation; very low quality evidence
5. Cardiac surgery with cardiopulmonary bypass	Avoid prophylactic transfusion	Weak recommendation; very low quality evidence
6. Intracranial hemorrhage on antiplatelet agents	Cannot recommend for or against platelet transfusion	Uncertain recommendation; very low quality evidence

The AABB guideline recommends against routine prophylactic platelet transfusion for patients who are not thrombocytopenic and have cardiac surgery with cardiopulmonary bypass (Answer A). If the time on the cardiopulmonary bypass circuit was short, platelets may not be dysfunctional. Evidence of bleeding should prompt consideration of transfusion (Answer B).

Patients are generally instructed to hold antiplatelet medications (Answer C) for a period of 7–10 days prior to surgery; thus, platelet transfusion is unnecessary. In general, the evidence for transfusion platelet in the presence of antiplatelets agents is mixed and controversial. There is no reason to transfuse platelets to reverse the anticoagulant effect (Answer D).

Please answer Questions 28–32 based on the following clinical scenario:

A 42-year-old woman requires a platelet transfusion. Her ABO/Rh type is A Rh negative.

28. From an ABO standpoint, which of the following ABO/Rh -types are most compatible with the patient?
A. Group O Rh negative whole blood-derived platelets pool
B. Group B Rh negative whole blood-derived platelets pool
C. Group AB Rh negative whole blood-derived platelets pool
D. Group B Rh negative apheresis-derived platelets
E. Group O Rh negative apheresis-derived platelets

Concept: Platelets have weak expression of A and B antigens on their surface. They do not express the Rh antigens. In the transfusion service, platelet compatibility typically refers to the infusion of plasma contained in the platelet (about 250–300 mL) that is incompatible with the ABO antigens on the patient's platelets and RBCs. Platelet compatibility guidelines, then, typically follow plasma compatibility guidelines (Table 8.6).

Answer: *C*—Because of the short shelf-life of platelets (5 days) and donor ABO distribution, it is not always possible to give an ABO type-compatible platelets to a patient. Infusion of ABO incompatible platelets may lead to a slightly (about 10%) lower increment than infusion of ABO compatible platelets. This has not been shown to result in a clinical difference (i.e., increased bleeding episodes) for the patient. Nonetheless, in a pediatric or neonatal patient with small blood volume, "reverse hemolysis" may be a real clinical concern if ABO incompatible platelets are transfused. Other choices (Answers A, B, D, and E) are incorrect based on Table 8.6.

TABLE 8.6 ABO Compatibility in Platelet Transfusion

Blood type	Platelets
A	A or AB
B	B or AB
O	Any
AB	AB

29. Which platelet unit from the list below is most likely to result in intravascular hemolysis if transfused to this patient?
A. Group O Rh negative whole blood-derived platelets pool
B. Group B Rh negative whole blood-derived platelets pool
C. Group AB Rh negative whole blood-derived platelets pool
D. Group B Rh negative apheresis-derived platelets
E. Group O Rh negative apheresis-derived platelets

Concept: Since platelet units contain plasma, they possess ABO antibodies. For example, a unit of type A platelets will contain anti-B, whereas a type O platelet will contain both anti-A and anti-B. On a rare occasion, an acute hemolytic transfusion reaction can result from platelet transfusion. When this occurs, it is most commonly the result of a type O apheresis platelet being infused into a group A or AB patient.

Answer: *E*—Empirically, group O individuals may have a high anti-A titer and, less often, a high anti-B titer. Group A and B individuals (Answers B, C, D) do not typically have high titer isoagglutinins. Because of this finding, some collection centers or transfusion service laboratories will test the isoagglutinin titers in their group O apheresis-derived platelet donors, and set policies to ensure that only group O recipients receive group O apheresis platelets with a high titer (most centers use the cutoff of 200 or below as low-titered products, although the data to support this number is weak). A group O whole blood-derived platelet pool (Answer A) may have a much lower risk of causing hemolysis, presumably because the high titer anti-A in one whole blood-derived platelet is diluted by other lower titer plasma in the pool.

30. What percentage of blood donors are considered universal plasma donors?
A. 1
B. 3
C. 10
D. 15
E. 20

Concept: Group AB is the universal donor type because these individuals do not have circulating isoagglutinins (e.g., anti-A or anti-B). Thus, there is no risk of causing hemolysis of the patient red cells upon transfusion of group AB plasma or platelets. The supply of group AB donors is entirely dependent on the frequency of group AB in the population, and the ability to recruit/accept these individuals into the donor pool. Blood centers work hard to build up the group AB plasma and platelet supply, but it can depleted quickly by a patient who is either group AB, or whose blood type is unknown.

Answer: *B*—Only 2.5%–3% of the US population is group AB. Other answer choices (Answers A, C, D, and E) are wrong. See Table 8.2 for more information on the ABO blood group distribution in Caucasians.

31. The above patient is group A Rh negative. The Blood Bank would like to give Rh negative platelets instead of Rh positive platelets because:

A. The patient has naturally occurring anti-D and thus, will clear the transfused platelets prematurely
B. The patient has naturally occurring anti-D and infusion of Rh positive platelets will result in hemolysis
C. The patient could form anti-D and be at risk of hemolysis with future transfusion of Rh positive RBCs
D. The patient could form anti-D and develop immune thrombocytopenia
E. Rh immune globulin is effective in preventing sensitization during pregnancy but not after a transfusion event

Concept: Unlike the antibodies that occur naturally with the ABO blood group system, antibodies of the Rh blood group system are not naturally occurring. Thus Rh negative individuals must be sensitized, usually through pregnancy or transfusion, in order to make anti-D. Although the D antigen is very immunogenic, the current immune status of the patient does play a role in whether the patient will make an anti-D. With platelet transfusion, the unit will contain small numbers of Rh positive red cells; platelets themselves do not express Rh antigens. In an apheresis platelet, the red cell dose is about 0.00043 mL/unit. In a whole blood platelet, the red cell dose is about 0.036 mL. Since the volume of RBCs in a platelet product is very small, administration of an adequate dose of Rh immune globulin (RhIg) will ensure reduction of anti-D formation. This is generally considered most important for the prevention of hemolytic disease of the fetus and newborn (HDFN) due to anti-D in an Rh negative female of child bearing potential.

In platelet shortage one may choose to issue an Rh positive platelet to a patient who has made anti-D, as the number of D-positive red cells in the platelet is very small and the patient would not be expected to have clinical manifestations from the resulting hemolysis.

Answer: *C*—If the patient forms an anti-D, there will not be an impact on the platelet transfusion, as platelets do not express the Rh antigen. However, once the anti-D is made, the patient cannot receive Rh positive RBCs, as hemolysis could result. Anti-D is not a naturally occurring antibody—exposure, such as transfusion and/or pregnancy, is required (Answers A and B). Anti-D cannot cause immune thrombocytopenia (Answer D).

An adequate dose of RhIg is effective in preventing sensitization both during pregnancy and after transfusion of Rh incompatible platelets (Answer E). Note that the dose of RhIg needed to prevent sensitization after RBCs transfusion is large, (and is usually not recommended due to the resulting massive extravascular hemolysis that can occur) compared to the dose needed to prevent sensitization after platelet transfusion (typically, 1 vial of RhIg protects for ~7 units of apheresis-derived platelets assuming each contains ~2 mL RBCs); however, in real life, apheresis-derived platelets contain about 0.00043 mL RBCs per unit.

32. If one is to administer RhIg following Rh positive platelet transfusion to an Rh negative individual, it should be given within what time frame following the transfusion?

A. 4 h
B. 12 h
C. 24 h
D. 48 h
E. 72 h

Concept: Findings from an old (prior to Human Subjects Review Boards) experiment involving Rh negative male prisoners report that RhIg administered up to 72 h after exposure to a small amount of D positive RBCs will prevent sensitization. The typical dosage is one 300 μg vial of RhIg per 30 mL of whole blood or 15 mL of RBCs.

Answer: *E*—RhIg should be given within 72 h of exposure to be fully effective. Other choices (Answers A, B, C, and D) are incorrect based on the recommendation.

End of Case

Please answer Questions 33–35 using the following clinical scenario.

A 27-year-old Rh negative woman is pregnant with her first child. She received a 300 μg dose of RhIg at 28 weeks gestation per standard protocols. Two weeks later, she is a restrained passenger involved in a moderate-speed motor vehicle crash.

33. At this stage of the pregnancy, if one wants to determine a quantitative amount of fetal cells in maternal circulation, which test should be ordered?

A. Rosette test
B. Kleihauer-Betke test
C. MCV of red cells in maternal circulation
D. ABO type of cells in maternal circulation
E. Rh type of cells in maternal circulation

Concept: At this stage of the pregnancy, the baby's ABO/D type is unknown. Periumbilical blood sampling (PUBS) would need to be done to determine the type; however, this is not usually readily available in the setting of trauma. The test to determine the presence of fetal cells (Rh negative or positive) is the Kleihauer-Betke (KB).

The KB stain uses the principle of differential resistance of fetal hemoglobin to acid. A blood film from mother's blood is made and then exposed to acid to remove adult hemoglobin from the cells; fetal hemoglobin is resistant to acid and will remain. Subsequent staining will make the fetal cells appear pink-red in color while the adult cells appear as ghost cells. 2,000 cells are counted and the ratio of fetal cells is determined from the count.

Of note, flow cytometry is more sensitive and specific than the KB stain for detecting fetal cells, but flow cytometry may not be available in all institutions and/or for emergent testing. In the setting of trauma, many other factors and tests (such as ultrasound, fetal monitoring, etc.) should also be considered in addition to the KB stain results, to determine further maternal-fetal management.

Answer: *B*—KB stain is used to determine the amount of fetomaternal hemorrhage (FMH).

Rosette test is performed when one knows the fetus is Rh positive. Further, it is only a qualitative test (Answer A). The KB test can be used as a quantitative measure of fetal hemorrhage, regardless of the fetal Rh type. Although it is true that the MCV of fetal cells is higher than that of adult cells, it is unlikely that the MCV result on the maternal sample will be informative enough to change management (Answer C). The amount of fetal blood in circulation may be too small to alter the MCV, yet the fetus may have lost a significant amount of blood.

While an ABO or Rh discrepancy may be seen in the setting of a very large fetal hemorrhage, the testing does not give a precise measure of fetal cells in circulation (Answers D and E). Thus it cannot be used to determine how much RhIg should be administered. And in the setting of trauma, there may already be an ABO discrepancy if the mother is group A, and has received uncrossmatched group O cells.

34. The KB result comes back showing zero fetal cells detected in maternal circulation. The recommended dose of RhIg according to the American College of Obstetricians and Gynecologists (ACOG) is:
A. 0 vials
B. 300 μg-vial
C. 600 μg-vial
D. 900 μg-vials
E. 1200 μg-vials

Concept: This question is a difficult one to answer, particularly because the patient received a 300 μg dose of RhIg 2 weeks prior. The half-life of IgG is approximately 24 days; thus, the effect of RhIg would still be present. Some transfusion medicine physicians might be comfortable making the recommendation to skip more RhIg. Others might aim for the smaller dose of RhIg (50 μg-vial). The majority of transfusion medicine physicians would likely recommend giving a full vial (300 μg-vial), since the usual formula for RhIg is to always add at least 1 vial of RhIg. The consequence of not clearing fetal D positive cells is significant, as the mother could make anti-D, with HDFN as the potential result for subsequent pregnancies. Nonetheless, the official recommendation from ACOG is to withhold further RhIg if a standard antenatal dose was administered within 3 weeks and there is no evidence of FMH.

Answer: *A*—Since this patient's KB test is negative, there is no evidence of excessive FMH. Thus, according to the ACOG guidelines, there is no need to administer another dose of RhIg since the last dose of RhIg was administered within 3 weeks. If the result of the KB demonstrated excessive FMH, then additional RhIg must be given. The dose is calculated in the usual way (see Question 35). The remaining choices (Answers B, C, D and E) are incorrect according to the ACOG guidelines.

35. The mother and fetus recover from the accident, and a healthy baby is delivered at 40 weeks gestation. The baby is Rh positive. What is the most appropriate plan at this point?
A. Check an antibody screen; if it is positive for anti-D, do not give RhIg
B. Give a (300 μg) vial of RhIg and plan to perform maternal antibody screen in 6 months to determine if the mother has made anti-D
C. Order a rosette test, with a reflex KB test if the rosette test is positive. Administer RhIg dose based on results
D. Order a rosette test and a weak D test on the cord blood sample
E. Estimate the number of vials of RhIg to give based on the amount of blood lost at delivery

Concept: Even if the mother received an additional dose of RhIg at 30 weeks, it is now 10 weeks since that dose. The testing should revert to the normal protocol for an Rh negative woman delivering an Rh positive baby. This entails performing a rosette test using a maternal sample. This is a qualitative test that reveals the presence of Rh positive fetal cells in maternal circulation. The Kleihauer-Betke test, or flow cytometry if available, is then needed to quantify the number of fetal cells in circulation, which leads to the appropriate dose of RhIg to administer.

Answer: *C*—A rosette test should be ordered. If the result is negative, 1 vial of RhIg should be given. If the result is positive, then a KB test should be performed. From the KB results, number of RhIg vials can be calculated as following:

Step 1: Calculate the fetal hemorrhage from maternal TBV and KB results (if no weight is given, then use 5,000 mL)

$$\Rightarrow Fetal\,hemorrhage\,(in\,mL) = \frac{(Maternal\,TBV) * (Fetal\,cells\,counted)}{Total\,cells\,counted}$$

Step 2: One 30-μg RhIg vial will cover 30 mL fetal whole blood. Therefore:

$$Number\,of\,RhIg\,vials\,required = \frac{Fetal\,hemorrhage\,(in\,mL)}{30}$$

Due to the inherent imprecision of this formula, recommendations for dosage adjustment are as follows:

- If the calculated dose to the right of the decimal point is ≥0.5, the number of vials should be rounded up to the nearest whole number plus one vial
- If <0.5, then round down plus one vial
- Example: If the calculated dose results as 1.5, the number of vials administered will be 3. If the result is 1.4, then 2 vials will be given

The rosette test is done on a maternal sample (Answer D). The weak D test should be performed on all baby samples that initially test Rh negative. The rosette test is considered invalid (could be falsely negative) if the child is weak D-positive, and should not be performed. If this is the case, then KB test should be performed. In the question above, the testing on the cord sample already showed the child to be Rh positive.

Given the sensitivity of some test methods, especially solid phase, anti-D from RhIg administered at 28 weeks may still be detectable on an antibody screen performed at the time of delivery. This should not dissuade one from determining the size of the fetal-maternal hemorrhage, and giving more RhIg (Answer A).

Neither giving a (300 μg) vial of RhIg and planning to perform maternal antibody screen in 6 months to determine if the mother has made anti-D (Answer B) nor estimating the number of vials of RhIg to give based on the amount of blood lost at delivery is completely protective against sensitization and is thus incorrect. It is preferable to quantitate the fetal-maternal hemorrhage and calculate the dose of RhIg. Small studies have shown that "eye-balling" the size of the hemorrhage can result in under-dosing of RhIg.

End of Case

36. A 35-year-old female with history of ITP refractory to corticosteroids and rituximab, is being evaluated to receive RhIg. What are the factors that need to be assessed before administering this drug?

A. The patient should have a red cell osmotic fragility test performed
B. The patient should be Rh positive and have a functional spleen
C. The patient should be tested for tuberculosis before receiving RhIg
D. The patient should be immunized against encapsulated organisms before receiving RhIg
E. RhIg is not indicated for treatment of ITP

Concept: RhIg has been proven to be effective in treatment of ITP, for patients who are Rh positive. Although the exact mechanism is unknown, it has been proposed that the RhIg binds to the Rh-antigen on the patient's RBCs, leading to immune-mediated clearance by the reticuloendothelial system, and presumably subsequent competition for clearance sights in the spleen is created, resulting in decreased clearance of antibody-coated platelets. Thus, a functional spleen is essential for RhIg to work in ITP.

The standard dose is 50–75 mcg/kg, depending on the patient's initial hemoglobin. It has been used since 1995 with a relatively low incidence of adverse events (0.03%–0.04%). Extravascular hemolysis is an expected complication, and is generally well tolerated. Importantly, patients should be monitored for at least 8 h for the occurrence of adverse events that may include hemolysis, hypersensitivity reactions, febrile reactions, and transfusion-related acute lung injury. The FDA placed a black-box warning indicating that intravascular hemolysis leading to death has been reported with the use of RhIg in the treatment of ITP. Thus, the patient needs to be monitored closely for signs, symptoms, and laboratory evidence of intravascular hemolysis in the healthcare setting for at least 8 h. Dipstick urinalysis to monitor hemoglobinuria and hematuria should be performed at baseline, at 2 h, at 4 h, and before the end of the monitoring period.

Answer: *B*—The patient should be Rh positive and have a functional spleen in order to receive RhIg as part of the treatment for ITP (Answer E). Most patients have normal red cell fragility, and RhIg would be expected to induce hemolysis by immune (rather than osmotic) mechanisms (Answer A). Furthermore, the risk of infection, or dissemination of infection, is not increased with the administration of RhIg (Answers C and D).

Please answer Questions 37 and 38 based on the following clinical scenario:

A 62-year-old multiparous woman has been recently diagnosed with AML. She is thrombocytopenic from chemotherapy. Platelet transfusion is ordered for a platelet count of 8,000/μL. The platelet transfusion is administered. Two hours later, the platelet count is drawn, and the count is 8,000/μL.

37. What is the most likely cause of this apparent lack of response to transfusion?
A. The patient is bleeding
B. The patient has splenomegaly
C. The patient has made alloantibodies against HLA antigen(s)
D. The patient had a fever just before the transfusion began
E. The patient is on an antifungal agent which is toxic to platelets

Concept: When it appears that the patient is not getting an adequate response to platelet transfusions, two broad reasons typically apply. The patient may have a clinical reason (such as bleeding, fevers, or be on medicine known to be toxic to platelets) for the refractoriness. Or the patient may have made alloantibodies against platelet receptor or HLA antigen(s). A good screening question before any testing is done is whether the patient has been pregnant or transfused in the past. If the answer is no, the probability that the patient has made platelet receptor or HLA alloantibodies is low. Immune thrombocytopenia (ITP) could also explain the lack of response; however it is unusual to have AML and ITP occurring together. Alloimmunization against platelet receptor or HLA antigens is much more likely.

Answer: *C*—Since this patient has immune-mediated refractoriness, most likely antibodies against the HLA antigens will be present.

Answer choices A, B, D, and E are all reasons for "clinical" refractoriness. One would expect that the platelet count will rise following transfusion, albeit not as high as the expected response for the dose administered.

38. What initial test would you recommend in order to determine if the platelet refractoriness is due to nonimmune causes versus immune mediated destruction?
A. HLA antibody screen
B. One hour posttransfusion platelet count
C. Patient's HLA type
D. Ultrasound of the spleen
E. No further testing needed; go ahead and order HLA matched platelets

Concept: When attempting to determine if the lack of response to transfusion is due to clinical reasons (i.e., nonimmune mediated), such as splenomegaly, or due to patient antibodies against HLA antigens (i.e., immune mediated), the initial test performed is the 1 h posttransfusion platelet count (Fig. 8.5). Ideally, this will be drawn 15 min to 1 hour after the platelet transfusion. The result test should help guide further testing or management:

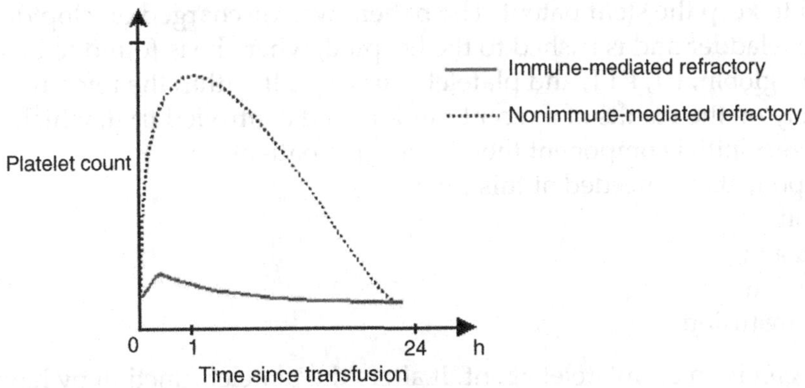

FIGURE 8.5 One-hour posttransplant platelet count interpretation.

If the platelet count increases, but not as high as one would expect, the refractoriness is more likely due to a clinical factor (e.g., splenomegaly). Management will be directed at either giving a larger dose of platelets, and/or giving platelets more frequently. The schedule of administration will be "titrated" to the desired platelet count for that patient.

Answer: *B*—The initial testing done in the algorithm to determine why the patient is refractory to platelet transfusion is the 1-h posttransfusion platelet count. The transfusion medicine physician will work to establish that this has been completed. Many times the clinician has a knowledge gap, and believes that a platelet count drawn up to 24 hours after the transfusion is still valid to detect immune refractoriness.

HLA antibody screen (Answer A) and ultrasound of the spleen (Answer D) are expensive compared to drawing a platelet count. Appropriate laboratory utilization is to perform the 1 h count first.

Order and issue HLA matched platelets, means that the patient HLA type must be known and a donor often be specifically recruited to donate the platelets, should not be done until the presence of HLA alloantibodies has been established (Answers C and E).

If there is no response to transfusion with the 1-h post-platelet count, testing should then be done to determine if the patient needs HLA matched platelets or crossmatched platelets. For HLA matched platelets, the baseline testing is the HLA antibody screen. If the patient has a very low platelet count and/or is at high risk of bleeding, the patient HLA type should be drawn at the same time. If the patient is not at high risk of bleeding or if the suspicion of alloimmunization is low, one could wait to obtain the patient's HLA type. HLA matching depends on HLA class I antigens (A and B), since only class I antigens are found on platelets.

An alternative test is the platelet crossmatch. This is done by testing the patient's plasma in a crossmatch with a select number of platelets for transfusion. The solid phase red cell adherence (SPRCA) test is the most widely used method. Compared with HLA matching, crossmatching can be easier and more cost effective. It also provides an advantage in facilitating the selection of platelets when platelet specific antibodies are present.

If the patient is at high risk of bleeding and the HLA antibody testing is in progress, one can recommend transfusion support with a pool of whole blood-derived platelets if available. This is based on the idea that the patient may lack an antibody to at least one of the platelets in the pool, and thus, get some response until an HLA matched or crossmatched platelets can be provided.

End of Case

39. A 70-year-old male patient was admitted to the cardiac catheterization laboratory after it was determined that he had acute occlusion of a coronary artery. A stent was placed, and clopidogrel

was administered to keep the stent patent. The patient was discharged on clopidogrel. Four months later, he falls from a ladder and is rushed to the hospital, where he is found to have an intracranial hemorrhage. Hemoglobin, PT, PTT, and platelet count are all within the reference range. If you base your decision solely on the results of a recent randomized controlled trial, which of the following represents the correct initial component therapy for this patient?

A. No blood components are needed at this time
B. RBC transfusion
C. Platelet transfusion
D. Plasma transfusion
E. Granulocyte transfusion

Concept: Clopidogrel is an antiplatelet agent. It alters the platelet function by having its active metabolite bind irreversibly to the $P2Y_{12}$ class of the ADP receptor; thus, it inhibits platelet activation and aggregation. This medication inhibits platelet function for the life-time of the platelets (7–10 days); thus, per the package insert of clopidogrel, if a surgical procedure is planned and antiplatelet therapy is not desirable, discontinuation of the medication for 5 days before the surgery is recommended.

In the past, since the half-life of the active metabolite of clopidogrel is short, platelet transfusion was recommended to be given if the patient is bleeding. This recommendation was given based on limited in vivo evidence suggesting that in a clinical scenario such as the one above, platelet transfusion may help to slow or reduce hemorrhage. The number of functioning platelets needed in order to be able to clot is not well known; older studies suggest that 20% functioning platelets will suffice. The dose of platelet transfusion may depend on the location of the bleeding or the urgency of the surgery. In one review, Sarode recommends transfusing 1 unit of apheresis derived platelets if the patient requires urgent surgery (not neuro- or ophthalmologic surgery) or presents with significant bleeding. If the patient presents with intracranial hemorrhage or requires neuro or ophthalmologic surgery, then 2 units of apheresis platelets may be necessary. Aspirin, irreversibly inhibits the cyclooxygenase of platelets already in circulation, but does not affect new platelets released into circulation. Therefore, some transfusion medicine physicians may only recommend transfusing 1 unit of apheresis platelets if the patient has intracranial bleeding or requires urgent neuro or ophthalmologic surgery.

However, the practice of platelet transfusion in attempt to reverse antiplatelet medications in patients with active bleeding may change in the near future based on the results of a recent phase III randomized open-label trial comparing platelet transfusion and standard of care vs. standard of care alone in patients with supratentorial intracerebral hemorrhage who had been on antiplatelet therapy for at least 7 days. The trial demonstrated that platelet transfusion in these patients "seem inferior to standard of care" and that "platelet transfusion cannot be recommended for this indication in clinical practice." Nonetheless, the trial has some limitations, such as small sample size and some imbalance in baseline prognostic variables.

Answer: *A*—Based on the results of a recent randomized trial, no blood component is required for this patient.

Since clopidogrel does not alter the function of red blood cells, coagulation factor, or white blood cells, red blood cells, plasma, or granulocyte transfusion (choices B, D, and E) is not clinically necessary. Platelet transfusion is the current clinical practice (Answer C); however, this practice is not supported by the result of a recent randomized clinical trial.

40. A 55-year-old male patient recently underwent coronary artery bypass surgery. The postoperative course was complicated by heart failure, requiring prolonged extracorporeal life support (ECLS). Daily platelet monitoring reveals a sudden drop in platelet count from 115,000 to 40,000/μL on postoperative day 9. The clinical team is concerned about the low platelet count in the context of the patient being anticoagulated with heparin for the ECLS circuit, and has requested

2 apheresis-derived platelet units. The patient has had a stable serum hemoglobin level for the past 8 h. How will you respond to their request?

A. Transfuse 1 dose of platelets, evaluate posttransfusion platelet count and, if lower than 50,000/μL, transfuse another dose
B. Send testing for heparin induced thrombocytopenia (HIT) and transfuse 1 dose of platelets
C. Send testing for HIT and avoid platelet transfusion
D. Start emergent therapeutic plasma exchange for thrombotic thrombocytopenic purpura (TTP) and avoid platelet transfusion
E. Start corticosteroids and intravenous immunoglobulin for immune thrombocytopenic purpura (ITP) and avoid platelet transfusion

Concept: Patients receiving large amount of heparin, especially in the context of surgery, are at increased risk of developing heparin induced thrombocytopenia (HIT). It is estimated that the frequency of HIT is ~0.76% in patients receiving therapeutic doses of unfractionated heparin, <0.1% in patients receiving subcutaneous unfractionated heparin, and ~0.2% in all exposed heparin patients. HIT is also more common in surgical patients versus medical patients (OR 3.25). Patients undergoing cardiac surgery have a high incidence of antibodies against platelet factor 4; however, the risk of true HIT in these patients may be similar to other surgical patients.

HIT is caused by antibodies against heparin that bind platelet factor 4 and result in platelet activation, with increased risk of thrombosis. Although there are immunoassays to test antibodies against PF4 and functional tests (serotonin release assay), HIT remains a clinical diagnosis. A 4-T score based on clinical parameters was developed to stratify these patients into low, intermediate, and high risk.

If HIT is clinically suspected, then heparin should be discontinued and the patient switched to other forms of anticoagulation, such as direct thrombin inhibitors (e.g., argatroban). Unless the patient is bleeding massively or bleeding into critical spaces (such as intracranial hemorrhage), then platelet transfusion should be avoided to reduce the risk of developing HIT with thrombosis (HITT) (so called "white clot syndrome" is a dreaded complication of this entity).

Answer: *C*—The clinical presentation suggests that HIT may be a possible diagnosis. This patient's 4T score is 5 (intermediated score) and is calculated as follows:

- thrombocytopenia = +2
- no evidence of thrombosis = +0
- timing of 9 days = +2
- may have other causes of thrombocytenia = +1

Thus, heparin should be stopped and testing for HIT should be initiated. Additionally, the clinical team should be encouraged to switch anticoagulation to a direct thrombin inhibitor while testing is underway. If the patient truly has HIT, not switching to an alternative agent could result in a thrombotic event even though the heparin has been discontinued.

The drop in platelet count to 40,000/μL, but not to <20,000/μL, is fairly typical of HIT, and is less typical of TTP or ITP. Furthermore, there is no evidence of hemolysis (hemoglobin stable); thus, TTP is unlikely (Answer choice D). ITP is a diagnosis of exclusion; thus, testing for HIT should be performed first (Answer E). When HIT is suspected, and the patient is not actively bleeding, then platelet transfusion should be avoided (Answers A and B).

41. A 29-year-old woman is undergoing her second cycle of chemotherapy for acute myelomonocytic leukemia. Granulocyte concentrates are requested by the clinical team. The typical indication for granulocyte transfusion in adults who have failed traditional antibiotics is which of the following?

A. Bacterial infection in a neutropenic patient expected to have relatively rapid marrow recovery
B. Viral infection in a neutropenic patient expected to have relatively rapid marrow recovery
C. Fungal infection in a lymphopenic patient expected to have relatively rapid marrow recovery
D. Viral infection in a lymphopenic patient expected to have relatively rapid marrow recovery
E. A neutropenic patient with acute myelomonocytic leukemia

Concept: The typical indication for granulocyte transfusion is to treat a bacterial and/or fungal infection in a neutropenic patient (absolute neutrophil count <500) who is expected to recover rapidly from the neutropenia and who has failed maximum antibiotic or antifungal dosage. However, fungal infection does not respond very well with granulocyte transfusion. Granulocytes transfusion is not indicated for patients with viral infection. In reality, clinical teams often request granulocyte transfusion as a last resort for a patient with neutropenia (possibly prolonged) and infection from a variety of pathogens. Factors to consider when this occurs include whether it will be possible to find granulocyte donors for a prolonged course of neutropenia. The patient must also be on appropriate antibiotics as the main treatment for the infection.

Answer: *A*—Granulocyte transfusion should only be given to neutropenic (and not lymphopenic, Answers C and D) patients with documented bacterial and/or fungal infection and with hope with marrow recovery and who has failed maximum conventional therapy. Granulocyte transfusion should not be used in patients with viral infection (Answers B and D) or in patients without neutropenia (Answer E).

42. Studies have shown that a main factor in the success of granulocyte transfusions is the dose of granulocytes administered to the patient. Because of this, the donor is usually given a one-time dose of steroids several hours prior to the collection, to enhance the demarginalization of the neutrophils away from the blood vessel walls. All of the following are relative contraindications to acceptance of a donor for granulocyte collection *except* which of the following?

A. Obesity
B. Diabetes
C. Cataracts
D. Severe hypertension
E. Peptic ulcer disease

Concept: Although the steroid dose administered to the donor is a one-time dose, these donors are often asked to give multiple donations, since the course of granulocytes is typically daily for at least a week. Thus donors are questioned ahead of time for the following: history of diabetes; severe hypertension; cataracts; and peptic ulcer disease. If the history is positive, the donor will likely be deferred from granulocyte donation.

Answer: *A*—Obesity (in the absence of diabetes) is not a contraindication to granulocyte donation. The remaining choices (Answers B, C, D, and E) represent conditions that can be exacerbated by steroid treatment.

43. A frequent modification made to granulocytes is which of the following?

A. Leukoreduction
B. Freezing
C. Irradiation
D. Washing
E. Pooling

Concept: Irradiation is required to prevent transfusion-associated graft versus host disease (TA-GVHD) if the donor is an HLA match or a blood relative. Because recipients are immunocompromised and the component is rich in T-lymphocytes (the cell responsible for TA-GVHD), a transfusion service may choose to irradiate all granulocytes. Such a policy can help to prevent process errors, if all granulocytes are handled the same way. Note that one would never leukoreduce a granulocyte product, as the efficacy of the product will be greatly compromised. If a CMV seronegative product is needed, the donor selected should be CMV seronegative.

Answer: *C*—Irradiation is typically performed on granulocyte products to prevent TA-GVHD. Leukoreduction should never be used in granulocyte product (Answer A). Freezing, washing, or pooling is not frequent product modification made to granulocyte (Answers B, D, and E).

44. One of the more common adverse consequences of granulocyte transfusion to the recipient is which of the following?

A. Transfusion associated graft versus host disease (TA-GVHD)
B. HLA alloimmunization
C. Chronic granulomatous disease
D. Immune thrombocytopenia
E. Cataracts

Concept: Besides transfusion reactions and transfusion-transmitted disease as potential complications, granulocyte transfusion has other risks. One of the rare adverse consequences is pulmonary toxicity. Moreover, because a patient requiring granulocyte transfusions typically requires not just one, but a week or longer course of daily transfusions, it is impossible to use only one granulocyte donor. This results in exposure to multiple HLA antigens from multiple different granulocyte donors, and subsequently leads to alloimmunization in patients. In the setting of a patient with a hematologic malignancy who is awaiting a hematopoietic progenitor cell transplantation, this can have significant consequences. For example, a patient who has aplastic anemia will have a less successful chance of engraftment if the patient is highly alloimmunized against HLA antigens.

Answer: *B*—HLA alloimmunization may be a risk for patients receiving granulocyte transfusions.

TA-GVHD could certainly be a complication of granulocyte transfusion if the product is not irradiated. Because of this, it is often standard practice to irradiate all granulocyte transfusions (see Question 43). CGD patient with a localized infection is a potential indication for granulocyte transfusions. It is not a potential adverse event of the transfusion (Answer C). Finally, immune thrombocytopenia or cataracts are not associated with adverse effects of granulocyte transfusion to the recipient (Answers D and E).

Suggested Reading

[1] AABB, Standards for Blood Banks and Transfusion Services, twenty-ninth, American Association of Blood Banks (AABB), 2016.

[2] S. Baraniuk, B.C. Tilley, D.J. del Junco, et al. Pragmatic randomized optimal platelet and plasma ratios (PROPPR) trial: design, rationale and implementation, Injury 45 (9) (2014) 1287–1295.

[3] J.L. Carson, B.J. Grossman, S. Kleinman, et al. Red blood cell transfusion: a clinical practice guideline from the AABB*, Ann. Int. Med. 157 (1) (2012) 49–58.

[4] M. Cushman, W. Lim, N.A. Zakai, 2011 Clinical Practice Guide on Anticoagulant Dosing and Management of Anticoagulant Associated Bleeding Complications in Adults, eighth ed., American Society of Hematology, Washington, DC, (2011).

[5] M. Desborough, S. Stanworth, Plasma transfusion for bedside, radiologically guided, and operating room invasive procedures, Transfusion 52 (2012) 20S–29S.

[6] J.M. Despotovic, M.P. Lambert, J.H. Herman, et al. RhIg for the treatment of immune thrombocytopenia: consensus and controversy, Transfusion 52 (5) (2012) 1126–1136.

[7] P.C. Hébert, G. Wells, M.A. Blajchman, et al. A multicenter, randomized, controlled clinical trial of transfusion requirements in critical care, N Eng. J. Med. 340 (6) (1999) 409–417.

[8] A. Holbrook, S. Schulman, D.M. Witt, et al. Evidence-based management of anticoagulant therapy: Antithrombotic therapy and prevention of thrombosis, 9th ed: american college of chest physicians evidence-based clinical practice guidelines, Chest 141 (2 Suppl.) (2012) e152S–e184S.

[9] R.M. Kaufman, B. Djulbegovic, T. Gernsheimer, et al. Platelet transfusion: a clinical practice guideline from the aabbplatelet transfusion: a clinical practice guideline from the AABB, Ann. Int. Med. 162 (3) (2015) 205–213.

[10] G.K. Lo, D. Juhl, T.E. Warkentin, C.S. Sigouin, P. Eichler, A. Greinacher, Evaluation of pretest clinical score (4 T's) for the diagnosis of heparin-induced thrombocytopenia in two clinical settings, J. Thromb. Haemost. 4 (4) (2006) 759–765.

[11] C.A. Mazzei, M.A. Popovsky, P.M. Kopko, Noninfectious complications of blood transfusion, in: M.K. Fung (Ed.), Technical Manual, American Association of Blood Banks, Bethesda, MD, 2014, pp. 665–696.

[12] M.H. Murad, J.R. Stubbs, M.J. Gandhi, et al. The effect of plasma transfusion on morbidity and mortality: a systematic review and meta-analysis, Transfusion 50 (6) (2010) 1370–1383.

[13] Services USDoHaH, Health NIo, National Heart L, Blood Institute, Evidence-based management of sickle cell disease: expert panel report, 2014. Available from: http://www.nhlbi.nih.gov/health-pro/guidelines/sickle-cell-disease-guidelines/.

[14] S.J. Slichter, R.M. Kaufman, S.F. Assmann, et al. Dose of prophylactic platelet transfusions and prevention of hemorrhage, N. Engl. J. Med. 362 (7) (2010) 600–613.
[15] E.P. Vichinsky, A. Earles, R.A. Johnson, et al. Alloimmunization in sickle cell anemia and transfusion of racially unmatched blood, N. Engl. J. Med. 322 (23) (1990) 1617–1621.
[16] S.T. Chou, T. Jackson, S. Vege, K. Smith-Whitley, D.F. Friedman, C.M. Westhoff, High prevalence of red blood cell alloimmunization in sickle cell disease despite transfusion from Rh-matched minority donors, Blood 122 (2013) 1062–1071.
[17] A.B. Docherty, R. O'Donnell, S. Brunskill, et al. Effect of restrictive versus liberal transfusion strategies on outcomes in patients with cardiovascular disease in a non-cardiac surgery setting: systematic review and meta-analysis, BJM 353 (2016) i1351.
[18] Z.M. Szczepiorkowski, N.M. Dunbar, Transfusion guidelines: when to transfuse, Hematol. Am. Soc. Hematol. Educ. Prog. 2013 (2013) 638–644.
[19] J.L. Carson, G. Guyatt, N.M. Heddle, et al. Clinical practice guidelines from the AABB: red cell transfusion thresholds and storage, JAMA 316 (2016) 2025–2035.
[20] H.P. Pham, M.C. Mueller, L.A. Williams III, et al. Mathematical model and calculation to predict the effect of prophylactic plasma transfusion on change in international normalized ratio in critically ill patients with coagulopathy, Transfusion 56 (2016) 926–932.
[21] American College of Obstetricians, GynecologistsACOG practice bulletin: prevention of Rh D alloimmunization, Int. J. Gynaecol. Obstet. 66 (1999) 63–70.
[22] R. Sarode, How do I transfuse platelets (PLTs) to reverse anti-PLT drug effect?, Transfusion 52 (2012) 695–701.
[23] M.I. Baharoglu, C. Cordonnier, R.A. Salman, et al. Platelet transfusion versus standard care after acute stroke due to spontaneous cerebral haemorrhage associated with antiplatelet therapy (PATCH): a randomised, open-label, phase 3 trial, Lancet 387 (2016) 2605–2613.
[24] C.K. Okerberg, et al. Cryoprecipitate AHF vs. fibrinogen concentrates for fibrinogen replacement in acquired bleeding patients–an economic evaluation, Vox sang 111 (2016) 292–298.
[25] A.W. Bryan Jr., E.M. Staley, T. Kennell Jr., A.Z. Feldman, L.A. Williams III, H.P. Pham, Plasma transfusion demystified: a review of the key factors influencing the response to plasma transfusion, Lab. Med. 48 (2) (2017) 108–112.
[26] AABB circular of information for the use of human blood and blood components. https://www.aabb.org/tm/coi/Documents/coi1113.pdf. Accessed 8-22-2017.

CHAPTER

9

Patient Blood Management

*Marisa B. Marques**, *Karen Dallas***, *Lance A. Williams, III**

*University of Alabama at Birmingham, Birmingham, AL, United States; **St. Paul's Hospital, Vancouver, BC, Canada

Blood management is a multidisciplinary initiative with roots in evidence-based medicine. Since Jehovah's Witnesses do not accept most blood products, we have known for many years that patients could be safely treated without transfusion. However, it was not until the publication of the transfusion requirements in critical care (TRICC) trial in 1999 that we had evidence to back up the theory that a restrictive transfusion approach was not inferior to a liberal approach and for some patients, was actually superior. Additionally, studies on the side-effects of transfusion (e.g., increased risk of sepsis and alloimmunization) provided more evidence for using the least amount of blood as possible for each and every patient. Thus, patient blood management programs (PBM) should be an essential part of any hospital's transfusion medicine service. This chapter focuses on the evidence behind PBM, the safe triggers that have been established, the many ways that hospitals can reduce blood usage and wastage, and the challenges involved in both academic and community-based PBM initiatives.

1. Which of the following statements is correct regarding patient blood management (PBM) and bloodless medicine?
   A. PBM is synonymous with bloodless medicine
   B. Bloodless medicine is practiced by surgeons of the Jehovah's Witness faith
   C. PBM and bloodless medicine are based on finding alternatives for transfusions
   D. PBM and bloodless medicine began during the HIV epidemic of the 1980s
   E. PBM targets the prevention of blood product use

   Concept: There is some confusion regarding these two terms and it is important to understand how they are similar and dissimilar. Both definitions have evolved, especially in the last 10 years, as new modalities to prevent and treat bleeding and anemia became available (e.g., recombinant factor concentrates and topical hemostatic agents). Further, they require a multidisciplinary team to carry out the plan for each individual patient.

   Answer: *C*—PBM and bloodless medicine are based on finding alternatives for transfusions. PBM is the combination of multidisciplinary and multimodal measures aimed at preventing or reducing the need for blood products with the goal of improving outcomes. If transfusion becomes necessary, a PBM approach does not prevent the use of blood products (Answer E). However, physicians and other healthcare professionals caring for such patients practice in ways that make transfusions less likely. PBM arose as a consequence of concerns related to the safety and the efficacy of blood products, as well as the risk of decreased supply and high cost (Answer D).

   In contrast to PBM, bloodless medicine (Answer A) relates to the care of patients who do not accept allogeneic blood products, either because of their religious beliefs, or for fear of the associated

Transfusion Medicine, Apheresis, and Hemostasis. http://dx.doi.org/10.1016/B978-0-12-803999-1.00009-2

risks. Bloodless medicine can be practiced by any physician, regardless of the patient's religious beliefs (Answer B).

2. Which of the following options describe the three pillars of patient blood management (PBM) in elective surgery?
   A. Optimize erythropoiesis, minimize blood loss, and manage anemia
   B. Suggest preoperative autologous donation, review medications, and optimize cardiac output
   C. Provide meticulous hemostasis, maximize oxygen delivery, and manage nutritional anemia
   D. Screen for drugs implicated in anemia, minimize iatrogenic blood loss, and optimize cardiac output
   E. Monitor bleeding, employ normovolemic hemodilution, and use evidence-based transfusion strategies

Concept: Elective surgical procedures offer several options to avoid intra- and postoperative transfusions, if patients are thoroughly evaluated preoperatively and their red cell mass is optimized. Furthermore, anesthesiologists and surgeons have a variety of tools to limit and to treat intra- and postoperative blood loss, without the need for allogeneic transfusions.

Answer: *A*—Optimize erythropoiesis, minimize blood loss, and manage anemia. These three pillars include a variety of activities that can be planned for the pre-, intra- and postop periods in order to avoid transfusions as seen below:

Preoperatively:
1. *Erythropoiesis optimization*: Identification, evaluation, and treatment of anemia; preoperative autologous blood donation; consideration of erythropoiesis-stimulating agents (ESA) if nutritional anemia is ruled out/treated; and referral for further evaluation, if necessary.
2. *Minimization of blood loss*: Identification and management of bleeding risk (past/family history); review of medications (antiplatelet, anticoagulation therapy); minimization of iatrogenic blood loss; procedure planning and rehearsal.
3. *Management of anemia*: Comparison between estimated blood loss with patient-specific tolerable blood loss; assessment/optimization of physiologic pulmonary and cardiac reserves; patient-specific management plan using appropriate blood conservation modalities to manage anemia.

Intraoperatively:
1. *Erythropoiesis optimization*: Timing of surgery with optimization of erythrocyte mass.
2. *Minimization of blood loss*: Meticulous hemostasis and surgical techniques; anesthetic blood conserving strategies (acute normovolemic hemodilution, cell salvage/reinfusion); pharmacologic/hemostatic agents.
3. *Management of anemia:* Optimization of cardiac output, ventilation and oxygenation; evidence-based transfusion strategies.

Postoperatively:
1. *Erythropoiesis optimization*: Management of nutritional/correctable anemia (e.g., avoid folate deficiency, iron-restricted erythropoiesis); ESA therapy, if appropriate; avoidance of drug interactions that can cause anemia (e.g., ACE inhibitor).
2. *Minimization of blood loss*: Monitoring and management of bleeding; maintenance of normothermia (unless hypothermia indicated); autologous blood salvage; minimization of iatrogenic blood loss; hemostasis/anticoagulation management; avoidance of adverse effects of medications (e.g., acquired vitamin K deficiency).
3. *Management of anemia:* Maximization of oxygen delivery; minimization of oxygen consumption; avoidance/treatment of infections promptly; evidence-based transfusion strategies.

Though the other answer choices (Answers B, C, D, and E) do describe activities that one might take pre-, intra-, and postoperatively, they do not describe the overarching pillars of perioperative blood management.

Of note, unless the patient has an antibody to a high frequency antigen for which locating donor blood would be nearly impossible or if they have a history of anaphylaxis due to severe IgA deficiency, autologous donation (Answer B) is typically not recommended. This is due to the low rate of actual use during the surgery, which leads to high rates of wastage of both the unit and the resources required to store the blood safely.

3. Iron-restricted erythropoiesis due to iron sequestration is a common mechanism of anemia in patients with chronic inflammatory diseases. What laboratory studies would typically characterize this condition?
   A. Increased transferrin saturation
   B. Low serum ferritin levels
   C. High bone marrow iron stores
   D. Macrocytic anemia
   E. High hepcidin levels

   Concept: In order to properly manage anemia, it is essential to understand the relationship between erythropoietin (EPO), iron, and erythropoiesis. In addition to absolute iron deficiency, iron-restricted erythropoiesis is a significant public health problem worldwide, which may be caused by functional iron deficiency and iron sequestration. Functional iron deficiency occurs during treatment with EPO or ESAs. Iron sequestration in enterocytes and macrophages occurs in patients with chronic inflammatory states, such as rheumatologic disorders, inflammatory bowel disease, malignancies, critical illness, and infections leading to anemia of chronic disease.

   Answer: *E*—High hepcidin levels. Hepcidin is a small hormone peptide mainly produced in the liver that is central to iron absorption in the gastrointestinal tract. In chronic inflammation, high hepcidin (probably cytokine-mediated) causes degradation of the iron exporter protein ferroportin, thereby preventing iron that needs to be used in erythropoiesis from leaving the reticuloendothelial system and enterocytes. Essentially, these patients do not respond to oral iron therapy, but do respond to IV iron therapy. Thus, these patients can be treated without exposing them to allogeneic blood products. In the future, hepcidin-antagonists may become an important tool in anemia management.

   Transferrin saturation (Answer A) is low in iron sequestration, while serum ferritin (Answer B) is normal to high. In anemia of chronic disease, the iron is typically sequestered in macrophages and enterocytes, not the bone marrow (Answer C) and this type of anemia is initially normocytic, but can progress to microcytic (Answer D).

   Table 9.1 summarizes laboratory parameters in the three types of iron-restricted erythropoiesis. Of note, anemia by two or more mechanisms may occur concurrently, and is not rare. One such example is chronic blood loss leading to absolute iron deficiency in patients with inflammatory bowel disease. In such patients, it is even more crucial to interpret laboratory parameters in light of the clinical information.

TABLE 9.1 Causes of Iron-Restricted Erythropoiesis

	Hepcidin RR: Not available	**Transferrin saturation (TSAT)** RR: 25%–35%	**Serum ferritin** RR: 33–350 ng/mL	**Serum iron** RR: 60–160 µg/mL
Iron deficiency anemia (IDA)	Low	Low(<20%)	Low (<30 ng/mL)	Low (<40 µg/mL)
Functional iron deficiency	Variable	Low (<15%)	Variable	Low
Iron sequestration	High	Low	Normal to High	Low

RR, Reference range.

4. A 67-year-old woman with rheumatoid arthritis complained of worsening fatigue and reports noticing melena for several months. Her complete blood cell count (CBC) showed a hypochromic microcytic anemia with a ferritin of 20 ng/mL and a transferrin saturation of 10%, despite oral iron

supplementation. Since she needs a partial colectomy for her colon cancer, her physician orders two units of red blood cells (RBCs) to correct her anemia. Which of the following statements is correct regarding this patient?

A. The dose of iron was probably insufficient and should be doubled for the next week
B. The units should be given immediately prior to the operation for the greatest benefit
C. The patient should donate autologous units to receive during the procedure
D. The patient likely needs intravenous iron to overcome decreased absorption
E. Her condition requires the addition of an ESA to the oral iron

Concept: While the prevalence of anemia in the elderly varies greatly among different studies, iron deficiency, as well as vitamin B12 and folate deficiencies, is thought to explain approximately one-third of cases. Another third of patients have anemia because of some degree of renal insufficiency and/or a chronic inflammatory condition. In the remaining, the etiology is not always clear. Decreased response to endogenous erythropoietin has been suggested to explain the so-called anemia of aging, while more practical reasons like medications, which suppress the bone marrow, should not be overlooked. In addition, there are many patients in which there is more than one cause of anemia. Combined comorbidities that contribute to anemia are likely to affect the response to treatment.

Answer: *D*—The patient needs intravenous iron to overcome impaired absorption. She probably also has anemia of chronic disease from her history of rheumatoid arthritis. The etiology of this common type of anemia was recently elucidated as a block in the intestinal absorption from high hepcidin, a hormone that controls the export of iron from enterocytes and macrophages into the circulation. For this reason, now that she has become severely iron deficient from chronic gastrointestinal bleed, she likely needs intravenous injections to replenish her iron stores and correct the anemia.

She has been on oral iron for several years. While increasing the dose may work, doing so could lead to increased side effects such as gastrointestinal (GI) upset (Answer A). Since this is a planned procedure, every effort should be made to avoid transfusion (Answer B). The patient is already anemic; therefore, autologous donation is not an option for her (Answer C). Since intravenous (IV) iron has not been given yet, placing the patient on an ESA (Answer E) might be too aggressive.

5. A 44-year-old woman with severe menorrhagia is being evaluated for a hysterectomy in 4 weeks. Her CBC reveals the following results: hemoglobin of 7.5 g/dL, a mean corpuscular volume (MCV) of 76 fL, and a platelet count of 350,000/μL. Which is the best treatment option?

A. Dietary modification
B. Oral iron supplementation
C. IV iron injections with concomitant vitamin C
D. IV iron injections
E. Erythropoietin and IV iron

Concept: The type of anemia and the patient's clinical history should help guide the differential diagnosis and the preoperative management of anemia. The World Health Organization (WHO) estimates that anemia affects 30% of nonpregnant women. Iron deficiency anemia (IDA) is very common in premenopausal women, and is often not given the necessary clinical attention. However, its recognition and proper treatment improve quality of life, as well as physical and cognitive performance. Furthermore, if not treated prior to hospitalization, IDA is associated with worse outcomes and a prolonged length of stay. Isolated IDA should be among the top differential diagnoses in a young female with microcytic anemia.

Answer: *B*—Oral (or PO) iron supplementation. The first choice to treat IDA is PO iron, due to its simplicity and low cost. Patients are typically not able to modify their diet sufficiently to make a difference in iron deficiency states (Answer A). Of note, only a few mg of iron may be absorbed by the GI tract each day, and not all patients tolerate PO iron due to its well-known GI side-effects (i.e., constipation). Thus, depending on the patient's preference and the degree of anemia (and iron deficit), it may be necessary to eventually use IV iron therapy (Answer D). IV iron may be particularly indicated to increase the hemoglobin enough to allow for a necessary surgical procedure to remove the source of blood loss, without the need for intra- or postop red cell transfusions. In patients that are particularly resistant to therapy, a combination of erythropoietin and IV iron (Answer E) may become necessary.

The following formula can be used to calculate a patient's iron deficit:

Cumulative iron deficit (mg) = body weight (kg) × [target hemoglobin (g/dL)—actual hemoglobin (g/dL)] × 2.4 + iron storage depot (mg). The iron storage depot for patients with body weight of 35 kg or higher is 500 mg.

If PO iron is the preferred course, the treating physician should be aware that approximately 150 mg of elemental iron is necessary to increase the hemoglobin by 1 g/dL, and ferrous iron (as opposed to ferric) formulations are preferred for their better absorption. The three available iron preparations, gluconate, sulfate and fumarate have increasing amounts of elemental iron per pill: 35 mg, 65 mg and 108 mg, respectively. Since the recommended daily dose to treat IDA is 150–180 mg, and absorption decreases with higher doses, patients must be advised to take iron pills 2–3 times a day. In addition, timing is crucial. Iron pills should be taken away from meals, or at least 1 h before or 2 h after eating to improve absorption.

Concomitant vitamin C (Answer C) may be beneficial to maintain stomach acidity, but is not required. Reticulocytosis within 1 week and increased hemoglobin level 2–3 weeks from the start of PO iron are signs of response to therapy.

6. The landmark Transfusion Requirements in Critical Care (TRICC) trial compared liberal and restrictive transfusion triggers and their effect on mortality in critically ill patients. Which of the following is the correct set of transfusion triggers used in the study and the main outcomes?

A. Liberal arm trigger = 10.0 g/dL; restrictive arm trigger = 7.0 g/dL; no difference in 30-day mortality in both arms
B. Liberal arm trigger = 9.0 g/dL; restrictive arm trigger = 7.0 g/dL; no difference in 30-day mortality in both groups
C. Liberal arm trigger = 9.0 g/dL; restrictive arm trigger = 7.0 g/dL; increased 30-day mortality in the restrictive arm
D. Liberal arm trigger = 10.0 g/dL; restrictive arm trigger = 7.0 g/dL; increased 30-day mortality in the restrictive arm
E. Liberal arm trigger = 10.0 g/dL; restrictive arm trigger = 6.0 g/dL; increased 30-day mortality in the restrictive arm.

Concept: The TRICC trial was the first prospective randomized trial and thus far, the largest, that compared a historically accepted hemoglobin transfusion trigger (liberal) with a lower hemoglobin alternative, in terms of outcomes in critically ill patients. The TRICC trial was designed as a multicentered noninferiority trial with the goal to test if a lower hemoglobin trigger would be as safe as the higher level.

Answer: *A*—Liberal arm trigger = 10.0 g/dL; restrictive arm trigger = 7.0 g/dL; equivalent 30-day mortality in both arms. In the TRICC trial, among the 838 patients enrolled, the mean hemoglobin and the percentage of patients transfused in each group were 10.7 g/dL and 99% for the liberal arm, and 8.5 g/dL and 67% for the restrictive arm, respectively. Deviation from protocol occurred in less than 4.5% in both groups, and overall, mortality at 30 days was 18.7% and 23.3% ($P = 0.11$) in the restrictive and liberal groups, respectively.

On the other hand, among patients with APACHE (acute physiology and chronic health evaluation) scores of ≤20 and those < 55 years of age, mortality was significantly lower in the restrictive transfusion approach arms (P values of 0.02). Based on these results, the authors concluded that a "restrictive strategy is at least as effective as and possibly superior to a liberal transfusion strategy in critically ill patients." Following the publication of the TRICC trial and several others with similar results, several medical societies have created clinical practice guidelines suggesting a hemoglobin trigger of 7 g/dL in critically ill patients without active or history of cardiac diseases.

None of the other choices (Answers B, C, D, and E) represent the correct triggers and/or interpretation of the trial.

7. A 62-year-old male presents to the emergency department with severe chest pain, and left-sided arm pain and numbness. He said the pain in his arm radiates up to his neck and he rates the pain a 6 out of 10. He has no signs or symptoms of bleeding and his vital signs are relatively stable. Initial lab tests reveal the following results:

Assay	Result
Hemoglobin	7.8 g/dL
Hematocrit	23%
Platelets	257,000/μL
Troponin	1.10 ng/mL

The ED physician orders 1 unit of RBCs for the patient. An astute medical technologist working in the blood bank knows about the TRICC trial and asks the blood bank director to call the physician to explain that transfusion is not indicated in this case. What is the best response to this request?

A. Agree—Transfusion is not indicated in this stable, nonbleeding patient
B. Disagree—Ask that she prepare 2 units of RBCs due to the patient's serious condition
C. Disagree—Ask that she prepare and release the unit, as requested by the physician
D. Agree—Transfusion is not indicated and the physician should receive a reprimand from the patient safety committee
E. Disagree—Ask that she prepare 1 unit of RBCs and 2 units of plasma due to the patient's serious condition

Concept: Different clinical conditions require various transfusion thresholds. Though the TRICC trial concluded that a restrictive transfusion policy was safe and effective in most patient populations, it did not have enough power to determine that the findings apply to patients with acute coronary syndromes (ACS). Therefore, most PBM programs recommend that patients with ACS be transfused at a hemoglobin trigger of <8 g/dL.

Answer: C—One unit of RBCs should be released, as requested, since this patient has ACS. Even though the patient is stable and not bleeding, his ACS makes him eligible for transfusion and the physician should not be reprimanded (Answers A and D). Since the patient's baseline hemoglobin is 7.8 g/dL, it is likely that 1 unit of blood will be sufficient to increase his hemoglobin above 8 g/dL; therefore 2 RBC units are not necessary, nor is any plasma warranted (Answers B and E).

8. A physician in the bone marrow transplant (BMT) unit orders a unit of platelets for a stable, nonbleeding patient with a current platelet count of 23,000/μL (range: 21,000–29,000/μL over the past 3 days). The patient is 14 days post-HPC (hematopoietic progenitor cell) transplant and has an absolute neutrophil count of 987/μL. There are no planned invasive procedures and the patient has no bruising or petechiae. This order comes in when the blood bank is critically short on platelets with only 8 left on the shelf and no platelets coming for 24 h (your minimum inventory is usually 15 or more). Since this blood bank is part of a large tertiary care center that has a busy trauma, transplant, and high-risk obstetric program, the blood bank technologist calls to the floor and explains the situation, but the physician insists that his patient receive priority. What is the appropriate response to his request?

A. Inform him that platelet counts >10,000/μL are safe in stable patients with no evidence of bleeding
B. Give him the platelets, since BMT patients are particularly susceptible to bleeding
C. Give him the platelets, since this will speed engraftment of the HPC transplant
D. Give him the platelets and call another blood supplier to help with critical shortage
E. Inform him that platelet counts of <5,000/μL are safe in stable patients with no evidence of bleeding

Concept: Similar to studies for safe thresholds for RBC transfusion, studies have also evaluated safe transfusion thresholds for platelets in patients with hematologic malignancies and post-HPC transplant. Though a multitude of clinical scenarios exists, studies suggest that stable, non-bleeding patients, that are not pre or postoperative do not require platelet transfusions until the platelet count is lower than 10,000/μL.

Answer: *A*—Platelet counts of >10,000/μL appear to be safe in stable nonbleeding patients; however, no studies demonstrate that platelet counts of <10,000/μL are safe (Answer E). The main study that demonstrated the safety of transfusing only when the platelet count drops below 10,000/μL was performed in BMT patients; thus, stable, non-bleeding BMT patients are no more susceptible to bleeding than any other patients (Answer B). The AABB has published guidelines that recommend various platelet thresholds. This patient has already engrafted (platelet count is >20,000/μL and ANC is >500/μL for three consecutive days without platelet and/or granulocyte transfusion) and transfusion of platelets will not "speed engraftment" of the HPC transplant (Answer C). Though you may call other blood suppliers to help with your critical shortage, it should not be done just to get this noncritical patient a unit of platelets that is not clinically necessary (Answer D).

An example of platelet transfusion thresholds in the author's facility is as follows but note that each hospital must decide on their own thresholds based on their unique patient population.

Indications for Apheresis Platelets—Adult Patients:

Platelet count of ≤10,000/μL for bleeding prophylaxis in hypoproliferative thrombocytopenias
Platelet count of ≤50,000/μL in the presence of bleeding
Platelet count of ≤100,000/μL *and* 12 h pre or postsurgery
Platelet count of ≤50,000/μL prior to a Cesarean-section or ≤20,000/μL for vaginal delivery
Massive transfusion (including autologous blood salvaged intra-operatively)
Open heart surgery with bleeding episode
ECMO as determined by patient-specific parameters
Bleeding due to dysfunctional platelets from antiplatelet therapy
Platelet count of ≤100,000/μL and intracranial hemorrhage

Preprocedure prophylaxis:

Lumbar puncture (elective): ≤50,000/μL or lower
Lumbar puncture (emergency, e.g., suspected meningitis): ≤20,000/μL
Transjugular liver puncture: ≤50,000/μL
Gastrointestinal endoscopy with biopsy: ≤20,000/μL
Bronchoscopy: ≤10,000/μL
Bronchoscopy with biopsy: ≤50,000/μL
Central venous catheter insertion: ≤10,000/μL
Epidural anesthesia: ≤80,000/μL
Spinal anesthesia: ≤50,000/μL

9. What statement reflects the purpose or mission of the Blood Utilization Committee (BUC)?
A. To decrease the hospitals' expenses with blood products
B. To ensure that blood product safety is maximized in all areas of the hospital
C. To reprimand physicians that use blood products outside of guidelines

D. To manage all physician decisions regarding blood transfusions
E. To punish technologists that make mistakes in patient testing

Concept: The BUC is a multidisciplinary committee ideally comprised of physicians, advanced practitioners, nurses, technologists, and other administrative leaders in the hospital. The mission of the BUC is to oversee the safety of blood product usage in the hospital. Such activities could include establishing evidence-based transfusion thresholds, reviewing root cause analysis of adverse events, and reviewing forms related to transfusion (e.g., informed consent for transfusion).

Answer: *B*—The BUC is tasked with ensuring that transfusion safety is optimized in all areas of the hospital, not just the blood bank. Though the BUC establishes safe transfusion thresholds, which may in turn reduce cost of blood products, the ultimate goal of such thresholds is patient safety and excellent quality of care (Answer A). Additionally, physicians that continually transfuse outside of guidelines may be reviewed by the BUC, but any reprimands are better received through their department chairs or the chief medical officer (Answer C). Similarly, though technologist errors may be discussed as part of a root cause analysis in a BUC meeting, any reprimands will come from their direct supervisors (Answer E). Random audits of blood product usage allow the medical director of the blood bank to provide the BUC with a general overview of blood transfusion patterns but the BUC is not tasked with clinical decisions (Answer D).

10. During a monthly audit of 50 transfusion orders, which of the following orders would warrant a letter to the physician from the medical director of the blood bank?
A. An order for one unit of RBCs to a stable, nonbleeding patient with a hemoglobin of 7.2 g/dL
B. An order for two units of RBCs to a stable, nonbleeding patient with a hemoglobin of 5.9 g/dL
C. An order for one unit of platelets to a stable, nonbleeding patient with a count of 37,000 /μL in preparation for an emergent coronary bypass procedure
D. An order for one unit of RBCs to a stable, nonbleeding patient with iron deficiency and a hemoglobin of 7.6 g /dL
E. An order for one unit of apheresis platelets to a patient on aspirin with mild bleeding and a platelet count of 162,000 /μL

Concept: A hospital's transfusion policy must take into account many different clinical scenarios for each blood product in the inventory. Regulatory agencies require that a certain percentage of transfusion orders be reviewed on a monthly basis to ensure that transfusion thresholds are being adhered to for most patients. It is the responsibility of the medical director of the blood bank to oversee this process and decide when a letter should be sent to the physician that transfused outside the guidelines, to inform him/her of transfusions that were deemed clinically inappropriate during the audit. In turn, it is expected that the ordering physician will provide an explanation for the decision to transfuse the specific patient. Oftentimes, there was a clinical reason to justify the order.

Answer: *D*—A stable, nonbleeding asymptomatic patient with iron deficiency anemia should be treated with iron therapy, not transfusion. Though, most hospitals have a transfusion threshold <7 g/dL for RBCs in stable nonbleeding patients, sending a letter to a physician that transfuses 1 unit at a hemoglobin of 7.2 g/dL or 2 units at a hemoglobin of 5.9 g/dL is not likely to be beneficial (Answers A and B). Sending out letters for transfusions close to the recommended triggers will likely lessen the overall effectiveness of the process. As mentioned earlier, many different clinical scenarios impact the decision to transfuse platelets. In these cases (Answers C and E), the decision to transfuse is appropriate given the patient's upcoming surgery, and the patient's recent use of an antiplatelet agent with clinical signs of bleeding. Although the benefit of platelet transfusion in bleeding patients who are on antiplatelet therapy is not conclusive, this action is not yet considered to be a major deviation from the standard of care by many experts.

11. During prolonged cold storage, RBCs undergo many structural and physiologic changes, including which of the following?

A. Decreased membrane rigidity
B. Decreased phospholipid vesiculation
C. Decreased 2,3-DPG (diphosphoglycerate) levels
D. Decreased extracellular hemoglobin
E. Decreased extracellular potassium

Concept: While in storage, RBCs undergo several different changes. The combination of such changes is referred to as the "storage lesion." Some changes are "visible," while others are biochemical. Overall, the changes affect the ability of the cells to carry and deliver oxygen, as well as to remain intact while traversing vessels of small caliber. In addition, the release of intracellular compounds into the supernatant, such as cytokines from contaminating leukocytes are implicated in febrile non-hemolytic transfusion reactions (FNHTRs). Leukoreduction (LRD) almost eliminates the risk of FNHTRs, but does not protect the unit from alterations in the red cells. The clinical significance of the storage lesion has been studied in some patient populations, with mixed results.

Answer: *C*—Decreased 2,3-DPG levels. The concentration of 2,3-DPG inside the RBCs determines the ability to deliver oxygen to the tissues. Lower 2,3-DPG increases the affinity of the cells for oxygen, thereby preventing relief of tissue hypoxia. Other changes in the cytoplasm and the membrane that occur during storage are listed in Table 9.2. The other choices (Answers A, B, D, and E) are incorrect (see Table 9.2).

TABLE 9.2 Changes in Red Blood Cells During Storage

RBC cytoplasm	RBC membrane	RBC medium (supernatant)
Decreased 2,3-DPG, ATP, and S-nitrosohemoglobin	Increased rigidity, phospholipid vesiculation, lipid peroxidation, protein oxidation, and oxidative stress	Decreased pH, increased free hemoglobin, potassium, sodium, lactate, glucose, cytokines (IL-1β, IL-6, IL-8, and membrane cofactor protein-1),

12. The ABLE (Age of Blood Evaluation trial in the resuscitation of critically ill patients) trial and the RECESS (Red Cell Storage Duration Study) trial were two large, multicenter prospective randomized clinical trials in critically ill patients and patients undergoing cardiac surgery, respectively, published in 2015. Which of the following answer choices represents one of the major conclusions of these trials?

A. Irradiated red cells prevented alloimmunization to HLA
B. A restrictive transfusion approach led to shorter stays in the ICU
C. Leukoreduction of red cells eliminated CMV transmission
D. The storage age of red cells was not associated with adverse outcomes
E. Red cell age was associated with increased risk of pneumonia

Concept: Level I evidence from well-designed clinical studies is necessary to guide transfusion decisions in many clinical scenarios. Since the TRICC trial in 1999, several studies have approached other aspects of red cell transfusions, such as the effect of the storage age of RBCs on patient outcomes.

Answer: *D*—The storage age of red cells was not associated with adverse outcomes. In the ABLE trial, the authors randomly assigned patients to receive units that had been stored for < 8 days (fresh-blood group) versus standard-issue red cells, defined as the oldest compatible units available throughout hospitalization. With an enrollment of more than 1200 patients in each group, they concluded that fresh red cells did not decrease 90-day mortality.

In the RECESS trial, patients 12 years of age or older undergoing complex cardiac surgery were randomly selected to be transfused with units that were no older than 10 days (shorter-term storage group) or with units that had been stored for at least 21 days (longer-term storage group) intra- and postoperatively. The primary outcome was change in the Multiple Organ Dysfunction Score (MODS) from the preoperative score to the highest within the first week, and the data from 1098 patients showed no difference between the groups. While these studies showed no effect of the storage lesion on patients' outcomes, other randomized trials are being conducted.

Irradiation of RBC units (Answer A) is used to prevent transfusion associated graft-versus-host disease (TA-GVHD), not alloimmunization. A restrictive approach has not been shown to lead to either shorter or longer stays in the ICU (Answer B). Though leukoreduction does decrease the transmission of CMV, it does not completely prevent it (Answer C). RBC age was not found to be associated with pneumonia (Answer E).

13. Patients of the Jehovah's Witness faith have various restrictions regarding transfusions. Which of the following products will they generally accept?

A. Freshly washed RBC units
B. Recombinant coagulation factor concentrates
C. Directed-donated RBCs from relatives
D. Apheresis platelets from other members of the church
E. Irradiated and CMV-safe RBCs

Concept: The ban on transfusions by the Watchtower Society started in 1945 and is based on the Bible passage in Genesis 9:4 that forbid blood consumption. Since then, their policies have evolved, taking into account the availability of blood derivatives and blood substitutes, in addition to blood products.

Answer: *B*—Recombinant coagulation factor concentrates. According to the church's policies, patients of the Jehovah's Witness faith may receive fractions of blood, such coagulation factor purified from human plasma, hemoglobin concentrates, as well as recombinant coagulation factors. They are also allowed to receive their blood collected during an operation by a cell saver device, as long as the tubing remains attached to the patient at all times. This process may also be called autotransfusion. However, even though the society generally accepts the above options, all treatment options should be discussed with the patient, so he/she can make the final decision that best fits with their faith and their optimal health. The other answer choices (Answers A, C, D, and E) are generally considered unacceptable by the church, but again, each patient and family must make the best decision for themselves.

14. Which of the following is a known risk of Hemoglobin-Based Oxygen Carriers (HBOCs)?

A. Anaphylaxis
B. Hypertension
C. Hypercoagulability
D. Malignant hyperthermia
E. Lung injury

Concept: HBOCs are the only alternative for patients whose anemia is life-threatening and allogeneic transfusions are not an option. These include patients of the Jehovah's Witness faith, those who refuse transfusions, and patients with multiple RBC alloantibodies for whom there are no compatible RBC units available. Nonetheless, at present, there is not any HBOC agent approved in the United States. This is most likely because of the risks observed during past randomized clinical trials carried out by eight manufacturers of HBOCs.

Answer: *B*—Hypertension. Vasoconstriction and hypertension occur because of scavenging of nitric oxide by HBOCs and have been associated with increased mortality, as well as cardiac, hepatic, renal, central nervous system, pancreatic, and gastrointestinal complications in various trials with volunteers or surgical or trauma patients. As of this writing, HBOCs are only used for "compassionate" use, when absolutely no other option is available for these patients. The other choices (Answers A, C, D, and E) are not known risks of HBOCs.

15. Many healthcare facilities face challenges when implementing PBM programs, including which of the following?
A. Decreased inventory of fresh RBC units
B. Longer length of stay for nontransfused patients
C. Difficulty finding evidence to guide transfusion guidelines
D. Lack of enthusiasm and/or resistance to change clinical practice
E. Decreased reimbursement from surgical procedures

Concept: PBM is defined as a multidisciplinary approach to the care of patients with anemia or at risk of developing anemia during hospitalization. In addition to physicians of various disciplines, nurses, laboratory professionals, hospital administration, pharmacists, must work in concert to prevent the development of anemia and/or treat it with alternatives to blood products. The ultimate goal of PBM is to avoid the risks of transfusions, increase safety, and improve patient outcomes.

Answer: *D*—Lack of enthusiasm and/or resistance to change clinical practice. For many practicing physicians, embarking on PBM requires a change in mindset, which in some cases has been solidified for many years. In terms of transfusion decisions, many physicians did not receive the necessary background training in medical schools in order to recognize that blood products may be similar to drugs, with associated benefits but also adverse effects. Furthermore, while the literature is abundant in support of PBM (Answer C), a busy clinician may not have the time to review publications outside of his/her specialty. Finally, the fear of loss of autonomy and a protocol-driven future practice must be acknowledged as barriers to successful PBM implementation.

PBM programs do not decrease the inventory of RBCs (Answer A). In fact, PBM programs may allow a hospital to decrease their standing order from blood suppliers as the program becomes more and more successful in decreasing utilization. There is no evidence that patients that are not transfused have longer hospital stays than those that are transfused (Answer B) and there is no effect on reimbursement for patients that are not transfused (Answer E).

16. Which of the following drugs decreases blood loss in orthopedic surgery by blocking the conversion of plasminogen to plasmin?
A. Tranexamic acid
B. Aprotinin
C. Recombinant activated factor VIIa
D. Prothrombin complex concentrate (PCC)
E. Recombinant thrombin

Concept: Surgical hemorrhage is associated with morbidity and mortality and should be both prevented and managed effectively when it occurs. Instead of using blood products during surgery, PBM proponents suggest the use of systemic or topical agents that improve hemostasis. Some agents, such as antifibrinolytics, may be used prophylactically preoperatively and intraoperatively or when increased blood loss ensues.

Answer: *A*—Tranexamic acid. Tranexamic acid (TXA) and ε−aminocaproic acid (EACA) are lysine analogs that inhibit the conversion of plasminogen to plasmin; thus, inhibiting fibrinolysis. Both have been shown to decrease bleeding in various orthopedic procedures, including complex spinal ones, without the risk of hypercoagulability. Aprotinin (Answer B) is another antifibrinolytic agent, but it is a serine protease that inhibits plasmin in addition to kallikrein.

Intravenous recombinant activated factor VIIa (Answer C) induces a thrombin burst on the platelet surface, but does not affect fibrinolysis. Prothrombin complex concentrates (Answer D) contain the vitamin K-dependent procoagulant factors (II, VII, IX, X) and anticoagulant proteins C and S. Recombinant thrombin (Answer E) is used topically for minor bleeding in raw surfaces, such as burns, adhesions, sutures, and muscle.

17. Patients with warfarin overdose accompanied by severe bleeding should receive which of the following treatments, if available?
A. Fresh frozen plasma
B. Cryoprecipitated AHF
C. Factor VIII concentrate
D. Recombinant activated factor VIIa
E. Prothrombin complex concentrate (PCC)

Concept: Millions of Americans take warfarin daily to prevent arterial and venous thromboembolic events, despite the availability of newer oral anticoagulants that directly inhibit the coagulation cascade. Warfarin is cheaper and has been in clinical use for almost 60 years. Conversely, it carries a risk of life-threatening hemorrhage. This risk is increased because its metabolism is affected by many common factors, such as diet and concurrent medications. For this reason, patients must monitor their International Normalized Ratio (INR) routinely to prevent under- and over-dosing. When overdosing with severe bleeding, such as intracranial hemorrhage, it is essential to treat the patient emergently to effectively reverse the drug action.

Answer: *E*—Prothrombin complex concentrate (PCC). Since the 2012 edition of the American College of Chest Physicians (ACCP) guidelines, the recommendation has been to use a 4-factor PCC (Kcentra) in these circumstances, along with IV Vitamin K (5–10 mg). PCCs provide a bolus of the deficient procoagulant factors in patients on warfarin (factors II, VII, IX, and X), without the large volume of plasma that would be required if plasma were used (Answer A). Furthermore, 4-factor PCCs also have natural anticoagulants, proteins C and S. A recent study confirmed the increased incidence of circulatory overload with plasma compared with PCC, in addition to the delay in restoring hemostasis when using plasma. The same authors provided dosage guidelines to treat patients with warfarin overdose and bleeding based on the INR as follows:

INR 2–<4: 25 IU of PCC/kg (maximum of 2500 units)
INR 4–6: 35 IU of PCC/kg (maximum of 3500 units)
INR >6: 50 IU of PCC/kg (maximum of 5000 units)

Cryoprecipitated AHF (Answer B) is typically used to correct hypofibrinogenemia. Factor VIII concentrate (Answer C) is used to treat hemophilia A patients. Though recombinant activated factor VIIa (Answer D) does help with hemostasis, it does not fully address other factors that are also deficient in warfarin overdose.

18. Among measures that have been shown to decrease transfusions in cardiac surgery, which one is widely accepted?
A. Use of autologous blood to prime the bypass circuit
B. Protocol-driven transfusion decisions
C. Prophylactic use of apheresis platelets prior to incision
D. Continuous infusion of tranexamic acid (TXA)
E. Intraoperative transfusion of RBCs and plasma in a 1:1 ratio

Concept: Patients undergoing open cardiac surgery are at significant risk of requiring allogeneic transfusions, especially during valve replacement or reoperations. As with other large operations, they benefit from optimization of their red cell mass by correcting anemia preoperatively, as well as by avoiding antiplatelet or anticoagulants that may be safely discontinued presurgery. Although some bleeding is expected in cardiac surgery, up to 10% of patients have an unusually high blood loss which necessitates the use of blood products. Unfortunately, the hemorrhage and the transfusions contribute to prolonged length of stay and worse outcomes, including death. Thus, initiatives which prevent increased bleeding intra- and postoperatively carry significant clinical benefit to the patient.

Answer: *B*—Protocol-driven transfusion decisions. Several studies have shown that using point-of-care testing (POCT) and preplanned triggers for transfusion decrease the overall need for blood products to manage bleeding in cardiac surgery. Since the etiology of bleeding is multifactorial, it is routine to use a variety of approaches to detect the hemostasis defect and restore it as promptly as possible by targeting the specific problem.

Examples of POCT include viscoelastic methods, such as thromboelastography (TEG) and rotational thromboelastometry (ROTEM), which use whole blood and are the only methods capable of detecting hyperfibrinolysis of the assays currently available. In addition, such methods provide concurrent information regarding platelet function and coagulation factor deficiencies, including fibrinogen. If hyperfibrinolysis is detected, the patient may respond to an antifibrinolytic, such as tranexamic acid, thereby, avoiding unnecessary exposure to blood products (Answer D).

With the availability of fibrinogen and PCC, these products may also be useful and more rapidly administered and effective than cryoprecipitated AHF or plasma. For platelet function, there are also dedicated POCT devices which may guide administration of desmopressin and/or platelet transfusions. POCT methods to measure the level of heparin are also beneficial to detect overanticoagulation that can be quickly reversed with protamine sulfate, since heparin does not respond to blood products. The other choices (Answers A, C, and E) are not consistent with a protocol-driven approach designed to minimize blood product usage, but rather might actually increase blood product usage.

19. Which of the following PBM initiatives applies to both medical and surgical hospitalized patients?

A. Avoidance of antiplatelet agents
B. Minimization of iatrogenic blood loss
C. Use of topical hemostatic agents
D. Acute normovolemic hemodilution
E. Autologous blood donation

Concept: PBM measures should take into account the risks and benefits of any intervention. Although certain classes of drugs affect hemostasis, increasing the risk of spontaneous hemorrhage, their benefit of decreasing life-threatening events, such as strokes and myocardial ischemia, outweigh their risk in most patients. On the other hand, certain medical practices that have become common and expected, may provide significant benefit if eliminated or modified.

Answer: *B*—Minimization of iatrogenic blood loss. Many studies have shown phlebotomy losses in hospitalized patients, both medical and surgical, are capable of inducing or worsening anemia to the point of having to transfuse the patient. This is particularly likely to happen in critically ill patients, whose ability to compensate for the blood collected for tests is diminished by a combination of their illness, decreased oral intake, and bone marrow suppression from various drugs. Another reason that patients in intensive care units are more susceptible to iatrogenic blood loss is that the frequency of laboratory testing is higher in that setting, often as a consequence of protocols, rather than a true clinical indication.

Practices changes that can affect a patient's risk of anemia from iatrogenic blood loss include eliminating the discard blood volume from arterial lines, collecting blood in small tubes as long as the laboratory can use them in their testing instruments, and limiting blood draws to only when it is clinically necessary. While the other choices (Answers A, C, D, and E) may be used as part of a patient's care, they do not apply to both surgical and medical patients.

20. In regards to alternatives to transfusion for patients with anemia and cancer, which of the following side-effects is sometimes associated with ESAs?

A. Increased platelet refractoriness
B. Altered mental status following infusion
C. Increased risk of venous thrombosis

D. Decreased response to certain chemotherapeutic drugs
E. Severe allergic reaction during rapid infusion

Concept: Patients with cancer often present with anemia at the time of the diagnosis, or develop anemia during the course of treatment. Common causes of anemia include iron deficiency from chronic blood loss, such as in colon cancer, marrow involvement by tumor, preexisting renal disease with erythropoietin deficiency, functional iron deficiency, iron sequestration, and myelosuppression caused by chemotherapy. In isolated IDA, iron supplementation alone may suffice, while in iron sequestration syndromes or renal insufficiency, ESAs are also needed. While ESAs have been shown to decrease transfusion requirements, their use has been limited, since safety concerns have arisen in the last 10 years.

Answer: *C*—Increased risk of venous thrombosis. In a metaanalysis of 38 studies with > 8000 patients with cancer and anemia, the incidence of venous thromboembolism in patients treated with ESAs was 7.5% compared with 4.9% (relative risk of 1.57) for those treated with placebo or standard of care. In addition, data from clinical trials in which ESAs were used to keep or reach hemoglobin levels in the reference range, raised concerns regarding the safety of these agents for this indication.

At the present time, it is suggested that the ESA dose to treat symptomatic anemia in patients with cancer be the minimal dose necessary to avoid transfusion, without raising the hemoglobin to >12 g/dL. Importantly, the decision to use an ESA should be individualized and made with the consent of the patient, taking into account the risks, benefits and alternatives. What remains to be determined is the mechanism by which ESAs increase the risk of venous thrombosis. The other choices (Answers A, B, D, and E) are not known side effects of ESAs.

21. A patient is undergoing a coronary bypass surgery that is expected to last 3–4 h. Which of the following is the correct maximum storage temperature and shelf-life for the blood product collected via acute normovolemic hemodilution (ANH)?
A. 30°C, 1 h
B. 21°C, 8 h
C. 30°C, 8 h
D. 21°C, 4 h
E. 10°C, 8 h

Concept: There are many available techniques to minimize allogeneic blood transfusion. During surgery, blood can be collected from the patient intravenously prior to the operation (e.g., ANH) or from the surgical field (e.g., intraoperative blood salvage). To ensure that bacterial growth and reinfusion do not occur, standards for storage time and temperature exist. Additionally, to prevent returning surgical field contaminants, microaggregate filters are used prior to reinfusion into the patient. Though not required to oversee perioperative PBM, the transfusion service should be involved in the development of the policies for the program.

For ANH, a percentage of the patient's total blood volume is collected shortly after the induction of anesthesia, and typically stored in approved plastic bags at room temperature. To maintain the patient's volume status, crystalloids, or colloids are then infused. The collected blood must be labeled with the patient's name, medical record number, the date and time of collection, and "Autologous Use Only." Once surgery is complete, the blood removed at the beginning of the operation is returned in the reverse order in which it was collected (i.e., unit collected last will be infused back first), as long as the shelf-life and/or storage temperatures were not exceeded.

For intraoperative blood salvage, blood that is lost during the operation is harvested in sterile bags and sometimes processed (i.e., washed) with saline, although this is not a requirement. Whether the blood is processed or not affects the allowable storage time.

Answer: *B*—Once collected from the patient using ANH, blood is typically stored in plastic bags at room temperature (~21°C) for up to 8 h. However, if the blood is stored at 1–6°C, it can be stored for up to 24 h from collection. The other choices (Answers A, C, D, and E) are incorrect for ANH. Other intraoperative blood salvage procedure storage temperatures and shelf-lives are as follows:

- Intraoperative blood recovery with processing—room temperature/4 h or 1–6°C/24 h
- Intraoperative blood recovery without processing—room temperature or 1–6°C/4 h
- Shed blood, no longer than 6 h regardless of the storage temperature

22. When discussing alternatives to transfusion for surgical patients, which of the following options correctly describes either autologous donation, ANH, or autologous blood cell salvage?

A. ANH may be employed in nonelective procedures
B. Autologous donation and intraoperative blood salvage are relatively contra-indicated in patients with cancer
C. Intraoperative blood salvage causes dilutional coagulopathy and should be accompanied by plasma and platelets
D. AABB offers protocols for ANH to guide its safe utilization
E. Preoperative autologous donations avoid the risk of clerical errors during blood administration

Concept: The multidisciplinary team engaged in PBM in the perioperative setting must be familiar with the various options to optimize erythropoiesis and minimize blood loss pre- and intraoperatively. Such options should be discussed with the patient during the planning stage, unless the procedure is emergent. The discussion should also be individualized, considering patient specific factors, such as religious beliefs, hemoglobin level, type of procedure, and likelihood of transfusion.

Answer: *A*—Acute normovolemic hemodilution (ANH) may be employed in nonelective procedures. Unlike presurgery donation, which requires several weeks of preparation, ANH may be used in emergent procedures. When large blood loss is expected, ANH lowers the hematocrit of the blood during the procedure, decreasing the absolute loss of red cell mass.

However, at present, standardized protocols for ANH are lacking (Answer D), which may explain its low acceptance among anesthesiologists and surgeons. Although intraoperative blood cell salvage (Answer C) may involve washing the patient's shed red cells prior to reinfusion, unless there is clinical coagulopathy, plasma and/or platelets are not indicated. Many surgeries with intraoperative blood salvage avoid allogeneic blood products altogether.

In order to avoid the infusion of malignant cells from the operative field back into the patient, autologous blood cell salvage is the only modality that is relatively contraindicated in patients with cancer (Answer B). Among the several limitations of autologous donations, units collected weeks prior to the scheduled operation often cause preoperative anemia and place the patient at higher risk of requiring a transfusion; furthermore, predonated units are not immune to errors in patient identification and administration (Answer E), and they are wasted if not used by the patient. Furthermore, the cells undergo the same storage changes as allogeneic units, such as depletion of 2,3 diphosphoglycerate (DPG) and adenosine triphosphate (ATP). Thus, autologous donations are the least favored among these three surgical alternatives to allogeneic red cells.

23. Since 2010, the WHO has urged the implementation of PBM programs in light of the mounting evidence of improved patient outcomes with an evidence-based approach to avoid and treat anemia without allogeneic blood products. Which of the following alternatives best describes the implementation of a PBM program?

A. PBM principles are universal and do not need to fit into each institution's culture
B. It is easier to implement PBM initiatives for all surgical patients at once than start with specific groups
C. The buy-in of local specialists that frequently order transfusions strengthens a PBM program
D. Educational materials and reminders to follow the PBM program are sufficient in most hospitals
E. If the Transfusion Services publishes guidelines for transfusions, there is no need for a PBM program

Concept: Medical decisions are among the most cherished privileges by physicians. For many, interference with their plan for each patient touches the core of what they believe to be protected. As a principle, many physicians believe that transfusions only benefit their patients, especially knowing the current negligible risk of infection transmission. The goal of a PBM program first and foremost should be to increase awareness of the subtle risks of transfusion and how to safely employ alternatives to avoid them. As with any change in culture, a PBM program must evolve over time.

Answer: *C*—The buy-in of local specialists that frequently order transfusions strengthens a PBM program. If possible, the program should be tailored to the practice setting (Answer A) and should start with initiatives that involve the medical or surgical services that routinely use blood products, such as cardiac surgery, anesthesiology, orthopedics, critical care, gastroenterology, and hematology-oncology (Answer B). In these settings, even single changes in practice, such as a protocol to treat preoperative anemia in patients undergoing hip arthroplasty or decreasing the hemoglobin trigger of hospitalized patients to 7 g/dL may yield significant results. Since every positive result of the PBM program should then be shared with the whole medical staff, early victories are likely to inspire others to follow. Without a consistent message and persistent effort to disseminate information a PBM program will likely not succeed (Answers D and E).

Please answer Questions 24-25 based on the following case scenario:

A 57-year-old male with a history of alcoholism presents to the emergency department after 2 weeks of dark tarry stools and a recent episode of bloody emesis. A stat CBC reveals the following laboratory results: hemoglobin 7.8 g/dL, platelets 275,000/μL, prothrombin time (PT) 12.9 s, and partial thromboplastin time (PTT) 36 s. Endoscopic examination reveals an upper GI bleed (Child-Pugh Class B). The patient's vital signs are currently stable.

24. Which statement is correct regarding transfusion of red blood cells in this patient?

A. Transfusions should be given for as long as necessary to fully correct the anemia
B. There is no recommendation because these patients are often critically ill
C. Transfusions are indicated to keep the hemoglobin at 10 g/dL, independent of other parameters
D. Every time the hemoglobin falls below the patient's baseline, a unit of red blood cells should be ordered
E. Transfusing red cells when the hemoglobin reaches 7 g/dL is associated with better outcomes than a trigger of 9 g/dL

Concept: Patients with massive GI hemorrhage are among the ones more likely to require massive transfusion, akin to trauma victims. Variceal bleeding can be life-threatening and laboratory tests are not always helpful to guide therapy. Since the liver synthesizes most coagulation factors, these patients often have both coagulopathic and anatomic etiologies for their bleeding. For these reasons, physicians tend to assume that they benefit from more, rather than fewer transfusions.

Answer: *E*—Transfuse only when the hemoglobin reaches 7 g/dL, in the absence of hemodynamic instability. Villanueva and colleagues performed a randomized controlled trial in patients with acute upper GI bleeding to compare the efficacy and safety of a restrictive transfusion strategy (transfusion when the hemoglobin level fell below 7 g/dL) with those of a liberal transfusion strategy (transfusion when the hemoglobin level fell below 9 g/dL). They found that patients in the restrictive-strategy group had better survival at 6 weeks (95% vs. 91%; P =0.02), and fewer adverse events: (40% vs. 48%; P = 0.02). Among patients with cirrhosis and Child-Pugh class A or B disease, the probability of survival was significantly higher (hazard ratio, 0.30), but not in those with cirrhosis and Child-Pugh class C disease (hazard ratio, 1.04). Thus, they suggested that limiting transfusions to when the hemoglobin reaches 7 g/dL is not only safe, but also associated with improved outcomes (Answer B). The other choices (Answers A, C, and D) represent a more aggressive transfusion strategy.

25. What other factors are important when deciding to transfuse this patient population?

A. Expanded blood volume and risk of rebleeding
B. Potential risk of acquiring hepatitis C
C. Risk of anaphylaxis from acquired IgA deficiency
D. Decreased factor VIII, increasing the risk of rebleeding
E. Increased susceptibility for transfusion-associated lung injury (TRALI)

Concept: Blood transfusions carry risks that can be classified as immediate and long-term. Among the former are the acute transfusion reactions such as hemolytic, febrile, allergic, infection from microbial contamination, transfusion-associated circulatory overload (TACO), and TRALI. While TRALI is the leading cause of death from transfusion in the United States according to the Food and Drug Administration, TACO, hemolysis, and microbially contaminated blood products are also implicated every year. Long-term consequences of transfusion include alloimmunization against red cell antigens or HLA, iron overload, and transfusion-related immunomodulation among the most common. For all these reasons, a PBM program aims at minimizing transfusions whenever possible, even in patients with cirrhosis who may otherwise appear to benefit from transfusions.

Answer: *A*—Expanded blood volume and risk of rebleeding. In the same study mentioned earlier (Villanueva et al.), the investigators noted a significant increase ($P = 0.03$) in the portal-pressure gradient in the patients that received more transfusions (liberal-strategy group) within the first 5 days of hospitalization. Furthermore, all patients in the same group had a higher incidence of rebleeding, while the rate of further bleeding in those with varices was 11% in the restrictive group compared with 22% in the liberal group. These data suggest that physicians should use caution when transfusing aggressively, since the volume transfused has major implications. Although many coagulation factors are lower in patients with cirrhosis than in normal individuals, factor VIII level is not affected because it is not exclusively produced by hepatocytes. The other choices (Answers B, C, D, and E) represent risks that any patient population might be susceptible to, but it is not specific for those with GI bleeding.

End of Case

Suggested Reading

[1] A. Shander, M. Javidroozi, S. Perelman, T. Puzio, G. Lobel, From bloodless surgery to patient blood management, Mt. Sinai. J. Med. 79 (2012) 56–65.
[2] L.T. Goodnough, A. Shander, Patient blood management, Anesthesiology 116 (2012) 1367–1376.
[3] L.T. Goodnough, Iron deficiency syndromes and iron-restricted erythropoiesis, Transfusion 52 (2012) 1584–1592.
[4] A. Shander, L.T. Goodnough, M. Javidroozi, M. Auerbach, J. Carson, W.B. Ershler, M. Ghiglione, J. Glaspy, I. Lew, Iron deficiency anemia—bridging the knowledge and practice gap, Transfus. Med. Rev. 28 (2014) 156–166.
[5] A.J. Friedman, A. Shander, S.R. Martin, R.K. Calabrese, M.E. Ashton, I. Lew, M.H. Seid, L.T. Goodnough, Iron deficiency anemia in women: a practical guide to detection, diagnosis, and treatment, Obstet. Gynecol. Surv. 70 (2015) 342–353.
[6] P.C. Hébert, G. Wells, M.A. Blajchman, J. Marshall, C. Martin, G. Pagliarello, M. Tweeddale, I. Schweitzer, E. Yetisir, A multicenter, randomized, controlled clinical trial of transfusion requirements in critical care. Transfusion requirements in critical care investigators, Canadian Critical Care Trials Group, N. Engl. J. Med. 340 (1999) 409–417.
[7] B. Cohen, I. Matot, Aged erythrocytes: a fine wine or sour grapes?, Br. J. Anaesth. 111 (2013) i62–i70.
[8] J. Lacroix, P.C. Hébert, D.A. Fergusson, A. Tinmouth, D.J. Cook, J.C. Marshall, L. Clayton, L. McIntyre, J. Callum, A.F. Turgeon, M.A. Blajchman, T.S. Walsh, S.J. Stanworth, H. Campbell, G. Capellier, P. Tiberghien, L. Bardiaux, L. van de Watering, N.J. van der Meer, E. Sabri, D. Vo, ABLE investigators; Canadian Critical Care Trials Group, age of transfused blood in critically ill adults, N. Engl. J. Med. 372 (2015) 1410–1418.
[9] M.E. Steiner, P.M. Ness, S.F. Assmann, D.J. Triulzi, S.R. Sloan, M. Delaney, S. Granger, E. Bennett-Guerrero, M.A. Blajchman, V. Scavo, J.L. Carson, J.H. Levy, G. Whitman, P. D'Andrea, S. Pulkrabek, T.L. Ortel, L. Bornikova, T. Raife, K.E. Puca, R.M. Kaufman, G.A. Nuttall, P.P. Young, S. Youssef, R. Engelman, P.E. Greilich, R. Miles, C.D. Josephson, A. Bracey, R. Cooke, J. McCullough, R. Hunsaker, L. Uhl, J.G. McFarland, Y. Park, M.M. Cushing, C.T. Klodell, R. Karanam, P.R. Roberts, C. Dyke, E.A. Hod, C.P. Stowell, Effects of red-cell storage duration on patients undergoing cardiac surgery, N. Engl. J. Med. 372 (2015) 1419–1429.
[10] http://www.jwfacts.com/watchtower/blood-transfusions.php.

[11] R.B. Weiskopf, T.A. Silverman, Balancing potential risks and benefits of hemoglobin-based oxygen carriers, Transfusion 53 (2013) 2327–2333.

[12] D.P. Fischer, K.D. Zacharowski, M.M. Müller, C. Geisen, E. Seifried, H. Müller, P. Meybohm, Patient blood management implementation strategies and their effect on physicians' risk perception, clinical knowledge and perioperative practice—the Frankfurt experience, Transfus. Med. Hemother. 42 (2015) 91–97.

[13] L.T. Goodnough, A. Shander, Current status of pharmacologic therapies in patient blood management, Anesth. Analg. 116 (2013) 15–34.

[14] R. Sarode, T.J. Milling Jr., M.A. Refaai, A. Mangione, A. Schneider, B.L. Durn, J.N. Goldstein, Efficacy and safety of a 4-factor prothrombin complex concentrate in patients on vitamin K antagonists presenting with major bleeding: a randomized, plasma-controlled, phase IIIb study, Circulation 128 (2013) 1234–1243.

[15] M.A. Refaai, J.N. Goldstein, M.L. Lee, B.L. Durn, T.J. Milling Jr., R. Sarode, Increased risk of volume overload with plasma compared with four-factor prothrombin complex concentrate for urgent vitamin K antagonist reversal, Transfusion 55 (2015) 2722–2729.

[16] B.L. Pearse, I. Smith, D. Faulke, D. Wall, J.F. Fraser, E.G. Ryan, L. Drake, I.L. Rapchuk, P. Tesar, M. Ziegenfuss, Y.L. Fung, Protocol guided bleeding management improves cardiac surgery patient outcomes, Vox Sang 109 (2015) 267–279.

[17] B. Hohmuth, S. Ozawa, M. Ashton, R.L. Melseth, Patient-centered blood management, J. Hosp. Med. 9 (2014) 60–65.

[18] D.P. Fischer, K.D. Zacharowski, P. Meybohm, Savoring every drop—vampire or mosquito?, Crit. Care 18 (2014) 306.

[19] J.A. Gilreath, D.D. Stenehjem, G.M. Rodgers, Diagnosis and treatment of cancer-related anemia, Am. J. Hematol. 89 (2014) 203–212.

[20] J.C. Oliver, R.L. Griffin, T. Hannon, M.B. Marques, The success of our patient blood management program depended on an institution-wide change in transfusion practices, Transfusion 54 (2014) 2617–2624.

[21] C. Villanueva, A. Colomo, A. Bosch, M. Concepción, V. Hernandez-Gea, C. Aracil, I. Graupera, M. Poca, C. Alvarez-Urturi, J. Gordillo, C. Guarner-Argente, M. Santaló, E. Muñiz, C. Guarner, Transfusion strategies for acute upper gastrointestinal bleeding, N. Engl. J. Med. 368 (2013) 11–21.

[22] A.W. Bryan Jr., E.M. Staley, T. Kennell Jr., A.Z. Feldman, L.A. Williams III, H.P. Pham, Plasma transfusion demystified: a review of the key factors influencing the response to plasma transfusion, Lab. Medicine 48 (2) (2017) 108–112.

CHAPTER

10

Perinatal, Neonatal, and Pediatric Transfusion—Principles and Practice

Edward C.C. Wong,**, Javi L. Hartenstine[†], Mrigender Virk[‡]*

*Children's National Medical Center, Center for Cancer and Blood Disorders, Washington, DC, United States; **George Washington University School of Medicine and Health Sciences, Departments of Pediatrics and Pathology, Washington, DC, United States; [†]University of California, Irvine, CA, United States; [‡]MedStar Georgetown University Hospital, Washington, DC, United States

The unique challenges of pediatric transfusion medicine include tailoring one's approach to treatment by taking into account many factors, such as the influence of patient age, developmental maturity on physiology and biology, and pathologic conditions involving the immunologic and hematologic systems. The main technical challenges include: (1) working with very small aliquots of blood components because of the inherently greater risk of volume overload and (2) the fact that risk of infection, product preservatives, and product storage breakdown can have an even greater effect in the pediatric population due to their unique physiology and biology. Research and evidence-based medicine in the field of pediatric transfusion is challenging due to the small number of experts and the challenge of designing research for such a vulnerable population. This chapter addresses core concepts in pediatric transfusion medicine, including: guidelines for the administration of blood products and component therapy; the associated relevant principles of immunology and hematology, and special considerations for certain pediatric populations.

1. The major IgG subclass antibodies implicated in hemolytic disease of the fetus and newborn (HDFN) include which of the following?
   A. IgG1 and IgG3
   B. IgG1 and IgG2
   C. IgG2 and IgG4
   D. IgG1 and IgG4
   E. IgG2 and IgG3

   Concept: HDFN, also known as erythroblastosis fetalis, is caused by destruction of fetal red blood cells (RBCs) by maternal antibodies. When fetal RBCs enter the maternal circulation, they may cause maternal alloimmunization; the frequency and severity of this is dependent on the antigens present on the RBCs. The mother may produce both IgM and IgG antibodies but only the IgG component can pass through the placenta and affect the fetus. Furthermore, IgG1 and IgG3 can fix complement and cause intravascular hemolysis. However, the correlation of IgG subclass with severity of disease is controversial.

   Answer: *A*—IgG1 and IgG3 can fix complement, and thus, can cause intravascular hemolysis. None of the other choices (Answers B, C, D, or E) have the correct combination.

Transfusion Medicine, Apheresis, and Hemostasis. http://dx.doi.org/10.1016/B978-0-12-803999-1.00010-9

2. HDFN due to ABO incompatibility is typically less clinically severe than Rh incompatibility. The best explanation for this is which of the following?
A. Mothers are given prophylactic anti-A and anti-B to prevent antibody formation
B. The trophoblasts contain a hydrolytic enzyme that destroys the ABO antibody
C. The newborn's red cells have well-developed ABO antigens
D. The newborn's plasma contains a blocking antibody for ABO antibody
E. Most of the antibody against the ABO antigen cannot cross the placental blood barrier

Concept: Since the use of Rh immunoglobulin (RhIG) to prevent HDFN due to anti-D, HDFN due to ABO incompatibility is now the most common cause of HDFN in the United States. Appropriate dosing and administration of RhIG has reduced the risk of an Rh negative mother being immunized by an Rh positive fetus from ~16% to <0.1%. ABO incompatibility occurs in approximately 15% of pregnancies, but results in HDFN in only 4% of these cases (0.6% of all pregnancies).

When a mother is exposed to fetal RBC antigens, she may develop IgG and IgM antibodies, of which IgG could cross the placenta and may cause HDFN, which occurs at varying rates and severities depending on the antigens involved. ABO antibodies are mostly IgM which do not cross an intact blood/placenta barrier. Nonetheless, since the amount of antibody is relatively small, it is usually not clinically significant (i.e., does not cause significant hemolysis).

Answer: *E*—Unlike anti-D, which is almost all IgG, most of the "naturally occurring" ABO antibodies are IgM, and thus, do not cross the placenta. However, those with blood type O have IgM anti-A and anti-B as well as IgG anti-A,B. These IgG antibodies may cross the placenta and bind to the fetal RBCs, and may give the fetus a positive direct antiglobulin test (DAT) at the time of birth. Nonetheless, since the amount of antibody is relatively small, it is usually not clinically significant (i.e., does not cause significant hemolysis). Therefore, the HDFN caused by ABO incompatible pregnancy is usually milder than the one caused by anti-D antibody and is more easily treated with phototherapy.

Other factors related to the decreased rate of hemolysis include variable titers of anti-A, or anti-B in the mother, weak A and B antigen expression on infant RBCs (Answer C), and absorption of antibody by the A and/or B antigen in plasma, vascular endothelium and other tissues in the infant and placenta. Absorption of the antibodies in the newborn's plasma is particularly true for those who are secretors. ABO HDFN occurs more often in the first child than HDFN due to anti-D because they are "naturally occurring;" thus, there is no need for a sensitizing exposure to form anti-A or anti-B. Unfortunately, there is no preventative treatment for ABO HDFN (Answer A). Neither trophoblasts (Answer B) nor the newborn's plasma (Answer D) contains substance that destroys antibodies.

3. Which of the following antibodies implicated in HDFN is most likely to suppress fetal erythropoiesis?
A. Anti-Fy^a antibodies
B. Anti-Jk^a antibodies
C. Anti-c antibodies
D. Anti-Rh17 antibodies
E. Anti-K1 (Kell) antibodies

Concept: The fetus affected by HDFN can have anemia due to hemolysis (i.e., RBC destruction) and/or production defects/suppression.

Answer: *E*—All the listed antibodies can cause HDFN; however, of those listed, only Anti-K1 (Kell) antibodies are known to suppress fetal erythroid precursors, causing severe anemia (without hemolysis) in the intrauterine and postpartum periods. Thus, a fetus affected by anti-K antibodies is more likely to have laboratory and clinical evidence of anemia, but not much evidence of hemolysis. Anti-Ge (Gerbich) antibodies can also suppress early fetal erythroid precursors, and cause similar signs, symptoms, and/or laboratory evidence of HDFN to anti-K1.

Antibodies against the Rh and Duffy antigens tend to cause severe hemolysis. Thus, anti-Fy^a, anti-c, and anti-Rh17 antibodies (Answers A, C, and D) are more likely to cause anemia due to hemolysis (and not suppress in production) and thus, will more likely to have prominent laboratory evidence of hemolysis (such as high indirect bilirubin). Kidd blood group antigens are not well developed during fetal development and at birth; thus, anti-Jk^a and/or anti-Jk^b rarely cause HDFN (Answer B). Additionally, antibodies from the Lewis blood group do not cause HDFN because the Lewis antigens are not well-expressed on fetal RBCs.

4. In addition to irradiation as well as being tested negative for Zika virus, which of the following is one of the product requirements for RBCs used for intrauterine exchange transfusions (IUTs) for a fetus with HDFN secondary to anti-C antibody present in the mother?

A. Washed 6 times
B. Saline resuspended with additional type AB plasma
C. C-antigen negative RBCs
D. EBV reduced risk
E. Rh positive RBCs

Concept: The use of prophylactic RhIG has significantly reduced the risk of HDFN due to anti-D in the United States. However, IUT remains essential for various causes of severe fetal anemia, including HDFN due to other antibodies, such as anti-C. IUT is considered as a therapeutic intervention for pregnancies between 18 and 35 weeks gestation. The RBC unit must be crossmatch compatible with maternal plasma. Irradiation is required to prevent transfusion associated graft-versus-host disease (TA-GVHD). To avoid fetal cytomegalovirus (CMV) infection, CMV reduced risk products must be given (either from a CMV seronegative donor or by leukoreduction). Only products that are negative for hemoglobin S should be transfused to prevent the possibility of fetal renal failure. Furthermore, since Zika virus has been linked with adverse effects on the fetus, the selected product should be tested negative for Zika virus.

Generally, middle cerebral artery peak velocity, which should be used before 35 weeks gestation, is useful but it has a false positive rate (12%) which could lead to unnecessary fetal blood sampling. If the velocity measurements are >1.5 multiple of the mean (MOM), then fetal blood sampling is warranted. Ultrasound guided fetal blood sampling, also known as percutaneous umbilical blood sample (PUBS), is not benign as there is an overall rate of 1%–2% of fetal loss. IUT is then performed if PUBS demonstrates a hemoglobin/hematocrit less than two standard deviations from the mean for gestational age.

Answer: C—If a mother has alloantibodies, the RBCs chosen must be negative for the corresponding antigens that may be attacked if transfused to prevent additional hemolytic burden to the fetus. The transfusion service usually attempts to provide fresh RBC units, but it may not be possible if the mother has alloantibodies to high frequency antigens or to multiple antigens. Washed units are not required (unless the unit is from the mother) because the plasma components pose no risk to the fetus (Answer A). The units should be volume reduced rather than saline resuspended in order to minimize volume delivered to the fetus (Answer B). No current product modifications can prevent Epstein Barr virus (EBV) transmission (Answer D). Rh positive units are not a requirement for IUT, but may be used in cases of HDFN caused by antibodies other than anti-D and when the fetus is known to be Rh positive (Answer E).

Please answer Questions 5 and 6 based on the following clinical scenario:

A neonatologist would like to perform a two blood volume exchange in a full-term infant weighing 3.2 kg with hyperbilirubinemia. The target hematocrit is 45%. The RBC unit that will be used during the reconstitution process has a hematocrit of 60%.

5. What is a typical exchange rate when performing a whole blood exchange in a neonate?

A. 5% of total blood volume every 15 min
B. 5% of total blood volume every 10 min
C. 5% of total blood volume every 5 min

D. 5% of total blood volume every 1–2 min
E. 10% of total blood volume every 1–2 min

Concept: Total blood volume in an average adult is equal to 70 mL/kg of body weight. This can vary between patients depending on body fat composition (obese patients have a total blood volume approximately equal to 50–60 mL/kg of body weight). Pediatric patients typically have higher blood volumes per kilogram body weight. The total blood volume for premature infants (<37 weeks gestation) can be estimated using 100 mL/kg of body weight and full-term infants would be estimated by using 85 mL/kg.

Answer: *D*—Whole blood exchange typically occurs at a rate of no more than 5% of total blood volume every 1–2 min. The baby weighs 3.5 kg and was born at full term, so the estimate total blood volume is 3.5 kg × 85 mL/kg = 297.5 mL. In this baby, the whole blood exchange would be no more than 5% (~15 mL) every 1–2 min. The other choices (Answers A, B, C, and E) are incorrect.

6. Potential complications related to whole blood exchange in neonates include which of the following?
A. Thrombocytosis
B. Hypercalcemia
C. Hypokalemia
D. Hyperthermia
E. Vascular insufficiency of lower limbs

Concept: Whole blood exchange may be used in the treatment of HDFN to remove bilirubin, maternal antibodies that cause the hemolysis, and the fetal RBCs that are being destroyed. Whole blood exchange is akin to massive blood transfusion. Thus, complications related to whole blood exchange are similar to those patients experiencing massive blood transfusion.

Answer: *E*—If the exchange is performed through an arterial umbilical catheter, vascular insufficiency of lower limbs may be a potential complication, due to the negative pressure exerted on the vessel during the exchange. Since whole blood reconstitution only uses RBCs and plasma, thrombocytopenia, not thrombocytosis, may occur, especially if the patient starts with a low platelet count (Answer A). Because this is similar to a massive blood transfusion, there is an increased risk of citrate toxicity, which could lead to hypocalcemia (Answer B). If given rapidly, hyperkalemia is a risk in infants (Answer C). Unless a blood warmer is used, hypothermia can occur during exchange transfusions (Answer D).

Other complications include hypo- and hyperglycemia. An inadequate glucose infusion rate can occur if other sources of glucose are discontinued during the exchange transfusion. In addition, hypoglycemic episodes occur more commonly with transfusion of citrate phosphate dextrose adenine-1 (CPDA-1) RBCs versus those in additive solution (AS), since AS-RBCs contain higher glucose concentrations. Although large-volume transfusions of AS-RBCs may cause transient hyperglycemia, this can be followed by rebound hypoglycemia from the insulin induced by the initial glucose load, especially if the transfusion is discontinued abruptly.

End of Case

7. A newborn presents with thrombocytopenia and abdominal petechiae and is presumptively diagnosed with NAIT. The mother is subsequently found to have antibodies against rare platelet antigens. Testing reveals that the maternal aunt lacks the platelet antigen and she donated platelets for the newborn, which were transfused emergently without any further product modifications. Two weeks later, the newborn presents with fever, diarrhea, elevated liver enzymes, and dermatitis. The cause of these symptoms is most likely which of the following?
A. Posttransfusion hepatitis C infection
B. Transfusion associated dyspnea
C. Bacterial infection due to bacterial contamination of platelets

D. Transfusion-associated graft versus host disease (TA-GVHD)
E. Delayed hemolytic disease of newborn (HDN)

Concept: TA-GVHD may occur when nonirradiated blood is transfused from one close blood relative to a related recipient due to similarities in their HLAs. The recipient is unable to recognize the lymphocytes as foreign; thus, there is not destruction of the donor T-lymphocytes, which can then attack the lymphoid tissue, bone marrow, and other organ systems of the recipient.

In homogenous populations, such as the Japanese or Israelis, sharing of HLA is reported. TA-GVHD usually affects the liver, skin, gastro-intestinal (GI) system, and bone marrow (late), leading to signs and symptoms of transaminitis, dermatitis, diarrhea, and pancytopenia (late). Unlike the GVHD complicating allogeneic HPC transplantation, TA-GVHD is almost universally fatal. Currently, there is no effective treatment for this condition. Therefore, it is essential for the transfusion service to identify the patients at risk for TA-GVHD and provide them with irradiated blood products in order to inactivate the donor T lymphocytes.

Answer: *D*—This neonate received platelets from a close blood relative. The signs and symptoms are suggestive of TA-GVHD. Documentation of donor T lymphocytes in the patient's circulation or tissues will confirm the diagnosis. Acute hepatitis C (Answer A) can result from transfusion; however, it does not typically present with dermatitis. Transfusion associated dyspnea (Answer B) and sepsis due to bacterial contamination (Answer C) from platelets are usually acute issues, occurring within minutes to several hours from the transfusion, not 2 weeks after the transfusion. Symptoms from GI tract, skin, and liver are not typical of HDN (Answer E).

8. In which of the following is platelet transfusion indicated in a newborn patient?
A. Stable infant, platelet count 30,000/μL
B. Unstable infant, not on ECMO, not bleeding, platelet count 65,000/μL
C. Infant with active bleeding, or invasive procedure, platelet count 41,000/μL
D. On ECMO, no bleeding, platelet count 150,000/μL
E. Stable neonate with NAIT and platelet count of 50,000/μL

Concept: Platelets may be transfused prophylactically or in the setting of active bleeding. Hemostasis and bleeding can differ between adults and children at similar platelet counts; therefore, specific guidelines have been developed for platelet transfusion in both populations.

In adults, platelet transfusions are used to prevent spontaneous bleeding when counts drop below 10,000/μL in a stable, nonbleeding patient. Higher thresholds may be used in cases of sepsis, fever, or coexisting coagulopathy. Varying platelet count thresholds are used in preparation for invasive procedures, from 20,000/μL for a central line placement up to 100,000/μL for neurosurgery. For pediatrics, platelet recommendations are as follows:

- Stable infant with platelet count < 20,000/μL
- Unstable infant, not on extracorporeal membrane oxygenation (ECMO), not bleeding with platelet count 30,000–50,000/μL
- Infant with active bleeding, or invasive procedure, with platelet count ≤50,000/μL
- On ECMO, not bleeding, with platelet count ≤100,000/μL

Similarly, RBC transfusion is not without risks and there are guidelines for both adult and pediatric patients. In order to improve the oxygen carrying capacity, some of the specific RBC transfusion indications in pediatric population are:

- Hematocrit <20% with low reticulocyte count and symptomatic anemia
- Hematocrit <30% with weight gain (<10 g/day observed over 4 days while receiving ≥100 kcal/kg/day)
- Hematocrit <35% and on >35% oxygen hood
- Hematocrit <35%–45% and on extracorporeal membrane oxygenation (ECMO) to maximize oxygen delivery to the tissues, since mixing occurs prior to systemic circulation of oxygenated

blood from the extracorporeal membrane and deoxygenated blood returning from the patient's circulation

Answer: *C*—Based on the above guidelines, infants with active bleeding, or invasive procedure and a platelet count ≤50,000/ μL should be transfused with platelets. The threshold of platelet transfusion for stable infant is 20,000/μL (Answers A and E). For unstable infants, not on ECMO and not bleeding, the threshold is between 30,000 and 50,000/μL (Answer B). Finally, the threshold is 100,000/μL nonbleeding infant on ECMO (Answer D).

9. In order to obtain an increment in the platelet count in a 5-kg infant by 60,000–100,000/μL, which of the following volumes of apheresis or whole blood derived platelets should be transfused?
A. 200 mL
B. 150 mL
C. 100 mL
D. 50 mL
E. 25 mL

Concept: Pediatric transfusion is challenging, partially due to the inability of the patient to tolerate large volumes of blood products. However, pediatric patients may not need large volumes of transfusion to raise the platelet count. For adults, each apheresis-derived platelet unit (~200–250 mL) is expected to raise the platelet count by ~30,000–60,000/μL. For pediatric patients, transfusing ~10 mL platelets/kg is generally expected to raise the platelet count by 60,000/μL to 100,000/μL. It is important to check the post transfusion count because the concentration of platelets in a platelet concentrate (apheresis or whole blood derived) can vary considerably.

Answer: *D*—Using the above formula, this infant needs 10 (mL/kg) × 5 kg or ~50 mL of platelet transfusion to raise the platelet count by approximately 60,000–100,000/μL. All of the other choices (Answers A, B, C, and E) are likely to be either under- or overdosed.

10. Which of the following is an accepted indication for leukoreduced blood components in neonates or infants?
A. Reduction of the incidence of hemolytic transfusion reactions
B. Prevention of alloimmunization to HLA Class I antigens
C. Reduction of CMV transmission
D. Prevention of TA-GVHD
E. Reduction of the incidence of bacterial transmission

Concept: Although leukocytes are reduced via prestorage or pretransfusion filtration, they are still present in cellular blood products and have been implicated in infectious disease transmission and other immune mediated sequelae, such as febrile non-hemolytic transfusion reactions (FNHTRs) and TA-GVHD. Each unit of whole blood contains ~2 to 5×10^9 leukocytes. Current leukoreduction filters leave a residual count of less than 5×10^6 [i.e., 3 log reduction, standard limit of WBCs set by the Food and Drug Administration (FDA)]. Leukoreduction may be performed at the time of collection (prestorage) or during transfusion at the bedside (poststorage). Prestorage leukoreduction is the method of choice because it has shown to decrease the risk of leukoreduction failures and it lessens the amount of cytokines that accumulate during storage.

Answer: *C*—The accepted indication for leukoreduction in neonates/infants is to provide CMV reduced risk products. Leukoreduced products also decrease febrile nonhemolytic transfusion reactions (FNHTRs) by decreasing the amount of WBCs available to release inflammatory cytokines. TA-GVHD can only be prevented by irradiation (Answer D); however, there are suggestions from newer studies that pathogen-inactivation techniques can also reduce TA-GVHD risk. Similarly, alloimmunization to HLA antigens is reduced, but not eliminated by leukoreduction (Answer B). Leukoreduction does not reduce the risk of bacterial transmission (Answer E).

11. Standard volume RBC transfusions (<20 mL/kg) for neonates need to be washed for which of the following reasons?

A. To avoid hyperkalemia
B. To prevent hemolytic transfusion reactions
C. To prevent T-activation
D. To remove harmful additives such as adenine and mannitol
E. Standard volume transfusions do not need to be washed

Concept: Washing is used to remove the plasma and/or supernatant in the RBCs and/or platelet products. Washing is labor intensive and also shortens the product shelf-life (i.e., washed RBCs must be used within 24 h and washed platelets must be used within 4 h after washing). Furthermore, the washing process results in cellular loss of both RBCs and platelets. It also activates the platelets and may result in platelet degranulation prior to the transfusion. Thus, currently, washing is only truly indicated for a few scenarios, such as in patients with severe IgA deficiency and with anti-IgA antibodies or patient with multiple severe/anaphylactic reactions.

Answer: *E*—Standard volume transfusion do not need to be washed for neonates. Hyperkalemia usually results after large volume of transfusion. The volume of plasma in a standard RBC transfusion (<20 mL/kg) does not usually place a child at risk for hyperkalemia (Answer A). Hemolytic transfusion reactions are often related to an incompatibility between infant's plasma and the RBCs given and thus, washing cannot prevent hemolytic reactions (Answer B). T-activation is the process by which the T-antigen is exposed due to an enzyme (typically a neuraminidase) which cleaves the terminal sugar residues on the surface of the patient's RBCs. Washing RBCs does not prevent T-activation (Answer C) and is therefore, not indicated, even though many neonatologists still request this modification. In standard volume transfusions, adenine and mannitol do not usually cause any problems; thus, washing is not indicated to remove the additives (Answer D).

12. Which of the following statements is true regarding complications of RBC transfusion in patients with sickle cell disease (SCD)?

A. Whole blood transfusions reduce the risk of stroke in patients with a high transcranial Doppler velocity
B. RBC transfusions should be avoided, even in patients with symptomatic anemia
C. RBC transfusion increases the risk of HLA alloimmunization
D. An RBC (stored in AS-1) transfusion contains approximately 10 g of iron for every mL transfused
E. Transfusing RBCs antigen matched for C, E and K antigens reduce the risk of HLA alloimmunization

Concept: Patients with SCD often require chronic transfusions to alleviate symptoms and prevent end-organ damage caused by chronic hemolysis and vasoocclusion. Transfusion of non-sickle RBCs increases the oxygen delivery to tissues; thus, reducing the sickling and microvascular thrombosis. For example, chronic transfusion therapy with RBCs to maintain hemoglobin S (HbS) level below 30% prior to the next transfusion reduces the risk of first stroke in the high-risk children with abnormal (high velocity) transcranial Doppler studies and recurrent stroke. However, transfusion does not come without risks. Besides transfusion reactions and transfusion transmitted infections, patients are at risk of other transfusion-related complications, such as iron over load and alloimmunization. This is mostly due to excess antigen exposure from multiple donors and their increased propensity to become alloimmunized. Therefore, besides giving these patients hemoglobin S negative products, many transfusion services elect to perform additional product modifications to reduce transfusion-related complications in these patients.

Answer: C—Even with leukoreduction, there is a risk of HLA alloimmunization with RBC transfusion, especially in patients requiring chronic transfusion. Because WBCs are still present in leukoreduced RBCs, HLA alloimmunization can occur, especially in SCD patients or those on chronic transfusion protocols. Previous studies report that use of prophylactic RBCs (antigen matched for C, E, and K-antigen) reduces the rate of RBC alloimmunization from 3% to 0.5% per unit (not necessarily HLA alloimmunization) (Answer E).

Red cell transfusions should be given as needed to SCD patients, especially those with symptomatic anemia (Answer B). Chronic RBC transfusion (and not necessarily whole blood) has been shown to reduce the risk of stroke in high risk patients (Answer A). One mL of "packed" RBCs contains approximately 1 mg of iron, so a transfusion of 1 mL of red cells (not packed because they are stored in additive solutions) contains approximately 0.5 mg of iron (because the AS-1 RBC unit has a hematocrit of ~55%–65%, Answer D).

13. In addition to be hemoglobin S negative, RBCs transfused to pediatric patients with SCD should have which of the following requirements?

A. Only ABO matched and crossmatch compatible
B. Leukoreduced in addition to being ABO compatible and crossmatch compatible
C. Parvovirus B19 negative in addition to being ABO compatible and crossmatch compatible
D. Fresh (<5 days old) and C-, E-, and K-antigen matched in addition to being ABO matched and crossmatch compatible
E. Irradiated in addition to being ABO matched and crossmatch compatible

Concept: Chronic transfusion patients, such as SCD patients, are at risk of developing multiple transfusion-related complications. Thus, many transfusion services elect to provide modifications to blood products to reduce the risk of these complications. For example, chronic transfusion therapy with RBCs has increased risk of alloimmunization of both RBC and HLA antigens. Therefore, measures to reduce this risk are recommended, including antigen matching and leukocyte reduction.

Answer: *B*—In addition to being ABO and Rh compatible and crossmatch compatible (Answer A), RBC units should be leukoreduced when given to patients requiring chronic transfusion to reduce the risk of RBC and HLA alloimmunization and FNHTRs.

Fresh units (Answer D) are not indicated based on the available evidence as of this writing, and may not be feasible if the patient has multiple alloantibodies. All patients should be transfused with crossmatch compatible RBCs. Leukocyte reduction also reduces risk of CMV transmission. CMV negative blood is not required. Irradiation (Answer E) is not typically required since T-cell function is not particularly compromised in patients with SCD.

Per previous studies, use of prophylactic RBCs antigen matched for C, E, and K reduces the rate of alloimmunization in RBC transfusion from 3% to 0.5% per unit; however, this is not universally required. Because WBCs are still present in leukoreduced RBCs, HLA alloimmunization can occur, especially in SCD patients on a chronic transfusion protocol. Parvovirus B19 suppresses marrow erythroid progenitor cells and thus, can be especially devastating in patients with reduced circulating RBC survival time. However, at this time, there is no recommendation of giving these sickle cell patients with Parvovirus B19 negative RBCs (Answer C). Of note, Parvovirus B19 is not part of the routine infectious testing for blood donors.

14. Reduction of RBC alloimmunization in pediatric patients with SCD can be accomplished by which of the following?

A. Providing ABO identical RBCs
B. Providing washed RBCs
C. Providing fresh (<5-day-old) RBCs
D. Providing irradiated RBCs
E. Providing phenotype matched units

Concept: Due to the increased frequency of RBC transfusions, patients with SCD have an increased risk of alloimmunization. Some studies suggest the alloimmunization risk becomes greater when RBCs are from Caucasian donors who have certain RBC antigens at higher frequencies compared to

African donors, though a recent study by Chou S.T. et al., has challenged this long held maxim [14]. In order to mitigate the risk of alloimmunization, prophylactic matching of certain RBC antigens (C, c, E, e, and K) is performed to reduce the rate of alloimmunization, but not completely eliminate alloantibody formation.

Answer: *E*—Only specific RBC antigen negative products can prevent red cell alloimmunization. Although studies have attempted to assess the best time to initiate antigen matching (prophylactically or after first antibody production) and the extent of antigens to match, most sickle cell programs now match, at a minimum, for RBC antigens (D, C, c, E, e, and K) after the patient begins a chronic RBC transfusion program. Studies have shown that extended phenotype matching (to include Jk, Fy, and other antigens) results in limited availability of compatible units and reduced cost effectiveness, without significantly reducing further alloantibody production. Furthermore, washed, fresh, irradiated RBC, or ABO identical units (Answers A, B, C, D) do not prevent or reduce RBC alloimmunization based on the currently available evidence.

15. Which of the following statements is true regarding patients with SCD with vasoocclusive crisis, marked anemia, and progression to splenic sequestration?
A. Patients can have a sudden increase in hemoglobin level due to splenic contraction
B. Over-transfusion is not a concern because they turn over red cells so quickly
C. ABO matched RBCs are sufficient and most patients do not require additionally antigen matching
D. They require fresh (<14-day-old) RBCs
E. Patients have a higher propensity to develop HPA1a alloantibodies

Concept: Patients with SCD often present to hospitals with acute painful episodes or "crisis." These episodes are due to microvascular occlusion (vasoocclusion) and subsequent tissue hypoxia. The main factor leading to these occlusions is polymerization of abnormal hemoglobin S followed by RBC sickling. In addition, many other factors play a role in the pathogenesis including leukocyte activation, endothelial adhesion molecules, vasoregulation, and blood viscosity. The main locations for acute vasoocclusive pain are long bones, abdomen, and chest. Frequent acute episodes and chronic occlusion can lead to severe hypoxia and ischemia with the potential for irreversible end-organ damage.

Answer: *A*—SCD patients can have a sudden contraction of the spleen resulting in auto-transfusion. Therefore, they should be carefully transfused to avoid raising the hematocrit >33% to avoid worsening the vasoocclusion. Iron overload is also a risk due to rapid red cell turnover and frequent RBC transfusion (Answer B).

SCD patients with inflammatory conditions have been shown to have increased RBC antigen alloimmunization risk, and frequent allogeneic red cell transfusions increases exposure to RBC antigens. Further compounding this risk is the fact that in the United States, the SCD patient population RBC genotype/phenotype tends to be very different from the RBC donor pool genotype/phenotype. Thus, currently, most SCD programs match, at a minimum, for RBC antigens (D, C, c, E, e, and K) after the patient begins a chronic RBC transfusion program (Answer C). There is no current evidence that these patients require fresher RBCs for transfusion (Answer D). HPA1a antigen is on platelet surface and thus, the risk of alloimmunization to HPA1a should not be increased with RBC transfusion (Answer E).

16. What is the minimum concentration of hemoglobin at which pediatric patients with beta thalassemia major, on a chronic transfusion program, can suppress ineffective hematopoiesis?
A. 10–11 g/dL
B. 11–12 g/dL
C. 12–13 g/dL
D. 13–14 g/dL
E. 14–15 g/dL

Concept: Beta thalassemia refers to a variety of genetic mutations, leading to impaired production of beta globin chains and a relative excess of alpha globin chains. The alpha globin chains precipitate within the cell and the degree of excess determines the clinical severity. Beta thalassemia major is the clinical term applied to patients with severely limited or absent beta globin chain production. They have transfusion dependent anemia, hepatosplenomegaly, bone marrow expansion, and are prone to infections and fractures. Beta thalassemia minor (beta thalassemia trait) refers to heterozygotes who have a single gene mutations, leading to reduced production. These patients are generally asymptomatic and discovered incidentally. Patients with an intermediate form of the disease, such as those who inherit a thalassemic variant on each gene, are said to have beta thalassemia intermedia. In addition, patients with beta thalassemia major (complete or near complete loss of beta globin production) have ineffective erythropoiesis, which results in a marked erythropoietic drive in the attempt to generate functional RBCs that can carry oxygen.

Answer: *D*—Only the transfusion of RBCs to a hemoglobin concentration between 13 and 14 g/dL can effectively suppress this erythropoietic drive (i.e., ineffective erythropoiesis). Patients with beta thalassemia major may be transfused to a higher Hct to suppress ineffective erythropoiesis, unlike SCD patients who should generally not be transfused to a Hct >33% due to concern over hyperviscosity, which can lead to vasoocclusive pain crisis or end organ ischemia/damage (e.g., stroke, severe intrahepatic cholestasis). All the other choices are incorrect (Answers A, B, C, E).

17. Which of the following is the best choice for treatment for a neonate with neonatal purpura fulminans secondary to homozygous protein C deficiency?
A. Cryoprecipitated AHF alone
B. Cryoprecipitated AHF and plasma
C. Plasma and purified protein C concentrate
D. Recombinant ADAMTS13 concentrate
E. Purified protein C concentrate or 4-factor prothrombin complex concentrate (4F PCC)

Concept: Neonatal purpura fulminans is a manisfestation seen in patients with homozygous congenital protein C or protein S deficiency. Protein C deficiency is a rare condition, occurring in ~1:500,000–1:1,000,000 births. Early lesions of purpura fulminans usually develop within 72 h of birth (in the perineal region, thighs, and abdomen) and they look similar to purpuric rashes; however, they will quickly progress to necrosis. Thrombocytopenia may also be present.

Answer: *E*—Treatment of neonatal purpura fulminans requires transfusion of products containing protein C. The products containing protein C include all plasma containing products and products derived from plasma, with the exception of cryoprecipitated AHF. Both purified protein C concentrate (Ceprotin) and four factor prothrombin complex concentrate (4F PCC, Kcentra) contain the highest concentrations of protein C.

For 4F PCC, each 500-unit vial contains ~420–820 IU protein C. This is important in order to avoid volume overload in pediatric patients. Thus, although plasma contains protein C, it is not the ideal product since protein C concentrate is available in 4F PCC (Answer C). Cryoprecipitated AHF only contains factor VIII, von Willebrand factor, factor XIII, fibrinogen, and fibrinectin. Hence, cryoprecipitated AHF does not contain protein C and should not be given in the treatment of neonatal purpura fulminans (Answers A and B). Recombinant ADAMTS13 has no role in the treatment of neonatal purpura fulminans (Answer D).

18. A 9-year-old male is currently on ECMO and is being anticoagulated with heparin. The ICU physician would like to follow the anti-Xa level to ensure that the patient is being properly anticoagulated. The anti-Xa level remains <0.1 IU/mL even after several dose increases and the ECMO technologist notices difficulty in the circuit flow. His antithrombin level is low at 25%. Of the options listed below, what is the best treatment for this patient?
A. Increasing the heparin dose
B. Plasma transfusion

C. 4F PCC infusion
D. Antithrombin concentrate infusion
E. Cryoprecipitated AHF transfusion

Concept: Congenital antithrombin deficiency is an autosomal dominant disorder. However, acquired antithrombin deficiency, is more common than the congenital form, and is usually due to liver disease or medications (e.g., L-asparaginase). Antithrombin deficiency usually presents with venous thrombosis. Often times, antithrombin deficiency is a reason for not being able to achieve therapeutic effect with heparin as the anticoagulant (as described in this example). Thus, antithrombin may be supplemented. Antithrombin is present in plasma products and plasma derivatives, with a half-life ranging from 2 to 4 days. Several antithrombin concentrates are now available.

Answer: *D*—Antithrombin concentrate, is FDA approved for congenital and acquired antithrombin III deficiency and among the choices above, is the most appropriate for treatment in this case. The dosing of antithrombin concentrate is similar to plasma based dosing for Factor VIII. However, specific antithrombin concentrate package inserts as well as transfusion medicine/hematology specialists should be consulted for proper dosing.

Plasma should only be reserved in emergencies when there is no antithrombin concentrate available (Answer B). 4F PCC may be contraindicated, as it could be pro-thrombotic in a patient with thrombophilia (Answer C). Cryoprecipitated AHF does not contain any substantial amount of antithrombin to be clinically useful (Answer E). Heparin resistance, as described in this case, is defined as the need of more than 35,000 U of heparin for 24 h to get the anti-Xa or PTT into therapeutic range, may be due to antithrombin deficiency, and thus, heparin is not a treatment for antithrombin III deficiency (Answer A).

19. Previously untreated patients with hemophilia A are typically treated with which of the following?
A. Cryoprecipitated antihemophilic factor (AHF)
B. Fresh frozen plasma (FFP)
C. Plasma derived, immunoaffinity purified, high purity factor VIII concentrates
D. Recombinant factor VIII concentrates
E. Plasma derived, intermediate purity factor VIII products with high von Willebrand factor levels

Concept: Hemophilia A (factor VIII deficiency) is an X-linked recessive disease that affects one in 5,000–10,000 males. The diagnosis is usually made within the first 2 years of life when an infant presents with easy bruising, joint/muscle bleeds, or CNS bleeding. Hemophilia is categorized by the baseline factor level as mild (5 to 40%), moderate (1%–5%), or severe (<1%). It has been demonstrated that primary prophylaxis with factor replacement reduces bleeding events and long-term sequelae as opposed to therapy limited for acute bleeding events. The choice for therapy is especially important in previously untreated patients because products have varying immunogenicity and the potential to incite factor inhibitor production in the patient.

Answer: *D*—Recombinant factor VIII products should be given to previously untreated patients with hemophilia A to reduce the theoretical risk of transfusion-transmitted infection. Interestingly, based on the result of a recent multicenter, randomized trial conducted in previously untreated patients (<6 years old) with severe hemophilia A (the SIPPET trial), recombinant factor VIII products may increase the risk of developing a factor VIII inhibitor [15].

Plasma derived factor VIII may be given if no recombinant factor is available or if the patient has developed an inhibitor versus the recombinant product (Answers C and E). While cryoprecipitated AHF does have a minimum of 80 IU of factor VIII per bag, it should not be used if any source of concentrated factor is available (Answer A). Plasma is not an option because the amount of factor VIII provided in plasma would require many liters of plasma to be transfused, causing potential fluid overload (Answer B).

20. Acute treatment for type I hereditary angioedema includes which of the following?
A. Specific factor concentrates containing antithrombin
B. Specific factor concentrates containing protein C

C. Specific plasma derived concentrates containing C1-inhibitor esterase inhibitor
D. Cryoprecipitated AHF
E. Platelets

Concept: Hereditary Angioedema (HAE) is a very rare (approximately one in 10,000 to one in 50,000) and potentially life-threatening genetic condition, in which episodic edema occurs in various body parts including the hands, feet, face and airway. Patients can often have excruciating abdominal pain, nausea, and vomiting that is caused by intestinal wall swelling. The specific gene defect results in either an inadequate production of C1 esterase inhibitor (C1-INH) protein or a non-functioning C1-INH protein. C1-INH helps to regulate the complex systems involved in the inflammatory response and coagulation. Three types of HAE can be differentiated with complement testing and, in the case of type III, genetic testing (Table 10.1).

TABLE 10.1 Different Types of HAE and Their Diagnostic Properties

	Type of hereditary angioedema		
	Type I	**Type II**	**Type III**
Key laboratory abnormalities	Low antigen level, normal function	Normal or elevated antigen levels, but dysfunctional	Normal antigen level, and normal function
C4/C2 level	Low	Low	C4 may be normal
C1q level	Normal	Normal	Normal
Other			Factor XII mutations may be present

In HAE types I and II, the treatment of choice in acute attacks consists of replacement with commercially available C1-INH concentrates or kallikrein inhibitor or if those are unavailable, FFP. In HAE type III, infusion of C1-INH has proven to be ineffective in which case, acute attacks of angioedema may be treated alteratively with bradykinin B2 receptor antagonists, kallikrein inhibitors, progesterone, danazol, and/or tranexamic acid, which are also used in the classic HAE types.

Answer: C—Factor concentrates with C1-INH can be used to treat hereditary angioedemia except in type III hereditary angioedemia. In cases where Cl-INH are unavailable, FFP may be tried as an emergent measure. Currently, the FDA has approved the C1-INH plasma derived concentrate (Berinert) for the treatment of acute abdominal and facial angioedema, largyngeal angioedemia attacks in adolescents and adults with HAE and the recombinant human C1-INH (rhC1-INH, Ruconest) to treat acute attacks of HAE in adolescents and adults.

In addition, the FDA has approved a recombinant kallikrein inhibitor, Ecallantide (Kalbitor), a potent, selective, reversible agent for the treatment acute HAE attacks in patients aged 16 years and older. This agent inhibits unregulated plasma kallikrein activity which results in excessive bradykinin generation, resulting in swelling during attacks. Finally, the FDA has also approved the selective bradykinin B2 receptor antagonist, Icatibant (Firazyr) for treatment of acute HAE attacks in adults. None of the products mentioned in the other choices (Answers A, B, D, and E) contain sufficient amounts of Cl-INH to be useful or has no significant C1 or kallikrein inhibitory ability.

21. Which of the following is a correct statement about pediatric massive transfusion protocols (MTPs)?
A. MTPs define the type and amount of blood products provided to the clinician
B. MTPs have been shown to decrease mortality in pediatric patients
C. MTPs have variable ratios of RBCs to plasma based on patient age and weight

D. MTPs require an active type and screen specimen
E. MTPs require that AB negative plasma be used

Concept: Massive hemorrhage often requires massive transfusion to maintain adequate circulation and hemostasis. In pediatric patients, massive hemorrhage can be defined as a patient with either of the following:

1. blood loss greater than one total blood volume within 24 h,
2. transfusion support to replace ongoing hemorrhage of greater than 10% total blood volume (TBV)/min, and
3. transfusion support needed to replace greater than 50% total blood volume within 3 h.

For optimal management of massively bleeding patients, regardless of etiology (trauma, obstetrical, surgical), effective preparation and communication between the laboratory, blood bank, and clinical teams is essential. A well-defined massive transfusion protocol (MTP) is a valuable tool to delineate how blood products are ordered, prepared, and delivered; determine laboratory algorithms to use as transfusion guidelines; and outline duties and facilitates communication between involved personnel.

Answer: *A*—A core concept of MTPs is that transfusion of RBCs, plasma and platelets be given in a predetermined ratio approximating the constitution of whole blood. To date, no studies have reported improved survival with the use of pediatric MTPs; however, this has been shown in adults (Answer B). Although AB plasma is ideal, it is not required (Answer E). MTPs do not require a blood bank specimen (Answer D); however, pretransfusion specimens should be drawn for type and screen as soon as possible because after large volume transfusions, it may be difficult to determine the patient's ABO type and alloantibody status. While the amount of RBCs and plasma given depends on the patient age and weight (Answer C), the ratio of RBCs to plasma and platelets is fixed in a MTP.

22. The following symptoms and signs are discovered in a pediatric patient receiving a large volume RBC transfusion: peaked T waves; widening QRS complex; nausea/diarrhea and bradycardia. Which of the following mechanisms is the most likely cause?
A. Hyperthermia
B. Citrate toxicity
C. Hypercalcemia
D. Hyperkalemia
E. Hyponatremia

Concept: Red cells undergo many changes during storage, which is termed "storage lesions". These include a decrease in both ATP and 2,3 DPG (2,3-diphosphoglycerate) and an increase in potassium. Further, RBC membrane loss can decrease the integrity of the RBCs, leading to hemolysis and formation of microparticles. These storage lesions may contribute to the adverse events that were reported from multiple retrospective studies. Nonetheless, results from recent three randomized clinical trials, including one in premature critically ill infants, did not show that fresh RBCs (i.e., RBCs that stored for <7 days) transfusion improved clinical outcome. However, large volume transfusions (such as in an MTP) may lead to metabolic derangement in addition to other complications in vulnerable populations, such as pediatric patients (Table 10.2).

Answer: *D*—Red cell storage lesions include decrease in adenosine triphosphate (ATP) content and function of ATP dependent ion channels, leading to potassium leakage out of cells. Thus, a large volume RBC transfusion could cause hyperkalemia with resultant cardiac conduction defects (peaked T waves and widening QRS complex) and gastrointestinal symptoms. Hypothermia, not hyperthermia, is a possible consequence of massive transfusion (Answer A). Typically a blood warmer is used to prevent this complication. Citrate toxicity is also a possibility with large volume

TABLE 10.2 Complications From Massive Transfusion

Adverse events	Comments and potential treatments
TRANSFUSION REACTIONS:	
Allergic	Allergic reactions can range from simple urticarial to anaphylaxis. Steroid and diphenhydramine might be given to patients with allergic transfusion.
Hemolytic transfusion reaction (acute and delayed)	Hemolytic transfusion might be reduced by giving O-RBC and AB plasma for emergency release of blood products.
Febrile nonhemolytic transfusion reaction	Febrile nonhemolytic transfusion reaction is a diagnosis of exclusion.
IMMUNOLOGIC REACTIONS:	
Transfusion-related acute lung injury (TRALI)	TRALI incidence could be reduced by transfusing male-only plasma. Fresh RBCs could be attempted.
Transfusion-related immunomodulation (TRIM)	TRIM may be responsible for increased risk of bacterial infection.
Transfusion-associated graft versus host disease (Ta-GVHD)	Irradiation of cellular blood products in patients at risk (such as neonates and immunosuppressed patients) to prevent Ta-GVHD.
Posttransfusion purpura (PTP)	May be treated with IVIG infusion, steroid, or plasma exchange.
METABOLIC COMPLICATIONS:	
Hypocalcemia[a]	Hypocalcemia is due to citrate overload from rapid transfusion of blood products. Neonates and patients with preexisting liver disease are at risk for hypocalcemia. Monitor ionized calcium level and correct if necessary.
Hypomagnesemia[a]	Hypomagnesemia is due to large volume of magnesium-poor fluid and citrate overload. Monitor ionized magnesium level and correct if necessary.
Hyperkalemia[a]	Hyperkalemia is due to hemolysis of RBC from storage and/or irradiation. Neonates and patients with preexisting cardiac and renal diseases are at risk for hyperkalemia. Monitor potassium level and correct if necessary. Fresh RBCs (<5–10 days old) and irradiated < 24 h prior to transfusion or washing may decrease risk.
Hypokalemia[a]	Hypokalemia is due to reentry into transfused RBCs, release of stress hormones, or metabolic alkalosis. Monitor potassium level and correct if necessary.
Metabolic alkalosis[a]	Metabolic alkalosis is due to citrate overload. Monitor acid-base status.
Acidosis[a]	Acidosis is due to hypoperfusion, liver dysfunction, and citrate overload. Monitor acid-base status.
Hypothermia[a]	Hypothermia is due to infusion of cold fluid and blood products, opening of body cavities, decrease heat production, and impaired thermal control. Neonates and infants are at increased risk. Blood warmer could be used.
OTHER ADVERSE EVENTS:	
Hemostatic defects[a]	Hemostatic defects result from complex mechanism (discuss in the pathophysiology section)
Infections	Infections can result from blood products or other resuscitated procedures, such as surgeries.
Transfusion associated circulatory overload (TACO)[a]	TACO should be differentiated from TRALI. Infants and patients with preexisting cardiac disease are at increased risk. Oxygen and diuresis can be used.
Air embolism	Air embolism is a rare fatal complications. Instructions and/or protocols on how to use rapid infuser must be followed.

[a]*Adverse events more likely to be due to rapid infusion of large amount of blood products in short period of time.*

Adapted from H.P. Pham, B.H, Shaz, Br. J. Anaesth. 11 (2013) i71–i82.

RBC transfusion, manifesting primarily as hypocalcemia, with common signs in older children and adults including perioral tingling and paresthesias (Answer B). In infants and those who cannot verbalize, citrate toxicity may manifest as jitteriness, apnea, cyanosis, poor feeding, lethargy, and seizures, with possible prolongation of the QT interval. Hyponatremia is not usually associated with large volume RBC transfusion, unless it results in volume overload [i.e., hypervolemic hyponatremia (Answer E)]. Hypercalcemia (Answer C) is not an adverse event associated with massive transfusion.

23. To minimize the risk of hyperkalemia-induced cardiac arrhythmias in infants who receive a rapidly infused, large volume transfusion (>20 mL/kg) of RBCs, one can use which of the following?
A. RBC units < 21 days old
B. RBCs units irradiated 20 days ago
C. Washed RBCs
D. Leukoreduced RBCs
E. A 170 μm filter with the transfusion

Concept: The plasma surrounding older RBCs has much higher amounts of potassium than fresher RBCs. In adults and infants receiving normal volumes of transfusion, this is usually not an issue, as the kidneys are able to excrete excess potassium. However, precautions must be taken in large volume transfusions for the pediatric population and in any transfusion for infants weighing less than 1200 g, since their renal function may not be able to handle excess potassium.

Answer: *C*—Potassium in stored RBCs (in mmol/L) increases linearly and is approximately equal to the number of days of RBC unit storage. Washed blood products can reduce the risk of hyperkalemia-induced cardiac arrhythmias. However, washed units expire within 24 h of washing. Irradiation results in further increases in potassium leakage from RBCs, which is why storage of RBCs after irradiation is limited to 28 days, or the time remaining until the initially issued expiration date (whichever is shorter) of the RBC product (Answer B). Leukoreduction does not affect the rate of potassium leakage from RBCs (Answer D).

Given the high potassium of older blood (typically greater than 7–10 days), only washing can remove the extracellular potassium that is present (Answer A). Standard blood transfusion filters (170 μm) have no effect on potassium levels (Answer E). Though a potassium absorbing filter that can decrease the amount of potassium transfused to the patient does exist, it is not widely used.

24. A pregnant patient with SCD is found to have an anti-G antibody. What other antibodies might this patient appear to have on their antibody panel?
A. Anti-Fy^a and Fy^b
B. Anti-C and D
C. Anti-Jk^a and Jk^b
D. Anti-C and S
E. Anti-S and s

Concept: The G antigen is part of the Rh blood group system, and is present on RBCs possessing both C and D antigens. Antibody panels in a patient with anti-G will initially appear to be consistent with the combination of anti-C and anti-D. This recognition by a transfusion medicine technologist or physician prompts further investigation to check for the presence of anti-G.

Answer: *B*—Antibodies against G antigen appear to react with the associated C and D antigens; however, anti-G does not react against RBCs without G antigen, containing only the C or the D antigens alone. Therefore, for the purpose of RhIG administration, it is very important in pregnant women to use serial adsorptions to identify the exact alloantibodies present (anti-G; anti-G and anti-D; anti-G and anti-C; or anti-G, anti-D, and anti-C) if their panel is consistent with anti-D and anti-C. If the women do not have anti-D, then they are candidates for RhIG at 28 weeks, after delivery, and/or at any time when feto-maternal hemorrhage (FMH) is suspected. Serial absorption and elution techniques are used to identify if this patient has anti-G alone, or if the patient has: anti-G; anti-G and anti-C; anti-G and anti-D; or anti-G, anti-D, and anti-C.

Other challenging antibody patterns are as follows: If the pattern on the antibody work-up demonstrates an antibody to a high prevalent antigen, then the technologist may suspect anti-Fy3 (if the plasma does not react with Fy^a and Fy^b negative cells) (Answer A), anti-Jk3 (if the plasma does not react with Jk^a and Jk^b negative cells) (Answer C), and anti-U (if the plasma does not react with S and s negative cells) (Answer E). There is no known antibody to a high prevalence antigen that mimics anti-C and anti-S on a standard antibody panel (Answer D).

25. Which of the following statements is true regarding paroxysmal cold hemoglobinuria (PCH)?

A. It rarely occurs after an antecedent viral infection
B. It occurs more often in adults than in children
C. The antibody in PCH is IgG
D. The antibody is frequently characterized as having anti-I specificity
E. The antibody is a monophasic hemolysin (i.e., bind and hemolyze at the same temperature)

Concept: PCH occurs when the body produces antibodies that hemolyze RBCs when the patient goes from cold to warm temperature. This disease entity was initially described as being associated with the late stage of syphilis. However, with the advancement of antibiotics for the treatment of syphilis, PCH is now more commonly seen as a transient syndrome in children with a recent history of viral illness. The patient will present with signs, symptoms, and/or laboratory evidence of hemolysis.

The autoantibody implicated in PCH is IgG and in most cases, it has specificity for the P antigen on RBCs. IgG-anti-P reacts with RBCs in colder areas of the body causing irreversible complement binding and imminent hemolysis of RBCs at warmer body temperature (biphasic hemolysis). The test for PCH consists of three patient specimens and three appropriate controls. The first tube is collected and kept at 4°C, the second tube is collected at 4°C and then warmed up to 37°C, and the third is kept at 37°C all the time. If hemolysis is observed in the second tube (and not in the first and third tube) and if the positive and negative controls work, then the result is consistent with a diagnosis of PCH.

Answer: C—The antibody in PCH is IgG. It hemolyzes cells in a biphasic manner, not monophasic (Answer E). In a patient with hemolysis of uncertain etiology PCH is part of the differential diagnosis, especially if laboratory testing reveals a positive DAT and negative elution. Currently, PCH occurs more often in children, usually after an antecedent viral syndrome (Answers A and B). Besides PCH, other typical differential diagnoses for a positive DAT with negative elution on antibody screen/panel include drug-induced antibodies and ABO incompatible transfusion. Typically, cold autoantibodies have I-specificity (in adults) and i-specificity (in children) (Answer D). These are typically IgM antibodies, and the antibody panel may show pan reactivity.

Please answer Questions 26–28 using the following clinical scenario

A full term 3.5-kg baby boy with a hemoglobin of 10 g/dL and hematocrit of 30% develops jaundice on day 2 of life, with a total serum bilirubin (TSB) of 25 mg/dL and indirect bilibrubin of 21 mg/dL. His blood type is O Rh positive, while mother is noted to be O Rh negative. The direct antiglobulin test (DAT) is 3+ for IgG only. The eluate reveals anti-D. Phototherapy is initiated but only reduces the TSB to 23 mg/dL in 24 h. The neonatologist taking care of the child would like to perform a two volume whole blood exchange.

26. Which of following statements is true about this case?

A. The neonatologist should continue to use phototherapy as this is the best way to reduce the total bilirubin to a safe level
B. Phototherapy utilizes a wavelength between 460 and 490 nm, which is in the ultraviolet spectrum

C. A two blood-volume exchange will require approximately 600 mL of reconstituted whole blood
D. The baby should be transfused with O Rh negative blood first before performing the exchange
E. Plasma exchange should be considered as the first option in this patient rather than whole blood exchange, because it will more effectively remove the bilirubin

Concept: Based on the immunohematology work-up, it seems that this neonate has HDFN secondary to anti-D from mother. Phototherapy was not effective, and thus, whole blood exchange is warranted. Typically, a two blood-volume is performed using reconstituted whole blood using RBCs and plasma. The Hct of this reconstituted blood is usually ~50%.

Answer: *C*—The most effective way to quickly reduce bilirubin levels in neonates with a high risk of kernicterus is via whole blood exchange, since it can both reduce the bilirubin and the Rh positive RBCs, which in this case is the target for anti-D causing hemolysis—not phototherapy or plasma exchange (Answers A and E). One has to calculate the total blood volume of full-term neonate (85 mL/kg), which equals approximately 300 mL. Thus, for a two volume exchange, we would need to prepare 600 mL.

If a whole blood exchange is going to be performed immediately, there is no need to transfuse the baby with RBCs, particularly O Rh negative blood (Answer D), unless needed in an emergency situation. In the case of phototherapy, blue-green light in the range of 460–490 nm is most effective (light in the ultraviolet spectrum spans from 'vacuum UV light' at wavelengths of 100 nm to UVA light reaching 400 nm, Answer B). The absorption of the blue-green light by normal bilirubin (4Z, 15Z-bilirubin) generates configuration isomers, structural isomers, and photooxidation products. The colorless photoisomerization products produced are readily excreted through the urine.

27. Given an initial hematocrit of 30%, what will be the approximate final Hct after a 2-blood-volume exchange? (You are using reconstituted whole blood with a hematocrit of 50%).
A. 47%
B. 52%
C. 57%
D. 62%
E. 65%

Concept: The final Hct based on the reconstituted whole blood Hct, patient's initial Hct, and the percent of blood exchanged can be estimated as following:

$$\text{Final } Hct \approx \frac{(\%\text{ exchanged } x\, Hct_{\text{reconstituted unit}}) + (100 - \%\text{ exchanged}) \times Hct_{\text{patient}}}{100}$$

Where
$Hct_{\text{reconstituted unit}}$ is the Hct of the reconstituted whole blood donor unit
Hct_{patient} is the initial Hct of patient
% exchanged: In general, a one volume exchange, results in a % exchange of 63% while a two volume exchange results in a % exchange of 85%.
Of note: whole numbers (and not decimal numbers) should be used in the above formula.

Answer: *A*—This patient has a two blood-volume exchange; thus, the percent exchanged is 85%. Using the above equation, yields:

$$\text{Final } Hct \approx \frac{(85 \times 50\%) + (100 - 85) \times 30\%}{100} = 47\%$$

28. Which of the following is a benefit of whole blood exchange?
A. Creatinine removal
B. Removal of the mother's antiplatelet antibodies
C. Removal of fibrinogen
D. Improved of oxygen delivery due to decreased whole blood viscosity
E. Improved oxygen delivery due to shifting oxygen dissociation curve to the right

Concept: In the course of whole blood exchange for severe hyperbilirubinemia, the mother's anti-D antibody and the affected RBCs are removed, but other benefits exist, too.

Answer: *E*—An additional benefit of whole blood exchange is that oxygen delivery is increased due to increased hematocrit. This causes an increase (not decrease) in whole blood viscosity (Answer D). In addition, because the patient's RBC's containing hemoglobin F is replaced by homologous RBC's containing hemoglobin A, the patient's overall oxygen dissociation curve is shifted to the right, allowing more off-loading of oxygen at lower pO_2. High bilirubin (and not creatinine) is the major risk factor for kernicterus in HDFN (Answer A). Since plasma is used as part of the whole blood reconstitution, fibrinogen as well as other coagulation factors are given back to the patient during the exchange (Answer C). However, thrombocytopenia should always be monitored, since platelets are not a large component of reconstituted whole blood products. Antiplatelet antibodies (Answer B) is not part of the pathogenesis of HDFN.

End of Case

Please answer Questions 29–32 using the following clinical scenario

A 3.5-kg term newborn female was born to a G2P2 previously healthy Caucasian mother and Caucasian father. Delivery was normal spontaneous vaginal with APGARs of 7 and 9 at 1 and 5 min, respectively. The infant was O Rh positive with a negative antibody screen. Mother had normal titers for Rubella, was positive for IgG anti-CMV only, and negative rapid plasma regain (RPR) with no history of herpes infection. Her past medical history was also unremarkable except for prior a pregnancy in which the older sibling of this child, with the same father, was noted to have low platelet counts and petechiae at birth. On physical exam, the newborn infant in this case was also noted to have petechiae on her lower and upper extremities and back. No other physical findings were noted. Ultrasonography did not detect any hemangiomas, and X-rays of the radius and ulna did not detect any abnormalities.

29. Based on the clinical findings above, which diagnosis should be MOST suspected in the differential diagnosis of thrombocytopenia for this patient?
A. Neonatal alloimmune thrombocytopenia (NAIT)
B. Toxoplasmosis, Other agents, Rubella, CMV, Herpes Simplex (TORCH) syndrome
C. Fanconi's anemia
D. Thrombocytopenia absent radii (TAR)
E. Kasabach-Merritt syndrome

Concept: The cause of neonatal thrombocytopenia can be grouped into three broad categories as a differential diagnosis (Fig. 10.1): immune thrombocytopenia, peripheral consumption, and decreased production. Immune thrombocytopenia may be autoimmune, alloimmune, or drug-induced. Peripheral consumption can be due to various etiologies that increase thrombosis or sequestration (hypersplenism, DIC, vWD). Lastly, there may be decreased production due to congenital disorders, infiltrative processes, and infections.

Answer: *A*—Given the clinical associations above as well as the unremarkable results from X-ray and ultrasound, the most likely diagnosis is neonatal alloimmune thrombocytopenia (NAIT). All of the earlier mentioned answers can be associated with isolated thrombocytopenia. Additionally,

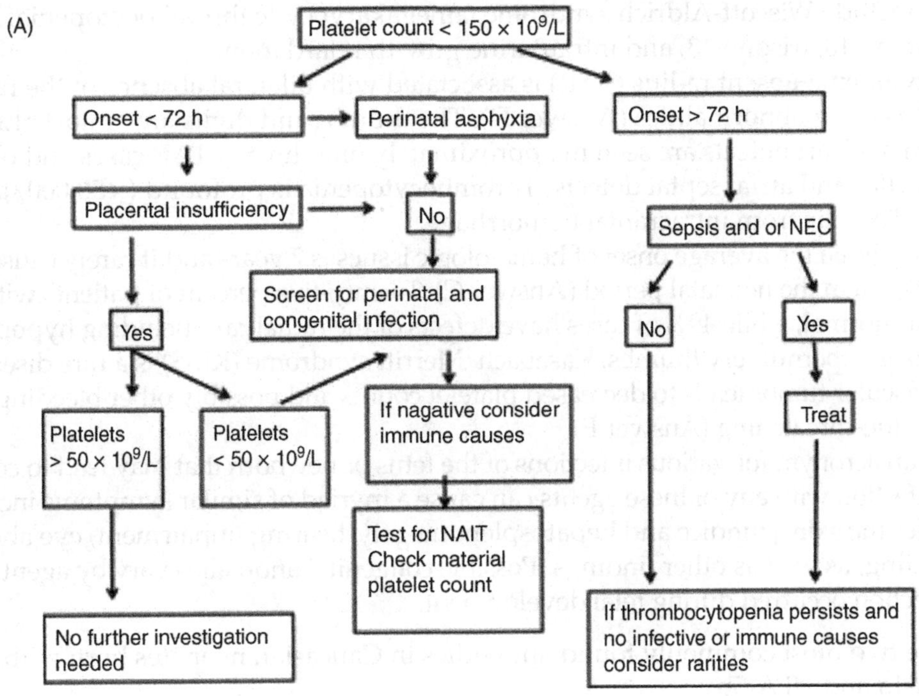

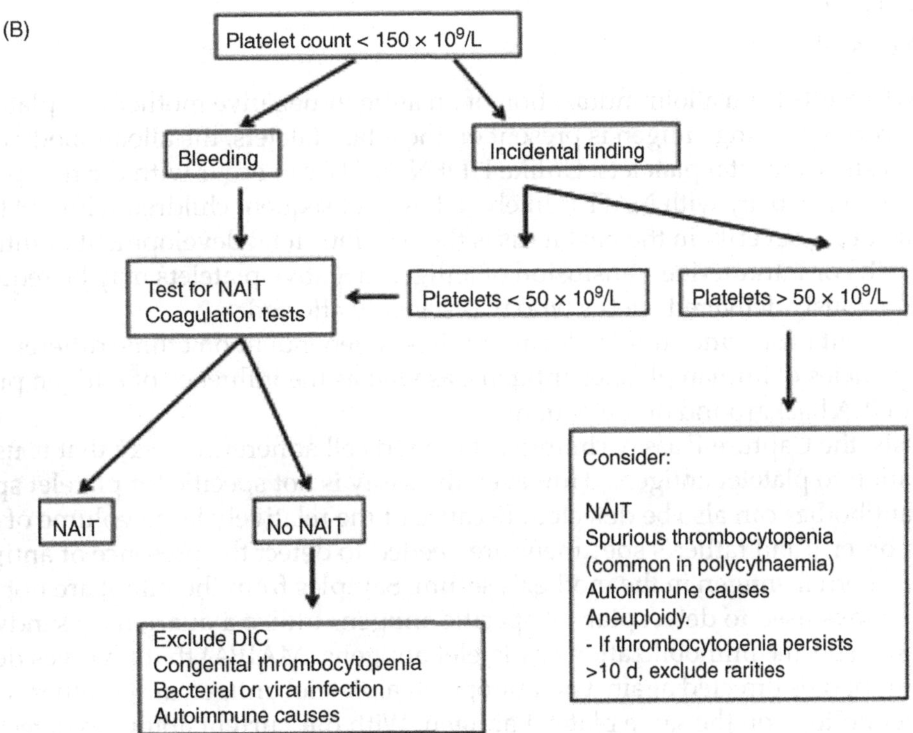

FIGURE 10.1 Diagnostic algorithm for approaching thrombocytopenia in (A) preterm and (B) term neonates. *Source: Adapted from S. Chakravorty, I. Roberts, Br. J. Haematol. 156 (2012) 155–162.*

other causes include Wiscott-Aldrich syndrome, amegakaryocytic thrombocytopenia, May-Hegglin anomaly, trisomy 18, trisomy13, and intrauterine growth retardation.

Thrombocytopenia-absent radius (TAR) is associated with bilateral absence of the radii in 100% of cases as well as ulnar abnormalities (Answer D). The thumbs and digits are almost always present in TAR patients. Heart defects are seen in approximately one-third of TAR cases and often include tetralogy of Fallot and atrial septal defects. Thrombocytopenia is profound (<50,000/μL) and 25% of patients with TAR die from intracranial hemorrhage.

In Faconi's anemia the average onset of hematologic issues is 7 years and it rarely causes isolated thrombocytopenia in the neonatal period (Answer C). Twenty-five percent of patients with Fanconi's anemia appear normal, while 49% of cases have defects of the radial ray, including hypoplasia or aplasia of the thumbs or supernumery thumbs. Kasabach–Merritt syndrome (KMS) is a rare disease where an underlying vascular tumor leads to decreased platelet counts and possibly other bleeding disorders, which may be life-threatening (Answer E).

TORCH is an acronym for various infections of the fetus or newborn that may lead to congenital anomalies. Infection with any of these agents can cause a myriad of similar symptoms including fever, bleeding under the skin, jaundice and hepatosplenomegaly, hearing impairment, eye abnormalities, difficulty feeding, as well as other findings. Possible congenital anomalies vary by agent and depend on when the infection occurred during fetal development.

30. Which are the two most commonly found antibodies in Caucasian neonates born with NAIT?
A. Anti-HPA-1a and HPA-5b
B. Anti-HPA-1a and HPA-3a
C. Anti-HPA-3a and HPA-5b
D. Anti-HLA class I
E. Anti-HLA class II

Concept: NAIT results from alloimmunization of an antigen-negative mother to a platelet-specific antigen. If the corresponding antigen is present on the fetal platelets, the alloantibody can cross the placenta and destroy the fetal platelets. Unlike HDFN, NAIT can occur with the first pregnancy. A mother who delivers a baby with NAIT is likely to have subsequent children with NAIT. Nonetheless, the only predictor for severity in the next fetus is the previous fetal development of intracranial hemorrhage. IVIG or intrauterine transfusion of antigen-negative platelets may be required if the woman has previously delivered a fetus who was severely affected.

The development of specific antiplatelet antibodies is dependent on ethnic differences in the antigenic frequencies of human platelet antigens, as well as the influence of antigen presentation in the context of HLA background of the patient.

For diagnosis, the Capture P assay (Immucor) is a red cell adherence assay that is used to detect alloimmunization to platelet antigens. However, the assay is not specific for platelet specific antigens, as anti-HLA antibodies can also be detected. Because of the relatively large volume of samples needed, the mother's and father's specimens are needed to detect the presence of antiplatelet antibody to a paternal antigen in the mother's serum. Samples from the infant are not needed.

The ELISA assays used to detect platelet specific antigens utilize a quartenary sandwich complex (i.e., monoclonal-specific immobilization of platelet antigens (MAIPA) that involves detecting mouse monoclonal antibodies directed against an epitope on a platelet antigen and mother's antibodies against another epitope on the same platelet antigen. With our current ability to detect a wide range of maternal antipaternal antigens, it is beginning to be appreciated that anti-HLA antibodies may be a cause of NAIT.

Answer: *A*—Anti-HPA-1a is the most commonly implicated antiplatelet antibody in NAIT in Caucasians. The second most common is anti-HPA-5b. However, in Asian populations, particularly the Japanese, the most common is anti-HPA-4b. The involvement of anti-HLA class I antibodies (Answer D) in the pathogenesis of NAIT is not fully elucidated; however, anti-HLA antibodies are not commonly found antibodies in neonates with NAIT. Platelets do not express HLA class II antigen (Answer E) on the surface. The remaining choices (Answers B and C) do not have the correct antibody combinations.

31. Several hours after birth, the patient was noted to have horizontal deviation of the eyes with tonic jerking of his extremities, eyelid blinking, and smacking mouth movements. A head ultrasound revealed intraventricular hemorrhage (IVH). A complete blood count (CBC) was ordered and showed the following:

White blood cell (WBC)	6,320/μL
Hgb	19.8 g/dL
Hct	58.3%
Plt	23,000/μL

What treatment, at a minimum, should be given to this patient at this time?

A. Platelets that are leukoreduced and ABO compatible
B. Platelets that are irradiated
C. Platelets that are volume-reduced
D. Platelets that are ABO compatible and less than 2 days old
E. Mother's platelets collected using apheresis technique

Concept: In order to treat suspected neonatal alloimmune thrombocytopenia (NAIT), antigen negative platelets should be given. However, finding such units may require time and thus, may delay treatment. If these are not available, then "stock or shelf" ABO compatible platelets should be immediately given as soon as possible, as studies have shown acceptable platelet increments to be possible in a large percentage of cases.

Answer: *A*—The quickest treatment in urgent situations (such as in this case when the patient has IVH) is to provide leukoreduced and ABO compatible platelets that are currently in the inventory. Although mother's platelets collected using apheresis are considered "ideal" (as they are negative for the antigen that the mother is producing antibodies against), this may take days to obtain and process (Answer E).

Irradiated, fresher (<2 days old) and/or volume reduced platelets do not need to be given (Answers B, C, D). Irradiated platelets are only needed in patients < 1200 g, or who have other risk factors for TA-GVHD. For example, if maternal platelets are transfused, then irradiation would be required because of the HLA similarities between the mother and newborn would put the newborn at risk of TA-GVHD. Additionally, the maternal platelets should be washed to remove the antibodies. Volume reduced platelets are only given to remove ABO incompatible plasma or to reduce the volume in patients who need to be volume restricted due to poor renal output or to avoid volume overload.

32. The mother would like to have another child in the future. Which of the following statements is true?

A. An effective antenatal treatment strategy is to give prophylactic IVIG
B. The degree of thrombocytopenia in the next child will be similar to the thrombocytopenia in the previous sibling who experienced gastrointestinal bleeding
C. This is an one-time event—there is no risk of NAIT in the future pregnancies
D. If there is a need for platelet transfusions, random donor platelets should be washed and irradiated
E. Treatment should not occur until there is evidence of intraventricular hemorrhage in the fetus

Concept: It is important that consultation with a perinatologist (obstetrician specializing in maternal-fetal medicine) be part of prenatal care of a mother with a history of a pregnancy that has been complicated by NAIT.

Answer: *A*—IVIG may be an effective treatment prenatally in mothers with a history previous children severely affect by NAIT. Unlike HDFN, NAIT can occur with the first pregnancy and severity may increase with each subsequent pregnancy (Answers B and C). However, having IVH in the previous baby is the only predictor for severity in the next fetus. Because intrauterine CNS hemorrhage can result from thrombocytopenia, antenatal treatment is often desired because of the risk of IVH in the current fetus (Answer E). Empiric use of IVIG can result in increased intrauterine platelet count sufficient to prevent intrauterine CNS hemorrhage. In the case that IVIG does not work, intrauterine platelet transfusions may be necessary, but these must be performed using antigen negative platelets that are leukoreduced (to reduce the risk of CMV infection) and irradiated (to prevent transfusion associated graft-vs-host disease) (Answer D).

End of Case

Acknowledgements

Edward Wong, MD is an employee of Quest Diagnostics Nichols Institute, Chantilly, VA.

Suggested Reading

[1] American College of Surgeons Trauma Quality Improvement Guidelines. Available from: https://www.facs.org/~/media/files/quality%20programs/trauma/tqip/massive%20transfusion%20in%20trauma%20guildelines.ashx.
[2] M.K. Fung, B.J. Grossman, C.D. Hillyer, C.M. Westhoff (Eds.), Technical Manual, eighteenth ed., AABB Press, Bethesda, MD, 2014.
[3] H.G. Klein, D.J. Anstee (Eds.), Mollison's blood transfusion in clinical medicine, twelfth ed., Wiley-Blackwell, Oxford, England, 2013.
[4] L.D. Petz, G. Garratty (Eds.), Immune hemolytic anemias, second ed., Churchill Livingstone, Philadelphia, PA, 2004.
[5] Standards for Blood Banks and Transfusion Services, thirtieth ed., AABB Press, Bethesda, MD, 2016.
[6] E.C.C. Wong, S.D. Roseff (Eds.), Pediatric Transfusion: A Physician's Handbook, fourth ed., AABB Press, Bethesda MD, 2015.
[7] E.C.C. Wong, W. Paul, Intrauterine, neonatal, and pediatric transfusion therapy, in: P.D. Mintz (Ed.), Transfusion therapy: clinical principles and practice,, third ed., AABB Press, Bethesda, MD, 2011, pp. 209–263.
[8] E.C.C. Wong, S.D. Roseff (Eds.), Pediatric Hemotherapy Data Card, fourth ed., AABB Press, Bethesda MD, 2015.
[9] G.P. Gravlee, R.F. Davis, M. Kurusz, J.R. Utley (Eds.), Cardiopulmonary bypass in infants and children, in: Cardiopulmonary Bypass Principles and Practice, second ed., Lippincott Williams & Wilkins, Philadelphia, PA, 2000 (Chapter 30).
[10] M.K. Fung, B.J. Grossman, C.D. Hillyer, Westhoff CM (Eds.), Technical Manual, eighteenth ed., AABB Press, Bethesda MD, 2014, pp. 590.
[11] R.M. Fasano, G.S. Booth, M. Miles, L. Du, T. Koyama, E.R. Meier, N.L. Luban, Red blood cell alloimmunization is influenced by recipient inflammatory state at time of transfusion in patients with sickle cell disease, Br. J. Haematol. 168 (2) (2015) 291–300.
[12] M.J. Maisel, A.D. McDonagh, Phototherapy for neonatal jaundice, N. Engl. J. Med. 358 (2008) 920–928.
[13] H.P. Pham, B.H. Shaz, Update on massive transfusion, Br. J. Anaesth. 11 (2013) i71–i82.
[14] S.T. Chou, et al. High prevalence of red blood cell alloimmunization in sickle cell disease despite transfusion from Rh-matched minority donors, Blood 122 (2013) 1062–1071.
[15] F. Peyvandi, et al. A Randomized Trial of Factor VIII and Neutralizing Antibodies in Hemophilia A, N. Engl. J. Med. 374 (2016) 2054–2064.

CHAPTER

11

Infectious Complications of Blood Transfusion

Evan M. Bloch, Cyril Jacquot**, Beth H. Shaz*[†]

*Johns Hopkins University, School of Medicine, Baltimore, MD, United States;
**Children's National Health System and George Washington University, Washington, DC, United States; [†]New York Blood Center, New York, NY, United States

Consequent to refined donor selection and laboratory screening, blood transfusion in the United States is remarkably safe. Nonetheless, emerging and reemerging pathogens may pose a risk to the blood supply, underscoring the need for the ongoing vigilance and development of innovative strategies to mitigate infectious risks. This chapter intends to provide a broad overview of the infectious complications of blood transfusion. This includes a description of the major transfusion-transmitted pathogens, associated preventative strategies in use (donor risk–based deferral and donor and product laboratory testing), and practical considerations surrounding how one responds to a suspected and/or confirmed transfusion-transmitted infections (i.e., quarantine, look-back, and notification).

1. Bacterial detection is performed on which of the following blood products?
   A. Granulocytes
   B. Red blood cells
   C. Plasma
   D. Cryoprecipitate
   E. Platelets

   Concept: Aseptic collection technique(s) and the storage conditions play an important role in bacterial growth in blood products. For instance, since platelets are stored at room temperature (with agitation) for up to 5–7 days, the risk of bacterial growth and potential bacterial sepsis after infusion increases with each day of storage.

   Answer: *E*—Platelets have the highest risk of bacterial contamination and concomitant risk of septic transfusion reactions because they are stored up to 5 days at room temperature (FDA will allow for 7 days with additional bacterial detection measures, such as point of release testing). Granulocytes (Answer A) are also stored at room temperature, but expire in 24 h; therefore, the chance of bacterial growth is minimal. Red blood cells (answer B) are refrigerated (storage at 1–6°C); plasma and cryoprecipitate (answer C and D, respectively) are stored frozen (–18°C).Thus, other products have lower bacterial growth and risk of septic transfusion reactions.

2. You are interviewing a new patient in your practice. This 47-year-old man reports a history of hepatitis B virus (HBV) 3 years prior. He is currently asymptomatic with no signs of liver disease. Which of the following combinations of results are consistent with resolution of his infection?
   A. HBV DNA negative, HBsAg negative, anti-HBs positive, total anti-HBc negative
   B. HBV DNA positive, HBsAg positive, anti-HBs negative, total anti-HBc positive (IgM positive)

Transfusion Medicine, Apheresis, and Hemostasis. http://dx.doi.org/10.1016/B978-0-12-803999-1.00011-0

C. HBV DNA positive, HBsAg positive, anti-HBs negative, total anti-HBc positive (IgM negative)
D. HBV DNA negative, HBsAg negative, anti-HBs positive, total anti-HBc positive
E. HBV DNA positive, HBsAg negative, anti-HBs negative, total anti-HBc negative

Concept: Both serology and nucleic acid testing (NAT) are available to evaluate HBV status in blood donors. The multitude of tests reflects the complexity and persistence of viral infection. Interpretation of results can be challenging and, in some cases, inconclusive. Currently, HBV screening of US blood donations consists of HBsAg (hepatitis B surface antigen), anti-HBc (hepatitis B core antibody) and HBV DNA testing.

Answer: *D*—HBV DNA negative, HBsAg negative, anti-HBs positive, total anti-HBc positive. The following table (Table 11.1) provides a guide to interpretation of HBV test results.

Notes: HBeAg may be used as a marker of infectivity given that its presence reflects active viral replication. HBcAg is the core form of HBeAg but unlike HBeAg, it does not circulate in the blood. Importantly, precore variants do not express HBeAg, thus potentially providing false reassurance if negative.

Acquisition of anti-HBs generally denotes immunity and disease resolution. In cases where interpretation is unclear, repeat interval testing may provide greater clinical insight. The other choices (Answers A, B, C, and E) are incorrect based on Table 11.1.

TABLE 11.1 Guide to interpretation of HBV test results

Condition	HBV DNA	HBsAg	Anti-HBs	Total anti-HBc	IgM anti-HBc
HBV vaccination (answer A)	Neg	Neg	Pos	Neg	Neg
Acute infection (answer B)	Pos	Pos	Neg	Pos	Pos
Chronic infection	Pos	Pos	Neg	Neg	Neg
Resolved infection (answer D)	Neg	Neg	Pos	Pos	Neg
1. "Window period" early infection (answer E) 2. False positive NAT	Pos	Neg	Neg	Neg	Neg
1. Resolved or resolving infection 2. False-positive anti-HBc	Neg	Neg	Neg	Pos	Pos or Neg
3. Occult infection	Pos	Neg	Neg	Pos	Pos or Neg

Neg, negative; Pos, positive.

3. In preparation for an elective knee replacement in an otherwise healthy 57-year-old man, an orthopedic surgeon asks you about the risk of transfusion-transmitted cytomegalovirus (TT-CMV). Which of the following blood product modifications is used to produce "CMV safe blood," which has reduced risk of TT-CMV?
   A. Leukoreduction
   B. Cell washing
   C. Irradiation
   D. Volume reduction
   E. Use of CPDA-1 anticoagulant

Concept: There is a high prevalence of CMV in both the general and blood donor populations (~50% of people are seropositive and this may be even higher in urban areas). It is not associated with serious morbidity for the majority of patients. Further, CMV is not a required test on donors based on FDA regulation. Although CMV is transfusion transmissible, seropositivity does not result in donor deferral (unlike other pathogens). Many seropositive recipients receive CMV untested blood without ill effect. However, a subset of patients who are CMV seronegative and immunocompromised (e.g., neonates, pregnancy, transplant patients) are at risk of severe TT-CMV; therefore they may benefit from receiving blood that is low risk for CMV, either through transfusion of blood from CMV seronegative donors or leukoreduction.

Answer: *A*—In donors who have been infected with CMV, the virus remains latent in leukocytes. Thus, leukoreduction ($<5 \times 10^6$ residual leukocytes) reduces the risk of TT-CMV. Transfusion of leukoreduced (CMV-safe) blood has been shown to have a similar risk profile to that of transfusion of CMV-seronegative units, both with a risk of approximately 1%–3%. Alternatively, pathogen inactivation can mitigate TT-CMV.

In regard to the other responses: cell washing (answer B) and volume reduction (Answer D) reduce the amount of plasma in an effort to reduce allergic reactions. Cell washing also lowers the potassium content in the unit. Irradiation (answer C) is indicated specifically to prevent transfusion-associated graft versus host disease. CPDA-1 (answer E) has no effect on infectious risk: CPDA-1 is an alternate red cell storage solution, which results in a shorter RBC shelf-life (35 days), more plasma, and higher potassium during storage.

4. Which of the following pathogens has the highest risk of transfusion transmission?
A. HIV
B. HTLV
C. Staphylococcus epidermidis
D. HBV
E. HCV

Concept: Both viral and bacterial agents may contaminate and be transmitted by blood product transfusion. With donor questionnaires and advanced testing methodologies, the risk of many viral infections, such as HIV, has dropped dramatically over the years. Bacterial contamination is more common than that with any of the viruses aforementioned, though bacterial contamination and subsequent infusion may not lead to any clinically significant effects.

Answer: *C*—Despite disproportionate focus on transfusion transmitted viruses, bacteria pose the highest exposure risk to transfusion recipients. Platelets are most commonly associated with bacterial contamination and septic transfusion reactions given that they do not undergo refrigerated storage. Skin floras (notably coagulase negative staphylococci) are most commonly implicated following introduction at time of collection. From the American Red Cross, the residual risk of posttransfusion sepsis following platelet transfusion is 9.4 per million (1:106 931). Multiple steps have been implemented to decrease the risk of bacterial contamination and related sepsis (most recent FDA reportable fatalities from 2014 demonstrated no bacteria related fatalities from apheresis platelets). The risks of the other named pathogens are comparatively low and estimated to be: 1:1,467,000 for HIV (Answer A), 1:1,148,000 for HCV (Answer E), 1:400,000 for HBV (Answer D), and 1:2,700,000 for HTLV (Answer B) products transfused.

5. Which of the following is true about syphilis screening and transfusion associated risk?
A. Screening is typically performed using a sensitive nontreponemal test (e.g., RPR or VDRL) followed by confirmatory testing with a highly specific treponemal test (e.g., TP-Microhemagglutin assay)
B. A donor who tests positive for syphilis (i.e., both screening and confirmatory testing are reactive), is not permanently deferred from blood donation
C. *Treponema pallidum*, the causative agent of syphilis has a high tolerance for refrigeration, processing and storage
D. The risk of transfusion transmitted syphilis in the United States remains high, despite routine laboratory based screening

E. Blood screening tests can distinguish between syphilis and other pathogenic treponemes, such as *T. pallidum* subsp. pertenue, endemicum, and carateum, the causative agents of yaws, bejel, and pinta, respectively.

Concept: Transfusion transmitted syphilis is extraordinarily rare in the United States; the last documented case was reported over 45 years ago. Reasons for the low incidence include a waning reservoir of syphilis in the United States, robust testing and an infectious agent that does not withstand routine blood banking conditions. Many have even questioned the value (cost-utility) of continued screening.

Answer: *B*—A donor positive for syphilis is deferred for 12 months after successful treatment. Despite being transfusion transmissible, *T. pallidum*, the causative agent of syphilis, is intolerant of refrigeration, processing, and cold storage (Answer C). The majority of laboratories that conduct transfusion screening in the United States, employ an inverse algorithm to that used in routine clinical practice.

Specifically, in clinical practice (i.e., patient testing) a sensitive nontreponemal test is used for primary screening followed by confirmation with a treponemal-specific test. In contrast, blood-donor screening uses the first line specific (treponemal test; e.g., MHA) test so as to avoid high numbers of false positives. A specific test is also used for confirmatory testing (e.g., IFA). A nonspecific test (e.g., RPR) may be used for the purpose of donor counseling and management (Answer A). A positive test result during transfusion screening prompts temporary (12 months)—rather than—permanent deferral from blood donation. New FDA guidance does not require retesting if the donor has been successfully treated, (i.e., the donor is eligible to donate after 12 months). The other choices (Answers D and E) are incorrect based on the previous information.

6. An asymptomatic, seemingly healthy adult presents to donate blood for the first time. A history of which of the following tick-borne disease renders them ineligible to donate blood?
A. *Anaplasma phagocytophilum*
B. *Borrelia burgdorferi*
C. *Babesia duncani*
D. *Ricketsia ricketssiae*
E. *Borrelia miyamotoi*

Concept: A history of Babesiosis confers indefinite deferral from blood donation. The rationale for this stems from the following: Babesia is readily transfusion transmissible, it has the ability to establish asymptomatic, persistent infection in immunocompetent adults and importantly there is currently no licensed testing of blood donors (although both serological and molecular assay development is underway). Studies suggest that Babesia can persist for up to 27 months following infection. Because many individuals are asymptomatic and infection persists for long periods (months to years), there is risk of donation by parasitemic donors, with subsequent risk of transfusion transmission.

Answer: *C*—Donor eligibility is based on a history of Babesiosis and does not discriminate between species. *Babesia duncani*, which is thought to be endemic in California and the Pacific Northwest, has only rarely been implicated in transfusion-transmitted babesiosis (as well as tick borne transmission). In regards to the other options, only Anaplasma (the causative agent of human granulocytic anaplasmosis) has been associated with transfusion although, currently there is no deferral policy specific to anaplasmosis (Answer A). None of the other named agents, *B. burgdorferi* (Lyme Disease) (answer B), *Ricketsia ricketssiae* (Rocky Mountain Spotted Fever) (Answer D) and *Borrelia miyamotoi* (*Borrelia miyamotoi* disease) (Answer E) have been shown to result in transfusion-transmitted infection. Of interest, A. *phagocytophila*, B. *burgdorferi* and B. *miyamotoi* share the same tick vector (*Ixodes scapularis*) as *Babesia microti*, the major cause of tick-borne and transfusion transmitted babesiosis (Fig. 11.1).

FIGURE 11.1 ***Questing Ixodid ticks*: (male—top and female—bottom) waiting on a blade of grass for a host to brush against them.** While the above are *I. pacificus* ticks, which were obtained in California (vectors for Lyme disease and anaplasmosis), the ticks are morphologically similar to *I. scapularis*, the vector for *B. microti*. *Source: Image courtesy of Anne Kjemtrup, California Department of Public Health.*

7. A 58-year-old Asian woman presents with fatigue and intermittent fevers for several weeks. She had previously been in good health. Past surgical history is notable for a cesarean section 30 years ago in Japan during which she required transfusion of two red blood cell units. On physical examination, she has hepatosplenomegaly and diffuse lymphadenopathy. Laboratory testing includes white blood cell count 280 × 10^9/L, hemoglobin 6.7 g/dL, platelets 140 × 10^9/L, calcium 12.8 mg/dL, and creatinine 1.2 mg/dL. Representative images of the peripheral blood are shown in Fig. 11.2A–B. Flow cytometry reveals the cells to be positive for CD2, CD3, CD4, CD5, CD25, and negative for CD7, CD8, CD19, CD20. T-cell receptor rearrangement studies are clonal.

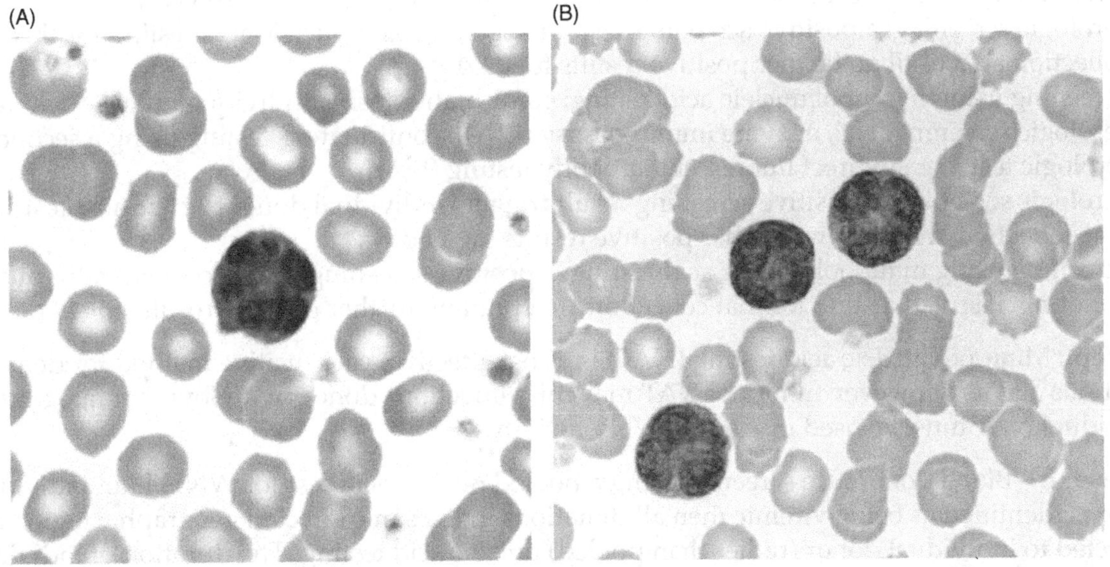

FIGURE 11.2 Photomicrographs A and B. *Source: Images courtesy of Dr. Yen-Chun Liu, Weill-Cornell Medical Center.*

Which agent is most likely implicated in this patient's disease?
A. HTLV
B. EBV
C. HIV
D. CMV
E. HPV

Concept: Several viruses have oncogenic properties and some may affect hematopoietic progenitor cells. The manifestations of these effects may be delayed by many years. These pathogens may be transmitted by blood transfusion. Allogeneic donor blood is screened for a subset of these viruses (HIV and HTLV) to reduce the risk of transmission.

Answer: *A*—Human T-cell lymphoma or lymphotropic virus (HTLV) is associated with development of adult T-cell lymphoma/leukemia (ATLL) as well as HTLV-associated myelopathy/tropical spastic paraparesis (HAM/TSP), a progressive neurological disease. HTLV is endemic in Southwestern Japan, the Caribbean and parts of Central Africa. Transmission by transfusion has been reported. Consequently, donor screening is undertaken in the United States using serological testing. The residual risk of HTLV transmission is estimated ~1:2,700,000. The infection may remain latent for many years (40–50 years in some cases) and the lifetime risk of developing ATLL in carriers is 3%–5%.

Patients with ATLL typically present with markedly elevated white blood cell counts, skin rash, lymphadenopathy, hypercalcemia and occasionally lytic bone lesions. The white blood cells often have characteristic distorted nuclei with "flower"-shaped morphology (Fig. 11.2). On flow cytometry, the tumor cells express T-cell-associated antigens (CD2, CD3, CD5), but usually lack CD7. CD25 is strongly positive in nearly all cases. Although most cases are CD4-positive/CD8-negative, CD4-negative/CD8-positive, or CD4/CD8 coexpression may also occur.

The other viruses listed are associated with some cancers. EBV (Answer B) is associated with B-cell neoplasms, such as Burkitt lymphoma and Hodgkin lymphoma. EBV can also cause B-cell lymphoproliferative disorders in immunodeficient patients (HIV, transplant recipients). However, EBV donor screening is not performed given the high prevalence in the general population (approximately 90% in US adults). CMV (Answer D) has been linked to epidermoid carcinoma of the salivary gland. HIV (Answer C) is associated with Kaposi's sarcoma, in part due to reduced immunosurveillance. HPV (Answer E) can cause cancers of the cervix, vulva, vagina, penis, and anus.

8. Which of the following best describes blood donor testing for West Nile Virus (WNV)?
A. Serologic screening; a positive screening result triggers pooled nucleic acid testing for that collection area until no further positive results for 14 days
B. Screening using minipool nucleic acid testing; confirmation using individual nucleic acid testing
C. Serologic screening [e.g., enzyme immunoassay (EIA)]; confirmatory testing using a second serologic test [e.g., indirect fluorescent antibody testing (IFA)]
D. Serologic screening; a positive screening result triggers individual donor nucleic acid testing for that collection area until no further positive results for 14 days
E. Screening using minipool nucleic acid testing; a positive screening result triggers individual donor nucleic acid testing for that collection area until no further positive results for 14 days

Concept: Minipool nucleic acid testing (NAT) is a cost effective, yet sensitive method for detecting infectious agents. However, minipool NAT may miss infectious donors with low viremia, thus individual NAT must be used in areas of WNV activity.

Answer: *E*—Blood donors are screened using a pooled nucleic acid test for WNV. If an individual donor is identified as being viremic then all donations in the same affected geographic region are subjected to individual donor (rather than pooled) nucleic acid testing. The duration of individual donor testing is informed by regional WNV activity. AABB maintains a listserv and a website (http://www.aabb.org/research/hemovigilance/Pages/wnv.aspx) identifying all presumed and confirmed viremic donations. For some blood centers, the region is typically defined as zip code and surrounding zip codes, while other blood centers use counties depending on the size of the region. Detriggering back to pooled testing occurs with no further activity for 14 days. The other choices (Answers A, B, C, and D) are incorrect based on standard protocols for WNV testing, as described previously.

9. Which methods are FDA approved to reduce bacterial contamination of platelet products?
A. Bacterial culture, gram stain
B. Bacterial culture, point of release testing
C. Point of release testing, gram stain
D. Bacteria culture, pH testing
E. Gram stain, pH testing

Concept: Historically, blood bankers would use imprecise methods to detect bacterial contamination of platelets, such as swirling the platelet bag and checking the pH level for changes. However, these methods are no longer considered acceptable to protect patients from contaminated platelets.

Answer: *B*—The FDA has approved a number of testing methods for detecting the presence of bacteria. These include culture based methods (bacT/ALERT, bioMerieux; eBDS Pall Corp), and point of release assays (Platelet Pan Genera Detection, Verax Biomedical; BacTx, Immunetics). Other countries have implemented pathogen inactivation technologies in place of bacterial culture methods. In addition to culture and testing, important steps to prevent bacterial contamination include use of an antiseptic arm scrub and a diversion pouch at the time of collection. The latter diverts the skin plug, a source of bacterial skin flora, into a separate pouch rather than into the collection bag. This diverted blood is used for donor testing, such as viral markers. The other choices (Answers A, C, D, and E) represent older methods of testing (e.g., Gram stain, and pH), which are no longer considered sufficient by the FDA.

10. Which of the following statements is *INCORRECT* in regards to a "lookback" procedure?
A. Lookback includes notification of consignees to quarantine in-date products from donors whose subsequent collection have screening reactive results
B. Allows for release of quarantined components if NAT or antibody confirmation is negative
C. It is performed in order to notify prior recipients of blood components that have been collected from donors who subsequently test confirmation positive for HIV
D. It is used to identify products that were collected and subsequently transfused when the donor may have been in the window period
E. It involves notification of the recipient with positive screening result prior to confirmation

Concept: The term "lookback" is used to refer to investigation of prior donations to determine risk of transfusion-transmitted infection.

Answer: *E*—A lookback is required when a blood component from a repeat donor tests repeatedly reactive by antibody or NAT for HIV and HCV. Lookbacks comprise the following components:
1) notification of consignees to quarantine in-date products,
2) further testing the donor using a current sample with notification of consignee of the test results,
3) release of units for transfusion if the subsequent result was NAT or antibody confirmation negative,
4) destruction of potentially infectious prior collections, and
5) notification of recipients or the recipient's physician of record if the subsequent donation tested NAT or confirmed antibody positive.

The other choices (Answers A, B, C, and D) are incorrect based on the definitions and explanation above.

Please answer Questions 11–14 based on the following clinical scenario:

A 46 year-old male from Scottsdale, Arizona was recently (5 weeks ago) transfused (5 units of RBCs, 4 units of plasma, 1 apheresis platelet and 1 pool of cryoprecipitate) following a motor vehicle accident. His medical history is significant for another motor vehicle accident 12 years prior, which had required emergency splenectomy following traumatic rupture. The patient is a smoker (20 pack year history). Over the course of the past month, the patient has grown increasingly tired and lethargic. During investigation of anemia, the following organism is noted on a Giemsa-stained peripheral blood smear examination (Fig. 11.3), which is shown to have been transfusion transmitted based on ancillary testing. Neither the donor (native to a town 45 min from Sacramento, California) nor the patient has a history of travel outside the United States.

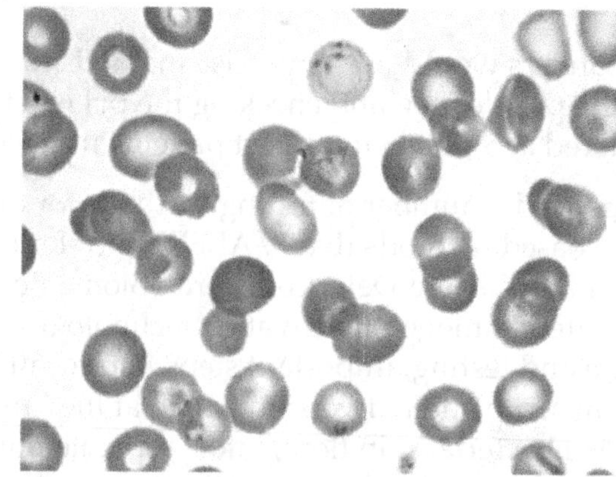

FIGURE 11.3 Oil immersion, 1000× magnification. *Source: Image courtesy of Anne Kjemtrup, California Department of Public Health.*

11. What is the organism that is depicted in the attached photomicrograph? (Fig. 11.3)
A. *Ehrlichia chaffeensis*
B. *Plasmodium falciparum*
C. *Babesia duncani*
D. *Trypanosoma cruzi*
E. *Plasmodium ovale*

Concept: Fig. 11.3 contains *Babesia duncani*-infected human Giemsa-stained red blood cells. Typical forms visible: dividing tetrad (top), ameboid form (middle), ring forms (bottom). *Babesia duncani*, unlike *B. microti*, has only rarely been implicated in human infection but both tick-borne and transfusion-transmission have been reported. Although the vector has not been identified, *B. duncani* is thought to be endemic in California and the Pacific Northwest. An important practical note: Maltese Cross/tetrad forms are considered pathognomonic of babesia infections but are very rarely identified in peripheral smears. The absence of tetrad forms does not exclude diagnosis. This diagnostic pitfall has accounted for multiple cases of delayed and missed diagnosis of babesiosis (i.e., misdiagnosis as malaria).

Answer: C—While virtually indistinguishable from *Plasmodium* spp. (malaria), microscopic features that may help to distinguish between malaria (Answer B and E) and babesia on peripheral blood smear examination include the presence of banana-shaped gametocytes (*P. falciparum*) and the pigment in malaria and the typical absence of extracellular forms in Babesiosis. Again, this is notoriously difficult and should be complemented by clinical history (e.g., travel) and laboratory diagnosis (serology ± PCR), both of which supported the diagnosis of babesiosis rather than malaria in this patient. *T.cruzi* (Answer D) and *E. chaffeensis* (Answer A) infection have trypomastigotes and characteristic cytoplasmic morula in peripheral blood leukocytes, respectively.

12. Which of the following factors may have contributed to the patient's risk of developing severe infection?

A. Age
B. Sex
C. Smoking
D. Contact with gasoline
E. Asplenia

Concept: Neonates, individuals >50 years old, the immunocompromised (cancer, HIV, immunosuppressant therapy) and/or those with apslenia are at high risk for severe Babesiosis. These subsets are notably over-represented among transfusion recipients.

Answer: *E*—The patient is asplenic, which rendered him susceptible to severe babesiosis. Age (answer A) is another risk factor for severe infection yet in this case applies to older patients (>50 years) given probable age-related decline in cellular immunity. None of the other responses (Answers B, C, and D) are known to be associated with severe babesia infection. Given that a high proportion of transfusion recipients are at risk of severe, complicated babesiosis, a strategy of maintaining a limited inventory of babesia-tested blood for use only in vulnerable patients is not viable (i.e., most recipients are "at risk").

13. Which of the blood products is most likely to be implicated with this transfusion transmitted infection?

A. Cryoprecipitate
B. Plasma
C. RBCs
D. Apheresis platelets
E. All of the above have been shown to transmit this infection efficiently

Concept: Babesia is a RBC based intraerythrocytic protozoan parasite and can be transmitted via any RBC containing blood product (e.g., packed RBC units and whole blood derived platelets).

Answer: *C*—Transfusion transmitted babesiosis (TTB) has not been reported following transfusion of plasma, cryoprecipitate or apheresis platelets (Answer A, B, and D). Most frequently it has been reported from RBC products and has been reported only rarely following whole blood derived platelet transfusion.

14. Which of the following blood banking procedures, has been shown to reduce transfusion-associated transmission of Babesia?

A. Leukoreduction
B. Irradiation
C. Refrigeration
D. Cell washing
E. None of the above

Concept: Babesia is very tolerant of standard storage and blood banking processing, as evidenced by multiple cases of TTB despite standard blood banking procedures.

Answer: *E*—All of the above procedures (i.e., leukoreduction, irradiation, refrigeration, and cell washing) (Answers A, B, C, and D) are ineffective at preventing TTB. TTB has even been reported following deglyerolization of frozen-stored red cells. Studies suggest that selected pathogen inactivation technologies (photochemical inactivation) are effective against babesia; however, none are -as yet- licensed for use in red blood cells or whole blood.

End of Case

15. For which of the following group of agents is nucleic acid testing used in blood donor screening?
   A. HTLV, WNV, HIV
   B. WNV, HCV, HBV
   C. HBV, HTLV, HIV
   D. HIV, HCV, HTLV
   E. HCV, HTLV, HBV

   Concept: Serological and/or nucleic acid testing methodologies are utilized to screen blood donors for infectious diseases.

   Answer: *B*—HTLV is screened using an enzyme-linked immunosorbent assay (ELISA). HIV and HCV are screened using NAT and antibody detection. HBV is screened using HBsAg, anti-HBc, and HBV NAT. WNV is screened using NAT. HCV, HBV, and HIV are tested together in a minipool using a multiplex assay. The other choices (Answers A, C, D, and E) are incorrect based on the information previously mentioned.

16. Which of the following donors may be eligible for reentry into the donor pool after being deferred for an infectious disease or a reactive screening test?
   A. Donor with history of treated malaria 2 years ago
   B. Donor with history of treated babesiosis 2 years ago
   C. Donor with multiple positive anti-HTLV tests
   D. Donor with positive HIV serology and confirmatory Western blot
   E. Donor with history of treated gonorrhea 14 months ago

   Concept: Both the donor health questionnaire and infectious disease screening tests are designed to defer donors with established infection or at risk of transfusion transmissible infection. In some cases, donors may be reinstated if sufficient time has elapsed after infection or if follow-up testing is negative (i.e., reactive test was a false positive). The policies differ widely between infectious agents as to whether reentry is allowable and how it is executed.

   Answer: *E*—A donor who has been diagnosed with gonorrhea is deferred for 12 months after completion of therapy. Gonorrhea is readily treatable and patients who undergo successful treatment are not considered to pose a risk to transfusion recipients. In regards to the other responses, donors with a history of babesiosis are indefinitely deferred from blood donation (Answer B), donors with a history of malaria who remain symptom free for 3 years while residing in nonendemic country may donate (Answer A) (http://www.fda.gov/BiologicsBloodVaccines/GuidanceComplianceRegulatoryInformation/Guidances/Blood/ucm365191.htm). Donors with repeat reactive HTLV (Answer C) serology and/or those with positive HIV (Answer D) serology and confirmatory Western blot are permanently deferred from donation given that both are chronically infected with transfusion transmissible pathogens. Of note, per AABB BBTS Standards, permanent deferral is defined as a deferral applied to a donor who will never be eligible to donate blood for someone else; in contrast indefinite deferral is defined as a deferral applied to a donor who is not eligible to donate blood for someone else for an unspecified period of time.

17. Which of the following combinations best describes current donor testing and deferral policy following a positive test result for *T. cruzi*?
   A. Chemiluminescent immunoassay screening; confirmation with alternative EIA; permanent deferral
   B. Individual donor nucleic acid primary screening/discriminatory NAT for confirmation; permanent deferral
   C. Screening with peripheral blood smear examination/radioimmunoprecipitation assay (RIPA) confirmation; 12-months deferral

D. Minipool nucleic acid primary screening/discriminatory NAT for confirmation; permanent deferral
E. Indirect fluorescent antibody (IFA) screening; manual peripheral smear examination for confirmation; 12-month deferral

Concept: Blood donor screening for *T. cruzi*, the causative agent for Chagas disease (American trypanosomiasis) is conducted using a serological assay chemiluminescent immunoassay that uses an epimastigote lysate to detect antibodies against *T. cruzi*. If the screening result is repeated reactive, confirmatory testing is conducted using an alternative serological test, for example, EIA. The RIPA was previously used for confirmation, but is no longer available.

Answer: *A—T. cruzi* testing was initiated in the United States in 2007 following historical reports of high transmissibility (12%–20%). A lack of evidence for incident or autochthonous transmission of *T. cruzi* led to a final recommendation of one time donor testing (i.e., donors are only tested once rather than at every donation). Testing is serology based; there is no NAT testing in use for transfusion screening (Answers B and D). *T. cruzi* carrier status may be life-long with those cases of transfusion transmission having occurred up to 40 years following departure from a *T. cruzi* endemic area; therefore, a seropositive result prompts mandatory permanent deferral from blood donation. The remaining choices (Answers C and E) do not contain tests used for *T. cruzi* screening.

18. Which of the following combinations of initiatives has successfully decreased the risk of HIV transmission via blood transfusion?
A. Donor education material, asking "are you feeling well", and advances in HIV donor testing
B. Donor education material, donor-risk behavior questions, and advances in HIV donor testing
C. Remuneration of blood donors, donor-risk behavior questions, and advances in HIV donor testing
D. Donor education material, donor-risk behavior questions, and advances in syphilis donor testing
E. Public health education materials, donor-risk behavior questions, and advances in HIV donor testing

Concept: The HIV epidemic in the early 1980's prompted the implementation of a more rigorous and multifaceted approach to decreasing TTI's.

Answer: *B*—The use of donor education material, specific deferral questions (e.g., IV drug use, male having sex with male, and commercial sex worker), and advances in HIV donor testing (e.g., HIV antibody assays, the previous use of p24 antigen assays, and subsequent NAT) have reduced the risk of HIV transmission from blood transfusion from about 1 in 2500 to currently about 1 in 1.47 million transfusions. In regards to some of the other responses, remunerated (i.e., paid or compensated) community blood donation was stopped several decades ago in the United States given the associated high risk of TTIs (Answer C). The question "are you feeling well enough to donate blood today" is a crude screen, and is ineffective for chronic asymptomatic infections, for example, HIV, HBV, HCV, Babesia, and so on (Answer A). Syphilis testing has—in the past been used as surrogate measure for high-risk behavior and TTIs (notably HIV). This is highly questionable given the low rates of HIV and syphilis coinfection in blood donors (Answer D). Finally, public health in general has been instrumental in reducing HIV incidence but is not directly targeted at blood donors (Answer E).

19. Which donor is currently eligible to donate?
A. 35-year old man who had symptomatic hepatitis at age 16, which subsequently resolved. He is unsure of the etiology but was evaluated by a physician at that time. He denies any recurrence and is now asymptomatic

B. 24-year old female college student who previously shared a house for 2 years with a friend with active hepatitis B. She moved out to her own apartment 9 months ago
C. 32-year-old male phlebotomist who received hepatitis B immune globulin 5 months ago following needle stick. Subsequent patient testing for HIV, HBV and HCV was negative
D. 59-year-old woman with history of hepatitis in the setting of ascending cholangitis. She completed treatment 7 months ago and is currently asymptomatic
E. 40-year-old man who reports one episode of IV drug use 18 years ago. He regularly sees his primary care physician and has no health issues

Concept: Throughout clinical medicine, there is a tendency to rely on laboratory data for diagnosis while underestimating the importance of the history and physical examination. Careful donor selection (i.e., history and examination) lessens reliance on laboratory screening to prevent transfusion-transmitted infection. In this regard, the donor health questionnaire serves to identify donors with a low risk profile for infectious diseases (e.g., hepatitis).

Answer: *D*—If donors report a history of hepatitis, further information should be obtained to determine the etiology and specifically whether a diagnosis of hepatitis B virus and/or C virus (HCV) has ever been established. The donor should be deferred during this investigation because the type of hepatitis is unclear (Answer A). Several of the questions on the donor health questionnaire are intended to gauge risk of viral hepatitis and carry a 1-year deferral if reported by the donor (e.g., close contact with a person who has active hepatitis, needle stick injury, and receipt of hepatitis B immune globulin administration) (Answers B and C). Donors with a history of nonviral hepatitis (e.g., gallstones, cholangitis) are eligible to donate if they have completed treatment and are feeling healthy (Answer D). FDA regulations changed from deferral of a donor with a history of viral hepatitis after the age of 11, to deferring donors who have a history of a relevant transfusion-transmitted infection, like hepatitis B or hepatitis C. Any use of IV drugs that have not been prescribed by a physician result in indefinite deferral from blood donation (Answer E) (https://www.federalregister.gov/articles/2015/05/22/2015-12228/requirements-for-blood-and-blood-components-intended-for-transfusion-or-for-further-manufacturing).

20. Which of the following vectors is correctly matched with the corresponding transfusion transmitted infection:
A. Anopheles mosquito—West Nile Virus (WNV)
B. *Ixodes scapularis*—Dengue virus (DENV)
C. Reduviid bug—*T. cruzi*
D. *Aedes albopictus*—*B. microti*
E. *Aedes aegypti*—*P. falciparum*

Concept: Knowledge of the vector for a given disease (i.e., its epidemiology, socio-behavioral risk factors, and geographic distribution) is critical to develop an effective mitigation strategy. Unfortunately, some agents have several vectors that may complicate prevention.

Answer: *C*—Reduviid bugs (Fig. 11.4A) are the vectors for *T. cruzi* (Fig. 11.4B), the causative agent of Chagas disease. Reduviid bugs are widely distributed in parts of Central and South America where they inhabit thatch roofs, thus primarily affecting rural or poor communities.

Female Anopheles (Answer A) mosquitoes transmit malaria (i.e., *Plasmodium* spp.); *Ixodes scapularis* (Answer B) transmits *B. microti*, *Aedes aegypti* (Answers D and E) and *Aedes albopictus* both transmit DENV; WNV has several mosquito vectors although the Culex mosquito is predominant in the United States.

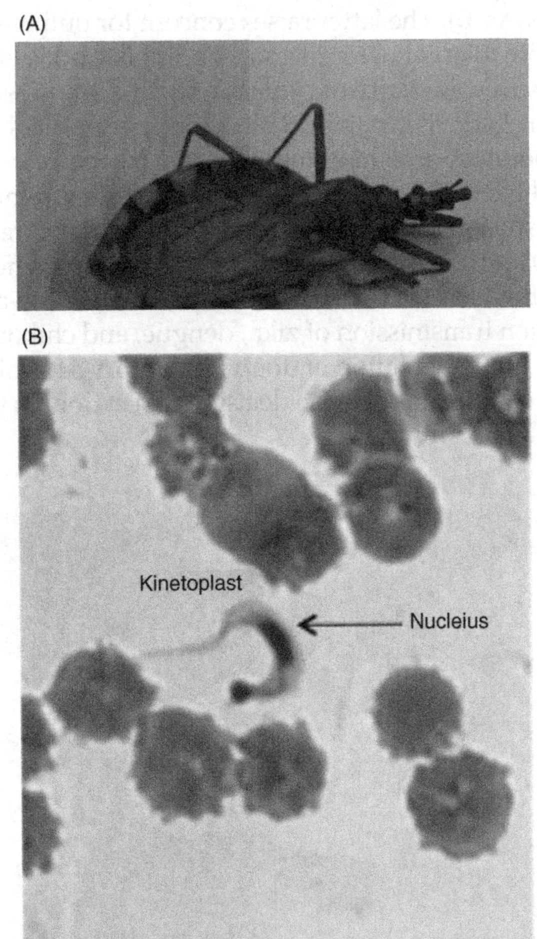

FIGURE 11.4 (A) *Triatoma infestans* "reduviid bug." (B) *T. cruzi* trypomastigote in a mouse blood smear stained with Giemsa (original magnification ×1000). *Source: Images courtesy of Lucia Maria Almeida Braz, PhD, Parasitology Lab—Instituto de Medicina Tropical da Universidade de São Paulo, São Paulo, SP, Brasil.*

21. Which of the following best qualifies the risk of Dengue virus (DENV) in regards to the US blood supply?

A. Despite evidence of transfusion transmission DENV has only rarely been shown to result in clinical disease following this route of transmission

B. DENV is primarily transmitted by Culex mosquitoes, raising concern for autochthonous transmission in the United States and concomitant risk to the blood supply

C. Despite an estimated 50–100 million new cases each year, DENV infection has never been shown to be transfusion transmissible

D. DENV is confined to the tropics and subtropics and does not pose risk to the US blood supply

E. Mandatory screening of blood donors has been in use throughout the continental United States since 2014 using NS1 antigen testing

Concept: Transfusion transmissibility and clinical penetrance are not the same: DENV is transfusion transmissible yet has only rarely been associated with clinical sequelae following transfusion transmission.

Answer: *A*—Recent studies suggest that DENV is transmissible via blood transfusion (37.5% of RNA positive donor units transmitted DENV to recipients); however, clinical penetrance is low (i.e., RNA is detectable in recipients yet does not typically result in signs or symptoms). Indeed, there are only ~5 described clinical cases of transfusion-transmitted dengue (Answer C) despite a high prevalence of Dengue in many parts of the world. The primary vectors for DENV are *A. aegypti* and *A. albopictus*, not the Culex mosquisto (Answer B). The latter raises concern for outbreaks in the United States given that the mosquito is more tolerant of temperate conditions and is already endemic in parts of the United States (Answer D). Despite rarity of clinically significant transfusion transmitted disease, DENV has prompted concern about risk to the US blood supply, particularly following emergence of DENV in Puerto Rico. No laboratory screening measures are currently in place (Answer E); if this were to be introduced, NAT would likely be the preferred strategy since viremia rather that seroreactivity is of foremost concern to transfusion transmission (http://www.aabb.org/programs/publications/bulletins/Documents/ab16-03.pdf). AABB recently released an Association Bulletin recommending self-deferral of donors who have traveled to Mexico, Caribbean, and Central and South America for 28 days to encompass risk of transfusion transmission of zika, dengue, and chikungunya viruses (Fig. 11.5). This was later amended following implementation of donor laboratory screening for Zika virus in 2016 first in the US territories and later continental US. The decision to test donors was prompted by uncertainty regarding clinical risk to transfusion recipients.

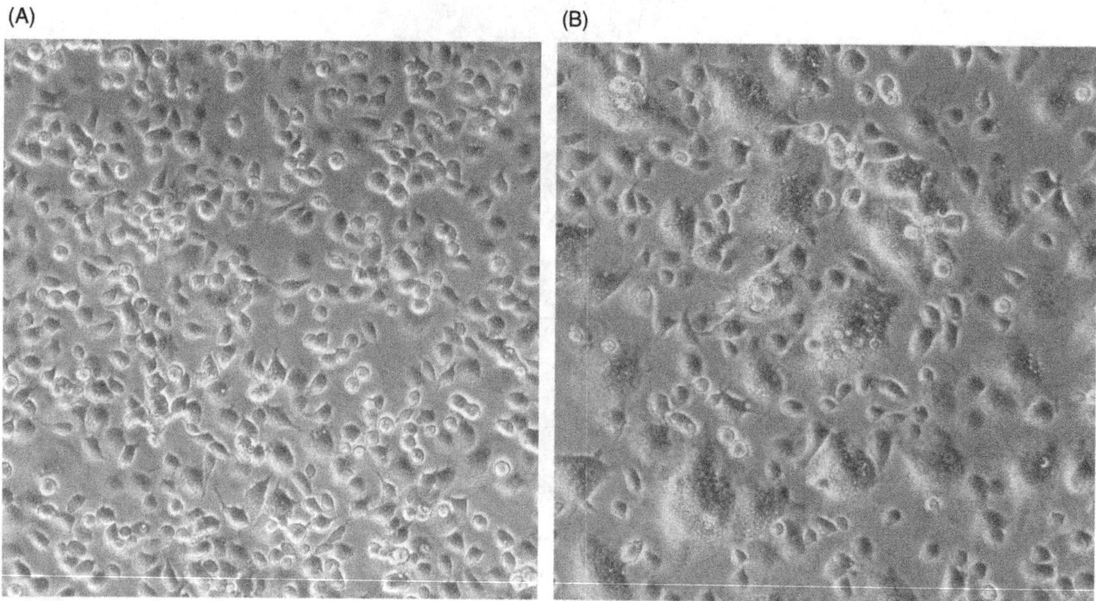

FIGURE 11.5 (A) Mock-infected *Aedes albopictus* clone C6/36 cells. The picture was taken at 6 days after inoculation. Optical microscopy. Magnification 320×. (B) DENV-2 (strain 46) infected *Aedes albopictus* clone C6/36 cells. Cytopathic effect, with syncytia or multinucleated giant cells, was seen on the 6th day after infection. Optical microscopy. Magnification 320×.

22. A 35-year-old avid hiker/backpacker in Minnesota complains of fever and flu-like symptoms for the past week. During his last outing 10 days ago, he noticed two ticks (measuring <1/4 inches, with black legs, and a black and red abdomen) on the back of his knee. He does not recall any recent mosquito bite. On examination, he has some erythema, induration and tenderness at the site. Laboratory testing shows white blood cell count 2.5×10^9/L with left shift, platelets 69×10^9/L, ALT 110 U/L and AST 97 U/L. A photomicrograph obtained from the associated peripheral blood smear is noted in Fig. 11.6.

Which of the following is true regarding this disease?

A. The causative organism is *Rickettsia rickettsia*
B. Transmission by transfusion is common where the disease prevalence is high
C. Doxycycline is an effective treatment
D. Most cases in the United States occur in the Southeast
E. Red blood cell inclusions are often seen on blood smear

Concept: Anaplasmosis is a tick-borne infectious disease caused by *A. phagocytophilum*. Patients typically present with thrombocytopenia, mild transaminitis (increased ALT and AST) and bacterial collections in leukocytes (appearing as basophilic morulae). Ticks in the *Ixodes persulcatus* complex are competent vectors for anaplasmosis, and can also spread *B. burgdorferi* (the agent of Lyme disease), *B. microti*, and Powassan virus.

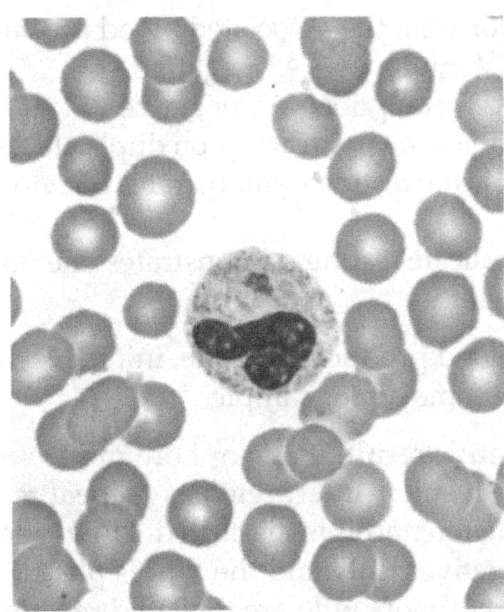

FIGURE 11.6 Peripheral blood smear stained with Giemsa showing granulocyte with Anaplasma morules (original magnification ×1000) *Source: Image courtesy of Dr. Bobbi Pritt, Mayo Clinic.*

Answer: C—*R. rickettsia* is the pathogen in Rocky Mountain spotted fever (Answer A), which is characterized by the classic triad of fever, centripedal maculopapular rash, and history of tick bite. Although transfusion-transmitted *A. phagocytophilum* has been reported, the rate is low even in endemic regions with high seroprevalence rates (Answer B). Currently there is no FDA guidance or AABB Standards regarding deferral. However deferral until after treatment is complete and symptoms have resolved or for 90 days in an untreated individual is prudent (https://www.aabb.org/tm/eid/Documents/169s.pdf). Anaplasmosis is most frequent in the upper Midwest and northeastern United States (Answer D). According to the CDC, 90% of cases occur in the following six states: New York, Connecticut, New Jersey, Rhode Island, Minnesota, and Wisconsin. Red blood cells show Inclusions are observed in cases of malaria and babesia (Maltese cross), but not in anaplasmosis (Answer E). Doxycycline may be used to treat both anaplasmosis and Lyme disease (Answer C).

23. Which of the following infectious disease testing is performed on source plasma donors but not on volunteer blood donors?

A. Syphilis
B. Parvovirus B19
C. EBV
D. Adenovirus
E. WNV

Concept: Source plasma donations have different infectious disease screening requirements from community blood donation due to distinct differences in processing. Since source plasma products are prepared from pooled plasma donations, even a small number of infectious donors can infect many recipients. Source plasma processing includes pathogen inactivation or removal methods, such as pasteurization, solvent/detergent (S/D) treatment, fractionation, and filtration.

In addition to rigorous donor selection, plasma processing procedures use several pathogen inactivation procedures, one of which is S/D treatment. Although S/D is highly effective against enveloped viruses, it is ineffective against nonenveloped viruses, such as parvovirus B19 and hepatitis A virus. Consequently, screening for Parvovirus B19 screening is routinely performed in source plasma donors, therefore screening is now routinely performed. Following institution of screening, parvovirus B19 transmission has not been reported.

Answer: *B*—Source plasma undergoes testing for parvovirus B19. All blood donations are screened for syphilis and WNV (i.e., these are not specific to source plasma) (Answers A and E). Currently, no blood product is screened for EBV or adenovirus (Answers C and D).

24. In which of the following situations can the associated blood products still be transfused?

A. Anti-HBc repeat reactive; HBV minipool NAT negative
B. Anti-HCV repeat reactive; HCV minipool NAT negative
C. Anti-HBc initial reactive, repeat testing nonreactive on duplicate testing; HBV minipool NAT negative
D. Anti-HBc initial reactive, duplicate testing demonstrates one nonreactive and one reactive, HBV minipool NAT negative
E. Anti-HCV initial reactive, duplicate testing demonstrates one nonreactive and one reactive, Hepatitis C minipool NAT negative

Concept: To provide the safest blood products possible, initial reactive serologic tests prompt duplicate repeat testing from the same donor sample.

Answer: C—After an initial positive result, if one or both of the second set of results are reactive during confirmatory testing, the result is interpreted as "repeat reactive"; consequently the unit is deemed unsuitable for transfusion and it is discarded. However, if the repeated tests are both negative, the result is "initial reactive" only and the unit is placed into inventory and transfused. In regards to the other responses, donors who are repeated reactive for Anti-HBc are placed on surveillance; nonetheless, their index donation is still discarded. If the donor is later reactive on a second donation, he/she is indefinitely deferred. Reentry is allowable after a minimum 8 weeks contingent on negative testing with anti-HBc, HBsAg, and HBV individual donor NAT.

Donors who are repeat reactive for anti-HCV but are HCV NAT nonreactive, will have their sample sent for confirmatory testing using an alternate FDA approved EIA. If negative they are indefinitely deferred; however, reentry may be allowable after 6 months with negative EIA and HCV individual donor NAT tests. The other choices (Answers A, B, D, and E) are incorrect based on the information above. Table 11.2, further describes testing methodologies, results, and interpretations in regards to donor products.

TABLE 11.2 Testing methodologies, results, and interpretations in regards to donor products

	Screening result			
Analyte	**Initial**	**Repeat**[a]	**Result interpretation**	**Suitable for transfusion**
HBc HBsAg	NR R	None N + N	NR IR	Yes Yes
	R	R + R or N + R	RR	No
HIV-1/2 plus O	NR	None	NR	Yes
HCV	R	N + N	IR	Yes
HTLV-I/II	R	R + R or N + R	RR	No
Chagas (*T. cruzi*)	R	R + R or N + R	RR	No
HIV-1 RNA (NHIV) HCV RNA (NHCV) HBV DNA (NHBV) WNV RNA (WNV)[b]	Mini-pool NEG Mini-pool POS	N/A Resolution POS	Nonreactive Reactive	Yes No
Syphilis	NEG POS/IND[c]	None POS/IND	NEG POS/IND	Yes No
Atypical RBC Antibody	NEG POS	None None	NEG POS	Yes RBC—yes if washed PLT/Plasma—no

IND, indeterminate; IR, initial reactive; Neg, negative; NR, non reactive; Pos, positive; R, reactive.
[a] *IR samples are retested in duplicate.*
[b] *NAT is performed in mini-pools of 16 samples for HIV, HCV, and HBV and for WNV. Positive pools are resolved to individual donations. Some selected samples may be tested initially as individual donations and not in mini-pools.*
[c] *Indeterminate syphilis results are treated as positive—HIV-1/2 plus O- detects antibody to HIV type 1 (groups M and O) and/or 2*

For Autologous Donations repeatedly reactive (RR) units receive a BIOHAZARD label regardless of confirmatory test results. The receiving hospital must agree to accept units with biohazard label prior to shipment. Units will be shipped upon the specific request of the Blood Bank Director and the treating physician.

25. Pathogen inactivation/reduction refers to a variety of approaches that address multiple infectious agents through global treatment of the blood product. In December 2014, the FDA approved the first pathogen inactivation system for both transfusable plasma and platelets in the United States (January 2013 for Octaplasma, pathogen inactivated plasma). A photochemical inactivation system (Intercept, Cerus, Concord, CA) has also obtained FDA approval in the USA. Which of the following organisms has showed the most resistance to pathogen inactivation?

A. HIV
B. *Plasmodium falciparum*
C. *Pseudomonas aeruginosa*
D. *Treponema pallidum*
E. HAV

Concept: Photochemical inactivation of blood products for works by intercalating light-activated compounds into DNA/RNA, preventing cellular replication. This has shown to be effective against a broad array of pathoegns (i.e., bacterial, viruses and protozoa alike). Two systems are available for use in transfusable plasma and platelets: the Mirasol Pathogen Reduction Technology System (riboflavin/UV light) system by TerumoBCT and the INTERCEPT Blood System (amotosalen/UV light) system by Cerus Corp. The latter is currently approved by the FDA although both are CE marked (European Union medical device approval).

Answer: *E*—In recently reported studies, hepatitis A virus showed a lower logarithmic reduction (0 on amotosalen platform, 1.8 on riboflavin platform) compared to the other listed pathogens (all greater than 3.2 log reduction) (Answers A, B, C, and D). The stated goal of the authors was to achieve a 3-log reduction with pathogen inactivation since this should eliminate the low viral loads that are associated with eclipse phase transmissions (i.e., NAT negative samples). Photochemical inactivation is also relatively ineffective against prions, some nonenveloped viruses and bacterial spores.

26. Which of the following is true about hepatitis E?

A. Infected patients are often symptomatic with marked liver enzyme abnormalities
B. Transmission is predominantly by IV drug use
C. The causative agent is a single-stranded DNA virus
D. Transfusion transmitted infection has not been reported
E. The main genotype in infected patients varies based on location

Concept: Five major hepatitis viruses have been described (Hepatitis A, B, C, D, and E). These vary with respect to mode of transmission, frequency of chronic infection, and treatment options.

Answer: *E*— Four major HEV genotypes have been described. Hepatitis E virus (HEV) infection is frequently asymptomatic (Answer A). 67%–98% of patients with HEV genotype 3 exhibit no symptoms. However, higher rates of severe disease and even death have been described in pregnant women and patients with underlying chronic liver disease. HEV is acquired primarily by feco-oral transmission (e.g., contaminated food) (Answer B). HEV is a single-stranded RNA virus (answer C). Cases of transfusion-transmitted infection have been reported (Answer D).

27. Historically, which of the following donor screening tests had the highest rate of false positive results?

A. HTLV serology
B. HIV NAT
C. HBsAg

D. Anti-HCV
E. Anti-HBc

Concept: Donor screening tests have a very high sensitivity and specificity. However, given that high volume testing is performed on a predominantly healthy (i.e., donor) population, false positive test results may exceed the number of true positive test results. This risks unnecessary donor deferral, incurring high cost to the blood center and provoking anxiety in the prospective donors.

Answer: *E*—In hepatitis B virus (HBV) infection, anti-HBc becomes detectable after Hepatitis B surface antigen (HBsAg) and before anti-HBs. In some cases, anti-HBc is the only positive detectable serological marker during the transition period between the disappearance of HBsAg and the rise in detectable antibody to surface antigen (anti-HBs); anti-HBc was useful before introduction of HBV NAT. However, anti-HBc persists in both chronic and resolved infection; therefore additional tests are necessary to determine if the donated blood is safe for transfusion. All the other choices (Answers A, B, C, and D) are incorrect based on the data.

28. Chikungunya (CHIKV) is another arboviral emerging infectious disease that has raised concern for transfusion-transmitted infection. Which of the following best describes the current mitigation strategy in the United States?
A. Laboratory based screening of blood donors for CHIKV is not routine in the United States
B. Given the high number of described cases of transfusion transmitted CHIKV (>100 cases in the recent Caribbean outbreak), CHIKV is the foremost priority for development of a mitigation strategy
C. CHIKV is spread, primarily, via Culex mosquitoes prompting concern given a shared epidemiology with West Nile Virus
D. Blood donors with a history of CHIKV are indefinitely deferred from blood donation
E. The major clinical feature of Chikungunya (both naturally acquired and transfusion transmitted) is neurological (meningoencephalitis), which is associated with a high fatality (~5%) in susceptible patients

Concept: To date, no cases of transfusion transmitted Chikungunya have been described.

Answer: *A*—Chikungunya is another arbovirus, whose rapid expansion and high infectivity in the general population has raised concern in regards to risk to the US blood supply. Nonetheless, no case of transfusion transmitted CHIKV has been described (Answer B) and there is no formalized donor deferral policy or laboratory based screening for CHIKV in place. An Association Bulletin from AABB has been adopted in blood centers whereby donors with a history of recent travel to the Caribbean with consistent symptoms and signs of CHIKV are temporarily deferred until 28 days (Answer D) after resolution of symptoms. The central feature of CHIKV is arthralgia (Answer E), which can be protracted and debilitating. However, CHIKV is not typically fatal. Like other arboviruses (e.g., DENV and WNV), the viremic period is very short rendering long-term deferral unnecessary. CHIKV has shared vectors with DENV (*A. aegypti* and albopictus) (Answer C) prompting concern for wider spread in the continental US, particularly given rapid expansion in the Caribbean, which has included rare cases in the United States. Pathogen inactivation has been shown to be effective, at least for use in platelets and plasma and was used during an outbreak of CHIKV in Reunion in 2005–06. It is debatable whether this was necessary given the absence of reported cases of transfusion transmission (even in CHIKV endemic countries where pathogen inactivation has not been available).

29. Which of the statements regarding transfusion-transmitted malaria in the United States is true?
A. History of malaria 1 year ago results in deferral from blood donation
B. Donors with a history of travel to a malaria endemic area in the past 12 months may be able to donate if a screening test (EIA or IFA) is negative
C. Transfusion transmitted malaria was common in the United States prior to initiation of routine laboratory-based donor screening

D. *P. malariae* is most frequently implicated in cases of transfusion transmitted malaria which is ascribed to its propensity for chronic asymptomatic parasitemia
E. Someone who has been a resident (i.e., has lived 5 consecutive years) in a malaria endemic country within the past 3 years may donate in the United States if laboratory screening is negative

Concept: Malaria is transfusion transmissible. Nonetheless, transfusion-transmitted malaria is rare in the United States with only seven cases reported since 2002.

Answer: *A*—There is NO laboratory-based donor screening for malaria in the United States (answers B and C). Instead, deferral is based on history of travel to or residency in a malaria endemic area. Nonresident travel to a malaria endemic area results in a 12-month temporary deferral from donation (irrespective of prophylaxis). Residency in an endemic country within the past 3 years (assuming no further travel to a malaria endemic area) results in temporary deferral from blood donation. Donors with a history of malaria that remain symptom free for the last 3 years while residing in a nonendemic country may donate (Answer E). Of interest, selective laboratory testing, using serological testing (EIA or IFA) is conducted donor selection in certain nonendemic countries, notably Australia, United Kingdom, and France. *P. falciparum* is most frequently implicated in transfusion transmitted malaria acquired from asymptomatic donors (Answer D); this may be surprising given that chronic parasitemia is more typically associated with the other species of malaria (e.g., *P. malariae*).

30. Which of the following statements regarding variant Creutzfeldt–Jakob disease (vCJD) is correct?
A. While no cases of transfusion transmitted vCJD have been reported, transfusion transmission is theoretically possible based in animal models
B. Susceptibility to vCJD is highest in individuals who are homozygous for a polymorphism at codon 129 of prion protein coding methionine rather than valine
C. Followed a high frequency of transfusion transmission in the1990s (over 450 cases described), an aggressive blood donor deferral policy was adopted in the United States, based on cumulative time spent in the United Kingdom and parts of Europe 1980–96
D. The risk of transfusion transmitted vCJD prompted implementation of leukoreduction and PrP^{Sc} screening, both of which have been shown to be successful in eliminating vCJD risk
E. Other prion diseases that have been associated with transfusion transmission include Scrapie and Chronic wasting Disease

Concept: vCJD is a prion disease that has been shown to be transfusion transmissible. However, to date this remains an extraordinarily rare infectious complication of blood transfusion with only ~4 cases described (3 with clinical sequelae and 1–2 with evidence of abnormal prion proteins).

Answer: *B*—Susceptibility to vCJD is highest in individuals who are homozygous for a polymorphism at codon 129 of prion protein coding methionine (MM) rather than valine (VV); subclinical disease has been questioned in heterozygous (MV) individuals. Donor selection is used to mitigate risk of vCJD in the United States. Specifically, the donor health questionnaire gauges risk using a series of questions about the cumulative time spent in United Kingdom and Europe. For example, anyone who has spent 3 months in the United Kingdom between 1980–96 or 5 consecutive years in Europe over the same time frame or has received a blood transfusion in Europe from 1980 to the present is indefinitely barred from blood donation. There are other laboratory-based strategies that have been investigated such as leukoreduction (already in routine use in most blood centers) and prion filters. The latter are designed to selectively bind the infectious protein (e.g., P-CAPT filters). Unfortunately, neither leukoreduction nor current filters have been shown to be completely effective in eliminating transfusion associated risk of vCJD (Answers A and D).

In regards to the other response options, Scrapie and Chronic Wasting Disease (answer E) are not known to affect humans, instead affecting sheep and elk/deer respectively. Other human prion diseases include Kuru and Gerstmann–Sträussler–Scheinker disease and fatal familial insomnia disease. There are no described cases of transfusion transmission with any of these other named prion diseases.

31. Which of the following statements is true about Ebola virus and the blood supply?

A. Ebola is a mosquito borne virus, which shares a similar geographic distribution to Zika virus
B. The incubation period of Ebola is 40–60 days
C. Ebola is transmitted through air borne exposure
D. People who have been exposed to a patient with Ebola virus should not donate for 28 days
E. Convalescent plasma is a first-line treatment in Ebola viral disease

Concept: Although there has not been a report of Ebola transmission via blood transfusion; nonetheless, transfusion transmission is theoretically possible.

Answer: *D*—Ebola virus is transmitted through direct contact with infected blood or body fluids and not through air (Answer C). The incubation period from the time of exposure to the manifestation of the signs/symptoms of the disease is 2–21 days (Answer B). Ebola patients have high fever and are acutely unwell, thus self deferring themselves from blood donation. Though there has not been a report of Ebola transmission via blood transfusion, given the infectivity via blood exposure, individuals who have potentially been exposed to Ebola should not donate for 28 days (Answer D). A recent review of the use of convalescent plasma (with unknown levels of neutralizing antibodies) in patients with acute Ebola infection reported no improvement in survival compared with historical controls (Answer E). Following a large pandemic in 2015–2016, Zika has been described in over 70 countries or territories. Notable outbreaks include Micronesia (2007), French Polynesia (2013–2014) and Brazil (2015–2016). Transfusion-transmitted Zika virus has recently been reported. Zika has a much broader geographic distribution than Ebola (Answer A).

32–35 matches the assays with their window periods;

32. HIV MP NAT	A. 50 days
33. HIV EIA	B. 9 days
34. HCV MP NAT	C. 11 days
35. HBV MP NAT	D. 7 days
	E. 21 days

Concept: A "window period" is the time between actual infection and disease detection by various testing methodologies. In recent years, the use of nucleic acid testing (NAT) has significantly reduced the window period in comparison to serologic based testing. Table 11.3 describes the window period for various diseases and the residual risk of transfusion transmission.

Answers: *32. B; 33. E; 34. D; 35. A*

TABLE 11.3 The window period for various diseases and the residual risk of transfusion transmission

Test	Window period (days)	Residual risk of transfusion
HIV MP NAT	9	1:1,467,000
HIV EIA	21	
HCV MP NAT	7	1:1,149,000
HCV EIA	51–58	
HBsAg	30–38	1:282,000–1:357,000
HBV NAT	40–50 (MP) and 15–34 (ID)	1:750,000
HTLV	80	1:2,993,000

36. You receive a call from a doctor who has been asked by his patient about the risk of *T. cruzi* transmission from blood transfusion. The patient lives in the United States and is concerned about an upcoming surgical procedure that will likely necessitate blood transfusion. In responding to the doctor, how would you best characterize the risk of transfusion transmitted *T. cruzi* in the United States?

A. High risk: high population seroprevalence; high rate of transmissibility and large numbers of transfusion associated cases (>150)
B. Intermediate risk: low population seroprevalence; high rate of transmissibility and modest numbers of transfusion associated cases (20–150)
C. Low risk: low-intermediate population seroprevalence; low/intermediate estimates of transmissibility with rare cases of transfusion transmission (<20)
D. No risk: low population seroprevalence; no documented transfusion associated cases
E. Unknown: insufficient data on seroprevalence, transmissibility and transmission in humans; testing initiated based on theoretical transmission model

Concept: There is very low risk of transfusion transmitted *T. cruzi* in the United States.

Answer: C—Although *T. cruzi* is known to be transfusion transmissible, it is very rare outside of endemic countries with only seven described cases in the United States and Canada over a 20 year period (Answers A, B, and D). The estimated rates of transmission (12%–20%) are based on historical data gleaned from South America (Answer E). There is a modest seroprevalence in the United States (largely in immigrants from Central and South America). Following initiation of screening the mean rate of donor seropositivity in the United States has been determined to be 1 in 25,000 to 1 in 30,000 donors. All documented cases of transfusion transmitted *T. cruzi* have followed whole blood or platelet (apheresis or whole blood derived) transfusions. A recent case of transmission following red cell transfusion was described in Belgium (its rarity merited publication). If indeed the high historical transmissibility estimates were true, the observed low rate of transfusion transmission in the United States remains unexplained. One possibility is underreporting given that most *T. cruzi* infection is subclinical. Another possibility is that the original estimates stemmed from endemic countries in the 1980's where fresh whole blood transfusion was common and high risk, paid or replacement donors were not unusual. In contrast, in the current day US exclusive reliance on voluntary, nonremunerated blood donors in the United States, and routine processing and refrigerated storage may have impacted transmissibility. In summary, one can inform the doctor that transfusion risk is extraordinarily low: anyone with a history of Chagas disease is indefinitely deferred and donors are screened the first time that they donate. Screening is not required on subsequent donations.

37. Which of the following combinations of variables, best describe criteria that merit implementation of screening for a given infectious agent?

A. High-clinical significance, low blood-borne phase, nonviable during storage
B. Tolerates storage, high prevalence, fecal-oral transmission
C. Transfusion transmissibility documented, minimal clinical significance, tolerates storage
D. >90% prevalence, tolerates storage, parental transmission
E. Asymptomatic blood-borne phase, tolerates storage, parental transmission

Concept: To maximize benefit and minimize cost, several variables are considered before implementing a screening program for an infectious agent.

Answer: *E*—There are several key variables that are considered prior to implementation of laboratory-based screening. Foremost, the agent needs to be associated with significant clinical disease in recipients. Second, the agent must be parenterally transmitted (i.e., outside of the GI tract) with the ability to tolerate storage and processing (e.g., refrigeration in the case of RBCs, and freezing in the case of plasma). Other variables that lend weight to laboratory-based screening include the presence of an asymptomatic chronic or acute blood-borne phase both of which increase the probability of an infected individual feeling sufficiently well to donate). Furthermore, the agent of concern should have a low prevalence in the recipient population. As one example, EBV is not screened given the high background prevalence in both donor and recipient alike (~95% of the population has been exposed) thereby, limiting the benefit of laboratory based screening. The other choices (Answers A, B, C, and D) are incorrect based on the information aforementioned.

38. What product has the highest risk of septic transfusion reaction?
A. Granulocytes
B. Fresh frozen plasma
C. Red blood cells
D. Platelets
E. Recombinant Factor VIIa

Concept: Product processing and storage times/conditions greatly influence the likelihood of bacterial contamination, transmission, and potential septic transfusion reactions.

Answer: *D*—The incidence of septic transfusion reaction by blood product is approximately 1 in 100,000 transfused apheresis platelets and 1 in 250,000 transfused RBCs. Septic reactions are rare following transfusion with frozen products and sterilized nonhuman clotting factors. RBCs are refrigerated (Answer C) and plasma is frozen; in contrast platelets are stored at room temperature accounting for the higher risk of septic reactions given bacteria survival and growth during storage. The majority of septic transfusion reactions from platelets are caused by aerobic Gram-positive species (e.g., *Staphylococcus aureus* and Coagulase negative *Staphylococcus* spp.). Platelets at 4–5 days after collection are more likely to cause reactions versus day 3 or less because of lag phase followed by exponential growth phase of the bacteria. The remaining choices (Answers A and E) have not been associated with septic transfusion reactions.

39. Which of the following statements about occult HBV is correct?
A. A portion of HBV RNA binds to hepatocyte proteins, forming a mini-chromosome
B. Anti-HBe is negative
C. HBsAg is positive
D. The viral load is too low for transmission by transfusion
E. Testing for HBV DNA in hepatocytes is more sensitive than testing in plasma in identifying occult disease

Concept: Occult hepatitis B infection (OBI) is a relatively new entity uncovered through the adoption of widespread HBV NAT testing. OBI is characterized by anti-HBc reactivity without detectable HBsAg and usually low, occasionally undetectable, levels of HBV DNA. The importance of OBI is that the virus is still transmissible during blood donation and that patients may still be at risk for disease progression.

Answer: *E*—The mechanism underlying OBI is the formation and persistence of a mini-chromosome of HBV DNA material in hepatocytes (Answer A). Patients with OBI may express anti-HBe (Answer B). OBI is defined by lack of HBsAg, illustrating that clearance of HBsAg does not necessarily reflect viral clearance (Answer C). In Australia, a low (0.2%–3.3%) rate of HBV transmission has been associated with OBI donors (Answer D). Even if HBV DNA is not detectable in plasma, viral material can be detected in hepatocytes (Answer E).

40. Which of the following responses is true regarding West Nile Virus?
A. WNV is found in the Northeastern United States
B. 20% of the cases are asymptomatic
C. WNV is an DNA virus
D. 5%–10% of NAT-reactive donations are negative on pooled NAT
E. WNV infects deer

Concept: West Nile virus (WNV), a single-stranded RNA flavivirus, is primarily spread by the mosquito.

Answer: *D*—WNV is a RNA virus (Answer C) that has a transmission cycle that involves birds (Answer E) and mosquitoes; humans are an incidental host and given acquisition via mosquito bites, the incidence of infection peaks in July through October. In 1999 WNV was first reported in the United States in New York and subsequently spread westward throughout the continental US (Answer A) where it caused (and continues to cause) significant seasonal epidemics. While ~80% of human WNV infections are asymptomatic (Answer B), symptomatic infections result in fever, myalgia, headcahe, nausea and vomiting. About one in 150 infected individuals will have severe disease resulting in meningitis and/or encephalitis (convulsion, coma, paralysis), and, less frequently, death.

41. Match the hepatitis virus with the correct family:
1. Hepatitis A
2. Hepatitis B
3. Hepatitis C
4. Hepatitis E

A. Flaviviridae
B. Picornaviridae
C. Hepadnaviridae
D. Hepeviridae

Concept: The viruses causing the most common viral hepatitis infections are members of different families.

Answers: 1-B, 2-C, 3-A, 4-D. Hepatitis A virus is a nonenveloped RNA virus. Hepatitis B virus is an enveloped DNA virus. Hepatitis C virus is an enveloped RNA virus. Hepatitis E virus is a nonenveloped RNA virus. Table 11.4 describes the hepatitis viruses and their routes of transmission.

TABLE 11.4 Hepatitis viruses and their routes of transmission

Virus	Hepatitis A	Hepatitis B	Hepatitis C	Hepatitis D	Hepatitis E
Family	Picornaviridae	Hepadnaviridae	Flaviviridae	Unassigned (subviral satellite requiring hepatitis B)	Hepeviridae
Type	Nonenveloped, RNA	Enveloped, DNA	Enveloped, RNA	Enveloped, RNA	Nonenveloped, RNA
Transmission	Fecal-oral	Blood	Blood	Blood	Fecal-oral
Association with chronic infection	No	Yes	Yes	Yes	Yes

42. Which of the following combinations best describes the geographic distribution and transfusion-associated risk of *B. microti*?

A. Endemic in Northeast and Upper Midwestern United States; transmissible via any red cell based product or whole blood platelet product; large numbers of transfusion associated cases with high all cause case fatality

B. Endemic in California and Pacific Northwestern United States; primarily transmitted via plasma; rare cause of transfusion transmission with no fatalities to date

C. Endemic in Southwestern United States; transmitted by granulocytes; rare cause of transfusion transmission with limited data on transfusion associated fatalities

D. Endemic throughout continental United States; transmitted by red cells and whole blood with seasonal (late Spring/Summer) transfusion transmission; rare cases of transfusion associated fatality
E. Nonendemic in the United States; rare cases of transfusion transmission ascribed to donor travel to Europe where the parasite is endemic; no cases of transfusion associated fatality in the United States but well-described in Europe

Concept: *B. microti* is endemic in the Northeast and Upper Midwestern United States. It is currently regarded as the foremost infectious risk to the US blood supply for which licensed donor screening is still unavailable.

Answer: *A*—There are over 100 species of Babesia that infect vertebrates yet only a few have been implicated in clinical human infection (babesiosis). *B. microti* is responsible for the overwhelming majority of both naturally-acquired as well as transfusion transmitted infection. Its geographic distribution is primarily focused in Massachusetts, Rhode Island, Connecticut, and New York (Answers C and D), but extends as far South as New Jersey and as far West as Minnesota. There have been over 200 cases of TTB with 32 all cause-related fatalities, most of which have been reported since 2005.

B. duncani, another Babesia species of clinical importance, is thought to be endemic to California and the Pacific Northwest (Answer B); it is comparatively rare with only a few described cases of TTB. *Babesia divergens* is endemic in Europe (Answer E) where it has resulted in naturally acquired human infection but not transfusion infection. Babesia is an intraerythrocytic parasite and is transmissible via any red cell containing product (e.g., RBCs, whole blood, whole blood-derived platelets).

No cases of TTB have been reported following plasma, cryoprecipitate or apheresis platelet transfusion. Unlike natural acquisition, TTB is neither seasonal nor geographically restricted given chronic asymptomatic parasitemia coupled with travel of both blood donors and blood products outside of endemic areas.

43. A repeat plasma apheresis donor is found to be reactive on HCV NAT following his donation on July 14, 2015. Concurrent HCV serology is negative. His previous donations occurred on June 17, 2014; September 14, 2014; November 19, 2014; February 14, 2015; and June 23, 2015. Infectious disease testing including anti-HCV and HCV NAT was negative at each of those prior donations. The plasma units associated with the most recent donation are currently still in the blood center while those from the June 2015 donation are in the inventory at a hospital transfusion service. The older units have all been transfused.

Which of the following measures should the blood center undertake first?
A. Defer the donor for 6 months
B. Retrieve components from the June 2015 donation
C. Report case to the public health department
D. Perform HCV testing on other members of donor's household
E. Notify transfusion service/recipients associated with the June 2014 donation

Concept: Donor "Lookback" is a process that is undertaken by blood centers when a donor tests reactive on infectious disease screening. Blood centers must review records to determine the disposition of previous donations in order to quarantine and/or destroy potentially infectious components. This process is intended to evaluate and minimize risk of infection and may entail notification of hospitals and/or transfusion recipients if deemed there is an increased risk of transfusion-transmitted infection.

Answer: *B*—According to the FDA guidance, hepatitis C lookback is indicated if there is a reactive NAT result even if concurrent serology is negative. NAT/serology discordant results raise the possibility of an acute HCV infection. In order to prevent transfusion of potentially infectious blood components, the blood center must retrieve and quarantine all in-date components immediately (Answer B). Although acute HCV infection may be reportable to the public health department depending on state law, this procedure can wait until the currently available blood products are quarantined (Answer C). As part of lookback, records of prior collections for the 12 months before the date of the NAT-reactive test must be reviewed and the blood bank at the receiving institution must be notified. After evaluation, the blood bank medical director may decide to notify the transfusion recipient's physician (and recipient). Since the June 2014 donation predated the positive NAT by more than 1 year, notification is not necessary (Answer E). However, the consignees of the more recent collections must be notified of the positive HCV NAT result. Once notified, the consignee must make a reasonable effort to notify the recipient or the recipient's physician of the need for counseling/testing within 12 weeks.

HCV infection results in an indefinite deferral rather than a 6 month deferral (Answer A). Finally, blood centers do not perform testing on household members of a newly reactive donor (Answer D). However, close contacts of the donor may themselves be deferred due to the infectious risk of hepatitis.

44. Which of the following statements is correct regarding Zika virus?

A. The first case of Zika virus infection in humans was reported in Brazil in 2015

B. Zika virus has gained recent attention for its propensity to cause severe symptoms and signs (specifically arthralgia and high fever) in the majority of those infected

C. Routes of acquisition include vector-borne (mosquito), perinatal, blood transfusion, and sexual transmission

D. Serological (antibody) testing is specific for Zika virus and is currently recommended as the diagnostic of in outbreak areas

E. Despite a rapid and expanding pandemic, Zika has not been implicated in long term clinical sequelae

Concept: Zika virus (ZIKV) is a mosquito-borne flavivirus that has gained recent attention for its rapid and expanding pandemic and the demonstration of serious neurological sequelae (e.g., microcephaly) following perinatal transmission. Given that there are already competent vectors (e.g., *Aedes albopictus*) in the United States, there is concern for autochthonous transmission; imported cases have been described in the United States and incident cases have been reported in Puerto Rico and parts of Central America and the Caribbean.

Answer: C—Although spread primarily via mosquitoes of the *Aedes* genus, many cases of perinatal transmission have been reported during the Brazilian ZIKV outbreak; one imported case of perinatal transmission and subsequent congenital microcephaly was reported in a women in Hawaii who spent time in Brazil during her pregnancy (Answer E). The first human case of ZIKV was reported in 1952 (Answer A), while the first large outbreak occurred on the Island of Yap in 2007. Cases of sexual, as well as transfusion transmission have also been described (Answer C). The latter has raised concerns for risk to the blood supply in areas where ZIKA is prevalent. A policy of donor deferral was adopted early during the emergence of Zika. Individual donor NAT screening of US donors was implemented in 2016 to contend with growing uncertainty surrounding the risk to the blood supply. While the detectable viremia in plasma is brief (1-2 weeks), the virus is detectable for 2-3 months whole blood. A total of four cases of transfusion transmitted Zika have been reported, all of which occurred in Brazil. All were based on detectable Zika RNA in the transfusion recipients rather than on signs of infection.

Serological testing both for clinical diagnostic use, as well as blood donor screening is suboptimal given significant cross-reactivity with other arbovirus infections (most notably DENV), several of which circulate in areas where ZIKV is being reported (Answer D). Symptoms and signs of ZIKV are mild and nonspecific in most patients (Answer B); rash is a prominent feature but is not pathognomonic for ZIKV infection. Nonetheless, severe complications (e.g., Guillain Barre syndrome and microcephaly) have been associated with ZIKV infection (answers B and E). Importantly, the clinical presentation of ZIKV infection is not easily distinguishable from DENV and CHIKV; therefore all three agents (ZIKV, DENV and CHIKV) should ideally be tested for in patents presenting in ZIKV outbreak areas with compatible symptoms and signs. Nucleic acid (RNA) testing should be pursued for diagnostic use where available; should ZIKV testing be introduced for blood screening, a NAT is likely to be the modality of choice. Pathogen reduction technology with photochemical activation has also been shown to be effective and was used to treat plasma and platelets during the French Polynesia outbreak. However PRT for red cell and whole blood treatment are not yet licensed for use.

Acknowledgments

The authors would like to thank Drs. Yen-Chun Liu (Weill-Cornell Medical Center), Bobbi Pritt (Mayo Clinic), Anne Kjemtrup (California Department of Health), José Eduardo Levi, Francielle T.G.S. Cardozo, and Lucia Maria Almeida Braz (Instituto de Medicina Tropical da Universidade de São Paulo, São Paulo, Brasil) for generously contributing images for use in the chapter, as well as Debra Kessler (New York Blood Center) for her critical reading of the chapter.

Suggested Reading

[1] B.H. Shaz, M. Desormeaux, Bacterial Detection Methods, in: B.H. Shaz, C.D. Hillyer, M. Roshal, C.S. Abrams (Eds.), Transfusion Medicine and Hemostasis, Clinical and Laboratory Aspects, second ed., Elsevier. 2013.

[2] L.M. Katz, R.Y. Dodd, Transfusion–transmitted diseases, in: B.H. Shaz, C.D. Hillyer, M. Roshal, C.S. Abrams (Eds.), Transfusion Medicine and Hemostasis, Clinical and Laboratory Aspects, second ed., Elsevier. 2013.

[3] C.D. Hillyer, B.H. Shaz, Overview of infectious disease testing, in: B.H. Shaz, C.D. Hillyer, M. Roshal, C.S. Abrams (Eds.), Transfusion Medicine and Hemostasis, Clinical and Laboratory Aspects, second ed., Elsevier. 2013.

[4] M.A. Kane, E.M. Bloch, R. Bruhn, Z. Kaidarova, E.L. Murphy, Demographic determinants of syphilis seroprevalence among US blood donors, 2011−2012, BMC Infect. Dis. 15 (2015) 63.

[5] R.O. Francis, D. Strauss, J.D. Williams, S. Whaley, B.H. Shaz, West Nile virus infection in blood donors in the New York City area during the 2010 seasonal epidemic, Transfusion 52 (2012) 2664–2670.

[6] FDA, Guidance for Industry: "Lookback" for Hepatitis C Virus (HCV): Product Quarantine, Consignee Notification, Further Testing, Product Disposition, and Notification of Transfusion Recipients Based on Donor Test Results Indicating Infection with HCV. Available from: http://www.fda.gov/BiologicsBloodVaccines/GuidanceComplianceRegulatoryInformation/Guidances/ucm238447.htm.

[7] Guidance for Industry Nucleic Acid Testing (NAT) for Human Immunodeficiency Virus Type 1 (HIV-1) and Hepatitis C Virus (HCV): Testing, Product Disposition, and Donor Deferral and Reentry. Available from: http://www.fda.gov/downloads/BiologicsBloodVaccines/GuidanceComplianceRegulatoryInformation/Guidances/Blood/UCM210270.pdf.

[8] E. Vannier, P.J. Krause, Human babesiosis, N. Engl. J. Med. 366 (2012) 2397–2407.

[9] R.J. Benjamin, S.L. Stramer, D.A. Leiby, R.Y. Dodd, M. Fearon, E. Castro, *Trypanosoma cruzi* infection in North America and Spain: evidence in support of transfusion transmission, Transfusion 52 (2012) 1913–1921 quiz 2.

[10] FDA. Revised Recommendations for Reducing the Risk of Human Immunodeficiency Virus Transmission by Blood and Blood Products. Draft guidance for industry. Available from: http://www.fda.gov/downloads/BiologicsBloodVaccines/GuidanceComplianceRegulatoryInformation/Guidances/Blood/UCM446580.pdf.

[11] R. Klamroth, A. Groner, T.L. Simon, Pathogen inactivation and removal methods for plasma-derived clotting factor concentrates, Transfusion 54 (5) (2014) 1406–1417.

[12] J. Kaiser-Guignard, et al. The clinical and biological impact of new pathogen inactivation technologies on platelet concentrates, Blood Rev. 28 (6) (2014) 235–241.

[13] M. Mungai, G. Tegtmeier, M. Chamberland, M. Parise, Transfusion-transmitted malaria in the United States from 1963 through 1999, N. Engl. J. Med. 344 (2001) 1973–1978.

[14] S. Zou, K.A. Dorsey, E.P. Notari, G. Foster, D. Krysztof, F. Musavi, R.Y. Dodd, S.L. Stramer, Prevalence, incidence, and residual risk of human immunodeficiency virus and hepatitis C virus infections among United States blood donors since the introduction of nucleic acid testing, Transfusion 50 (2010) 1495–1504.

[15] C.R. Seed, et al. Infectivity of blood components from donors with occult hepatitis B infection—results from an Australian lookback programme, Vox Sang 108 (2) (2015) 113–122.

[16] D. Kessler, WNV screening, in: B.H. Shaz, C.D. Hillyer, M. Roshal, C.S. Abrams (Eds.), Transfusion Medicine and Hemostasis, Clinical and Laboratory Aspects, second edition, Elsevier. 2013.

[17] B.L. Herwaldt, J.V. Linden, E. Bosserman, C. Young, D. Olkowska, M. Wilson, Transfusion-associated babesiosis in the United States: a description of cases, Ann. Intern. Med. 155 (2011) 509–519.

[18] AABB, ed. AABB Standards for Blood Bank and Transfusion Services. 29th ed. Bethesda; 2014.

[19] M.K. Fung, B.J. Grossman, C. Hillyer, C.M. Westhoff (Eds.), AABB Technical Manual, 18th ed, AABB, Bethesda, 2014.

[20] D. Musso, T. Nhan, E. Robin, C. Roche, D. Bierlaire, K. Zisou, et al. Potential for Zika virus transmission through blood transfusion demonstrated during an outbreak in French Polynesia, November 2013 to February 2014, Euro Surveill 19 (14) (2014).

[21] L.R. Petersen, M.P. Busch, Transfusion-transmitted arboviruses, Vox Sang. 98 (4) (2010) 495–503.

[22] Centers for Disease Control and Prevention: Zika Virus. Available from: http://www.cdc.gov/zika, 2015.

CHAPTER

12

Noninfectious Risks of Transfusion

Jeffrey S. Jhang, Suzanne Arinsburg

Icahn School of Medicine at Mount Sinai, New York, NY, United States

The goal of blood component therapy is to achieve the desired effect without adverse outcomes. The desired effect may be to increase oxygen carrying capacity, correct coagulation factor deficiencies, or provide cellular components. However, adverse outcomes due to errors, accidents, or transfusion reactions cannot be completely avoided. Despite risk mitigation, the rate of transfusion reactions is 2.4 reactions per 1000 units transfused. Transfusion reactions can be just an annoyance (e.g., urticarial), but more severe reactions can be life threatening (e.g., anaphylaxis). The Transfusion Medicine specialist must be able to recognize, treat, and prevent future reactions by providing a comprehensive consultation for all reported reactions. The blood banking community has been very successful at reducing the risk of transfusion transmitted infections and is continually improving transfusion safety and risk mitigation. The purpose of this chapter on noninfectious risks of transfusion is to enhance the ability of the user to recognize, treat, and prevent these transfusion reactions.

1. A 55-year-old man (patient type O) with myelodysplastic syndrome is scheduled to receive a red blood cell (RBC) unit at the same infusion center as another 72-year-old man (patient type B) with chronic lower gastrointestinal bleeding who is also scheduled to be transfused. The two patients have the same last name and similar sounding first names. The type O patient receives the RBC unit that was intended for the type B patient. Subsequently, the patient experiences a fulminant intravascular acute hemolytic transfusion reaction and dies of multiorgan failure 3 days later.

In which of the following ways should the Food and Drug Administration (FDA), Center for Biologics Evaluation and Research (CBER) be notified of this transfusion-related fatality?

A. Notification by phone, fax, or email as soon as possible
B. Notification by phone, fax, or email within 24 h
C. Notification by phone, fax, or email within 7 days
D. Written notification as soon as possible
E. Written notification within 24 h

Concept: Transfusion related fatalities must be reported to the FDA as soon as possible via phone, fax, or email, followed by a written report within 7 days (see link http://www.fda.gov/BiologicsBloodVaccines/SafetyAvailability/ReportaProblem/TransfusionDonationFatalities/). In addition, state departments of health may also require reporting.

Answer: *A*—The code of federal regulations requires that the CBER branch of the FDA, Office of Compliance and Biologics Quality (OCBQ), be contacted as soon as possible after confirming a fatal complication of blood collection or transfusion. The report can be by phone, fax, or, email. The regulation also requires the reporting facility to submit a written report of the investigation within 7 days. The other choices (Answers B, C, D, and E) do not reflect the appropriate notification method or time frame.

Transfusion Medicine, Apheresis, and Hemostasis. http://dx.doi.org/10.1016/B978-0-12-803999-1.00012-2

2. Which of the following lists the three most common FDA reported transfusion related fatalities from 2010 to 2014?
   A. ABO HTR (hemolytic transfusion reaction), non-ABO HTR, TACO (transfusion-associated cardiac overload)
   B. Anaphylaxis, TRALI (transfusion-related acute lung injury), TACO
   C. Microbial infection, TRALI, GVHD (graft-versus-host disease)
   D. Non-ABO acute hemolytic reaction, microbial infection, TACO
   E. TRALI, TACO, non-ABO HTR

   Concept: The FDA compiles and publishes the statistics on the reported adverse events secondary to the transfusion of blood products. This allows the blood banking community to monitor trends. For example, the recognition of increasing reports of TRALI and related fatalities led to the efforts to mitigate the risks of TRALI by increasing donor antihuman-leukocyte antibody (anti-HLA antibody) testing and moving to a plasma supply that is ~90% from male donors.

   Answer: *E*—The lay public and many health care providers may mistakenly think that transfusion transmitted infectious disease and ABO incompatible acute hemolytic transfusion reactions (ABO-HTR) are the most common adverse transfusion events. However, the most common transfusion related fatality reported to the FDA during the period 2010–14 was transfusion-related acute lung injury (TRALI), representing 42 of 176 (41%) of the fatalities reported, followed by transfusion-associated circulatory overload (TACO) (22%), non-ABO-HTR (14%), and microbial infection (8%). Overall, however, the most common transfusion related adverse events (including nonfatal) reported by transfusion services are febrile nonhemolytic transfusion reactions, allergic reactions, and delayed serologic reactions (Health and Human services NBCUS 2011 survey; 51,000 total reported to transfusion services). The other choices (Answers A, B, C, and D) do not represent the correct combination of adverse events.

3. Urticarial reactions are one of the most common reactions seen with the transfusion of blood components. Which of the following is the best way to prevent urticarial reactions in patients with no history of such reactions?
   A. Premedicate with acetaminophen
   B. Premedicate with diphenhydramine
   C. Premedicate with prednisone
   D. Premedicate with acetaminophen and diphenhydramine
   E. No treatment is necessary

   Concept: Urticarial reactions are the mildest form of allergic transfusion reactions, and patients typically present with epidermal wheals or hives and pruritus. Symptoms may develop at any time during or shortly after a transfusion. The hives may occur on any part of the body and can vary in size and severity. More severe cases may be associated with angioedema or dermal swelling most typically around the eyes and mouth and rarely includes the throat and tongue. These reactions respond well to treatment with antihistamines with or without steroids. These reactions occur when preformed IgE antibody reacts with proteins in the transfused blood product. The IgE-protein complexes bind to Fc receptors on mast cells and basophils leading to degranulation and release of bioactive mediators including histamine, leukotrienes, and prostaglandins that cause vasodilation and smooth muscle contraction (type I hypersensitivity).

   Answer: *E*—Randomized controlled trials have failed to demonstrate that premedication decreases the risk of allergic (diphenhydramine) or febrile nonhemolytic transfusion reactions (acetaminophen). Premedication is generally not indicated for patients with no history of allergic reactions. Premedication with antihistamines or corticosteroids may be used in patients with recurrent urticarial reactions or severe allergic reactions. If necessary, diphenhydramine should be administered 30 min prior to transfusion if given orally and 10 min prior to transfusion if given intravenously. The optimal timing and dosage for prophylactic steroid administration has not been determined. The other choices (Answers A, B, C, and D) are incorrect because premedication is not necessary in patients without history of allergic reactions to blood products.

4. A 13-year-old girl underwent an allogeneic hematopoietic progenitor cell (HPC) transplant 1 week ago. She has received multiple units of platelets to keep the platelet count >10,000/μL. Today, after transfusion of a unit of platelets, she developed an urticarial rash on her chest without fever, wheezing, stridor, hypotension, or facial edema. Which of the following would be the best treatment of this reaction?

A. Stop the transfusion immediately and administer epinephrine 0.3 mg subcutaneously
B. Stop the transfusion, administer diphenhydramine, and restart the unit if the urticarial reaction resolves
C. Stop the transfusion, administer diphenhydramine, and administer a unit of washed red blood cells
D. Continue the transfusion and administer diphenhydramine
E. Continue the transfusion and administer diphenhydramine and acetaminophen

Concept: Nonhemolytic reactions (allergic and febrile nonhemolytic) are the most common transfusion reactions reported to transfusion services. Allergic reactions are usually due to reactions against plasma proteins. Allergens can be passively transmitted by transfusion in a dependent fashion; the allergen binds to antibodies, the Fc fragment binds to Fc receptor on mast cells, and then mast cell degranulation leads to histamine release. Mild to moderate allergic reactions consist of urticaria, erythema, and pruritus. Moderate to severe reactions (anaphylactic) consist of varying degrees of angioedema, laryngeal edema with stridor, wheezing, and hypotension.

Answer: *B*—At most institutions, mild allergic reactions can be treated by pausing the transfusion, administering diphenhydramine, and then continuing the transfusion if the symptoms resolve. Future transfusions can be managed with premedication (e.g., diphenhydramine, steroids, etc.), although as stated in Question 3, this strategy is not clearly proven to prevent allergic reactions. Although, corticosteroids and H2 antagonist can be administered, it is not clear that there is any significant benefit. If a patient has severe reactions that are refractory to pharmaceutical premedication, then washed products may be provided. In addition, IgA-deficiency and/or haptoglobin deficiency (in Japanese patients) should be ruled out when severe reactions are experienced.

Of note, mild allergic transfusion reactions, such as the one described here, are the only transfusion reaction for which you may restart the transfusion, if the reaction resolves after administering diphenhydramine. For any other transfusion reactions, you must stop the transfusion and complete a transfusion reaction workup. Although the patient experiences mild allergic reaction, the transfusion must be stopped and the patient should be observed for symptoms resolution before transfusion can be continued (Answers D and E). The patient does not experience any laryngeal edema and has stable vitals; thus, epinephrine is not necessary at this time (Answer A). Furthermore, since this is only a mild allergic reaction, washed RBCs are not indicated (Answer C).

5. A 42-year-old male is receiving a unit of RBCs prior to a below the knee amputation due to a presurgery hemoglobin of 6.9 g/dL. One minute into the transfusion, the patient becomes severely dyspneic, his blood pressure drops to 60/25, and he passes out. Intubation and resuscitation with epinephrine is required for him to regain consciousness. When the patient is stabilized, he reports that he has had a similar reaction in the past with plasma transfusion, but the cause was never identified. Which of the following is the most likely cause of his symptoms?

A. IgM deficiency
B. IgE deficiency
C. IgA deficiency with anti-IgA present
D. Epinephrine
E. Unrelated to the transfusion

Concept: IgA deficiency is the most common primary immunodeficiency, but the prevalence varies widely among different populations. Relative IgA deficiency is most common in Caucasians with a prevalence of about 1/500 individuals; however, absolute IgA deficiency is rare. Absolute IgA deficiency is defined as IgA levels <0.05 mg/dL. Patients with absolute IgA deficiency can form class specific anti-IgA antibodies that are thought to potentially cause anaphylactic transfusion reactions. Patients with relative IgA deficiency can occasionally form subclass specific anti-IgA antibodies that have been associated with mild allergic transfusion reactions, but not anaphylactic reactions as seen earlier. Patients with a history of anaphylactic transfusion reactions and documented absolute IgA deficiency with anti-IgA antibodies should receive plasma-containing components (such as plasma or platelets) from absolute IgA deficient donors, since washing these products (e.g., plasma) is not possible or causes activation (e.g., platelets). RBCs can often be washed 6 times to prevent anaphylactic reactions in patients with anti-IgA. Patients with haptoglobin deficiency (usually in people of Japanese origin) or deficiencies in complement components, usually C4, have also been reported to form antibodies that can cause anaphylactic transfusion reactions.

Answer: *C*—Allergic/anaphylactic transfusion reactions are most commonly due to the patient's response to the transfusion of donor proteins to which the recipient is presensitized. These reactions are type 1 hypersensitivity reactions that are not fully understood but are due to patient factors, donor factors, and component storage factors. Anaphylactic reactions due to anti-IgA antibodies in IgA deficient patients occur but are uncommon. Anaphylactic reactions tend to happen at the very beginning of the transfusion. It is important to note that such anaphylactic reactions can even happen in "packed" RBCs because there is still a small amount of plasma that is not removed. IgM and IgE deficiency (Answer A and B) are not known to lead to anaphylactic reactions. Epinephrine is not known to cause allergic reaction (Answer D). Indeed, it is a mainstay of treatment for severe allergic/anaphylactic reactions. This reaction, especially with the patient's history of a similar reaction, does appear to be related to the transfusion (Answer E).

6. A 57-year-old woman receiving chemotherapy for breast carcinoma has a platelet count of 13,000/μL and requires insertion of a tunneled central venous catheter. The surgeon would like the platelet count to be greater than 20,000/μL for the procedure. Fifteen minutes after receiving an irradiated, single donor apheresis platelet unit, the patient's temperature rises from 36.5 to 37.6°C. The patient is otherwise well appearing and has no complaints. The blood pressure is 120/70 mmHg, heart rate 77 bpm, and respiratory rate 16/min, and are relatively unchanged from pretransfusion vitals. Her temperature returned to normal in 30 min without intervention. The platelet unit had been discarded and was not returned to the blood bank. Which of the following interventions would most likely decrease the incidence of this presumed transfusion reaction?

A. Premedication with diphenhydramine
B. Using diversion pouch during collection
C. HLA-matching
D. Irradiation
E. Leukoreduction

Concept: Febrile nonhemolytic transfusion reactions (FNHTRs) and mild to moderate allergic transfusion reactions account for the majority of adverse reactions reported to hospital transfusion services (FNHTR ~1:1000 transfusions). FNHTRs are generally thought to be due to cytokines that are released by granulocytes during storage or due to anti-leukocyte antibodies in the recipient that activate leukocytes in the blood product.

Per the National Healthcare Safety Network (NHSN) guidelines, febrile nonhemolytic transfusion reactions (FNHTRs) are defined as follows:

Occurs during or within 4 h of cessation of transfusion

And Either

Fever (greater than or equal to 38°C/100.4°F oral and a change of at least 1°C/1.8°F) from pretransfusion value)

Or

Chills/rigors are present

It is not life threatening and is not associated with hemolysis. However, since fever may be the only initial sign of an acute hemolytic transfusion reaction, it is important to stop the transfusion and perform a thorough transfusion reaction workup [which usually includes at a minimum of a clerical check, visual check for hemolysis, and direct antiglobulin test (DAT)]. FNHTRs can be treated with acetaminophen and if rigors are present, meperidine may be useful. Patients with a history of repeated FNHTRs may benefit from premedication with acetaminophen and in cases of severe cytokine sensitivity, washing of RBCs may be necessary to proceed with transfusion.

Answer: *E*—FNHTRs occur when recipient anti-leukocyte or anti-HLA antibodies activate infused donor white cells or cytokines are released during storage and passively transfused into the recipient. Since donor leukocytes are implicated in these reactions, leukoreduction either at the bedside or prestorage leukoreduced products can reduce the rate of FNHTRs. Randomized controlled trials have failed to demonstrate that premedication decreases the risk of allergic (diphenhydramine) or febrile nonhemolytic transfusion reactions in patients with no history of repeated FNHTRs or allergic reactions (Answer A). The use of diversion pouch (Answer B) reduces the incidence of bacterial contamination and not FNHTR. HLA matching (Answer C) is used in patients with platelet refractoriness and is not related to FNHTR. Irradiation (Answer D) does not reduce FNHTR—it only prevents transfusion-associated graft-versus-host disease (TA-GVHD).

7. A 55-year-old female status post-HPC transplant for acute myeloid leukemia is admitted to the Bone Marrow Transplant unit with suspected pneumonia. Throughout the past 2 days her temperature has ranged from 96.3 to 101.4°F. Today she received 1 unit of RBCs for a hemoglobin of 6.8 g/dL. She develops chills approximately 20 min after the completion of the transfusion. Her vital signs are as follows: Pretransfusion: BP 123/75, HR 67, RR 15, T 98.5°F; Intratransfusion: BP 123/75, HR 67, RR 15, T 100.6°F. She is treated with acetaminophen with no change in her symptoms and her temperature continues to fluctuate throughout the night. Which of the following is the most likely cause of her peritransfusion signs and symptoms?

A. Septic transfusion reaction
B. Febrile nonhemolytic transfusion reaction
C. Delayed hemolytic transfusion reaction
D. Acute hemolytic transfusion reaction
E. Unrelated to transfusion

Concept: FNHTRs are considered a diagnosis of exclusion, as all other causes of fever in a transfusion recipient must be ruled out. Allergic transfusion reactions and FNHTRs are the most common reactions reported to transfusion services, both occurring at a rate of approximately 1%. FNHTRs are thought to be due to the recipient's cytokine response to the interaction of recipient antileukocyte antibodies to leukocytes or platelets in the transfused blood component or the recipient's reaction to the presence of cytokines that have accumulated in the transfused blood component. Prestorage leukoreduction has greatly reduced, but not eliminated the incidence of these reactions. These reactions are treated with antipyretics. If severe rigors or shaking chills are present, meperidine is given.

Answer: *E*—The fever in this case should still be worked up, but is most likely due to her underlying condition and is therefore, unrelated to the transfusion. Hemolytic transfusion reactions, septic reactions, transfusion-related lung injury, nonimmune hemolysis along with signs and symptoms related to the patient's underlying disease including new onset infection or sepsis must be ruled out prior to diagnosing a patient with FNHTR. If repeated, severe FNHTRs occur, washing of blood products may be necessary to clear the cytokines from the products and allow the patient to receive clinically necessary transfusions. Septic reaction (Answer A) would be a more severe and immediate reaction than the symptoms described. There is no laboratory evidence of hemolytic reaction (Answers C and D). Since the patient has underlying infections and was febrile prior to the transfusion, FNHTR (Answer B) is certainly a possibility, but the most likely explanation for the peritransfusion temperatures is the patients underlying condition.

8. A 37-year-old woman G2P1001 underwent an elective C-section for a placenta accreta with a 1.6 L estimated blood loss. The patient was typed as O Rh positive and was transfused 2 units of O Rh positive RBCs and was no longer bleeding. The following day, laboratory tests showed that the hemoglobin rose only 1 g/dL, the total bilirubin was 1.7 mg/dL, and the direct bilirubin was 0.3 mg/dL. The patient was afebrile and denied back pain. The urine was straw colored. The blood bank was consulted for a possible transfusion reaction. The pre- and the posttransfusion DAT were negative. The pretransfusion sample was not hemolyzed, but the posttransfusion sample was visually slightly pink. The patient and a segment from the transfused units typed O Rh positive, and the patient's antibody screen was also negative. Which of the following would most likely be responsible for the pink plasma?
A. Infusion of 0.9% sodium chloride solution with the unit
B. Infusion of the unit through a 16 G intravenous catheter
C. Infusion of the unit through a blood warmer set to 44°C
D. Infusion of type O plasma with the unit
E. Infusion of 5% albumin with the unit

Comment: Nonimmune hemolysis may occur when RBCs lyse prior to reaching the patient. The signs, symptoms, and sequelae can closely resemble the presentation of an immune mediated acute hemolytic transfusion reaction (AHTR). Hemolysis may be due to the following reasons:
- mechanical lysis when RBCs are transfused through a small-bore needle (e.g., 21 gauge);
- during surgeries where a bypass pump is used or with intraop blood salvage devices and also with critically ill patients on extracorporeal membrane oxygenation (ECMO) circuits;
- infusion of hypoosmotic fluids, such as 0.45% normal saline or D5W with the RBCs; infusion of incompatible plasma with the RBCs (such as transfusion of O plasma to a patient with blood type A);
- thermal injury from blood warmers exceeding the safe temperature (>42°C) or inappropriate use of a microwave to warm the unit.

Answer: C—Hemolysis due to a blood warmer set to a temperature that is too high (44°C) is the likely cause of hemolysis in this case. 0.9% sodium chloride (Answer A), 5% albumin (Answer E), and ABO compatible plasma (Answer D) should not cause red cell lysis. However, it is best to not infuse any fluids or medications other than normal saline with blood components. Hemolysis can occur due to infusion of hypoosmotic fluids, such as 0.45% normal saline of D5W. The red cells lyse in the tubing and the hemolysate is infused. This can resemble in signs and symptoms an AHTR. A 16 G intravenous catheter (Answer C) is not expected to cause mechanical hemolysis.

9. Washed products are indicated for which of the following patients?
A. Patient with paroxysmal nocturnal hemoglobinuria
B. Patient with paroxysmal cold hemoglobinuria
C. Patient with selective IgA deficiency
D. Patient with recurrent severe allergic transfusion reactions
E. Patients receiving granulocyte transfusions

Concept: Severe allergic transfusion reactions are uncommon, but may be associated with severe morbidity and mortality. Symptoms of allergic transfusion reactions range from mild urticarial reactions to severe anaphylactic reactions with hypotension, laryngeal edema, and shock. Severe reactions may also be associated with gastrointestinal symptoms including nausea, vomiting, and diarrhea. These reactions are typically treated with antihistamines, corticosteroids, and supplemental oxygen. Epinephrine and β2-agonists may be administered in severe cases. Premedication is recommended in patients with a history of these types of reactions.

Answer: *D*—Washing cellular products removes plasma proteins, which decreases the risk of severe allergic transfusion reactions. Washing is no longer recommended in patients with paroxysmal nocturnal hemoglobinuria (Answer A). If the patient only has IgA deficiency without anti-IgA antibody (Answer C) and/or a history of anaphylactic reaction, nonwashed products may be provided. Other indications for washing include repeated, severe FNHTRs and for transfusions of blood products to neonates in order to reduce the risk of hyperkalemia. Washing is not indicated in patients with paroxysmal cold hemoglobinuria (Answer B) or for those receiving granulocyte transfusion (Answer E).

10. A 47-year-old woman received 2 units of RBCs for vaginal bleeding 9 years ago. The patient is now admitted for a hysterectomy for uterine fibroids. The patient is type O Rh positive and has a negative red cell antibody screen. During the procedure, the patient receives 2 units of O Rh positive crossmatch-compatible RBCs. After the surgery, the patient's hemoglobin is 8.9 g/dL, but 12 days later decreases to 7.2 g/dL and she is transfused an additional unit of RBCs. After another 2 days, the patient's red cell antibody screen is positive, anti-K antibodies are identified in the patient's plasma, and the autocontrol is positive. The DAT is positive with polyspecific and anti-IgG, but negative for C3 (complement). An eluate prepared from the patient's red cells shows an anti-K antibody. The patient is otherwise clinically stable. The haptoglobin, urinalysis, and total and direct bilirubin are all within the reference range. The hemoglobin appropriately rose from 7.2 to 8.4 g/dL after transfusion. Which of the following treatments should the patient receive?

A. Eculizumab
B. Close observation
C. Rh immune globulin
D. Intravenous immune globulin (IVIG)
E. RBC exchange transfusion

Concept: Not all delayed hemolytic transfusion reactions (DHTR) result in detectable clinical and/or laboratory findings of hemolysis. These DHTRs are more accurately called delayed serologic transfusion reactions (DSTR). As with DHTRs, the patient has an initial exposure to a RBC antigen by pregnancy or prior transfusion/transplantation and the antibody becomes undetectable by conventional blood banking methods. Future transfusions are subject to an anamnestic response and the resulting antibody production and binding to the corresponding RBC antigen may or may not be clinically significant. A DHTR can result when intra- or extravascular hemolysis results in decreasing hemoglobin, jaundice, increasing bilirubin, and hemoglobinuria. However, if there are no clinical changes or laboratory studies suggesting hemolysis, then the reaction is limited to a new blood bank/serologic finding, and is called DSTR.

Answer: *B*—Since there are no clinical or laboratory findings associated with the antibody response (i.e., hemolysis) in this case, the patient is experiencing a DSTR and not a DHTR. Therefore, close clinical monitoring (e.g., jaundice) and monitoring of laboratory studies (e.g., hemoglobin, haptoglobin, bilibrubin, LDH, urinalysis) may be all that is necessary.

In the event of significant hemolysis, treatments, such as intravenous immune globulin (IVIg) (Answer D) and RBC exchange (Answer E) can be considered. Eculizumab (Answer A) blocks the formation of the membrane attack complex and decreases complement-mediated RBC lysis. It is used in the treatment of paroxysmal nocturnal hemoglobinuria and atypical hemolytic uremic syndrome (aHUS), but it is not well studied in hemolytic transfusion reactions (although there have been a few case reports of using eculizumab in the treatment of ABO incompatible hemolytic transfusion reactions).

11. A 54-year old man with hepatitis C cirrhosis is brought in by ambulance to the emergency department for massive upper gastrointestinal bleeding, most likely from esophageal varices. He has mental status changes, hypotension, and tachycardia. Emergency, uncrossmatched O Rh positive RBCs are requested from the blood bank, since there is no type and screen sample available yet. By the time 6 units of RBCs are transfused, the pretransfusion sample is typed as O Rh positive and the antibody screen is negative.

An endoscopic procedure is performed, the bleeding ceases, and the patient is stable with a hemoglobin of 7–8 g/dL. By the fifth hospital day, the hemoglobin has dropped to 5.8 g/dL, the total bilirubin and direct bilirubin are 2.1 and 0.6, respectively, the haptoglobin is <7 mg/dL (reference range 20–200 mg/dL), and the urine is noted to be slightly pink with blood detected via dipstick, with no red cells seen on microscopic exam. The patient now has a positive antibody screen and the direct antiglobulin test is 2+ for IgG and w+ for C3/complement. Which of the following red cell alloantibodies is most likely to be identified in the patient's plasma?

A. Anti-K
B. Anti-Ch
C. Anti-Jk^a
D. Anti-Le^a
E. Anti-P1

Concept: DHTRs occur more than 24 h after the transfusion event. They are almost always due to an anamnestic response to a RBC antigen that the patient has been previously exposed to by pregnancy or previous transfusion or transplantation. During initial testing in our patient, antibody levels were too low to be detected by traditional blood banking methods. However, after re-exposure to the antigen by transfusion, the antibody is rapidly produced. Although most DHTRs are associated with extravascular clearance of red cells, intravascular hemolysis can occur. Classically, anti-Kidd (anti-Jk^a and anti-Jk^b) antibodies can behave in this manner, termed evanescence. Intravascular hemolysis is due to complement-mediated RBC lysis that is associated with decreasing hemoglobin; spherocytes on peripheral blood smear; increasing indirect bilirubin; low to absent haptoglobin; and hemoglobinuria. However, DHTRs most often cause extravascular hemolysis where RBCs are cleared by the reticulo-endothelial system. With extravascular hemolysis, red cell survival is decreased, but the laboratory changes of intravascular hemolysis are usually milder or undetectable.

Answer: C—This DHTR with intravascular hemolysis is most likely due to an anti-Kidd antibody, which classically becomes undetectable after an initial exposure. Upon a second exposure, there is a robust amnestic response that leads to extravascular hemolysis and in the case of anti-Kidd antibodies, intravascular hemolysis as well. Anti-K (Answer A) usually leads only to extravascular hemolysis. Anti-Chido, -Lewis, and -P1 antibodies (Answers B, D, and E) are usually not clinically significant.

12. Which of the following is the most common presenting symptom of an acute hemolytic transfusion reaction?

A. Fever
B. Hypotension
C. Oliguria
D. Flank pain
E. Rigors

Concept: Acute hemolytic transfusion reactions accounted for 26% (8 out of 30 total fatalities) of transfusion-related mortalities reported to the FDA for fiscal year 2014. 13% of these were due to ABO incompatibility and 13% were due to non-ABO causes including non-ABO antibodies, unidentified antibodies, or hyperhemolysis. Administration of ABO incompatible blood components may lead to rapid hemolysis and severe morbidity or mortality; however, in the majority of cases, only minor morbidity occurs. The severity of the reaction is due to multiple factors including the volume of incompatible red cells or plasma transfused, the ABO group of the product and the patient, and the strength and titer of the recipient or donor antibody.

Answer: *A*—Fever is the most common presenting symptom of acute hemolytic transfusion reactions. This is the reason that FNHTRs can never be ignored as being "just a fever" and you must complete a transfusion reaction workup to rule out acute hemolysis. Chills, rigors, and flank or back pain (Answers D and E) may also be seen in mild acute hemolytic reactions, while hypotension, disseminated intravascular coagulopathy, red urine, renal failure, and shock (Answers B and C) may be seen in more severe reactions.

Please answer Questions 13–18 based on the following clinical scenario:

An 80-year-old woman on the antiplatelet medication, clopidogrel, trips and falls and is brought into the local emergency department. She did not experience loss of consciousness. A small subdural hematoma is noted on CT scan. Complete blood count is within normal limits. Two doses of apheresis platelets are ordered. The patient is A Rh positive, but due to a shortage of platelets, O Rh positive units are prepared for her. Thirty minutes into the second unit of platelets, the patient complains of back and flank pain.

13. Which of the following should be done at this time?

A. Continue the transfusion

B. Stop the transfusion and draw a complete blood count, type and screen, and complete metabolic panel

C. Continue the transfusion, but treat with diphenhydramine to resolve her symptoms

D. Stop the transfusion, draw a complete blood count, type and screen, haptoglobin, and complete metabolic panel and send a urine sample for urinalysis

E. Stop the transfusion, draw a complete blood count, type and screen, haptoglobin, and complete metabolic panel, send a urine sample for urinalysis, and send the bag and tubing to the blood bank

Concept: All blood banks and transfusion services should have processes and procedures for the administration of blood components including the recognition, evaluation, and reporting of adverse events. These procedures should delineate the monitoring of the patient during the transfusion and when the transfusion should be discontinued. The clinical indications for pausing a transfusion along with the signs and symptoms of a potential transfusion reaction should be described. The patient should be evaluated clinically to determine if the transfusion should be discontinued and the reaction reported to the blood bank. The label on the blood component container should be compared with patient records to determine if an error occurred. The results of a complete blood count, repeat type and screen, haptoglobin, complete metabolic panel, and urinalysis will be helpful in determining if a hemolytic transfusion reaction occurred and the severity of the reaction.

Answer: *E*—Whenever a patient experiences an adverse event or change in vital signs during a transfusion, the transfusion should be stopped immediately. With the exception of urticarial reactions, the reaction should be reported to the blood bank and a full transfusion reaction workup should be completed. A posttransfusion blood sample and the blood component bag, even if empty, along with any tubing and attached intravenous lines or solutions should be sent back to the blood bank. If a patient has an urticarial reaction, it is permissible to treat the reaction and restart the transfusion if all signs and symptoms abate. All the other choices (Answers A, B, C, and D) do not represent a full transfusion reaction workup.

14. A posttransfusion sample was sent to the blood bank as part of a transfusion reaction workup. The sample is positive for hemolysis on visual inspection, as opposed to the pretransfusion sample which had no evidence of hemolysis. Which of the following is at the top of your differential diagnosis?

A. Delayed hemolytic transfusion reaction (DHTR)

B. Transfusion-related acute lung injury (TRALI)

C. Nonimmune hemolysis
D. Febrile nonhemolytic transfusion reaction (FNHTR)
E. Acute hemolytic transfusion reaction (AHTR)

Concept: A reaction that occurs within 24 h of transfusion is considered an acute transfusion reaction. Delayed transfusion reactions occur 24 h to days or weeks posttransfusion. When a transfusion reaction that may be hemolytic is reported to the blood bank, a clerical check, repeat type and screen on the pretransfusion and posttransfusion samples, a DAT, and a visual inspection on the posttransfusion plasma are performed. Cultures may be performed if bacterial contamination is suspected. Ideally, additional blood components are not transfused until the transfusion reaction workup is completed and the cause of the reaction is identified.

Answer: *E*—Fever, flank pain, and back pain are common presenting symptoms of an AHTR. The patient must be treated as if an AHTR has occurred until such a reaction can be ruled out. The signs, symptoms, and timing of the reaction are not consistent with a DHTR (Answer A), TRALI (Answer B), or FNHTR (Answer D). AHTR due to an immune cause must be ruled out before attributing the reaction to a nonimmune cause of hemolysis (Answer C).

15. According to the AABB Standards for Blood Banks and Transfusion Services, which of the following is one of the steps in the laboratory evaluation of a suspected hemolytic transfusion reaction?
A. Inspect a posttransfusion reaction sample for hemolysis and compare with a pretransfusion sample
B. Perform autoadsorption on the pretransfusion sample
C. Perform an indirect antiglobulin test on the posttransfusion sample and compare to the pretransfusion sample if positive
D. Review and provide an interpretation of all laboratory results by the supervising technologist
E. Perform bacterial cultures on the suspected unit

Concept: The AABB Standards for Blood Banks and Transfusion Services states that blood banks or transfusion services must have policies, processes, and procedures for the evaluation and reporting of suspected transfusion reactions. Additionally, there must be a process for the administration of blood and blood components that delineate the recognition, evaluation, and reporting of suspected transfusion reactions and adverse events. After a transfusion is discontinued, blood component labels and patient records must be checked to ensure that the blood component and patient were properly identified. The blood component, all tubing, any attached intravenous solutions, and a posttransfusion sample obtained from the patient must be sent to the blood bank.

Answer: *A*—For suspected hemolytic reactions, the following steps must be included in the laboratory evaluation:
- Perform a clerical check to make sure that the patient received the product and blood type intended for them
- Inspect a posttransfusion reaction plasma or serum sample for evidence of hemolysis and compare with a pretransfusion sample
- Perform a direct antiglobulin test on the posttransfusion sample and compare to the pretransfusion sample if positive
- Repeat ABO group determination on the posttransfusion sample
- Have processes to define which circumstances require additional testing and what tests should be performed
- Review and provide an interpretation by the medical director

Other choices (Answers B, C, D, and E) are incorrect based on the AABB Standards.

16. A transfusion reaction workup is ordered by the blood bank. The results are as follows:
Clerical check—No clerical error detected.

Type

	Anti-A	Anti-B	Rh	A cells	B cells
Pretransfusion	4+	0	4+	0	4+
Posttransfusion	4+	0	4+	0	4+

Antibody screen

	Pretransfusion	Posttransfusion
	Gel	Gel
Cell I	0√	0√
Cell II	0√	0√

Direct Antiglobulin Test (DAT)

	Pretransfusion	Posttransfusion
Polyspecific	Neg	4+
Anti-IgG	Neg	2+
Anti-C3d	Neg	4+

Eluate—nonreactive on all panel cells

What is the next step in the workup of this patient?

A. Repeat all the above testing
B. Order a complete metabolic panel and haptoglobin to confirm the presence of hemolysis
C. Send the sample to a reference laboratory to identify a possible antibody to a low incidence antigen
D. Run the eluate against A cells and B cells
E. Request information regarding the patient's transfusion and pregnancy history

Concept: A DAT is a routine test in the evaluation of a transfusion reaction. The red blood cells in the posttransfusion sample are washed and then tested against a polyspecific antihuman globulin to look for the presence of bound IgG or complement. If the polyspecific test is positive, two monospecific tests, anti-IgG, and anti-C3d, are performed to identify the presence of bound IgG or IgM, respectively. A positive result posttransfusion with a negative result on the pretransfusion sample indicates that an incompatible product may have been transfused. If the monospecific anti-IgG test is positive, the bound antibody is eluted from the RBCs and reacted against a panel of group O RBCs to determine the specificity of the bound antibody. Transfusion of group O platelets to a group B or more commonly a group A patient can result in an acute hemolytic transfusion reaction when the donor has high titers to group A or group B antigens.

Answer: *D*—The eluate is always tested against a panel of group O RBCs, as are most standard antibody identifications. However, if anti-A or anti-B is the cause of a reaction, it will not be detected, and the eluate must be run against group A cells and group B cells. A positive test result against group A or group B red blood cells indicates ABO incompatibility between the transfused product and the patient. The controls (i.e., check cells) demonstrated that the test results are valid, and thus, repeated testing (Answer A) may not be indicated at this time. Transfusion and pregnancy history (Answer E) are always important in transfusion testing but further history may not be helpful in this case. Visual check for hemolysis is the most sensitive test for intravascular hemolysis and order more serum markers for hemolysis (Answer B) may not be helpful to solve the problem here. There is no indication of a low incidence antibody in this case (Answer C).

17. The eluate was run against A and B cells. The result is as follows:

	Gel
A1 cell	4+
B cell	0√

Additionally, an anti-A antibody titration was performed on the remainder of the second platelet unit and found to have an anti-A titer of 256.
What is the pathophysiology of this reaction in this patient?
A. Preformed IgG in the transfused component binds to the patient's RBCs
B. Preformed IgG and IgM in the transfused component binds to the patient's RBCs
C. Preformed IgG in the patient binds to the donor's RBCs
D. Preformed IgG and IgM in the patient binds to the donor's RBCs
E. Nonimmune hemolysis of the transfused blood component due to small gauge needle used for transfusion

Concept: AHTRs occur when either incompatible RBCs are transfused to a patient with preformed antibodies or incompatible plasma with preformed antibodies is transfused to a patient with the cognate antigen. The preformed IgG or IgM binds to the incompatible antigen, complement is activated, thereby, resulting in acute intravascular hemolysis. Opsonization by IgG can lead to rapid phagocytosis of red blood cells and cytokine activation further aggravating the effects of intravascular hemolysis. Complement activation and cytokine release leads to neutrophil activation, release of anaphylatoxins, activation of the coagulation cascade, platelet activation, free radical formation, vasodilation, and smooth muscle contraction. This leads to flank and back pain, gastrointestinal symptoms, disseminated intravascular coagulopathy and thrombi formation, end organ damage due to ischemia and free radical injury, hypotension, and shock. The free hemoglobin from the hemolyzed red blood cells is directly nephrotoxic.

Answer: *B*—The patient in this case is type A Rh positive. The transfused platelets were group O with a high titer of anti-A antibody. This is confirmed from the eluate result—the antibody has A specificity. Group O individuals can make both anti-A and anti-A,B antibodies which are IgM and IgG isotype antibodies, respectively. The severe morbidity in this patient is likely due to high titers of anti-A IgM isotype antibodies. Many institutions may decide to only transfuse low-titer O (anti-A and anti-B titer <50) platelets to non-O patients to prevent reverse hemolysis. Furthermore, due to the small blood volume, it is usually recommended to transfuse ABO identical platelets to neonates to prevent this type of AHTR. The other choices (Answers A, C, D and E) are incorrect based on the explanation above.

18. The patient is now vomiting and develops red urine. Her temperature increases to 101.5°F, and she continues to have flank and back pain. What is the next step in the treatment of this patient?
A. Aggressively infuse intravenous fluids
B. Start hemodialysis
C. Transfuse a new dose of apheresis platelets
D. Transfuse emergency release RBCs
E. Aggressively diurese the patient

Concept: The treatment of an AHTR includes prompt recognition of the reaction and discontinuation of the transfusion and infusion of saline to maintain renal blood flow and urine output. Low dose dopamine and diuretics, such as furosemide may be used to maintain renal function. Supportive care should be used to manage hypotension, disseminated intravascular coagulation, and electrolyte abnormalities. Consulting a nephrologist may be beneficial in managing the renal status of the patient and any electrolyte abnormalities.

Answer: *A*—Aggressive infusion of intravenous crystalloid fluids is the mainstay of treatment for this patient at this time. Later, if the patient becomes fluid overloaded, diuretics, such as furosemide may be considered. If the patient is determined to have acute kidney injury, dialysis (answer B) may be initiated. Aggressively diuresing this patient (Answer E) may be harmful initially since it may add more insults to the kidney by reducing the blood flow to the kidneys that are already exposed to the toxic free hemoglobin from the hemolysis (i.e., cause prerenal injury on top of renal injury). Unless the patient develops signs and symptoms of anemia and thrombocytopenia from the disseminated intravascular coagulation (DIC), RBCs and platelets (Answers C and D) are not needed at this time.

End of Case

Please answer Questions 19–21 based on the following clinical scenario:

A 55-year-old woman with liver cirrhosis presents to the emergency department with a severe upper gastrointestinal bleed. Her hemoglobin is 4.7 g/dL and her vital signs are HR 98 BP 90/60 RR 16 T 37.0°C. The physician orders 2 units of RBCs to be infused prior to endoscopy. The first unit was administered over 1 h without complications, but after 25 mL of the second unit was infused, the patient developed shortness of breath and the oxygen saturation dropped to 86%. Vital signs at this time were HR 111 BP 88/58 RR 22 T 37.2°C. The patient was subsequently intubated and frothy fluid was observed in the endotracheal tube. Her chest radiograph was clear prior to transfusion, but now she has bilateral infiltrates. An echocardiogram shows an ejection fraction of 67% and normal left atrial size. Her brain natriuretic peptide (BNP) level is 35 pg/mL (reference range: <100 pg/mL).

19. Which of the following measurements in the *recipient* is part of the diagnostic criterion for transfusion-related acute lung injury (TRALI) according to the National Health Safety Network's (NHSN) Hemovigilance Protocol?
A. Anti-human neutrophil antigen (HNA) antibodies are detected
B. Anti-heparin/platelet factor 4 antibodies are detected
C. Increased BNP
D. Normal troponin levels
E. $PaO_2/FiO_2 \leq 300$ mmHg

Concept: Transfusion-related acute lung injury (TRALI) presents as respiratory distress occurring within 6 h of administration of any plasma containing product (e.g., RBCs, platelets, plasma, cryoprecipitated AHF). Donor anti-HLA (human leukocyte antigen) or anti-HNA (human neutrophil antigen) antibodies that are passively transfused to the recipient through plasma containing blood components are thought to be the cause of TRALI; however, the exact mechanism is not known. During certain activated states in the recipient or amid the presence of bioactive response materials, the patient's leukocytes can be activated causing damage to pulmonary endothelium. The resulting pulmonary leakage results in an acute respiratory distress-like clinical picture.

For the definitive diagnosis of TRALI, the following criteria should be met:

No evidence of acute lung injury (ALI) prior to transfusion
And
Onset of symptoms within 6 h of cessation of the transfusion
And
Hypoxemia defined by any of these methods:

- $PaO_2/FiO_2 \leq 300$ mmHg
- O_2Sat $< 90\%$ on room air
- Other clinical signs of respiratory distress, such as increased respiratory rates

And
Radiographic evidence of bilateral infiltrates
And
No evidence of left atrial hypertension or circulatory overload (e.g., normal BNP levels)

The differential diagnosis of TRALI may include cardiogenic pulmonary edema (TACO), bacterial contamination, and anaphylaxis; however, TRALI is usually a diagnosis of exclusion.

Answer: *E*—Although laboratory testing of the donor (anti-HLA or anti-HNA antibodies, Answer A) and recipient (HLA antigen typing) may be supportive, TRALI is a clinical diagnosis that requires hypoxemia, which can be measured by the PaO_2 on arterial blood gas. Increased BNP levels (Answer C) are usually observed in patients with transfusion-associated circulatory overload (TACO). Patients with TRALI usually have normal BNP level. Troponin levels (Answer D) have no role in the diagnosis of TRALI. Anti-heparin/platelet factor 4 antibodies (Answer B) are detected in patients with heparin induce thrombocytopenia, not TRALI.

20. Which of the following would be the best treatment for this transfusion reaction?
A. Broad spectrum antibiotics
B. Epinephrine 0.3 mg subcutaneously
C. Diuresis with furosemide
D. Oxygen and mechanical ventilation
E. Therapeutic plasma exchange

Concept: The pulmonary injury caused by TRALI does not have a specific therapy, but patients clearly benefit from supportive therapy with oxygen and mechanical ventilation or positive airway pressure, as needed. Most of the reactions resolve over 48–72 h (~80%), but others may have clinically worsening disease. The overall mortality rate for TRALI is estimated to be somewhere in the range of 5%–10%, but with recent mitigation strategies (see Question 21), that number is decreasing.

Answer: *D*–Treatment for TRALI is mostly supportive. Oxygenation and mechanical ventilation or positive airway pressure support may be necessary. Distinguishing TRALI from cardiogenic pulmonary edema is important because diuretics (Answer C) are therapeutic for TACO, but potentially contraindicated in TRALI. For TACO, diuretics relieve the intravascular fluid overload. However, with TRALI, the fluid is localized to the lungs, thus giving diuretics risks that intravascular volume will be decreased unnecessarily with resulting hypotension. Corticosteroids are often administered, but have not been proven to be beneficial. Antibiotics (Answer A) may play a role in septic reaction. Epinephrine (Answer B) is an important treatment modality in anaphylactic reaction. Neither antibiotics and epinephrine nor plasma exchange (Answer E) has a role in the treatment of TRALI.

Case Continued…

After 24 h of supportive care with mechanical ventilation, the patient improves and is extubated. There has been significant improvement in the chest radiograph and the infiltrates are nearly resolved. You immediately report the adverse event to the blood collection facility. The donor center requests a human leukocyte antigen (HLA) type of the patient, which comes back as HLA-A2, 28; B8, 35; DR15, 17.

21. Which of the following laboratory results found in the blood *donor*(s) would best support a diagnosis of TRALI?
A. Anti-HLA-A2 antibodies
B. Anti-HLA-B37 antibodies
C. Anti-HPA-1a antibodies
D. Elevated platelet count
E. Leukopenia

Concept: Donor antibodies against recipient leukocytes have been implicated in the pathogenesis of TRALI. However, antibodies are not always detected in all cases of clinically diagnosed TRALI. The most common antibodies that are detected in donors implicated in TRALI are anti-HLA antibodies and anti-HNA (human neutrophil antigen) antibodies. The donor population at the greatest risk of having alloantibodies is multiparous women. Therefore, one TRALI mitigation strategy is to defer multiparous women from donating plasma and platelets. Another, more expensive option is to test all female donors for anti-HLA/HNA antibodies. This strategy increases the testing costs, but avoids indiscriminately disqualifying women from donating plasma products. The 30th edition of *Standards for Blood Banks and Transfusion Services* from the AABB states the plasma and whole blood for allogeneic transfusion must be from male donors, nulliparous females, or female donors who have tested negative for HLA antibodies.

Answer: *A*—The patient expresses the HLA-A2 antigen and the donor's plasma contains an anti-HLA-A2 antibody. Therefore, this antibody is the most likely cause of TRALI in this case. Although the donor also has an anti-HLA-B37 antibody (Answer B), the patient does not express this HLA antigen, which decreases the suspicion that it is pathogenic. Going forward, this donor will likely be permanently deferred from donating plasma products, since they have been proven to be the source of TRALI in our patient. Human platelet antigen (HPA) antibodies (Answer C) have not been implicated in the pathogenesis of TRALI. The patient (not the donor) may have leukopenia (Answer E) at the time the signs and symptoms of TRALI occur. Thrombocytosis (Answer D) is not part of the TRALI pathogenesis.

End of Case

22. A 78-year old woman takes warfarin daily due to a past medical history of atrial fibrillation, myocardial infarction (status post coronary artery bypass grafting), and congestive heart failure. Her admission BNP level is 336 pg/mL (reference range: <100 pg/mL). She presents to the ED after a fall at home and is diagnosed with an intracranial bleed. She is transfused with 3 units plasma (group A, type specific). Midway through the third unit of plasma, the patient becomes acutely dyspneic and her oxygen saturation drops from 99% to 68% on room air. Her other vital signs are as follows: HR 112, BP 153/98, RR 29, T 97.8°F. A chest radiograph reveals bilateral infiltrates and a posttransfusion BNP level is 912 pg/mL. A transfusion reaction workup revealed no evidence of hemolysis or incompatibility. Which of the following is the most likely cause of her symptoms?

A. Anaphylactic transfusion reaction
B. TRALI
C. AHTR
D. TACO
E. Unrelated to the transfusion

Concept: TACO is defined as acute pulmonary edema due to intravascular fluid overload. Definitive diagnosis of TACO is aided by following the NHSN guidelines:
New onset or exacerbation of three or more of the following within 6 h of cessation of transfusion:

- Acute respiratory distress (dyspnea, orthopnea, cough, hypoxemia)
- Elevated brain natriuretic peptide (BNP)
- Elevated central venous pressure (CVP)
- Evidence of left heart failure
- Evidence of positive fluid balance (e.g., hypertension, tachycardia, jugular venous distension)
- Radiographic evidence of pulmonary edema, typically bilateral

Elderly patients, infants, and patients with a history of congestive heart failure are at greatest risk of TACO; however, patients receiving large volumes of blood products, especially at high flow rates, are also at risk. Treatment includes supplemental oxygen and diuretics to decrease intravascular volume. Slow transfusion rates, splitting blood products, close monitoring, and use of diuretics in patients at risk may be useful in prevention of TACO.

Answer: *D*—Hypoxemia, hypertension, tachycardia, and an elevated BNP after transfusion of multiple units of plasma in an elderly female with a history of congestive heart failure strongly supports a diagnosis of TACO. The presence of an elevated BNP as evidence of left atrial hypertension is especially useful in cases of respiratory distress after transfusion, as such a finding argues against a diagnosis of TRALI. Use of prothrombin complex concentrates in patients on warfarin with intracranial bleeding can rapidly correct coagulation factor deficiencies without risk of volume overload and should be considered in these patients. The clinical signs and laboratory values in this case do not support a diagnosis of an anaphylactic reaction (Answer A), TRALI (Answer B), or acute hemolysis (Answer C). Answer E is wrong because the patient clearly has evidence of a reaction related to the transfusion.

23. TACO is commonly associated with which one of the following?
A. Anti-HNA antibodies
B. Anti-IgA antibodies
C. Leukopenia
D. Electrocardiographic changes
E. Pulmonary edema

Concept: TACO is volume overload associated with the transfusion of blood components. Patients receiving large volumes of blood components at high transfusion rates are at increased risk of developing TACO, but TACO may also be seen with transfusion of small volumes of blood products even at slow transfusion rates in patient populations at risk. Infusion volumes of other fluids and medications along with the cardiac and renal status of the patient are important contributing factors. Patients with congestive heart failure are particularly susceptible to developing TACO. Elderly patients and infants are at greatest risk, but TACO has been reported in all age groups across many different patient populations. TACO is most frequently reported after transfusion of red blood cells but is also commonly seen with plasma transfusions. TACO is thought to be underrecognized and underreported; however, recent advances in hemovigilance programs have led to increased awareness of this complication.

Answer: *E*—Conceptually, TACO is acute pulmonary edema caused by transfusion-related intravascular volume overload. Signs and symptoms may include dyspnea, orthopnea, jugular venous distention, elevated brain natriuretic peptide, hypertension, and chest radiograph changes including a widened cardiothoracic ratio. This must be distinguished from TRALI, which is a noncardiogenic pulmonary edema, because the treatment of the two conditions is very different. Anti-HNA (Answer A) is involved in TRALI, as explained earlier. Anti-IgA (Answer B) can be responsible for anaphylactic transfusion reactions. Leukopenia (Answer C) is not associated with TACO, but can be seen within 4 h of TRALI, if a CBC is drawn within that time frame. This is due to leukocyte sequestration within the lung parenchyma. EKG changes (Answer D) are not typically seen in TACO.

24. A 33-year-old woman with acute myeloid leukemia underwent matched unrelated peripheral blood hematopoietic progenitor cell (HPC) transplant 1 week ago. Which of the following blood product types/modifications should be provided to prevent transfusion associated graft versus host disease (TA-GVHD)?
A. CMV-negative
B. HLA-matched
C. Irradiated
D. Leukoreduced
E. Washed

Concept: Chronic graft versus host disease (cGVHD) is often seen after allogeneic transplantation and is treated with increased immunosuppression. Unlike cGVHD, TA-GVHD is a nearly uniformly fatal complication of transfusion. TA-GVHD is an immunologic complication of transfusion that can occur if an immunosuppressed patient is transfused T lymphocytes. Immunocompromised patients are not able to destroy the transfused lymphocytes, which leads to TA-GVHD. The first sign is usually a skin rash, but later the lymphocytes attack the patient's skin, gastrointestinal tract, liver, and bone marrow. Inactivation of lymphocytes by irradiation of blood products containing lymphocytes (e.g., red cells, platelets, granulocytes) is used to prevent TA-GVHD. Irradiation is most commonly performed using radioactive cesium or X-ray irradiators, but amotosalen/UV irradiation can also achieve the same result.

The following are guidelines for irradiation indications and nonindications:

Well-documented Indications

- Patients with congenital cellular immunodeficiency disorders, such as DiGeorge Syndrome
- Neonates from families with previous child known or suspected of having inherited immunodeficiency
- Intrauterine transfusions
- Low birth weight neonates < 1200 g or prematurity
- Hematologic malignancies and solid tumors (neuroblastoma, sarcoma, Hodgkin Disease)
- Autologous and allogeneic HPC transplant recipients
- Post-natal transfusions up to 4 months after intrauterine transfusions
- Recipients of directed donations from biologic relatives
- HLA-matched or crossmatch-compatible platelet donations
- Current or past exposure to fludarabine, nelarabine, cladribine, clofarabine, and deoxycoformycin
- Patients treated with drugs that affect T-cell number or function, such as Campath (anti-CD52)
- Granulocyte transfusions

Recommended Indications

- Aplastic anemia
- Patients with solid tumors receiving intensive chemotherapy and/or radiation

Nonindications

- Solid organ transplant recipients
- Healthy, full-term neonates
- Patients with HIV/AIDS

Answer: C—TA-GVHD is prevented by inactivating lymphocytes in the blood component (e.g., red cells, platelets, granulocytes) by gamma irradiation or amotosalen/UV irradiation. If irradiation is used, standards require that the blood product receive at least 2500 cGy at the center and at least 1500 cGy at any part of the bag. Leukoreduction (Answer D) or washing (Answer E) is not sufficient to reduce the number of white cells to a level that eliminates the risk of TA-GVHD. Although CMV-negative products (Answer A) may be indicated for some allogeneic HPC transplant patients, such products provide no protection against TA-GVHD. HLA-matched products (Answer B) require irradiation but it is not a product modification to reduce the risk of TA-GVHD.

25. A 55-year-old man with hepatitis C cirrhosis is undergoing a deceased donor liver transplant. After the transplanted liver is reperfused, the patient starts massively bleeding and a massive transfusion protocol (MTP) is activated. The patient receives 18 units of RBCs, 12 units of plasma, 3 units of single donor platelets, and 5 units of cryoprecipitated AHF. Which of the following electrolyte disorders would be expected to result specifically from the rapid infusion of citrate in the anticoagulant?

A. Acidosis
B. Hypermagnesemia
C. Hypocalcemia

D. Hypernatremia
E. Increased zinc

Concept: A patient undergoing resuscitation for massive trauma faces injury from two sources. First, the trauma itself leads to tissue damage, coagulopathy, and acidosis. Second, the rapid infusion of blood components can lead to hyperkalemia, hypokalemia, hypothermia, and accumulation of citrate. Citrate is the anticoagulant present in blood components and when transfused in moderate amounts, the liver rapidly metabolizes the citrate. However, when blood components are rapidly infused (e.g., in MTP), citrate can bind a significant amount of calcium, zinc, and magnesium. Severe hypocalcemia can lead to paresthesias, muscle spasms, tetany, seizures, cardiac changes with decreased ejection fraction and hypotension, and hypocoagulability. Since stored red cells leak potassium as they get older, massive transfusion can lead to hyperkalemia. In addition, hypokalemia can also occur as potassium depleted donor cells reaccumulate potassium intracellularly. Monitoring and correction of calcium, potassium, and pH during an MTP are important.

Answer: *C*—During a massive transfusion, large volumes of blood components and the accompanying anticoagulant, citrate can be infused. Citrate chelates calcium, magnesium, and zinc leading to hypocalcemia and hypomagnesemia. Hypocalcemia can result in spasms, seizures, cardiac changes, and impaired coagulation. The other choices (Answers A, B, D, and E) are not associated with citrate toxicity.

26. Which of the following patient populations is most at risk of developing hyperkalemia from massive transfusion?
A. Trauma patients
B. Patients in liver failure
C. Elderly patients
D. Newborns
E. Patients with diabetes

Concept: Red cell storage lesions are changes to a RBC unit over the storage period. Glucose, adenine triphosphate (ATP), and 2,3-diphosphoglycerate (2,3-DPG) levels decrease while potassium levels increase due to extracellular leakage into the storage solution. The amount of extracellular potassium increases over the life of the RBC unit. This rarely causes any clinical effect in transfused patients, as the relative quantity of extracellular potassium is low. It is thought that the free potassium is rapidly diluted in the patient's bloodstream and is redistributed both within the patient's own RBCs and the transfused RBCs, as ATP is repleted and the sodium-potassium pumps are restored. Additionally, in patients with normal renal function, the extra potassium is easily excreted in the urine.

Answer: *D*—Newborns and premature infants receiving massive or exchange transfusion, fetuses receiving intrauterine transfusions, and patients with renal failure receiving massive transfusions are at the greatest risk for developing hyperkalemia, due to their inability to process excess potassium. Additionally, rapid transfusion of large volumes of RBCs through a central venous catheter can cause transitory hyperkalemia that may cause cardiac arrhythmias and resultant electrocardiographic changes. Patients with acute liver failure (Answer B) are at risk of citrate toxicity, not hyperkalemia. Neither trauma patients (Answer A) alone (without receiving massive transfusions) nor patients with diabetes (Answer E) or elderly patients (Answer C) without renal failure are at risk of hyperkalemia.

Please answer Questions 27 and 28 based on the following clinical scenario:

A 55-year-old white woman, G3P3003, without significant past medical or surgical history underwent a hysterectomy for a prolapsed uterus. During the procedure, there was unanticipated bleeding and her hemoglobin fell from 11.5 g/dL preoperatively to 6.9 g/dL. She was transfused 2 units of RBCs without incident. On the postoperative day #7, she presented to her primary care physician (PCP) because she developed a purpuric rash over her lower extremities and moderate epistaxis. A CBC revealed her platelet count had dropped from 234,000/μL pretransfusion to 4,000/μL. Her aPTT and fibrinogen were all within the reference range. There were no schistocytes or platelet clumping observed on the peripheral blood film. Her PCP ordered a transfusion of 1 unit of apheresis platelets, but there was no rise in her platelet count. A heparin-induced thrombocytopenia (HIT)-PF4 ELISA assay performed the same day was negative. One day later, an antiplatelet alloantibody was identified and the patient was diagnosed with posttransfusion purpura.

27. Which of the following platelet antigens is the antibody most likely directed against?

A. HPA-1a
B. HPA-1b
C. HPA-3a
D. HPA-4b
E. HPA-5a

Concept: Posttransfusion purpura (PTP) is a rare and potentially life threatening thrombocytopenia that occurs 5–10 days after a transfusion, usually of RBCs. It is more common in multiparous women who may have preformed alloantibodies against platelet antigens, such as HPA-1a (60% of cases), but also against HPA-3a, -3b, -4a, -5a, and -5b. Interestingly, the alloantibody in PTP leads to not just clearance of transfused platelets, but of native platelets as well. The mechanism for this bystander effect is not known. PTP can be mistaken for other underlying causes of thrombocytopenia, such as medications, sepsis, disseminated intravascular coagulation, and heparin induced thrombocytopenia. Therefore, there must be a high suspicion for PTP. Testing of the patient will show that the patient has an anti-HPA-1a alloantibody and lacks the implicated platelet antigen HPA-1a (i.e., the patient is homozygous HPA-1b).

Answer: *A*—The majority of PTP is caused by an anti-HPA-1a antibody in a patient who is HPA-1b/b. This represents about 60% of cases. Less common antibodies implicated in PTP include anti-HPA-3a, -3b, -4a, -5a, and -5b alloantibodies. The allelic frequency of platelet antigens is dependent on the population being studied. For example, virtually no Asian patients are HPA-1b/b whereas 2% of Caucasians are HPA-1b/b. The other choices (Answers B, C, D, and E) are incorrect due to the explanation earlier.

28. If the patient now has bleeding, which of the following is the best treatment option?

A. Transfusion of platelet from random donors daily
B. Intravenous immune globulin
C. Therapeutic plasma exchange
D. Intravenous Rh immunoglobulin infusion
E. Corticosteroids

Concept: Posttransfusion purpura (PTP) is a rare complication of transfusion that can occur when a patient with preformed human platelet antigen (HPA) alloantibodies receives a blood component containing platelets expressing the cognate antigen. The patient's own platelets along with the transfused platelets are destroyed by an unknown mechanism. These patients typically present with profound thrombocytopenia and wet purpura approximately 10 days posttransfusion. PTP most commonly occurs after transfusion of RBCs or whole blood, but may also be seen after transfusion of plasma or platelets. Antibodies against HPA-1a are identified in the majority of cases. The associated thrombocytopenia is usually self-limited and resolves within 2 weeks; however, in cases with severe bleeding, intravenous immune globulin (IVIG) is the treatment of choice. Platelet transfusions in these patients are not harmful, but are also unlikely to be helpful because the transfused platelets will be quickly destroyed. Steroids and therapeutic plasma exchange have also been used as second-line therapies. Platelets lacking the HPA-1a antigen are often requested for these patients, but HPA-1a negative platelets are often difficult to obtain and recurrence of PTP has rarely been reported.

Answer: *B*—This patient demonstrates a classical presentation of PTP. Watchful waiting or IVIG are the treatments of choice for these patients. In non-bleeding patients, watchful waiting is sufficient; however, in bleeding patients, more aggressive measures are warranted. An increase in the platelet count is usually seen within a few days after IVIG treatment. Platelet transfusion (Answer A) is of limited benefit. If the patient is massively bleeding or bleeding into critical space (such as intracranial bleeding), then HPA-1a negative platelets may be used for transfusions. Plasma exchange (Answer B) or corticosteroids (Answer E) are not first line therapies in PTP. Rh immunoglobulin (Answer D) is used to prevent anti-D formation and is not part of the management of PTP.

End of Case

29. An 18-year-old man was diagnosed with β-thalassemia major during his first year of life and has been receiving chronic red cell transfusion (1 RBC unit every 2 weeks) since childhood. Which of the following is the best estimate of the iron he receives per RBC unit?
A. 10–20 mg
B. 50–100 mg
C. 100–150 mg
D. 200–250 mg
E. 500–550 mg

Concept: Humans shed approximately 1 mg of iron per day, mostly in the form of epithelial cells shed in the gastrointestinal tract and the same amount of iron is absorbed from food. Therefore, in the absence of bleeding, there is typically no net gain or loss of total body iron. Menstruating women, however, will lose iron if not compensated by increased absorption from food or supplements. In contrast, patients with beta-thalassemia major, sickle cell disease, myelodysplastic syndrome, congenital anemias (e.g., Diamond-Blackfan), and aplastic anemia often require chronic transfusions and can become iron overloaded. Each RBC unit contains 1 mg of iron for every 1 mL of "packed" RBCs. Therefore, approximately 200–250 mg of iron is present in a typical RBC unit.

The rate of iron accumulation depends on the indication for transfusion, frequency of transfusion, duration of transfusion dependence, and whether the patient has been receiving and being compliant with iron chelation therapy (e.g., deferoxamine). This iron will accumulate in macrophages, but can also accumulate in hepatocytes, cardiac myocytes, pancreatic cells, and pituitary cells. This can lead to the long-term complications of hypertrophic or dilated cardiomyopathy with congestive heart failure, delayed puberty, diabetes, and liver cirrhosis. The other choices (Answers A, B, C, and E) do not represent the correct amount.

Answer: *D*—Each RBC unit contains approximately 200–250 mg of iron. Since there is no other method for iron to be unloaded in a nonbleeding patient, a chronically transfused patient receiving 2 RBC units per month could accumulate 400–500 mg of iron. Other choices (Answers A, B, C, and E) are not correct.

30. An 8-year-old boy with sickle cell disease (SCD) had an ischemic stroke at 5 years of age, which was acutely treated with a RBC exchange. Since then, he has been receiving simple transfusion every 2–4 weeks for secondary stroke prophylaxis at another hospital and has not been treated with hydroxyurea or chelation therapy. His ferritin measured prior to the first visit with his new hematologist was 2660 ng/mL. Which of the following is the next best step in his treatment?
A. Continue transfusions and add deferoxamine
B. Continue transfusions and monitor ferritin every 2 weeks
C. Discontinue transfusions and add deferoxamine
D. Discontinue transfusions and add hydroxyurea
E. Discontinue transfusions, add hydroxyurea, and begin phlebotomy

Concept: Pediatric patients with (SCD) can develop cerebrovascular infarcts, most commonly between 3 and 15 years of age. Without treatment, the recurrence rate for pediatric stroke in SCD is approximately 60%. However, this risk can be significantly reduced by chronic transfusion therapy, either by simple transfusion or red blood cell exchange (RBCx). One of the main complications of chronic transfusion is iron overload, which can lead to cardiomyopathy, liver cirrhosis, diabetes, and growth retardation. Monitoring a patient's iron status is important and can be monitored using serum ferritin, iron content of liver measured after biopsy, or by T2*MRI measurement of liver iron concentration. Patients benefit from early chelation therapy to prevent iron overload when receiving chronic transfusions.

Answer: *A*—Chronic transfusion is necessary to reduce the risk of a recurrent stroke. Patients on long term transfusion therapy benefit from early chelation therapy to prevent irreversible organ damage. Although discontinuing transfusions and replacing it with hydroxyurea therapy (Answer D) and phlebotomy (Answer E) has been studied, it is unlikely to be better than chronic transfusion with chelation therapy. Chronic RBCx is another option to decrease the risk of iron overload, but increases the number of units that the patient is exposed to over time, thus, may increase the chance of forming alloantibodies. The ferritin level is already very high and it will increase with continued transfusion (Answer B).

31. Which of the following adverse reactions is associated with granulocyte transfusions?
A. Increased risk of Zika virus transmission in Zika-negative patients
B. Increased risk of transfusion-associated circulatory overload in adult patients
C. Increased risk of pulmonary reactions in patients with pulmonary aspergillosis
D. Increased risk of graft rejection in patients receiving HLA matched transplants
E. Increased risk of acute hemolytic transfusion reactions in pediatric patients

Concept: Granulocyte transfusions are indicated for severely neutropenic patients who are expected to have bone marrow recovery and with documented bacterial or fungal infections that are refractory to conventional antibiotic therapy. Patients should have an absolute neutrophil count less than 500/μL.

The resolving infection in people with neutropenia with granulocytes (RING) study began in 2008, but failed to enroll enough subjects to make any conclusions about the effectiveness of granulocyte transfusions. The advances in antibiotic therapy along with the lack of evidence have contributed to the decreased use of granulocyte transfusions in recent years.

Doses less than 1×10^{10} granulocytes are considered to be ineffective. Granulocytes should be ABO-compatible to prevent hemolytic reactions and irradiated to prevent transfusion-associated graft versus host disease. Cytomegalovirus (CMV)-seronegative donors are recruited to donate for CMV-negative patients. Ideally, units are preferred to be transfused within 6 h of collection, but must be transfused within 24 h of collection for optimal efficacy. Patients are at risk of developing anti-human leukocyte antigen (HLA) antibodies despite being severely immunocompromised. Evidence supports the theory that the development of anti-HLA antibodies in response to granulocyte transfusions can lead to an increased risk of rejection of HLA-mismatched allogeneic HPC transplants.

Answer: *C*—Pulmonary reactions are the most frequent reactions reported with granulocyte transfusions. These reactions are seen more commonly in patients with pulmonary infections, such as aspergillosis. It is thought that the granulocytes aggregate in the lungs in these patients and leads to adverse reactions. Reports of severe pulmonary reactions were reported with concomitant administration of amphotericin and granulocytes, and temporal separation by a minimum of 12 h is recommended. Zika virus (Answer A) does not only infect white blood cells; thus, the risk for Zika transmission is not necessarily higher with granulocyte transfusion compared to other blood products. Similarly, granulocyte transfusion does not confer higher risk for graft rejection (Answer D), AHTR (Answer E), or TACO (Answer B).

32. One day ago, a 67-year-old man underwent an uneventful aortic valve replacement that did not require transfusion of blood components. His hemoglobin has been slowly drifting down and he now requires a RBC transfusion. Which of the following RBC modifications or testing should be used to reduce transfusion related immune modulation (TRIM)?

A. C-, E-, and K-antigen matched
B. CMV negative
C. Irradiated
D. Leukoreduced
E. Sickle negative

Concept: Since the 1960s–70s, studies on the negative and positive effects of transfusion have reported outcomes that suggest posttransfusion immunosuppression. These immunomodulatory effects are collectively termed TRIM. Though the exact mechanism is unclear, the infusion of foreign WBCs in blood products leads to alterations in both humoral and cellular immunity in the transfusion recipient. The reported negative effects of TRIM include increased posttransfusion infections, cancer metastasis, and alloimmunization.

In general, the more WBCs present in the blood product, the greater the immunomodulation. To decrease the effects of TRIM, blood products undergo leukoreduction to decrease, but not eliminate WBCs.

Other benefits of leukoreduction include:

- Decrease in the incidence of febrile nonhemolytic transfusion reactions
- Decrease in CMV transfusion transmitted infections
- Decrease in the rate of HLA alloimmunization

Of note, AABB standards state that leukoreduced products must contain $< 5 \times 10^6$ WBCs. Thus, leukoreduction does not completely eliminate WBCs from a blood product and will not protect against TA-GVHD.

Answer: *D*—Providing leukoreduced RBC units decreased TRIM-related consequences of transfusion, in addition to the other three benefits mentioned earlier. Consequently, providing prestorage leukoreduced RBCs is now common practice worldwide. The other choices (Answers A, B, C, and E) do not have a known effect on TRIM.

Please answer Questions 33–36 based on the following clinical scenario:

A 78-year-old man presented to the emergency department with a lower gastrointestinal bleed. His blood pressure was 82/50 mmHg, heart rate 125 bpm, and hemoglobin 6.2 g/dL. The physician ordered 1 unit of RBCs from the blood bank. At the same time, there was a 24-year-old man with SCD who presented to the emergency department with fever, pain, shortness of breath, O_2 saturation of 82% on room air and 93% on 4 L nasal cannula, and a new right lower lobe infiltrate on chest X-ray. His hemoglobin was reported at 5.8 g/dL for which his physician ordered 1 unit of RBCs while awaiting bed assignment. The blood bank servicing the ED has a two-sample policy; the two samples confirmed that the 78-year-old man, named James Harrison was A Rh positive with a negative antibody screen and the 24-year-old man named John Harrison was O Rh positive with a negative antibody screen. The correct group A and group O red cell units were issued to the emergency department for James and John Harrison, respectively.

The nurse started the administration of the unit for John Harrison and about 5 min into the transfusion (~ 30 mL transfused), the patient complained of worsening pain in his knees, abdomen, and back, his temperature increased to 101.2°F, and his urine turned dark red.

33. Which of the following was the most likely cause of the patient's signs and symptoms?

A. Acute hemolytic transfusion reaction
B. Delayed hemolytic transfusion reaction
C. Delayed serologic transfusion reaction
D. Febrile nonhemolytic transfusion reaction
E. Sickle cell disease with pain crisis

Concept: In the event of an ABO incompatible red cell transfusion, the clinical presentation is variable depending on the volume of incompatible red cells transfused and unique patient characteristics. The most common sign/symptom of an ABO acute hemolytic transfusion reaction (AHTR) is fever. Patients may also complain of back pain, chills, and a feeling of impending doom. Symptoms may be accompanied by hypotension and hemoglobinuria. A late manifestation may be the development of disseminated intravascular coagulation and renal failure. Laboratory studies may show an ABO discrepancy between the pre- and posttransfusion samples, a positive DAT (may be negative if all transfused red cells are hemolyzed), and hemoglobinemia. The following changes may also be seen: elevated lactate dehydrogenase, undetectable haptoglobin, increased indirect bilirubin, and urinalysis with positive blood but no red cells identified on microscopic examination.

Answer: *A*—The clinical findings suggest the presence of an acute hemolytic transfusion reaction (AHTR) because of the sudden onset of fever and hemoglobinuria. The timing goes against a DHTR (Answer B) which would more likely occur in 5–10 days rather than during the first 5 min of the transfusion. A delayed serologic transfusion reaction (DSTR, Answer C) is detected on serology, but is not associated with the clinical signs and symptoms and laboratory findings seen with an AHTR or DHTR. A febrile nonhemolytic transfusion reaction (FNHTR, Answer D) can present with fever and chills, but the hemoglobinuria and pain are not consistent with a FNHTR. Pain crisis in sickle cell disease (Answer E) does not usually happen abruptly, especially within 5 min of transfusion. An AHTR must always be ruled out before attributing a reaction to a FNHTR or underlying disease condition.

34. The transfusion was immediately stopped and the bag clamped and returned to the blood bank with a posttransfusion blood sample. Urinalysis and blood samples for basic metabolic panel, hepatic profile, and lactate dehydrogenase were submitted to the main laboratory.

While waiting for laboratory results to be reported, which of the following is the most important treatment to initiate?

A. Corticosteroids
B. Eculizumab
C. Mannitol
D. Hydration with IV fluids
E. Intravenous immunoglobulin (IVIG)

Concept: The first step in the treatment of an AHTR should be immediate cessation of the transfusion and evaluation of a transfusion reaction by the blood bank. The blood bank investigation will include a clerical check (verification of the compatibility label, container label, and the issued product) and visual inspection of the returned unit and a posttransfusion sample for hemolysis. A repeat blood type and antibody screen may be performed on the pretransfusion and posttransfusion sample. A DAT will be performed and if positive, an eluate may be prepared and used to identify antibody specificity.

Answer: *D*—If an AHTR is suspected, especially in the presence of hemoglobinuria, fluids should be initiated as soon as possible and infused at the fastest rate tolerated by the patient. If the reaction is severe, mannitol (Answer C) and dobutamine may also be considered. Although some physicians will administer intravenous immune globulin or steroids (Answers A and E), there are no definitive studies to show that these interventions are effective. A developing strategy in the treatment of AHTRs is the use of eculizumab (Answer B), which is a monoclonal antibody against the complement C5 component. A few case reports have documented the success of using eculizumab in the treatment of ABO incompatible transfusions, but this is not yet common place.

35. The blood bank reports the transfusion reaction to the blood bank director, who immediate shares the results with the ED physician. The clerical check did not pass. The red cell unit intended for James Harrison (type A) was incorrectly administered to John Harrison (type O). The posttransfusion sample confirmed that the patient was O positive, the antibody screen was negative, and visual hemolysis was significant (cherry red). The direct antiglobulin test was positive with polyspecific reagent, anti-IgG, and anti-C3; the eluate agglutinated against B red cells, but not A1 red cells. The returned unit was type A positive and the compatibility label showed the patient names of James Harrison. Which of the following was the most likely cause of this mistransfusion?

A. Failure to identify the patient when collecting the blood for type and screen
B. Failure of the nurse to identify the patient and the unit at the time of blood administration
C. Failure to inspect the unit for discoloration
D. Mistyping of the red cell unit at the blood donor center
E. Technologist incorrectly reading the blood type of the patient

Concept: Approximately 1:76,000 red cell transfusions is associated with an AHTR. Of these reactions, very few are fatal. The blood banking community has greatly safeguarded against ABO type mistransfusions so that ABO AHTRs are now only the fourth leading cause of transfusion-related fatalities reported to the FDA. AHTRs due to ABO incompatibility are most commonly caused by errors in drawing the specimen for type and screen [aka wrong blood in tube (WBIT)], errors in performing the test and entering results, and failure to identify the compatibility label with the patient or improper patient identification.

Proper identification of the patient should always be performed at the time that specimens are drawn from the patient. The patient should be asked their name and date of birth, which should be compared to the patient's identification bracelet and the specimen label. The specimen label should be labeled at the bedside and signed by the phlebotomist to certify that they properly followed the specimen collection guidelines. The blood bank should perform patient and specimen/product identification during all steps of testing, result reporting, and issuing of blood products. At the time of blood administration, identification of the patient and the unit is critical for ensuring patient safety. The patient should be identified by asking the name and date of birth and these should match the patient identification bracelet. The blood container label contains the donor identification number, expiration date, and blood type, which should match the compatibility label attached to the container. At last, the information on the compatibility label should match the patient identification bracelet. A second person should independently verify the same identifying information prior to initiate a transfusion.

In the event that the sample was not actually drawn from the correct patient and the patient does not have a historical blood type on record, the incorrect blood type may be assigned to the patient. To prevent this kind of error, institutions have implemented bar code patient identification with label printing at the bedside and/or policies requiring a second confirmatory sample to be submitted prior to issuing type specific blood. This confirmatory sample should be drawn at a different time and preferably by a different person. Barcode verification of the patient and blood component has also been instituted to improve transfusion safety. This electronic verification can replace the check by the second person or it can be used to enhance an established two-person verification step.

Answer: *B*—The most likely cause of this AHTR is failure to identify the patient at the bedside or failure to match the compatibility label on the bag with the patient identification bracelet. Most likely, the unit intended for James Harrison was accidentally picked up and erroneously administered to John Harrison because of the similar last names. If patient and unit identification is not performed according to policy, similar names may easily be missed. Although identification errors when collecting samples or labeling samples for type and screen do occur, the implementation of a two-specimen policy significantly reduces the risk of erroneous assignment of blood type. Although mistyping at the blood collection facilities do occur, the standard of practice in blood banks is to confirm all or a subset of units accepted into the inventory. The other choices (Answers A, C, D, and E) are all possibilities in this case, but do not represent the most likely cause.

36. After an hour of hydration with normal saline, the patient's symptoms resolved. The urine was slightly yellow colored and clear and another blood sample showed that the plasma was straw-colored and clear.

Which of the following should be performed as soon as possible?

A. Performance of a root cause analysis
B. Preparation of a defense against a lawsuit
C. Establish a policy for type O Rh negative red cells for all patients in the ED
D. Termination of the nurses involved in transfusing the unit
E. Establish a policy that all same name or similar sounding names be given an alias

Concept: An immediate investigation should be initiated after any error in transfusion to prevent immediate recurrence and to initiate a long-term analysis and corrective action plan. The conduct of investigation should not be punitive, but instead be a measure to discover the events and determine how to improve the process. Following the investigation, a root causes analysis team synthesizes the information and establishes all the causes that contributed to the error(s). After the root causes are identified, a team puts together a corrective action plan to prevent future events. In the event of negligence or breaking of standard of care, disciplinary action may be needed for the participants in the errors.

Answer: *A*—A root cause analysis should always be performed when an error results in patient harm or potentially could have caused patient harm. It is best to follow a series of defined steps to outline the causes of the error and then address corrective actions for each root cause. The root cause(s) is usually due to a systematic error and is rarely the result of a single error made by a single person. Usually a series of failures leads to an error. Preparing a defense against a law suit (Answer B) is likely to be premature at this time. Transfusing all patients in the ED with O Rh negative RBCs (Answer C) is not necessary and may be unsafe (i.e., the patient may have unexpected red cell alloantibodies). Terminating the nurses involved (Answer D) does not solve the problem since the process likely needs to be examined and improved. Answer E may be one of the solutions; however, a root cause analysis should be performed as the first step.

End of Case

Please answer Questions 37–40 based on the following clinical scenario:

A 27-year-old male with acute myeloid leukemia received a prophylactic dose of single donor apheresis platelets for a platelet count of 7,000/μL. Pretransfusion vital signs were heart rate 75, blood pressure 110/70, respiratory rate 18, and temperature 37.2°C. The patient complained of chills and rigors 40 min into the transfusion. Vital signs at that time were heart rate 110, blood pressure 80/50, respiratory rate 24, and temperature 39.4°C.

37. Which of the following is the most likely cause of this patient's signs and symptoms?

A. FNHTR
B. Anaphylactic transfusion reaction
C. Septic transfusion reaction
D. TRALI
E. TACO

Concept: Transfusion of a contaminated blood product can lead to no reaction, minor fever, sepsis, or death depending on the species and concentration of the organism transfused and the clinical status of the recipient. Septic transfusion reactions typically present with an increase in temperature by at least 2°C above the pretransfusion temperature or a temperature greater than 39°C. Hypotension, rigors, chills, tachycardia, nausea and/or vomiting, and dyspnea may also be seen. Mild reactions may only present with fever and chills while more severe reactions may present with septic shock. The majority of platelet contaminants are Gram positive skin flora and generally cause no reaction to only mild symptoms. These reactions are often misdiagnosed (frequently as febrile nonhemolytic transfusion reactions) and underrecognized. Gram negative contaminants are associated with more severe reactions due to the presence of preformed endotoxin.

If a septic reaction is suspected, the transfusion should be immediately stopped. The bag and all tubing should be sent to the blood bank for Gram stain and culture. Gram stain and culture should be performed on the remaining product in the bag and not the segments due to the high rate of false negative results. If the culture is positive, additional testing must be performed to identify the organism as per AABB Standards. Blood cultures should also be drawn from the patient, ideally from the opposite arm. Identification of the same organism in both the patient and the implicated component supports the diagnosis of a septic transfusion reaction.

Answer: *C*—An increase in temperature by greater than 2°C or temperature greater than 39°C accompanied by hypotension, rigors, and tachycardia in response to a platelet transfusion is highly suggestive of a septic transfusion reaction. A FNHTR (Answer A) would be part of the differential diagnosis, but the severity of the signs and symptoms would more likely indicate a septic transfusion reaction. TRALI and TACO (Answers D and E) would present with more respiratory symptoms. Anaphylactic reactions (Answer B) don't typically present with fever but may present with profound hypotension, tachycardia, and respiratory distress.

38. Which of the following products is most commonly associated with fatalities due to septic transfusion reactions?

A. Red blood cells
B. Platelets
C. Plasma
D. Cryoprecipitated AHF
E. Prothrombin complex concentrates

Concept: Platelet products are most commonly contaminated with bacteria due to storage conditions. Platelets are stored at room temperature in a protein rich, aerobic environment. This environment facilitates bacterial growth and proliferation. After an initial lag phase, bacteria reach an exponential phase of growth; therefore, severe septic reactions typically occur on day 4 or 5 of storage. Red blood cells are stored at 1–6°C. Cold temperatures lead to poor viability and inhibit bacterial growth. Gram negative, psychrophilic bacteria are the most commonly isolated contaminants in red blood cell units. *Pseudomonas* species, *Yersinia enterocolitica*, and *Serratia* species are most commonly implicated in septic transfusion reactions after red blood cell transfusion. Due to preformed endotoxin, these reactions are frequently fatal. Plasma components and cryoprecipitated AHF are stored at −18°C or colder leading to poor viability and inhibiting growth of bacterial contaminants. Contamination of these products generally occurs during thawing in a contaminated water bath.

As of this writing, the AABB Standards requires that transfusion services have methods to limit and detect or inactivate bacteria in platelet products. Transfusion services and collection centers have taken measures to decrease the risk of bacterial contamination including implementation of improved skin decontamination techniques, diversion pouches to collect deep-seated bacteria and skin plugs, and culturing of platelets prior to release. As a result of these measures, the estimated residual risk of sepsis after transfusion is about 1:108,000 for single donor apheresis platelets and about 1:25,000 for whole blood derived (i.e., random donor) platelets. The estimated residual risk of a fatal septic transfusion reaction is approximately 1:500,000 and 1:250,000 for single donor apheresis platelet components and whole blood derived platelets, respectively. The institution of point-of-issue testing for bacterial contamination of platelet components is expected to further decrease the risk.

Answer: *B*—Platelet transfusions, including both pooled platelet products and apheresis platelets, are most commonly implicated in fatal septic transfusion reactions reported to the FDA. Red blood cell transfusions (Answer A) may also be implicated in fatal septic transfusion reactions, but less commonly so compared to platelet components. Transfusion of plasma, cryoprecipitated AHF, and prothrombin complex concentrates (Answers C, D, and E) are unlikely to lead to a septic transfusion reaction due to other storage conditions or processing.

39. The blood bank called the blood center to notify the director of the reported transfusion reaction. What is the next step performed by the collection facility for this case?
A. Recall the donor for blood cultures
B. Culture any in-date products from the implicated donor
C. Recheck infectious disease markers on the implicated donor
D. Recall all remaining blood components from that donation
E. Perform a lookback on previous donations from that donor

Concept: If a septic transfusion reaction is suspected, the blood supplier should be contacted immediately to prevent transfusion of other components from that donation. The blood supplier will quarantine all products on-site and notify transfusion services that received these products. It is imperative that these actions occur to prevent additional septic transfusion reactions in other patients.

Contamination of blood products can occur after improper disinfection technique at the needle puncture site. Complete sterilization of the needle puncture site is not possible. The first few milliliters of blood are collected in a diversion pouch to prevent product contamination by deep-seated bacteria in hair follicles, sebaceous glands, and the skin plug. Another source of bacterial contamination is transient bacteremia or subclinical infection in asymptomatic donors. All donors are questioned regarding signs and symptoms of infection, as well as recent antibiotic use. Donors with any sign of infection or those currently on antibiotics for a recent infection are deferred.

Answer: *D*—It is imperative that the blood center is notified of a potential septic transfusion reaction to either prevent distribution or transfusion of any other components from the same donation. The other choices (Answers A, B, C, and E) do not reflect the immediate action required by the collection center.

40. Which of the following organisms is most likely to cause the symptoms seen in this case after platelet transfusion?
A. *Staphylococcus* species
B. *Klebsiella pneumoniae*
C. *Babesia microti*
D. *Yersinia enterocolitica*
E. *Pseudomonas* species

Concept: Bacterial contamination of platelet components is a significant issue affecting approximately 1 in 5000 platelet components, but is estimated to be higher for whole blood derived platelets. Transfusion of a contaminated product may result in no symptoms, mild symptoms, sepsis, or death. Gram positive skin flora, including *Staphylococcus* species and *Streptococcus* species are most commonly isolated but aerobic and anaerobic diphtheroid bacilli and Gram positive bacilli are also commonly identified. *Staphylococcus epidermidis, Echerichia coli,* and *Klebsiella pneumoniae* are most commonly associated with fatal septic transfusion reactions.

All apheresis platelet products are held for at least 24 h after collection before being cultured. Only aerobic cultures are routinely performed by collection facilities as anaerobic cultures have high false positive rates leading to wastage. The platelets are released to transfusion services if cultures remain negative for 12–24 h. The culture bottles are incubated until the expiration date of the unit or until the culture bottle becomes positive. Slow growing bacteria may be missed but are unlikely to reach exponential phase of growth and cause a septic transfusion reaction.

Answer: *A*—*Staphylococcus* species and other skin flora are the most common cause of bacterial contamination of platelet products and septic transfusion reactions. *Klebsiella* (Answer B) has been implicated in fatal septic transfusion reactions, but is not the most common causative agent.

Yersinia enterocolitica and *Pseudomonas* species (Answers D and E) have been implicated in fatal septic transfusion reactions with red blood cell transfusion. *Babesia microti* (Answer C) is a parasite that may be transmitted by transfusion of red blood cells.

End of Case

Suggested Reading

[1] Food and Drug Administratrion Center for Biologics Evaluation and Research. Transfusion/Donation Fatalities, 2015. Available from: http://www.fda.gov/downloads/BiologicsBloodVaccines/SafetyAvailability/ReportaProblem/TransfusionDonationFatalities/UCM459461.pdf.

[2] P.W. Ooley (Ed.), Standards for Blood Banks and Transfusion Services, thirtieth ed., AABB, Bethesda, MD, 2016.

[3] M. Goldman, K.E. Webert, D.M. Arnold, J. Freedman, J. Hannon, M.A. Blajchman, TRALI consensus panel. Proceeding of a consensus conference: towards an understanding of TRALI, Trans. Med. Rev. 19 (1) (2005) 2–31.

[4] National Healthcare Safety Network. National Healthcare Safety Network Biovigilance Component Hemovigilance Module Surveillance Protocol, 2016. Available from: http://www.cdc.gov/nhsn/pdfs/biovigilance/bv-hv-protocol-current.pdf.

[5] M.K. Fung, B.J. Grossman, C.D. Hillyer, C.M. Westhoff (Eds.), Technical Manual, eighteenth ed., AABB, Bethesda, MD, 2014.

CHAPTER

13

Hemostasis and Thrombosis—Laboratory Diagnosis and Treatment

Marisa B. Marques, George A. Fritsma***

*University of Alabama at Birmingham, Birmingham, AL, United States; **The Fritsma Factor, Your Interactive Hemostasis Resource, Birmingham, AL, United States

Transfusion medicine physician and technologists are confronted daily with hemostasis and thrombosis-related concerns as they select and administer blood components, coagulation factor concentrates, anticoagulants, and agents to manage anticoagulant therapy. This chapter provides an introduction and overview of hemostasis. Primary hemostasis focuses on platelet function and their interaction with the vasculature, endothelium, and the coagulation mechanism. Secondary hemostasis focuses on the coagulation cascade and is subdivided into the extrinsic, intrinsic, and common enzymatic pathways. Coagulation also includes control systems (e.g., thrombomodulin) and clot dissolution (aka fibrinolysis). Depending on the defect, hemostasis disorders may be congenital or acquired, resulting in hemorrhage and/or thrombosis. The questions in this chapter will explore normal hemostasis, disorders of hemostasis, and the laboratory assays that predict, identify, and monitor treatment of hemorrhage and thrombosis.

1. What are the necessary reagents or components of the prothrombin time (PT) assay?
A. $CaCl_2$, negatively charged particle suspension, phospholipid
B. Negatively charged particle suspension, phospholipid
C. $CaCl_2$, tissue factor, phospholipids
D. $CaCl_2$, tissue factor, thrombin
E. $CaCl_2$, thrombin

Concept: The PT and activated partial thromboplastin time (APTT) are commonly used clot-based assays. They are often ordered as part of a standard admission screen that may also include a complete blood count (CBC) and urinalysis. Though they are ineffective as screening assays, surgeons insist on ordering them to reduce the perceived risk of intraoperative bleeding, and internists request them to reduce the risk of drug-related hemorrhage. PT reagents are designed to trigger the "extrinsic" and "common" coagulation pathways and are used to detect coagulopathies and monitor warfarin therapy. The APTT reagents are used to trigger the "intrinsic" and common pathways and are used to detect coagulopathies, to monitor unfractionated heparin therapy, and to screen for lupus anticoagulant (LA) (Fig. 13.1).

Answer: C—The components of the PT assay are $CaCl_2$, tissue factor, and phospholipids. Tissue factor triggers coagulation at the level of coagulation factor VII, and the phospholipid mixture supports assembly of the Va–Xa coagulation complex "prothrombinase," which activates prothrombin to form thrombin (Fig. 13.2). Because coagulation testing specimens are collected in 3.2% sodium citrate, a calcium ion chelation compound, it is necessary to replace the bound calcium with $CaCl_2$.

Transfusion Medicine, Apheresis, and Hemostasis. http://dx.doi.org/10.1016/B978-0-12-803999-1.00013-4

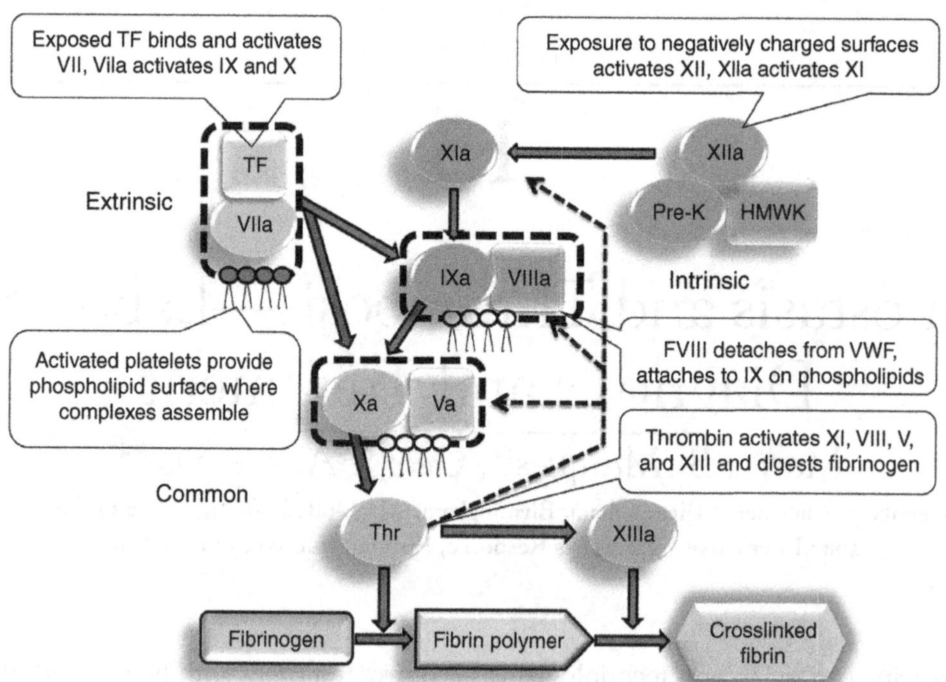

FIGURE 13.1 Overview of hemostasis and coagulation. *Thr*, thrombin; *TF*, tissue factor; *HMWK*, high molecular weight kininogen; *Pre-K*, prekallekrein; *VWF*, von Willebrand factor. *Source: Courtesy of George Fritsma.*

The thrombin time (TT) assay employs thrombin (Answer E), which tests the penultimate step, conversion of fibrinogen to fibrin to form the clot. No coagulation reagent is composed of both tissue factor and thrombin (Answer D). APTT is a two-step assay; the first reagent is a negatively charged particulate suspension, such as ellagic acid, celite, or kaolin, mixed with phospholipids (Answer B). The negatively charged particulate suspension triggers activation of factor XII to initiate the intrinsic pathway. The phospholipid mixture supports the VIIIa–IXa complex that activates factor X and the Xa–Va complex that activates prothrombin. The $CaCl_2$ is separate and is added after an incubation period to trigger clotting (Answer A).

2. Which of the following coagulation factors is needed for a normal prothrombin time (PT), activated partial thromboplastin time (APTT), and thrombin time (TT)?
A. Fibrinogen
B. Prothrombin
C. Factor V
D. Factor X
E. Prekallikrein

Concept: In vitro clot-based assays, such as the PT, APTT, and thrombin time, are employed to test for deficiencies of one or more of the plasma procoagulant factors. Deficiencies of each factor affect the clotting times of each of these tests differently, depending on the factor's role in the cascade.

Answer: *A*—In clot-based assays for example, the PT, APTT, and TT, thrombin cleaves fibrinogen to form fibrin, which is detected by the coagulation instrument (coagulometer). In the PT and APTT, thrombin is generated by activation of upstream reactions; however, in the TT, thrombin is added to the patient's plasma to act upon fibrinogen to generate a fibrin clot. Thus, prothrombin and all other procoagulant factors are not needed to form a clot in the TT (Answers B, C, D, and E). The PT depends on factors VII, X, V, prothrombin, and fibrinogen, while the APTT requires all coagulation factors except factors VII and XIII. Of note, though the normal reference range for fibrinogen is typically 150–400 mg/dL, the PT/APTT is not prolonged until the level reaches <100 mg/dL.

3. A patient with history of delayed postoperative bleeding has a normal PT, APTT, TT, and fibrinogen. Which of the following deficiencies/conditions is the patient likely affected by?
A. Factor XIII deficiency
B. Vitamin C deficiency
C. Hemophilia C
D. VWD subtype 2N
E. Hemophilia A

Concept: While several bleeding disorders may be detected or identified by testing the patient with the PT, APTT, TT, and fibrinogen assay, these assays do not detect qualitative platelet deficiencies, thrombocytopenia, vascular disorders, factor XIII deficiency, or mild coagulation factor deficiencies.

Answer: *A*—Factor XIII deficiency has a prevalence of 1 in 5,000,000. Factor XIII "crosslinks" fibrin, stabilizing the clot. In factor XIII deficiency, hemostasis is usually achieved soon after a hemostatic challenge, such as surgery. However, bleeding may ensue a few hours later when the initial fibrin clot deteriorates. Along with hypofibrinogenemia, factor XIII deficiency is currently one of the few remaining indications for cryoprecipitate transfusion, since commercial factor XIII concentrates only became available in the United States in 2011. Delayed bleeding is also seen in patients with hyperfibrinolysis, such as those with deficiencies of fibrinolysis control proteins like plasminogen activator inhibitor-1 (PAI-1), plasminogen activator inhibitor-2 (PAI-2), and α_2-antiplasmin.

Patients with vitamin C deficiency (Answer B) have normal PT, APTT, TT, and fibrinogen, but their bleeding is mucocutaneous, such as gum bleeding and bleeding around the hair follicles, and is not typically delayed. Patients with hemophilia A or C (Answers C and E) have a prolonged APTT due to decreased factors VIII or XI, respectively. Von Willebrand disease (VWD) subtype 2N (Answer D) is caused by a mutation in the VWF binding site for factor VIII, causing a moderate deficiency of factor VIII, which may also be detected by a prolonged APTT. APTT reagents are calibrated to become prolonged when the factor VIII or factor IX level is below 30%–40%; thus, the APTT may not become consistently prolonged in mild VWD.

4. What point of care (POC) assay may be used to assess fibrinolysis, while treating a patient with trauma-induced coagulopathy?
A. PT and APTT
B. Plasminogen activity assay
C. Thromboelastograph (TEG)
D. Plasminogen activator inhibitor-1 (PAI-1) assay
E. Euglobulin lysis time

Concept: Trauma-induced coagulopathy (TIC) is multifactorial and manifests as either hemorrhage or thrombosis, as well as increased fibrinolysis. While fibrinolysis inhibitors, such as tranexamic acid (TXA) may be useful, detecting hyperfibrinolysis is not possible with the clot-based hemostasis assays.

Answer: *C*—Two instruments that perform thromboelastography are the thromboelastograph (TEG, Fig. 13.2) and the rotational thromboelastometry (ROTEM). Both may be used in a central laboratory or at the bedside and use fresh or anticoagulated whole blood. The result is depicted as a tracing of the clot formation through primary and secondary hemostasis, and fibrinolysis. Abnormally rapid fibrinolysis, a component of TIC pathogenicity, is easily detected by these tests, offering an opportunity to treat the patient early with antifibrinolytics (e.g., TXA), unless clinically contraindicated. The PT and APTT (Answer A) may be offered as POC tests of coagulation but give no information about fibrinolysis. The plasminogen activity assay (Answer B), PAI-1 assay (Answer D), and the obsolete euglobulin lysis time (Answer E) are referral tests that are generally available from tertiary care or reference facilities.

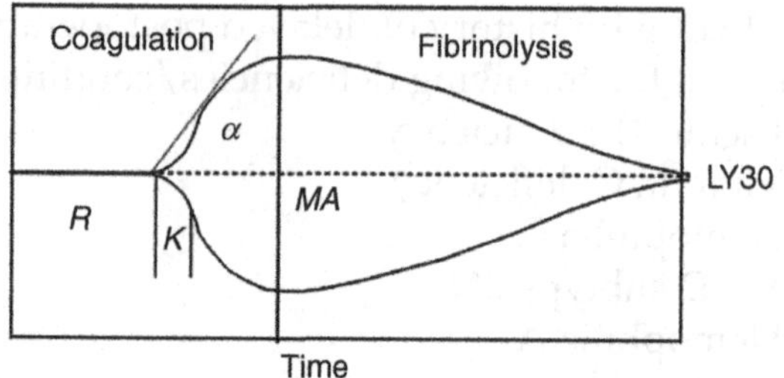

FIGURE 13.2 **TEG Parameters.** *R*, clot onset; *K* and α, clot k`inetics; *MA*, maximum amplitude; *LY 30*, fibrinolysis. *Source: Modified from Figure 2 in Pham HP, Shaz BH. Update on massive transfusion. Br. J. Anaesth 2013 Dec;111 Suppl 1:i71–82. doi: 10.1093/bja/aet376.*

5. Why is the PT/INR (international normalized ratio) employed to monitor warfarin therapy?
A. The PT reagent is abundant and cheap
B. The PT is most sensitive to prothrombin deficiency
C. The PT is most sensitive to factor VII deficiency
D. The PT is most sensitive to factor X deficiency
E. The PT is most sensitive to factor XII deficiency

Concept: Warfarin is a vitamin K antagonist that controls coagulation by suppressing the posttranslational modification of both pro- and anticoagulation vitamin K-dependent factors (factors II, VII, IX, X, protein C, S, and Z, Table 13.1).

Answer: C—The PT is most sensitive to factor VII deficiency. Because factor VII has the lowest plasma concentration and possesses the shortest half-life of the coagulation factors (Table 13.2), it is the first of the vitamin K-dependent factors whose activity becomes deficient subsequent to warfarin therapy initiation. Consequently, the PT becomes prolonged within approximately 6 h after the first warfarin dose. Prothrombin and factor X have longer half-lives and higher plasma concentration (Answers B and D), and factor XII is part of the intrinsic pathway whose deficiency does not affect the PT (Answer E). The PT reagent is abundant and relatively cheap, but this is not a primary reason

TABLE 13.1 Bleeding Disorders That Affect the PT/APTT

PT and APTT values	Acquired disorder	Congenital disorder
PT prolonged, APTT normal	Early liver disease Vitamin K deficiency[a]	Factor VII deficiency[b]
PT normal, APTT prolonged	Factor VIII or IX inhibitor[c]	Factor VIII, IX or XI deficiency[d]
PT and APTT prolonged	Disseminated intravascular coagulation (DIC), rat poison (superwarfarin), advanced liver disease, amyloidosis, LA-induced factor deficiency[e]	Fibrinogen[f], prothrombin, factor V or X deficiency[g]
PT and APTT normal	Thrombocytopenia, platelet function disorders, acquired von Willebrand disease (VWD), scurvy, hyperfibrinolysis (iatrogenic or amyloidosis)	Mild factor deficiency, mild VWD, factor XIII deficiency[h]

Use this table when there is bleeding and the patient is receiving no anticoagulant. Rule out heparin using the thrombin time, which typically prolongs to >21 s when heparin is present.

[a] *To distinguish* liver disease *from* vitamin K deficiency, *assay factors V and VII. If only VII is reduced, suspect vitamin K deficiency but if both are low, suspect liver disease. In liver disease, vitamin K therapy may be unsuccessful.*

[b] *Congenital factor VII deficiency is rare and causes mild to moderate childhood bleeding that correlates poorly with FVII activity. Some patients may present with thrombosis.*

[c] *Factor VIII inhibitor in a non-hemophiliac is a rare but life-threatening autoimmune disease called "acquired hemophilia." The inhibitor is detected using APTT mixing studies and the factor VIII assay. The inhibitor is measured using the Nijmegen Bethesda assay and reported in Nijmegen Bethesda units (NBU).*

[d] *Factor VIII and IX deficiencies, hemophilia A and B, are X-linked and are diagnosed in childhood, unless mild. Factor XI deficiency, hemophilia C, is autosomal recessive and is most prevalent in Ashkenazi Jews, though it is found in any ethnic group. Intensity of hemophilia C bleeding correlates poorly with factor XI activity.*

[e] *LA is seldom associated with bleeding but often associated with thrombophilia. It may cause bleeding when it induces rare deficiencies of factors, such as prothrombin and factor X. LA is detected and confirmed with a series of assays and mixing studies.*

[f] *Fibrinogen deficiency usually only prolongs PT and APTT when the concentration is <100 mg/dL. When suspecting afibrinogenemia or hypofibrinogenemia, perform the fibrinogen assay; available from all acute care laboratories.*

[g] *Congenital deficiencies of prothrombin, factor V and factor X are rare.*

[h] *Establish factor XIII deficiency using a factor XIII immunoassay available from a reference laboratory or tertiary care facility.*

TABLE 13.2 Half-Lives of the Coagulation Factors and Their Plasma Concentrations

Factor	Half-life	Plasma level
Fibrinogen	100–150 h	300 mg/dL
Prothrombin	60 h	100 μg/mL
V	24 h	7 μg/mL
VII	6 h	0.5 μg/mL
VIII	12 h	0.1 μg/mL
IX	24 h	5 μg/mL
X	48–52 h	10 μg/mL
XI	48–84 h	6 μg/mL
XIII	150 h	290 μg/mL
VWF	12 h	40 μg/mL

for its use in monitoring warfarin therapy (Answer A). Of note, the full effect of warfarin may not be reflected in the PT/INR until 2–3 days after initiation of therapy, therefore, dosage adjustments are not recommended before this time period in order to avoid giving increased doses unnecessarily.

6. The following presurgical screen coagulation testing results were found in a patient with no bleeding history. The CBC included a manual blood film examination, a portion of the field is illustrated in Fig. 13.3. What result is abnormal, and what may be the cause?

Assay	Result
Prothrombin time (PT)	13.9 s
Activated partial thromboplastin time (APTT)	31 s
Thrombin time (TT)	18 s
Fibrinogen	347 mg/dL
Platelet count	75,000/μL

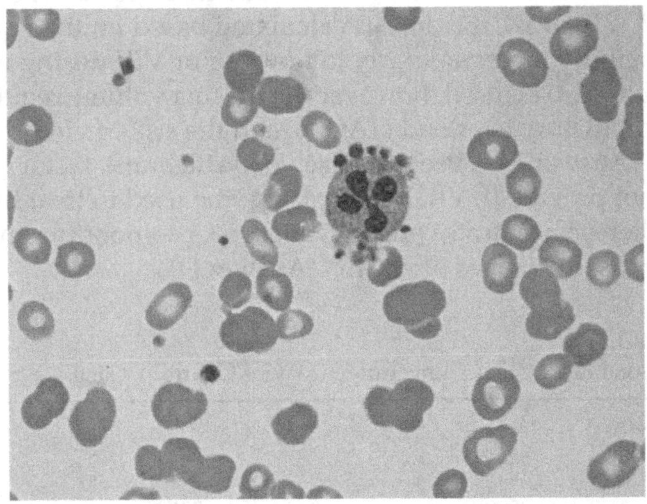

FIGURE 13.3 **Wright-stained blood film.** *Source: Image courtesy of Dr. Vishnu Reddy, University of Alabama at Birmingham.*

A. The PT; patient has factor VII deficiency
B. The APTT; patient has a potential factor XII deficiency
C. The thrombin time, the patient has a fibrinogen deficiency
D. The platelet count, this could be a spurious thrombocytopenia
E. The patient is on heparin therapy

Concept: Although they continue to be used, presurgical coagulation screening profiles do not predict intraoperative hemorrhage. The false positive rate (type 1 or α error) is high, and false positives often delay surgery, trigger the need for additional laboratory assays, and create patient anxiety.

Answer: *D*—The platelet count is abnormal and could be the result of a spurious or pseudothrombocytopenia. The blood film reveals platelet satellitism, a benign condition seen in blood films prepared from whole blood collected in EDTA anticoagulant (lavender closure tube). An EDTA-dependent plasma antibody causes platelets to bind neutrophils in a satellite pattern, falsely reducing the platelet count. To obtain an accurate platelet count, the laboratory scientist requests a fresh blood specimen using a sodium citrate (blue closure) tube and repeats the CBC. The CBC results are multiplied by 10/9 (or 1.11, a correction factor) to compensate for citrate dilution. Answers A, B, and C are incorrect, since only the platelet count is abnormal per the laboratory report. Answer E is incorrect because heparin therapy would likely prolong both the APTT and the TT.

7. What readily available therapy can be used to treat a warfarin-induced deficiency of coagulation factors II, VII, IX, and X prior to surgery or during bleeding episodes without transmitting viruses or causing transfusion-associated circulatory overload?
A. Plasma
B. Desmopressin
C. Cryoprecipitate
D. Cryoprecipitate reduced plasma (i.e., cryo-poor plasma)
E. Four-factor prothrombin complex concentrate (PCC)

Concept: Congenital factor II, VII, or X deficiencies are rare. Each deficiency has a prevalence of less than 1 in 1,000,000. However, all three factors become reduced in vitamin K deficiency, liver disease, and warfarin therapy.

Answer: *E*—Prothrombin complex concentrates (PCCs) are plasma-derived products available since 1980 that provide factors II, VII, IX, and X. PCCs are screened for viruses and may be used to reverse bleeding in liver disease and warfarin overdose. PCC (Kcentra, CSL Behring, King of Prussia, PA) dosing for patients on warfarin is provided in Table 13.3. For patients with isolated deficiencies of factors II, VII, or IX, the dose is calculated based on the patient's plasma volume and the desired factor activity target similarly to how factor VIII dosing is determined.

Plasma (Answer A) may also be used, however the plasma volume required carries a high risk of transfusion associated circulatory overload (TACO), and the risk of viral transmission exceeds that of PCCs. Desmopressin (Answer B) is used to raise von Willebrand factor in VWD. Cryoprecipitated AHF (Answer C) does not provide II, VII, IX or X; it may be used to treat hypofibrinogenemia and factor XIII deficiency where concentrates are unavailable. Cryo-poor plasma has factors II, VII, IX and X, but has the same disadvantages of plasma (Answer D).

TABLE 13.3 Four-Factor Prothrombin Complex Concentrate (PCC, KCentra) Dosing

Pretreatment INR	2– < 4	4–6	>6
Units of PCC factor IX/kg	25	35	50
Maximum dose of PCC factor IX/kg	2500	3500	5000

Please answer Questions 8–10 based on the following clinical scenario:

A hemostasis laboratory technologist performs a presurgical hemostasis screen for a 56-year-old female with no history of bleeding who is scheduled to have a breast reconstruction and obtains the following results:

Assay	Result
PT	11.9 s
APTT	43 s
Thrombin time (TT)	20 s
Fibrinogen	450 mg/dL
Platelet count	230,000/μL

8. Based on the results, the surgical resident for the case orders 2 units of plasma to give preoperatively. He postulates that factor XII deficiency might be the cause of the APTT prolongation. You tell him this is not likely the correct reason, due to which of the following?
A. Factor XII deficiency causes a very prolonged APTT
B. Factor XII deficiency is mostly X-linked and the patient is a female
C. Factor XII deficiency is very rare in patients without bleeding
D. Factor XII deficiency usually manifests in childhood as a cause of hemarthroses
E. Factor XII deficiency is often accompanied by other hemostatic abnormalities

Concept: Factor XII, prekallikrein (PK, Fletcher factor) and high-molecular weight kininogen (HMWK, Fitzgerald factor) are contact factors that initiate the intrinsic pathway of coagulation tested by the APTT. Deficiencies in any of them can prolong the APTT, but do not lead to clinically significant bleeding.

Answer: *A*—Factor XII deficiency often causes a markedly prolonged APTT (i.e., ~120–200 s) because it is the first factor to become activated when the patient's plasma is exposed to a contact activator, such as glass, silica, or kaolin. However, since none of the contact factors are needed for hemostasis in vivo, their deficiencies (autosomal) are asymptomatic and would not present in childhood (Answers B, C, and D). Factor XII deficiency is present in 1%–3% of the general population and is usually an isolated deficiency (Answer E).

9. Given these results, what is the most logical next step to identify the cause of the APTT prolongation?
A. Perform a coagulation factor XI activity assay
B. Performs coagulation factor VIII activity assay
C. Perform a lupus anticoagulant (LA) profile
D. Perform an APTT mixing study
E. Perform a factor VIII inhibitor assay

Concept: An isolated prolonged APTT from an outpatient plasma specimen may indicate an acquired or congenital coagulation factor deficiency of the intrinsic system, factors VIII, IX, XI, or XII, PK, or HMWK. It could also signal the presence of a lupus anticoagulant (LA); a specific factor inhibitor typically antifactor VIII; or anticoagulant therapy, such as the direct thrombin inhibitor dabigatran. An outpatient is unlikely to be receiving intravenous unfractionated heparin therapy, but could be taking dabigatran, both of which prolong the APTT. However, the normal TT rules out dabigatran and heparin. Lupus anticoagulant appears in 1%–2% of unselected individuals and is more likely to prolong the APTT than the PT. An acquired specific inhibitor, such as auto-anti-factor VIII (acquired hemophilia A), reduces the factor VIII activity level and prolongs the APTT. However, an acquired specific factor inhibitor usually causes bleeding, unlike LA, which can lead to thrombosis, (Fig. 13.4).

Answer: *D*—The APTT mixing study is available in all coagulation laboratories. Patient plasma is mixed 1:1 with reagent pooled normal plasma (NP) and an APTT is performed on the mix. If the APTT on the mix is prolonged beyond 10% of the NP APTT (or an alternative limit as locally defined), the result is reported as "uncorrected"; presumptive evidence for the presence of LA. If the result is corrected, a second 1:1 mix is incubated 2 hours and the mixture APTT is compared to incubated NP. If the APTT of the incubated mix is prolonged beyond 10%, the result is reported as presumptive evidence for a specific factor deficiency, usually anti-factor VIII. If both the unincubated and incubated mix results are corrected, the result is reported as presumptive evidence of a coagulant factor deficiency such as VIII or IX. Factor VIII and IX testing (Answers A and B) are used in follow-up to a mixing study that indicates a possible coagulation factor deficiency. The LA profile (Answer C) is a follow up to an uncorrected unincubated mix result and a factor VIII inhibitor (Answer E), is a follow-up to an uncorrected incubated mix result.

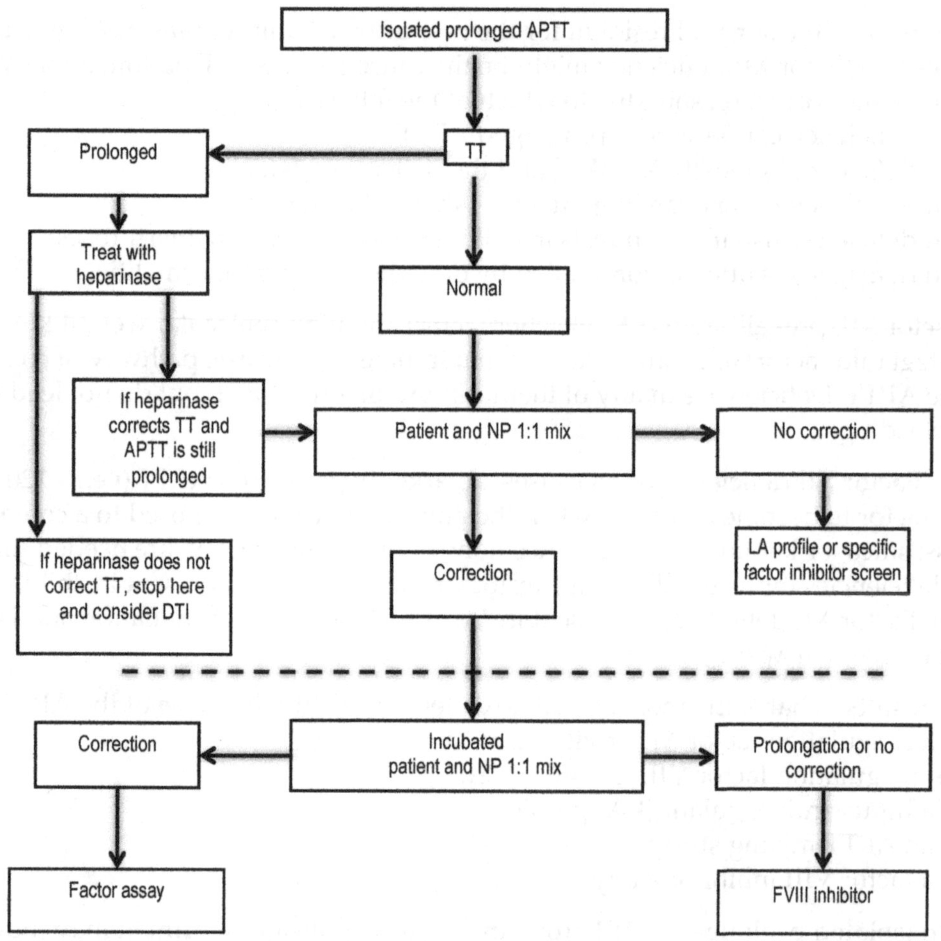

FIGURE 13.4 **Mixing study algorithm.** *DTI*, direct thrombin inhibitor; *NP* normal plasma. *Source: Courtesy of George Fritsma.*

10. The surgery resident is told of the possible explanations for the prolonged APTT and is confused by the term "lupus anticoagulant." What is the best explanation?

A. LA is an alloantibody to endothelial cells in patients with systemic lupus erythematosus
B. LA is a plasma protein that causes bleeding in patients with lupus vulgaris
C. LA is an autoantibody to phospholipid-binding proteins mostly associated with thrombosis
D. LA is a rare cause of postoperative bleeding in patients undergoing general surgery
E. LA is an alloantibody to plasma proteins in patients with systemic lupus erythematosus

Concept: LAs are heterogeneous IgG autoantibodies that bind plasma phospholipid-binding proteins, such as β2 glycoprotein I or prothrombin. Despite the term "anticoagulant," LA is typically associated with thrombophilia (and not bleeding) by a poorly understood mechanism that involves endothelial activation. In uncommon instances, prothrombin-specific LAs may be associated with bleeding (hypoprothrombinemia LA) due to the fast clearance of prothrombin.

Answer: C—LA is an autoantibody to phospholipid-binding proteins mainly associated with thrombosis, though in a few situations LAs may cause bleeding (Answers B and D). LAs are suspected and detected because they interfere with and cause prolongation of clot-based assays, whose reagents are prepared with reduced phospholipid, such as the PTT-LA and the dilute Russell viper venom test (DRVVT). The other choices (Answers A and E) do not describe the LA target.

End of Case

11. The hemostasis laboratory technologist performs a mixing study on a plasma specimen from a 13-year-old Caucasian female of European descent with complaints of easy bruising, menorrhagia, and frequent epistaxis. She has no signs of deep tissue or joint bleeding. Her results are provided as follows:

Assay	Result	Comment
Initial APTT	45 s	Reference interval: 25–35 s
Reagent normal plasma (NP)	30 s	From package insert, validated by QC manager
1:1 immediate mix APTT	32 s	Within 10% of NP = correction
37°C 2h incubated NP APTT	34 s	For comparison
37°C 2h incubated 1:1 mix APTT	36 s	Within 10% of incubated NP = correction

Of the following choices, which test should be performed next?
A. A lupus anticoagulant (LA) profile
B. Coagulation factor VIII activity
C. Coagulation factor XI activity
D. Coagulation factor XII activity
E. Anticardiolipin antibody assay

Concept: An APTT mixing study is the logical first step to evaluate an isolated prolonged APTT, as it distinguishes between a coagulation factor deficiency (coagulopathy) and an inhibitor. In a mixing study, the patient's plasma that produced a prolonged APTT result is mixed 1:1 with reagent normal plasma (NP) and an APTT is immediately performed on the mixture.

If the initial mix result indicates correction, the assay is repeated after a 1–2 h incubation at 37°C. This incubation allows for inhibitors to bind and prolong the previously "corrected" APTT. Most specific inhibitors are of IgG isotype and are time- and temperature-dependent. For example, this incubation step could detect antifactor VIII, which may not be detected in the immediate mix.

Correction may be defined as an APTT mix result within 10% of the NP APTT result; however, some laboratory directors prefer to use the upper limit of the reference interval as the comparison limit, while others calculate the Rosner index. In either event, for this case, it does not matter because the immediate and incubated mixing study results indicate correction, implying a factor deficiency.

Answer: *B*—Coagulation factor VIII activity should be measured, since VWD disease should be suspected in this case based on the clinical signs/symptoms and the mixing study results. LA and specific inhibitors are ruled out because the mixing study results indicate immediate and incubated correction (Answer A). Since the patient is a female, hemophilia A and B (factor VIII and IX deficiency), are unlikely. However, low factor VIII may indicate VWD, whose prevalence is 1%–2% of an unselected population. Furthermore, the patient's symptoms are suggestive of VWD.

Factor XI deficiency, hemophilia C, also named Rosenthal syndrome, is a possibility, but its prevalence is one in a million, and half of all Rosenthal sufferers are of Jewish parentage. Thus, if the patient is of Jewish parentage, a factor XI asay could be performed (Answer C). The prevalence of factor XII deficiency is 1%–3% (Answer D), but it has no association with bleeding. It is rare to test for the contact factors XII, PK, or HMWK, since they have little clinical consequence, though it may be useful to assay them only to establish why the APTT is prolonged and no explanation is found with initial testing. Anticardiolipin antibody assay would not be contributory in a bleeding patient, but is sometimes performed in patients with thrombosis (Answer E).

12. A patient is suspected of having LA, and a LA profile is performed. The results are shown as follows:

Assay	Result	Reference interval
PTT-LA	67 s	34–50 s
Staclot-LA; hex-phase phospholipid APTT reagent	54 s	
Difference	13 s	8 s difference
DRVVT	47.5 s	30.9–41.5 s
DRVVT with high phospholipid	33 s	
DRVVT ratio	1.4	< 1.2

What is the correct interpretation of these patient's results?

A. Inconclusive
B. Positive for LA
C. Negative for LA
D. Possible specific factor inhibitor
E. Possible coagulation factor deficiency

Concept: Owing to LA's heterogeneous test reaction patterns, LA detection and confirmation requires two different phospholipid-dependent assays, as specified in the 2009 International Society on Thrombosis and Haemostasis (ISTH) update of guidelines for LA detection (Table 13.4). The assays most often used are the dilute Russell viper venom time (DRVVT) and a form of the APTT, such as the PTT-LA, in which the reagent phospholipid concentration is reduced, rendering the assay sensitive to LA. The DRVVT triggers coagulation at the level of factor X, bypassing potential interference by intrinsic pathway factor deficiencies or inhibitors, whereas the PTT-LA initiates coagulation at the level of contact activation, factor XII, and may be affected by several factor deficiencies, excluding only factor VII deficiency.

TABLE 13.4 Guidelines for LA Testing

International Society on Thrombosis and Haemostasis (ISTH) update of guidelines for LA detection provides the following sequence of assays
1. Prolonged phospholipid-dependent clot formation using an initial screen assay, such as a low phospholipid APTT or DRVVT
2. Failure to correct the prolonged clot time when mixing with normal platelet-poor control plasma and repeating the test (i.e., mixing study not corrected)
3. Shortening or complete correction of the prolonged screen assay result by addition of a reagent formulated with excess phospholipids[a]
4. Exclusion of other coagulopathies

[a]*Reagents with excess phospholipids include the hexagonal phase phospholipid reagent (StaClot-LA for the APTT-based system) or the high phospholipid reagent for the DRVVT-based system. The two assay systems are not confirmatory, meaning that a positive result in either one is diagnostic for LA.*

Answer: *B*—These results indicate that the patient is positive for LA. Though limits of positivity for the Staclot-LA should be established in each laboratory, a difference of 8 s or more separating the APTT-based assays, or a ratio of 1.2 or greater between the DRVVT-based assays conclusively identify LA (Answers A and C).

In the event that the APTT-based assay confirms LA and the DRVVT-based assay does not, the laboratory scientist performs factor assays, usually VIII and IX, to ensure the originally prolonged APTT was not caused by a specific factor deficiency or inhibitor (Answers D and E), such as antifactor VIII.

Chronic LA is associated with antiphospholipid syndrome (APS), whose clinical manifestations include venous and arterial thrombosis and repeated spontaneous abortions. Positive LA results are confirmed with a repeat assay 12 weeks or more subsequent to the original primary result to establish LA chronicity, as transient elevated C-reactive protein, an acute phase reactant that is not associated with thrombosis, interferes with the Staclot-LA assay, mimicking LA.

13. Which immunoassay is the most reliable in supporting the diagnosis of antiphospholipid antibody syndrome?
A. Antiphosphatidylserine antibody
B. Anti-β2-glycoprotein I antibody
C. Antinuclear antigen antibody
D. Antiheparin-PF4 antibody
E. Anticardiolipin antibody

Concept: Lupus anticoagulants (LAs) are members of the antiphospholipid antibody (APL) family, which may be implicated in venous and arterial thrombotic events, a condition called antiphospholipid antibody syndrome (APS). In addition to clot-based LA testing, enzyme immunoassays are available to test for several disparate members of the APL family. In all cases, clot-based LA testing and immunoassays for APLs must be positive twice at least 12 weeks apart, to rule out transient APLs.

Answer: *B*—Anti-β2-glycoprotein I antibody assay is the best choice among the given options. Although the anticardiolipin assay (Answer E) has been available longer than the anti-β2-glycoprotein I assay and is still used, the latter is more closely associated with APS. Antiphosphatidylserine (Answer A) testing is available from specialty reference laboratories for rare instances when APS is suspected and all other assays are negative. Antinuclear antigen antibodies (Answer C) are associated with systemic lupus erythematosus, not LA, and the antiheparin-PF4 antibody (Answer D) is the cause of heparin-induced thrombocytopenia (HIT).

14. Which anticoagulant therapy is most often used during pregnancy to maintain fetal viability for women with APS?
A. Aspirin
B. Warfarin
C. Rivaroxaban
D. Enoxaparin
E. Dabigatran

Concept: A diagnostic criterion of APS is recurrent first or second trimester fetal loss. Once diagnosed, fetal loss is minimized using appropriate anticoagulant therapy.

Answer: *D*—Enoxaparin is a low molecular weight heparin, as are dalteparin and tinzaparin. These preparations are effective and safe, and may be self-administered subcutaneously daily throughout pregnancy. Aspirin does not appear to be effective in APS (Answer A), and warfarin (Answer B) is contraindicated during pregnancy because it is teratogenic. The direct oral anticoagulants (DOACs), rivaroxaban and dabigatran (Answers C and E) have neither been approved for management of APS nor for use in pregnancy.

15. A 43-year old male with a history of alcohol abuse arrives in the emergency department with acute upper gastrointestinal (GI) bleeding. His initial hemostasis laboratory results are shown here.

Assay	Result
PT	18.0 s
APTT	39.5 s
TT	20 s
Fibrinogen	475 mg/dL
Platelet count	189,000/μL

What additional hemostasis assays may be helpful?

A. D-dimer
B. Platelet aggregometry
C. Factors V and VII
D. Factor VIII
E. Factors X and V

Concept: The two most common acquired coagulopathies are liver disease and vitamin K deficiency. In liver disease, most coagulation factor levels are reduced. Fibrinogen and factor VIII levels may rise, as they are acute phase reactants. In vitamin K deficiency, provided the liver is otherwise healthy, prothrombin (II), VII, IX, and X are reduced, as are coagulation control proteins C, S, and Z.

Answer: C—Factor V and VII activity assays distinguish between liver disease and vitamin K deficiency. Both are reduced in liver disease, but only factor VII is reduced in vitamin K deficiency. Factor VII is chosen over II, IX, or X (Answer E) because factor VII has the shortest half-life and the lowest plasma concentration. Thus, it is the first to be affected by warfarin therapy, vitamin K deficiency, or liver disease. Factor V is chosen because it is not vitamin K-dependent, and its activity level diminishes in liver disease but not in vitamin K deficiency.

Factor VIII (Answer D) is not a reliable indicator of liver disease, as it is an acute phase reactant whose level may actually rise early in liver disease. The platelet count or platelet functional assays (Answer B) and D-dimer (Answer A) are also unreliable since changes may be related to a variety of disorders. D-dimer levels may become useful if there is a clinical suspicion of DIC, as when the patient has an infection with a Gram-negative organism, such as spontaneous bacterial peritonitis, a common complication of cirrhosis.

16. A patient with adenocarcinoma has the following hemostasis test results:

Assay	Result
PT	17.5 s
APTT	36 s
TT	18 s
Fibrinogen	125 mg/dL
Platelet count	52,000/μL

Which additional hemostasis assay(s) is/are indicated?

A. Platelet aggregometry
B. Factors VIII, IX, and XI activity
C. Factors V and VII activity
D. Factor VIII inhibitor
E. D-dimer

Concept: Tumors often secrete coagulation-triggering molecules that may activate coagulation, consuming coagulation factors, and depositing systemic microthrombi, called DIC. Microthrombi are gradually degraded through fibrinolysis, in which plasmin digests the fibrin clot, forming an array of fragments called fibrin degradation products, labeled D, E, X, and Y. Another product is D-dimer, which initially forms during the factor XIII crosslinking reaction and is released during fibrinolysis. Several manufacturers have developed kits that employ D-dimer-specific monoclonal antibodies. D-dimer assays may be used to rule out venous thromboembolic disease in low or moderate clinical suspicion and to detect and monitor chronic and acute DIC.

Answer: *E*—D-dimer is the best follow-up test. Adenocarcinomas secrete procoagulant substances that may trigger chronic or acute DIC. As coagulation factors are consumed, the PT, APTT, and TT become prolonged; platelets become activated and are consumed, resulting in thrombocytopenia. Although most coagulation factors are diminished, the most consistent abnormal laboratory result is the D-dimer level that may be elevated to 50–100 times the reference range limit, indicating ongoing fibrinolysis. The D-dimer assay is a mainstay of DIC detection and management.

Platelet function, as measured by aggregometry (Answer A) may be unaffected in DIC, and while factor activities are likely to be decreased, individual factor assays are impractical as initial markers (Answers B and C). Platelet aggregometry and factor assays are high-complexity tests, whereas the D-dimer is an automated assay available in acute care facilities with a relatively quick turnaround time. There is no reason to suspect a factor VIII inhibitor in this case, given the PT, APTT, and TT are all prolonged. A factor VIII inhibitor would prolong the APTT without affecting the PT or TT (Answer D).

17. A patient being prepared for a valve replacement undergoes a platelet aggregometry test with the following results:

Agonist	**Percent aggregation**	**Reference interval**
Arachidonic acid	18%	
Thrombin	85%	> 65% for all agonists
Collagen	41%	
Ristocetin	71%	

What is the most likely reason for these results?

A. Storage pool disease
B. VWD
C. Bernard-Soulier syndrome
D. Glanzmann thrombasthenia
E. Aspirin or a nonsteroidal antiinflammatory drug (NSAID)

Concept: In platelet aggregometry, several agonists (activators) distinguish among platelet functional disorders depending on their response tracings (normal tracings, Figs. 13.5 and 13.6). Arachidonic acid is a substrate for the eicosanoid synthesis pathway, ADP and collagen bind their respective platelet membrane receptors, and ristocetin reduces the negative surface charges that cause platelets to repel each other when suspended in plasma. Thrombin is the most avid agonist; it also binds a specific platelet membrane receptor and triggers full aggregation and platelet secretion. Epinephrine is rarely used in current whole blood platelet aggregometry (Fig. 13.7).

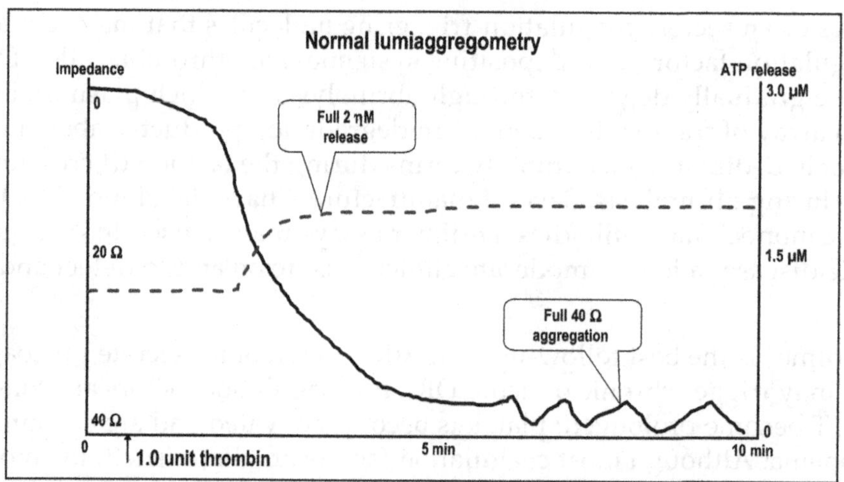

FIGURE 13.5 **Normal lumiaggregometry response to 1 U/mL thrombin, illustrating full 40 Ω aggregation and full 2 ηM ATP secretion (release).** *ATP*, Adenosine triphosphate. *Source: Courtesy of George Fritsma.*

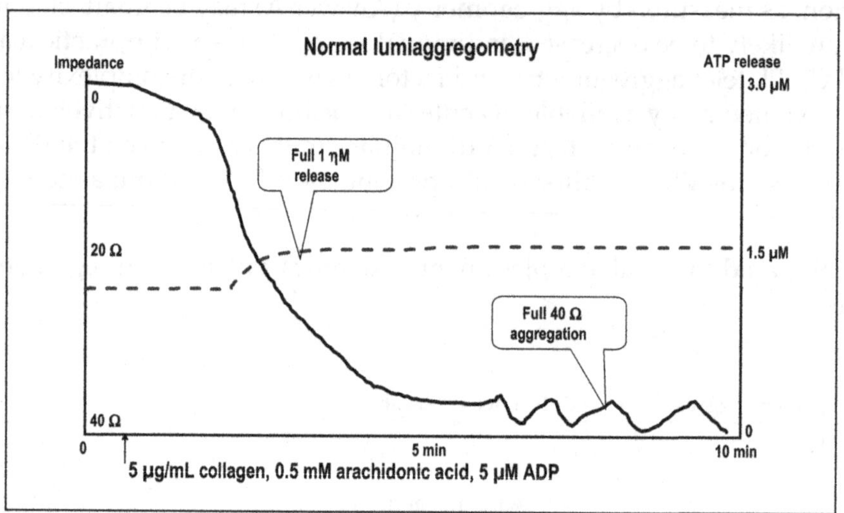

FIGURE 13.6 **Normal lumiaggregometry response to 5 μg/mL collagen, 0.5 mM arachidonic acid, or 5 μM ADP, illustrating full 40 Ω aggregation and full 1 ηM ATP secretion (release), approximately one-half of the thrombin-generated ATP secretion response.** *ADP*, Adenosine diphosphate. *Source: Courtesy of George Fritsma.*

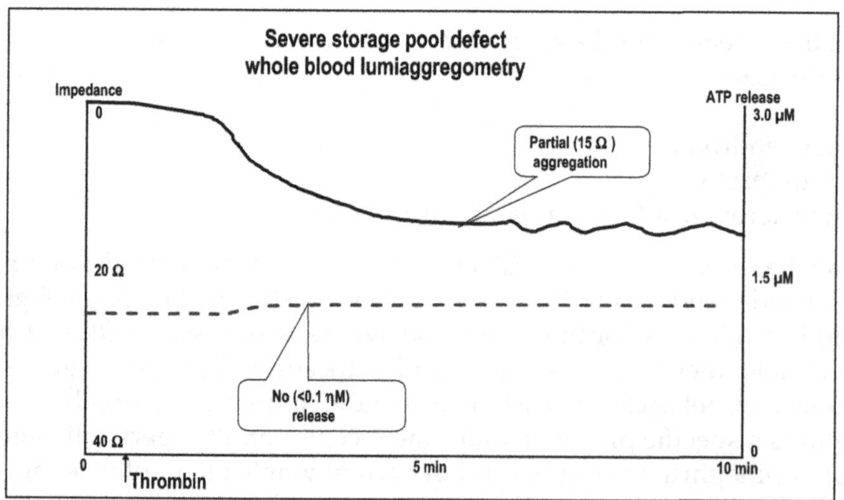

FIGURE 13.7 **Platelet lumiaggregometry response to 1 U/mL thrombin in severe storage pool disease: slight aggregation but no secretion (release).** *Source: Courtesy of George Fritsma.*

Answer: *E*—The patient took aspirin or a nonsteroidal antiinflammatory drug (NSAID). Because aspirin irreversibly acetylates platelet cyclooxygenase, the first of several enzymes in the eicosanoid synthesis pathway, arachidonic acid fails to be converted to thromboxane A2, a platelet-activating product, reducing aggregation. When drug ingestion is excluded, the clinician may diagnose the uncommon platelet secretion (aspirin-like) disorder. This is a hereditary reduction of one of the eicosanoid synthesis pathway enzymes that suppresses platelet activation and reduces platelet secretions. The aggregation tracing patterns are similar to those that appear with aspirin or NSAID ingestion, so platelet secretion disorder is nicknamed "aspirin-like" disorder. The effect of aspirin is also partially expressed in a decreased response to collagen. The ristocetin response is normal, ruling out VWD or Bernard-Soulier syndrome (Answers B and C), and the normal response to thrombin rules out storage pool disease (Answer A). Since the results show response to thrombin and collagen, it is unlikely to be Glanzmann thrombasthenia (Answer D) (Figs. 13.8 and 13.9) (Table 13.5).

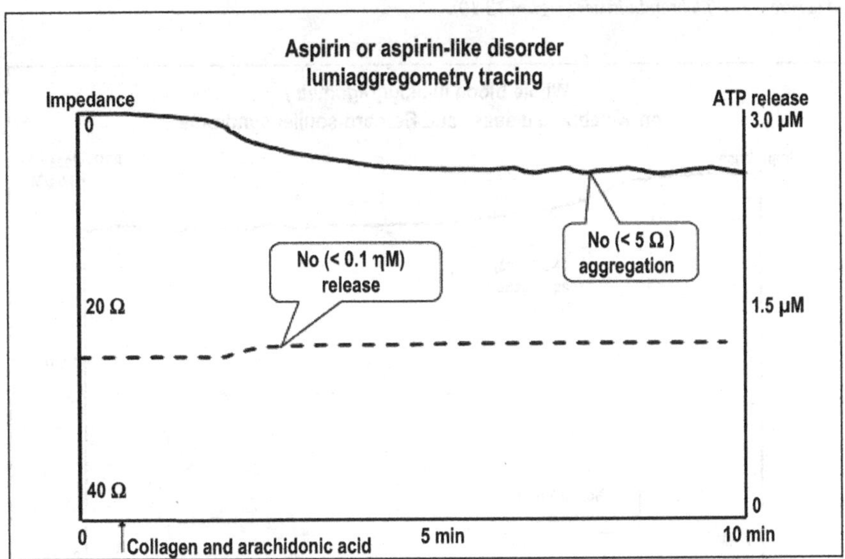

FIGURE 13.8 **Platelet lumiaggregometry response to 1 μg/mL collagen, 0.5 mM arachidonic acid shows no aggregation or secretion (release).** *Source: Courtesy of George Fritsma.*

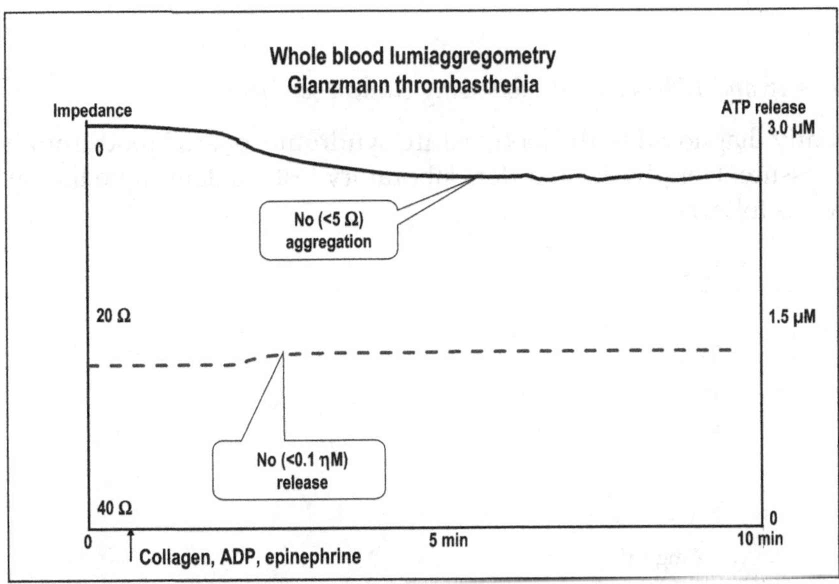

FIGURE 13.9 **Platelet lumiaggregometry response to 5 μg/mL collagen and 5 μM ADP in Glanzmann thrombasthenia (loss of glycoprotein IIb/IIIa), illustrating minimal aggregation and secretion.** *Source: Courtesy of George Fritsma.*

TABLE 13.5 Expected Aggregation and Secretion Results in Various Conditions

Agonist	Aspirin or release defect	Membrane defect	Storage pool deficiency
Thrombin	Aggregation and secretion normal	Aggregation and secretion normal	Aggregation & secretion reduced (Fig. 13.7)
Arachidonic acid	Aggregation and secretion reduced	Aggregation and secretion normal	Aggregation & secretion reduced (Fig. 13.8)
ADP & collagen	Aggregation and secretion vary by agonist concentration	Aggregation and secretion reduced (Fig. 13.9); Glanzmann thrombasthenia)	Aggregation & secretion reduced
Ristocetin[a]	Normal	Reduced (Fig. 13.10: Bernard-Soulier syndrome)	Normal

[a] *Ristocetin induces* agglutination, *not aggregation, and no secretion, as ristocetin does not activate platelets. Ristocetin-induced agglutination is reduced in Bernard-Soulier syndrome and severe von Willebrand disease (Figure 13.10).*

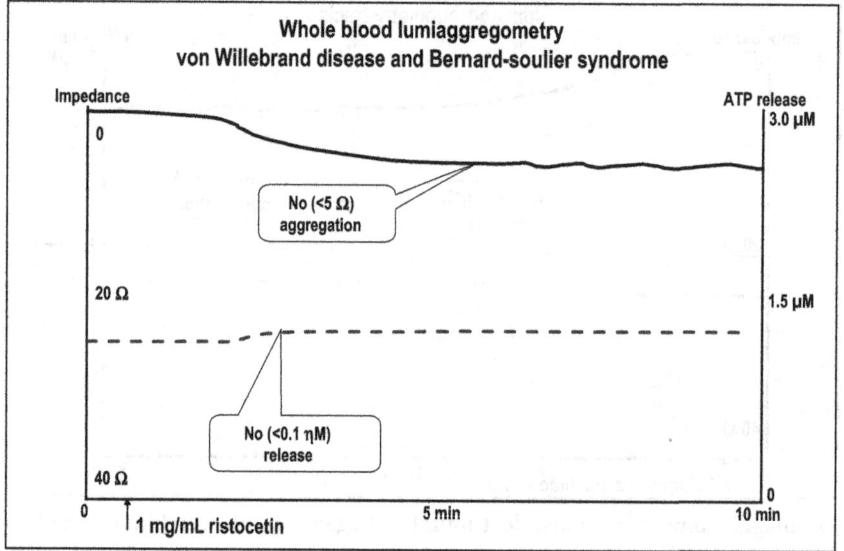

FIGURE 13.10 Platelet lumiaggregometry response to 1 mg/mL ristocetin in Bernard-Soulier syndrome (loss of glycoprotein Ib/IX/V) or severe VWD, illustrating minimal agglutination and no secretion. *Source: Courtesy of George Fritsma.*

Please answer Questions 16 and 17 based on the following clinical scenario:

A 65-kg patient, recently diagnosed with Goodpasture syndrome, oozes blood from her IV line insertion point despite local pressure. Her physician orders laboratory tests to determine the cause of her bleeding. The results are shown as follows:

Assay	Result
PT	16.5 s
APTT	34 s
TT	19 s
Fibrinogen	567 mg/dL
D-Dimer	550 ng/mL
Platelet count	234,000/µL
Creatinine	3.7 mg/dL

18. Since her PT is prolonged, the physician orders 2 units of plasma. The transfusion service technologist prospectively reviews the order and calls the pathology resident. What is the best advice to provide to the clinician?

A. The degree of PT prolongation requires 15 mL/kg of plasma, corresponding to 3–4 units for a 65 kg patient
B. Her bleeding is more likely due to an inherited platelet defect, and platelet transfusion is indicated
C. Goodpasture syndrome causes platelet dysfunction that can be ameliorated by dialysis or DDAVP
D. Order a PT mixing study to determine if a nonspecific inhibitor is present
E. There is no evidence of hemostatic abnormality and her oozing is likely due to a localized problem

Concept: The type of bleeding and the patient's clinical history are often more informative in the construction of the differential diagnosis than the results of the initial coagulation assays. While patients sometimes have unexpected bleeding without an abnormal laboratory test result, others have laboratory abnormalities that are unrelated to the bleeding.

Answer: *C*—Goodpasture syndrome (or Goodpasture disease) is an acquired fulminant autoimmune disorder of unknown origin characterized by the development of antiglomerular basement membrane collagen autoantibodies that cross-react with the collagen of pulmonary alveoli. Goodpasture syndrome causes acute glomerulonephritis with renal failure and pulmonary hemorrhage. As in many kidney disorders, the uremia of Goodpasture syndrome causes an incompletely characterized nonhereditary platelet dysfunction (Answer B) that is partially ameliorated by dialysis, or DDAVP. Decreased platelet function causes a characteristic mucocutaneous bleeding diathesis that may require systemic treatment (Answer E). Treating this patient with plasma will not improve the acquired platelet dysfunction (Answer A) and risks transfusion-associated circulatory overload (TACO). Since the APTT is normal and the PT is only minimally prolonged, a mixing study is not likely to be beneficial, especially in light of the patient's diagnosis of Goodpasture syndrome (Answer D).

19. As a follow-up, you suggest a complete blood count for the Goodpasture syndrome patient. The hematocrit is 18%, with a mean corpuscular volume (MCV) of 90 fL. Which statement is correct?

A. The anemia is probably due to iron deficiency and should be treated with oral iron supplementation
B. Low hematocrit is known to decrease platelet adhesion in renal failure and RBC transfusion may help reverse the bleeding
C. She has a significant risk of hemoptysis and should be given a platelet transfusion to prevent bleeding
D. The correlation between the hematocrit and uremic platelet dysfunction suggests that increasing the platelet count is useful
E. She should be transfused RBCs, plasma, and platelets in a 1:1:1 ratio

Concept: Rheology is the study of fluids in motion and is applicable when discussing the influence of the hematocrit in hemostasis, especially the ability of platelets to adhere to the endothelium.

Answer: *B*—Reduced hematocrit is known to decrease platelet adhesion in renal failure and RBC transfusion may help reverse bleeding. In anemia, increasing the RBC mass is more beneficial to platelet function than transfusing platelets to increase the platelet count. Additionally, donor platelets will become dysfunctional by the same mechanism (incompletely described) as the patient's own platelets, so platelet transfusions are contraindicated (Answers C and D) unless there is life-threatening bleeding in the absence of response from all other measures. Iron deficiency anemia is ruled out by the normal MCV—the MCV of IDA is consistently < 80 fL (Answer A). With no evidence of massive bleeding, transfusing in a 1:1:1 ratio is not indicated (Answer E).

End of Case

Please answer Questions 20 and 21 based on this clinical scenario:

A 2-day-old baby boy with congenital heart disease is being evaluated for cardiovascular surgery. Preoperative laboratory tests reveal the following results:

Assay	Result	Newborn reference interval
PT	18.5 s	10.1–15.9 s
APTT	53.7 s	31.3–54.3 s
HCT	46%	48%–68%
Platelet count	300,000/μL	150,000–450,000/μL

20. How should these results be interpreted?

A. The prolonged PT is of no clinical consequence

B. The prolonged PT may signal a need for vitamin K therapy

C. Factor assays are needed to explain the prolonged PT

D. Intraoperative plasma and cryoprecipitate will correct his coagulopathy

E. The mother was receiving unfractionated heparin

Concept: Preop laboratory tests are unnecessary in the absence of a bleeding history, though they are often ordered. Since congenital heart defects often cause hepatic congestion and the liver produces most of our coagulation factors, testing this child is warranted. Newborns' expected factor levels differ from adults, making it essential to interpret their test results with age-matched reference ranges.

Answer: *B*—Vitamin K therapy may be indicated in this case. Because the source of vitamin K is dietary, newborns are often vitamin K-deficient, so their vitamin K-dependent coagulation factor activities (II, VII, IX, and X) are reduced for the first 6 months of life. This is most often reflected in a prolonged PT because the PT responds first to factor VII deficiency, and factor VII has the smallest plasma concentration and the shortest half-life. This can be corrected by administering oral vitamin K.

The prolonged PT cannot be ignored (Answer A), and there may be some value in performing coagulation factor assays (Answer C) to understand why the PT is prolonged and the APTT is borderline; however, this would require excessive blood specimen volume from a tiny newborn, and would delay surgery. Conversely, the administration of vitamin K is safe and simple. Cryoprecipitate is unlikely to provide the necessary coagulation factors and plasma brings the risk of circulatory overload (Answer D). Unfractionated heparin does not cross the placental barrier (Answer E).

21. The surgeon consults with the pediatric hematologist who orders various factor activity level assays. The results are as follows:

Assay	Patient	Newborn reference interval
Factor V	57%	34%–105%
Factor VII	29%	28%–104%
Factor VIII	188%	50%–178%
Factor IX	29%	15%–91%
Fibrinogen	429 mg/dL	167–399 mg/dL

How do you interpret these results?

A. He may have congenital mild factor VII deficiency and should receive preoperative plasma

B. He may have mild hemophilia B and should receive preoperative factor IX concentrate

C. His liver is affected by severe congestion and he should receive preoperative plasma

D. These factor levels are expected for a newborn and he is not at increased risk of bleeding

E. The high factor VIII and fibrinogen levels may be a thrombosis risk factor

Concept: When interpreting a newborn's laboratory test results, take into account their age, since factor levels change over time until they reach adult levels by approximately 6 months.

Answer: *D*—These factor levels are expected for a newborn and he is not at an increased risk of bleeding. Even factors that are lower than what would be expected in adults, such as factors V, VII, and IX, are at hemostatic levels and not at the level seen in congenital deficiencies (Answers A and B). The vitamin K-dependent factors VII and IX are low normal, reflecting vitamin K deficiency, but do not provide evidence of liver congestion (Answer C). Although chronically elevated factor VIII and fibrinogen levels may be associated with thrombophilia in adults, both are acute phase reactants. Their slight elevation in this case may be an acute response to inflammation, but do not signal thrombotic risk in a newborn (Answer E).

End of Case

22. The parents of a 1-year-old boy report that he has begun to experience episodes of severe pain in his ankles and knees since beginning to walk. His laboratory results are provided as follows:

Assay	Result	Reference interval
PT	14.1 s	12.6–14.6 s
APTT	63 s	25–35 s
Fibrinogen	267 mg/dL	220–498 mg/dL
Platelet count	320,000/μL	150,000–450,000/μL

His APTT mixing studies revealed immediate and 2-h incubated mix correction. What is the logical next step?

A. Perform a platelet aggregation study
B. Assay coagulation factors VIII and IX
C. Assay coagulation factors XII, PK, and HMWK
D. Test for a specific inhibitor, such as antifactor VIII
E. Test for a lupus anticoagulant (LA)

Concept: The patient's symptoms reflect anatomic (joint and soft tissue) bleeding characteristic of coagulation factor deficiencies. Hemophilia, a congenital single coagulation factor deficiency, and VWD are considerations; however, VWD bleeding is typically mucocutaneous, not anatomic. The immediate and 2-h incubated mixing study corrections indicate a factor deficiency.

Answer: *B*—Assaying coagulation factors VIII and IX is the best follow-up step. The most likely diagnosis in this patient is hemophilia A, a congenital factor VIII deficiency, which occurs in 1 in 5000 male births and comprises about 80% of hemophiliacs. Hemophilia B, a congenital factor IX deficiency, affects 1 in 30,000 males. Hemophilia C, Rosenthal syndrome, or factor XI deficiency, is an autosomal recessive disorder with mild bleeding that affects one in a million males and females, half of who are of Jewish parentage.

Among hemophilia A and B cases, about 30% arise from de novo mutations with no family history and the most common genetic abnormality is factor VIII intron 22 inversion. Most hemostasis laboratories are equipped to assay factors VIII, IX, and XI using automated coagulometers. Deficiencies of the contact factors XII, PK, and HMWK (Answer C) prolong only the APTT, but have no clinical consequence, as they are not necessary for normal hemostasis. This patient's clinical symptoms are not typical for a platelet disorder, which usually manifests with mucocutaneous bleeding (Answer A), as compared to his joint bleeding. Young patients do not tend to form specific inhibitors (Answer D). Since the patient has a bleeding disorder, not a clotting disorder, LA testing would not be necessary (Answer E).

23. An adult male with severe hemophilia A (factor VIII < 1%) presents to the emergency room while on vacation in another state. He is diagnosed with a joint bleed requiring factor VIII concentrate therapy. He weighs 70 kg and his hematocrit is 40%. His dosing for factor concentrate was computed using the following formula to achieve a factor VIII plasma level of at least 80%:

Blood volume (BV, mL) = weight (kg) × 70 mL/kg
Use 60 mL/kg for obese, BMI 25–30
Use 50 mL/kg for morbidly obese, BMI >30
Plasma volume (PV, mL) = BV × (100%–HCT%)
Determine IU (%) of FVIII concentrate desired based on factor VIII plasma level
Units of FVIII required = (desired FVIII in IU/mL—initial IU/mL) × PV (mL)

In our example:
BV = 70 kg × 70 mL/kg = 4900 mL
PV = (100%–40%) = 60% × 4900 mL = 2940 mL
Units of FVIII required = (0.8–0 IU/mL) × 2940 = 2352 IU to achieve 80% factor VIII activity.

The patient was given two vials (1200 IU each) by intravenous push. To monitor efficacy, a specimen was collected for a factor assay 15 min after the concentrate was administered, with the following results:

Plasma dilution	Seconds (s)	Raw factor VIII activity (%)	Computed factor VIII activity (× dilution) (%)
1:10 (undiluted)	95	10	10
1:20	99	8	16
1:40	107	5	20
1:80	108	4	32
1:160	108	4	32

What do the factor VIII assay results indicate?

A. Adequate response to factor VIII concentrate
B. Erroneously low factor VIII concentrate dose
C. Specific factor VIII inhibitor
D. Lupus anticoagulant
E. Accompanying factor II deficiency

Concept: The coagulation factor computation shown here is generic and may be used for factor VIII concentrate dosing, unless the manufacturer provides an alternative in the accompanying package insert.

Coagulation factor assays are performed using four simultaneous dilutions. Because the calibrator is diluted 1:10, the first patient's plasma dilution is 1:10 and is paradoxically considered "undiluted" for computation purposes. Subsequent dilutions are usually serial (20, 40, 80, 160, …). If all the dilutions generate results within 10% of the first result after multiplying by their respective dilution factors, the results are deemed "parallel," the assay is validated, and there is no inhibitor (Table 13.6).

TABLE 13.6 Example of parallel dilution results.

Plasma dilution	Seconds (s)	Raw factor VIII activity (%)	Computed factor VIII activity (× dilution)
1:10 "undiluted"	90	20	20 %
1:20	105	10	20 % (parallel)[a]
1:40	107	5.5	22 % (parallel)
1:80	110	2.6	20.8 % (parallel)

[a] *<10% difference from undiluted indicates parallelism, no inhibitor*

Answer: C—The results in this case likely represent a specific factor VIII inhibitor, an IgG alloantibody that partially neutralizes the infused factor VIII. The anticipated plasma factor level was 80 units/dL or 80%, but the highest measured level was only 32%, consistent with a factor VIII inhibitor. The identical results for dilutions 1:80 and 1:160 indicate that the inhibitor has been "diluted out." The operator typically extends the dilutions until consecutive tubes generate results that duplicate within 10%. Based on the results, this is not an adequate response to the dose given (Answers A and B). Lupus anticoagulant (Answer D) may generate similar nonparallel dilution results and must be considered in the laboratory differential identification; however, the patient's diagnosis is long-established. These results do not imply the possibility of a second factor deficiency (Answer E).

24. A laboratory technologist performs a VWD profile on a specimen from a 23-year-old male who complains of easy bruising and frequent epistaxis. His results are as follows:

Assay	Patient (%)	Reference interval (%)
VWF:RCo (activity)	25	50–166
VWF:Ag (concentration)	28	50–249
Factor VIII	27	55–186

VWF:RCo, von Willebrand factor ristocetin cofactor; VWF:Ag, von Willebrand factor antigen.

Based on these results and the clinical history, what is the most likely diagnosis for this patient?
A. Hemophilia A
B. Bernard-Soulier syndrome
C. VWD type 1
D. VWD type 2
E. VWD type 3

Concept: VWD is a mucocutaneous (systemic) bleeding disorder with a worldwide incidence of 1%–2%, caused by von Willebrand factor (VWF) deficiency. VWD may be a quantitative (types 1 and 3) or qualitative (type 2) deficiency. Type 1 implies a moderate deficiency, while type 3 VWD implies complete VWF absence. VWF is necessary for platelet adhesion to injured arterial or arteriolar walls, which is why VWD is associated with platelet-type mucocutaneous bleeding. VWF is also the carrier molecule for coagulation factor VIII, as free coagulation factor VIII has a half-life of seconds. Consequently, factor VIII activity levels reflect VWF concentrations. Because VWF is markedly deficient in type 3 VWD, factor VIII activity is proportionally deficient and the patient suffers both mucocutaneous and soft-tissue (hemophilia-like) bleeding.

A standard VWD profile includes a VWF activity (ristocetin cofactor, VWF:RCo) assay, VWF antigen concentration immunoassay (VWF:Ag), and a coagulation factor VIII activity assay. The FVIII level more closely parallels VWF:Ag than VWF:RCo. The VWF:RCo is a commonly performed VWF activity assay, however, its precision is poor, exceeding coefficient of variation levels of 30% in external quality control assays.

Although the reference interval remains consistent for an unselected (random) population, VWF activity varies by blood group, as shown in the table below. Consequently, a disproportionate incidence of people with "low VWF" have blood type O.

Blood group	VWF:RCo (activity) (%)
O	36–157
A	48–234
B	57–241
AB	64–238
Overall	50–166

Answer: C—VWD type 1 is the most likely diagnosis. Approximately 80% of VWD patients possess the type 1 quantitative variant, in which the VWF:RCo to VWF:Ag ratio is expected to be ≥ 0.7. Type 2 VWD (Answer D) is qualitative, implying the VWF:Ag level is near normal or mildly decreased but the activity is decreased, producing a VWF:RCo to VWF:Ag/ratio of < 0.7. Type 3 (Answer E) implies a complete absence of VWF. The ratio limit is established locally and may vary.

VWD is diagnosed when VWF activity is <30%. Levels of 30%–50% may be associated with mild intermittent mucocutaneous bleeding, but are termed "low VWF" rather than VWD. Bernard-Soulier syndrome (BSS, Answer B) is a rare inherited platelet membrane receptor abnormality whose symptoms may resemble VWD symptoms, but with normal VWD profile results. A low FVIII activity level with normal VWF activity and concentration would be consistent with hemophilia A (Answer A).

25. What is a common initial therapeutic option for Type 1 VWD?
A. Tranexamic acid
B. Cryoprecipitate
C. DDAVP inhaler
D. VWF concentrate
E. Plasma

Concept: Type 1 VWD is a moderate quantitative deficiency of VWF in which there may exist relatively abundant VWF stored in the Weibel-Palade bodies of endothelial cells. Increasing the plasma activity of VWF prior to surgical procedures may be indicated to avoid significant blood loss.

Answer: C—A DDAVP inhaler (desmopressin or Stimate) can induce VWF secretion from endothelial cell storage. In many instances, DDAVP is sufficient to prevent or control bleeding in patients with type 1 VWD. The antifibrinolytic drug tranexamic acid (answer A) may be added if bleeding continues. Cryoprecipitate and plasma (Answers B and E) are not recommended, except in situations where no other preparation is available, as they are blood products and confer a risk of infectious disease transmission.

The physician may employ a presurgical DDAVP challenge, in which the baseline VWF:Ag level is measured, the physician asks the patient to inhale DDAVP, and the VWF:Ag is measured again. If the level rises appropriately, DDAVP may be used prior to surgery. If not, tranexamic acid or VWF concentrate are used. VWF concentrates are indicated if bleeding is not controlled by DDAVP or tranexamic acid.

26. A patient appears to have type 2 VWD because the VWF:RCo to VWF:Ag ratio is <0.7. The laboratory technologist performs the ristocetin-induced platelet agglutination (RIPA) assay using customary (1.0 mg/mL) and reduced ristocetin concentrations: 0.5, 0.25, and 0.1 mg/mL, respectively, a procedure called the ristocetin response curve. The technologist observes platelet agglutination at all concentrations of ristocetin. What VWD subtype is most likely based on these results?
A. VWD type 1
B. VWD subtype 2A
C. VWD subtype 2B
D. VWD subtype 2M
E. VWD type 3

Concept: Type 2 VWD is a qualitative VWF deficiency. The VWF:Ag-based concentration in type 2 VWD may be normal or moderately reduced, but VWF:RCo activity is diminished to ≤ 30%. In type 2 VWD, the VWF:RCo to VWF:Ag ratio is <0.7, and the concentration of large VWF multimers, the molecules necessary to support platelet adhesion, is diminished, as demonstrated in VWF multimeric electrophoresis. Type 2 VWD is classified into subtypes 2A, 2B, 2M, and 2N (Table 13.7).

TABLE 13.7 VWD Types and Subtypes

Type	Prevalence	Bleeding	Pathology	Assay
1	75%–80% of VWD	Mild-moderate mucocutaneous	Various dominant VWF mutations throughout gene	VWF:RCo/VWF:Ag ≥0.7
Subtype 2A	10%–15% of VWD	Moderate-severe mucocutaneous	Dominant or recessive mutation in ADAMTS13 cleavage site of VWF	VWF:RCo/VWF:Ag <0.7, large multimers absent in electrophoresis
Subtype 2B	5% of VWD	Mild-moderate mucocutaneous	Dominant gain of function mutation in GP Ib/IX/V binding site of VWF	Agglutination present in ristocetin response curve with low ristocetin concentrations
Subtype 2M	Infrequent	Moderate mucocutaneous	Dominant or recessive loss of function mutation in GP Ib/IX/V binding site of VWF	VWF:RCo/VWF:Ag <0.7, large multimers present in electrophoresis
Subtype 2N	Infrequent	Mild soft tissue	Recessive mutation in factor VIII binding site of VWF	VWF:RCo/VWF:Ag ≥0.7, normal electrophoresis, FVIII decreased
3	Rare	Severe mucocutaneous and soft tissue	Recessive deletion with decreased mRNA expression	VWF absent
Acquired	Rare	Variable, depending on the type	Anti-VWF antibody, increased VWF consumption, reduced production	VWF decreased

ADAMTS13, A disintegrin-like and metalloprotease domain with thrombospondin motifs-13; GP, glycoprotein; RIPA, ristocetin-induced platelet agglutination; VWD, von Willebrand disease; VWF, von Willebrand factor; VWF:Ag, von Willebrand factor antigen; VWF:RCo, von Willebrand factor ristocetin cofactor assay.

In VWD subtype 2A, large VWF multimers are diminished due to mutations that enhance VWF cleavage by ADAMTS13 (VWF-cleaving protease) in plasma.

VWD subtype 2B is caused by a "gain of function" mutation in which large multimers possess increased platelet membrane avidity. The multimers bind and activate circulating platelets, causing mild thrombocytopenia and reduction in large multimers.

In VWD subtype 2M, the multimer distribution appears normal, but a mutation results in poor platelet binding, the opposite of subtype 2B. Subtype 2M may often be mistaken for subtype 2A, except that the electrophoretic pattern of VWF is normal.

In VWD subtype 2N, the coagulation factor VIII binding site is defective, owing to a gene mutation. Also called Normandy type VWD or "autosomal hemophilia A," the symptoms appear as soft tissue (anatomic), and not mucocutaneous hemorrhage, and affect both genders.

Answer: C—VWD subtype 2B is most likely. In subtype 2B, the VWF mutation causes a VWF multimer gain of function. The large multimers are abnormally avid for platelet receptors, and bind platelets in vivo. The coated platelets respond to ristocetin at both normal and reduced concentrations. The ristocetin-induced platelet agglutination (RIPA) assay resembles the VWF:RCo assay except that patient's own platelets are used in lieu of reagent platelets. While VWF:RCo is quantitative, RIPA is qualitative. Patient platelet agglutination at ristocetin concentrations that are < 1.0 mg/mL confirms VWD subtype 2B.

This is an important clinical distinction. DDAVP is relatively contraindicated in type 2B as it may release hyperfunctioning VWF, leading to thrombocytopenia. VWF concentrate is usually the preferred treatment for VWD subtype 2B.

In VWD types 1 and 3 (Answers A and E), and in subtypes 2A and 2M (Answers B and D), patient's platelets do not respond to ristocetin, or only slightly to 1.0 mg/mL.

27. Which of the following mechanisms is associated with increased risk for venous thromboembolism?
A. Methylene tetrahydrofolate reductase (MTHFR) mutation
B. Aspirin resistance
C. Elevated LDL-cholesterol
D. Reduced HDL-cholesterol
E. Factor V Leiden (FVL) mutation

Concept: Venous thromboembolism (VTE) includes deep venous thrombosis (DVT) and pulmonary embolism (PE). Most VTEs are provoked by acquired risk factors, such as venous stasis from prolonged immobilization, cancer, or intravenous indwelling catheters. Additional risk factors for VTEs can be abnormalities in the coagulation mechanism control systems, such as antithrombin, protein C, and protein S deficiencies. Conversely, arterial thrombotic events, such as acute coronary syndromes (ACS) and stroke are typically associated with inflammatory conditions and lipid abnormalities and, less likely, with coagulation mechanism imbalances.

Answer: *E*—Factor V Leiden mutation increases the risk of venous thromboembolism. The factor V Leiden (FVL) mutation renders coagulation factor V partially resistant to activated protein C, a control protein whose function is to digest and decrease the activity of activated factor V as it complexes with activated factor X. Therefore, FVL results in increased thrombin generation and thus, increases the risk of VTE. MTHFR mutation (Answer A) was once implicated in VTE, but recent outcomes studies have shown no relationship between MTHFR polymorphisms and thrombosis. Aspirin resistance and lipid abnormalities are most often associated with arterial thrombosis (Answers B, C, and D).

28. A patient presents for follow-up and warfarin monitoring by INR 2 months after an episode of venous thrombosis. At this visit, his physician decides to also run a thrombophilia profile to determine the etiology of the thrombosis. He receives laboratory results that report concurrent protein C and protein S activity deficiencies, but no other inherited thrombophilia indications. What is the best explanation for these results?
A. Protein C and S assays were performed while the patient was taking warfarin
B. These results may indicate interference caused by antithrombin deficiency
C. The patient may have an LA
D. The patient is a double heterozygote, and is at high risk for pulmonary embolism
E. Protein C and S activity levels remain reduced for several months after a thrombotic event

Concept: The typical laboratory profile to detect congenital thrombosis risk factors includes antithrombin, protein C and protein S activity assays, the activated protein C resistance ratio, the prothrombin G20210A mutation molecular assay, and an LA profile that may include the anticardiolipin antibody and the anti-β2-glycoprotein I antibody. Thrombophilia profiles should be performed at least 14 days after warfarin therapy is discontinued.

Answer: *A*—Protein C and S levels are most likely low because they were performed while patient was taking warfarin. Warfarin is a vitamin K antagonist that induces the liver to produce inactive vitamin K-dependent factors, II, VII, IX, and X; and control proteins C, S, and Z. Testing patients while on warfarin results in factitiously low levels of protein C and protein S, which do not indicate thrombosis risk factors. The probability of inherited protein C and S deficiency, a double heterozygote (Answer D), is less than one in a million and should be discounted clinically until proven otherwise.

Antithrombin activity levels are reduced by heparin and active thrombosis, but do not affect protein C or protein S activity levels (Answer B). Although there is no longer active thrombosis 2 months since the episode, similar to the anticoagulant guidelines, the thrombophilia profile should only be ordered 14 or more days subsequent to a period of active thrombosis (Answer E). LA has no effect on Protein C and S levels (Answer C).

29. Which finding is *not* part of the "4T" scoring system for heparin-induced thrombocytopenia (HIT)?
A. A 30%–50% decrease in platelet count
B. Thrombosis and skin lesions
C. Timing of platelet count decrease
D. No alternate explanation for platelet count decrease
E. Positive antiheparin:PF4 immunoassay

Concept: Between 1% and 5% of patients who receive unfractionated heparin (UFH) for more than 5 days develop an IgG antibody to a heparin-PF4 complex detectable in an immunoassay. A small percentage of these antibodies activate platelets and cause microvascular arterial and/or venous thrombi. In thee cases, HIT may cause pulmonary emboli, limb gangrene, stroke, and myocardial infarction.

Answer: *E*—A positive antiheparin:PF4 immunoassay is not part of the 4T score. Often, if a patient has a 4T score in the intermediate to high range, heparin is discontinued, and an enzyme-linked immunoassay (ELISA) is performed as the initial test for HIT. The other choices (Answers A, B, C, and D) represent parts of the 4T scoring system (Table 13.8).

TABLE 13.8 Pretest HIT Scoring: The 4 Ts

4 Ts	2 points	1 point	0 points
Thrombocytopenia	Platelet count fall >50% and platelet count nadir ≥ 20,000/μL	Platelet count fall 30%–50% and platelet count nadir 10–19,000/μL	Platelet count fall <30% and platelet count nadir < 10,000/μL
Timing of platelet count fall	Clear onset days 5–10 or fall in ≤1 day with prior heparin exposure ≤30 days	Onset unclear or fall in ≤1 day with prior heparin exposure 30–100 days	Fall <4 days without recent heparin exposure
Thrombosis or other sequelae	Confirmed new thrombosis, skin necrosis, or acute systemic reaction	Progressive, recurrent, or suspected thrombosis, or nonnecrotizing lesions	None
Other thrombocytopenia causes	None apparent	Possible	Definite
Clinical probability score	High: 6–8 Intermediate: 4–5 Low: ≤ 3		

30. Which assay is typically used to confirm HIT?
A. Serotonin release assay (SRA)
B. Lateral flow colorimetric immunoassay
C. Platelet aggregometry using heparin as the agonist
D. Enzyme immunoassay for antiheparin-PF4
E. Washed platelet suspension aggregometry using heparin as the agonist

Concept: Because the immunoglobulin antiheparin-PF4 is the basis for HIT, it seems logical that a standard ELISA or a platelet function assay would define the disorder. Most assays, however, have high sensitivity to detect the antibodies, but poor specificity for the diagnosis of HIT.

Answer: *A*—The serotonin release assay (SRA) is available from selected reference laboratories that possess radioactive isotope licenses. Platelet aggregometry-based assays (Answers C and E) tend toward specificity with reduced sensitivity, producing unacceptable false negative results. Immunoassays (Answers B and D) produce high false positive results due to their high sensitivities.

A typical HIT diagnostic sequence starts with application of the clinical 4T algorithm, followed by an immunoassay, which, if positive, requires confirmation using the SRA. The lateral flow colorimetric immunoassay method offers a short turn-around, but requires confirmation. In 2017, IL-Werfen began to offer a 30 minute turnaround immunoassay with improved sensitivity and specificity, however, the initial response to HIT relies on the 4T approach, with discontinuation of heparin if clinical suspicion is high.

31. When HIT is suspected, which of the following anticoagulant therapies may be substituted for UFH?
A. Warfarin
B. Dabigatran
C. Argatroban
D. Rivaroxaban
E. Enoxaparin

Concept: When HIT is suspected, UFH therapy is withdrawn immediately; however, a fast-acting anticoagulant must be started in its place to manage the ongoing thrombosis.

Answer: *C*—Argatroban, administered by infusion is the approved option. Each 2.5 mL vial contains 250 mg of concentrated argatroban, which is diluted 100-fold with normal saline prior to infusion. Argatroban is associated with a higher risk of hemorrhage than unfractionaled heparin so patients must be observed closely. Warfarin (Answer A) requires 5 days to reach anticoagulant efficacy, and may be thrombogenic during early administration, leading to warfarin skin necrosis. The DOACs dabigatran and rivaroxaban (Answers B and D) are not approved for use in HIT. Low molecular weight heparin, such as enoxaparin (Answer E) cross-reacts with the HIT antibody, and confers a slight (~1% of UFH prevalence) risk of de novo HIT.

32. In accordance with trauma management guidelines developed after 2005, in addition to red blood cells, what is the best choice for emergency fluid replacement in trauma patients with major bleeding and the signs and symptoms of shock?
A. Intravenous saline
B. Intravenous crystalloid solution
C. Thawed group AB or A plasma
D. Prothrombin complex concentrates
E. Cryoprecipitate

Concept: Hemorrhage from blunt or penetrating trauma may trigger shock with associated hypotension, hypothermia, and acidosis (base excess). Emergency medical personnel attempt to control shock with blood pressure stabilization, body temperature control, and base excess control using electrolyte solutions. Managing coagulopathy is key to survival. Studies reported between 2005 and 2012 have led to balanced blood product therapy, 1:1:1 RBCs, plasma, and platelets, to improve trauma-induced coagulopathy management.

Answer: *C*—Thawed group AB or group A plasma is the best choice in this specific situation. Saline, crystalloids, and colloids (Answers A and B) dilute plasma coagulation factors and intensify the coagulopathy of trauma and shock. Crystalloids and colloids also promote fluid imbalance and edema. Prior to 2005, initial use of crystalloids and colloids for fluid resuscitation in trauma patients was recommended over plasma. However, current massive transfusion protocols recommend early coagulation management with group AB or A plasma, with a subsequent 1:1:1 balanced blood product administration of plasma, RBCs, and platelets. Coagulation factor concentrates, such as fibrinogen (or cryoprecipitate) or the prothrombin complex concentrate (Answers D and E), may be used later in the resuscitation when laboratory assays indicate specific deficiencies.

33. The transfusion service receives an order for 4 units of plasma to be administered to a patient in the emergency room with massive gastrointestinal bleeding and a past medical history significant for atrial fibrillation. She is presumed to be taking an oral anticoagulant to prevent stroke. While the plasma units are thawing, you review her coagulation test results and note a markedly prolonged TT and APTT, and a mildly prolonged PT. What anticoagulant do you suspect the patient is taking and what is the best treatment strategy at this time?

A. Warfarin, and you recommend intravenous vitamin K and plasma
B. Dabigatran, and you recommend a dose of idarucizumab
C. Rivaroxaban, and you recommend dialysis STAT
D. Apixaban, and you recommend 10 units of cryoprecipitate
E. Edoxaban, and you recommend 6 units of plasma instead of 4

Concept: The direct oral anticoagulants (DOACs) dabigatran, rivaroxaban, apixaban, betrixaban and edoxaban have been approved for the prevention of stroke in atrial fibrillation or VTE following orthopedic surgery. While one of their main benefits is that there is no required laboratory monitoring, all prolong the clot-based clotting tests PT, APTT, and/or TT to varying degrees. For the PT and the APTT, their effect depends on the reagent source and coagulometer being used, and is not easy to predict.

Answer: *B*—As of 2017, dabigatran, a direct thrombin inhibitor (DTI) is the only DOAC that prolongs the thrombin time (TT) and the APTT, and may slightly prolong the PT. None of the other DOACs, nor warfarin (Answer A), prolong the TT. For reversal, the patient should receive an infusion of idarucizumab, a specific monoclonal antibody that neutralizes dabigatran approved in 2015. The recommended dose is 5 g, provided as two separate 50-mL vials, each containing 2.5 g.

Rivaroxaban, apixaban, edoxaban and betrixaban (Answers C, D, and E) are direct factor Xa inhibitors that are not dialyzable and do not respond to even large doses of plasma or cryoprecipitate. These agents may require treatment with activated coagulation factor concentrates to overcome their inhibitor effect. Andexanet alfa, a direct anti-Xa inhibitor reversal agent, is under US FDA consideration (Table 13.9).

34. A 65-year-old female with nonvalvular atrial fibrillation arrives in the emergency department (ED) experiencing severe hematuria. Her husband reports she takes rivaroxaban. The ED physician suspects an overdose. If properly calibrated, what laboratory assay is available to detect rivaroxaban?

A. PT
B. APTT
C. Factor X activity
D. Plasma-diluted thrombin time
E. Chromogenic anti-Xa

Concept: Rivaroxaban is a direct factor Xa antagonist. Others in the same group are betrixaban, apixaban and edoxaban. An additional DOAC, dabigatran, is a direct thrombin inhibitor (DTI). The DOACs have emerged as an alternative to warfarin for oral anticoagulation because they do not require routine laboratory monitoring.

Answer: *A*—Though regular monitoring of rivaroxaban is not required or recommended, the PT has a semiquantitative dose-response relationship with the plasma rivaroxaban level, if calibrated. As of 2017, the PT is the only FDA-approved method to measure direct anti-Xa anticoagulants. The PT response depends on the reagent and coagulometer, and must be calibrated to rivaroxaban at each laboratory using plasma with rivaroxaban. Without this calibration, the results are less meaningful.

TABLE 13.9 Example of an Anticoagulant Reversal Protocol

Generic drug	Elimination half-life	Removal by dialysis	Reversal strategy
Apixaban Class: Factor Xa inhibitor	8–15 h (longer in renal impairment)	No	Drug activity can be assessed with an anti-Xa activity assay (awaits FDA approval as of 2017) • If ingested within 2 h, administer activated charcoal • Consider prothrombin complex concentrate (Kcentra), 50 IU/kg Kcentra may partially correct PT/APTT but will not affect anti-factor Xa activity assay results and will not increase drug clearance; correlation of shortening PT/APTT with reduction in bleeding risk is unknown and should not be used, since it does not correlate with clinical response
Argatroban Class: Direct thrombin inhibitor	40–50 min	~20%	• Discontinue infusion • APTT may be used to check presence of the drug in the plasma
Bivalirudin (Angiomax) Class: Direct thrombin inhibitor	25 min (up to 1 h in severe renal impairment)	~25%	• Discontinue infusion • Degree of anticoagulation can be assessed with APTT
Dabigatran Class: Direct thrombin inhibitor	14–17 h (up to 34 h in severe renal impairment)	~65%	• Thrombin time markedly prolonged • If ingested within 2 h, administer activated charcoal • Infuse idarucizumab 5 g in two 2.5 g/50 mL vials • Consider dialysis or idarucizumab re-dosing for refractory bleeding after initial administration • Administer no coagulation factor concentrates
Dalteparin and Enoxaparin Class: Indirect thrombin inhibitor: low molecular weight heparin	3–5 h (longer in renal impairment)	~20%	• Use protamine sulfate for partial neutralization (~60%) • Degree of reversal can be assessed with anti-Xa activity Time since last dose — Protamine dose for each 100 units of drug administered < 8 h — 1 mg (alternative: 50 mg fixed dose) 8–12 h — 0.5 mg (alternative: 25 mg fixed dose) >12 h — Not likely to be useful (alternative 25 mg fixed dose)
Fondaparinux Class: Factor Xa inhibitor	17–21 h (significantly longer in renal impairment)	No	• Fondaparinux levels can be assessed by anti-Xa activity (using different calibrator from the heparin anti-Xa test) • Consider FEIBA 20 IU/kg or rFVIIa 30 μg/kg FEIBA may partially correct PT/APTT but will not affect anti-factor Xa activity results and will not speed drug clearance; correlation of shortening PT/APTT with reduction in bleeding risk is unknown, and should not be used since it does not correlate with clinical response
Heparin Class: indirect thrombin inhibitor	30–90 min (dose dependent)	Partial	• Use protamine sulfate for neutralization (100%) • Degree of reversal can be assessed with anti-Xa activity Time since last dose of heparin — Dose of protamine sulfate for each 100 units of heparin Immediate — 1 mg (alternative: 50 mg fixed dose) 30 min–2 h — 0.5 mg (alternative: 25 mg fixed dose) >2 h — 0.25 mg (alternative: 10 mg fixed dose)

TABLE 13.9 Example of an Anticoagulant Reversal Protocol (*cont.*)

Generic drug	Elimination half-life	Removal by dialysis	Reversal strategy
Rivaroxaban, edoxaban, betrixaban Class: direct factor Xa inhibitor	Healthy: 5–9 h >70 years old: 11–13 h (longer in renal impairment)	No	• Drug activity assessed with anti-Xa activity (Awaiting FDA approval in 2017) • Consider Kcentra. Andexanet alfa, a direct anti-Xa inhibitor reversal agent, was under US FDA consideration in 2017 Kcentra may partially correct PT/APTT but will not affect anti-factor Xa activity results and will not speed drug clearance; correlation of shortening PT/APTT with reduction in bleeding risk is unknown, and should not be used since it does not correlate with clinical response

WARFARIN: VITAMIN K ANTAGONIST

INR	Clinical scenario	Management
Any INR	Life-threatening bleeding (e.g., intracranial hemorrhage)	• Hold warfarin • Give phytonadione (vitamin K) 10 mg IV for 10–30 min (repeated if necessary depending on the INR) • Consider 4-factor PCC (Kcentra) for all life-threatening bleeding o INR 2–<4; 25 units/kg; not to exceed 2500 units o INR 4.0–6.0; 35 units/kg; not to exceed 3500 units o INR >6.0; 50 units/kg; not to exceed 5000 units • If Kcentra is not available or contraindicated, alternative treatment is recommended over no treatment: Treatment with plasma 10–15 mL/kg IV along with vitamin K 10 mg IV over 10–30 min
Any INR	Serious bleeding	• Hold warfarin • Vitamin K 10 mg IV over 10–30 m, repeated if necessary depending on the INR • May supplement with plasma or Kcentra depending on the urgency of the situation (see dosing schedule above)
<5	No significant bleeding	• Lower the warfarin dose and monitor more frequently or hold warfarin and resume therapy at an appropriately adjusted dose when INR is in the therapeutic range
5–9	No significant bleeding	• Hold warfarin for 1 or 2 doses, monitor more frequently, and resume therapy at an appropriately adjusted dose when INR is in the therapeutic range • May consider vitamin K 1–2.5 mg PO × 1 if the patient is at increased risk of bleeding
	More rapid reversal required due to urgent surgery needed	• Administer vitamin K ≤ 5 mg PO × 1 with the expectation that a reduction in INR will occur in 24 h • If INR remains elevated, administer an additional dose of vitamin K 1–2 mg PO × 1
>9	No significant bleeding	• Hold warfarin and administer vitamin K 2.5–5 mg PO × 1 with the expectation that the INR will be reduced substantially in 24–48 h • Monitor INR more frequently, administer additional vitamin K if necessary, and resume therapy at an appropriately adjusted dose when INR is in the therapeutic range

The chromogenic anti-Xa provides a quantitative measurement when specific calibrators are used; however, in 2017, calibrators for the DOACs do not have FDA approval; therefore, results can be released as "research use only" (RUO, Answer E). The APTT does not prolong with rivaroxaban, apixaban, betrixaban, or edoxaban (Answer B). Coagulation factor X activity is not affected by any DOAC (Answer C). The plasma-diluted thrombin time (Answer D) provides reproducible linear results for dabigatran, but awaits FDA clearance in 2017.

Please answer Questions 35 and 36 based on the following clinical scenario:

A 59-year-old woman with a history of handling rat poison in her attic comes to the emergency room with bruises all over her torso and neck. Her PT is >100 s, INR >12, and her APTT is 86 s. Since she has congestive heart failure, the ED resident is concerned that giving her plasma will cause transfusion-associated circulatory overload, so he orders cryoprecipitate instead.

35. After reviewing this case, the transfusion medicine resident calls the ED resident to explain that cryoprecipitate is not indicated because of which of the following reasons?
A. Each dose would expose her to 10 donors with high risk of disease transmission
B. Cryoprecipitate units are of small volume and she appears hypovolemic
C. Vitamin K-dependent factors are not present in cryoprecipitate
D. Because it takes too long to thaw cryoprecipitate and her bleeding is serious
E. The transfusion service is out of cryoprecipitate of her blood group

Concept: Rat poison (brodifacoum, "superwarfarin"), a vitamin K antagonist, causes a coagulopathy that is not always identified because it is based heavily on history. When patients report exposure, the PT/INR is likely to be prolonged. Depending on the PT prolongation, which reflects the degree of deficiency of vitamin K-dependent factors II, VII, and X, the APTT is also likely to be prolonged because it requires factors II, IX, and X. In this patient with a PT >100 s (INR > 12), her APTT is also expected to be significantly prolonged. In order to treat a patient with brodifacoum poisoning, it is essential to know which blood product contains the vitamin K-dependent factors.

Answer: *C*—Vitamin K-dependent factors are not present in cryoprecipitate.

Cryoprecipitate contains factors VIII, VWF, fibrinogen, factor XIII, and fibronectin, but does not contain vitamin K-dependent factors. Fibrinogen and factor XIII deficiency are the main indications for cryoprecipitate because factor VIII and VWF concentrates are readily available. Fibrinogen and factor XIII concentrates are also available, but are not commonly used in the United States.

As an untreated plasma product, cryoprecipitate confers a modest risk of viral transmission (Answer A). Given that she suffers from congestive heart failure, hypervolemia is a greater risk than hypovolemia (Answer B).

As each unit of cryoprecipitate is typically only 10–15 mL, the units thaw rapidly (Answer D), and cryoprecipitate's ABO blood type does not need to be matched for the patient's blood type as RBCs or plasma (Answer E).

36. The physician decides to administer PCC at 50 IU/kg based on the initial INR, which controls the bleeding and shortens the PT. On the next day, her hemoglobin starts to decrease again, and the PT and APTT become more prolonged than before PCC therapy. What is the explanation for this?
A. The PCC dose was not enough for the severity of the poisoning
B. Rat poison has ~120-day half-life and must be managed for months
C. Platelets should have been given in conjunction with plasma for this indication
D. Vitamin K would have been a better therapeutic choice
E. Activated PCC (such as FEIBA) would have been more effective

Concept: Understanding the half-life of rat poison helps predict the effect of treatment and plan for subsequent needs of the patient. PCC increases the plasma levels of the vitamin K-dependent factors temporarily, but their production continues to be blocked by residual brodifacoum.

Answer: *B*—Brodifacoum's half-life is long and its effects must be managed for months. Rat poisons, also known as superwarfarins or long-acting anticoagulant rodenticides (LAARS), include brodifacoum, difenacoum, and chlorphacinone. Unlike warfarin, which has a half-life of 20–60 h, the various LAARs have half-lives of 16–270 days. Thus, treatment must extend until the PT returns to normal.

The PCC dose was appropriate based on the patient's INR (Answer A), but brodifacoum requires prolonged therapy (several doses over time). Brodifacoum does not affect platelets (Answer C). Vitamin K (Answer D) should have been administered with the PCC, but the PCC was necessary to control bleeding, which vitamin K would not accomplish alone. Activated PCC is not indicated, as it confers a risk of thrombosis (Answer E).

Standard brodifacoum therapy is PCC and vitamin K. PCC provides all four procoagulant vitamin K-dependent factors II, VII, IX, and X, and the control proteins C and S. PCC is given acutely to raise the levels of such factors when the patient is experiencing bleeding. At the same time, intravenous vitamin K should be administered, to allow the liver to start synthesizing these factors. While one 10 mg dose should correct an overdose of warfarin, repeated vitamin K doses are necessary to treat rat poisoning due to the very long half-lives of the LAARs.

End of Case

Suggested Reading

[1] L.T. Da Luz, B. Nascimento, A.K. Shankarakutty, et al. Effect of thromboelastography (TEG®) and rotational thromboelastometry (ROTEM®) on diagnosis of coagulopathy, transfusion guidance and mortality in trauma: descriptive systematic review, Crit. Care 18 (2014) 518.

[2] M. Franchini, G. Lippi, Prothrombin complex concentrates: an update, Blood Transfusion 8 (2010) 149–154.

[3] V. Pengo, A. Tripodi, G. Reber, et al. Update of the guidelines for lupus anticoagulant detection. Subcommittee on Lupus Anticoagulant/Antiphospholipid Antibody of the Scientific and Standardisation Committee of the International Society on Thrombosis and Haemostasis, J. Thromb. Haemost. 7 (2009) 1737–1740.

[4] S.A. Just, M. Nybo, H. Laustrup, et al. Single test isolated lupus anticoagulant positivity is associated with increased plasma levels of inflammatory markers and dyslipidemia, Lupus 25 (2016) 241–247.

[5] G.A. Fritsma, Thrombotic disorders and laboratory assessment, in: E.M. Keohane, L.J. Smith, J.M. Walenga (Eds.), Rodak's Hematology: Clinical Principles and Applications, fifth ed., Elsevier, 2016, pp. 689–712 Chapter 39.

[6] L.D. Brace, Qualitative disorders of platelets and vasculature, in: E.M. Keohane, L.J. Smith, J.M. Walenga (Eds.), Rodak's Hematology: Clinical Principles and Applications, fifth ed., Elsevier, 2016, pp. 739–759 Chapter 41.

[7] P.F. Fogarty, C.M. Kessler, Hemophilia A and B, in: C.S. Kitchens, C.M. Kessler, B.A. Konkle, Consultative Hemostasis and Thrombosis, third ed., Saunders, Philadelphia, 2013 (Chapter 4): 45–59.

[8] M.B. Marques, G.A. Fritsma, Quick Guide to Hemostasis, third ed., AACC Press, Washington, DC, (2015).

[9] E.J. Favaloro, Diagnosing von Willebrand disease: a short history of laboratory milestones and innovations, plus current status, challenges, and solutions, Semin. Thromb. Hemost. 40 (2014) 551–570.

[10] National Heart, Lung, and Blood Institute. The diagnosis, evaluation, and management of von Willebrand disease. NIH Publication 08-5832. Bethesda, MD, US Department of Health and Human Services, National Institutes of Health. 2009.

[11] E.J. Favaloro, I. Bodó, S.J. Israels, S.A. Brown, Von Willebrand disease and platelet disorders, Haemophilia 20 (Suppl. 4) (2014) 59–64.

[12] J.A. Heit, Thrombophilia: clinical and laboratory assessment and management, in: C.S. Kitchens, C.M. Kessler, B.A. Konkle, Consultative Hemostasis and Thrombosis, third ed., Elsevier, 2013 (Chapter 14): 205–239.

[13] E.J. Favaloro, R.C. Wong, Antiphospholipid antibody testing for the antiphospholipid syndrome: a comprehensive practical review including a synopsis of challenges and recent guidelines, Pathology 46 (2014) 481–495.

[14] S.M. Bates, Thrombophilia in Pregnancy, in: C.S. Kitchens, C.M. Kessler, B.A. Konkle, Consultative Hemostasis and Thrombosis, third ed., Elsevier, 2013 (Chapter 35): 627–650.

[15] J.B. Holcomb, E.E. Fox, X. Zhang, et al. Cryoprecipitate use in the PROMMTT study, J. Trauma Acute Care Surg. 75 (1 Suppl. 1) (2013) S31–S39.

[16] N. King, M.H. Tran, Long-acting anticoagulant rodenticide (superwarfarin) poisoning: a review of its historical development, epidemiology, and clinical management, Transf. Med. Rev. 29 (2015) 250–258.

[17] R. Palla, F. Peyvandi, A.D. Shapiro, Rare bleeding disorders: diagnosis and treatment, Blood 125 (13) (2015) 2052–2061.

[18] J.C. Chapin, K.A. Hajjar, Fibrinolysis and the control of blood coagulation, Blood Rev. 29 (2015) 17–24.

[19] V. Baca, G. Montiel, L. Meill_on, et al. Diagnosis of lupus anticoagulant in the lupus anticoagulant-hypoprothrombinemia syndrome: report of two cases and review of the literature, Am. J. Hematol. 71 (2002) 200–207.

[20] S. Kankirawatana, R. Berkow, M.B. Marques, A neonate with bleeding and multiple factor deficiencies, Lab. Med. 37 (2006) 5–7.

[21] V. Koenig-Oberhuber, M. Filipovic, New antiplatelet drugs and new oral anticoagulants, Br. J. Anaest. 117 (S2) (2016) ii74–ii84.

CHAPTER

14

Therapeutic and Donor Apheresis

Jill Adamski, Tina S. Ipe**, Theresa Kinard**

*Mayo Clinic Arizona, Phoenix, AZ, United States; **Houston Methodist Hospital, Houston, TX, United States

Apheresis machines separate whole blood into cellular and plasma fractions. This allows for therapeutic interventions through removal of pathologic substances from these fractions (e.g., therapeutic plasma exchange to remove acetylcholine receptor antibodies in patients with myasthenia gravis) or by isolation and treatment of a specific blood component (e.g., collection and photoactivation of WBCs to induce immunomodulation by extracorporeal photopheresis). Apheresis procedures are also used to collect blood products for transfusion and hematopoietic progenitor cells for transplant. This chapter focuses on the basic principles of apheresis procedures, indications for therapeutic apheresis that are supported by evidence-based guidelines, and the management of adverse events associated with these procedures.

1. Centrifugation is the principal method of some apheresis instruments to perform cell separation by specific gravity to allow selective cell or plasma removal. What is the correct order of blood components, from the lowest to highest specific gravity?

A. Red blood cells, platelets, lymphocytes, granulocytes, plasma
B. Red blood cells, granulocytes, lymphocytes, platelets, plasma
C. Plasma, lymphocytes, platelets, granulocytes red blood cells
D. Plasma, platelets, lymphocytes, granulocytes, red blood cells
E. Plasma, red blood cells, granulocytes, lymphocytes, platelets

Concept: Centrifugation of whole blood causes blood components to separate based on differences in specific gravity. Separation depends on the centrifugal force (determined by the radius of the centrifuge and revolutions per minute) and dwell time in the centrifuge. The specific gravity of each component overlaps, thus 100% separation cannot be achieved. This may have clinical implications during donor or therapeutic apheresis procedures due to contamination of adjacent blood components. For example, hematopoietic progenitor cells (HPCs) collected by apheresis may have significant contamination with red blood cells (RBCs) and platelets. High levels of RBCs may require product modifications to prevent potential hemolytic transfusion reactions in an allogeneic recipient. Patients, who may already be thrombocytopenic, may also lose a significant amount of platelets and require platelet transfusion following a HPC collection procedure.

Answer: *D*—The components with the highest specific gravity are furthest away from the rotating axis. From lowest to highest specific gravity, plasma can be found closest to the axis, followed by platelets, white blood cells (lymphocytes, monocytes), granulocytes (basophils, eosinophils, neutrophils), and last are the red blood cells. All other answer choices (Answers A, B, C, and E) have the incorrect order of components.

Transfusion Medicine, Apheresis, and Hemostasis. http://dx.doi.org/10.1016/B978-0-12-803999-1.00014-6

Please answer Questions 2–4 based on the following clinical scenario:

A 33-year-old man presents to the emergency department (ED) with a 1 week history of blood in his urine and this afternoon, he began coughing up blood. A computed tomography (CT) scan of his chest confirms that the patient has diffuse alveolar hemorrhage (DAH). The ED physician is concerned about Goodpasture syndrome and calls the apheresis team for a consultation. The following are his laboratory results:

Hemoglobin (Hgb) = 9.7 g/dL
Hematocrit (Hct) = 31%
Prothrombin Time (PT) = 12 s
Activated partial thromboplastin time (aPTT) = 29 s
Fibrinogen = 205 mg/dL
Creatinine = 3.7 mg/dL
Aspartate amino transferase (AST) = 89 IU/L
Amino alanine transferase (ALT) = 112 IU/L

2. According to the 2016 (seventh ed.) American Society for Apheresis (ASFA) Guidelines for Therapeutic Apheresis, Goodpasture syndrome/antiglomerular basement membrane disease (anti-GBM disease) is considered an ASFA category III indication (i.e., role of apheresis is not established) for therapeutic plasma exchange (TPE) if which of the following condition is present?
A. Serum creatinine >6.6 mg/dL
B. Anti-GBM antibody titers <1024
C. Dialysis independence
D. Diffuse alveolar hemorrhage
E. O_2 saturation <90%

Concept: Goodpasture syndrome is due to antibasement membrane antibodies that are against type IV collagen in the lungs and kidneys. Severe damage can lead to organ dysfunction and life-threatening bleeding. Whether TPE is effective often depends on the degree of organ destruction that has already occurred.

Answer: *A*—Several studies report that most patients who are dialysis dependent or have creatinine levels > 6.6 mg/dL at the time of TPE initiation will not recover renal function (Answer A). Thus, TPE is less beneficial in these settings and is considered a category III indication.

It is important that TPE be initiated early in the disease process in order to increase the likelihood of renal recovery. In the setting of dialysis independence or DAH (regardless of dialysis requirement) TPE is considered a category I treatment (Answers C and D). Anti-GBM titers (Answer B) and oxygen saturation (Answer E) are not taken into consideration when deciding on the urgency of the procedure or the likelihood of success with TPE. Table 14.1 contains the definitions of the four different ASFA categories for therapeutic apheresis.

TABLE 14.1 American Society for Apheresis Indication Categories

Category	Definition
I	Apheresis, alone or in conjunction with other therapies, is considered a *first-line* intervention for these indications.
II	Apheresis, alone or in conjunction with other therapies, is considered a *second-line* intervention for these indications.
III	Role of apheresis therapy has not been established for these indications. Decisions should be made on case-by-case basis.
IV	Apheresis therapy is ineffective or harmful. IRB approval should be sought if apheresis is performed for these indications.

3. The decision is to perform an urgent TPE on this patient based on the overall clinical picture and the possible presence of DAH. What replacement fluid will be used for the first procedure?
A. 5% albumin
B. 5% albumin and normal saline
C. 5% albumin and cryoprecipitated AHF
D. Cryoprecipitated AHF and plasma
E. Plasma

Concept: During any apheresis procedure, a part of the blood is typically removed and replaced by an appropriate fluid. For example, in patients with sickle cell disease (SCD), sickled RBCs are removed and replaced with normal donor RBC units. In TPE, plasma is removed from the patient because there is a pathologic substance within the plasma.

For most diseases, 5% albumin is an ideal replacement for the plasma that is removed. Albumin is used because, unlike normal saline, it will stay mostly intravascular and maintain oncotic pressure. A potential risk of using albumin as a replacement fluid is that you are not replacing any coagulation factors that are also present in plasma. Thus, if the coagulation factor levels are decreased too much (after repeated procedures) or if a patient is already at high risk of bleeding, the bleeding risk may dramatically increase after the procedure. Thus, in select patient populations, plasma may be a better replacement fluid option, to either treat the coagulopathy or to decrease the risk of intra- or post-TPE bleeding.

Answer: *E*—In patients with Goodpasture syndrome and DAH, use of plasma replacement is the preferred option. Even though the patient does not have an evident coagulopathy, removal of his plasma, hence the nonselective removal of the pathologic antibodies and coagulation factors, may place him at risk of intra- or post-TPE worsening of his DAH. After the DAH has resolved, most institutions will switch to 5% albumin (Answer A) or 5% albumin and saline (Answer B) as the replacement fluid(s), as long as the patient does not have signs of being coagulopathic. Since the patient does not have any evidence of a fibrinogen deficiency, cryoprecipitated AHF is not necessary at this time (Answers C and D).

In these cases, the urgency of the procedure often depends on whether the patient has DAH. If DAH is confirmed or highly suspected, ASFA considers this a category I indication for therapeutic plasma exchange (TPE) and most practitioners would consider this an emergent case, since the condition carries a ~50% risk of mortality without treatment. A slightly less urgent category I indication is Goodpasture syndrome with no DAH, but for a patient who is not yet dialysis dependent. This is because there is still a probability of renal function recovery. However, if a patient is already dialysis dependent, the ASFA category drops to category III. The ASFA guidelines recommend daily to every other day treatment for Goodpasture syndrome for at least 10–20 days, since anti-GBM antibodies typically fall to undetectable levels within 2 weeks of TPE and immunosuppressive therapy.

4. The apheresis unit typically uses citrate as its anticoagulant. Since you are also using plasma as your replacement fluid, which of the following adverse events may be observed during TPE in this patient?
A. Hyperkalemia
B. Anemia
C. Leukopenia
D. Hypertension
E. Hypocalcemia

Concept: Though heparin can be used, citrate is the more commonly used anticoagulant for TPE at most apheresis centers. Citrate works an anticoagulant by binding calcium in the blood; thus, preventing certain coagulation factors from completing secondary hemostasis. Although citrate is used to prevent blood from clotting in the extracorporeal circuit, a small amount is returned to the patient. Therefore, the patient is at risk of hypocalcemia during and shortly after the procedure.

There are three physiologic ways that the body manages the infused anticoagulant as follows: (1) Dilution due to a large volume of distribution; (2) Metabolism by the liver and skeletal muscle; (3) Excretion by the kidneys. Thus, as in this case, patients with impaired liver and kidney function are at higher risk for citrate toxicity.

Answer: *E*—Hypocalcemia secondary to citrate toxicity is a common adverse event in TPE. In this patient, not only is he exposed to procedure-related citrate, but citrate is also used as an anticoagulant in plasma storage (replacement fluid during the procedure), thereby, increasing the risk of citrate toxicity even more. This patient also has renal insufficiency, preventing normal metabolism and excretion of citrate. Therefore, continuous supplementation of calcium by IV is recommended in this setting and/or one could consider adjusting the whole blood to anticoagulant (WB:AC) ratio to accommodate for the potential for citrate toxicity.

Hypokalemia, not hyperkalemia is a risk during TPE (Answer A). Some centers add potassium into their albumin bottles to mitigate hypokalemia. Hypertension is not a common side effect of TPE (Answer D). While anemia can be a consequence of long-term apheresis, it is not a concern in the acute setting (Answer B). Since your machine is set to remove mostly plasma, leukocytes should not be significantly decreased during TPE (Answer C).

End of Case

5. A 43-year-old female is undergoing a series of TPE to treat antibody mediated rejection of her renal allograft. She had a renal biopsy last night and laboratory results are as follows: PT = 16 s; fibrinogen = 87 mg/dL. Therefore, you elect to use plasma as the replacement fluid during the exchange to minimize the risk of bleeding from the recent biopsy site. Shortly after the procedure starts, the patient complains of weakness, headache, confusion, and chest pain. You stop the procedure and order a STAT ionized calcium level, which returns as 1.21 mmol/L (reference range: 1.1–1.35 mmol/L). You remember the patient is taking cyclosporine, which causes you to suspect which of the following additional side effects of citrate anticoagulation?
   A. Hypomagnesemia
   B. Hypercalcemia
   C. Hyponatremia
   D. Hypermagnesemia
   E. Hyperkalemia

Concept: Anticoagulation is required during apheresis procedures to prevent the extracorporeal circuit from clotting. Heparin and citrate are commonly used options for this purpose. The most common adverse event associated with TPE is hypocalcemia due to chelation of calcium by citrate. The symptoms of hypocalcemia are normally reversible by slowing down the rate of the procedure or pausing the procedure to allow the physiologic mechanisms for calcium homeostasis (e.g., mobilization from albumin and bone). However, in some cases, the patient will not recover quickly and will require supplementation with exogenous calcium (oral or IV). Citrate also chelates other divalent cations, including magnesium. The symptoms of hypomagnesemia are similar to those seen in hypocalcemia; however, they are not quickly reversible by physiologic processes or magnesium supplementation.

Answer: *A*—Patients who are taking cyclosporine often have hypomagnesemia that will be exacerbated in the presence of citrate. Therefore, low magnesium levels should be corrected prior to initiating TPE to minimize the risk of prolonged, symptomatic hypomagnesemia in the patient. Apheresis procedures tend to lead to a decrease in minerals; thus hypercalcemia, hypermagnesemia, and hyperkalemia (Answers B, D, and E) are incorrect. Hyponatremia (Answer C) is not a known side effect of apheresis.

6. Which of the following is a Food and Drug Administration (FDA) approved indication for extracorporeal photopheresis (ECP)?
   A. Graft-versus-host disease (GVHD)
   B. Heart transplant rejection

C. Lung transplant rejection
D. Cutaneous T-cell lymphoma (CTCL)
E. Crohn's disease

Concept: ECP is an immunomodulatory apheresis procedure often used to treat CTCL, GVHD, and heart and lung transplant rejection. During ECP, white blood cells are separated by centrifugation and routed into a treatment bag. After enough cells have been collected, 8-methoxypsorolen (8-MOP) is injected into the bag, which is then placed into an ultraviolet (UV) light chamber. The UV light photoactivates the 8-MOP which intercalates with the white blood cell (WBC) deoxyribonucleic acid (DNA) to induce apoptosis or programmed cell death. The cells are then returned to the patient to be processed by the immune system (e.g., spleen, lymph nodes, etc.).

Answer: *D*—As of this writing, the only FDA approved indication for ECP is CTCL, which was approved as an ECP indication by the FDA in 1987 based on the results of a multicenter clinical trial that reported positive responses in >70% of patients with refractory CTCL. ASFA deems CTCL a category I indication for ECP therapy. For off-label use of ECP in other conditions, such as GVHD, heart transplant, and under certain conditions, lung transplant rejection (Answers A, B, and C), Centers for Medicare and Medicare Services (CMS) have issued Medicare National Coverage Determination (NCD) provisions that extend coverage to these indications. However, as of this writing, autoimmune conditions (e.g., Crohn's disease) (Answer E) are not eligible for ECP reimbursement in the United States.

7. You are asked to explain the risks of ECP to a patient who is considering ECP for treatment of chronic graft-versus-host disease (cGVHD), secondary to a hematopoietic progenitor cell (HPC) transplant for acute myeloid leukemia (AML). She is fearful of the long-term consequences of undergoing ECP since she read on the internet that the precise mechanism for how ECP works to reverse GVHD is unknown. You tell her that the only known long-term complication of repeated ECP treatments is which of the following?
A. Increased risk of infection
B. Increased risk of malignancy
C. Increased risk of iron-deficiency anemia
D. Increased risk of diabetes
E. Increased risk of osteopenia

Concept: GVHD is a potential consequence of HPC transplants. In fact, approximately 30% of HPC transplant patients will experience some degree of chronic GVHD. Symptoms can include skin lesions and tightness, dry/irritated eyes, gastrointestinal upset, difficulty breathing, and other symptoms in many different organ systems. Essentially, the donor lymphocytes recognize the recipient as foreign and begins attacking tissues all over the body. Typically, conventional immunosuppression regimens are used as first line treatment for GVHD. In cases where GVHD is unresponsive to conventional therapy, ECP is an option that bone marrow transplant physicians will often consider adding to the patient's treatment plan.

Answer: *C*—Iron deficiency anemia is certainly a risk for this patient since she will likely receive many ECP treatments in the next 6 months. Patients who undergo long-term ECP treatments at short intervals (e.g., 1–2 week intervals) are at risk of developing iron deficiency anemia. At the end of each procedure the extracorporeal blood is returned to the patient; however, it is not possible to return all of the RBCs and there may be a cumulative loss over multiple procedures, eventually leading to iron deficiency anemia.

Because ECP is immunomodulatory, not immunosuppressive, there does not appear to be increased risks of complications seen with the typical immunosuppressive regimens, such as increased risk of infection, malignancy, diabetes, and osteopenia (Answers A, B, D, and E).

8. A 23-year-old female with a history of sickle cell disease (SCD) is brought by ambulance to the ED. She developed acute onset of left-sided paralysis an hour ago. Her husband tells you that she is 12 weeks pregnant and she has had no SCD complications in 2 years, but she did receive a few blood transfusions when she was younger. You are consulted to perform a red blood cell exchange (RBCx) for acute stroke. The ED physician reports that a CT scan of her head confirms the presence of an acute ischemic event. How would you advise the ED physician regarding the possibility of RBCx in this patient?

A. RBCx is not indicated to treat acute stroke in sickle cell disease
B. RBCx is contraindicated due to her pregnancy
C. RBCx should be performed as soon as possible due to the acute stroke
D. RBCx cannot be performed until after the hemoglobin S level is determined by hemoglobin electrophoresis
E. RBCx should be performed as soon as possible due to her pregnancy

Concept: Patients with SCD are at a very high risk of ischemic events due to ongoing chronic inflammation and the sickle shape of their RBCs. Hemoglobin S (HbS) RBCs are misshaped and thus, are more prone to get stuck in small vessels. Additionally, this damage leads to a chronic inflammatory state that causes sickled cells in narrow vessels to stick together, leading to even more blockages. These blockages can manifest as mild pain episodes, to acute chest syndrome (ACS), to acute strokes. In severe cases, RBCx is used to quickly reduce the load of HbS cells, while increasing normal donor cells that have good oxygen carrying capacity.

Answer: C—Acute stroke in patients with sickle cell disease is considered a category I indication for RBCx according to the 2016 American Society for Apheresis (ASFA) guidelines (Answer A). Though a definitive threshold for the percent of sickle cells that puts a patient at increased risk of ischemic events has yet to be agreed upon, most practitioners believe that levels above 30% represent an increased risk. Thus, the goal of most RBCx procedures for acute stroke is a HbS level of <30%.

In this case, the patient has a remote history of blood transfusion; therefore, the assumption can be made that her HbS levels are nearly 100% and waiting for confirmation of that percentage is not a reason to delay the procedure (Answer D). The goal of RBCx is to reduce the HbS level to below 30% so the instrument can be programmed with a FCR (fraction of cells remaining) of 30%. The FCR is calculated by dividing the desired HbS by the current HbS then multiplying by 100%. In this case, the calculation is as follows: $(30/100) \times 100\% = 30\%$.

Pregnancy, is neither an indication, nor a contraindication for RBCx (Answers B and E).

Additional notes: Before starting an RBCx, a type and screen must be performed to avoid acute or delayed hemolytic transfusion reactions. ABO/Rh matched or compatible units will be selected. If the patient's phenotype is known, then the transfusion service may elect to provide the units that are matched for D-, C-, c-, E-, e-, and K-antigen. If the antibody screen identifies alloantibodies, the patient must receive antigen negative RBCs for those corresponding antigens. A qualitative screen for HbS is performed on the RBC units to ensure that they are negative for HbS (i.e., sickle cell negative); however, if this testing is unavailable, it should not delay the procedure.

9. A 21-year-old female with homozygous SCD presents to the emergency room complaining of diffuse joint and bone pain. She has been ill recently with nausea and vomiting. She has frequent pain crises and has required numerous transfusions in the past, but has never had an RBCx. Her last transfusion was 3 months ago. Her vital signs are as follows: heart rate 107, respiratory rate 16, blood pressure 118/79, temperature 98.6°F. Her O_2 saturation on room air is 98%. Her physical exam, which included a neurologic and respiratory evaluation was unremarkable. She is mildly jaundiced and overall ill appearing. Her hemoglobin is 7.2 g/dL, white blood cell count (WBC) is 15,000/μL, and platelet count is 250,000/μL. Her typical baseline hemoglobin is 8.2 g/dL.

Hemoglobin electrophoresis results will not be available until the next morning, but her last results from a month ago were reported as HbS = 79%. What would be your response to a consult for RBCx in this patient?

A. RBCx is not indicated for pain crisis
B. RBCx is indicated on a nonemergent basis
C. RBCx is not indicated due to the patient's low hemoglobin
D. RBCx should not be performed until hemoglobin electrophoresis results are reported
E. RBCx should be avoided to reduce the risk of iron overload

Concept: As discussed previously in Question 8, patients with SCD undergo constant vasoocclusive events. The most common clinical complaint associated with vasoocclusion is pain crisis, which is treated with analgesics, hydration, and oxygenation addressing the underlying cause (e.g., infection), with or without simple transfusion. Only in more severe complications is RBCx indicated (e.g., acute stroke, ACS, multiorgan failure, and hepatic or splenic sequestration).

Answer: *A*—Pain crisis alone is not a currently accepted indication for RBCx. This applies to both acute and nonacute settings (Answer B). The ASFA guidelines give a category I recommendation for RBCx in cases of acute stroke and a category II for ACS. Multiorgan failure, hepatic/splenic sequestration, intrahepatic cholestasis are category III. Though controversial as of this writing, priapism is also a category III indication.

SCD patients who are chronically transfused are at increased risk of iron overload, which can lead to organ failure, such as hepatic cirrhosis, due to iron deposition. Iron overload can be treated with iron chelation therapy, which may be poorly tolerated by some patients. Alternatively, RBCx can be performed with net red cell deficit over time to reduce iron overload. The patient's hemoglobin would not preclude RBCx (Answer C). During acute events, such as acute strokes, lack of hemoglobin electrophoresis results should not prevent RBC exchanges (Answer D), but again, in simple pain crisis, RBCx is not indicated.

Unless suppressed with chemotherapy (e.g., hydroxyurea), simple transfusion, and/or RBCx, homozygous SCD patients will have approximately 85%–90% HbS levels. Lastly, patients that undergo long-term simple transfusions are at increased risk of iron overload because the body does not have a natural mechanism to excrete iron. If iron chelation therapy fails, the patients are often placed on RBCx maintenance programs to avoid iron overload (Answer E).

10. You are consulted to perform a course of TPE for a patient with myasthenia gravis (MG). She is currently in the intensive care unit (ICU) due to a myasthenic crisis that developed after a recent respiratory infection. She is not yet intubated, but the clinical team states her respiratory status is declining and they think that she might require intubation soon. Upon reviewing her chart, you discover that she has seronegative myasthenia gravis. How would you advise the primary team regarding TPE in this patient?

A. Patients with seronegative MG do not respond to TPE
B. It is too late to perform TPE because the patient is already in crisis
C. The patient needs to be retested for acetylcholine receptor antibodies before you perform the exchange
D. TPE should be performed emergently to prevent her from being intubated
E. The initial TPE regimens for seropositive and seronegative myasthenia gravis are identical

Concept: MG is a neuromuscular disease caused by antibodies against the receptors at the neuromuscular junction. The majority of patients with MG have antibodies against the acetylcholine receptors (anti-AChR). However, approximately 50% of patients without anti-AChR will have antibodies against muscle specific receptor tyrosine kinase (anti-MusK). There are other known antibodies to receptors such as low-density lipoprotein receptor-related protein (LRP4). Patients without evidence of the above antibodies are classified as "seronegative," but likely have other autoantibodies that are acting at the neuromuscular junction to cause MG signs and symptoms. TPE works on many antibody related diseases by nonselectively removing antibodies from the patient's plasma, as is the case for all forms of MG.

Answer: *E*—Patients with seronegative MG likely have autoantibodies against other important proteins found at the neuromuscular junction or they may have low affinity anti-AChR or anti-MusK that are not detected with conventional assays. Thus, the initial TPE treatment regimens for all forms of MG are the same—a typical induction regimen consists of processing 225 mL/kg over a period of up to 2 weeks (which is approximately 1 plasma-volume exchange every other day for five treatments, Answer E). Approximately 10%–15% of patients with seronegative MG will respond to TPE (Answer A).

The ASFA guidelines rate MG as a category I indication for TPE. This is especially the case for patient in crisis (Answer B). While TPE will not prevent a patient from being intubated (Answer D), it may shorten the time needed for successful extubation. It is not necessary to check antibody levels in a patient who has already been diagnosed with MG (Answer C).

11. A 46-year-old female is admitted to the hospital with a severe headache (8 of 10 on pain scale) and shortness of breath with a respiratory rate of 26 respirations per minute and an O_2 saturation of 86%. She reportedly takes a daily multivitamin and daily 81 mg aspirin therapy. A chest X-ray (CXR) show interstitial infiltrates without evidence of pneumonia. A head CT reveals no evidence of acute ischemic or hemorrhagic damage. Shortly after her radiologic examination, she becomes confused and agitated. A CBC reveals a WBC count of $187 \times 10^9/L$. Which of the following is the most likely explanation for her clinical symptoms?

A. Thrombotic stroke
B. Infection with *Neisseria meningitidis*
C. Leukostasis
D. Dementia
E. Medication overdose

Concept: WBC counts of more than $100 \times 10^9/L$ can lead to end-organ damage secondary to microvascular obstruction by WBCs (e.g., leukostasis). The pulmonary and central nervous systems are the most common organ systems affected by leukostasis. This entity typically occurs in patients with leukemia of either myeloid or lymphocytic lineage. Severely symptomatic leukostasis typically occurs in patients with WBC/blast counts $>100 \times 10^9/L$ in AML patients or $>400 \times 10^9/L$ in ALL patients. However, in M4 or M5 AML variants, leukostasis might occur at lower counts.

Answer: *C*—Leukostasis can cause clinical symptoms in multiple organ systems. Cerebrovascular symptoms present in the patient include headaches, visual impairments, confusion, somnolence, delirium, coma, and other focal neurologic deficits. Though these symptoms can often be due to stroke, infections (e.g., meningitis), dementia, or medication overdoses (e.g., warfarin), the patient in this case is not at significant risk for these conditions (Answers A, B, D, and E). Additionally, the elevated WBC count confirms leukostasis as the most likely cause of her symptoms. Pulmonary symptoms include dyspnea, tachypnea, and hypoxemia. Radiographic findings are pertinent for interstitial or alveolar infiltrates without evidence of pneumonia.

Symptomatic leukostasis secondary to AML or ALL is an ASFA category II indication for leukocytapheresis and is typically considered a medical emergency. The goal of most procedures is to decrease the WBC/blast count to $< 100 \times 10^9/L$. This intervention is often used as a bridge to chemotherapy and may help increase short term survival; however, long term survival in patients receiving leukocytapheresis is not improved.

Similar interventions are performed for thrombocytosis. Symptomatic thrombocytosis secondary to essential thrombocythemia or polycythemia vera, is also considered a ASFA category II indication for thrombocytapheresis. In this setting, platelet counts often exceed 1,000,000/μL putting patients at risk for thrombotic complications, with bleeding complications (secondary to an acquired vonWillebrand disease) occurring at even higher counts. The goal of thrombocytapheresis is to normalize platelet counts while definitive cytoreduction therapy takes effect.

12. According to the ASFA guidelines, prophylactic leukocytapheresis has an impact on which of the following outcomes?
    A. Symptom severity
    B. Tumor lysis syndrome
    C. Chemotherapy response
    D. Long-term survival
    E. Overall mortality

    Concept: While the definitive treatment is chemotherapy, leukocytapheresis is indicated to decrease symptoms of leukostasis. Multiple cohort studies report no impact of leukocytapheresis on early or late mortality, long-term survival, response to chemotherapy, or tumor lysis syndrome. However, decreasing the WBC burden decreases the clinical symptoms associated with leukostasis. Thus, prophylactic leukapheresis is a category III indication in these cases.

    Answer: *A*—Prophylactic leukocytapheresis is an ASFA category III indication and is sometimes used in cases of hyperleukocytosis because decreasing the WBC burden decreases the symptom severity associated with leukostasis. However, as of this writing, there is no known benefit of performing a leukapheresis on tumor lysis syndrome, chemotherapy response, long-term survival, or overall mortality (Answers B, C, D, and E).

13. A healthy donor is undergoing granulocyte collection at your center. Shortly after the collection, the donor complains of edema in his arms and legs. Which of the following is the most likely cause of the edema?
    A. Granulocyte stimulating factor (G-CSF)
    B. Antibiotics
    C. Corticosteroids (CS)
    D. Erythrocyte-sedimenting agent (ESA)
    E. 5% albumin

    Concept: Therapeutic granulocyte transfusions may be beneficial for neutropenic patients with severe bacterial or fungal infections. Granulocytes can be collected from healthy donors without stimulating agents, but the polymononuclear (PMN) cell dose from an unstimulated donor is small in magnitude (doses up to 10^9 cells). When stimulated with G-CSF, corticosteroids (CS) alone, or a combination of G-CSF/CS, a larger collection of PMNs is possible (doses up to 10^{11} cells). ESAs are also used in granulocyte collections to enhance separation of granulocytes from red blood cells.

    Answer: *D*—An ESA is frequently used to enhance the separation of granulocytes from RBCs during the leukocytapheresis procedure. ESA agents, such as hydroxyethyl starch (HES) are associated with side effects, such as allergic reactions, edema, and weight gain. In most centers, healthy donors are stimulated with G-CSF alone, CS alone, or a combination of G-CSF/CS to achieve larger collections of PMNs (Answers A and C). None of the side effects mentioned earlier are typical acute reactions with those agents. Antibiotics (Answer B) would not be used to treat a granulocyte donor, since being infectious would disqualify them from donating and because antibiotics will not increase the collection yield. 5% albumin (Answer E) is not typically used during these types of procedures because fluid replacement if unnecessary, and even if it were used, it would not cause the side-effect mentioned in this case.

14. A 59-year-old male was admitted to the hospital for acute ascending paralysis. He recently had gastroenteritis and a preliminary diagnosis of acute inflammatory demyelinating polyneuropathy (AIDP) or Guillain-Barré Syndrome (GBS) is made. He has bilateral upper and lower extremity weakness and is intubated for respiratory distress. The neurologist consults you for urgent TPE, and asks you if you can perform the procedure through a peripherally inserted central catheter (PICC), since the patient has poor peripheral access. What is your response?
    A. Yes, TPE can be performed through a PICC line
    B. No, peripheral vascular access will be used since the patient will only receive two TPE procedures

C. No, the patient should get a temporary nontunneled central venous catheter (CVC) immediately
D. No, a subdermal port should be placed for TPE
E. No, because TPE is not indicated for the treatment of GBS

Concept: Therapeutic apheresis procedures may be performed using a variety of vascular access options. However, there are many considerations for each patient, such as duration of treatment, risk of infection, and the setting in which the treatment will take place. Additionally, the diameter of the line must be large enough to allow sufficient flow rates and the tensile strength must be able to resist the negative pressure forces of the apheresis machine.

Answer: C—CVCs offer the advantage of having a single access point, sustained flow rates, and the ability to be shared for other patient care needs, but carry the associated risks of inserting the line, maintenance for adequate use, and complications that may arise while the line is in place (infection, thrombosis, bleeding, etc.). Temporary, nontunneled central lines may be considered for patients with poor vascular access, who need multiple procedures, and whose course of treatment is temporary (less than 7–10 days). These temporary CVC are inadequate for prolonged therapies and outpatient treatments due to the risk of infection.

If treatments will occur in an outpatient setting or prolonged inpatient setting, a tunneled CVC is often more appropriate. Ports may be considered for patients who require potentially lifelong apheresis treatments (Answer D). Peripheral intravenous (IV) catheters, including peripherally inserted central lines (PICC) are usually inadequate for apheresis procedures due to the inability to sustain high flow rates (Answer A). Important factors that determine flow rates are the catheter diameter and length. Doubling the diameter size increases flow rate by 16-fold. Flow rate is also inversely proportional to the length of the catheter. These factors are described by Hagen-Poiseuille's law.

TPE is an ASFA category I indication for AIDP or GBS, and is therefore, indicated in this case (Answer E). Peripheral venous access is acceptable if the patient has adequate veins, muscular tone, and the ability to accommodate large gauge steel needles (16–18 gauge). While this is a preferred option due to minimal maintenance, immediate availability, and avoidance of risks associated with central venous catheters, peripheral access may not be optimal due to clinical factors, such as number of procedures, patient comfort, and the ability of the veins to sustain adequate blood flow. Unfortunately, this patient's clinical condition precludes successful TPE to be performed by peripheral access (Answer B) and multiple TPE procedures will be necessary for clinical improvement. While typical treatment may require 5–6 treatments over the course of 2 weeks, a temporary nontunneled catheter can be requested, which can be performed quickly at the bedside.

Please answer Questions 15 and 16 based on the following clinical scenario

The apheresis service is consulted to evaluate a 25-year-old female who was recently diagnosed with familial hypercholesterolemia after suffering a myocardial infarction. Lipid lowering agents have been unable to reduce her low-density lipoprotein (LDL) sufficiently and she is experiencing significant side effects from the medications. Her referring physician would like her to start LDL apheresis and you request more details of the patient's medical history.

15. Which of the following statements is most accurate?
A. The patient is ineligible for LDL apheresis insurance coverage if she is homozygous for familial hypercholesterolemia
B. If she has homozygous familial hypercholesterolemia, her LDL cholesterol level should be higher than 1000 mg/dL before considering LDL apheresis therapy
C. She should have a trial of lipid lowering agents for at least 12 months before considering LDL therapy
D. The goal of LDL apheresis is to reduce time-average total cholesterol by more than 50% and LDL by more than 60% from baseline

E. If the patient is heterozygous for familial hypercholesterolemia, they should only be considered for LDL apheresis if her LDL is consistently above 500 mg/dL

Concept: Familial hypercholesterolemia (FH) is an autosomal dominant condition that causes decreased production of apolipoprotein-B (apo-B). The lack of apo-B leads to insufficient hepatic removal of low density lipoprotein cholesterol (LDL-C). The condition places patients at risk for advanced atherosclerosis and early death due to myocardial infarction. LDL apheresis received a category I recommendation for FH homozygotes and a category II recommendation for heterozygotes.

Answer: *D*—Cholesterol recovers to baseline within 14 days following an LDL apheresis procedure. While the reduction appears transient, the goal of LDL apheresis is to reduce time-average total cholesterol by more than 50% and LDL by more than 60% from baseline, which requires more than 65% reduction in total cholesterol or greater than 70% reduction of LDL-C following a procedure.

LDL apheresis is performed to acutely remove LDL-C from plasma in patients who are at high risk for coronary heart disease for whom diet or drug therapy has been ineffective for at least 6 months or when drug therapy is not tolerated (Answer C). Maximum tolerated drug therapy is defined as a trial of drugs from at least two separate classes of hypolipidemic agents, such as bile acid sequestrants, HMG-CoA reductase inhibitors, fibric acid derivatives, or niacin/nicotinic acids. LDL apheresis using a selective column effectively reduces LDL-C without reduction in HDL, albumin, or immunoglobulins.

Patients with heterozygous hypercholesterolemia who have failed maximal drug therapy for 6 months may consider LDL apheresis if LDL remains >300 mg/dL, or 160 mg/dL if they have a documented history of coronary heart disease, and they can be initiated on LDL apheresis every other week (Answer E). Patients with familial hypercholesterolemia may have 2–3 times acceptable levels of LDL-C. LDL-apheresis may be considered for patients with homozygous (Answer A). Patients with familial hypercholesterolemia who have LDL-C >500 mg/dL are eligible for therapy and may need LDL apheresis once a week (Answer B).

16. The patient has been undergoing LDL apheresis every 2 weeks for the previous 8 months. Today, shortly after initiation of the procedure, the patient complained of feeling faint and began to sweat profusely. Her blood pressure dropped from 138/80 mmHg to 64/44 mmHg and the patient lost consciousness. The rapid response team was called and the patient was resuscitated. The apheresis provider calls the patient's cardiologist to inform him of the event and the cardiologist shares a change in the patient's medical history that may account for the adverse event during the procedure. Which of the following represents the patient history that most likely explains the adverse event?

A. The patient was recently started on an angiotensin receptor blocker (ARB) for hypertension
B. The patient was recently started on an angiotensin converting enzyme (ACE) inhibitor for hypertension
C. The patient was recently started on a beta-blocker for hypertension
D. The patient was recently started on a calcium channel blocker for hypertension
E. The patient started taking St. John's wort for treatment of hypertension

Concept: ACE inhibitors mediate antihypertensive effects by preventing the conversion of angiotensin I to angiotensin II, a potent vasoconstrictor, by the angiotensin converting enzyme. The angiotensin converting enzyme also breaks down bradykinins, potent vasodilators, to prevent hypotension. During LDL apheresis, plasma is exposed to negatively charged surfaces that activate the contact pathways (factor XII, kallikrein-kinin) creating high levels of bradykinin. Normally, the angiotensin converting enzyme rapidly degrades bradykinin to minimize vasodilation. However, in the presence of an ACE inhibitor, bradykinin concentrations can rapidly increase causing an "anaphylactiod" response, characterized by vasculature collapse and shock.

Answer: *B*—The patient was recently started on an ACE inhibitor for the treatment of hypertension therefore bradykinin rapidly increased in the patient's plasma during the procedure causing cardiovascular collapse. Other classes of antihypertensive drugs have not been associated with anaphylactoid responses (Answers A, C, and D). Although St. John's wort is not known to interfere with bradykinin metabolism, it is important to consider alternative therapies when investigating adverse reactions (Answer E).

End of Case

17. Which of the following statements about medication removal by TPE is correct? (VD = volume of distribution)

A. A drug with a high VD and high protein binding will be removed more effectively than a drug that has low VD and high protein binding
B. Heparin has a low VD and is not removed during TPE
C. In the event of bleeding and overdose, rivaroxaban can be effectively removed by TPE
D. Drugs with a low VD will be concentrated in the intravascular space
E. Drugs with low VD should be dosed immediately before TPE

Concept: VD is a pharmacokinetic parameter used for drug dosing regimens and indicates the extent of drug distribution among the intravascular and extravascular space. Protein and lipid binding affect a drug's VD. For example, a drug with high lipid binding ability will move out of the intravascular space and into the tissues, resulting in a high VD.

For an average 70-kg adult, the estimated total body water is estimated to be ~40 L. Drugs with low VD (< 20 L) are assumed to be distributed mainly to the intravascular space. An intermediate VD (20–40 L) means the drug is distributed into all body water (i.e., intra- and extravascular). A high VD (> 50 L) means the drug is concentrated in tissue outside intravascular space.

Answer: *D*—Drugs with a low VD (less concentration in the tissues) have a higher probability of being removed with TPE, especially if the drug is highly protein bound (Answer D). Administration of drugs with a low VD should be dosed or timed appropriately in relation to TPE (i.e., given after TPE is completed) (Answer E). Drugs with a high volume of distribution will be difficult to remove with TPE (Answer A). Heparin has a very low VD and can potentially be removed during TPE. Therefore, periprocedural monitoring of the patient's therapeutic range is recommended (Answer B). Rivaroxaban has a moderate VD (50 L) and is not effectively removed by TPE (Answer C).

18. A 35-year-old female in previously good health presents to the emergency room complaining of extreme fatigue. She also reports that she is currently having unusually prolonged and heavy menstruation for the past 5 days. The patient is pale and tachycardic. You notice the patient has diffuse patches of petechia and bruising. Her pertinent laboratory values are as follows:

	Patient value
White cell count	*13 × 10⁹/L*
Hemoglobin	*6.7 g/dL*
Hematocrit	*19%*
Platelet	*7 × 10⁹/L*
LDH	*900 U/L*
PT	*12 s*
PTT	*21 s*
Fibrinogen	*225 mg/dL*
Creatinine	*1.0 mg/dL*

A peripheral blood film shows reticulocytosis and moderate schistocytes. Which of the following steps are appropriate for this patient's management?

A. Order 2 units of platelets for bleeding and thrombocytopenia
B. Order 1 unit of platelets to transfuse during catheter placement for TPE
C. Order 4 L of 5% albumin from the pharmacy to be used as the replacement fluid during TPE
D. Prepare to perform a 3-plasma volume TPE every 8 h until laboratory and clinical improvement
E. Collect a sample for ADAMTS13 testing prior to initiating emergent TPE

Concept: Acquired autoimmune thrombotic thrombocytopenic purpura (TTP) is a form of thrombotic microangiopathy (TMA). TTP results from accumulation of von Willebrand factor (vWF) ultralarge multimers due to the lack of or loss of function of the metalloproteinase, ADAMTS13 (*a disintegrin and metalloproteinase with a thrombospondin type 1 motif, member 13*). Low levels of ADAMTS13 may be congenital (i.e., Upshaw Schulman Syndrome), or acquired (ADAMST13 inhibitor/autoantibody). Accumulation of ultra-large vWF results in activation of platelets, formation of intravascular microthrombi, thrombocytopenia, and microangiopathic hemolytic anemia (MAHA).

TTP was historically defined as a pentad: anemia, thrombocytopenia, fever, renal failure, and altered mental status; however, fever and altered mental status may not manifest until late in the disease process and renal failure is associated with hemolytic uremic syndrome (HUS) and not a common finding in TTP. Evidence of thrombocytopenia and MAHA (e.g., schistocytes on the blood film, elevated LDH, elevated bilirubins, etc.) is sufficient to initiate TPE prior to confirming TTP. TPE should continue daily until TTP can be ruled in or out with an ADAMTS13 activity assay. ADAMTS13 levels should be drawn prior initiating TPE. Drawing a level after TPE could lead to a false negative result due to the additional ADAMTS13 provided by using donor plasma during the exchange. Very low ADAMTS13 activity (<5%–10%) is diagnostic of TTP. At diagnosis, most patients with acquired TTP will have evidence of an inhibitor. ADAMTS13 levels may also be low due to other conditions (sepsis, postoperative, hepatic cirrhosis, chronic inflammation), including other microangiopathies (i.e., disseminated intravascular coagulation); however, ADAMTS13 deficiency is usually less profound (10%–40%) in those conditions.

TPE for TTP is an ASFA category I indication and requires at least 1 plasma volume exchange daily to remove the pathologic inhibitor. Plasma is used as replacement fluid to give the patient normal exogenous ADAMTS13. TPE should be continued until clinical and laboratory improvement and stabilization (normal platelet counts and LDH for 2–3 days). The benefits of tapering TPE treatments after clinical improvement have been studied, but not proven to be of definitive benefit. Nonetheless, some centers will taper patients after initial stabilization by performing TPE every other day for 2–3 procedures.

Answer: *E*—Daily TPE should be initiated for this patient immediately for a presumptive diagnosis of TTP, not every 8 h (Answer D). The patient is presenting with thrombocytopenia and petechia. She also has laboratory evidence of MAHA (anemia, schistocytes on peripheral smear, elevated LDH). Normal coagulation parameters make DIC less likely. Albumin should not be used as the sole replacement fluid because it does not contain any ADAMTS13, unlike donor plasma (Answer C). Prior to the recognition of TPE as an effective treatment, the mortality of thrombotic thrombocytopenic purpura (TTP) was as high as 90%.

Platelet transfusions are relatively contraindicated, based on anecdotal reports and theoretical risk of thrombotic complications (Answers A and B). While some providers may be hesitant to insert a central line, it can be performed safely in most patients with thrombocytopenia without platelet transfusions. Alternatively, TPE can be performed by peripheral IV access if the patient is mentally alert and has satisfactory antecubital veins.

19. A 20-year-old woman is admitted to the hospital for work-up of ocular pain and reduced vision in her right eye. She states that she has experienced occasional bladder incontinence in the previous week. The patient also reports weakness in her left leg. A MRI shows numerous enhancing lesions in her spinal cord. Serologic testing is ordered but the results won't be available for several days. The neurologist asks you to perform TPE. What is the most likely diagnosis/indication for TPE in this patient?

A. Neuromyelitis optica (NMO)
B. Myasthenia gravis (MG)
C. Chronic inflammatory demyelinating polyneuropathy (CIDP)
D. Guillain-Barre syndrome (GBS)
E. Acute disseminated encephalomyelitis (ADEM)

Concept: NMO is an autoimmune disease that is characterized by demyelinating lesions in the optic nerve and spinal cord. These lesions are associated with vision loss, paresthesia, paraplegia, radiculopathy, and ocular pain. Aquaporin-4 autoantibodies are pathogenic in NMO.

Answer: *A*—The symptoms and radiographic findings in this patient are most consistent with a diagnosis of NMO. The pathogenic autoantibody associated with NMO, antiaquaporin-4, can be removed with TPE and therefore, it is postulated that TPE has a role in managing NMO (Answer A). ASFA lists NMO as a category II indication (adjunct or second line therapy) for TPE.

Plasma exchange is indicated for all of the autoimmune neurologic conditions listed in the other choices (Answers B, C, D, and E). CIDP is presumably caused by an autoimmune attack on the peripheral nerves (Answer C) and is characterized by progressive weakness and impaired sensory function. MG is associated with lesions at the neuromuscular junction (Answer B) and is associated with muscle weakness and bulbar symptoms. GBS is characterized by symmetrical weakness and parasthesia that spreads proximally (Answer D). ADEM is an acute demyelinating disease that predominantly affects the white matter of the brain and spinal cord (Answer E). This entity is most commonly associated with CNS findings (nausea/vomiting, headache, confusion, seizures) but patients often have weakness and may experience optic neuritis.

20. A 30-year-old man presents to the hospital with hypertensive urgency. On admission, his blood pressures were >180 systolic and >100 diastolic. Laboratory findings are listed as follows:

	Patient value
White cell count	*8.2 × 10⁹/L*
Hemoglobin	*7.0 g/dL*
Hematocrit	*22%*
Platelet	*65 × 10⁹/L*
LDH	*856 U/L*
Haptoglobin	*6.5 mg/dL*
Creatinine	*5.0 mg/dL*

A peripheral blood film shows 1+ schistocytes.

The apheresis service is consulted to perform TPE for a low suspicion of acquired autoimmune TTP. The ADAMTS13 activity assay is performed locally and before the TPE is initiated, the results are reported. The ADAMTS13 activity is 50% (normal >70%) and this result and the other presenting laboratory findings are not consistent with a diagnosis of acquired TTP. TPE is not initiated.

Additional samples are sent to reference laboratories and the following results were obtained several days after admission: PCR testing for Shiga toxin was negative; however, PCR testing for complement mutations was positive for the *A304V* mutation in the membrane cofactor protein (MCP). The apheresis provider is consulted again to perform TPE on this patient. What is the best response to this request?

A. Yes, because the ADAMTS13 was lower than the reference range at presentation and TPE should have been started several days ago

B. Yes, because Shiga toxin producing *Escherichia coli* is not the only bacteria that causes this condition
C. Yes, because this patient is experiencing uncontrolled activation of alternative complement system
D. No, because the mutated protein, MCP, is only present on the cell membrane and not found in the plasma
E. No, because this condition is caused by malignant hypertension

Concept: This patient has a thrombotic microangiopathy (TMA). TMAs are characterized by MAHA and thrombocytopenia, as seen earlier in the TTP case. There are numerous TMA etiologies, including acquired autoimmune TTP, typical HUS, atypical HUS, malignant hypertension, autoimmune disorders, and cancer. This patient has atypical HUS due to a mutation in the MCP protein.

Answer: *D*—Atypical hemolytic uremic syndrome (aHUS) due to mutations in MCP are considered to be ASFA category IV (i.e., not beneficial or potential for harm) indications for TPE. The MCP protein is associated with the cell membrane and is not found in plasma. For conditions such as these, the first line therapy is treatment is eculizumab, which is FDA approved. However, TPE may be indicated for other complement mutations that cause uncontrolled activation of the alternative pathway (Answer C) but only as a temporizing intervention until definitive therapy can be initiated (e.g., eculizumab and addressing the underlying cause that is triggering uncontrolled complement activation).

Patients with TMA that present with serum creatinine >2.3 mg/dL and platelet counts $>30 \times 10^9/L$ are unlikely to have acquired autoimmune TTP. Although the patient's ADAMTS13 activity is lower than "normal," it's important to remember that reference ranges are developed using healthy controls. ADAMTS13 levels suggestive of acquired autoimmune TTP are typically <10% (Answer A). Infection associated HUS may be caused by non-shiga toxin producing bacteria (e.g., *Streptococcus pneumonia*) but the role for TPE is unclear in this setting (Answer B). Malignant hypertension can cause TMA and the treatment for this condition is blood pressure control (Answer E).

21. ADAMTS13 autoantibodies have been identified in which of the following drug-induced TMA?
A. Clopidogrel
B. Cyclosporine
C. Gemcitabine
D. Tacrolimus
E. Ticlopidine

Concept: Several drugs, such as ticlopidine, clopidogrel, cyclosporine, tacrolimus, gemcitabine, quinine can be associated with a drug-induced TMA via different mechanisms. The ASFA guidelines provide recommendations for each of these drugs.

Answer: *E*—Ticlopidine is the most common drug associated with TMA and it is the only drug for which ASFA has deemed TPE as a first line therapy for TMA (category I). This drug can cause immune mediated TMA through the induction of autoantibodies directed against ADAMTS13, mimicking acquired autoimmune TTP. Patients with acquired autoimmune TTP and ticlopidine induced TMA have similar response rates to TPE, 85% and 87%, respectively.

The ASFA guidelines indicate that there is not enough evidence for or against the use of TPE for TMAs associated with clopidogrel, cyclosporine, or tacrolimus; thus, they are category III indications (Answers A, B, and D). TPE does not appear to be effective for treating gemcitabine and quinine induced TMA and ASFA states that TPE should not be used for these indications (category IV) (Answer C).

22. A 56-year-old man has a remote history of heparin-induced thrombocytopenia (HIT). A heart allograft is now available and the transplant is scheduled in the next four h. The transplant surgeon wants to anticoagulate the cardiopulmonary bypass (CPB) circuit with heparin during the transplant and asks you if the patient needs TPE immediately before the operation. A HIT antibody screening test (PF4:heparin enzyme-linked immunosorbent assay (ELISA) was performed yesterday and was negative. What is your recommendation?

A. TPE is not indicated because once a patient has a diagnosis of HIT they should never be exposed to heparin again
B. TPE is not indicated because the HIT screening test was negative for HIT antibodies
C. TPE is indicated prior to transplantation because the antibody is still present but below the level of detection of the HIT ELISA
D. TPE should be performed immediately after transplant due to the well-known amnestic antibody response when patients are reexposed to heparin
E. TPE should be performed in tandem with CPB during the transplant

Concept: HIT is caused by antibody directed against heparin-PF4 (platelet factor 4) complexes. The antibody binds to the complexes and the heavy chain engages the Fc receptor on the platelet surface, leading to platelet activation, release of prothrombotic microparticles, which can lead to clot formation. The most common clinical presentation of HIT is thrombocytopenia (at least a 50% decrease from baseline since the initiation of heparin therapy). However, the most serious and sometimes deadly consequence is the formation of venous or arterial thromboses that can lead to death or limb amputations due to tissue ischemia. Clinical pretest probability scoring systems, such as Warkentin's 4T score, help to support clinical suspicious of HIT. In the laboratory, ELISA testing is very sensitive for HIT antibodies, and the SRA (serotonin release assay) is used as the confirmatory test because it is very sensitive and specific.

Answer: *B*—This patient has a remote history of HIT and a recent negative ELISA result. Since, the HIT antibody ELISA has nearly 100% sensitivity and the patient has a remote (>100 day) history of HIT, TPE is not indicated in this case since the operation will happen within 24 h of a negative HIT ELISA test. If the patient had clinically significant HIT antibodies and required heparin reexposure, the ASFA guidelines recommend preoperative TPE procedures until the antibody titer is negative by the method your lab is using. Recent studies have shown that HIT ELISAs often remain strongly positive after the SRA turns negative during a course of TPE; thus some clinicians believe that the SRA is more representative of the true risk. Typically, intra- and postop TPE is not necessary (Answers D and E).

The American College of Chest Physicians (ACCP) released guidelines for anticoagulating patients with a history of HIT who must undergo surgery with cardiopulmonary bypass. The recommendation states, "...in patients with a history of HIT in whom heparin antibodies have been shown to be absent who require cardiac surgery, we suggest the use of heparin (short-term use only) over non-heparin anticoagulants (Grade 2C)" (Answer A). HIT antibodies are typically present for up to 100 days and reexposure to heparin after the antibodies have disappeared has not been associated with amnestic antibody responses (Answer D). If the patients have HIT antibodies, the ACCP recommends that "...for patients with a history of HIT in whom heparin antibodies are *still present* who require cardiac surgery, we suggest the use of non-heparin anticoagulants" (Answer C).

23. A 54-year-old male patient with Waldenstrom macroglobulinemia presents with headache, confusion, blurred vision, dyspnea, and epistaxis. What statement about TPE for this patient is most correct?

A. TPE should only be performed if the serum viscosity is elevated
B. TPE should be performed every other day for 2 weeks in order to see symptomatic improvement
C. TPE should be performed and should provide symptomatic improvement after 1–3 procedures performed 24 h apart
D. TPE should not be performed because the patient is at risk for post-TPE pulmonary edema
E. TPE should not be performed because the patient most likely has metastatic disease to the brain

Concept: Waldenstrom macroglobulinemia is a lymphoplasmacytic lymphoma in the bone marrow. A consequence of the disease is an excess of IgM paraprotein in the bloodstream. Patients with Waldenstrom or other gammopathies (e.g., multiple myeloma) are at risk for developing plasma hyperviscosity due to the excess immunoglobulins. Left untreated, this can cause significant morbidity.

Answer: *C*—Elevated levels of monoclonal IgM immunoglobulins are the cause of hyperviscosity in Waldenstrom's macroglobulinemia. IgM is primarily located in the intravascular compartment and therefore, a significant amount of the paraprotein can be removed with 1 TPE. Because of the efficiency of IgM removal, typically only 1–3 procedures are needed for significant relief from hyperviscosity symptoms.

A sample for viscosity measurements can be drawn before initiating TPE. However, viscosity measurements do not always correlate with clinical symptoms. Therefore, treatment should not be delayed while waiting for results (Answer A). Hyperviscosity due to IgG or IgA will require more TPE procedures at every-other-day frequency because these antibodies are distributed between the intra- and extravascular spaces (Answer B). Patients with hyperviscosity are at some risk for pulmonary edema following TPE due to the acute decrease in the intravascular oncotic pressure resulting fluid shifts into the extracellular compartment (Answer D). Though any patient with lymphoma may have metastatic disease, this should not prevent or delay TPE for symptomatic relief from hyperviscosity (Answer E).

24. A 39-year-old female with recurrent focal segmental glomerular sclerosis (FSGS) in a transplanted kidney is undergoing 2 TPEs per week to prevent graft failure. The patient has a mature fistula in her left arm that has been used for years without difficulty. During her last TPE, the patient developed a hematoma when a nursing student tried unsuccessfully to access her fistula and the procedure was aborted after the apheresis nurse noticed swelling around the fistula. The patient reports for her next scheduled TPE and the ultrasound reveals a thrombosis within the fistula. The patient has one peripheral vein in her right arm that is suitable for access pressures required during the apheresis procedures. What is the best course of care for this patient?

A. Cancel all future TPE until the patient undergoes a thrombectomy of the fistula
B. Cancel all future TPE until the patient gets a tunneled central venous catheter
C. Arrange for emergent placement of a temporary central venous catheter to allow for ongoing TPE
D. Place the access and return lines in the same peripheral vein to allow for ongoing TPE
E. Program the apheresis instrument to perform a discontinuous procedure using a single access site

Concept: Most therapeutic apheresis procedures are performed using a continuous flow that requires separate access and return sites. Continuous procedures are faster than discontinuous (or intermittent) procedures because the separation chamber is not emptied until the end of the process. Discontinuous procedures may be performed with a single access site that also serves as the return site. Once a volume of blood is separated and plasma is transferred to the waste bag (for TPE), the separation chamber is emptied and then filled again to repeat the separation process (cycle) again.

Answer: *E*—Because this patient has good peripheral access in her right arm, discontinuous/single needle procedures can be performed until the fistula is functional or an alternative access is established.

Without maintenance of TPE, this patient is at risk of losing function of her transplanted kidney due to recurrent FSGS. If at all possible, procedures should not be delayed to minimize risk of allograft loss (Answers A and B). Some apheresis instruments have dual programming that allows for continuous (double-needle) or discontinuous (single-needle) procedures. Although a central line is another option for this patient, a long-term tunneled line, not a temporary catheter (Answer C) should be placed to minimize infectious complications in this immunocompromised patient. Access and return needles should not be placed in the same vein to prevent recirculation and damage to the vasculature (Answer D).

25. A 21-year-old female from Connecticut seeks evaluation in the emergency room for recurrent fevers, chills, night sweats, fatigue, and altered mental status. She had been seen by a community physician

2 weeks ago, who diagnosed her with the common cold. Her laboratory work up reveals anemia (hemoglobin 6.2 g/dL), leukocytosis with a left shift, a normal platelet count, and impaired renal function (creatinine 3.9 mg/dL). Incidentally, a peripheral smear was performed on the patient's blood and the laboratory technologist observed intracellular parasites. The pathology resident on-call identifies the parasites as being consistent with *Babesia* and estimates that the parasite load is approximately 15%. The apheresis service is consulted to perform TPE. Which of the following statements is accurate?

A. TPE should be performed to remove the parasites
B. RBC exchange is not beneficial for treating severe Babesiosis
C. RBC exchange should be performed to lower the parasitemia
D. Transfuse 2 units of RBC to correct anemia, no other interventions are needed
E. TPE should be performed to address the patient's renal dysfunction

Concept: It is very important to understand disease mechanisms and procedure mechanisms when deciding which apheresis procedure to perform. For example, while TPE is very efficient at removing antibodies from a patient with myasthenia gravis, it would be ineffective in treating a disease affecting the RBCs, such as sickle cell disease or infection with *Babesia*.

Answer: *C*—Red blood cell exchange is considered an ASFA category II indication for Babesiosis and is indicated in this case (Answer B). Most patients receive antibiotics, such as azithromycin, as a first line treatment. However, in cases that do not respond to antibiotics, or cases with high parasite loads (>10%) or with significant comorbidities, such as significant hemolysis, disseminated intravascular coagulation, pulmonary, renal, or hepatic compromise, red blood cell exchange of 1–2 blood volumes is indicated. TPE is not indicated for treatment of Babesiosis because it does not address the infected RBCs (Answers A and E). While transfusing RBC units would provide uninfected RBCs to the patient and may treat the symptomatic anemia, this would be insufficient to address the severe parasitemia that causes other symptoms (Answer D).

Please answer Questions 26–30 based on the following clinical scenario:

You have been consulted by the transplant nephrology service to perform TPE on a 46-year-old female with antibody mediated acute kidney rejection. Her kidney biopsy shows histologic changes consistent with antibody mediated rejection and positive complement staining (C4d positivity). The transplant physician reports high levels of donor specific antibodies (DSA).

26. The transplant nephrology physician would like you to perform 10 daily TPE. What is your response and recommendation?

A. Daily TPE as requested with close monitoring of coagulation factors, and replace with plasma as necessary to prevent bleeding complications
B. Daily TPE as requested with close monitoring of immunoglobulins, and replace with intravenous immunoglobulins as necessary to prevent infectious complications
C. 2–3 daily TPE, followed by TPE every other day until clinical improvement or appropriate decrease of donor specific antibodies
D. 2 TPE procedures, performed on consecutive days, every week until clinical improvement or appropriate decrease of DSA
E. A limit of 5 total TPEs because additional procedures will provide no additional benefit

Concept: Organ transplantation is a balance between HLA matching and immunosuppression. Even closely matched donor and recipient pairs may have histocompatibility mismatches that could lead to rejection of the transplanted organ. While immunosuppression can be successful in preventing development of donor specific antibodies, antibody production can be the result of medical noncompliance or other immunologic triggers, such as infection.

Renal transplantation complicated by antibody mediated rejection is an ASFA category I indication to perform TPE. A single one-volume plasma exchange is estimated to exchange only ~63% of the patient's original plasma due to recirculation throughout the procedure. In other words, 63% of the intravascular protein (including pathogenic antibodies) can be removed with a one-volume plasma exchange. The relationship between plasma removal and plasma volumes processed is not linear because of recirculation during the procedure. Therefore, increasing the plasma exchange to 1.5 or 2 plasma volumes only increases plasma removal to 72% and 86%, respectively.

After an exchange, approximately 30%–40% of a pathogenic antibody can remain in the patient's plasma or be distributed in the tissues of the extravascular space. After the exchange, the antibody may slightly increase due to continued production by the patient or due to tissue redistribution. In severe or acute cases, daily exchanges may be performed initially, especially if the pathogenic antibody is in a high concentration. However, after the initial procedure, it is more common to perform every other day TPE to allow for a more efficient removal over time and to allow the patient to naturally recover coagulation factors that are also removed with TPE.

Since the implicated antibodies are IgG, they are present in plasma and within tissues; thus, TPE is unable to efficiently remove total antibody with multiple daily procedures. Additionally, consecutive daily TPE increases the likelihood of needing plasma as a replacement fluid to supplement coagulation factors. Initial daily procedures may be reasonable if concentration of the targeted substance to be removed is high. However, subsequent procedures with longer interprocedure intervals allow for reequilibration of the pathologic target from the extravascular space to the intravascular space. The benefits of long-term TPE are determined clinically by the patient's reduction in antibody levels and the ability to recover renal function.

Answer: C—For a patient with acute renal rejection and high DSAs, it is reasonable to perform 2–3 daily TPE procedures followed with TPE every other day, or 3 times a week, with routine monitoring of antibody levels, coagulation factors, and renal function. Since TPE therapy alone is not sufficient to treat antibody-mediated rejection, immunosuppression therapies should be coordinated with the referring transplant physician and administered accordingly with TPE.

The strategies in the other choices are either too aggressive and may harm the patient due to coagulopathy and infections, respectively (Answers A and B) or are insufficient and may not help the patient fully recover (Answers D and E).

27. The patient's HLA type and results of the patient's HLA antibodies are as follows. MFI = Mean fluorescence intensity.

HLA	A	B	Bw	C	DRB1	DRB3/4/5	DQB1	DQA1	DPB1
Patient	11	34	6	4	10	53	6	05	17:01
	31	48	5	7	12	53	6	05	88:01
Donor	11	8	6	5	10	52	2	05	04:01
	4	44	5	7	12	53	6	05	04:01

MFI	<2000 (Low)	2000–5000 (Intermediate)	>5000 (High)
HLA Antibodies	*A1*	*B8*	*DR52*
	A4	*B44*	*DQ2*
	A24	*B43*	
	B76		
	C5		
	*DRB1*04:01*		

What are your instructions to your team for drawing specimens for antibody testing?

A. Through the apheresis access line at the end of a TPE procedure

B. 24–48 h after the most recent TPE

C. 15 min after intravenous immunoglobulin infusion (IVIG)

D. 30 min after starting TPE
E. One week after completion of the TPE treatment series

Concept: The timing of collection of blood samples for laboratory tests should be coordinated appropriately in relation to the scheduled TPE and any other treatments (e.g., IVIG). Laboratories drawn immediately following an exchange are not reflective of the patient's steady state due to inadequate reequilibration, the selection of replacement fluid used, and potentially, the extracorporeal anticoagulation.

The patient has many HLA antibodies detected with varying MFIs. However, the antibodies of interest for monitoring antibody-mediated rejection are the antibodies specific for antigens on the donor kidney. For this patient, HLA antibodies to *A4, B8, B44, C5, DR52, and DQ2* are directed to the donor kidney and should be monitored for treatment efficacy. This patient also has other HLA antibodies not directed against the kidney donor, which could be explained by previous exposures, such as blood transfusion and/or pregnancy. The presence of multiple antibodies, or antibodies to a common HLA antigen, may increase the percentage panel reactive antibody (PRA), but only antibodies specific for the donor are mediating the rejection.

Answer: *B*—This question aims to avoid drawing laboratories too soon in proximity to a TPE procedure. This patient's antibody levels should be drawn at least 24–48 h following a procedure, which may be immediately prior to the start of the next TPE procedure when performing TPE at every-other-day intervals. While there is no official recommendation on when to recheck HLA antibody levels during the course of TPE treatment, most practitioners would retest after 4–5 TPEs. However, waiting 1 week following the last procedure is likely too long to wait and may create unnecessary delays in detecting persistent high DSAs that may require further TPEs (Answer E).

Drawing laboratories 30 min after starting TPE (Answer D) or at the end of a TPE (Answer A) would likely result in falsely low results, since the role of TPE is to remove antibodies. Depending on the methodology of detecting and quantifying DSAs, high doses of IVIG may interfere with the interpretation of the results; thus, levels should be drawn prior to infusing IVIG (Answer C).

28. The patient will be receiving TPE at the outpatient clinic. She has a left arm fistula, which was created for dialysis prior to transplant, but has not been used in the past 4 months. What type of vascular access should be used for TPE?
A. The fistula can be used for TPE
B. To avoid multiple needle sticks, the patient can get a nontunneled temporary central venous catheter
C. To reduce the risk of infection, the patient can get a PICC line
D. Since the patient will require multiple procedures, the patient should get a tunneled central venous catheter
E. Since the patient will require multiple procedures, the patient should get an implantable port

Concept: Therapeutic apheresis procedures may be performed via a variety of vascular access options. Important considerations for choosing access are patient comfort and safety, sufficient flow rates, and durability of the access for the planned TPE series. Peripheral venous access is acceptable if the patient has adequate veins, muscular tone, and the ability to accommodate large gauge steel needles (16–18 gauge). While this is a preferred option due to minimal maintenance, immediate availability, and avoidance of risks associated with central venous catheters, peripheral access may not be optimal due to clinical factors, such as the number of procedures, patient comfort, and the ability of the veins to sustain adequate blood flow.

Central venous catheters (CVCs) offer the advantage of having suitable access for all procedures, sustained flow rates, and the ability to be shared for other patient care needs, but carry the associated risks of inserting the line, maintenance for adequate use, and complications that may arise while the line is in place (infection, thrombosis, bleeding, etc.). Temporary central lines may be considered for patients with poor vascular access, who need multiple procedures, and whose course of treatment is temporary (less than 7–10 days). These temporary CVCs are inadequate for prolonged therapies and outpatient treatments due to the risk of infection. If treatments will occur in an outpatient setting or prolonged inpatient setting, a tunneled CVC is more appropriate. Ports may be considered for patients who require potentially lifelong apheresis treatments. Peripheral IV catheters, including peripherally inserted central lines (PICC) are inadequate for apheresis procedures.

Answer: *A*—Therapeutic apheresis procedures may be safely performed via a functional fistula, thus, this is a viable option for TPE access in this patient. As with performing therapeutic apheresis procedures by other forms of access, the apheresis operator must be competent to perform the procedure via a fistula, and must be specially trained to access, deaccess, and manage fistula associated complications. As long as the fistula is functional, the other options are unnecessary in this patient (Answers B, C, D, and E).

29. Half-way through her first TPE, she starts complaining of perioral tingling, which progresses to "vibrations" in her arms and chest. Her symptoms are most likely due to which of the following adverse reactions?
A. Hypocalcemia, due to citrate toxicity
B. Hypocalcemia, due to antirejection medication
C. Metabolic acidosis, due to citrate toxicity
D. An allergic reaction to the plasma used for the replacement fluid
E. A hemorrhagic stroke caused by reduced clotting factor and extracorporeal anticoagulation

Concept: See Question 4 earlier for a general explanation of citrate use during TPE. During TPE with citrate anticoagulation, patients may develop symptoms of hypocalcemia. Typical symptoms include the following: paresthesia, perioral tingling, body vibrations, and possibly tinnitus. Severe hypocalcemia can progress to tetany and cardiac dysrhythmias. Laboratory assessment of ionized calcium is not necessary before treating the patient's symptoms if hypocalcemia is suspected, but may be useful to prove that the symptoms were due to hypocalcemia and to guide future treatment planning. Interventions include pausing the procedure, slowing the inlet flow rate, increasing whole blood to anticoagulation ratio, and/or providing supplemental calcium (oral or IV). Patients who need multiple procedures and have recurrent symptoms may have IV calcium added to subsequent procedures routinely.

Answer: *A*—This patient most likely has symptoms due to hypocalcemia and should be presumptively treated as such. Recognition and early treatment (e.g., stopping the procedure, giving calcium supplementation, and restarting the procedure at a slower inlet flow rate) can prevent progression to more serious complications.

Metabolic alkalosis (not acidosis) results from citrate toxicity (Answer C). Calcineurin inhibitors can cause hypomagnesemia (which can mimic hypocalcemia) by suppressing reabsorption of magnesium from renal tubules (Answer B). The symptoms of this patient are not consistent with an allergic reaction (Answer D) and the patient has not lost significant clotting factors nor is she receiving significant systemic anticoagulation at this time point during her first procedure (Answer E).

30. She completes her 8th and final TPE without incident. Since her preprocedure fibrinogen level was 153 mg/dL, 5% albumin was used as replacement fluid. She had a follow-up appointment with her nephrologist the following morning. Her renal function tests are improved and the physician decides to perform a kidney biopsy. No immediate complications were noted; however, she develops acute pain and swelling at the site of the biopsy shortly thereafter. She goes to the emergency room where they discover a large perinephric hematoma. What is the most likely cause of this postprocedure complication?
A. Puncture of the renal artery during the kidney biopsy
B. Coagulopathy due to decreased clotting factors following multiple plasma exchanges
C. Uremia from renal failure

D. Thrombocytopenia from therapeutic plasma exchange
E. Residual systemic citrate anticoagulation from therapeutic plasma exchange

Concept: There are a variety of options for replacement fluid for TPE. The most common fluids are 5% albumin, donor plasma, and occasionally normal saline. Selecting a particular replacement fluid, or a combination of fluids, is dependent on the patient's clinical need. TPE is a nonselective process of removing plasma proteins, so beneficial clotting factors are also removed along with the pathogenic substance unless plasma is used as a replacement fluid. Since fibrinogen has a long recovery time (this protein is primarily located in the intravascular space and baseline synthesis is slower than other clotting factors), plasma concentration of fibrinogen can be used as a surrogate marker for factor replacement with plasma during the exchange. Patients who will be completing TPE with recent or anticipated surgical procedures may be at bleeding risk if clotting factors are low.

Answer: *B*—Bleeding following her kidney biopsy may have been due to a variety of reasons; however, in light of her recent multiple TPEs with 5% albumin as replacement; coagulopathy is the most likely explanation. Assuming that 65% of her fibrinogen was removed during the procedure, she was likely depleted to a level of ~50 mg/dL at the end of her TPE. Therefore, while not at risk of spontaneous hemorrhage, she would be at risk of hemorrhage during traumatic events or during surgical procedures.

Had the apheresis team known about the post-TPE biopsy, they could have used plasma for part of the replacement fluid to supplement clotting factors or they could have recommended delaying the biopsy until the clotting factors recovered. This emphasizes the importance of communication between the apheresis and the clinical team. Anatomically, puncture of the renal artery from a percutaneous biopsy is unlikely (Answer A). While she may have had renal failure from her acute rejection, her renal function tests were improved, ruling out a uremic bleed (Answer C). TPE does not cause significant thrombocytopenia (Answer D), and the citrate should have been metabolized by the time the biopsy was performed on the following day (Answer E).

End of Case

Please answer Questions 31–34 based on the following clinical scenario:

A 47-year-old male with multiple myeloma will receive an autologous HPC transplant.

31. The goal for this autologous collection is 6×10^6 CD34+ cells/kg. The patient begins mobilization with G-CSF at a total dose of 10 μg/kg per day. On the 4th day of mobilization, his preharvest peripheral cluster of differentiation (CD34) count is only 3 cells/μL. By institutional protocol, the patient will receive plerixafor today and autologous stem cell collection will start the next morning. Which of the following statements regarding this patient's mobilization is correct?
A. Once a patient begins plerixafor therapy, G-CSF must be discontinued
B. Plerixafor is dosed at 0.24 mg/kg, regardless of renal function
C. Plerixafor is contraindicated for patients with myeloma
D. Plerixafor administration should be administered approximately 10–12 h prior to his scheduled HPC collection
E. Plerixafor may be administered for a total of 6 days, if needed to collect the appropriate number of stem cells

Concept: Plerixafor is a HPC mobilizing agent and is indicated in combination with G-SCF for peripheral HPC collection in patients with nonHodgkin's lymphoma and multiple myeloma. It is a CXCR4 chemokine receptor inhibitor and blocks binding of stromal cell-derived factor-1α (SDF-1α), which normally anchors the stem cells to the marrow matrix and prevents it from reaching the peripheral blood. Plerixafor use results in an increase of circulating HPC available for peripheral harvesting. Plerixafor can be considered if the patient has failed mobilization with G-CSF alone after 4 days, and can be administered for up to 4 days for collection.

Answer: *D*—This patient has not responded to G-CSF alone and can begin plerixafor. Plerixafor is a subcutaneous injection that should be administered approximately 11 h prior to initiation of HPC collection. Plerixafor can be considered if the patient has failed mobilization with G-CSF alone after 4 days, and can be administered for up to 4 days for collection (Answer E). The dosage is weight-based and also depends on the patient's renal function (Answer B). In patients with adequate renal function, dosing is 0.24 mg/kg of actual body weight. If creatinine clearance is ≤50 mL/min, the dose is 0.16 mg/kg of actual body weight. Plerixafor is given in combination with G-CSF (Answer A) and is not contraindicated for patients with myeloma (Answer C).

32. After receiving plerixafor, the patient is now ready for HPC collection. On the morning of his 1st day of collection, the patient's white cell count is 78×10^9/L, hemoglobin is 11.4 g/dL, platelet count is 35×10^9/L, and preharvest peripheral CD34 count is 28 cells/μL. He completes his first collection without complications. Which laboratory test(s) should be performed before discharging the patient?
A. Check his PT and aPTT, transfuse 2 units of plasma if his clotting times are prolonged to prevent bleeding complications
B. Check a hemoglobin and hematocrit, transfuse 2 units of RBCs if his hemoglobin is <10 for his history of myeloma
C. Check his peripheral CD34 count to estimate the number of HPCs collected in the product bag to let the patient know if he will have to return for another day of collection tomorrow
D. Check his peripheral CD34 count, administer plerixafor if the CD34 count is less than 50 cells/μL
E. Check a platelet count, transfuse 1 unit of platelets if the platelet count is less than 10,000/μL

Concept: Any apheresis procedure has the potential to remove components other than those desired. Reduction in platelet count and hemoglobin may be anticipated following a HPC collection, and the degree of reduction depends on the collection instrument and the operator. Many RBCs and platelets may be collected with the harvest; therefore, depending on the patient's values prior to collection, the patient may need a postcollection CBC for assessment. Reductions may be minimal or may be dramatic enough to warrant treatment, such as transfusion of blood components.

Answer: *E*—This patient was moderately thrombocytopenic prior to collection, and could potentially require a platelet transfusion after the procedure to prevent bleeding complications. The patient's hemoglobin was normal preprocedure and should not be a problem postprocedure (Answer B). For parameters that are not abnormal preprocedure, caution should be used in testing for abnormalities or treating any laboratory values prophylactically. For example, the patient's PT and aPTT may be falsely prolonged due to extracorporeal anticoagulation post-TPE. If the patient is not coagulopathic prior to starting collection, the collection procedure itself will not result in clinically significant depletion of coagulation factors, but may result in abnormal laboratories (e.g., PT and aPTT) (Answer A). The patient's postcollection peripheral CD34 cell count cannot reflect or estimate the number of stem cells collected in the product (Answer C). After the HPC product is enumerated for CD34, if the patient needs additional collections to achieve the collection goal, he will continue receiving plerixafor injections for subsequent collections (Answer D).

33. After 1 day of collection, the patient is 0.3×10^6 CD34/kg short of his collection goal (6×10^6 CD34+ cells/kg). However, he has gained 5 kg since starting mobilization and complains of shortness of breath and difficult breathing while laying flat. His heart rate is 98 bpm, he is hypotensive, and his oxygen saturation is 88%. Auscultation reveals bilateral pulmonary crackles. He has peripheral edema and abdominal distension, but he does not have elevated jugular venous pressure. In addition to symptomatic support with supplemental oxygen, the next best step in his management is which of the following?
A. Order a stat CT because you suspect capillary leak syndrome
B. Order a stat CT because you suspect an air embolism
C. Order a stat echocardiogram because you suspect acute heart failure
D. Order furosemide and postpone his collection until tomorrow
E. Continue with the HPC collection, but ask the patient to sit during the procedure rather than lay down

Concept: Mobilization with G-CSF may rarely cause serious and life-threatening side effects that must be identified and treated promptly. Serious complications include splenomegaly (with risk of splenic rupture), capillary leak syndrome, retinal hemorrhage, acute iritis, or thrombotic complications. Capillary leak syndrome is characterized by the development of edema, ascites, and multiorgan dysfunction which includes noncardiogenic pulmonary edema that may be associated with pleural effusions. The exact mechanism by which G-CSF mobilization causes capillary leak syndrome is not completely known, but may be due to leukocyte activation and production of inflammatory mediators that disrupt the vascular endothelium. Common, less severe, adverse symptoms associated with G-CSF include local injection site reactions, bone pain, myalgia, headache, flu-like symptoms, or loss of appetite.

Reactions to plerixafor include: diarrhea, nausea, fatigue, injection site reactions, headache, arthralgia, dizziness, and vomiting. Rare but serious reactions include: allergic reactions (including anaphylaxis), vasovagal reaction, orthostatic hypotension, and syncope. Plerixafor may also contribute to splenomegaly. Both mobilization treatments may cause extreme leukocytosis, and therefore, patients/donors are at risk for leukostasis.

Answer: *A*—The differential is broad considering his history of malignancy and associated cardiac involvement. His findings on physical exam suggest excess fluids (crackles, edema, abdominal distension), but without elevated JVP or elevated blood pressure, it is suggestive of intravascular depletion. Capillary leak syndrome should be considered and his pulmonary symptoms can be quickly evaluated with a CT scan, which may reveal pericardial effusion, pulmonary edema, pleural effusion, or ascites in cases of capillary leak syndrome or signs of thrombus formation. If the CT scan is unremarkable, additional testing should be performed before proceeding with the HPC collection (Answers C, D, and E). If capillary leak syndrome is ruled out, other causes of dyspnea, such as a pulmonary embolus or heart failure, should be ruled out with further testing. Air embolism (Answer B) is certainly a cause of shortness of breath, but is not a known side effect of G-CSF.

34. A stat CT reveals a moderate pleural effusion, mild pulmonary edema, ascites, and a small pericardial effusion, confirming your suspicion of capillary leak syndrome. The patient's first day of collection yielded a CD34 cell dose of (5.7×10^6)/kg, based on his actual weight of 122 kg. The patient's ideal body weight is 71 kg. What is the next appropriate step in this patient's management?

A. Postpone the second day of collection until tomorrow due to his emergency thoracentesis today
B. Stop mobilization and collection, especially in light of this patient's clinical condition
C. Transfer the patient to the intensive care unit and proceed with the collection
D. Cancel the second collection and notify the transplanting hematologist that the patient has failed HPC collection
E. Administer furosemide after his thoracentesis and proceed with the collection in the apheresis unit

Concept: The current minimum number of cells needed for transplantation is accepted to be 2×10^6 CD34 cells/kg, although higher doses may be desired in certain clinical situations. CD34 cell dosing has been historically based on actual body weight, yet there has been great interest and research about the effects of engraftment when using alternative dosing weights for overweight patients, such as the recipient's ideal or adjusted body weight. New literature reports that using the recipient's ideal or adjusted body weight is a good predictor of engraftment, and can lead to successful engraftment with no adverse effect on engraftment, while reducing the number of collection days.

Answer: *B*—The patient's adverse reaction to mobilization was severe, and with the number of cells already collected based on the patient's actual weight, the risks currently outweigh the benefits of continuing the collection to achieve the requested goal. Therefore, the collection should be discontinued. Proceeding with the collection in any setting or with any additional treatment (Answers A, C, D, and E) is not worth the risk to the patient.

End of Case

Suggested Reading

[1] N. Aqui, U. O'Doherty, Leukocytapheresis for the treatment of hyperleukocytosis secondary to acute leukemia, Hematol. Am. Soc. Hematol. Educ. Program 2014 (1) (2014) 457–460.
[2] S.R. Cataland, H.M. Wu, How I treat: the clinical differentiation and initial treatment of adult patients with atypical hemolytic uremic syndrome, Blood 123 (16) (2014) 2478–2484.
[3] R.L. Edelson, Mechanistic insights into extracorporeal photochemotherapy: efficient induction of monocyte-to-dendritic cell maturation, Transfus. Apher. Sci. 50 (3) (2014) 322–329.
[4] J.N. George, C.M. Nester, Syndromes of thrombotic microangiopathy, N. Engl. J. Med. 371 (7) (2014) 654–666.
[5] R.B. Ibrahim, C. Liu, S.M. Cronin, et al. Drug removal by plasmapheresis: an evidence-based review, Pharmacotherapy 27 (11) (2007) 1529–1549.
[6] A. Kaplan, Complications of apheresis, Semin. Dial 25 (2) (2012) 152–158.
[7] S.Y. Lee, V. Sanchorawala, D.C. Seldin, J. Mark Sloan, N. Andrea, K. Quillen, Plerixafor-augmented peripheral blood stem cell mobilization in AL amyloidosis with cardiac involvement: a case series, Amyloid 21 (3) (2014) 149–153.
[8] L.A. Linkins, A.L. Dans, L.K. Moores, et al. Treatment and prevention of heparin-induced thrombocytopenia: Antithrombotic Therapy and Prevention of Thrombosis, 9th ed: American College of Chest Physicians Evidence-Based Clinical Practice Guidelines, Chest 141 (2 Suppl.) (2012) e495S–e530S.
[9] K. Matevosyan, C. Anderson, R. Sarode, Isovolemic hemodilution-red cell exchange for prevention of cerebrovascular accident in sickle cell anemia: the standard operating procedure, J. Clin. Apher. 27 (2) (2012) 88–92.
[10] B.C. McLeod, Therapeutic apheresis: history, clinical application, and lingering uncertainties, Transfusion 50 (7) (2010) 1413–1426.
[11] N. Ratcliffe, N.M. Dunbar, J. Adamski, et al. National Institutes of Health State of the Science Symposium in Therapeutic Apheresis: scientific opportunities in extracorporeal photopheresis, Transfus. Med. Rev. 29 (1) (2015) 62–70.
[12] J.A. Reese, D.W. Bougie, B.R. Curtis, et al. Drug-induced thrombotic microangiopathy: experience of the Oklahoma Registry and the Blood Center of Wisconsin, Am. J. Hematol. 90 (5) (2015) 406–410.
[13] B.S. Sachais, J. Katz, J. Ross, D.J. Rader, Long-term effects of LDL apheresis in patients with severe hypercholesterolemia, J. Clin. Apher. 20 (4) (2005) 252–255.
[14] F.A. Sayani, C.S. Abrams, How I treat refractory thrombotic thrombocytopenic purpura, Blood 125 (25) (2015) 3860–3867.
[15] J. Schwartz, A. Padmanabhan, N. Aqui, R.A. Balogun, L. Connelly-Smith, M. Delaney, N.M. Dunbar, V. Witt, Y. Wu, B.H. Shaz, Guidelines on the use of therapeutic apheresis in clinical practice-evidence-based approach from the writing Committee of the American Society for Apheresis: the Seventh Special Issue, J. Clin. Apher. 31 (3) (2016) 149–162.
[16] H.P. Pham, L.A. Williams III, Plasma vs. cryoprecipitate for fibrinogen replacement in therapeutic plasma exchange procedures, J. Clin. Apher. 30 (6) (2015) 382–383.
[17] H.P. Pham, J. Schwartz, How to approach an apheresis consultation using the American Society for Apheresis guidelines for therapeutic apheresis procedures, ISBT Sci. Ser. 10S (2015) 79–88.
[18] H.P. Pham, J. Schwartz, How do we approach a patient with symptoms of leukostasis requiring emergent leukocytapheresis, Transfusion 55 (10) (2015) 2306–2311.
[19] L.A. Williams, M.B. Marques, Pathology consultation on the diagnosis and treatment of thrombotic microangiopathies (TMAs), Am. J. Clin. Pathol. 145 (2) (2016) 158–165.
[20] P.K. Bendapudi, S. Hurwitz, A. Fry, M.B. Marques, S.W. Waldo, A. Li, L. Sun, V. Upadhyay, A. Hamdan, A.M. Brunner, J.M. Gansner, S. Viswanathan, R.M. Kaufman, L. Uhl, C.P. Stowell, W.H. Dzik, R.S. Makar, Derivation and external validation of the PLASMIC score for rapid assessment of adults with thrombotic microangiopathies: a cohort study, Lancet Haematol. 4 (4) (2017) e157–e164.
[21] C.H. Kim, S.C. Simmons, L.A. Williams III, E.M. Staley, X.L. Zheng, H.P. Pham, ADAMTS13 test and/or PLASMIC clinical score in management of acquired thrombotic thrombocytopenic purpura: a cost-effective analysis, Transfusion (2017) doi: 10.1111/trf.14230 [Epub ahead of print].
[22] T.S. Ipe, H.P. Pham, L.A. Williams III, Critical updates in the 7th edition of the American society for apheresis guidelines, J. Clin. Apher. (2017) doi: 10.1002/jca.21562 [Epub ahead of print].

CHAPTER

15

Special Clinical Scenarios in Transfusion Medicine and Hemostasis

D. Joe Chaffin, Holli M. Mason**, Chelsea Hayes****

*Blood Bank Guy Web Site (BBGuy.org), LifeStream Blood Bank, San Bernardino, CA, United States; **Harbor-UCLA Medical Center, Torrance, CA, United States; ***Cedars-Sinai Medical Center, Los Angeles, CA, United States

The majority of transfusions are performed in settings that could be considered "routine." However, many unusual circumstances may arise that call for "nonroutine" thinking and practice on the part of the Transfusion Medicine team. These special clinical scenarios include examples such as urgent transfusions, massive transfusions, autoimmune hemolytic anemia, intrauterine transfusion, and sickle cell disease. The questions and cases in this chapter will discuss those issues and more, and will emphasize recognition and management of patients and problems that are out of the norm in transfusion medicine.

1. A 43-year-old female with acute myelogenous leukemia has a hemoglobin (Hgb) level of 6.6 g/dL, with weakness and mild tachycardia. Her physician orders a unit of red blood cells (RBCs) for transfusion. Your experienced blood bank technologist selects a compatible unit that expires in three days, instead of an identically compatible unit that expires in 4 weeks. Her trainee asks, "What is different between the older unit and the younger unit?" Which of the following is the most accurate response?

A. "The older unit has a higher ATP level than the younger unit"
B. "The older unit has more 2,3-DPG than the younger unit"
C. "The older unit has a lower potassium level than the younger unit"
D. "The older unit has less free hemoglobin than the younger unit"
E. "The older unit has a lower pH than the younger unit"

Concept: In the United States, RBC units may be stored in different anticoagulant preservative solutions. RBCs stored in additive solutions are approved for storage for up to 42 days. However, during storage, the RBCs undergo several changes collectively termed "storage lesions."

Answer: *E*—pH levels drop substantially during RBC storage due to accumulation of lactic acid. All of the other choices are exactly opposite of what occurs during storage, as ATP and 2,3-diphosphoglycerate (2,3-DPG) levels decline dramatically (Answers A and B), and potassium and free hemoglobin levels increase (Answers C and D). Numerous well-described storage lesions occur in the product during storage, including decrease in 2,3-DPG, pH, ATP levels, nitric oxide levels, and RBC membrane flexibility. On the other hand, other substances increase during storage, including potassium, free hemoglobin, and various membrane breakdown products (sometimes called "biological response modifiers"). These changes are indisputable, but their effect is unclear. Although some retrospective studies suggest that receiving "older blood" may negatively impact outcomes,

Transfusion Medicine, Apheresis, and Hemostasis. http://dx.doi.org/10.1016/B978-0-12-803999-1.00015-8

several recent large prospective randomized studies did not demonstrate any significant adverse clinical outcome from receiving "older blood." This allows transfusion services to continue to use the traditional "first in-first out" strategy (i.e., transfusing the oldest compatible unit in inventory to avoid wastage) for most patient populations.

2. A 60-year-old female (blood group A Rh-positive) with lymphoma had an allogeneic hematopoietic progenitor cell (HPC) transplant from an A Rh-negative donor 3 weeks ago. Her attending physician is requesting HLA-matched platelets because the patient "is not responding" to platelet transfusions. Which of the following is the most appropriate next step?
   A. Search your inventory for appropriately HLA-matched platelets
   B. Issue a unit of "fresh" group O Rh-negative platelets from inventory
   C. Issue a unit of washed, irradiated platelets
   D. Check to see if the patient has had a 1 hour posttransfusion platelet count
   E. Perform a platelet crossmatch procedure

   Concept: The management of a patient who does not respond to platelet transfusion (i.e., someone who is "refractory" to platelet transfusion) is challenging, time-consuming, and strains resources of both the transfusion service and supporting blood donor center. There are two types of refractoriness—immune- (i.e., antibody mediated) and nonimmune-mediated (e.g., fever, medication, and/or splenomegaly). Before embarking on the process of managing immune-mediated platelet refractoriness, transfusion services should ensure that the patient is truly immune refractory. The simplest way to evaluate for refractoriness is to evaluate the patient's response to platelet transfusion with a platelet count result drawn anywhere from 10 min to 1 h posttransfusion. This time frame is important because beyond 1 h, consumptive processes may quickly decrease the platelet count, mimicking immune refractoriness.

   Answer: *D*—Accurate evaluation of the patient's response to transfusion is the best first step in a platelet refractoriness workup. HLA-matched platelets (Answer A) are not commonly found in the inventory of a transfusion service; they are usually collected by a blood center from specific donors and are not immediately available. Group O, Rh-negative platelets (Answer B) are not a "universal" product like their RBC counterparts, and would not be expected to give a greater response in any patient (they may even cause acute hemolysis of the recipient's own RBCs if they contain high-titered anti-A or anti-B antibodies against the patient's RBCs). Washing platelets (Answer C) does nothing to help refractory patients; in fact, the loss of up to one-third of the platelets during washing may give the patient even less of a response. Platelets will also be activated and release granules during the wash process; thus, functionally, they may be inferior as well. This patient should receive irradiated platelets due to her recent HPC transplant, but not because of refractoriness (irradiation is indicated to prevent Transfusion-associated Graft-versus-Host Disease, or TA-GVHD, in susceptible recipients). Finally, performing a platelet crossmatch (Answer E) prior to establish the presence of refractoriness is premature.

3. A 46-year-old G5P5005 female has relapsed acute myelogenous leukemia (AML) and recently completed consolidation chemotherapy. She has anti-D detected in her plasma. Her current platelet count is 5,000/µL, and she has epistaxis and numerous petechia on her arms. However, her count was also 5,000/µL yesterday, and the team is concerned that she is refractory to platelet transfusion due to the presence of alloantibodies against platelet and/or HLA antigens. If this is proven true, which of the following strategies is likely to offer the most rapid availability of a specialized platelet product useful to treat this patient?
   A. Crossmatched platelets
   B. HLA-matched platelets
   C. ABO-identical platelets
   D. HLA antigen-negative platelets
   E. Direct-donated platelets (collected from family members)

Concept: There are three main strategies available for patients who are truly refractory to platelet transfusion due to formation of anti-HLA antibodies: Platelet crossmatching, HLA matching (see Table 15.1 for the degree of matching), and HLA antigen-negative products (i.e., negative for the antibodies the patient has against the HLA antigen). All have strengths and weaknesses. The platelet crossmatch procedure, available in some transfusion services and many immunohematology reference laboratories, is designed to rapidly predict in vitro compatibility between patient's serum and donor's platelets. The patient's serum can be tested against many different donors, and units showing no incompatibility are chosen. This can occur with or without a known patient or donor HLA type, and is generally the fastest route to a compatible product for a refractory patient. Serum from highly alloimmunized patients; however, may be incompatible with all donors, and crossmatching becomes a less effective strategy. Of note, Rh antigens only present on RBCs and are not on platelets; thus, although the patient has anti-D, Rh-negative platelets will not help in this case.

TABLE 15.1 HLA Match Grades, Abbreviated

Grade	Definition
A	All four antigens (HLA-A and HLA-B) identical to recipient
B1U	Three antigens present in donor, all identical to recipient
B1X	Four antigens present in donor; three identical and one cross-reactive
C	One incompatible (not crossreactive) antigen present
D	Two incompatible (not crossreactive) antigens present

Answer: *A*—Crossmatched platelets can be rapidly acquired (within 24–48 h), and offer results essentially equivalent to HLA matched and HLA antigen-negative platelets. HLA matched platelets (Answer B) require knowledge of the patient's HLA-A and -B type (i.e., an untyped patient will delay the process), and the blood donor center usually must recruit a matched donor for collection (another potential delay). Blood donor centers may also struggle to find a "perfect" HLA match if the patient's type is uncommon, and the "matched" platelets may not really be perfectly matched (see Table 15.1 for the HLA match grade system). If a patient has developed anti-HLA antibodies, HLA testing laboratories can specifically identify those antibodies. Once the identity of the patient's antibodies is known, platelets can be chosen that lack the target HLA antigens (Answer D) for those antibodies. This strategy, known as antigen-negative platelet selection, is conceptually similar to the transfusion service practice of choosing an RBC unit negative for an RBC antigen for a patient who has an RBC alloantibody (K-antigen negative units for patients with anti-K, for example). The strategy usually allows more flexibility in donor choice than the HLA matching strategy, and may allow faster access to compatible platelets than HLA matching. While ABO-identical platelets (Answer C) may be the fastest and easiest to acquire, and many institutions will provide such platelets as a first step in a patient being evaluated for alloimmunization, once the presence of anti-HLA antibodies is established, no additional benefit is conferred. Lastly, collecting platelets from family members does not guarantee those platelets will either be a perfect HLA match, lack the antigen(s) against which the patient has antibodies, or be crossmatch compatible (Answer E).

4. A 42-year-old Caucasian female is seen in the Emergency Department complaining of a "blood blister" in her mouth. On examination, she has a 1.3-cm hemorrhagic bulla on the mucosa of her lower lip, and several smaller similar-appearing lesions elsewhere in her oral mucosa. She has numerous petechia on her arms and legs, and a 3-cm area of purpura on her left upper arm. When questioned, she reports that she has experienced occasional epistaxis and has seen similar skin changes that resolved in a few weeks many times in the past. She denies previous oral mucosa lesions. Her physical examination, including a neurological examination, is otherwise normal. Her platelet count is 21,000/μL; all other parameters on her CBC are within normal limits. The peripheral smear confirms the thrombocytopenia, but the platelets appear normal. Which of the following is likely the best next step for her management?

A. Emergency splenectomy
B. Urgent transfusion of 2 units of apheresis platelets
C. STAT request for platelet antibody testing from the laboratory
D. Corticosteroids
E. Romiplostim

Concept: Immune thrombocytopenic purpura (ITP) is an often-chronic condition caused by platelet autoantibodies. In young to middle age adults, there is a significant female predominance. ITP is a diagnosis of exclusion, as many other diseases can cause thrombocytopenia. However, once the diagnosis of ITP is established, unless the patient is bleeding significantly, platelet transfusion usually plays no role in therapy. The transfused platelets will be quickly destroyed by the autoantibody, as are the native platelets. Primary treatment options include corticosteroid therapy and/or intravenous immunoglobulin (IVIG). Steroids are inexpensive and increase platelet counts in most cases, while IVIG is much more expensive, but has a much more rapid effect on platelet counts. Other options (typically secondary choices) include splenectomy, rituximab, or thrombopoietin agonists (e.g. romiplostim). If the patient is Rh-positive and has not had a splenectomy, then Rh immune globulin (RhIG) may be used as part of the therapy. However, due to the Food and Drug administration's (FDA) black box warning about the risk of hemolysis with large doses of RhIG, caution is warranted.

Answer: *D*—Corticosteroids are typically first-line therapy in newly diagnosed adult ITP patients, especially those presenting with platelet counts below 30,000/μL. IVIG may be used if the response to steroids is not as rapid as desired. Splenectomy and thrombopoietin agonist treatment (Answers A and E) are generally reserved for second-line treatment in the event of failure of initial efforts to increase the platelet count. Platelet transfusions (Answer B) are only indicated in the face of significant bleeding. Identification of platelet autoantibodies is not required to diagnose ITP (Answer C), and such testing is not readily available in most hospital laboratories.

Please answer Questions 5 and 6 based on the following clinical scenario:

A 26-year-old Asian female gives birth to her first child, a female, with an estimated gestational age of 39 weeks. The baby has APGAR scores of 6 at 1 min and 6 at 5 min, and is transferred to the NICU for evaluation. The cord hemoglobin is 10.5 g/dL, and the NICU physician requests a transfusion.

5. Which of the following is true regarding pretransfusion testing for this baby?
A. Rh typing is not necessary
B. ABO reverse grouping may be performed using maternal serum
C. The antibody screen may be performed on maternal serum
D. ABO type must be determined using anti-A, anti-B, and anti-A,B
E. Pretransfusion testing is not necessary if the baby is given O, Rh-negative RBCs

Concept: Pretransfusion testing for infants below the age of 4 months, as for adults, includes ABO grouping, Rh typing, and screening for unexpected (non-ABO) RBC antibodies. However, the testing may be abbreviated or modified in these patients in several ways. First, ABO cellular grouping (forward grouping or front-typing) is required, using at minimum anti-A and anti-B reagents, but serum grouping (i.e., reverse grouping or back-typing) is not required. Babies do not typically form their own ABO antibodies until approximately 4 months or more after birth. Moreover, the antibody screen may be performed using either the mother's or baby's serum.

Note: These rules for pretransfusion testing extend until the baby is 4 months of age or at the end of the neonate's hospital admission, whichever is sooner. As a result, many blood bankers inaccurately believe that the "neonatal" period extends from birth to age 4 months, but by official definition, the neonatal period ends at 1 month.

Answer: *C*—The mother's serum is perfectly acceptable as the source for the antibody screen. Rh typing (Answer A) is required for infant pretransfusion testing. Reverse grouping (Answer B) is not required at all for these patients, so there would be no reason to use the maternal serum to perform the test, nor would it always be accurate, as the mother and child are not necessarily ABO-identical. Anti-A,B reagent (Answer D) is not required for ABO grouping. In routine scenarios, using O Rh-negative RBCs to avoid pretransfusion testing should not be encouraged (Answer E).

6. The neonate is determined to be blood group A. No significant unexpected antibodies are identified. The baby is expected to be in the hospital for a maximum of 3 weeks. Which of the following is true for this and future transfusions for this infant?
A. The baby may receive group A RBCs with no serologic crossmatch
B. Every 3 days, a repeat pretransfusion testing is required
C. The antibody screen should be repeated once the infant is 1-month old
D. A serologic crossmatch is required before group O RBCs are given
E. Group B RBCs are acceptable, since the antibody screen is negative

Concept: The rules for pretransfusion testing during the period from birth to 4 months are different from those for adults.

Answer: *A*—Once an infant has an ABO and Rh type, if the antibody screen is negative, a serologic crossmatch is not required before any transfusion of ABO identical or group O RBCs (Answer D), until the age of 4 months or at the end of the neonate's hospitalization, whichever is sooner. However, if the baby is to receive RBC units during that period that is NOT group O and not compatible with the maternal ABO group; then the baby's serum must first be evaluated for the presence of anti-A or -B (most likely originating from the mother before birth), using the sensitive antihuman globulin methods.

The normal adult rule requiring a new pretransfusion testing workup every 3 days in those who have been transfused within the last 3 months (the "3-day rule") (Answer B) does not apply to infants up to age 4 months (Answer C) because the original test is valid for the whole time frame. Group B RBCs (Answer E) are not acceptable for a group A baby, even if the baby has no anti-B (the antibody screen is irrelevant, as it detects non-ABO antibodies).

End of Case

7. A 65-year-old male presented to the emergency department with a 5-day history of fever and cough. He stated that his urine "starting to turn dark in color" recently. In the ED, he was found to be mildly hypoxic, febrile to 38°C, and chest X-ray showed bilateral infiltrates suggestive of pneumonia. Blood cultures drawn were positive for *Streptococcus pneumonia*. The urinalysis was positive for hemoglobin. Laboratory results were notable for the following: Hgb 7.4 g/dL and total bilirubin 5.2 mg/dL. Blood bank testing was unremarkable (Group A, Rh-positive, negative antibody screen), but further testing by a curious technologist showed that the patient's RBC agglutinated strongly in the presence of plasma from all random donors. Which of the following is the most likely recommendation for transfusing this patient?
A. Avoid transfusion of all blood products
B. Unmodified red cells, platelets, and plasma
C. Irradiated red cells; unmodified platelets and plasma
D. Unmodified red cells; washed platelets and plasma
E. Washed red cells and platelets; avoid plasma

Concept: Polyagglutination is a structural modification of red cell surface oligosaccharides that exposes underlying "cryptantigens." This change renders them susceptible to agglutination by virtually all plasma from adults, and occasionally, leads to hemolysis, classically in the setting of an infection. "T activation" (also called "T polyagglutination") is the most common infection-related polyagglutination, associated with organisms, such as *Streptococcus pneumonia*, *Clostridium* *perfringens*, and *Vibrio cholera*, as well as the influenza virus. Many other types of polyagglutination exist, and reference laboratories distinguish them by testing the patient's RBCs with a panel of lectins (seed extracts with agglutinating activity, Table 15.2).

TABLE 15.2 Lectin Reactions for Various Types of Polyagglutination

Lectin	T	Th	Tk	Tx	Tn	Cad	HEMPAS
Arachis hypogaea	+	+	+	+	–	–	–
Solichos biflorus	–	–	–	–	+	+	–
Salvia sclarea	–	–	–	–	+	–	–
Glycine soja	+	–	–	–	+	–	–

Answer: *B*—Currently, in adult patients, even if hemolysis is present, the majority of transfusion services do not recommend transfusion with washed cellular products (Answer E). In pediatric patients, reports of severe hemolysis receiving plasma-containing blood products have led some pediatricians to request washed RBCs and platelet products for patients with polyagglutination; however, the final decision should be a collaborative effort between the clinical team and transfusion medicine based on the latest evidence. Plasma itself should be avoided if at all possible (Answer D), but it is not necessary to avoid transfusion of all blood products (Answer A). Irradiation has no impact on polyagglutination (Answer C).

8. A 31-year-old group O, Rh-negative, G4P3 female has a strong anti-D (4+ reaction) and is currently pregnant (approximately 31 weeks estimated gestational age). Monitoring of the child indicates a strong possibility of severe hemolytic disease of the fetus and newborn (HDFN), confirmed by a cordocentesis sample with fetal hematocrit 18%, fetal blood type A, Rh-positive, and a 4+ direct antiglobulin test (DAT). Due to the severity of the anemia and hemolysis, an intrauterine transfusion is planned. In addition to being hemoglobin S negative, which of the following is the best choice of blood product for this transfusion?
A. Group A, Rh-positive, washed RBCs
B. Group O, Rh-positive, irradiated, leukoreduced RBCs
C. Group A, Rh-negative, washed, CMV-seronegative RBCs
D. Group O, Rh-negative, irradiated, leukoreduced RBCs
E. Group O, Rh-negative, washed, CMV-seronegative RBCs

Concept: Intrauterine transfusion is uncommonly performed, as the incidence of severe HDFN due to anti-D has declined with the widespread use of RhIG in developed countries. Blood for these transfusions should be: (1) group O; (2) negative for the antigen targeted by the mother's antibody; (3) irradiated in an effort to avoid transfusion-associated graft versus-host disease (TA-GVHD); (4) "CMV-safe" (functionally, to most neonatologists, this means CMV-seronegative, but many will also accept leukoreduced RBCs as equivalent); and (5) Hemoglobin S negative to avoid complications in a low-oxygen environment. Many institutions will also give fresh units (<7 days old) and may hyperconcentrate units to a hematocrit >80%, if the unit is preserved in AS solution.

Answer: *D*—Since the mother carries anti-D, the transfused units must be Rh-negative, not Rh-positive (Answers A and B). Though the fetus is blood group A (Answers A and C), group O is generally the best choice of RBCs, to avoid any ABO-compatibility issues. Irradiation and CMV-safe status are likewise important (note, as earlier, that equivalence of leukoreduced and CMV-seronegative for prevention of CMV transmission is usually not accepted by many neonatologists). There is no indication for washing in this case (Answers C and E).

9. A 65-year-old Caucasian male with no significant medical history presents to his physician complaining of decreased exercise tolerance and yellow discoloration in his eyes. His physical examination is significant for several enlarged axillary lymph nodes and prominent scleral icterus. His hemoglobin is 5.2 g/dL, and his total bilirubin is 3.6 mg/dL (direct bilirubin 2.9 mg/dL). Scattered spherocytes are seen on his peripheral smear. His chest X-ray suggests mediastinal lymphadenopathy. Which of the following is the most likely finding in the transfusion service when his blood is sent for compatibility testing?
A. ABO discrepancy, with interference with cellular grouping
B. Positive antibody screen, with panagglutinin pattern
C. Positive direct antiglobulin test with anticomplement only
D. Positive Donath-Landsteiner test
E. Negative antibody screen, with no ABO discrepancy

Concept: Approximately half of patients with warm autoimmune hemolytic anemia (WAIHA) have no associated clinical disease. However, the most common clinical association when one is actually found is a hematologic malignancy, such as chronic lymphocytic leukemia or non-Hodgkins lymphoma. Patients may actually present with hemolytic anemia months prior to the malignant diagnosis. Warm autoantibodies typically react against all screening cells at or near body temperatures, as well as all cells on an antibody panel (panagglutinin), and usually have a positive direct antiglobulin test (DAT) with anti-IgG. The eluate is also usually panreactive with all tested cells.

Answer: *B*—The physical findings are suspicious for a hematologic malignancy, such as lymphoma. Blood bank testing will most likely reveal a positive antibody screen (Answer E) with panagglutination to all cells. This patient has a fairly classic clinical and laboratory presentation of a warm-reacting autoantibody causing extravascular hemolysis, consistent with WAIHA. Warm autoantibodies do not usually interfere with ABO testing (Answer A), and most do not have a positive DAT with anticomplement only (though that pattern may be seen in 10% of cases) (Answer C). The Donath-Landsteiner test (Answer D) is used in the diagnosis of paroxysmal cold hemoglobinuria (PCH) and is not relevant to the diagnosis of WAIHA.

10. A 76-year-old female with systemic lupus erythematosis (SLE) is transferred to your medical center from a small rural hospital. She tells her nurse that she has a 5 year history of chronic anemia, and insists that she has been warned that her blood "is not compatible with anyone else's blood!" She has not been transfused in over a year, however. The transfusion service discovers a panreactive pattern, with all testing cells reacting with microcolumn (gel) technology. The autocontrol and DAT are both positive (with both anti-IgG and anticomplement positive in the DAT) and her eluate is also panreactive. Her hemoglobin is 6.1 g/dL, and the clinician would like to transfuse RBCs. Which of the following tests or procedures is the next best step to find blood for this patient?
A. Warm autoadsorption
B. Incubating patient serum with rabbit erythrocyte stroma reagent
C. Choosing "least incompatible" RBCs
D. Drug adsorption testing
E. Perform crossmatches of random RBC units until a compatible one is found

Concept: WAIHA is associated with autoimmune disorders, such as SLE, or with an underlying malignancy, as in the previous question. Selecting blood for patients with WAIHA is challenging because essentially all donor RBCs will be incompatible due to the panreactivity of the warm autoantibody. This can make it difficult to detect underlying alloantibodies that the patient may also have formed (e.g., antibodies, such as anti-K, anti- Jk^a, and anti-E). Most autoantibodies do not cause substantial hemolysis when donor RBCs are transfused, but the underlying alloantibodies can cause severe hemolysis if they are not detected. Blood banks use several strategies to choose the safest blood for these patients when transfusion cannot be avoided. Warm autoadsorption can be performed when the patient has not been recently transfused. This procedure removes the autoantibody from the serum by incubating the patient's own RBCs with the serum to "soak up" the autoantibody. The serum left behind ('adsorbed serum') may then be tested for alloantibodies. If an alloantibody is identified, the patient should receive blood that is negative for the corresponding antigen. If the patient is recently transfused, then alloadsorption, which is a more complicated procedure, is necessary. Alternatively, transfusion services can give donor RBCs that are phenotypically "matched" for all major blood group antigens with the patient (Rh, Kell, Kidd, Duffy, and Ss) to avoid exposing the patient to antigens that could be targeted by possible alloantibodies that the patient may have formed. The matching can be done by serologic methods or by molecular genotyping.

Answer: *A*—This patient has a clinical and laboratory presentation consistent with chronic WAIHA and management must include a way to detect or manage any underlying or hidden alloantibodies. Since the patient has not been recently transfused, warm autoadsorption is the best choice among those listed (molecular or serologic phenotyping would have also been equally good choices if listed). Without adsorption, the patient will react against all RBCs that you attempt to crossmatch; thus, performing crossmatches on random RBC units to find compatible ones is not useful (Answer E). Elution involves removing an antibody from the surface of a red cell and determining its specificity, but this is less useful than autoadsorption in most WAIHA cases, as the eluate will simply react against all test RBCs. Rabbit erythrocyte stroma (RESt) (Answer B) is used to adsorb cold autoantibodies, which are not evident in this clinical or laboratory presentation. Incubating drugs (Answer D) with test RBCs is a strategy useful in the workup of drug-induced hemolytic anemia, which is in the differential diagnosis in this case, but is less likely. The "least incompatible" option (testing patient serum vs. numerous donor RBCs and choosing blood from the donor against which there is the least reactivity) is not really a strategy at all. Rather, it is simply a way to choose which unit to give after alloantibodies have either been ruled out or managed by one of the earlier mentioned strategies (Answer C). Blood banks typically opt to choose the least incompatible (or "most compatible") unit among the units crossmatched, but there is no real evidence that this strategy does anything to reduce hemolysis.

11. A 27-year-old male presents to the Emergency Department by ambulance, following a motor vehicle accident. The patient has a history of sickle cell disease with multiple previous hospital admissions secondary to pain crises. The patient's blood pressure is 105/75, pulse 96, oxygen saturation 99% on 2 L. A CBC reveals hemoglobin of 6.8 g/dL. Ultrasound identifies a possible hepatic laceration. When two RBC units are requested for transfusion, review of the prior blood bank work-up reveals a history of alloantibodies to E, S, Fy^a, Fy^b, and Jk^b. The blood bank calls to notify you that no RBC units lacking all of the appropriate antigens are in inventory. What is the most reasonable next step?

A. Request antigen negative units from a blood center
B. Assess the urgency of the need for transfusion with the clinician
C. Issue 2 units of O, Rh-negative red cells, uncrossmatched
D. Issue 2 units of O, Rh-negative red cells, least incompatible
E. Request directed donations from family members

Concept: In urgent transfusion scenarios, communication with the clinical team caring for the patient is extremely important. This is especially true when there may be delays in the provision of compatible blood for transfusion. If the patient is hemodynamically stable, it may be most appropriate to hold off on transfusion and request antigen negative units from a blood center. If the patient is hemorrhaging or showing signs of shock (such as hypotension, tachycardia, cool extremities, and significantly decreased hemoglobin), transfusion is urgently needed. In these situations, transfusion of uncrossmatched or least incompatible units, lacking as many of the appropriate antigens as possible, may be necessary.

Answer: *B*—This patient has a possible bleeding source on imaging, and the patient's hemoglobin is below normal. However, the patient has a history of sickle cell disease and his usual baseline hemoglobin level is not known. In this situation, it would be most appropriate to speak with the team caring for the patient to assess clinical status. If the patient's hemoglobin level is at or near his baseline and there is no evidence of hemodynamic and/or oxygen-extraction compromise, delaying transfusion while waiting for antigen-negative units to arrive from an outside facility is the best option (Answer A). Note that this patient's antibody combination is compatible with less than 1% of donors. If there is clinical concern for active bleeding and the patient is showing signs of hypovolemia, delaying transfusion would be inappropriate and uncrossmatched or least incompatible (Answers C and D) units lacking as many of the appropriate antigens as possible should be transfused and the patient should be monitored for signs of hemolysis. If the patient will require long-term transfusion support, family members may be screened for directed donation. Blood relatives (Answer E) may be more likely than the general donor population to lack the antigens corresponding to the patient's alloantibodies. However, coordinating a directed donation takes time and is not the first choice in this setting.

12. A 60-year-old female with alcoholic cirrhosis arrives in the Emergency Department with hematemesis and melena. The patient is hypotensive and her hemoglobin is 4.6 g/dL. A type and crossmatch sample is sent to the blood bank, but due to the urgent situation, 4 uncrossmatched O Rh-positive RBC units are immediately transfused. Multiple bleeding variceal vessels are identified and clipped. The patient's blood pressure stabilizes and her posttransfusion hemoglobin is 8.3 g/dL. The patient is admitted in stable condition. The blood bank calls to notify you that the type and screen was positive, the antibody identification revealed an anti-E alloantibody, and 1 of the 4 units transfused was positive for the E antigen. What is the most appropriate next step?

A. No action is needed; Anti-E antibodies are not reported to cause hemolytic transfusion reactions
B. Notify the clinician caring for the patient, obtain a posttransfusion DAT, and monitor for signs of hemolysis
C. Notify the clinician caring for the patient and immediately administer fluids and diuretics to keep urine output >1 mL/kg/h
D. Transfuse two O-negative red cell units that lack the E-antigen
E. Begin immediate preparations to perform an RBC exchange

Concept: Patients who require emergent transfusion may receive uncrossmatched red blood cells that are later found to be incompatible due to the presence of preformed alloantibodies in the patient's plasma. When this occurs, the blood bank must assess the risk of developing a clinically significant acute hemolytic transfusion reaction (AHTR), communicate that risk to the clinical team caring for the patient, and obtain laboratory tests to adequately monitor the patient for signs of hemolysis. The severity of the reaction (if any) varies depending on the antibody involved. Such patients should be carefully monitored, but urgent intervention is often unnecessary.

Answer: *B*—This patient was administered a single uncrossmatched type O, Rh-positive RBC unit that was later found to be incompatible due to the presence of anti-E in the patient's plasma. Antibodies to the E antigen typically do not activate complement and have only rarely been associated with severe AHTRs, but can cause delayed hemolytic transfusion reactions (DHTRs) (Answer A). Following transfusion, the patient's hemoglobin increased appropriately and she appears clinically stable; thus, a severe AHTR appears unlikely. Despite this, it is appropriate to notify the clinician and monitor the patient for signs of hemolysis (Answer B). Urgent administration of fluids (Answer C) and diuretics to ensure renal blood flow is not indicated and an RBC exchange (Answer E) is also not indicated based on the current clinical status. Given the posttransfusion hemoglobin of 8.3 g/dL (which means the patient appropriately responded to 4 transfused units), additional transfusions are likely not needed at this time (Answer D).

13. A 37-year-old woman is admitted with end stage renal disease and severe, symptomatic anemia. She is short of breath and complains of fatigue. Vital signs are as follows: blood pressure 90/60, pulse 110, and respirations 28. Her hemoglobin is 4.9 g/dL. She says that blood transfusions are "against her beliefs," as she is a Jehovah's Witness. Which of the following is the most appropriate response to this scenario?

A. Ask her spouse for permission to transfuse despite the patient's wishes
B. Seek a court order for emergency transfusion
C. Discuss her individual limitations on blood transfusion
D. Counsel her that refusing transfusion means she is ineligible for kidney transplant
E. Treat her renal failure with hemodialysis instead of peritoneal dialysis

Concept: An adult patient has autonomy to make decisions about his or her healthcare and may refuse any therapy. Jehovah's Witnesses generally refuse blood transfusions based on Biblical references (such as in Lev. 17: "...the life of all flesh is the blood thereof: Whoever eat it shall be cut off"). This prohibition has been in place since 1945. Despite this, congregants may vary in the strictness of their interpretation of this prohibition. Officially, the Watchtower Bible and Tract Society states that the church's official position is a prohibition on the acceptance of whole blood, red blood cells, platelets, and plasma. The use of "fractions" of blood (such as clotting factors, fibrinogen, albumin, or immunoglobulins) is left to the congregant's personal choice.

Answer: *C*—It is important to fully discuss what is and is not acceptable to a Jehovah's Witness patient. Some will accept fractions of blood as listed earlier, some will actually accept some primary components (such as plasma), while others will not accept any part of the blood. If the patient is an adult with the capacity to make decisions regarding her care and treatment, attempting to circumvent the patient's wishes with legal (Answer B) or family maneuvers is not generally appropriate (Answer A). Other appropriate steps include restricting the patient to bedrest to minimize oxygen demand and providing supplemental oxygen, minimizing blood draws, and considering starting erythropoietin and iron therapy. Congregants must decide for themselves if they will accept tissue/organ transplant, so refusal of transfusion has no bearing on eligibility for renal transplant (Answer D). Finally, hemodialysis may waste up to 175 mL of blood per session, whereas peritoneal dialysis will have no blood loss (Answer E).

14. The neonatal intensive care unit (NICU) team is called to the nursery to consult on a jaundiced newborn. The baby's bilirubin is 37 mg/dL at 12 h of age. The DAT is positive and the mother's antibody screen is positive. The NICU team recommends an exchange transfusion. However, the parents are Jehovah's Witnesses and they refuse to give consent for the exchange transfusion. The most appropriate next step is:

A. No action; respect the autonomy of the parents to make this decision
B. Request permission to give a series of simple transfusions
C. Perform a plasma exchange since that treatment would be acceptable to the parents
D. Seek an emergency injunction to treat from the court
E. Treat the baby with ultraviolet light

Concept: Courts recognize the right of parents to consent to treatment for their children. They also recognize that these rights are not absolute, and must promote the welfare of the children. When a baby's life is in imminent danger of death or permanent disability, legal action is indicated. Policies on specific approach to these cases vary, but in general a knowledgeable individual is available to assist with emergent requests to the court for injunctions to treat a minor. Most of these decisions are based on a United States Supreme Court case, *Prince v. Massachusetts* 321 US 158 (1944), in which the court held that the government has broad authority to regulate the treatment of children. Parents cannot make decisions that may permanently harm their children.

Answer: *D*—Parents have autonomy over their own treatment, but cannot make decisions that may result in permanent harm to their child (Answer A). Therefore, urgent legal intervention is appropriate in this case in order to protect the child from harm. Simple transfusions (Answer B) would be ineffective because an exchange transfusion removes passive antibody, as well as the antigen positive cells, cell debris, and excess bilirubin as the donor blood is infused. Additionally, the parents are unlikely to consent to simple transfusions for the same reason, they are refusing consent for the exchange transfusion. Plasma may be acceptable to some Jehovah's Witnesses; however, a plasma exchange (Answer C) would be a poor substitution for a whole blood exchange and is technically challenging in a newborn. Ultraviolet lights (Answer E) are used for mild physiologic jaundice, but would not be effective for the severe hyperbilirubinemia seen in this case.

15. A 38-year-old African American female with sickle cell disease (SCD) is admitted with symptoms of acute pain crisis, including severe pain in her lower back and abdomen. She has a history of chronic anemia with multiple transfusions, and her current hemoglobin is 6.8 g/dL. After admission, she is diagnosed with necrotic bowel, a probable pelvic abscess is suspected, and an exploratory abdominal surgery is planned. Which of the following is the most appropriate reason for this patient to receive a simple transfusion of RBCs?

A. In preparation for major surgery
B. As a result of her infection and possible sepsis
C. Because of the acute pain episode
D. Because of her chronic anemia
E. To prevent iron overload by suppressing hematopoiesis

Concept: There is good evidence to recommend transfusion in SCD patients prior to major surgery. In studies randomizing SCD patients to aggressive transfusion, conservative transfusion, or no transfusion, patients in both the transfusion arms had fewer complications; however the aggressive transfusion arm had more alloimmunization. The National Heart, Lung, and Blood Institute (NHLBI) recommend transfusing to a hemoglobin of 10 g/dL prior to major surgery. However, the NHLBI's report also states that if the pretransfusion hemoglobin is >8.5 g/dL, a hematologist should be consulted as to the best management strategy. Depending on multiple factors, RBC exchange may not be clinically necessary for preoperative preparation.

Answer: *A*—This patient is likely to require abdominal exploration for removal of a segment of necrotic bowel. Simple transfusion of anemic SCD patients prior to major surgery is appropriate. There is no indication for transfusing for infection unless the patient is also significantly anemic (Answer B). Acute pain episodes (Answer C) are treated with hydration and analgesia; transfusion is not recommended unless there are other indications. Chronic anemia (Answer D) is generally well tolerated by patients with SCD and in the absence of other indications, is not an indication for transfusion. While simple transfusion may suppress hematopoiesis (Answer E), it does not appreciably slow iron accumulation, since RBCs contain about 200–250 mg of iron per unit. Red cell exchange may be preferable to simple transfusion in reducing iron overload and in acute SCD complications, such as acute stroke.

16. A 46-year-old African-American male with SCD presents with avascular necrosis of the right femoral head due to bone infarctions related to his SCD. An orthopedic surgeon strongly recommends a hip replacement in 2 weeks. In preparation for surgery, the surgeon requests 3 units of crossmatched RBCs. The transfusion service has no previous records for this patient, though he reports many previous transfusions elsewhere. However, none of those were transfused within the past 3 months. While preparing to choose the optimal blood product, which of the following is the most important step for the transfusion?

A. Inform the patient that he should ask relatives for directed donation
B. Inform the patient that he should donate autologous blood
C. Reserve the "freshest" units in inventory, preferably less than 5 days old
D. Perform RBC phenotype
E. Prepare to wash the units that are chosen for transfusion

Concept: Patients with SCD are at high risk of alloimmunization to red blood cell antigens (20%–25% form such antibodies, vs. about 2% in the general population). Prior to transfusion, the transfusion service should establish the presence or absence of such antibodies by contacting any other facilities at which the patient receives transfusion, performing an antibody screen, and determining the patient's RBC antigen phenotype. The phenotype may not be valid if the patient was recently transfused. If the patient has no alloantibodies, limited antigen matching between recipient and RBC donor for at least D, C/c, E/e, and K antigens is generally accepted as a method to prevent alloimmunization. In the presence of alloantibodies, RBCs that are more completely phenotypically matched (to include all major blood group systems) is recommended to prevent further alloantibody formation. In general, sickle cell patients should receive leukocyte-reduced blood that is also negative for hemoglobin S.

Answer: *D*—In a chronically transfused SCD patient who does not have a known immunohematologic history at your facility, obtaining any alloantibody history from outside facilities is essential because antibodies can disappear (or evanesce) over time and thus, would not be identified during a blood bank workup. However, if the patient has been to many different facilities, a simple history check is insufficient to safely provide units for the patient. Thus, evaluation for alloantibodies and assessment of the patient's RBC phenotype is a vital part of the in-house patient evaluation. The antibody screen is important, not only to avoid the corresponding antigen on donor cells, but to aid in the decision to limit phenotypic matching, or to extend it. The phenotype may be performed using traditional serologic methods (requiring a pretransfusion sample from a patient who has not been recently transfused) or by using molecular genotyping (Answer D) methods (which may be performed despite recent transfusion).

Attempting to obtain direct donated RBCs units (Answer A) is premature at this point given that the transfusion service does not know what antibodies, if any, this patient has. Furthermore, direct donated units have higher risk of transfusion transmitted infection comparing to units from volunteer donors. The use of autologous blood collection (Answer B) for SCD patients is contraindicated. Blood from these patients will not increase oxygen carrying capacity or decrease the fraction of hemoglobin S. Storage conditions may also induce sickling in SCD red cells. There is no need to provide freshest blood (Answer C) for sickle cell anemia patients for simple transfusion. In general, blood does not need to be washed (Answer E) unless the patient has experienced repeated allergic reactions or is truly IgA deficient with anti-IgA antibodies.

17. A 16-year-old African-American male with a history of SCD and chronic transfusion presents to the Emergency Department with headache, severe fatigue, and dyspnea. His most recent transfusion was approximately 1 month ago. His CBC has the following results: WBC 6.4 × 10^9 cells/L, hemoglobin 4.5 g/dL, HCT 13.5%, and platelets 90,000/μL. His reticulocyte count is <1%. Which of the following is the most likely diagnosis?
A. Hyperhemolysis syndrome
B. Parvovirus B19 infection
C. Acute chest syndrome
D. Pneumonia due to an encapsulated bacteria
E. Acute pain crisis

Concept: Parvovirus B19 (which causes the childhood infection "erythema infectiosum" or "Fifth Disease") may also cause profound, transient red cell aplasia by directly invading the red cell precursors ("Transient Aplastic Crisis", TAC). The cytotoxic effect of the virus becomes more obvious in patients with SCD because of the shortened red cell life span. The nonred cell counts are usually within normal parameters, although pancytopenia can be seen.

Answer: *B*—The findings are consistent with, but not exclusive for TAC caused by Parvovirus B19. IgM and IgG serology tests for Parvovirus B19 would confirm the diagnosis. The extremely low reticulocyte count in the face of profound anemia indicates a lack of production rather than increased destruction.

Hyperhemolysis (Answer A) (severe, accelerated red cell hemolysis) is unlikely because the reticulocyte count would be expected to be very high (up to 20%) in response to overwhelming red cell destruction. Additionally, hyperhemolysis following transfusion typically occurs within 7 days. Hyperhemolysis is typically a delayed hemolytic transfusion reaction whereby donor cells are hemolyzed and bystander native cells get caught up in the immune reaction, causing the hemoglobin level to drop lower than the pretransfusion level. Acute chest syndrome, pneumonia due to an encapsulated bacterium, and acute pain crisis are all conditions to which patients with SCD are prone (Answers C, D, and E). However, none cause selected red cell aplasia. This patient has dyspnea, but this can be explained by the severe anemia causing an oxygen deficit.

18. A 29-year-old Caucasian female with systemic lupus erythematosus was admitted with progressive jaundice. She was found to have autoimmune hepatitis with resulting end-stage liver disease and treatment with high-dose steroids was initiated. She developed progressive thrombocytopenia with a nadir platelet count of 25,000/μL, and empiric IVIG was administered to treat presumed idiopathic thrombocytopenic purpura (ITP). Over the next week, the patient subsequently developed progressive anemia (hemoglobin 6.9 g/dL and haptoglobin <6 mg/dL). In the transfusion service, the patient tested as group A Rh-positive, with a negative antibody screen. The direct antiglobulin test (DAT) was positive with a monospecific anti-IgG reagent, and an eluate prepared from the patient's RBCs did not show reactivity with any of the cells on the antibody identification panel. What is the appropriate next step?

A. Perform flow cytometric testing for paroxysmal nocturnal hemoglobinuria (PNH)
B. Evaluate for reactivity between the patient's eluate and group A RBCs
C. Evaluate for reactivity between the patient's eluate and cefotetan-treated RBCs
D. Perform the Donath-Landsteiner test
E. No further work-up is indicated

Concept: Intravenous immunoglobulin (IVIG) may be used to treat immune thrombocytopenic purpura (ITP), among many other diseases. This product contains concentrated human immune globulins, and may occasionally contain high titers of ABO antibodies, especially anti-A. Less commonly, ABO antibodies in IVIG may directly cause hemolysis in recipients. In the serologic evaluation of such cases, a positive DAT with a negative eluate is generally seen when the antibody in question is directed against an antigen that is not present on the antibody identification panel cells. It is important to remember that all cells on the antibody identification panel are type O, and would therefore, produce no reactivity when tested against an eluate containing anti-A or anti-B antibodies. If anti-A or anti-B antibody is suspected, the eluate should be tested against type A or type B RBCs, respectively.

Answer: *B*—The most appropriate next step would be to evaluate for reactivity between the patient's eluate and group A RBCs. PNH (Answer A) results from reduced or absent glycosylphosphatidylinositol (GPI)-anchored proteins, including complement regulatory proteins, and is characterized by a negative DAT. The Donath-Landsteiner test (Answer D) is used to diagnose paroxysmal cold hemoglobinuria (PCH), a rare sporadic hemolysis that usually presents following a viral infection in young children. PCH is generally characterized by DAT positivity with anti-C3 reagent and a negative eluate. Although drug-induced hemolytic anemia (Answer C) may be characterized by a positive DAT and negative eluate, there is no mention of cefotetan in the current case. Although up to 15% of hospitalized patients have a positive DAT, most of which are due to clinically insignificant, nonimmunologic adsorption of IgG, other causes of positive DAT positivity should be ruled out before assuming a clinically insignificant false positive test (Answer E).

19. A 25-year-old Rh-negative pregnant female presents for her initial prenatal appointment at approximately 10 weeks gestation. This is her second pregnancy. She received Rh immunoglobulin (RhIG) during her first pregnancy and delivered a healthy Rh-positive infant. Currently, her routine prenatal laboratory work, including complete blood count, thyroid-stimulating hormone, and HbA1c

levels are within normal limits. Her antibody screen is positive. Anti-C and anti-D alloantibodies are identified. The obstetrician is concerned and calls the blood bank for help and to ask about whether the patient should receive RhIG later in the pregnancy. How do you respond?
A. Yes, since all Rh-negative pregnant women require prenatal RhIG
B. Possibly, since this may represent anti-f and not anti-D
C. Possibly, since this may represent anti-G and not anti-D
D. No, because once anti-D is formed, RhIG is contraindicated
E. No, because once any Rh system antibody is formed, RhIG is contraindicated

Concept: The G antigen is a member of the Rh blood system and is present on RBCs containing either C or D antigens. A G-negative individual is at risk of forming an anti-G antibody if exposed to blood with *either* the C or D antigen. Antibodies to G react like a combination of anti-C and anti-D, and can only be distinguished through adsorption and elution studies. In women of childbearing potential, it is important to distinguish anti-G from anti-D and/or anti-C, because RhIG is unnecessary in pregnant D-negative women who have made anti-D. If anti-G is present without anti-D, the patient should be protected from D exposure by administering prophylactic RhIG.

Answer: *C*—It is unlikely that the patient in the current case has developed a true anti-D because she appropriately received Rh immunoprophylaxis during her first pregnancy. It is possible that her first child was positive for C and as a result, also positive for G, thus, exposing her to the G antigen and potential anti-G formation, which mimics the combination of an anti-C and anti-D. After confirmation that she has anti-G only, she should receive prenatal RhIG at the appropriate time to prevent D alloimmunization, similar to all Rh-negative pregnant women who have not yet formed an anti-D (Answers A, D, and E). The f antigen occurs when both c and e antigens are present on the same Rh protein, such as in individuals with either the Dce (R^0) or dce (r) haplotypes (Answer B).

20. A 60-year-old male with Crohn's disease, currently being treated with mesalamine, presents to the Emergency Department with a gastrointestinal bleed. The patient also has a past medical history of refractory multiple myeloma, hypertension, and diabetes. His medication list includes daratumumab, lisinopril, and glipizide. The blood bank receives an order to crossmatch 2 RBC units for transfusion. Using solid-phase testing methodology, the patient's plasma shows weak panreactivity with all cells on the antibody identification panel. The DAT is positive with monospecific anti-IgG reagent, and an eluate prepared from the patient's RBCs also shows panreactivity. The plasma continues to show panreactivity with all panel cells following autoadsorption. What is the most appropriate next step?
A. Treat reagent RBCs with dithiothreitol (DTT)
B. Treat reagent RBCs with papain
C. Repeat testing using the microcolumn (gel) platform
D. Perform a computer crossmatch only
E. Provide least incompatible units

Concept: Daratumumab is an anti-CD38 monoclonal antibody approved in the United States in 2015 for the treatment of refractory multiple myeloma. CD38 is expressed on the surface of all RBCs; as a result, the drug in patient plasma binds to all reagent cells causing weak agglutination on all cells in the antibody identification panel (panreactivity). The autocontrol and DAT may also be positive. Unlike samples with warm autoantibodies, autoadsorption fails to eliminate the daratumumab-induced panreactivity and antihuman globulin (AHG) crossmatching is typically positive with all RBC units tested. ABO and Rh typing is not affected. Dithiothreitol (DTT) destroys the CD38 antigen on reagent RBCs and eliminates the panreactivity by removing the drug's target antigen. DTT also destroys Kell system antigens, so K-negative units should be provided unless the patient is known to be K-positive.

Answer: *A*—The serologic work-up is consistent with daratumumab interferenced. Reagent RBCs should be treated with DTT to eliminate panreactivity, then the patient serum should be tested to rule out the presence of alloantibodies. Daratumumab-induced panreactivity cannot be eliminated using papain-treated cells (Answer B) and occurs regardless of testing platform, including tube, gel, and solid phase methods (Answer C). Unless emergent transfusion is necessary, it would be inappropriate to provide computer crossmatch or least incompatible units (Answers D and E) because an underlying clinically significant alloantibody may be masked by the panreactivity.

21. A 60-year-old Ashkenazi Jewish female is admitted for the evaluation of syncope. Her workup reveals severe aortic stenosis, and she is scheduled for aortic valve replacement 2 days from now. She reports a prior history of appendicitis with subsequent open appendectomy 30 years ago, complicated by intraoperative and postoperative bleeding, as well as mild bleeding following a childhood tooth extraction and heavy periods until menopause 10 years ago. She reports her mother experienced bleeding following surgery, as well, but she does not know her deceased father's bleeding history (both parents were of Ashkenazi Jewish descent). Her INR is 1.1 and aPTT is 52 s. Mixing with normal pooled plasma corrects her aPTT to 35 s immediately and it remains 36 s after 2-h incubation. Her fibrinogen is 318 mg/dL and her platelet count is 125,000/μL. Factor levels are ordered and reveal the following: factor VIII 92%, factor IX 103%, and factor XI 6%. Which of the following is the best advice regarding this patient's management with the available information?

A. She should receive Factor VIII concentrate prior to surgery
B. She should receive cryoprecipitated AHF transfusion prior to surgery
C. She should receive plasma transfusion prior to surgery
D. She should receive platelet transfusion prior to surgery
E. Blood product administration prior to surgery is not indicated

Concept: Factor XI (FXI) deficiency, also known as hemophilia C, is an inherited bleeding disorder that is rare in the general population but has a relatively high prevalence in Ashkenazi Jews (up to 8%). Inheritance is autosomal, so incidence is similar between males and females. Bleeding manifestations are widely variable, and most have none at all. Some with FXI deficiency may bleed significantly with only mild deficits, while others with similar levels have no bleeding. Epistaxis, menorrhagia, or bleeding after routine procedures is most common in those with bleeding, often with initial diagnosis occurring later in life. Unlike hemophilia A and B, spontaneous bleeding and hemarthrosis are rare. FXI deficiency leads to isolated prolongation of aPTT with correction on immediate and incubated mixing studies. Patients with factor XI deficiency do not need factor replacement for routine daily activities, but they may require treatment prior to surgery or in trauma. FXI concentrates are not available in the United States and preoperative prophylactic replacement generally is done with plasma infusion.

Answer: C—The patient's history, including her Ashkenazi Jewish heritage, personal and family history of bleeding following surgery, and prolongation of aPTT with correction on mixing studies, is suggestive of a diagnosis of FXI deficiency, which is confirmed by the factor assays. Given her prior history of postoperative bleeding, FXI replacement by plasma transfusion prior to surgery is the best choice (Answer E). The half-life of factor XI in plasma is ~2 days; thus, if necessary, maintenance doses may be given every other day. The clinical history and the patient's female gender are not generally consistent with a diagnosis of Factor VIII deficiency (hemophilia A), so Factor VIII concentrate is not indicated (Answer A). Fibrinogen levels are normal and platelets are only slightly decreased; therefore, cryoprecipitated AHF and platelet transfusions are not necessary (Answers B and D).

22. A pharmacist calls you to discuss a patient who presented with a lower extremity deep vein thrombosis (DVT). Despite the initiation of a heparin drip 2 days ago, the patient's aPTT remained subtherapeutic (therapeutic range depends on the laboratory, but is typically between 70 and 110 s). The pharmacist increased the heparin dose to 35,000 units per day but the aPTT still did not respond appropriately. The anti-Xa was measured, and it is less than 0.1 IU/mL (therapeutic range is 0.3–0.7 IU/mL). What do you recommend to the pharmacist?

A. Increase heparin dose
B. Switch to a direct thrombin inhibitor

C. Switch to a LMWH
D. Measure antithrombin activity
E. Immediately measure protein C and S activity

Concept: Antithrombin (AT, formerly "antithrombin III") is a natural anticoagulant that inhibits factor IIa and factor Xa in the coagulation cascade. Heparin and low molecular weight heparin (LMWH) cause anticoagulation by potentiating AT activity. AT deficiency may lead to increased thrombotic risk, as well as insensitivity to heparin therapy (patients have a lack of therapeutic aPTT prolongation in the presence of more than adequate heparin administration). AT deficiency should be considered in any patient demonstrating heparin resistance. In patients with suspected AT deficiency and venous thromboembolism, antithrombin activity can be measured and if low, antithrombin concentrate can be provided. Of note, some facilities use the anti-Xa test to monitor heparin therapy.

Answer: *D*—The absence of aPTT prolongation in the current case, even after the heparin dose was increased, suggests AT deficiency and antithrombin activity levels should be checked. If low, supplementation with antithrombin concentrate should allow for heparin to reach therapeutic range. If antithrombin activity testing or antithrombin concentrate are not readily available, then anticoagulation may be changed to a direct thrombin inhibitor, such as argatroban, which does not require functional AT (Answer B). Further increasing heparin dose (Answer A) and switching to a LMWH (Answer C), which also relies on AT activity, could delay necessary anticoagulation. Deficiency of protein C and S (Answer E) can infer an increased risk of clotting, but do not affect the function of heparin.

23. A 22-year-old G1P0000 female has blood drawn following her first prenatal visit to her obstetrician. The ABO and Rh results (using gel technology) are as follows:

	Anti-A	Anti-B	A1 RBCs	B RBCs	Anti-D
Patient RBCs	4+	0	N/A	N/A	1+
Patient serum	N/A	N/A	0	4+	N/A

Note: All controls were valid and appropriate.

Given these results, which of the following is the best strategy?
A. Issue a report identifying the mother as group A, Rh-positive
B. Perform a weak D test
C. Recommend RhIG injection to prevent the mother from forming anti-D
D. Consider genetic testing to detect D variant
E. Notify the obstetrician that the fetus is at high risk for HDFN from the patient's strong anti-B

Concept: D-antigen testing reagents are designed to result in strongly positive reactions when a Rh-positive person is testing using reagent anti-D (4+ strength). When a "weaker than expected" ($\leq 2+$) positive reaction occurs when testing a patient or pregnant mother, transfusion services may elect to manage the individual as Rh-negative. This may result in unnecessary RhIG injections and use of scarce resources. A committee from the College of American Pathologists proposed an alternative in 2015.

Answer: *D*—No discrepancies are seen in the ABO testing. The results are typical for a group A person, with the RBCs reacting strongly with reagent anti-A and the serum reacting strongly against group B reagent RBCs (the anti-B in a group A person is predominantly IgM, so HDFN is unlikely even if the fetus is group B; Answer E). The D testing, however, shows a much weaker than expected reaction of the patient's RBCs with anti-D (serologic weak D), and suggests an altered, or "variant" D antigen. In general, D variants fall into one of two categories: First, those that have less D antigen on the RBC surface, and second, those that have a form of D antigen that is altered or missing parts. Those with the quantitative problem (traditionally called "weak D," or the older term "Du") are at minimal risk for making anti-D when exposed to normal D antigens, while those with the altered or missing parts (traditionally called "partial D") are usually at risk for making anti-D when exposed. Distinguishing between the two types is not possible with routine serologic testing, however, as both may show variable reactions with anti-D, from completely negative to weakly positive.

Typically, the Rh testing is done at immediate spin; thus, if a person with a D variant has a *negative* test with anti-D, an antihuman globulin (AHG) test using anti-D will usually be positive; in this circumstance, the AHG test is called a "weak D test." However, when the anti-D test is *weakly positive*, the weak D test would only confirm the presence of the D antigen, and is not useful (Answer B). Issuing a result that ignores the weaker-than-expected reaction (Answer A), as roughly half of transfusion services would reportedly do (based on a 2014 survey), may lead to formation of anti-D, as a pregnant female carrying a D-positive infant may be exposed during pregnancy or at delivery. For that reason, in accordance with some testing kit package inserts, many transfusion services treat a weak reaction as D negative (and might recommend RhIG injection at 28 weeks' gestational age; Answer C).

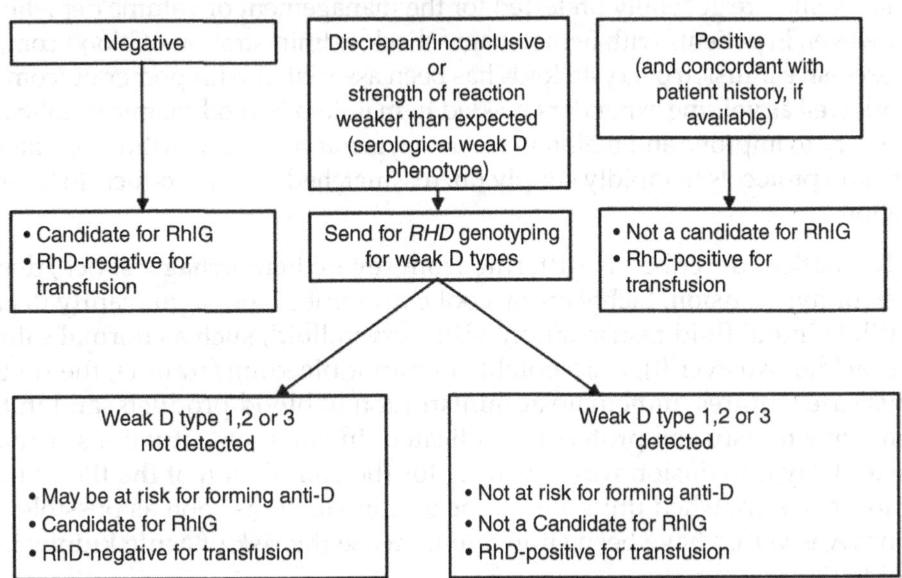

FIGURE 15.1 Algorithm for resolving serological weak D phenotype. *Source: Adapted from S.G. Sandler, et al., Transfusion 55 (2015) 680–689.*

Recognizing that practice was not uniform in this area, the College of American Pathologists Transfusion Medicine Resource Committee recommended recently that patients with either serologic weak D or discrepant results (for example, being called Rh-positive as a blood donor and Rh-positive as a recipient or pregnant female) undergo *RHD* genotyping (Answer D) to help separate those with weak D types 1, 2, or 3 from all other forms of weak D and all forms of partial D. If the patient is weak D type 1, 2, or 3 (as 75% will be), they may then be treated as Rh-positive, with an extremely low risk of forming anti-D from either transfusion of Rh-positive RBCs or pregnancy exposure to a Rh-positive baby. In other words, they may receive Rh-positive RBCs, and if pregnant, would not require RhIg injections. The remaining 25% that are not weak D types 1, 2, or 3 would be treated as Rh-negative, and at risk for forming anti-D with transfusion or pregnancy exposure to the D antigen (Fig. 15.1). Thus, this group of patients would receive Rh-negative products and are eligible for RhIg prophylaxis.

Please answer Questions 24–29 based on the following clinical scenario:

Paramedics bring an unconscious 30-year-old male into the emergency department with a gunshot wound to the abdomen. An IV line running normal saline is in place. The on-call physician observes agonal breathing and intubates the patient for airway protection. He is hypotensive, tachycardic, unresponsive, and his extremities are cool to the touch. Intravenous access is established, laboratories are drawn, and a type and crossmatch sample is sent to the blood bank. The STAT hemoglobin level is 4.8 g/dL.

24. When considering the patient's volume and hemodynamic status, what is the next appropriate step?
A. Continue volume resuscitation with rapid infusion of normal saline
B. Switch to Ringer's Lactate solution
C. Switch to hydroxyethyl starch
D. Initiate a massive transfusion protocol
E. Stop infusion and wait for blood bank to provide crossmatch compatible RBC units

Concept: Initial management of patients in hemorrhagic shock includes maintaining intravascular volume; however, the optimal type of fluid resuscitation is the subject of debate. There are three main classes of replacement fluids, including crystalloid solutions, such as normal saline and Ringer's lactate, colloid-containing solutions, such as albumin and hyperoncotic starch, and blood products. In general, the type of fluid that is being lost should determine the choice of replacement fluid. Crystalloids, primarily normal saline, are generally preferred for the management of volume depletion not due to bleeding. However, in patients with hemorrhagic shock, administration of blood components is indicated and excessive infusion of crystalloids has been associated with poorer outcomes. While crystalloids work well as volume expanders needed to maintain hemodynamic stability, transfusion of RBCs is necessary to improve and maintain tissue oxygenation. Many institutions have created a massive transfusion protocols to rapidly supply uncrossmatched blood products to bleeding patients in emergent situations.

Answer: *D*—The patient described is suffering from severe hemorrhagic shock, demonstrated by the presence of hypotension, tachycardia, cool extremities, and significantly decreased hemoglobin. While initial fluid resuscitation with a crystalloid, such as normal saline (Answer A) or Ringer's lactate (Answer B), is acceptable in minor bleeding/trauma, the severity of the situation in this case requires immediate administration of blood products, and initiation of the institution's massive transfusion protocol is indicated. In emergent situations such as this, it is inappropriate to delay transfusion while waiting for the completion of the Blood Bank work-up (Answer E) and uncrossmatched units should be administered as soon as possible. Hyperoncotic starch solutions (Answer C) have been shown to increase the risk of acute kidney injury and should be avoided.

25. The patient becomes bradycardic with heart rate dropping to 40 bpm. The blood bank immediately releases blood products to the emergency department. At what point does the transfusion meet the criteria of "massive transfusion"?
A. One RBC unit transfused over 1 h
B. Two RBC units transfused over 4 h
C. 8–10 RBC transfused over 24 h
D. One blood volume replaced over 72 h
E. Only when an RBC exchange has been initiated

Concept: Massive transfusion is a situation where an amount of blood potentially equal to the patient's blood volume is transfused in a relatively short period of time. It is usually defined as transfusion of 8–10 red blood cell units in 24 h. Additional definitions include exchange transfusion in a neonate, 4–5 RBCs transfused over 1 h, and 1 total blood volume replaced over 24 h. Massive transfusion is most commonly required following trauma, but may also result from other clinical situations, including liver disease, as well as obstetric and surgical hemorrhage.

Answer: *C*—The most common definition of massive transfusion is administration of 8–10 RBC units in 24 h. The volume of blood in the other choices is too small to be considered massive transfusion. Transfusion of 1–2 RBC units over 1–4 h would be considered routine (Answers A and B). To be defined as massive transfusion, a total blood volume would need to be replaced over 24 h, not 72 h (Answer D). RBC exchanges in adults are often planned procedures, performed in the absence of hemorrhage, to treat blood disorders, such as sickle cell anemia and severe babesiosis (Answer E).

26. Which of the following is considered a complication of massive transfusion?
A. Hyperthermia
B. Metabolic acidosis
C. Hypernatremia
D. Hypercalcemia
E. Hyperkalemia

Concept: There are numerous potential complications associated with massive transfusion.

Answer: *E*—Hyperkalemia is a reported complication of massive transfusion. During blood storage, potassium "leaks" out of the red cells and into the supernatant at a rate of approximately 1 mEq/day. This is usually clinically insignificant because of the small volume of supernatant. After routine transfusion, free potassium moves quickly back into RBCs. However, delivery of larger quantities of free potassium with massive transfusion may cause problems in patients with preexisting hyperkalemia, renal disease, or if long-stored RBC units are rapidly transfused directly into the central circulation through a central venous catheter. Fresher units or (less commonly) washed red cell units may avoid this complication.

Rapid delivery of a large number of "cold" blood products can lead to a decrease in core body temperature (or hypothermia, (Answer A). Hypothermia can result in cardiac arrhythmias, as well as coagulopathy caused by diminished coagulation factor activity and decreased platelet adhesion to von Willebrand factor. The use of a blood warmer when delivering a large number of blood products reduces the risk of hypothermia. In addition to hypothermia-related coagulopathy, volume replacement with crystalloid or colloid solutions may lead to coagulopathy secondary to dilution of platelets and clotting factors.

A large dose of citrate from blood anticoagulant-preservative solution is delivered to patients during massive transfusion. Citrate exposure can lead to both hypocalcemia (Answer D) and metabolic alkalosis (Answer B). Citrate binds to calcium and decreases plasma free calcium concentration, which may result in hypocalcemia. In most patients, the liver rapidly metabolizes citrate and the resulting hypocalcemia is transient. However, some individuals may develop clinically significant hypocalcemia leading to paresthesia or cardiac arrhythmia. Thus, monitoring calcium level, such as ionized calcium level, is important during massive transfusion. Additionally, citrate is metabolized to bicarbonate, which may result in metabolic alkalosis (Answer B). This is especially true in the presence of renal ischemia or underlying renal disease, which prevents bicarbonate excretion in urine. Although metabolic acidosis (Answer B) may be seen in many patients receiving massive transfusion, this is most likely caused by an increase in lactate production secondary to impaired tissue oxygenation resulting from the patient's underlying disease and is not directly related to the transfusion. There is no association between altered sodium levels and massive transfusion (Answer C).

27. Which of the following is the main benefit of providing a predetermined mixture of plasma, platelets, and red cells (i.e., a massive transfusion protocol) to a hemorrhagic patient?
A. Prevention of dilutional coagulopathy
B. Prevention of metabolic alkalosis
C. Prevention of hypocalcemia
D. Prevention of hyperkalemia
E. Prevention of hypokalemia

Concept: Studies in military trauma settings suggest that transfusing plasma, platelet, and red cell units to massively bleeding patients in a 1:1:1 ratio may lead to improved survival, though the largest civilian study to date did not show a large benefit. Blood volume resuscitation using crystalloid fluids only (e.g., saline) will quickly lead to dilution of coagulation factors, platelets, and hematocrit.

Answer: *A*—A predetermined mixture of plasma, platelets, and red cells given to a hemorrhagic patient may prevent dilutional coagulopathy. All three components are delivered in equal amounts to maintain an adequate hematocrit and hemostatic levels of coagulation factor and platelets. A 2015 randomized trial (PROPPR) failed to show a 24 h or 30 day survival difference between severely injured patients transfused with a 1:1:1 ratio versus a 1:1:2 ratio. Nevertheless, the 1:1:1 target is considered the current standard of care. Although metabolic alkalosis, hypocalcemia, hyperkalemia, and hypokalemia (Answers B, C, D, and E) are associated with massive transfusion, altering the ratio of product administration will not prevent the development of these complications.

28. A blood bank technologist calls to let you know that the patient's ABO and Rh typing is delayed and the inventory of O, RhD-negative RBC units is running low. The technologist asks if you will consider allowing her to provide Rh-positive units to this patient. What is the most appropriate response?

A. Yes, Rh type never matters in an emergency
B. Yes, this is acceptable in all patients except females of childbearing potential and those without a history of anti-D
C. Yes, but only if the decision is first discussed with the clinical team caring for the patient
D. Yes, but only if we have a historical negative antibody screen
E. No, it is never acceptable to release potentially incompatible blood

Concept: In emergent situations, an incomplete blood bank work-up should never delay blood administration; immediate transfusion of uncrossmatched units may be necessary. With multiple bleeding patients or a single prolonged massive transfusion event, the inventory of Rh-negative units may become depleted, forcing the blood bank to decide which patients should receive D-negative units and which should receive D-positive units. Although the risk of anti-D formation in this setting is equal in male and female blood recipients (according to modern studies that risk is approximately 22%), the consequence of the antibody is far greater in young females because future pregnancies are at risk of the complication of HDFN. For this reason, females of childbearing potential should receive Rh-negative RBC units whenever possible.

Answer: *B*—In an emergency, it is acceptable to supply Rh-positive blood to men and postmenopausal women prior to completion of an antibody screen unless the patient is known to have anti-D (Answer E). If at all possible, Rh-positive units should be avoided in females of childbearing potential to avoid alloimmunization and subsequent HDFN in future pregnancies. Because the current patient is male and emergently requires blood administration, the blood bank should make the decision to supply Rh-positive units (Answer A). During the current trauma, the treating physicians should not be interrupted to discuss the Rh type of blood being sent to the patient (Answer C). However, the possibility of Rh switching during a massive transfusion event should be discussed during the establishment of the institution's massive transfusion policy. Although a prior negative antibody screen may seem reassuring (Answer D), it does not reflect the patient's current antibody status and is not particularly useful in deciding whether or not to supply Rh-positive units in an emergency. If uncrossmatched blood is administered and later, after completion of the blood bank work-up, found to be incompatible due to the presence of a non-ABO alloantibody, the patient's physician should be notified and the patient should be monitored for signs and symptoms associated with hemolysis.

29. The patient is brought to the operating room for abdominal exploration. Approximately 2 L of blood is present in the abdominal cavity, and damage to the right iliac artery is identified and repaired. The patient's blood pressure stabilizes and he is transferred to the intensive care unit. He continues to ooze blood from his abdominal incision and a thromboelastography (TEG) study is performed (Fig. 15.2). Based on the TEG tracing, which of the following blood products should be administered?

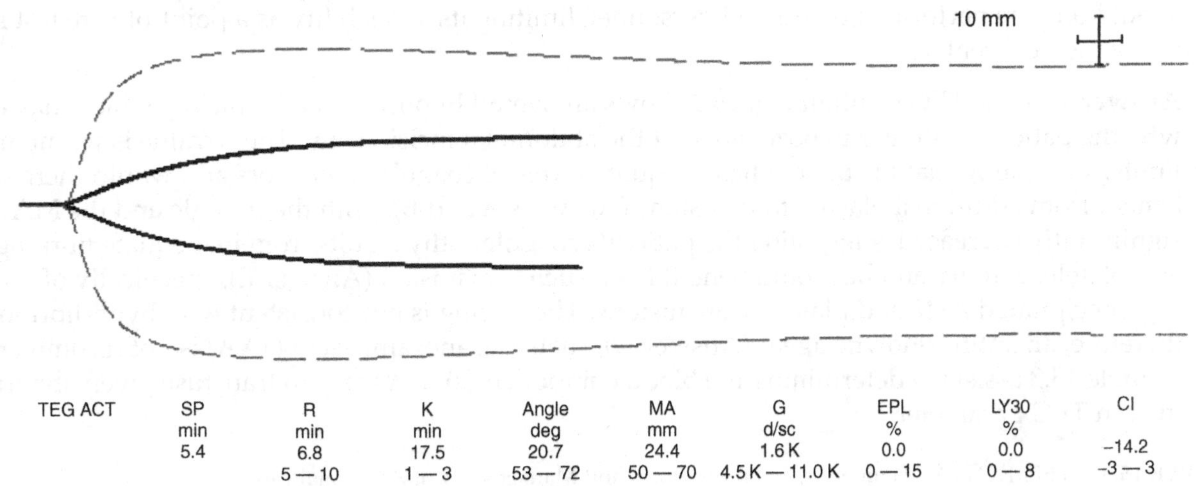

FIGURE 15.2 Patient's TEG result.

A. Plasma only
B. Plasma and cryoprecipitated AHF
C. Cryoprecipitated AHF and platelets
D. Antifibrinolytic agent
E. No additional blood products should be given

Concept: Routine coagulation tests, such as INR and aPTT, are fragmented and only offer a snapshot of the coagulation process. Thromboelastography (TEG) and rotational thromboelastometry (ROTEM) are similar real-time, point-of-care analyzers that evaluate global clotting mechanisms by measuring the rate of clot formation as an interaction between platelets and coagulation factors. Both methods work by measuring the difference in oscillation between a cup containing the patient's whole blood and a pin suspended in the blood. Results are presented as a graphic representation of clot formation and lysis. TEG tracings contain four main components; (1) the R-value reflects clotting factor activation via initial thrombin generation, (2) the α-angle reflects the amount of fibrinogen being converted to fibrin, (3) the maximum amplitude (MA) represents the platelet contribution to the clot; and (4) the LY30 represents fibrinolysis. Fig. 15.3 shows a result from a normal TEG tracing.

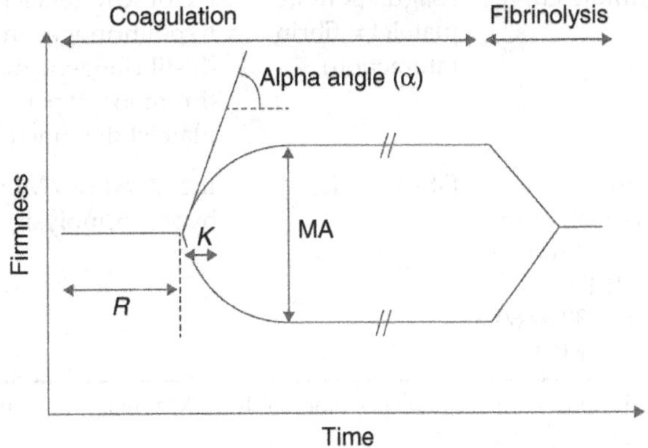

FIGURE 15.3 Unremarkable TEG tracing. *Source: Modified from Figure 2 in H.P. Pham, B.H. Shaz, Update on massive transfusion. Br. J. Anaesth. 13 (2012) 171–182.*

The TEG tracing provides immediate information that may be used to aid in decisions regarding blood product transfusion. The use of TEG to successfully guide blood product transfusion in both cardiac surgery and liver transplantation is well documented. TEG requires daily calibration and should only be performed by trained personnel, limiting its availability as a point-of-care test in many medical centers.

Answer: C—The TEG result in Fig. 15.2 shows an overall hypocoagulable tracing, which may explain why the patient continues to ooze blood at the abdominal incision site. The R-value is within normal limits, suggesting that the patient has adequate levels of coagulation factors and would therefore not benefit from additional plasma transfusions (Answers A and B). Both the α-angle and the MA are significantly decreased, suggesting the patient's coagulopathy results from inadequate fibrinogen and platelet activity and he would benefit from such transfusion (Answer E), specifically of cryoprecipitated AHF and platelet transfusions. The tracing is not consistent with hyperfibrinolysis; therefore, an antifibrinolytic agent (Answer D), such as tranexamic acid (TXA) is not recommended.

Table 15.3 assists in determining the blood component(s) necessary to transfuse given abnormality in each TEG parameter.

TABLE 15.3 TEG/ROTEM Parameters, Interpretations, and Management of Abnormalities

TEG parameter	ROTEM parameter	Definition	Hemostatic phase	Etiologies for abnormalities	Potential management
R	CT	Time from the begin of the test till the first sign of clot formation	Initiation of coagulation	Prolonged R/CT: factor deficiencies or anticoagulants Shortened R/CT: plasma hypercoagulability	Plasma for prolong R/CT
K	CFT	Time from the begin of clot formation to the time when the curve reaches amplitude of 20 mm.	Amplification of coagulation	Prolonged K/CFT: factor deficiencies, hypofibrinogenemia, dysfibrinogenemia, thrombocytopenia, or platelet dysfunction	Cryoprecipitated AHF
α	α	Angle between the baseline and the tangent to the curve through the starting point of coagulation	Propagation of coagulation (i.e., thrombin burst)	Low α: factor deficiencies, hypofibrinogenemia, dysfibrinogenemia, thrombocytopenia, or platelet dysfunction	Cryoprecipitated AHF
MA	MCF	Amplitude measured at the maximum curve width	Propagation of coagulation (i.e., platelet = fibrin interaction)	Low MA/MCF: factor XIII deficiency, hypofibrinogenemia, dysfibrinogenemia, thrombocytopenia, or platelet dysfunction	Platelets Might consider plasma or cryoprecipitated AHF for factor XIII deficiency if ongoing bleeding and persistently low MA/MCF
LY	ML	Reduction in area under curve (LY) or in amplitude (ML) from time MA/MCF is achieved until 30 or 60 min after MA/MCF	Fibrinolysis	Increased LY/ML: hyperfibrinolysis	Antifibrinolytic medication

α, Alpha angle; CFT, clot formation time; CT, clotting time; K, kinetics time; LY, lysis; MA, maximum amplitude; MCF, maximum clot firmness; ML, maximum lysis; R, reaction time.

Modified from Table 5 in H.P. Pham, B.H. Shaz, Update on massive transfusion. Br. J. Anaesth. 13 (2012) i71–i82.

End of Case

Please answer Questions 30–33 based on the following clinical scenario:

A 34-year-old G3P2002 female presents to the Emergency Department in labor. She is approximately 37 weeks gestation and has chosen to forgo prenatal care because her first two pregnancies were uneventful.

The patient is admitted to Labor and Delivery. All hematology and chemistry laboratory parameters are within normal limits. The blood bank serologic workup reveals her to be group O negative with a positive antibody screen. The patient denies any RhIG injections during this pregnancy. The antibody identification work up is performed, and an anti-K alloantibody is present.

30. Considering the information earlier, for which of the following is the unborn baby most at risk?
A. Hemolytic disease of the fetus/newborn due to Rh mismatch
B. Red cell aplasia due to maternal anti-K
C. Intrauterine growth retardation due to lack of prenatal care
D. Development of anti-D because no RhIG was administered
E. Severe hyperbilirubinemia

Concept: Though anti-D is the most well-known cause, HDFN can occur with antibodies other than anti-D. Of note, HDFN due to ABO antibodies is currently more common than HDFN due to anti-D due the widespread use of RhIG in the United States.

Approximately 9% of Caucasians and 2% of African-Americans carry the K antigen. In K-negative individuals, exposure to the antigen commonly leads to formation of anti-K. Unlike other alloantibodies, anti-K predominantly acts on fetal early red cell precursors and inhibits development of normal, mature, red blood cells. This can lead to severe anemia/red cell aplasia in K-positive infants born to mothers with anti-K. Anti-K may be destructive at lower maternal serum titers than other alloantibodies (typically at 8 compared to 16 with anti-D, for example) and can cause very severe anemia, as early as 20 weeks gestation.

Answer: *B*—The antibody screen is an important part of prenatal care and any antibody should be a cause for increased surveillance of the pregnancy, but anti-K is of even greater concern in early pregnancy. The mother is Rh-negative and is at risk of forming anti-D, but she has not done so at this point (Answers A and D). As a result, the baby is not at risk for anti-D HDFN. Lack of prenatal care does not in and of itself, put the baby at increased risk of growth retardation (Answer C). The mother is at risk for developing anti-D because she did not receive her prophylactic dose of RhIG at 28 weeks. However, receiving the injection postpartum can mitigate this risk if she has not been alloimmunized. As noted earlier, anti-K exerts its effect primarily on early red cell precursors, causing red cell aplasia. Since anti-K prevents the formation of red cells, there are fewer destroyed and less build-up of red cell breakdown products, such as bilirubin (Answer E), as would be seen in HDFN due to other antibodies (e.g., anti-D).

31. Which of the following is the most appropriate next step for the blood bank to take in order to prepare for delivery?
A. Mix a unit of group O negative, K-negative, red cells with AB plasma for a neonatal whole blood exchange
B. Immediately issue a vial of RhIG to Labor and Delivery
C. Crossmatch 1–2 units of compatible blood for the mother
D. Alert the local blood center of possible need of rare blood in case of maternal hemorrhage
E. Perform a titer on maternal anti-K to assess the fetal risk

Concept: When a pregnant female is preparing for delivery and has a clinically significant red cell alloantibody, such as anti-K, it is prudent for a transfusion service to prepare for possible transfusion by obtaining antigen-negative RBCs. Although the risk of hemorrhage requiring transfusion is relatively low, it is significant enough to plan ahead in a patient with known antibodies to avoid the need for emergency uncrossmatched blood that may be positive for the K antigen.

Answer: *C*—Crossmatching 2 units of K-negative blood is a reasonable step to prepare for possible transfusion of a pregnant female in labor. Since the likelihood of finding K-negative units is ~91%, these units should be relatively easy to locate (Answer D). Since the infant is not at risk for the typical HDFN, but if affected, may be at risk for anemia, simple transfusion rather than exchange transfusion (Answer A) is usually the preferred treatment for red cell aplasia secondary to anti-K. Red cell exchanges are most often performed for severe hyperbilirubinemia, which we would not expect in this situation. At present, the patient has not formed anti-D. RhIG should be administered, but there is no harm in waiting until after the delivery (Answer B). At that time, a full work up will determine the baby's Rh type and the degree of fetal-maternal hemorrhage in order to administer the proper dose of RhIG within 72 h from delivery. This patient did miss her prophylactic dose of RhIG typically administered at 28 weeks estimated gestational age in the United States; however, it is unlikely that urgent administration of RhIG will change the risk of developing an anti-D. At this late stage, the anti-K titer is of little use (Answer E). Anti-K tends to be significant at a lower titer and acts earlier in pregnancy than other alloantibodies. Fetal monitors and clinical indications are more useful predictors of outcome at this point.

32. After several hours of labor, a baby girl is born via normal spontaneous vaginal delivery. The baby's APGAR scores are 7/10 at 1 min and 9/10 at 5 min. The baby appears healthy and vigorous. A cord blood sample is collected and sent to the Blood Bank. An average amount of blood is lost (approximately 300 mL) at parturition. Traction is applied to the umbilical cord in order to deliver the placenta. Following delivery of the placenta, the nurse notes increased blood loss. After an hour, the cumulative blood loss is quantified as approximately 1000 mL. The mother's heart rate has increased to 110 bpm, and her blood pressure has decreased to 95/50. Standard obstetric measures to decrease postpartum bleeding are employed, including external uterine massage, misoprostol administration, and warming blankets. The most appropriate next step in this patient's management should be which of the following?

A. Order a CBC and wait for the results
B. Transfuse 2 units of apheresis platelets
C. Begin to thaw plasma in anticipation of use
D. Infuse Lactated Ringers solution at maximum rates until vitals stabilize
E. Transfuse 2 units of RBCs and alert the blood bank to be prepared for possible massive transfusion protocol activation

Concept: Mothers with significant postpartum hemorrhage and symptomatic changes should undergo early treatment. Evidence of decreased oxygen carrying capacity along with hemorrhage is best treated by red blood cell transfusion.

Answer: *E*—Despite an average blood loss during childbirth, this mother has now lost 1 L of blood in the postpartum period. Her increased heart rate and low blood pressure indicate that she is hypovolemic. She needs volume, but she also needs increased oxygen carrying capacity. Available RBC units on hand for her should be transfused now and, critically, the blood bank should be told that a massive transfusion may be needed so appropriate resources may be prepared. All transfusing facilities should consider having a massive transfusion protocol, but such a protocol is essential at trauma centers, facilities performing surgery, and those with obstetric units. Depending upon the size of the facility and the frequency of massive transfusions, plasma may be prethawed or urgently thawed when an alert warning is received. Similarly, platelets may be readily available at large institutions, but a small hospital may not have platelets on the shelf at all times. Regardless of the logistics involved, the process should be formalized well before a massive transfusion event takes place. In this case of a known anti-K antibody, K-negative blood should be selected and transfused. It is not necessary to use AB plasma in this patient who is known to be group O.

It would not be appropriate to wait for laboratory confirmation of the clinical picture demonstrated here (Answer A), since the estimated blood loss is approximately 1 L. Loss of 1 L of blood is significant and the mother's altered vital signs confirm the seriousness of the situation.

While it may be necessary to give platelets, should the patient progress to require massive transfusion or if evidence of disseminated intravascular coagulation (DIC) were present, their use would not be appropriate as first line therapy (Answer B). The cause of bleeding in a postpartum woman is most likely to be due to uterine atony, retained placenta, or coagulopathy. Platelets alone would do little to improve any of these three problems.

Though plasma transfusion (Answer C) is not indicated at this time, early thought should be given to thawing that product, which is the reason for having a massive transfusion protocol. In the absence of a massive transfusion protocol, it may be prudent to order plasma at this point in conjunction with a request for the red cells already crossmatched.

Since this patient has lost a significant amount of blood, replacement with IV fluid (Answer D) may temporarily reverse the derangement of vital signs due to hypovolemia. However, simply replacing volume without replacing red cells that carry oxygen to tissues will cause a significant oxygen deficit in the tissues. Additionally, normalizing volume causes the peripheral circuits to open up, bypassing the physiologic protection and prioritization of vital organs (in a hypovolemic state, blood is shunted to core vital organs at the expense of the periphery to prioritize oxygen delivery). This may lead to more rapid end-organ damage. IV fluid also dilutes coagulation factors needed to stop bleeding and drives the system toward coagulopathy. Over time, osmotic pressures act to equalize intravascular and interstitial fluids, which could create significant tissue edema combined with new hypovolemia and possibly cardiovascular collapse.

33. Despite obstetric interventions, including a dilation and curettage (D and C), once it is noted that the placenta is incomplete, the patient continues to bleed with a total quantitative blood loss of over 2000 mL. The patient's vital signs continue to remain unstable. A massive transfusion protocol is activated. What is the primary goal of an obstetrical massive transfusion protocol?

A. To reverse coagulopathy
B. To reverse hyperfibrinolysis
C. To provide a red cell to plasma ratio of 2:1
D. To avoid platelet transfusions
E. To use only crossmatched blood products

Concept: Dilution and consumption of coagulation factors, acidemia (exacerbated by the low pH of anticoagulated donor blood), and hypothermia may all quickly push the patient toward coagulopathy (Fig. 15.4). Massive transfusion protocols are designed to combat the coagulopathy by providing blood products at near-physiologic levels (i.e., provide reconstituted "whole blood").

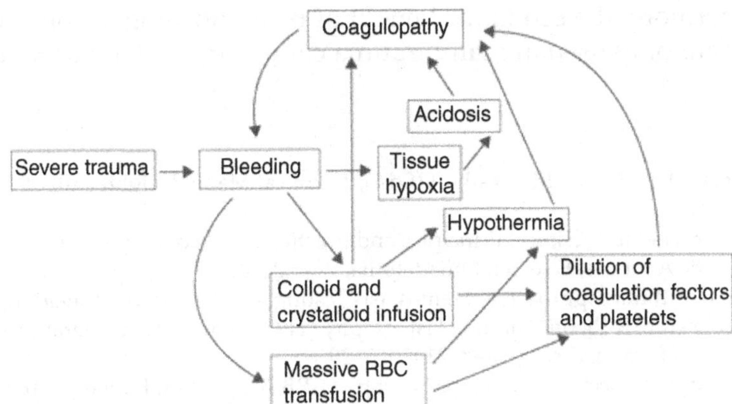

FIGURE 15.4 **Hemostatic abnormalities in massive transfusion.** Dilutional coagulopathy, activation of inflammatory mediators, hyperfibrinolysis, thrombocytopathy, and metabolic abnormalities (hypothermia, hypocalcemia, and acidosis) all contribute to the pathogenesis of the hemostasis abnormality in massive hemorrhage. *Source: Modified from Figure 1 in in H.P. Pham, B.H. Shaz, Update on massive transfusion. Br. J. Anaesth. 13 (2012) i71–i82.*

Answer: *A*—The early addition of red cells and plasma at a near 1:1 ratio (Answer C) is given with the direct goal of reversing coagulopathy. Additional measures to prevent a spiral toward coagulopathy include blood warmers, patient warming systems, such as specialized blankets with warm air blown into an interior space, and corrections of acid/base and metabolic derangements.

Platelet transfusions (Answer D) are generally recommended as part of the massive transfusion protocol. For trauma patients, the Army Surgeon General has recommended 1:1:1 red cell to plasma to platelet ratio (Answer C). In this recommendation, the "1 unit of platelets" refers to 1 unit of concentrated platelets derived from whole blood in a pool of 6. Because of this recommendation, many transfusion services use a 6:6:1 ratio, which refers to 6 RBC units, 6 plasma units, and 1 apheresis-derived platelet unit (equivalent to a pool of 6 whole blood-derived platelet concentrates). Hyperfibrinolysis may occur in trauma patients; however, the main goal of a massive transfusion protocol is not to reverse hyperfibrinolysis (Answer B).

Massive transfusion protocols typically use uncrossmatched blood (Answer E). The seriousness of the situation and urgency to provide blood quickly makes the time spent to crossmatch multiple units unacceptably long. In this scenario, where the patient has a known antibody, it is prudent to provide K-negative RBC as much as possible. Most blood donors (91% of Caucasians) are negative for K. If the patient's antibody status is unknown, it is generally felt that the risk of uncrossmatched blood is lower than the risk of serious harm due to massive blood loss.

The incidence of red blood cell antibodies in the population at large and hospitalized patients is generally about 2% and 4%, respectively. In other words, 96%–98% of patients will lack antibodies and any ABO compatible unit will be acceptable. Of those who have an antibody, there is a moderate likelihood that the donor will lack the antigen, depending upon the antigen/antibody pair. Finally, if the patient has the misfortune of having an antibody and being transfused an antigen positive unit of blood, the result will most likely be a clearance of those donor cells over the next days to weeks (i.e., delayed hemolytic transfusion reaction). This is because the antibodies to minor red cell antigens are most commonly IgG and the result will be extravascular hemolysis facilitated by splenic clearance. Acute hemolytic transfusion reactions have been reported, but are rare.

Note: The utility of cryoprecipitated AHF in massive transfusion protocols is not well studied, but some studies point to the benefit of maintaining sufficient levels of fibrinogen to participate in the clotting cascade. Therefore, the use and timing of cryoprecipitated AHF can vary from hospital to hospital. In centers that do use cryoprecipitated AHF during trauma, it is typically used with the second MTP cooler and every other cooler thereafter, unless the TEG tracing dictates more frequent usage is required. Fibrinogen concentrate may be used instead of cryoprecipitated AHF in an effort to reduce the high rate of cryoprecipitated AHF wastage; however, there is no head-to-head study regarding the efficacy between fibrinogen concentrates and cryoprecipitated AHF in adult patients. Furthermore, the economic benefit of using fibrinogen concentrate is unclear; thus, currently, it is not commonly used at many trauma centers in the United States.

Suggested Reading

[1] M.K. Fung, B.J. Grossman, C.D. Hillyer, C.M. Westhoff (Eds.), Technical Manual, eighteenth ed., AABB Press, Bethesda, MD, 2014.
[2] P.D. Mintz (Ed.), Transfusion Therapy: Clinical Principles and Practice, third ed., AABB Press, Bethesda, MD, 2011.
[3] L.D. Petz, Immune Hemolytic Anemias, Elsevier, Philadelphia, PA, (2004).
[4] K. Kottke-Marchant, An Algorithmic Approach to Hemostasis Testing, CAP Press, Northfield, IL, (2008).
[5] L. Teodoro da Luz, B. Nascimento, S. Rizoli, Thrombelastography (TEG): practical considerations on its clinical use in trauma resuscitation, Scand. J. Trauma Resusc. Emerg. Med. 21 (2013) 29.
[6] Mitigating the Anti-CD-38 Interference with Serologic Testing. AABB Association Bulletin0 #16-02, 2016.
[7] M.L. Panico, G.Y. Jenq, U.C. Brewster, When a patient refuses life-saving care: issues raised when treating a Jehovah's witness, Am. J. Kidney Dis. 58 (4) (2011) 647–653.
[8] United States Supreme Court, Prince v. Massachusetts 321 US 158, (1944).
[9] A. Babinszki, R.H. Lapinski, R.L. Berkowitz, Prognostic factors and management in pregnancies complicated with severe kell alloimmunization: experiences of the last 13 years, Am. J. Perinatol. 15 (12) (1998) 695–701.

[10] A. Lyndon, D. Lagrew, L. Shields, E. Main, V. Cape. Improving Health Care Response to Obstetric Hemorrhage (California Maternal Quality Care Collaborative Toolkit to Transform Maternity Care) Developed under contract #11-10006 with the California Department of Public Health; Maternal, Child and Adolescent Health Division; Published by the California Maternal Quality Care Collaborative, 3/17/15. Available from: https://www.cmqcc.org/resources-tool-kits/toolkits/ob-hemorrhage-toolkit.

[11] E.P. Vichinsky, C.M. Haberkern, L. Neumayr, et al. A comparison of conservative and aggressive transfusion regimens in the perioperative management of sickle cell disease. The Preoperative Transfusion in Sickle Cell Disease Study Group, N. Engl. J. Med. 333 (1995) 206–213.

[12] National Institutes of Health, National Heart, Lung, and Blood Institute, Division of Blood Diseases and Resources. The Management of Sickle Cell Disease, fourth ed., 2002. NIH Publication No. 02-2117.

[13] S.D. Sosler, B.J. Jilly, C. Saporito, et al. A simple, practical model for reducing alloimmunization in patients with sickle cell disease, Am. J. Hematol. 43 (1993) 103–106.

[14] J.B. Holcomb, B.C. Tilley, S. Baraniuk, et al. Transfusion of plasma, platelets, and red blood cells in a 1:1:1 vs a 1:1:2 ratio and mortality in patients with severe trauma: The PROPPR randomized clinical trial, JAMA 313 (5) (2015) 471–482.

[15] S.G. Sandler, W.A. Flegel, C.M. Westhoff, et al. It's time to phase in RHD genotyping for patients with a serologic weak D phenotype, Transfusion 55 (2015) 680–689.

[16] H.P. Pham, B.H. Shaz, Update on massive transfusion, Br. J. Anaesth. 13 (2012) i71–i82.

[17] C.K. Okerberg, L.A. Williams III, M.L. Kilgore, C.H. Kim, M.B. Marques, J. Schwartz, H.P. Pham, Cryoprecipitate AHF vs. fibrinogen concentrate for fibrinogen replacement in acquired bleeding patients—an economic evaluation, Vox Sang. 111 (3) (2016) 292–298.

[18] S.G. Sandler, S.D. Roseff, R.E. Domen, B. Shaz, J.L. Gottschall. Policies and procedures related to testing for weak D phenotypes and administration of Rh immune globulin: results and recommendations related to supplemental questions in the Comprehensive Transfusion Medicine survey of the College of American Pathologists, Arch. Pathol. Lab. Med. 138 (5) (2014) 620–625.

C H A P T E R

16

Clinical Histocompatibility Testing

Michael D. Gautreaux, John Schmitz***

*Wake Forest University School of Medicine, Winston-Salem, NC, United States; **University of North Carolina School of Medicine, Chapel Hill, NC, United States

Histocompatibility testing, also known as "HLA" testing, is among the highest complexity testing performed in clinical laboratories. Most HLA laboratories utilize molecular methodologies for typing patients at several HLA loci and a variety of testing methodologies for HLA antibody screening and crossmatching. These methodologies can range from 1960s-era complement-dependent cytotoxicity to modern multilaser platforms. Different transplant programs and HLA laboratories have different criteria for testing their patients. As a result, each HLA laboratory is mandated to have an agreement with each of the transplant programs that they support. These criteria are mandated by various regulatory and accrediting agencies. This chapter will provide a review and overview of histocompatibility testing for transplantation and other clinical purposes.

1. Which of the following accrediting organizations have oversight over clinical histocompatibility (HLA) testing for solid organ transplantation in the United States?
   A. Centers for Medicare and Medicaid Services (CMS)
   B. The Department of Health and Human Services (HHS)
   C. American Society for Histocompatibility and Immunogenetics (ASHI)
   D. Foundation for the Accreditation of Cellular Therapy (FACT)
   E. American Association of Blood Banks (AABB)

   Concept: All clinical laboratories that are in the United States (US) or those that test samples for US hospitals fall under the jurisdiction of the Clinical Laboratory Improvement Amendments (CLIA). CLIA is administered by the Centers for Medicare and Medicaid Services (CMS). CMS deems various accrediting organizations (AOs) to accredit laboratories for various specialties or subspecialties under CLIA. Additionally, the US Department of Health and Human Services regulates organ transplantation through its Organ Procurement and Transplantation Network (OPTN). Established by the National Organ Transplantation Act (NOTA), the OPTN is to be administered by an independent contractor, whose contract is put out for bids every 3 years.

   Since NOTA came out, the only contractor to bid for the OPTN contract has been the United Network for Organ Sharing (UNOS). This arrangement is commonly referred to as OPTN/UNOS, which also designates AO's to assist in ensuring the quality and safety of the laboratory.

   Answer: C—Two of the main AO's, deemed acceptable by both CLIA/CMS and OPTN/UNOS, for the specialty of histocompatibility, are the American Society for Histocompatibility and Immunogenetics (ASHI) and the College of American Pathologists (CAP). Most laboratories will designate either ASHI or CAP as its accrediting body, but many have both.

   Though clearly involved in the process of maintaining high quality in laboratory testing, CMS and HHS (Answers A and B) are regulatory, not accrediting agencies. FACT and AABB

Transfusion Medicine, Apheresis, and Hemostasis. http://dx.doi.org/10.1016/B978-0-12-803999-1.00016-X

(Answers D and E) are accrediting agencies, but are not involved in HLA laboratory accreditation. FACT focuses on all aspects of hematopoietic progenitor cell (HPC) transplant and AABB focuses on blood collection and transfusion services.

2. Which of the following cells does not express HLA Class I antigens?
   A. Nucleated red blood cells
   B. Platelets
   C. Precursor red blood cells
   D. Mature red blood cells
   E. Lymphocytes

   Concept: Conceptually, HLA antigens [aka major histocompatibility complexes (MHCs)] function to help our bodies recognize self versus nonself, and to assist in antigen presentation. The major classes of HLA are class I and II.

   Answer: *D*—HLA class I antigens are present on most cells/tissues in the body, especially nucleated cells. However, they are not present on mature red blood cells (RBCs). Once the RBC expels its nucleus during maturation, HLA class I expression diminishes. Thus, nucleated RBCs (Answer A) and precursor RBCs (Answer C) do express class I antigens. Other cells that express class I antigens include platelets (Answer B), lymphocytes (Answer E), monocytes, granulocytes, and cells/tissues that make up solid organs.

   HLA class II antigens are also present on a variety of cell types, especially antigen presenting cells, such as B-cells, activated T-cells, monocytes, macrophages, and dendritic cells. They are not present on resting T-cells.

3. The antigens presented by HLA molecules are of intracellular or extracellular protein origin. Which of the following class I MHC related molecules is involved in the presentation of lipid antigens to T-cells?
   A. CD-1
   B. MICA
   C. MICB
   D. FcRn
   E. HFE

   Concept: There are several MHC-related molecules, some of which present antigen, but most do not.

   Answer: *A*—Cluster of differentiation 1 (CD-1) is a family of glycoproteins expressed on the surface of various human antigen-presenting cells. They are related to the class I MHC molecules, and are involved in the presentation of lipid antigens to T cells. There are five isoforms of CD-1 (CD1a, b, c, d, e), all of which present different microbial lipid antigens. For example, CD-1a presents lipids from mycobacteria, such as mycobactin, a lipopeptide produced by *Myobacterium tuberculosis*, while CD1b can present mycolic acid also from *M. tuberculosis*.

   MHC-class I polypeptide-related sequence A (MICA) and sequence B (MICB) (Answers B and C) are both stress-induced ligands for the NK-cell receptor NKG2DL, an activation receptor also found on some T-cells. MICA and MICB cellular expression is upregulated in situations of cellular stress (infection, malignancy) and mark the cell for destruction by NKG2D expressing NK and CD8 T-cells.

   Another MHC-related molecule is newborn Fc receptor (FcRn) (Answer D). The FcRn gene is found on chromosome 19 and is important in absorbing IgG from mother's milk in the gut of newborns. In addition to providing passive transfer of IgG in neonates, FcRn protein also plays a role in the recycling of IgG, significantly extending IgG's half-life to over 20 days when compared to other immunoglobulin isotypes.

   Human hemochromatosis protein (HFE) (Answer E), initially and incorrectly called HLA-H, is encoded on chromosome 6 near the HLA region. Like class I HLA, HFE associates with β_2-microglobulin. This gene helps regulate iron uptake through interaction with the transferrin receptor. HFE is prominent in small intestinal absorptive cells, gastric epithelial cells, tissue macrophages, monocytes, granulocytes, and the syncytiotrophoblast, an iron transport tissue in the placenta. Defects in the HFE gene lead to hereditary hemochromatosis.

4. A patient is awaiting a kidney transplant and appears to be well-matched for a deceased donor. The patient has a calculated panel reactive antibody (cPRA) of 85%. The HLA typings are listed as follows:

	HLA-A	HLA-B	HLA-Bw	HLA-C	HLA-DR	HLA-DR 51, 52, 53	HLA-DQ	HLA-DP
Patient	2, 24	27, -	4,-	2,-	7, 11	52, 53	2,7	4, -
Donor	2, 24	27, -	4,6	2,-	7, -	53	2,-	4, -

A crossmatch between the patient and the donor is positive for both T- and B-cells. Which of the following might be the best reason that such a well-matched transplant has a positive crossmatch?

A. The patient is reacting to the Bw6 epitope
B. The patient is reacting to the DR11 antigen
C. The donor is reacting to the Bw6 epitope
D. The donor is reacting to the DR11 antigen
E. The crossmatch is falsely positive

Concept: Prior to kidney transplants, all patients and donors must be crossmatched. In the lymphocyte crossmatch, a donor's lymphocytes are isolated from either peripheral blood or tissue, such as a lymph node or a piece of spleen. The B-cells and T-cells are considered separately. The reason is that T-cells express class I HLA (HLA-A, B, and –C) and B-cells express both class I and class II HLA. If a patient's serum is reactive with both T-cells and B-cells, this is an indication that the sera are reacting with a class I antigen. If the patient's serum is reactive with only B-cells, that indicates reactivity to class II (HLA-DR, -DQ-, and -DP, see Table 16.1).

TABLE 16.1 Crossmatching Interpretation

T-cell	B-cell	Interpretation
Negative	Negative	No antibody
Positive	Positive	Class I ± Class II antibody
Negative	Positive	Class II antibody ± low-titer class I antibody
Positive	Negative	Non-HLA antibody or technical error ⇒ repeat test

Answer: *A*—Since the crossmatch in this scenario is both T-cell positive and B-cell positive, this leads to the initial consideration that this is primarily a response to a class I antigen. The only mismatch at class I is the Bw6 epitope. This result indicates that the transplant from this donor to this recipient is contraindicated. *Note*: This is the epitope, not the antigen. An epitope is the specific part of the antigen that can be recognized by the immune system and may be bound by an associated antibody.

The Bw6 epitope, along with the Bw4 epitope, resides on various HLA-B antigens. These are highly exposed epitopes, found at amino acid positions 80–83, and are part of the peptide binding region of the HLA molecule. Bw4 is the major ligand for the KIR3DL1 receptor of NK cells. The Bw6 epitope has no known ligand function in humans; however, it can interact with KIR from macaques. This suggests that human KIR have only recently lost the ability to bind to Bw6.

As mentioned earlier, all HLA-B molecules have either Bw4 or Bw6, never both on the same molecule. HLA-B27 is an oddity in that most forms of HLA-B27 (such as the HLA-B27:01 allele) express Bw4, while one particular form (HLA-B27:08) expresses the Bw6 epitope. Since most typing for solid organ transplantation is only for the "antigenic" level, particular alleles can't usually be differentiated. It is apparent that this donor has the B*27:08 allele, expressing Bw6, along with one of the other alleles of B27, such as B*27:01, B*27:05, etc., all of which express Bw4. The recipient does not express the HLA-B*27:08 allele, as determined by the lack of Bw6 epitope.

Additionally, as shown by the patient's PRA, they are highly sensitized, probably to Bw6. As a matter of fact, antibodies to Bw6 alone will give a patient a cPRA of 85% as determined by the OPTN/UNOS cPRA calculator (https://optn.transplant.hrsa.gov/resources/allocation-calculators/cpra-calculator/). Patients lacking Bw6 on their own HLA can make strong antibody response to Bw6 from a donor and vice versa.

The crossmatch detects recipient antibody against donor cells so the DR11 antigen (Answer B), which is a self-antigen is incorrect. Answers C and D are incorrect because they are asking about donor response to antigens and this assay measures recipient response to antigens. Answer E could theoretically be possible; however, given the documented presence of donor specific antibody identified in the pretransplant antibody testing, this probability is very low. This is one of the reasons that histocompatibility laboratories are very scrupulous in their quality control and quality assurance testing.

5. A renal transplant candidate has been tested for the presence of HLA antibodies using single antigen bead array. With the Class I bead set, an antibody to HLA-A*02:01 was detected. The median fluorescent intensity (MFI) of the bead was 1100. Your laboratory uses a value of 1000 MFI as a positive cutoff. You have been asked to recommend a crossmatch method that might determine if this antibody is donor specific. Which of the following crossmatch methods would you recommend?

A. Solid phase crossmatch
B. Basic NIH CDC crossmatch
C. Amos modified CDC crossmatch
D. Antihuman globulin augmented crossmatch
E. Flow cytometric crossmatch

Concept: Crossmatch methods use donor lymphocytes incubated with candidate serum to detect the presence of donor specific antibodies. A number of different crossmatch methods have been developed and used clinically. One of the major distinctions of these methods is their sensitivity. Modifications such as addition of a wash step, and antihuman globulin increase the sensitivity of the crossmatch.

Answer: *E*—The most sensitive crossmatch method (i.e., the one that can detect the lowest level of donor specific antibody) is the flow cytometric crossmatch. In this case, the amount of antibody in question (A*02:01) is relatively low and may not even be detected by a flow crossmatch. The complement-dependent cytotoxic (CDC) methods (Answers B and C) would be less likely to detect the donor specificity of this serum. In fact, the original NIH CDC crossmatch method is the least sensitive of the methods listed.

The AMOS crossmatch (a wash step is added to the NIH method) has improved sensitivity but not as high as the flow crossmatch. Antihuman globulin (AHG) augmented crossmatch (Answer D) is an improvement on the AMOS crossmatch, but is still a complement-based cellular cytotoxic assay, all of which are not as sensitive as flow cytometric crossmatches. Studies have shown that even the sensitive AHG crossmatch is less sensitive than the flow crossmatch. Solid phase crossmatches (Answer A) rely upon HLA molecules affixed to the surface of a plate or to microbeads. When HLA antigens are removed from the surface of a cell and placed onto a solid surface, there is always a certain amount of degradation. This lessening of the potential antigens for antibody to bind to, as well as the lack of reliable vendors for this method, have caused it to not be considered as reliable as flow cytometry.

6. A 50-year-old patient is presenting with malaise, fatigue, and flu-like symptoms 5 years after receiving a heart transplant. A biopsy and ultrasound indicate that the patient is suffering from graft arteriosclerosis. Which of the following types of rejection are most associated with these findings and clinical history?

A. Acute antibody-mediated rejection
B. Acute cellular rejection
C. Chronic antibody-mediated rejection
D. Chronic cellular rejection
E. Quilty phenomenon

Concept: Transplanted organs can undergo both antibody mediated and cellular rejection. In the past, it was thought that only cellular mediated rejection occurred in the years after transplant; however, it has been recently recognized that antibody-mediated rejection can also occur many years post-transplant.

Transplant rejection can be broadly broken down into two different types, acute and chronic. Originally, the differences between the two were based on the amount of time from the transplant that the rejection occurred (acute—within a year of transplant; chronic—greater than 12 months posttransplant). Later, as better understanding of the mechanisms of rejection became apparent, it was determined that acute rejection was primarily driven by differences between the donor and recipient, as reacted to by differences in intact MHC found on the donor cells. Chronic rejection, by contrast, is driven by donor MHC, primarily, that has been processed by recipient antigen presenting cells. Chronic rejection often resembles the wound-healing process, where organ parenchymal tissue is replaced by scar tissue.

Rejection can also be described as antibody-mediated or cellular. Developed through exposure to the non-self-antigen, a transplant recipient can have specific antibody reacting with the donor tissue upon a transplant event (a secondary exposure) or through a de novo process if there has been no prior exposure. Cellular rejection occurs when allo-reactive killer T-cells, also called cytotoxic T lymphocytes (CTLs), have CD8 receptors that dock to the transplanted tissue's MHC class I molecules, which display the donor's self-peptides, such presentation of self-antigens helped maintain self-tolerance. Thereupon, the T-cell receptors (TCRs) of the killer T cells recognize their matching epitope, and trigger the target cell's programmed cell death by apoptosis.

Answer: C—This case most closely represents chronic antibody mediated rejection.

Graft arteriosclerosis, with or without atherosclerosis, is significantly associated with features of chronic antibody mediated rejection (Answer C), such as endothelial activation, complement component C4d deposition and donor-specific anti-HLA antibodies (DSA). In particular, anti-class II HLA DSAs appears to be more prevalent than anti-class I HLA DSAs. Importantly, unrecognized antibody-mediated rejection may be observed years before allograft failure. Consequently, it may be important to test posttransplant patients for the presence of DSA prior to the presence of symptoms.

The Quilty phenomenon (Answer E) is a subendocardial infiltration of lymphocytes found in a biopsy of a transplanted heart. For a long time, this was thought not to have a deleterious effect. More recently, it has become apparent that the presence of Quilty is associated with graft failure. Acute rejection, regardless of antibody-mediated (Answer A) and cellular (Answer B) happens too quickly for such an indolent process. Chronic cellular rejection (Answer D) is not found to happen in blood vessels, it is much more likely to occur in graft parenchyma.

7. A 32-year-old female is being assessed for a living donor renal transplant. HLA antibody testing demonstrated a high level of allosensitization (cPRA = 99%) likely due to multiple pregnancies. The decision to desensitize (remove/decrease the level of HLA antibody) was made in order to achieve a crossmatch negative transplant. The patient received multiple rounds of therapeutic plasma exchange (TPE) using 5% albumin as the replacement fluid and post-TPE intravenous immune globulin (IVIg). Antibody analysis with single antigen multiplex bead array documented a decline in the level of donor specific antibodies over several treatments (Fig. 16.1). Paradoxically, one donor specific HLA antibody demonstrated an increase in MFI value over this same time frame (Fig. 16.2).

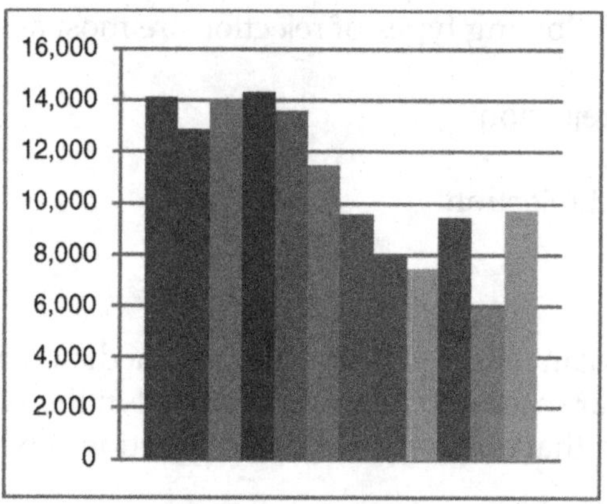

FIGURE 16.1 **HLA-DR15 reactivity with longitudinal sera.**

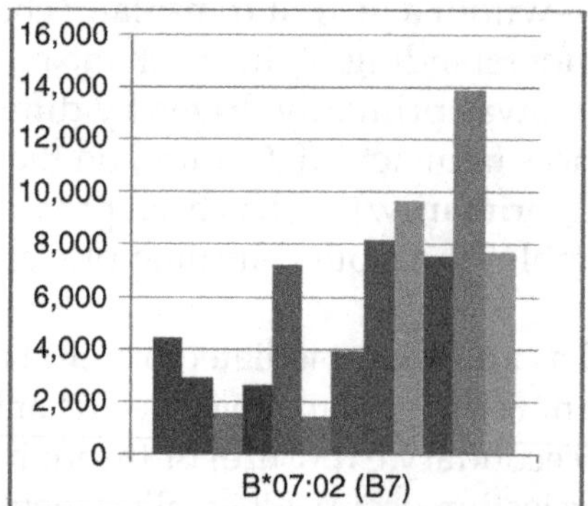

FIGURE 16.2 **HLA-B7 reactivity with longitudinal sera.**

To assess paradoxical finding the laboratory performed a single antigen multiplex bead array test with the serum diluted 1:10 prior to testing. This resulted in an increase in the MFI value of the HLA-B7 bead (Fig. 16.3).

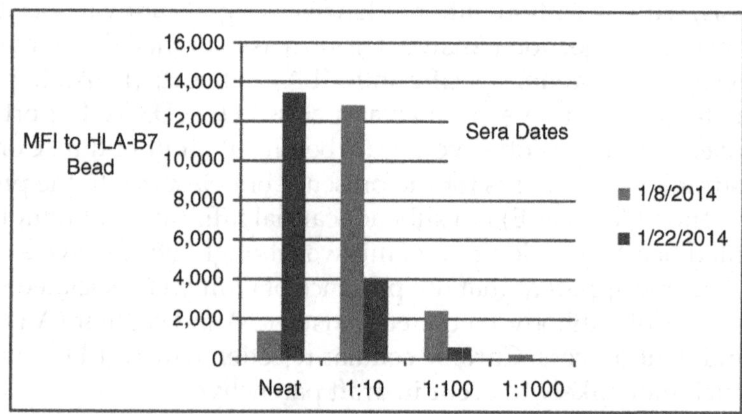

FIGURE 16.3 **HLA-B7 reactivity after serial dilution.**

What is the term that best defines this phenomenon?

A. Prozone effect
B. Plasma exchange interference
C. IVIg interference
D. Postzone effect
E. Ouchterlony interference

Concept: Single antigen bead analysis allows the identification of antibody specificities in a reliable and sensitivity fashion. However, the assay is subject to some interferences that can falsely reduce reactivity, one example being interference by residual complement proteins in the serum.

Answer: *A*—High levels of antibodies to HLA antigens can lead to a prozone phenomenon (or Hook effect), also known as antibody excess. In this instance, the level of antibody detected is falsely low. This phenomenon can be uncovered by testing the serum sample in a diluted fashion, as opposed to the usual neat (undiluted) analysis. This phenomenon may also be due to the effects of residual complement components in the serum being tested.

Postzone effect (Answer D) is where there is antigen excess. There is no such entity as Ouchterlony interference (Answer E). Ouchterlony double inmmunodiffusion is a laboratory technique that is used to quantify antibodies and antigens via immunodiffusion methods. Though the passive transfer of antibodies are possible in both TPE and IVIG (Answers B and C), they are not likely the cause of this specific antibody in this case. TPE with albumin replacement would not add any foreign antibodies into the patient's system. IVIg may be associated with hemolysis due to the passive transfer of anti-A and anti-B antibodies, but other antibodies are not commonly encountered.

8. Which of the following cross-ABO transplants (i.e. ABO-incompatible transplants) has been implemented into the national kidney allocation system in the United States, as of December 2014?
A. O to A
B. O to B
C. O to A_2
D. A_2 to B
E. A_2 to O

Concept: While it may not be always appreciated, the ABO system is indeed a histocompatibility antigen system. Transplantation across the ABO blood group barrier was long considered a contraindication for transplantation, but in an effort to increase donor pools; specific regimens for ABO-incompatible (ABOi) transplantation have been developed and are now widely used as an available treatment option.

Answer: *D*—Blood group B transplant candidates are at a historical disadvantage in finding suitable donors because of the relative scarcity of B donors. Changes to the kidney allocation system in 2001 allowed for more B candidates to be transplanted, the composition of the B waitlist was 16% which was above the transplant rate. In 2014, a change to the Kidney Allocation System was changed to allow for donors having A_2 or A_2B to be considered for donation to B recipients.

A_2 is more properly known as non-A_1. A1 and A2 are the two most commons subgroups of the A blood type. A2, encoded by the ABO*A201 gene, is almost identical to A1, encoded by the ABO*0101 gene, except for a substitution mutation at one site and a single deletion at another site. The result is a disruption of the stop codon and an A-transferase product with an extra 21 amino acid (AA) residue at the C-terminus. Consequently, RBCs bearing A2 have only a quarter of the A antigens per RBC.

The lower level of A blood group antigens of the surface of A_2 RBCs allows for B patients with a sufficiently low anti-A titer (usually less than 4) to receive an organ from, such as an A_2 donor. The results of the change in allocation has allowed for blood type B patients to have a better chance of receiving a transplant. All of the other choices are incorrect (Answers A, B, C, and E).

9. Which of the following must be expressed on the surface of a pre-T-cell in the thymic cortex in order to successfully undergo positive selection?
A. CD-25
B. β_2-microglobulin
C. Intact HLA molecules
D. T-cell receptor
E. CD-19

Concept: Up to 10% of a person's T-cells are capable of responding to a donor's different HLA molecules, even though the usual immune response only responds that strongly after an initial sensitization. The reason for this extreme reaction is because T-cells are selected in the thymus (after positive, then negative selection) so that their T-cell receptors can recognize self-HLA with foreign peptide and a donor organ can have foreign HLA with self-peptides. In essence, this reaction is a crossreaction of what a T-cell is educated to react to.

Answer: *D*—A pre-T-cell must have an intact (even potentially autoreactive) T-cell receptor in order to go through positive selection in the thymus. A pre-T-cell is formed in the bone marrow and travels to the thymus. In the thymus, these cells begin rearranging their T-cell receptor genes in an attempt to make a functional receptor. This receptor must be able to bind to self-HLA in a process called positive selection. The pre-T-cell then moves to another area of the thymus to undergo negative selection.

In negative selection, the T-cell receptor must not be able to bind too strongly to the self HLA. If it does, the cell is put on the pathway to apoptosis. Only after undergoing positive selection followed by negative selection can a pre-T-cell emerge from the thymus as a fully functional T-cell. CD25 (Answer A) is a marker of regulatory T-cells and CD19 (Answer E) is commonly used as a B-cell marker. Recognition of HLA antigens (Answer C) as self or non-self is an important function of our immune system, but does not play a role in positive selection in the thymus. β_2-microglobulin (Answer B) is a component of MHC Class I molecules that are important for development of CD8 T-cells, among many other functions. Increased β_2-microglobulin levels can be associated with lymphoma and multiple myeloma.

10. Some consider pregnancy as the most successful form of allogeneic transplant in existence. One of the reasons for success is that there is no HLA-A, -B, or -C expressed at the maternal/fetal interface. Which of the following HLA genes in involved in protecting fetal tissue against attack by maternal NK cells?
A. HLA-E
B. HLA-F
C. HLA-G
D. HLA-DQ
E. HLA-DP

Concept: Natural killer cells are a type of cytotoxic lymphocyte that recognizes a variety of activating and inhibitory receptors. NK cells can act in a manner that is reflective of both innate and adaptive immunity. NK cells have a multitude of receptors some of which can activate the NK cells and some of which inhibit NK cells. Most receptors perform either the stimulatory or the inhibitory function. HLA-G, which binds to the NK cell receptor KIR2DL4, can perform stimulatory functions or inhibitory function depending on where the receptor is engaged.

If KIR2DL4 is expressed on the NK-cell's surface, along with the FcRγ receptor, interaction with HLA-G causes the NK cell to become activated. If KIR2DL4 is found on the internal membrane of an endosome, without FcRγ, and binds to a solubilized HLA-G (free from the cell surface), the NK cell is inhibited. This appears to be one of the most important, although not exclusive, ways in which the female is prevented from immunologically rejecting a fetus.

Answer: *C*—HLA-G when solubilized and internalized within an endosome of a NK cell causes the NK cell to be inhibited when interacting with fetal tissue at the maternal/fetal interface. When bound to CD94/NKG2C, HLA-E (Answer A) inhibits NK cell activity. The function of HLA-F (Answer B) still remains unclear. It mainly resides intracellularly and rarely reaches the surface of the cell. HLA-DQ and DP (Answers D and E) function as cell surface receptors on antigen presenting cells.

11. A 29-year-old female on the renal transplant waitlist has had monthly serum samples submitted for cPRA determination for 2 years. All samples to date have been negative for antibodies to HLA Class I and II antigens with the use of a flow cytometric bead screening assay. On the most recent serum, however, a positive result was obtained with the flow cytometric screening assay. Because of this positive screening result, the sample was tested further using multiplex bead array (Luminex) for HLA Class I and II antibodies. This testing revealed multiple class I and II HLA antibodies corresponding to a cPRA value of 85%. What is your recommendation as to listing of unacceptable antigens for this patient?

A. List all antigens that are positive as unacceptable
B. Request a new serum to confirm the unexpected result
C. Request a clinical evaluation for sensitizing events
D. List only HLA Class I antibodies
E. Repeat the test to confirm the initial finding

Concept: Pretransplant antibody testing is performed to determine HLA antibody status. This testing is performed in order to detect the presence of HLA antibodies and determine the specificity, so that unacceptable antigens can be defined for candidates. An unacceptable antigen is classified as a donor antigen to which the candidate has specific antibody. Because of the detrimental effect of donor specific antibodies, donors with these corresponding antigens are avoided. In addition, this testing is performed longitudinally pretransplant in order to identify changes in sensitization status that may lead to changes in cPRA values over time. The frequency of testing is defined by the transplant center (e.g., monthly or quarterly testing).

Answer: C—Sensitizing events that influence HLA antibody production include pregnancy, transfusion, and transplantation. In this case, an unexpected positive antibody test is encountered after years of negative results. A request for information on sensitizing events in order to provide clinical correlation with the result is an important approach for interpretation of this sample and future samples and may be the preferred initial approach. If no sensitizing events are documented than retesting and/or requesting a new sample would be appropriate follow up strategies.

If no sensitizing events are uncovered, it would be reasonable to retest in order to confirm that there were no problems with the assays (Answer E). Alternatively, a request for a new sample might be reasonable to address the possibility of a sample mislabeling (Answer B). Simply listing the newly positive antigens as unacceptable (Answer A), as defined in the test, or listing only HLA Class I antibodies (Answer D) is questionable given the long history of negative results.

12. A 34-year-old female has been listed for her second kidney transplant. Her first transplant was lost to chronic allograft nephropathy. Her most recent cPRA determination is 54%. She received an offer for an immunologically compatible kidney from a deceased donor based upon her unacceptable antigen repertoire (i.e., no donor specific antibodies were identified in her most recent serum tested). Given the lack of donor specific antibody, the transplant surgeon requested a crossmatch with the most recent serum, which was collected 45 days ago. What should you recommend to the surgeon regarding this request?

A. Recommend the crossmatch be performed, as requested
B. Recommend that the crossmatch not be performed due to high risk of immunologic complications
C. Recommend that a crossmatch wait until a current serum can be collected
D. Recommend that the transplant proceed due to the lack of donor specific antibody
E. Recommend that a virtual crossmatch be performed

Concept: Pretransplant crossmatches are performed to confirm the absence of donor specific antibody prior to transplantation.

Answer: C—The crossmatch should wait for a current serum to be collected for use. In a sensitized patient, a crossmatch, physical or virtual is mandatory (Answer B). In order to ensure the most accurate and current assessment of donor reactivity a serum from the day of transplant should be used for crossmatching. This is most important in a candidate that has preexisting HLA antibodies.

In this case, one can't exclude the possibility of a sensitizing event (e.g., infection or transfusion) having occurred in the candidate that might elevate existing antibody levels or induce the formation of new, donor specific antibodies. Therefore, a crossmatch, physical or virtual (a prediction of the physical crossmatch based on recipient HLA antibody profile and donor HLA type) should be performed with a current specimen (most centers consider serum to be current at different timelines, the standard appears to be no greater than 30 days).

The development of donor specific antibodies would alter the risk in this transplant scenario, which is critical information that needs to be provided to the clinical transplant team, thus, doing the crossmatch with the most recent specimen (45 days old) is not the best option (Answer A). While a virtual crossmatch (Answer E) would be possible, most centers would consider it a risky proposition since the last serum tested is relatively old and, if a sensitizing event occurred in the interim, would not be an accurate picture of the patient's immune status. Likewise, (Answer D) is also relying on an "out-of-date" serum to make a determination is, in actuality, a virtual crossmatch and is also not optimal.

13. A 46-year-old male, with ESRD secondary to granulomatosis with polyangiitis (GPA), is on the renal transplant waitlist. He was recently placed on rituximab therapy for this condition. Two weeks after starting this therapy, a preliminary flow cytometric crossmatch with a potential living kidney donor was performed. The crossmatch results were T-cell negative and B cell positive. However, he has no history of any sensitizing events (e.g., transfusion, prior transplant). What is the most likely explanation for the positive B cell crossmatch?

A. The B cell crossmatch is nonspecific
B. The rituximab therapy caused the positive B cell result
C. The underlying disease, GPA, caused the positive B cell crossmatch
D. The patient neglected to report a blood transfusion
E. The flow crossmatch was incorrectly performed

Concept: The flow cytometric crossmatch is the most sensitive test to detect donor specific HLA antibody. However, the test does have limitations and is susceptible to interference. In particular, certain biologic therapies may cause false positive results.

Answer: *B*—The secondary antibody is binding to the rituximab, mimicking a positive crossmatch. Rituximab is a B-cell specific, chimeric monoclonal antibody. Because it includes a human Fc region, it can bind to donor B cells and be detected with the antihuman IgG-FITC antibody used in the crossmatch, resulting in a positive result. GPA (Answer C), while an autoimmune disease, is associated with autoantibodies to neutrophils and not lymphocytes, so this disease state in and of itself would not be expected to result in false positive donor specific crossmatch results. The patient's history is verified by his records (Answer D). Given the history of rituximab therapy, Answers A and E are less likely explanations. Donor specificity of the B cell crossmatch could be further explored by performing an autocrossmatch (recipient lymphocytes + recipient serum). All crossmatches are performed with positive and negative control sera, so the correct performance of the test is verified with each run, as a result the other choices (Answers A and E) are incorrect.

14. A patient has developed a strong antibody mediated rejection to a mismatch of a previous donor, where the only mismatch was in HLA-A2. A family member is well-matched to the patient except for having an HLA-B57. The crossmatch between the patient and the family member is positive for both T-cell and B-cells, and all controls were appropriate. What may be the cause of such an unexpected result?

A. The "well-matched" family member may not be a biological relative
B. Possible reactivity to untyped class II antigens
C. Inter-locus cross-reactivity of shared epitopes
D. Antibodies to the B-cell marker CD-19
E. A crossover event in the HLA-B locus has occurred

Concept: Antigens are usually defined as the biological targets of antibody binding. However, that explanation can be expanded. The antigen binding site on the immunoglobulin is called the idiotope and the binding site on the antigen is called an epitope. If epitope is shared amongst several antigens, then an antibody that binds to one epitope will usually bind to the same epitope on another antigen.

This brings to mind the concept of crossreactivity. HLA antigens share epitopes through the processes that generate polymorphism. Some new alleles are the result of point mutations arising in previously existing antigens. Many arise from genetic recombination or gene conversion. These processes extend the range of antigens. The polymorphisms that arise can cause changes in the conformation of the HLA, which can then affect which peptides which can bind in the peptide binding groove of the HLA molecule.

There are several examples of new alleles that arise from homologous recombination. HLA-A69 arose from having the 5′ end of HLA-A68 with the 3′ end of HLA-A2. As a result, A2, A68, and A69 share several epitopes. They can be said to belong to the same crossreactive epitope group or CREG. There are many CREGs, such as 1C, 2C, 4C, 5C, 6C, 7C, 8C, 10C, and 12C. Usually, these CREGs are found within the same locus. For example, the 2C CREG comprises many A antigens, such as A2, 23, 24, 68, 69, etc.

Many CREGs are further subdivided, if the particular epitope has been defined. The 28p epitope is a 4 amino acid sequence of TTKH found at amino acid residues 142–145. This epitope is found on HLA-A2, 68 and 69. Another epitope common in the 2C CREG is 17p. This epitope is unique in that it is found on HLA-A2 and HLA-B57 and B58. This epitope, being share across two different HLA loci, is referred to as an interlocus epitope and is the reason for interlocus crossreactivity.

Answer: C—The positive T and B cell crossmatch is most likely due to interlocus crossreactivity. Since the test controls are reacting appropriately, Answers B and D are incorrect. The biologic relatedness of the donor is irrelevant as far as epitope reactivity is concerned, as is the presence of a cross over event. Thus, the other choices (Answers A and E) are incorrect because the relevant factor is the presence of the shared epitope on the HLA molecule present.

15. A mixed-race patient (Caucasian father and African-American mother) and his family are HLA typed to see if one of the family members could be a potential bone marrow donor for the patient. The transplant program requires a 10 out of 10 match at the following HLA loci: HLA-A, -B, -C, -DRB1, and -DQB1. The HLA typings are listed as follows:

	HLA-A	HLA-B	HLA-C	HLA-DR	HLA-DQ
Patient	1, -	8, -	7, -	17, -	2,
Father	1, 3	7, 8	7, -	15, 17	2, 6
Mother	1, 30	8, 42,	7, 17	17, 18	2, 4
Sibling 1	3, 30	7, 42	7, 17	15, 18	4, 6
Sibling 2	1, 3	8, -	7, -	17, -	2, -

Based *solely* on the HLA typings shown above, which of the following is probably true?

A. The father of the patient is not the father of Sibling 2
B. A crossover event in the HLA-A locus has occurred in Sibling 2
C. Sibling 1 is probably not a biological sibling to the patient
D. Sibling 2 is probably not a biological sibling to the patient
E. A crossover event in the HLA-C locus has occurred in Sibling 1

Concept: This question brings up a major concept that personnel in the HLA laboratory have to deal with on a consistent basis, the haplotype. The term haplotype is a combination of the terms "haploid genotype" and is defined as the genes present on a particular chromosome, as inherited from each parent. Hence, there is in each chromosomal pair, a maternal haplotype and a paternal haplotype.

Based on Darwinian genetics, there are 4 haplotypes in a pair of parents. By convention, we label the 2 haplotypes of the father as A and B, while we label the mother's haplotypes as C and D. Also, by convention, we label the patients haplotypes as A and C (A from the father and C from the mother). Studying the typings in a sufficient number of family members, at minimum the patient and both parents, allows us to sort out the haplotypes. In the example above, we can sort the haplotypes as follows:

	A	B	C	D
HLA-A	1	3	1	30
HLA-B	8	7	8	42
HLA-C	7	7	7	17
HLA-DR	17	15	17	18
HLA-DQ	2	6	2	4

Reading the columns from top to bottom allows one to see the HLA alleles assigned to each haplotype. Based on this family study, we can see that the mother and father appear to share a similar haplotype (A1, B8, C7, DR17, DQ2). Consequently, the patient appears to be homozygous at all HLA loci. In all likelihood, the patient has two of the same allele, one from each parent. It is common to designate an allele that appears to be homozygous as a number with a dash, hence, HLA-A1,-.

It is important to keep in mind that certain haplotypes, given the way that chromosomes are inherited, tend to be found more common in certain racial groups. However, this is not always the case. Racial designations can be self-identified in an individual even though the HLA haplotypes from an individual's racially-identified ancestors are no longer present. However, in the HLA typings of this question, the two most common haplotypes found in Caucasians and African-Americans were used. It is interesting to note that the most common haplotype in Caucasians (HLA-A1, -B8, -C7, -DR17, -DQ2) is the second most common haplotype found in African Americans.

As can be seen in the typing of Sibling 2, there appears to be a haplotype that doesn't appear to be from the father. When an additional haplotype is present it is called "E" by convention. In this case, it appears that some recombination may have occurred. This allows us to consider another major concept that personnel in the HLA laboratory have to deal with on a consistent basis, the crossover event.

A crossover event is a manifestation of homologous recombination. Homologous recombination is a type of genetic recombination in which nucleotide sequences are exchanged between two similar or identical molecules of DNA. It is most widely used by cells to accurately repair harmful breaks that occur on both strands of DNA, known as double-strand breaks. In cells that divide through mitosis, homologous recombination repairs double-strand breaks in DNA caused by ionizing radiation or DNA-damaging chemicals. Left unrepaired, these double-strand breaks can cause large-scale rearrangement of chromosomes in somatic cells, which can in turn lead to cancer.

In addition to repairing DNA, this process helps produce genetic diversity when cells divide in meiosis to become specialized gamete. It does so by facilitating chromosomal crossover, in which regions of similar but not identical DNA are exchanged between homologous chromosomes.

Answer: *B*—In the scenario of this question, it appears that a crossover event occurred where the HLA A3 of the father's B haplotype was attached to the HLA-A locus on the chromosome carrying haplotype A. Based on a careful analysis of the HLA typings in the family, Answer A is unlikely due to the shared "A" haplotype in sibling 2 with the father. The remaining choices (Answers C, D, and E) are less likely due to the identified HLA alleles present.

16. The patient is a 40-year-old Caucasian in need of a renal transplant. He has four siblings, all of whom agreed to be assessed as potential donors. As part of the workup, the patient and all four siblings were HLA typed with the following results:

Patient: HLA-A1, 24; B57, 62; DR7, 17
Sibling 1: HLA-A1, 2; B57, 44; DR7, 4

Sibling 2: HLA-A11, 2; B7, 44; DR15, 4
Sibling 3: HLA-A11, 2; B7, 44; DR15, 4
Sibling 4: HLA-A1, 24; B57, 62; DR7, 17

Based on the HLA types provided, sibling 4 was the preferred donor. What is the correct genotype for this patient and donor?

A. A1; B57; DR7 and A24; B62; DR17
B. A1; B62; DR7 and A24; B57; DR17
C. A1; B57; DR17 and A24; B62; DR7
D. A1; B62; DR17 and A24; B57; DR7
E. A1; B44; DR7 and A24; B57; DR17

Concept: HLA loci are linked on the short arm of chromosome 6 and inherited as haplotypes. In the absence of high resolution (allele level) HLA typing, determination of haplotypes can be used to determine HLA identity. One can infer haplotypes by assessing inheritance of antigens from parents. Alternatively, if sufficient siblings with a representation of 4 haplotypes are available, one can infer haplotypes (and HLA identity) via examination of siblings HLA types. In the absence of four unambiguous haplotypes in a group of siblings, one would need to perform high resolution HLA typing to determine HLA identity due the presence of multiple common alleles within an allele group (antigen equivalent) at the various HLA loci.

Answer: *A*—Using a simple matching strategy, the genotypes of the patient and donor are A1; B57; DR7 and A24; B62; DR17. All of the other choices (Answers B, C, D, and E) have the incorrect combination and/or location of the antigens.

17. A 53-year-old male with end stage renal disease is being activated on the UNOS waitlist. His HLA type (reported with serologic equivalents of a molecular HLA typing) is as follows:
HLA-A1, 33; B8 (Bw6), 78 (Bw6); Cw7, -; DR1, 17; DR53, DQ2, 5
His most recent serum submitted for antibody analysis demonstrated a cPRA of 100% with the following specificities identified using a single antigen bead multiplex array:

A3, 11, 24, 25, 26, 29, 32, 34, 43, 66, 68, 69, 74
B7, 13, 27, 42 48, 55, 56, 57, 58, 60, 61, 63, 67, 81
DR103, 4, 7, 8, 9, 11, 12, 13, 16, 51
DQ4, 6, 7 8, 9
DQA1*03
DP2, 3, 4, 6, 9, 11, 14, 17, 18, 20, 23, 28

The patient was crossmatched with a potential deceased donor using this same serum. The deceased donor's HLA phenotype was as follows:
HLA-A1, -; B8, -; Cw7, -; DR17, 10; DQ2, 5
However, the flow cytometric crossmatch results were T-cell negative and B-cell positive. Given that the patient had no apparent donor specific antibody (0 antigen mismatch donor), what is the first step would you recommend to determine the cause of this unexpected positive B-cell crossmatch?

A. Perform DQA and DPB typing on the donor
B. Retype the donor to determine if they were mistyped
C. Repeat the crossmatch
D. Repeat the HLA antibody analysis on the recipient
E. Report the result as a false positive B-cell crossmatch

Concept: The calculated PRA (cPRA) is determined to assess the likelihood of finding a crossmatch compatible donor. In this case, a cPRA of 100% suggests a small chance of finding such a donor. However, since patients with a 100% cPRA are eligible for national sharing, the chance does increase. Currently, cPRA values are based upon unacceptable antigens at the HLA-A, B, C, DRB1/3/4/5 and DQB1 loci. As such, as of this writing, one cannot list unacceptable for candidates with DQA or DPB1 antibodies. While it is likely that the HLA-DQA1 type of the donor is the same as the patient due to linkage disequilibrium, the same cannot be said for HLA-DPB1. Studies have documented a discordance between unrelated donors and recipients of up to 80% for DPB1 in pairs that are 10 of 10 allele matched at HLA-A, B, C, DRB1, and DQB1.

Answer: *A*—The allocation system does not currently take into account all antigens to which a recipient may be sensitized, therefore, performing DQA and DPB typing on the donor may explain the positive crossmatch. Though the other choices (Answers B, C, D, and E) may be eventually completed, they would not be the first step.

18. The patient, a 51-year-old male with end stage renal disease secondary to diabetes mellitus type II received a living donor kidney transplant in 1996. The transplant failed after 15 years, with evidence of transplant glomerulopathy. Donor and recipient HLA types were:

Recipient: HLA-A2, 29, B44 (Bw4), 51(Bw4); Cw5, 16; DR7, 13; DR52, 53; DQ2, 6
Donor: HLA-A28, 31; B*51, 60; DR4, 7; DR53; DQ2, 8 (C locus not typed)

The patient was subsequently listed for a deceased donor kidney transplant. HLA antibody analysis at the time of listing demonstrated HLA Class I and II antibodies with flow cytometric screening beads. Single antigen multiplex bead array analysis demonstrated the following specificities:

HLA-A3, 11, 25, 30, 31, 32, 33, 34 36, 68, 69, 74
HLA-B7, 8, 13, 27, 38, 39, 41, 42, 45, 47, 48, 50, 55, 56, 60, 61, 63, 67, 73, 81, 82
HLA-C2, 6, 9, 10, 15, 17
HLA-DR1, 4, 8, 10, 11, 14, 15, 51
HLA-DQ2, 4, 6, 7, 8, 9 (possible DQA reactivity noted)

A deceased donor kidney offer was received for this patient. Flow cytometric crossmatching was T-cell negative and B-cell negative with donor cells that HLA typed as follows:

HLA-A1, 29; B8, 44; Cw7, 16; DR13, 17; DR52; DQ2, 6.

The patient's posttransplant course was complicated by biopsy proven rejection 2 months posttransplant. No C4d staining was noted. HLA antibody analysis demonstrated the presence of the following antibodies:

HLA-A3, 11, 32, 33, 34 36, 68, 69
HLA-B7, 13, 27 38, 39, 41, 42, 47, 48, 55, 60, 61, 67, 73, 81, 82
HLA-Cw2, 6, 9, 10, 17
HLA-DR1, 4, 13
HLA-DQ2, 4, 6, 7, 8, 9 (possible DQA reactivity noted)

The nephrologist contacted the HLA laboratory to question the presence of a DQ2 antibody (MFI = 6707) and questioned why a recipient that HLA typed as DQ2 had developed a donor specific DQ2 antibody. How would you answer his/her question?

A. The posttransplant antibody test was falsely positive for DQ2 antibody
B. The flow crossmatch with the deceased donor was falsely negative (i.e., should have been positive with the DQ2 antibody)
C. The patient has an allele specific antibody to DQ2
D. The DQ2 antibody should be ignored as it is impossible for a person to make a self-reactive antibody that is relevant to transplant rejection
E. HLA-DQ2 has been "split" into other antigens

Concept: The HLA system is the most polymorphic genetic system known. There are thousands of allelic variants at the various classical transplant loci (HLA-A, B, C, DRB1/3/4/5, DQB1).

Answer: *C*—The patient has an antibody to a specific allele of HLA-DQ2. It is not uncommon for a donor and recipient with the same low resolution HLA type (i.e., DQ2) to have differing allelic variants that could be immunogenic and lead to HLA antibody formation (Answer A). In this case, neither false negativity of the test, nor an antigen split is likely (Answers B and E) and higher resolution HLA typing is necessary to explore the findings in this case. High resolution HLA typing demonstrated that the recipient possessed the HLA-DQB1*02:02 allele, as did the living donor (first transplant).

The deceased donor, however, possessed the HLA-DB1*02:01 allele. While HLA-DQ2 antibody was detected prior to the deceased donor transplant, it was at a very low level, insufficient to cause a positive crossmatch. In this case, a moderate level of HLA DSA was detected to the mismatched HLA-DQB1*02:01 allele, suggesting an allogeneic immune response to the mismatched allele. While in theory it is possible for a person to make a self-reactive antibody (Answer D), is not appropriate in this case as the donor and recipient do not possess the same DQ2 allele, reinforcing that the level (resolution) of HLA typing typically performed for solid organ transplantation may not always be sufficient to resolve antibody reactivity patterns seen in practice.

19. A renal transplant candidate is being listed for deceased donor transplant. HLA antibody testing indicates allosensitization. Among the antibodies detected is an A*11 antibody. Examination of the single antigen bead assay results demonstrates that 2 alleles of A*11 are represented in the panel, A*11:01 and A*11:02. The candidate has exclusive reactivity to the A*11:02 bead and no reactivity to the A*11:01 bead. You make a recommendation that A*11 not be listed as an unacceptable antigen in the UNET system. What is the rationale for this decision?

A. The A*11:02 reactivity is falsely negative
B. The A*11:02 allele is very infrequent in the population and thus less likely to be present in a potential donor
C. The A11 antigen is rare in the donor population and thus you are not concerned
D. Reactivity to A11 is not associated with adverse events such as rejection episodes
E. The A*11:01 reactivity is falsely positive

Concept: Listing an antigen (allele) as unacceptable eliminates any donor with that antigen (allele) from donation to the candidate in question. In order to maximize the opportunity for transplant, one may not want to list an antigen as unacceptable if it is rare (or uncommon) in the population and the candidate antibody is specific to that rare (or less common) allele.

Answer: *B*—Since the HLA-A*11:01 is much more common, listing A11 as unacceptable based on HLA antibody to the less common HLA-A*11:02 variant will only serve to deny the patient from consideration with a more common, safe antigen. However, the lab personnel should make sure that the transplant team understands that the crossmatch must be performed prior to transplant. In this case, the A*11:02 allele is much less common than A*11:01. Therefore, not listing A*11 as unacceptable is less likely to result in the offer of a crossmatch incompatible donor kidney. This possibility is additionally controlled for by the performance of the final crossmatch to document lack of donor specific HLA antibody.

Since multiple sera are typically tested for transplant candidates due to wait time that must be accrued for organ allocation, there is typically sufficient replicate testing performed to determine the "true" presence of an alloantibody (Answers A and E). A11 (Answer C) is not a rare allele group and is no more or less likely to serve as a target of an allogeneic response compared to most other HLA antigens (Answer D).

20. A kidney patient at a dialysis center received a one unit transfusion of red blood cells due to a hemoglobin level of 6.7 g/dL over a period of 2 h. Approximately 4 h later, the patient appears in the emergency department, apparently suffering from acute respiratory distress. Her vital signs and laboratory results are as follows:

Heart rate: 77 BPM
Blood pressure: 80/60
Respiratory rate: 32 RPM
O_2Sat: 78%
Temperature: 100.2°F
Hemoglobin: 8.2 g/dL
WBC: 5,230/μL
Brain natriuretic peptide (BNP): 75 pg/mL (reference range: < 100 pg/mL)
Chest X-ray (CXR): Complete white out of lungs bilaterally (no CXR is available for comparison)

The patient's vital signs continue to deteriorate and eventually she requires intubation and admission to the ICU. She recovers fully in approximately 48 h. Which of the following diagnoses is most likely?
A. Transfusion-associated graft-versus-host disease
B. Transfusion-related acute lung injury
C. Febrile nonhemolytic transfusion reaction
D. Anaphylactic reaction
E. Transfusion-associated circulatory overload

Concept: Antibodies to leukocyte antigens do not only occur in potential transplant recipients. They can occur in donors of blood products, potentially at very high titers, and, if transfused, can actually be passively transferred to a recipient of blood products that contain serum. There are three transfusion reactions in which HLA or antibodies to HLA play a significant role: transfusion-related acute lung injury, transfusion associated graft-versus-host disease, and febrile nonhemolytic transfusion reactions.

Answer: *B*—Transfusion-related acute lung injury, better known as TRALI, is a leading cause of transfusion related mortality. Called by numerous terms throughout medical history, consensus on a definition of TRALI did not occur until 2004. TRALI is caused by antileukocyte antibodies from the donor cross reacting with white cells in the patient. TRALI can occur in reverse where the antibodies are in the patient, but less likely and less common, especially with leukoreduction. Antibodies are either anti-HLA or antigranulocyte. TRALI is characterized by sequestration and neutrophil priming in the lungs which leads to pulmonary capillary leakage. TRALI usually occurs within 6 h of a transfusion and is not temporally related to a competing etiology of acute lung injury.

Transfusion associated circulatory overload (TACO) (Answer E) is the main differential diagnosis linked to TRALI. TACO is often caused by transfusion of a large amount of blood products in a short time period and is often seen more commonly in elderly patients with decreased cardiac function. Signs and symptoms of TACO include increased blood pressure, increased heart rate, and respiratory distress. The BNP level is increased in TACO due to the increased intravascular fluid overload. BNP levels are not increased in TRALI because the fluid overload is restricted to the lungs.

Transfusion associated graft-versus-host (TA-GvHD) (Answer A) occurs when viable lymphocytes in the transfused unit react to the recipient's HLA. TA-GvHD, usually occurs in immunosuppressed patients, typically becomes evident 8–10 days posttransfusion and is almost uniformly fatal. Irradiating blood units with 25 Gray of radiation has almost completely eliminated TA-GvHD where it is performed. Leukoreduction is another method that can reduce or eliminate TA-GvHD.

Febrile nonhemolytic transfusion reactions (Answer C) are some of the most common reactions in transfusions. Characterized by chills, rise in body temperature of 1 or 2°C within 30–60 min, patients also can have rigor, flushing, tachycardia, nausea, and vomiting. While unpleasant, this condition can be treated easily with antipyretics, like acetaminophen. Antibodies in the recipient are reacting to leukocytes or platelets in the blood product. Antibody/antigen complexes may directly activate WBCs to produce pyrogenic cytokines to cause fever. The antigens that are most reacted to are HLA, although antibodies to human neutrophil and human platelet antigens are also implicated. In fact, study of these transfusion reactions in the late 1950s and early 1960s led J.J. van Rood in the Netherlands and Rose Payne in the United States to the discovery of the antigen system that eventually became HLA.

Though an anaphylactic reaction (Answer D) can cause hypotension, fever, and changes in the CXR are uncommon.

21. A patient presents to an oncologist for treatment of multiple myeloma and possible bone marrow transplant. One of the drugs discussed with the patient is milatuzumab, a humanized monoclonal antibody against CD74. CD74 is also known as "invariant chain." Invariant chain is important in the assembly of which of the following HLA molecules?
A. HLA-A
B. HLA-B

C. HLA-C
D. HLA-DR
E. HLA-DM

Concept: Chemotherapy agents and monoclonal antibodies often target cellular functions in order to cause or prevent cell death.

Answer: *D*—Invariant chain is a protein that acts as a chaperone to facilitate the passage of HLA-DR from the endoplasmic reticulum to the endosome where it will receive its peptide. This chaperone function prevents peptide from prematurely filling the peptide groove. It is often found on the surface of hematological tumors, while not being present in normal tissue. Invariant chain expressed on the cell surface interferes with Fas-mediated apoptosis, which allows such cells escape a possible signal to undergo programmed cell death.

Invariant chain or CD74 does not interact with class I HLA (HLA-A, -B, and -C) (Answers A, B, and C), as class I derives its peptides from intracellular sources. HLA-DM (Answer E) does not express peptides on the cell surface, but does interact with invariant chain to facilitate the placement of peptide into the peptide binding groove of HLA-DR, -DQ-, and -DP.

22. A 43-year-old male needs HPSC transplantation for acute myeloid leukemia. His HLA type is A*30:01, A*74:01; B*07:06, B*35:01; C*04:01; C*15:05; DRB1*11:02, DRB1*15:02. There are five potential eligible allogeneic unrelated donors for this patient with the following HLA types:

Donor	HLA type
1	A*30:01, A*30:02; B*07:06, B*35:01; C*04:01; C*15:05; DRB1*11:02, DRB1*15:02
2	A*30:01, A*74:01; B*07:02, B*35:01; C*04:01; C*15:05; DRB1*11:02, DRB1*15:02
3	A*30:01, A*30:02; B*07:02, B*35:01; C*04:01; C*15:05; DRB1*11:02, DRB1*15:02
4	A*30:01, A*30:02; B*07:02, B*35:01; C*03:02; C*15:05; DRB1*11:02, DRB1*15:02
5	A*30:01, A*74:01; B*07:02, B*35:01; C*04:01; C*15:05; DRB1*13:02, DRB1*15:01

Based on the HLA typing alone, which of the above five donors is best for the patient?
A. Donor 1
B. Donor 2
C. Donor 3
D. Donor 4
E. Donor 5

Concept: HLA is part of the immune regulation system. In transplantation, the difference in HLA types between patient and donor may provoke the immune system and lead to graft rejection and failure. In HPC transplantation, the closer the match between the patient and the donor, the higher the chance of transplantation will be successful. High resolution HLA matching for HLA-A, -B, -C, and DRB1 is associated with best survival. However, many patients may not find a perfect match; thus, the transplant physician must decide on the donors with the closest match possible.

In Asian or African-American patients, it may be difficult to find an HLA matched donor through the registry or among the family members. Thus, after weighing the risks versus the benefits, transplant physician may decide to use the HPC source from a haploidentical donor (i.e., a half-matched related donor, either from the patient's parent, sibling, or child). In the past, in haploidentical HPC transplantation, host and donor T-cells reactive to HLA antigens may result fatal graft rejection and/or severe graft-versus-host-disease (GvHD). Recent advances in immunosuppressive therapy (such as high dose posttransplantation cyclophosphamide) and/or reduced intensity conditioning regimen has improved the efficacy and safety of haploidentical HPC transplantation.

Answer: *B*—None of the above five donors has complete HLA match with the patient at HLA- A, -B, -C, and -DRB1 loci. Thus, the one with the closest match (donor #2) should be chosen. Studies have shown that a single mismatch at -B or -C locus is better tolerated than a mismatch at -A or DRB1 locus. Each additional mismatch is associated with ~9%–10% decrease in absolute survival. Further, a single HLA-DQB1 mismatches appear to be better tolerated compared to HLA-A, -B, or -C (increased risk of acute GVHD) and HLA-A, -C, or -DRB1 (transplant related mortality).

Based on the given HLA phenotype, donor 1 has 1 mismatch at A locus, donor 2 has 1 mismatch at B locus, donor 3 has 2 mismatches at both -A and -B loci, donor 4 has 3 mismatches at -A,-B, and -C loci, and donor 5 has 2 mismatches at -B and -DRB1 loci. Given that more mismatches associated with worse outcome, donors 1 and 2 are better than donors 3, 4, and 5 (Answers C, D, and E). Also, given that a single mismatch at -B or -C locus is better tolerated than a mismatch at -A or DRB1 locus, donor 2 is a better donor than donor 1 (Answer A).

23. Which of the following HLA alleles is tested to see if abacavir is contraindicated for the treatment of HIV infection in a newly-diagnosed patient?
A. B*57:01
B. B*15:02
C. B*58:01
D. DRB1*01:01
E. DQB1*02:01

Concept: Certain HLA alleles are known to be involved in drug hypersensitivity. Screening for HLA alleles implicated in adverse drug reactions are an increasing part of testing performed in HLA laboratories.

Answer: *A*—HLA-B*57:01 is routinely tested for in newly positive HIV-infected individual to ensure that Stevens-Johnson Syndrome (SJS)/Toxic Epidermal Necrolysis (TEN) does not occur in recipients of abacavir. It is well known that the anti-HIV drug abacavir can interact negatively with HLA-B*57:01 to the point that testing for the presence of the allele before prescribing abacavir, as part of antiretroviral therapy, is recommended by the Department of Health and Human Services Panel of Antiretroviral Guidelines for Adults and Adolescents.

Other HLA alleles known to be involved in drug hypersensitivity syndromes are HLA-B*15:02 (Answer B) in carbamazepine induced SJS/TEN, HLA-B*58:01 (Answer C) in allopurinol induced SJS/TEN. Amoxicillin-induced drug-induced liver injury (DILI) is associated with HLA-DRB1*15:01. All routinely tested HLA loci (HLA-A, -B, -C, -DR, DQ, and -DP) (Answers D and E) have been associated in some way with adverse drug reactions.

24. A 2-year-old child presents to a pediatric gastroenterologist with diarrhea, malnutrition, and a general failure to thrive. The gastroenterologist suspects celiac disease (CD) and orders an HLA typing in order to determine if the patient has the risk alleles for CD. Which of the following HLA genes are associated with CD?
A. HLA-DRB1
B. HLA-B27
C. HLA-DQA1
D. HLA-DR3
E. HLA-DQ2

Concept: Celiac disease is an increasingly common autoimmune disease of the small intestine that occurs in susceptible persons by the ingestion of gluten proteins from cereal grains. Gluten, a composite of glutenin and gliadin, is abundant in western diet and is well-tolerated by most individuals. Glutenin and gliadin have very high proline content. Highly proline-rich areas of such proteins are resistant to proteolytic degradation by proteases in the human intestine; but, are good substrates for transglutaminase, which can impart negative charges to proline-rich gluten peptides.

Gluten peptides, modified by transglutaminases, can bind especially effectively to HLA-DQ, especially HLA-DQ2.5, -DQ2.2, and DQ8. HLA-DQ-2.5 and alternative designation for the heterodimer consisting of the products of the HLA-DQA1*05:01 combined with HLA-DQB1*02:01. HLA-DQ2.2 consists of the combination of HLA-DQA1*02:01 with HLA-DQB1*02:02, while DQ8 consists of the combination of HLA-DQA1*03 with DQB1*03:02. Individuals homozygous for HLA-DQ2.5 are at the highest risk for CD compared to heterozygous individuals. The American College of Gastroenterology has issued guidelines which recommend HLA-DQ2/DQ8 genotyping testing be used to effectively rule out the disease in selected clinical situations.

Answer: *E*—HLA-DQ2 and DQ8 genes are most associated with celiac disease. The incorrect answer choices have other common HLA associations as follows: Sjögren's = HLA-DRB1 (Answer A); ankylosing spondylitis = HLA-B27 (Answer B); multiple sclerosis = DQA1 (Answer C); and myasthenia gravis (MG) = DR3 (Answer D), which has a positive association with early onset MG, but a negative association with late onset MG.

25. A patient of Turkish descent presents to a physician with painful oral mucous membrane ulcerations and tender nodules on both shins. The physician suspects Behçet's disease. A typing for a particular HLA-B antigen is ordered. Which of the following HLA typings has been associated with Behçet's disease?

A. HLA-B35
B. HLA-B51
C. HLA-B53
D. HLA-B57
E. HLA-B61

Concept: The pathogenic role of HLA-B51 in Behçet's disease has yet to be fully determined. However, the inhibitory effect of HLA-B51 binding to killer Ig-like receptor DL1 (KIR3DL1) on NK cells appears to be most important. This interaction is the normal default response for NK cells, where as long as cells express normal levels of HLA-B on their surface, they are protected from cytotoxicity. A reduction in HLA class I can be induced by tumor or viruses, leading to recognition and killing. Alternatively, increased MICA expression via an activating signal on binding to NKG2D can overcome the inhibitory effect of HLA class I.

Answer: *B*—Behçet's disease is associated with HLA-B51. Other choices (Answers A, C, D, and E) are wrong.

Suggested Reading

[1] J.P. Vivian, R.C. Duncan, R. Berry, et al. Killer immunoglobulin receptor 3DL1-mediated recognition of human leukocyte antigen B, Nature 479 (7373) (2011) 401–405.

[2] M. Maiers, L. Gragert, W. Klitz, High-resolution HLA alleles and haplotypes in the United States population, Hum. Immunol. 69 (2) (2008) 141.

[3] C.J. Brown, C.V. Navarrete, Clinical relevance of the HLA system in blood transfusion, Vox Sang. 101 (2011) 93–105.

[4] N. Heddle, Pathophysiology of febrile nonhemolytic transfusion reactions, Curr. Opin. Hematol. 6 (6) (1999) 420.

[5] R. Pavlos, S. Mallal, E. Phillips, HLA and pharmacogenetics of drug hypersensitivity, Pharmacogenomics 13 (11) (2012) 1285–1306.

[6] Guidelines for the Use of Antiretroviral Agents in HIV-1-Infected Adults and Adolescents. Panel's Recommendations for HLA-B*5701 Screening. Available from: https://aidsinfo.nih.gov/guidelines/html/1/adult-and-adolescent-arv-guidelines/7/hla-b--5701-screening.

[7] E. Vivier, D. Raulet, A. Moretta, et al. Innate or adaptive immunity? The example of natural killer cells, Science 331 (6013) (2011) 44–49 2011.

[8] S. Rajagopalan, E.O. Long, KIR2DL4 (CD158d): an activation receptor for HLA-G, Front. Immunol. 3 (2012) 258.

[9] R. Carapito, S. Bahram. Genetics, genomics, and evolutionary biology of NKG2D ligands. Immunol. Rev. 267 (1) (2015) 88–116.

[10] N. Goebl, C. Babbey, A. Datta-Mannan, D. Witcher, V. Wroblewski, K.W. Dunn, Neonatal Fc receptor mediates internalization of Fc in transfected human endothelial cells, Mol. Biol. Cell. 19 (2008) 5490–5505 2008.

[11] S. Parkkila, A.K. Parkkila, A. Waheed, R.S. Britton, X.Y. Zhou, R.E. Fleming, S. Tomatsu, B.R. Bacon, W.S. Sly, Cell surface expression of HFE protein in epithelial cells, macrophages, and monocytes, Haematologica 85 (2000) 340–345.
[12] W.J.H. Griffiths, A.L. Kelly, S.J. Smith, T.M. Cox, Localization of iron transport and regulatory proteins in human cells, QJM 93 (2000) 575–587.
[13] S. Porcelli, M. Brenner, J. Greenstein, J. Terhorst, S. Balk, P. Bleicher, Recognition of cluster of differentiation 1 antigens by human CD4 − CD8− cytolytic T lymphocytes, Nature 341 (1989) 447–450.
[14] S. Markus, S.M. Behar, The role of group 1 and group 2 CD1-restricted T cells in microbial immunity, Microbes Infect. 7 (2005) (2005) 544–551.
[15] J. Stamnaes, L.M. Sollid, Celiac disease: autoimmunity in response to food antigen, Semin. Immunol. 27 (2015) 343–352.
[16] A. Rubio-Tapia, I. Hill, C. Kelly, A. Calderwood, J. Murray, Diagnosis and management of celiac disease, Am. J. Gastroenterol. 108 (2013) 656–676.
[17] R. Roches, P. Cresswell, Invariant chain association with HLA-DR molecules inhibits immunogenic peptide binding, Nature 345 (1990) 615–618.
[18] Z. Berkova, S. Wang, X. Ao, et al. CD74 interferes with the expression of Fas receptor on the surface of lymphoma cells, J. Exp. Clin. Cancer Res. 33 (1) (2014) 80.
[19] A. Loupy, C. Toquet, P. Rouvier, et al. Late failing heart allografts: pathology of cardiac allograft vasculopathy and association with antibody-mediated rejection. Am. J. Transplant. 16 (2016) 111–120.
[20] N.E. Hiemann, C. Knosalla, et al. Quilty indicates increased risk for microvasculopathy and poor survival after heart transplantation, J. Heart Lung Transplant. 27 (2007) 289–296.
[21] G.R. Wallace, HLA-B*51 the primary risk in Behçet disease, PNAS 111 (2014) 8706–8707.
[22] M.S. Leffell, The calculated panel reactive antibody policy: an advancement improving organ allocation, Curr. Opin. Organ Transplant. 16 (4) (2011) 404–409.
[23] J.E. Locke, et al. Proinflammatory events are associated with significant increases in breadth and strength of HLA-specific antibody, Am. J. Transplant. 9 (9) (2009) 2136–2139.
[24] L. Rees, J.J. Kim, HLA sensitisation: can it be prevented?, Pediatr. Nephrol. 30 (4) (2015) 577–587.
[25] H.M. Gebel, R.A. Bray, P. Nickerson, Pre-transplant assessment of donor-reactive, HLA-specific antibodies in renal transplantation: contraindication vs. risk, Am. J. Transplant. 3 (12) (2003) 1488–1500.
[26] R.A. Bray, et al. Clinical cytometry and progress in HLA antibody detection, Methods Cell. Biol. 103 (2011) 285–310.
[27] A. Picascia, T. Infante, C. Napoli, Luminex and antibody detection in kidney transplantation, Clin. Exp. Nephrol. 16 (3) (2012) 373–381.
[28] B.D. Tait, et al. Review article: Luminex technology for HLA antibody detection in organ transplantation, Nephrology (Carlton) 14 (2) (2009) 247–254.
[29] K. Fleischhauer, et al. Effect of T-cell-epitope matching at HLA-DPB1 in recipients of unrelated-donor haemopoietic-cell transplantation: a retrospective study, Lancet Oncol. 13 (4) (2012) 366–374.
[30] A.A. Zachary, et al. HLA antibody detection and characterization by solid phase immunoassays: methods and pitfalls, Methods Mol. Biol. 882 (2012) 289–308.
[31] E. Schwaiger, et al. Complement component C3 activation: the leading cause of the prozone phenomenon affecting HLA antibody detection on single-antigen beads, Transplantation 97 (12) (2014) 1279–1285.
[32] G.A. Böhmig, A.M. Farkas, F. Eskandary, T. Wekerle, Strategies to overcome the ABO barrier in kidney transplantation, Nat. Rev. Nephrol. 11 (2015) 732–747.
[33] A. Seltsam, M. Hallensleben, A. Kollmann, R. Blasczyk. The nature of diversity and diversification at the ABO locus. Blood 102 (8) (2003) 3035–3042.
[34] http://www.owenfoundation.com/Health_Science/Blood_Group_A_Subtypes.html.
[35] C.F. Bryan, W.S. Cherikh, D.A. Sesok-Pizzini, A2 /A2 B to B renal transplantation: past, present, and future directions, Am. J. Transplant. 16 (1) (2016) 11–20.

CHAPTER

17

Cellular Therapy

Ronit Reich-Slotky, Yen-Michael S. Hsu*, Joseph Schwartz***

*New York Presbyterian Hospital, New York, NY, United States; **New York-Presbyterian Hospital—Columbia University, New York, NY, United States

Cellular therapy is quickly becoming an integral part of transfusion medicine practice. Therefore, transfusion medicine professionals are required to be familiar with the practice of manufacturing therapeutic cellular products for clinical use. This chapter is designed with practical questions allowing for comprehension of the fundamental knowledge and important clinical issues in cellular therapy. The topics covered by this chapter include evaluating the manufacturing of various hematopoietic progenitor cells (HPCs) products, reviewing donor eligibility and suitability for HPCs collections, and understanding the unique challenges of HPC mobilization/collection. The chapter also provides relevant clinical scenarios, such as specific transfusion support for patients receiving HPC transplantation with ABO-incompatible stem cell graft. Last, the chapter highlights the relevant FDA regulations that are critical in the field of cellular therapy.

1. Human cells, tissue, and cellular and tissue-based products (HCT/P) are regulated solely under section 361 of the Public Health Service Act and Chapter 21 of Code of Federal Regulations (CFR) part 1271, if they meet certain criteria. Which one of the following is one of those criteria:
A. The HCT/P is more than minimally manipulated
B. The HCT/P is intended for nonhomologous use
C. The manufacture of the HCT/P involves the combination of the cells or tissues with another article (except for water, crystalloids, or a sterilizing, persevering, or storage agent)
D. The HCT/P does not have a systemic effect and is not dependent upon the metabolic activity of living cells for its primary function
E. The HCT/P has a systemic effect or is dependent upon the metabolic activity of living cells for its primary function and is intended for allogeneic use from unrelated donor

Concept: The FDA published a set of three rules in order to implement its proposed approach to the regulation of HCT/Ps. These rules (listed below) established a comprehensive regulatory framework applicable to HCT/Ps, including donor eligibility requirements, good tissue practice regulations, and appropriate enforcement provisions.

- The first rule on registration and listing was finalized on January 19, 2001. By March 29, 2004, FDA expected all establishments that recover, process, store, label, package, or distribute HCT/Ps, or that screen or test the donor of the HCT/Ps, to be registered with FDA and list their HCT/Ps.
- The donor eligibility rule was the second rule that was finalized on March 25, 2004.
- The third rule regarding current good tissue practice was published on November 24, 2004.
- The donor eligibility rule and the current good tissue practice rule became effective on May 25, 2005.

Transfusion Medicine, Apheresis, and Hemostasis. http://dx.doi.org/10.1016/B978-0-12-803999-1.00017-1

Answer: *D*—In the United States, HCT/P products are regulated under the 21 CFR 1271 Good Tissue Practices (GTP) regulations and section 361 of the Public Health Service Act if they meet all of the following criteria:

1. The HCT/P is minimally manipulated
2. The HCT/P is intended for homologous use only
3. The manufacture of the HCT/P does *not* involve combination of the cell or tissues with another article (water, crystalloids, or a sterilizing, persevering or storage agent are an exception)
4. Either: The HCT/P does not have a systemic effect and is not dependent upon the metabolic activity of living cells for its primary function; or
5. The HCT/P has a systemic effect or is depend upon the metabolic activity of living cells for its primary function and is for autologous use; for allogeneic use in first-degree or second degree blood relative; or is for reproductive use.

If one or more of the above criteria is not met, the HCT/P is regulated as a drug, device, and/or biologic product under section 351 of the Public Health Service Act and other applicable regulations in title 21 of the Code of Federal Regulations (Answers A, B, C, and E).

2. Which of the following hematopoietic progenitor cell (HPC) donors does not need a donor eligibility determination (testing and screening)?
A. Sibling allogeneic donor
B. Unrelated allogeneic donor
C. Autologous donor
D. Cord blood donor
E. Mother donating for her child for a haploidentical transplantation

Concept: The FDA donor eligibility rule was finalized on March 25, 2004 and became effective on May 25, 2005. This is subpart C of 21 CFR part 1271. The FDA requires tissue-based product (HCT/P) establishments to screen and test cell and tissue donors for risk factors for and clinical evidence of relevant communicable disease agents and diseases (RCDADs).

Answer: *C*—Exceptions to the donor eligibility rule are provided in section 1271.90. Donor eligibility determination is not required for cells and tissues for autologous use. Another exception is reproductive cells or tissue donated by a sexually intimate partner of the recipient for reproductive use. All allogeneic donors (Answers A, B, D, and E) must undergo donor eligibility determination.

3. Donor eligibility determination in the context of HPC collection is primary geared toward which of the following?
A. Reducing the risk of transmitting relevant communicable diseases from the donor to the recipient
B. Reducing the risk of transmitting relevant communicable diseases from the recipient to the donor
C. Protect the safety of the donor during the collection procedure
D. Protecting the safety of the collection team
E. Assess the recipient's eligibility to receive the transplantation

Concept: Certain diseases, such as hepatitis C, or HIV, can be transmitted from infected donors through the transplant of HCT/Ps derived products. To prevent the introduction, transmission, or spread of such diseases, it is necessary to take appropriate measures to prevent the use of cells or tissues from infected donors. Thus, before the use of most HCT/Ps, the eligibility of cell or tissue donor must be evaluated for relevant communicable diseases via donor questionnaire screening and infectious disease testing.

Answer: *A*—The HCT/P regulations are intended to reduce the risk of transmitting communicable diseases from an allogeneic HCT/P donor to the transplant recipient. In most cases, a donor who tests reactive for a particular disease, or who possesses clinical evidence of or risk factors for such a disease, would be considered ineligible. Though donor and their own safety during collection (Answer C) is a concern of the collection staff and protecting the safety of the collection team

(Answer D) is important, they are not a focus of donation eligibility regulations. Donor eligibility determination does not involve assess the recipient's eligibility for transplantation (Answer E).

4. Which of the following statements regarding donor eligibility determination is correct?
A. Ineligible donors are never accepted for donation
B. Positive results for CMV antibodies makes a donor ineligible to donate
C. Positive results for Hepatitis B surface antigen (HBsAg) makes a donor ineligible to donate
D. Donor screening by the questionnaire is not as important as donor testing
E. Donor screening by testing is not as important as the results of the donor history questionnaire

Concept: This final donor eligibility rule requires establishments to make donor-eligibility determinations for cell and tissue donors, based on the donor history questionnaire screening and by reviewing the test results for relevant communicable disease agents and diseases (RCDADs).

Answer: C—The term "relevant communicable disease agent or disease" (RCDADs) is defined in Chapter 21 of Code of Federal Regulations (CFR) part 1271.3(r) along with a list of agents that are considered RCDADs (Table 17.1 based on 21 CFR 1271.75, 21 CFR 1271.80, 21 CFR 1271.85, and http://www.fda.gov/downloads/BiologicsBloodVaccines/GuidanceComplianceRegulatoryInformation/Guidances/Tissue/UCM091345.pdf for full details of the eligibility determination criteria). Hepatitis B is part of the RCDADs per the FDA.

TABLE 17.1 RCDADs Screening and Testing Requirements

Agent	Required for	Screening questions[a]	Testing[b]
HIV 1 and 2	All HCT/Ps	x	x
Hepatitis B	All HCT/Ps	x	x
Hepatitis C	All HCT/Ps	x	x
Syphilis	All HCT/Ps	x	x
Human transmissible spongiform encephalopathy, including Creutzfeldt-Jacob disease	All HCT/Ps	x	
West Nile virus	All HCT/Ps	x	
Sepsis	All HCT/Ps	x	
Vaccinia (virus used in smallpox vaccine)	All HCT/Ps	x	
HTLV-I and II	Viable, Leukocyte-Rich HCT/Ps	x	x
CMV	Viable, Leukocyte-Rich HCT/Ps		x

[a] *Screening is not required for autologous donors. Please refer to the FDA donor eligibility rule for further details surrounding autologous donations.*
[b] *More than one test may be necessary to adequately and appropriately test for a single RCDAD (e.g., anti-HCV and HCV NAT for Hepatitis C).*

CMV is not considered a RCDAD. Thus, positive testing for CMV antibodies doesn't make the donor ineligible to donate (Answer B). However, donors of viable leukocyte-rich HCT/Ps must be tested for CMV. Facilities must have SOPs governing the release of an HCT/P from a donor who tests positive for CMV [21 CFR 1271.85(b)(2)]. SOPs should describe how the CMV test results are communicated to the physician using the product. This final donor eligibility rule also is not prohibited to use an HCT/P from an ineligible donor (Answer A) or a donor who has not yet been determined eligible. The rule provides provisions of how to address those situations. This is a different process from deferral of blood donors.

Different from blood donors, HCT/Ps from ineligible donors may be used (Answer A) if:

- It is for allogeneic use in a first or second-degree blood relative;
- It consists of reproductive cells or tissue from a directed donor as defined in 21 CFR.3(I); or
- There is documented urgent medical need.

HCT/Ps from ineligible donors must have unique identification, storage, labelling, and accompanying record requirements. The physician using the product must be notified of the results of screening and testing. This notification must be documented.

Similar to blood donation, both donor questionnaire screening and testing (Answers D and E) are important process in donor eligibility determination.

5. What is the difference between HPC/P donor eligibility and donor suitability?
   A. There is no difference as both terms are used interchangeably
   B. Donor eligibility refers to a HPC/P donor that that is deemed fit to process to the collection process based on medical evaluation from a health care provider, while donor suitability refers to a donor meets all the donor screening and testing requirement related to infectious disease transmission
   C. Donor eligibility refers to a HPC/P donor that meets all the donor screening and testing requirement related to infectious disease transmission, while donor suitability refers to a donor that is deemed fit to proceed to the collection process based on medical evaluation from a health care provider
   D. Donor eligibility is based on the product content while donor suitability is based on the HLA matching
   E. Donor eligibility is based on the HLA matching while donor suitability is based on the product content

Concept: The purpose of the regulations in the HPC/P donation process is to protect both the donors and recipients. As explained in Question 3, donor eligibility process is employed to prevent the transmission of RCDADs from the donor to the recipient. On the other hand, donor suitability is a process to ensure the safety of the donor before, during, and after the collection process.

Answer: *C*—Donor eligibility refers to a HPC/P donor that meets all the donor screening and testing requirement related to infectious disease transmission while donor suitability refers to a donor that is deemed fit to process to the collection process based on medical evaluation from a health care provider. The suitability evaluation process must account for the entire collection process from the initial evaluation, mobilization (if applicable), to collection, and postcollection care. Answer B is thus incorrect.

"Eligibility" and "suitability" refer to independent determination processes, and thus, should not be used interchangeably (Answer A). It has nothing to do with the HLA matching and/or product content (Answers D and E). A donor may be eligible but not suitable and vice versa. For example, a donor with hepatitis B may be well enough to proceed with the donation (suitable for donation) but is not eligible based on the donor questionnaire screening. On the other hand, a donor with a hip fracture may be eligible for donation but not suitable for donating bone marrow (but may be suitable for donating HPC collection by apheresis).

6. A 15-year-old patient with the diagnosis of acute myeloid leukemia (AML) requires allogeneic HPC transplantation. The patient has three full siblings who are willing to be the HPC donor. What is the approximate probability of identifying an HLA-identical (six antigen match) among the siblings?
   A. 58%
   B. 50%
   C. 44%
   D. 25%
   E. 0%

Concept: HLA is present on the short arm of chromosome 6. Each parent has two copies of chromosome 6. Thus, based on Mendelian genetics rules—the chance to have two siblings with identical HLA is 1 in 4 (i.e., 25%). Vice versa, the probability of having two siblings that do not have an identical HLA type is 75%. Therefore, the formula to determine the probability of finding a sibling with identical HLA is: $1-(0.75)^n$, in which n represents the number of siblings.

Answer: *A*—Based on the formula, the probability of finding a sibling with identical HLA is $1-(0.75)^3= 0.578$ (~58%). The other choices (Answers B, C, D, and E) are incorrect based on the calculation above.

7. Which of the following is the most reliable predictor of peripheral blood progenitor cell collection yields from mobilized donors?
   A. The nucleated cell count (WBC) in the peripheral blood
   B. The absolute neutrophil counts (ANC) in the peripheral blood
   C. The number of CD34+ cell in the peripheral blood
   D. The number of CD3+ cell in the peripheral blood
   E. The number of CD133+ cell in the peripheral blood

Concept: HPC products transplant dose is defined by the number of CD34+ cells per recipient weight in kilogram (typically 2–5 × 10^6 CD34+ cells/kg as it has been shown to associated with good outcomes). Many programs use the measurement of the level of peripheral blood circulating CD34+ cells prior to collection (preharvest PB CD34+ count) as a predictor of successful stem cell collection. There is no established formula to calculate collection efficiency, but many centers use the following:

$$\textit{Collection efficiency} = \frac{\textit{Product CD34+dose}}{[\textit{Preharvest PB CD34+cells(in } mL^{-1})] \times [\textit{Volume processed(in mL)}]}$$

Thus, mathematically, if the average collection efficiency of the facility is known, then the above equation may be rearranged to estimate the volume needed to process in order to collect the target product CD34+ dose.

$$\textit{Product CD34+dose} = [\textit{Collection efficiency}] \times [\textit{Preharvest PB CD34+(in } mL^{-1})] \times [\textit{Volume processed(in mL)}]$$

Example:
Peripheral blood CD34+ cell = 27/μL (or 27,000/mL)
Total apheresis procedure blood volume processed = 18 L (or 18,000 mL)
CD34+ cells in final product = 2.4 × 10^8

$$\textit{Collection efficiency} = \frac{2.4\times10^8}{(27\times10^3)\times(18\times10^3)} = 49.4\%$$

Answer: C—The most reliable predictor of the apheresis collection yield is the peripheral blood CD34+ cell count as determined by flow cytometry. The test is typically performed the day before or the day of the scheduled collection. A preharvest PB CD34+ cell concentration of ≥20/μL is considered by many transplants centers to be a good predictor of a collection that will yield a product with a minimal CD34+ cell dose of 2 × 10^6/kg. This cut off of preharvest PB CD34+ of 20/μL should be used as a general guideline and not as a strict criterion to start the collection process by apheresis. Since the collection process is not linear and there is limit to the daily volume that can be processed, donors with low peripheral blood CD34+ count might require additional collection day(s). CD13 and CD133 (Answers D and E) are not markers of HPCs. WBC (Answer A) is not the best predictor of the apheresis collection yield and HPCs are not neutrophils; thus, ANC (Answer B) is not a surrogate marker for HPCs.

8. Which of the following factors can be associated with inadequate CD34+ cell mobilization?
   A. Age <50 years
   B. No prior radiotherapy
   C. Male gender
   D. Female gender
   E. Minimal bone marrow involvement with disease.

Concept: In order to increase the number of circulating HPCs collection by apheresis and to collect an adequate cell dose for HPC transplantation, mobilization agents, such as G-CSF and/or plerixafor, are given to the donor. The incidence of inadequate CD34+ cell mobilization varies between 5-30% (based on the definition and collection goal).

Answer: *D*—Factors associated with inadequate CD34+ cell mobilization includes the following:

- Age > 65 years
- Female gender
- Thrombocytopenia at the time of mobilization
- Disease status, including the presence and extent of bone marrow involvement with diseases, as well as bone marrow damage due to prior radiation therapy and the use of stem cell toxins as part of chemotherapy cycles, such as dacarbazine, platinum analogs, fludarabine, and lenalidomide.

Based on the earlier mentioned information, the other choices (Answers A, B, C, and E) are wrong.

9. A type B Rh positive AML patient is getting a HPC product collected from a donor with a blood type of A Rh positive. The above scenario is an example of which of the following?

A. Full match transplantation
B. Minor mismatch only
C. Major mismatch only
D. Partial mismatch only
E. Bidirectional mismatch

Concept: ABO incompatibility (ABOi) between the blood types of the donor and the transplant recipient could result in similar types of adverse reactions as seen in blood transfusion since HPC products contain variable amount of donor RBCs and plasma. The degree of mismatch is graded in the following categories:

- Minor mismatch is defined when the donor has compatible RBCs but incompatible plasma to recipient (e.g., Group O donor to Group A recipient).
- Major mismatch is defined when the donor has compatible plasma to recipient but incompatible RBCs (e.g., Group A donor to Group O recipient).
- Bidirectional mismatch is defined when the donor has incompatible RBCs and incompatible plasma to recipient (e.g., Group A donor to Group B recipient, shown in the question). This may potentially result in acute/delayed hemolysis when excessively large amount of incompatible RBCs and/or incompatible hemagglutinin are infused.

Answer: *E*—Per definition earlier, this is an example of bidirectional mismatch; thus, the other choices (Answers A, B, C, and D) are incorrect.

10. A 46-year-old AML male patient with Group A Rh positive blood type was scheduled to receive a fresh bone marrow derived product from a matched unrelated donor (MUD) with an AB Rh positive blood type. The patient weighs 80 kg. The volume of the collected bone marrow product was 1,000 mL and it contains 2.3×10^8 TNC/kg, 2.8×10^6 CD34+/kg with a hematocrit of 36%. What processing procedure should be performed on the bone marrow product before issuing it for transplant?

A. Plasma reduction
B. Red cell reduction
C. Cryopreservation
D. Anti-B titer
E. Issue for transplant without further manipulations and infuse slowly

Concept: ABOi HPC transplantation may associate with adverse events. There are several methods that may be employed to reduce the severity of these complications.

A major ABO incompatibility implies that the recipient has circulating isoagglutinin antibodies against the donor's red blood cells (RBCs), which can cause acute RBC hemolysis, delayed red cell engraftment, and even red cell aplasia. RBC reduction can significantly reduce the RBC volume in the HPC product and reduce the likelihood of acute hemolysis at the time of infusion. There are no specific regulations regarding the RBC volume allowed in an HPC product, and many laboratories use self-determined thresholds. A threshold of ≤0.4 mL/kg or total 20-30 mL RBC volume is commonly used. Therapeutic plasma exchange pretransplant may be performed in the patient to reduce the recipient anti-donor's issoagglutinin.

In minor ABO incompatible HPC transplantation, since the donor has compatible RBC to recipient, but incompatible plasma, the patient may experience acute and/or delayed hemolysis. Thus, plasma reduction may be considered to reduce the infused isoagglutinin amount. Passenger lymphocyte syndrome is one of the recently recognized posttransplantation complications involving competent graft lymphocytes sensitized against host RBC antigens (including ABO and other RBC antigens). This could lead to delayed hemolysis within 1 to 2 weeks of graft infusion. The incidence of this complication can be affected by the pretransplantation conditioning regimen, GVHD prophylaxis treatment, and the types of stem cell product infused. In a severe and life-threatening situation, RBC exchange can be considered to mitigate the extensive hemolysis.

For bidirectional mismatch, a combination of procedures and techniques used in major and minor mismatched may be considered.

Answer: *B*—This is a case of a major ABO incompatible transplant of a HPC product with a high excess of incompatible donor RBCs. The cellular content of the HPC product in the question (2.8×10^6 CD34+/kg) will require administration of the entire dose based on the minimum number of cells typically considered necessary for successful engraftment.

Anti-B titer on the recipient should be done in this case, but will not address the need to reduce the exposure to RBCs with different ABO antigen (Answer D). Because the hematocrit of the product is 36%, the RBC volume is 1000 × 0.36 or 360 mL (or 4.5 mL/kg recipient weight). RBC reduction should be performed on this product prior to transplant. RBC reduction can be achieved either manually by centrifugation or sedimentation with hydroxyethyl starch. Alternatively, automated cell washing devices can be used to manufacture a buffy coat, thus reducing RBC volume.

Plasma reduction (Answer A) is performed on the HPC product in case of minor mismatch or bidirectional mismatch. Cryopreserved of the products is done when the patient is not ready to be transplanted at the time the product is collected (Answer C). None of other choices (Answers A, C, D, and E) directly address the issue of excess RBCs in the HPC content (which may lead to acute hemolysis).

11. If transfusion is necessary during the posttransplantation period for a patient with bidirectional, ABOi HPC transplantation, what ABO type of RBCs and platelets/plasma should be provided?

A. Group A RBC with Group B plasma/platelets
B. Group O RBC with Group A plasma/platelets
C. Group B RBC with Group AB plasma/platelets
D. Group O RBC with Group AB plasma/platelets
E. Group AB RBC with Group O plasma/platelets

Concept: In the setting of ABOi HPC transplantation, it is critical to provide appropriate blood products, and have established policies in place to guarantee that the correct ABO-typed product is dispensed by the blood bank. There are three distinct phases of transfusion support for stem cell transplant patients. Phase 1 is the time when the patient undergoes preparation for HPC transplantation. Phase 2 initiates with the induction of chemotherapy, and extends until donor isoagglutinins and the recipients RBCs are undetectable. Phase 3 occurs after full donor conversion. Table 17.2 can be used as a guidance to select the blood products of appropriate blood group for patients at each of the three phases of ABOi HPC transplantation blood transfusion support. There is no consensus on which Rh type to give to patients when there is a discrepancy between patient's and donor's.

TABLE 17.2 Transfusion Support for Patients With ABOi HPC Transplantation

		Phase 1	Phase 2			Phase 3
Recipient	Donor	All products	RBCs	Platelets	Plasma	All products
O	A	Recipient	O	A; AB; B; O	A, AB	Donor
O	B	Recipient	O	B; AB; A; O	B, AB	Donor
O	AB	Recipient	O	AB; A; B; O	AB	Donor
A	O	Recipient	O	A; AB; B; O	A, AB	Donor
A	B	Recipient	O	AB; A; B; O	AB	Donor
A	AB	Recipient	A	AB; A; B; O	AB	Donor
B	O	Recipient	O	B; AB; A; O	B, AB	Donor
B	A	Recipient	O	AB; B; A; O	AB	Donor
B	AB	Recipient	B	AB; B; A; O	AB	Donor
AB	O	Recipient	O	AB; A; B; O	AB	Donor
AB	A	Recipient	A	AB; A; B; O	AB	Donor
AB	B	Recipient	B	AB; B; A; O	AB	Donor
		Pretransplant	Transplant→ Donor conversion			Donor conversion

PLTs should be selected in the order selected.

Adapted with permission from E.M. Staley et al., An update on ABO incompatible hematopoietic progenitor cell transplantation. Transfus. Apher. Sci. 54 (2016) 337–344.

Answer: *D*—Group O RBC with Group AB plasma/platelets should be given during the peritransplant period for a bidirectional mismatch. After the full hematologic recovery, then blood products with ABO and Rh type may be given. Based on the table mentioned, the other choices (Answers A, B, C, and E) are incorrect.

12. Compared to mobilized peripheral blood HPC products; a typical bone marrow-derived HPC product contains which of the following combinations of cells?
A. Less CD34+ content, more T-cells, less mesenchymal cells, and higher hematocrit
B. Less CD34+ content, less T-cells, more mesenchymal cells, and higher hematocrit
C. Less CD34+ content, less T-cells, more mesenchymal cells, and lower hematocrit
D. Less CD34+ content, less T-cells, less mesenchymal cells, and lower hematocrit
E. Higher CD34+ content, less T-cells, more mesenchymal cells, and higher hematocrit

Concept: There are three common HPC products: bone marrow (HPC(M)), peripheral blood from mobilized donor (HPC(A)), and umbilical cord blood (HPC(C)). The cellular composition of these HPC products is significantly different (Table 17.3). These differences can contribute to short term and long-term transplant outcomes and side effects.

TABLE 17.3 HPC Products General Characteristics

	HPC, Marrow	HPC, Apheresis	HPC, Cord
Volume	~1000–1500 mL	~300–600 mL	~20–50 mL
Hematocrit	~25%–50%	~ 2%–8%	NA
HPC cells (CD34+)	Adequate	High	Low
Mature T-cells	Low	High	NA
Engraftment time	Medium	Fast	Slow
GVHD risk	Medium	High	Low

Answer: *B*—Compared to HPC(A); HPC(M) products usually contain two to fivefold less CD34+ cells and 10-fold less T-cells. On the other hand, HPC(M) products contain significantly more mesenchymal stromal cells and other microenvironmental cells. It also contains bone spicules, fat, clots, and anticoagulant; thus, filtration may be necessary during collection or prior to processing to remove unwanted particles. Due to the large product volume and the procurement process, the hematocrit of bone marrow products is significantly higher. The use of umbilical cord blood for HPC transplant is steadily increasing. The advantage of HPC(C) is that they are readily available and pose no risk to the donor. Also, cord blood T-cells are naïve and have lower risk of GVHD. On the other hand, due to their limited size (typically, 25–50 mL), cord blood products have lower TNC and progenitor cells content, thus resulting in much slower engraftment time, especially if used for adult patients. Based on the table mentioned earlier, the other choices (Answers A, C, D, and E) are incorrect.

13. The accuracy of the automated WBC measurement of bone marrow and umbilical cord products can be overestimated due to the presence of which of the following?
A. Megakaryocytes
B. Nucleated RBC
C. Platelets
D. Bone spicules
E. Mesenchymal stem cells

Concept: The traditional method for counting cells is electrical impedance, which is used in almost every hematology analyzer. The change in impedance is proportional to cell volume and can be used to differentiate between the different leukocytes subpopulations. This technique does not work as well for bone marrow due to the presence of immature myeloid cells and nucleated red blood cells (NRBC) that have similar volume as lymphocytes.

Answer: *B*—Unlike peripheral blood, bone marrow HPC products contain nucleated RBCs, which are identified as lymphocytes by many hematology analyzers, thus overestimating the true TNC content. New hematology analyzers correct automatically the WBC count when NRBCs are detected, but if this function is not available, it is important to indicate that the reported WBC is uncorrected. It is also possible to manually correct the WBC count by reviewing a smear of the product and assessing the percentage of NRBC. Megakaryocytes (Answer A) is larger than lymphocytes, and thus, do not interfere with the count. On the other hand, platelets (Answer C) are smaller than lymphocytes, and thus, should not interfere with the WBC count. Neither bone spicules (Answer D) nor mesenchymal stem cells (Answer E) should interfere with the WBC.

14. CD34 is expressed on the surface of bone marrow (BM) and peripheral blood (PB) WBC cells in a steady state (without mobilization) in low concentrations. These low concentrations are represented correctly by which of the following?
A. CD34 is expressed on 0.01%−0.1% of PB WBC cells and on 1%–3% of BM cells
B. CD34 is expressed on 0.01%− 0.1% of BM cells and on 1%–3% of PB WBC cells
C. CD34 is expressed on 0.05%−0.5% of PB WBC cells and on 3%–4% of BM cells
D. CD34 is expressed on 0.05%−0.5% of BM cells and on 3%–4% of PB WBC cells
E. CD34 is expressed on 1% of PB WBC cells and on 5% of BM cells

Concept: While there is no unique and unambiguous surface marker that identifies HPCs, some surface markers are used as surrogate markers. CD34, a glycoprotein found on the surface of early HPCs and other cell types (e.g., endothelial cells), is frequently used to guide collection and transplantation dose, as it has shown to correlate with neutrophil and platelet engraftment.

Answer: *A*—In steady state, CD34 is expressed on about 1%−3% of BM cells. The large majority of these cells are committed progenitors. CD34 is expressed on 0.01%−0.1% of PB white blood cells. Thus, mobilization regimens are used to increase the level of circulating CD34+ cells in order to have more cells to be collected via apheresis. Based on the information above, the other choices (Answers B, C, D, and E) are incorrect.

15. A 27-year-old healthy man is going to donate HPCs for his sister's HPC transplantation. He receives G-CSF as part of the mobilization regimen. What is the most common side effect of granulocyte colony-stimulating factor (G-CSF) using for HPC mobilization?
A. Prostate cancer
B. Fever
C. Bone pain
D. Nausea
E. Anorexia

Concept: Over the past decade, there has been a major shift from using HPC(M) to HPC(A) from allogeneic donors. It started with related donors and has extended to healthy volunteers. Granulocyte colony-stimulating factor (G-CSF) is the cytokine most commonly administered to healthy donors as part of the mobilization regimen for HPC(A) collection and has been administered to thousands of related and unrelated donors worldwide. The mechanism of G-CSF on HPC mobilization is not completely understood. Part of the mechanism is that G-CSF increases the number of bone-resorbing osteoclasts which secrete high levels of the mobilizing chemokine IL-8, the proteolytic enzymes MMP-9, and cathepsins that cleave SDF-1, resulting in the release of HPCs into the circulation. Donors usually take the medication (~10 μg/kg) for 4–5 days with the collection happens on day 4 or 5.

Answer: *C*—Studies have shown that the most common adverse event in HPC donors taking G-CSF is bone pain, peaking on day 4 after 3 days of administration. It occurs most frequently in the axial skeleton (back, hip, and chest/sternum). Female donors are more likely to have pain compared to male donors. Acetaminophen or ibuprofens are usually effect for pain relief. Most of the bone pain decreases significantly after G-CSF termination. Fever, nausea, and anorexia (Answers B, D and E) may occur, but are not the most common. There is no known increase in cancer risk in the HPC donor receiving G-CSF comparing to the general population (Answer A). Rarely, splenic rupture due to enlargement of the spleen has been reported.

16. In which of the following clinical scenarios should G-CSF be avoided?
A. Patient with lymphoma in need for autologous HPC collection
B. Healthy donor donating HPC by apheresis collection for her sibling with AML
C. Healthy donor donating HPC by apheresis collection for an unrelated patient with ALL
D. Donor with hemoglobin SC disease who is donating to his sibling with sickle cell disease.
E. Donor with sickle cell trait who is donating to his sibling with sickle cell disease.

Concept: As mentioned in the previous question, G CSF is relatively safe when used as mobilization agent. Nonetheless, life-threatening complications have been reported in donors with sickle cell disease or complex hemoglobinopathies. Thus, voluntary accreditation organizations requirements include the evaluation for the risk of hemoglobinopathy prior to administration of the mobilization regimen. Although it is one of the acceptable methods, testing for hemoglobinopathy is not required.

Answer: *D*—Donor with Hemoglobin SC disease should not be mobilized with G-CSF. There are multiple reports of donors with sickle cell disease or complex hemoglobinopathies, such as hemoglobin SC or hemoglobin S-β thalassemia, who developed severe symptoms after receiving mobilization with G-CSF, including severe vasoocclusive crises, multiorgan failure, and death. While the mechanism is not fully understood, it might relate to the activation of the neutrophils by the G-CSF. In these patient populations, G-CSF is contraindicated. Nonetheless, these donors may be suitable to donate bone marrow. Thus far, there is no data to support the risk of G-CSF as a mobilizing agent in patients with sickle trait (and no other coexisting hemoglobinopathy) (Answer E). Since the donors in Answers A, B, and C do not have any hemoglobinopathy, G-CSF may be used as mobilization agent.

Please answer Questions 17–19 based on the following clinical scenario:

A 75-year-old patient requires an autologous HPC transplantation. He recently read about plerixafor (Mozobil, Genzyme Corp, Cambridge, MA) and would like to know more information.

17. The mechanism of action for plerixafor is which of the following?
A. Direct inhibition of the CXCR-4 receptor on the surface of the HPCs
B. Direct inhibition of the stromal cell-derived factor-1 (SDF-1α) on the surface of BM stroma cells
C. Stimulates stem cell proliferation
D. CD38 receptor antagonist
E. Inhibitor of CD34-proliferation-inhibitory factor

Concept: HPCs are normally found in very small concentrations in the peripheral blood (~0.01%–0.1% of WBC in the peripheral blood). Thus, mobilization agents are used in order to achieve efficient stem cell collections. Commonly used HPC mobilizing methods include the use of chemotherapy and growth factors, such as G-CSF. Recently, a CXCR-4 antagonist, plerixafor, a novel mobilization agent with a different mechanism of action, was approved as a mobilization agent.

Answer: *A*—Plerixafor is an inhibitor of the CXCR4 chemokine receptor and blocks binding to its cognate ligand SDF-1α. SDF-1α is also known as CXCL12. By blocking the CXCR4-SDF-1α interaction, the HPCs are released from the bone marrow into the peripheral blood for a period of 12–24 h. The other choices (Answers B, C, D, and E) do not correctly describe the mechanism of plerixafor.

18. Plerixafor is FDA approved for which of the following?
A. Healthy allogeneic HPC donors
B. Patients with AML who failed mobilization with G-CSF
C. Patients with multiple myeloma as a first line sole mobilization agent
D. Patients with multiple myeloma and Hodgkin's lymphoma who failed mobilization with G-CSF
E. Patients with multiple myeloma and non-Hodgkin's Lymphoma, in combination with G-CSF

Concept: Plerixafor, an inhibitor of the CXCR4 chemokine receptor and blocks binding to its cognate ligand SDF-1α, is a novel mobilization agent. The efficacy and safety of plerixafor in conjunction with G-CSF in non-Hodgkin's lymphoma (NHL) and multiple myeloma (MM) adult patients was evaluated in two randomized, double-blind, placebo-controlled, multicenter phase three studies. The results showed that for NHL patients—a goal of 5×10^6 CD34+/kg in ≤ 2 collections was achieved in 59.3% and 19.6% patients mobilized with plerixafor and G-CSF, respectively (P-value < 0.001). For MM patients, a goal of 6×10^6 CD34+/kg in ≤ 4 collections was achieved in 71.6% and 34.4% patients mobilized with plerixafor and G-CSF, respectively (P-value < 0.001). If plerixafor is used as mobilization agent (at a dose of 0.24 mg/kg), then peripheral blood HPC collection by apheresis must be initiated approximately 10–14 h postadministration due to the pharmacokinetics (i.e., its peak mobilization of CD34+ cells occurring ~10–14 h following injection and its effect dramatically decrease after 24 h).

Answer: *E*—Based on those phase III clinical trials, plerixafor is FDA approved for usage in combination with G-CSF to mobilize HPCs for the HPC for collection by apheresis and subsequent autologous transplantation in patients with NHL and MM. It may be used off-labeled in patients who failed mobilization with G-CSF and/or chemotherapy (Answers B and D), but this is not the FDA indication. Plerixafor is used in combination with G-CSF and not as the only mobilizing agent (Answer C). At this time, plerixafor is not approved to be used in healthy donor (Answer A).

19. The most common side effect of plerixafor (Mozobil) is which of the following?
A. Fever
B. Bone pain
C. Diarrhea
D. Restless leg syndrome
E. Sore throat

Concept: Clinical safety data for plerixafor comes primarily from the two randomized, double-blind, placebo-controlled, multicenter phase III studies as described in the previous question and is supported by other uncontrolled studies in 543 patients.

Answer: *C*—Based on those phase three trials and other studies, the most common adverse events (≥10%) associated with Plerixafor were: diarrhea, nausea, fatigue, injection site erythema, headache, arthralgia, dizziness, and vomiting. Bone pain (Answer B) is the most common side effect associated with G-CSF mobilization. If the patient experiences fever or sore throat (Answers A and E), he or she must be evaluated carefully for any signs of infection and/or sepsis because the infection may contaminate the HPC products. Restless leg syndrome (Answer D) is not usually associated with plerixafor administration.

20. A 30-year-old female donor is undergoing a large volume peripheral blood HPC collection procedure (3–5 total blood volumes). In the middle of the procedure, she complains of paresthesia, headaches, and nausea. What would be one of the immediate steps to mitigate the donor symptoms?
A. Decrease citrate infusion rate by slowing down the procedure
B. Increase citrate infusion rate by speeding up the procedure
C. Decrease citrate infusion rate by supplementing with potassium
D. Proceed the collection procedure as it is until completed as the symptoms are not related to the procedure
E. Abort the procedure

Concept: There are known risks and adverse reactions that are associated with an apheresis collection procedure. The apheresis procedure requires the use of anticoagulant, typically anticoagulant-citrate-dextrose formula A (ACD-A). Hypocalcaemia is a known citrate-related toxicity, and its symptoms include headaches, nausea, and paresthesia. If untreated, it can result in tetany, cardiac arrhythmia, and death.

Answer: *A*—Due to their smaller blood volume, female and pediatric donors, who undergo large volume peripheral blood HPC collection, have higher citrate-related toxicity adverse events, especially if citrate is used as the sole anticoagulant. Citrate is metabolized in the liver. Thus, patients with liver failure are also at increased risk for citrate toxicity. Slowing down the procedure will decrease the citrate infusion rate and thus, answer B is wrong. Other options include oral or IV calcium based on symptoms and local standard operating policies. Although the procedure does not need to be terminated at this time (Answer E), the procedure should be slow down and/or the patient should be treated since the symptoms are related to the procedure (Answer D). Although citrate can induce hypokalemia and metabolic alkalosis, the symptoms are not consistent with hypokalemia (Answer C).

21. What HPC modification is performed prior to transplantation to reduce GVDH risk?
A. T-cell reduction
B. Cryopreservation
C. Red cell reduction
D. Slow infusion rate
E. CD34+ cell reduction

Concept: HLA incompatibility is a major immunological barrier to a bone marrow transplant both pre- and posttransplant due to the risk of both acute and chronic GVHD. GVHD is a known side effect of allogeneic transplant and is the result of donor immune cells recognition of recipient foreign antigens. The occurrence of GVHD can be minimized by reduction of the number of T-cells in the product. T-cell reduction can be achieved by either positive or negative selection, using specific antibodies to CD34 or CD3 epitopes, respectively.

Answer: *A*—Reduction of T-cells may reduce GVHD risk. Cryopreservation (Answer B) is usually done when the transplant occurs after the HPC collection. Red cell reduction (Answer C) is performed for major ABOi and bidirectional incompatible HPC transplants. CD34+ cells are critical for the engraftment, and thus, HPC products should be carefully handled and manipulated to prevent the loss of CD34+ cells (Answer E). Slow infusion rate does not reduce GVHD (Answer D).

22. A potential consequence of T-cell reduction is:
A. Increased adverse events during infusion
B. Loss of the graft versus leukemia (GVL) effect
C. Increased risk of GVHD
D. Decreased adverse events for major ABO incompatible products
E. Shorter time to neutrophils engraftment

Concept: While removal of T-lymphocytes from a graft helps reduce the occurrence of GVHD, it also carries a major risk. T-cell depletion from the graft results in the loss of the potential immunologic antitumor effect of donor's T-lymphocytes, or the potential GVL effect. It increases the risk of disease relapse. It can also increase the risk of infections due to the potential engraftment delay of T-cell depleted graft.

The above concept is the reason that HPC(M) is the preferred choice of product for HPC transplantation for nonmalignant hematologic conditions, such as sickle cell disease. This is because HPC(M) contains less T-cells than HPC(A), which leads to lower risk of GVHD. Furthermore, the patients receiving HPC transplantations for nonmalignant hematologic conditions do not need GVL effects.

Answer: *B*—Removal of T-cells from the graft can result in loss of the potential GVL effect of allogeneic grafts. T-cell reduction of the HPC products decreases risk of GVHD (Answer C). It does not affect the adverse events during infusion or for major ABOi products (Answers A and D). T-cell depleted HPC products may lead to delayed engraftment and graft failure (Answer E).

23. For a cellular product from an allogeneic donor that contains red blood cells at the time of infusion, which of the following is true?
A. One ABO group and Rh typing is enough to ensure patient safety.
B. Confirmatory ABO group and Rh typing should be available from samples collected at the same time.
C. Confirmatory ABO group and Rh typing should be available from two independently collected samples.
D. ABO group and Rh typing is not relevant to the success of HPC transplant
E. ABO matching is more important than HLA typing

Concept: To avoid the unintentional use of ABO incompatible products that can result in a transfusion reaction, allogeneic donors and recipients should be tested for ABO group and Rh type. The testing should be done from two independently collected samples. The test can be performed on peripheral blood or on the cellular therapy product. Recipients of incompatible ABO group products should also be screened for red cell antibodies.

Answer: *C*—Confirmatory ABO group and Rh typing should be available from two independently collected samples. Also, in HPC transplantation, HLA matching is more important for engraftment and survival than ABO compatibility (Answer E). There are no uniform recommendations to what needs to be done, if at all, when alloantibody is detected in the donor's plasma. This will be up to the local policy of the transplant center. Based on the information above, the other choices (Answers A, B, and D) are incorrect.

24. The minimally required tests that a laboratory has to perform to determine quantity and quality of autologous HPC products are which of the following?
A. Sterility, total nucleated cell (TNC) count, CD34+ cells count, viability
B. Sterility, TNC count, CD34+ cells count, CD3+ cells count

C. Endotoxin, CD34+ cells count, viability, colony forming unit (CFU)
D. Blood typing, sterility, viability, TNC count
E. Blood typing, CD34+ cells count, viability, TNC count

Concept: To comply with current good clinical practice (cGMP) and voluntary accreditation agencies, prior to their release, HPC products have to be tested for safety and quality. Autologous products have to be tested for microbial contamination and viability. They should also be assessed for their TNC and hematopoietic stem cells content (representing by the CD34+ cells). For allogeneic products, CD3+ cells content, HLA and ABO typing of the donor and/or the products are also required in addition to the tests required for autologous HPC products.

Answer: *A*—The minimally required testing of HPC products from autologous donor includes: microbial testing (sterility), total nucleated cell (TNC) count, and viability, and enumeration of CD34+ cells. Viability is commonly performed by methods, such as trypan blue exclusion (viability = ratio of cells not stained blue over total cells counted), or 7-AAD staining with flow cytometry.

Additional testing is required for products from allogeneic donors. ABO group and Rh typing (Answers D and E) is required for allogeneic donors only. CD3+ (Answer B) is a marker of T lymphocytes and is only required for allogeneic products. Endotoxin testing and potency assays, such as CFU (Answer C), are typically required for more than minimally manipulated investigational products, CD34+ enriched, or expanded products.

25. Which of the following statements regarding HPC cryopreservation is incorrect?
A. DMSO is FDA approved for clinical use
B. DMSO can cause side effects such as nausea, vomiting, and hypertension
C. Thawed HPCs should be infused as soon as possible
D. When possible, cryopresrved HPC should be stored at less than –150°C.
E. DMSO concentration in thawed HPC products can be reduced by washing

Concept: Cryopreservation of living cells requires the use of cryoprotectants to reduce cells freezing damage. DMSO is the standard cryoprotectant used for cryopreservation of HPC, and it is usually used at a concentration of 5%–10% combined with isotonic solution and serum albumin. Post thaw DMSO can be toxic to HPC cells in the liquid products. Immediate diluting and washing the product can reduce this toxicity. If the product is to be thawed and infused, without washing the exposure time to DMSO should be minimized. DMSO is most commonly associated with gastrointestinal side effects; such as nausea and vomiting that appear shortly after administration. Also common are cardiovascular side effects such as hypertension and electrocardigraphic abnormalities. There are no specific guidelines regarding the maximal DMSO volume allowed for infusion. Clinical programs are required to self-determine the appropriate volume of DMSO load. The circular of information for the use of cellular therapy products recommends not to exceed 1 mL of DMSO per kilogram of recipient weight per day of administration.

Answer: *A*—In the USA, DMSO is not approved by the FDA for clinical use. The use of DMSO as a drug was halted by the FDA in 1965, and it is currently approved to be used in human only for interstitial cystitis. Since there is no equivalent reagent for cryopreservation, it is used as the standard HPCs cryoprotectant. It can be used under approval by the appropriate agencies (i.e., investigational device exemption [IDE]). Like other reagents with no clinical use approval, a lot-to-lot functional qualification has to be performed on each new DMSO lot prior to use. Other choices (Answers B, C, D, and E) are wrong.

26. The dose of donor lymphocytes infusion (DLI) is based upon what type of cells?
A. CD3+ cells
B. CD19+ cells
C. CD34+ cells

D. CD45+ cells
E. CD133+ cells

Concept: DLI is used for treatment of patients with relapse disease postallogeneic HPC transplant, using cells from the same donor. It is also used to target posttransplant infectious diseases, such as cytomegalovirus (CMV) and Epstein-Barr virus (EBV). The principle behind DLI is to use the immunologic antitumor effect of the donor's T-cells, and it is used primarily in patients with relapsed chronic myeloid leukemia (CML), AML and chronic lymphocytic leukemia (CLL). The most common cell source for DLI is either stimulated or unstimulated peripheral blood collected by apheresis.

For dosage, typical treatment dosing options exist. A probable scheme will be 1×10^7 T-cells/kg recipient weight initially. If remission does not occur, the dose can be increased to 5×10^7 T-cells/kg recipient weight and then 1×10^8 T-cells/kg recipient weight. Higher DLI dose associates with higher risk of GVHD. Transient pancytopenia and an infectious predilection are also associated with DLI infusion.

Answer: *A*—The dosage of DLI is usually based on the lymphocyte content and the number of CD3+ cells (a marker of T lymphocytes), which is measured by flow cytometry enumeration. The other choices (Answers A, C, D, and E) are not markers for T-cells.

27. Engraftment after autologous HPC transplant is determined by:
A. Donor chimerism
B. Nucleated RBC recovery
C. Absolute neutrophil count (ANC) and platelets recovery
D. Circulating HLA antibodies
E. Donor lymphocyte percentage

Concept: Short-term engraftment for both autologous and allogeneic transplant is determined by peripheral blood ANC and platelets recovery. Although ANC recovery in allogeneic transplant is most likely to result from the donor, this cannot be proven without performing donor chimerism tests, especially for nonmyeloablative or reduce intensity regiments. Chimerism tests use molecular testing to measure the frequency of donor cells compared to the frequency of recipient. Chimerism test is required for all allogeneic transplants.

Answer: *C*—Engraftment post autologous HPC transplant is determined by the recipient peripheral blood ANC and platelets recovery. For ANC recovery, it is defined as the first day of three consecutive laboratory values of $\geq 0.5 \times 10^9$/L obtained on different days. For platelets recovery, it is defined as the first day of three consecutive laboratory values of $\geq 20 \times 10^9$/L obtained on different days, with no platelets transfusion support for seven consecutive days. Although donor chimerism (Answer A), and donor lymphocyte percentage (Answer E) are important information, they are not part of the definition of engraftment post autologous HPC transplant. The other choices (Answers B and D) are incorrect based on the information earlier mentioned.

28. Assuming that recipient and donor HLA matching is similar, which of the following statements is most accurate regarding time to neutrophils engraftment and the source of the HPCs:
A. PB-derived HPCs engraft faster than BM- and UCB-derived products
B. BM-derived HPCs engraft faster than PB-derived but slower than UCB-derived products.
C. UCB-derived HPCs engraft faster than BM- and PB-derived products
D. PB-derived HPCs engraft faster BM-derived products but slower than double UCBs
E. BM-derived HPCs engraft faster than PB- and UCB-derived products

Concept: On average, BM and Umbilical Cord-derived HPC products contain substantially less total nucleated cells and progenitor cells (CD34+ cell) than mobilized peripheral blood products. HPC(A) product time to engraftment is faster than the other two cell sources.

Answer: *A*—The average time to achieve an absolute neutrophil count (ANC) greater than 0.5×10^9/L is usually between 9 and 11 days for mobilized peripheral blood products, while the average neutrophil engraftment time for bone marrow and umbilical cord blood products is 16–20 days and greater than 20 days, respectively. Therefore, other choices (Answers B, C, D, and E) are incorrect.

29. Umbilical cord bloods (UCB) have been used as an alternative source for HPC transplantation for more than 3 decades. What are some of the main advantages of using UCB as a HPC source compare to the peripheral blood or the bone marrow?

A. Easy procurement and relatively high CD34 cell dose.
B. High TNC and fast engraftment.
C. Easy procurement and reduced GVHD.
D. Fast immune-reconstitution and reduced transplant-related mortality (TRM).
E. Higher dose of alloreactive T-cells and donor cells availability for posttransplant cellular immunotherapy.

Concept: Umbilical cord blood transplantation is an alternative option for patients who do not have fully matched related or unrelated donors. Cord bloods are used for both malignant and nonmalignant diseases. The easy procurement with relatively low risk for the donor (mother and baby) is of great advantage and allows increasing the diversity of available grafts. Other advantages include the lower level of alloreactive T-cells due to the immature immune system, a factor that contributes to lower incidence and severity of graft versus host disease (GVHD). This also allows the use of less stringent donor and recipient HLA-match criteria, and the HLA matching is usually based on the three loci HLA-A, HLA-B, and HLA-DRB1.

Answer: *C*—The main advantage of using UCB for transplant is the easy procurement and availability, and the reduced alloreactive T-cells in the graft that result in lower GVHD rates and intensity.

The major disadvantages of using a cord unit are its low cell dose (Answer A) and the inability to recollect the donor, which prevents the possible use of posttransplant immunotherapy used for treat infections and relapse (Answer E). The relatively CBU lower cell dose is the major cause of delayed neutrophil engraftment (Answer B) and immune-reconstitution that contributes to increased TRM (Answer D). A dose of 2.5–3 $\times 10^7$ nucleated cells/kg recipient weights is typically used for patients with malignant diseases. In recent years, new approaches were developed to overcome the limitation in dosage with relatively good efficacy and safety outcomes. These include double umbilical cord transplantation, ex vivo cord blood expansion, and combined transplantation with additional cells from different sources.

Please answer Questions 30–32 based on the following clinical scenario:

A 48-year-old male patient with multiple myeloma was scheduled for peripheral blood autologous HPC collection after G-CSF based mobilization. A double-lumen dialysis catheter was placed by an interventional radiologist on the morning of the scheduled collection, in preparation for a 24 L HPC-A collection. Toward the end of collection, the patient complained of chills and weakness. His temperature was 100.2°F. The patient received antipyretic medication with symptom resolution. The collection was complete and the product was transported to the processing laboratory.

The processing laboratory received and characterized the product for TNC and CD34+ content. The product was plasma reduced, cryopreserved, and stored for future transplant. Sterility testing was performed on the product twice: (1) upon receipt from apheresis and (2) at processing completion prior to product placement in the freezer.

Two days later, the laboratory was informed by the microbiology department that the first culture bottle (upon receipt from apheresis) was positive for *S. epidermidis*. The second culture bottle (at processing completion) was negative.

30. How should cryopreserved autologous products with positive microbial culture be stored?
A. There are no special storage requirements for products with positive microbial culture
B. All cellular therapy products with positive microbial culture should be quarantined
C. Only products from donor with positive infectious disease test (i.e., HTLV, HIV, etc.), but not positive microbial culture, have to be quarantined
D. Autologous products with positive microbial culture are exempt from quarantine
E. Products with positive microbial culture should be discarded

Concept: The FDA and voluntary accreditation agencies require that processing laboratory will test HCT/P product for sterility. All products should be tested at a minimum after processing, but many laboratories test their products prior to processing. That allows capturing contaminations that originated in a septic donor or where introduced to the donor blood stream during line placement. Products with positive microbial testing should be stored in a way that minimized cross-contamination of other products and wrong distribution. Further, to reduce the risk of cross contamination and to provide appropriate warning on contaminated products, some voluntary accreditation bodies recommend that products with a known positive culture be labeled similarly to products from donors with a positive infectious disease test result and requires the use of "biohazard" label for contaminated products.

Answer: *B*—Cellular therapy products with positive infectious disease test results for relevant communicable disease agents and/or positive microbial cultures have to be quarantined. Quarantine does not necessarily require physical separation of these products, but the laboratory is expected to store it in a way that minimizes potential cross-contamination, such as using vapor phase liquid nitrogen freezers and proper labeling. Based on the above information, the other choices (Answers A, C, D, and E) are incorrect.

31. Since the second and final culture (at processing completion) is negative (no bacterial growth detected), the product sterility is determined as which of the following?
A. Negative for microbial culture
B. Positive for microbial culture
C. Positive for microbial testing only at infusion
D. Since it is an autologous product, microbial testing results do not apply
E. Since sterility testing is required only post processing, this product is negative for microbial culture

Concept: Processing facilities are required to validate the microbial testing procedure and demonstrate that it is capable of identifying organisms within acceptable time. However, the ability to correctly identify clinically significant microbial contamination in products depends on the specific assay's sensitivity, the product volume that is used for testing, and the type of contamination (i.e., aerobic, anaerobic bacteria). It is also important to take into account that the processing itself, such as volume reduction and the addition of cryoprotectant, can reduce the sensitivity of the detection. Additionally, products with positive microbial testing require an investigation of all possible causes and should include at least a root cause analysis of the collection and processing procedures. The goal is to identify any possible breach of sterility, finding evidence of a septic donor, and ruling out the possibility of false positive test result. The laboratory is also required to audit and monitor trends of products contamination and recommend change of practice if necessary.

Answer: *B*—Once a product's sterility test is positive for microbial contamination, it should be considered to have positive microbial cultures even if additional tests do not show any growth, and the product labeling should not change. Although positive microbial testing, and especially skin flora bacteria, such as *S. epidermitis*, can sometimes be the result of the testing technique and not the product itself, it is in the best interest of the recipient to consider the product as positive and treat the patient accordingly during and after transplant. It is also required for the laboratory to investigate all aspects of collection and processing and determine if procedures were not followed properly at any point in time. In this case, since the donor's temperature did spike toward the end of collection, there is a reasonable chance that the contamination source is a septic donor. Therefore, the laboratory should investigate if blood cultures were performed on the donor at the day of or the day after collection and see if those came positive for any bacteria. Based on this explanation the other choices (Answers A, C, D, and E) are incorrect.

32. Two months after product collection, the patient is schedule to receive his cryopreserved HPC product for transplant. Are there any specific requirements before releasing of products with microbial positive culture?

A. There are no special disposition requirements for products with positive microbial culture
B. There are no special disposition requirements for products with positive microbial culture from autologous donor only
C. Only the processing laboratory medical director should review and approve the use of the products with positive microbial culture
D. The laboratory medical director and the clinical physician should approve the use of the product with positive microbial culture and the recipient should be notified and provide consent for its use
E. As long as the appropriate antibiotic is used prior to transplant, there are no additional requirements to release products with positive microbial culture

Concept: Since the processing facility is typically the first one to receive the microbial culture results, it is the processing facilities' responsibility to notify the collection facility and the clinical program of any microbial contamination of products. This notification should be done as soon as possible and preferably prior to product release for infusion. Freshly infused products usually do not have available microbial testing results prior to infusion, but cryopreserved products should at least have preliminary results. If final microbial testing results are not available prior to release, it can still be infused providing documentation of approval for release by both the processing facility and the transplant medical directors. If microbial testing is reported positive after infusion, it is important to have in place a method to alert the transplant physician and treat the patient with antibiotics.

Answer: *D*—Infusion of contaminated products require the approval by both the processing laboratory and the clinical medical director. The clinical program is usually responsible for informing the recipient and document an urgent medical need prior to transplant. Based on this explanation the other choices (Answers A, B, C, and E) are incorrect.

End of Case

Please answer Questions 33–35 based on the following clinical scenario:

A 57-year-old male with AML is being evaluated for a possible HPC transplantation.

33. The patient's sister is deemed to be eligible and suitable to donate HPC(A) for the patient. The patient's blood type is O Rh positive and her blood type is A Rh positive. For this type of ABO incompatible HPC transplantation, what is the most severe engraftment-related hematologic complication?

A. Immediate hemolysis
B. Delayed hemolysis
C. Pure red blood cell aplasia
D. Neutropenia
E. Thrombocytopenia

Concept: Both immediate and delayed hemolysis can be severe complication of ABO incompatible HPC transplantation. However, these are not considered engraftment-related outcomes.

Myelopoiesis and megakaryopoiesis are not affected by ABO incompatibility between donors and recipient. Blood group ABO antigens are expressed in the early erythroblast stage of erythropoiesis. Therefore, a high titer of isoagglutinins against incompatible donor RBCs (e.g., major mismatch or bidirectional mismatch) may be associated with the suppression of graft erythropoiesis. Pure red blood cell aplasia (PRCA) may then require prolonged red blood transfusion, while other hematopoietic lineages (platelets and WBCs) are recovered normally. While pretransplantation isoagglutinin titer reduction with donor-type RBC transfusion is associated with reduced posttransplantation RBC transfusion need, the direct causal relationship between isoagglutinin titer and PRCA incidence is not well established. If PRCA does not self-resolve, additional pharmacologic agents can be used, such as corticosteroids, cyclophosphamide, cyclosporine A, and antithymocyte globulin. More recently, rituximab (anti-CD20) and alemtuzumab (anti-CD52) have been shown to induce remission in refractory PRCA. Medical procedures, such as splenectomy and plasmapheresis can be considered as well.

Answer: *C*—The ABO incompatible between the patient and his sister can be classified as major mismatch. Thus, PRCA may be a complication for this type of transplant. Both delay and acute hemolysis are not considered as engraftment-related complication (Answers A and B). Myelopoiesis and megakaryopoiesis are not affected by ABO incompatibility; thus, Answers D and E are not correct.

34. The patient underwent the transplant using the sister's HPC(A), and is now having acute GVHD. Which of the following organs are three most commonly affected in acute GVHD?

A. Skin, gastrointestinal (GI) tract, liver
B. Skin, GI tract, bone marrow
C. Skin, liver, lung
D. GI tract, liver, lung
E. Liver, lung, bone marrow

Concept: Classical acute GVHD is usually diagnosed within the first 100 days of transplantation or persistent, recurrent, or late acute GVHD (occurring past 100 days after transplantation) without chronic GVHD manifestations. It remains one of the most significant barriers to successful HPC transplantation. Its pathogenesis involves the interaction between donor T-cells and host tissues. In a simple model, chemotherapy leads to damage host tissues (mostly in the GI tract). This results in the translocation of bacteria and endotoxin into the bloodstream, as well as the release of inflammatory cytokines. Major histocompatibility complex (MHC) antigen expression is then upregulated on host antigen-presenting cells resulting in the recipient antigen presenting to the donor T-cells. The final phase of the GVHD responses involves T-cell cytotoxic effects against host tissues. Besides the dose of T-cells in the HPC product, other risk factors associated with acute GVHD include: HLA and/or gender mismatch, unrelated donor, multiparity donor, older donor, ABO mismatch, and myeloablative regimen. Symptoms of acute GVHD include maculopapular erythromatous rash, nausea, vomiting, diarrhea, and elevated liver enzymes. First line of treatment for acute GVHD is corticosteroids. If the patient fails steroid treatment, other medications, such as cyclosporine, tacrolimus, mycophenolate mofetil, and antithymocyte globubin, may be attempted.

Answer: *A*—The skin is the most commonly affected organ in acute GVHD, following by GI tract and liver. Lung and bone marrow are not typically affected in acute GVHD. This is different from transfusion associated GVHD caused by transfusing of nonirradiated cellular products to immunosuppressive patients, where the bone marrow is severely affected in addition to the skin, GI tract, and liver. Based on this explanation the other choices (Answers B, C, D, and E) are incorrect.

35. The patient returns to the clinic presenting with signs and symptoms suggesting chronic GVHD. What is the first line of treatment for chronic GVHD?

A. Corticosteroids
B. Extracorporeal photopheresis

C. Therapeutic plasma exchange
D. Pentostatin
E. Rituximab

Concept: Chronic GVHD is a major cause of nonrelapse morbidity and mortality after allogeneic HPC transplantation. It is due to suppression and dysregulation of the immune system, resulting in increased risk of infection and multi-organ dysfunction. Chronic GVHD usually involves the skin, eyes, mouth, liver, GI tract, vagina, and lungs. In the past, any GVHD developed after 100 posttransplantations was classified as chronic GVHD. However, this is no longer true. The recent NIH guidelines classify chronic GVHD as classical chronic GVHD (signs and symptoms consistent with chronic GVHD only) and overlap syndrome (features both acute and chronic GVHD). The diagnosis of chronic GVHD requires the following:

- At least one diagnostic sign (i.e., manifestation that establishes the presence of chronic GVHD without the need of further testing, such as poikiloderma in skin, esophageal web, vaginal stenosis, etc.) or one distinctive sign (i.e., manifestation highly suggestive of chronic GVHD but insufficient alone to establish the diagnosis, such as skin depigmentation or keratoconjunctivis sicca) confirmed by biopsy, laboratory test, or radiology in the same or other organ
- Distinction from acute GVHD (maculopapular erythromatous rash, nausea, vomiting, diarrhea, and elevated liver enzymes)
- Exclusion of other diagnoses, such as infection, drug-induced, or secondary malignancy

Answer: *A*—Corticosteroids are the main first-line therapy for patients with chronic GVHD. For the patients who do not respond after 3 months on steroids or demonstrate progressive GVHD, second line therapies include cyclosporine, tacrolimus, sirolimus, mycophenolate mofetil, extracorporeal photopheresis (Answer B), pentostatin (Answer choice D), and rituximab (Answer E). Therapeutic plasma exchange (Answer C) is not typically part of the regimen for chronic GVHD.

End of Case

Suggested Reading

[1] Guidance for Industry: Eligibility Determination for Donors of Human Cell, Tissues, and Cellular and Tissue Based Products (HCT/Ps), 2007. Available from: http://www.fda.gov/downloads/biologicsbloodvaccines/guidancecomplianceregulatory-information/guidances/tissue/ucm091345.pdf

[2] H.V.C. Schoemans, Cellular biology of hematopoiesis, in: R.B.E Hoffman, S.J. Shattil (Eds.), Hematology: Basic Principles and Practice, sixth ed., Elsevier, Philadelphia, PA, 2009, pp. 200–212.

[3] T.F. Wang, S.H. Wen, R.L. Chen, et al. Factors associated with peripheral blood stem cell yield in volunteer donors mobilized with granulocyte colony-stimulating factors: the impact of donor characteristics and procedural settings, Biol. Blood Marrow Transplant. 14 (11) (2008) 1305–1311.

[4] W. Bensinger, F. Appelbaum, S. Rowley, et al. Factors that influence collection and engraftment of autologous peripheral-blood stem cells, J. Clin. Oncol. 13 (10) (1995) 2547–2555.

[5] C. Chabannon, A.G. Le Corroller, F. Viret, et al. Cost-effectiveness of repeated aphereses in poor mobilizers undergoing high-dose chemotherapy and autologous hematopoietic cell transplantation, Leukemia 17 (4) (2003) 811–813.

[6] J.F. DiPersio, E.A. Stadtmauer, A. Nademanee, et al. Plerixafor and G-CSF versus placebo and G-CSF to mobilize hematopoietic stem cells for autologous stem cell transplantation in patients with multiple myeloma, Blood 113 (23) (2009) 5720–5726.

[7] J.F. DiPersio, I.N. Micallef, P.J. Stiff, et al. Phase III prospective randomized double-blind placebo-controlled trial of plerixafor plus granulocyte colony-stimulating factor compared with placebo plus granulocyte colony-stimulating factor for autologous stem-cell mobilization and transplantation for patients with non-Hodgkin's lymphoma, J. Clin. Oncol. 27 (28) (2009) 4767–4773.

[8] Cellular Therpy: A Physician's Handbook, first ed., AABB, Bethesda, MD, 2004.

[9] G.S. Kao, Assesment of collection quality, in: A.E. M., K. Loper (Eds.), Cellular Therapy: Principles, Methods, and Regulations, AABB, Bethesda, MD, 2009, pp. 291–302.

[10] R.D. Sutherland, M. Keeney, Enumeration of CD34+ cells by flow cytometry, in: E.M. Areman, K. Loper (Eds.), Cellular Therapy: Principles, Methods, and Regulations, AABB, Bethesda, MD, 2009, pp. 538–550.

[11] K. Gutensohn, M.M. Magens, P. Kuehnl, W. Zeller, Increasing the economic efficacy of peripheral blood progenitor cell collections by monitoring peripheral blood CD34+ concentrations, Transfusion 50 (3) (2010) 656–662.

[12] M. Croskell, K. Loper, D. McKenna, Basic cellular therapy manufacturing procedures, in: E.M. Areman, K. Loper (Eds.), Cellular Therapy: Pronciples, Methods, and Regulations, AABB, Bethesda, MD, 2009, pp. 303–313.

[13] C.A. Tormey, J.E. Hendrickson, Transfusion support for hematopietic stem cell transplant recipients, in: M.K. Fung, B.J. Grossman, C.D. Hillyer, C.M. Westoff (Eds.), Technical Manual, eighteenth ed., AABB, Bethesda, MD, 2014, pp. 631–644.

[14] E.M. Staley, J. Schwartz, H.P. Pham, An update on ABO incompatible hematopoietic progenitor cell transplantation, Transfus. Apher. Sci. 54 (3) (2016) 337–344.

[15] A. Hubel, Cryopreservation of Cellular Therapy Products, in: E.M. Areman, K. Loper (Eds.), Cellular Therapy: Principles, Methods, and Regulations, AABB, Bethesda, MD, 2009.

[16] International Standards for Hematopietic Cellular Therapy Product Collection, Processing, and Administration: Accreditation Manual, sixth ed., Foundation for Accreditation of Cellular Therpay (FACT), 2015.

[17] Standards for Cellular Therapy Services, seventh ed., AABB, Bethesda, MD, 2015.

[18] M. Körbling, P. Anderlini, Peripheral blood stem cell versus bone marrow allotransplantation: does the source of hematopoietic stem cells matter?, Blood 98 (10) (2001) 2900–2908.

[19] M.L. Beck, Red blood cell polyagglutination: clinical aspects, Semin. Hematol. 37 (2) (2000) 186–196.

[20] A. Srivastava, H. Pearson, J. Bryant, et al. Acidified chloroquine treatment for the removal of class I HLA antigens, Vox Sang. 65 (2) (1993) 146–150.

[21] A.H. Filipovich, D. Weisdorf, S. Pavletic, G. Socie, J.R. Wingard, S.J. Lee, P. Martin, J. Chien, D. Przepiorka, D. Couriel, E.W. Cowen, P. Dinndorf, A. Farrell, R. Hartzman, J. Henslee-Downey, D. Jacobsohn, G. McDonald, B. Mittleman, J.D. Rizzo, M. Robinson, M. Schubert, K. Schultz, H. Shulman, M. Turner, G. Vogelsang, M.E. Flowers, National Institutes of Health consensus development project on criteria for clinical trials in chronic graft-versus-host disease: I. Diagnosis and staging working group report, Biol. Blood Marrow. Transplant 11 (12) (2005) 945–956.

[22] M.H. Jagasia, H.T. Greinix, M. Arora, K.M. Williams, D. Wolff, E.W. Cowen, J. Palmer, D. Weisdorf, N.S. Treister, G.S. Cheng, H. Kerr, P. Stratton, R.F. Duarte, G.B. McDonald, Y. Inamoto, A. Vigorito, S. Arai, M.B. Datiles, D. Jacobsohn, T. Heller, C.L. Kitko, S.A. Mitchell, P.J. Martin, H. Shulman, R.S. Wu, C.S. Cutler, G.B. Vogelsang, S.J. Lee, S.Z. Pavletic, M.E. Flowers, National Institutes of Health Consensus Development Project on Criteria for Clinical Trials in Chronic Graft-versus-Host Disease: I. The 2014 Diagnosis and Staging Working Group report, Biol. Blood Marrow Transplant. 21 (3) (2015) 389–401.

C H A P T E R

18

Human Tissue Banking and Hospital-Based Surgical Tissue Management

Annette J. Schlueter, Beth M. Alden

University of Iowa, Iowa City, IA, United States

Many steps are required to ensure safe and efficacious autologous and allogeneic human tissue transplants. These steps start with appropriate donor screening and testing, proceed through tissue recovery, processing, packaging/labeling, storage, distribution, and implantation. Tissue tracking through this sequence is critical to ensure the ability to respond appropriately to tissue recalls and adverse events. The entities that perform the various steps typically include an organization that recovers tissue from the donor, a tissue bank that processes the tissue, and a tissue service that manages the steps that take place at the implanting facility. This chapter provides a review of important concepts and regulations related to this process, with a focus on elements that are important for hospital-based tissue services.

1. Which of the following human cells, tissues, and cellular and tissue-based products (HCT/Ps) are regulated solely as "361 products" by the Food and Drug Administration (FDA)?
   A. Liver for organ transplant
   B. Cultured cartilage cells
   C. Peripheral blood hematopoietic progenitor cells (HPC) from unrelated donors
   D. Amniotic membrane
   E. Cultured nerve cells

   Concept: HCT/Ps are highly regulated products. The FDA regulates human tissue based on perceived risk to the recipient. "361 products" are considered to be lower risk than "351 products." "361 products" are defined as articles containing or consisting of human cells or tissues that are intended for implantation, transplantation, infusion, or transfer into a human recipient. To be a 361 HCT/P, the product must meet all four of the following criteria:
   - It is minimally manipulated.
   - It is intended for homologous use.
   - Its manufacture does not involve combination with another article, except for water, crystalloids, or a sterilizing, preserving, or storage agent.
   - It does not have a systemic effect and is not dependent upon the metabolic activity of living cells for its primary function, or if it has such an effect, it is intended for autologous use or allogeneic use in close relatives or for reproductive use.

   Table 18.1 lists examples of products regulated under section 361 and 351.

Transfusion Medicine, Apheresis, and Hemostasis. http://dx.doi.org/10.1016/B978-0-12-803999-1.00018-3

TABLE 18.1 Examples of "351 Products" and "361 Products"

Products regulated under section 361	Products regulated under section 351
Bone	Cultured nerve cells
Cartilage	Cultured cartilage cells
Cornea	Lymphocyte immune therapy
Fascia	Gene therapy products
Ligament	Human cloning
Pericardium	Human cells used in therapy involving transfer of genetic material
Hematopoietic stem/progenitor cells derived from peripheral blood or cord blood if for autologous use or use in a first or second degree blood relative	Hematopoietic stem/progenitor cells derived from peripheral blood or cord blood for allogeneic use not in a first or second degree blood relative
Sclera	Unrelated allogenic stem/progenitor cells derived from peripheral blood or cord blood
Skin	
Tendon	
Vascular graft	
Heart valve	
Dura mater	
Reproductive cells and tissues (e.g., semen, oocytes, embryos, amniotic membrane)	

FDA expressly excludes the following from regulation as HCT/Ps:

- vascularized human organs for transplantation
- whole blood or blood components or derivative products already regulated as biologics under 21 CFR Parts 607 and 207
- secreted or extracted human products except semen (e.g., milk collagen and cell factors)
- minimally manipulated bone marrow for homologous use (and not combined with another article except for water, crystalloids, or a sterilizing, preserving, or storage agent)
- cells, tissues, and organs derived from animals other than humans

Answer: *D*—Amniotic membrane is an unmodified human tissue implant. Organ transplants (Answer A) are regulated under the Health Resources and Services Administration in the United States. Cultured tissue (Answer B) and unrelated peripheral blood HPC (Answer C), and cultured nerve cells (Answer E) are regulated as 351 products.

2. The FDA requires registration of establishments that perform various functions related to human cells, tissues, and cellular and tissue-based products HCT/Ps. A hospital based tissue service must register if they perform which of the following functions?

A. Receive corneas from a distributor and dispense for implantation within the same institution
B. Store autologous skin for implantation during the same admission
C. Process corneas for an in vitro research study on treatment of blindness
D. Recover semen for immediate transfer to a sexually intimate partner
E. Ship autologous cranial flaps for reimplantation at another institution

Concept: FDA registration is required for all institutions that recover, screen, test, package, process, store, label, and/or distribute HCT/Ps. However, they have allowed some exceptions that are considered to not significantly affect the safety of the product for the patient who ultimately receives it. These exceptions generally are in circumstances where the exempted function does not significantly increase the risk for the introduction, transmission or spread of communicable disease.

Answer: *E*—Shipment of cranial flaps to another institution is considered distribution by the FDA, and therefore, requires registration. Storage within an institution solely for implantation, transplantation, infusion, and/or transfer within that same facility is exempt from registration (Answers A and B), as is any processing of tissue for nonclinical research (or educational) purposes (Answer C). Recovery of reproductive cells for immediate transfer into a sexually intimate partner is also exempt (Answer D). Other exemptions to registration include (1) removal of HCT/P's from an individual and reimplantation of the HCT/P's into that same individual during the same surgical procedure, (2) carriers who accept, receive, carry, or deliver HCT/P's in the usual course of their business as a carrier, and (3) individuals under contract with a registered establishment that are engaged solely in recovering cells or tissues for that establishment.

3. Your hospital is considering whether their tissue service should be centralized or decentralized. Advantages to a decentralized service include which of the following?
A. Faster access to the tissues by the operating room/clinic teams
B. Smaller inventory of tissues across the institution
C. Preferred pricing from tissue vendors
D. Less coordination required for responding to tissue recalls
E. Less monitored storage units required

Concept: Management of transplantable tissues is a complex process. There are two different models that are used in hospitals, centralized and decentralized. Centralized tissue services have one site or section within the organization responsible for most or all of the functions of the tissue service, including vendor qualification, ordering, inspection, storage, distribution, and tracking of tissues. In decentralized services, these functions are the responsibility of the surgeon or surgical service. However, since management by multiple services could lead to poor follow-up, The Joint Commission (JC) requires that in decentralized models, one person is responsible for keeping track of the individual surgical specialties. Though no regulatory agency mandates a particular model to be used, the AABB recommends the centralized model and encourages the transfusion medicine service to be the centralized hub of tissue service management.

Answer: *A*—Faster access to the tissues by the operating room/clinic teams is an advantage of a decentralized service.
The other choices (Answers B, C, D, and E) represent advantages of a centralized tissue service, which also include the following:

- Decreased overall inventory at the facility
- Less storage units to manage
- Improves chance of obtaining preferred vendor pricing
- Easier to track the identity of all patients who have received grafts in the event of a recall

4. The purpose of qualifying a tissue vendor is which of the following?
A. Ensure a reliable supply of safe, efficacious tissues
B. Identify the least expensive source of a particular tissue
C. Understand the relationship of the vendor to other tissue distributors
D. Able to track a tissue from source to recipient
E. Find a location for autologous tissue storage

Concept: Tissue vendor qualification is the process of evaluating suppliers of tissue for their ability to provide a quality product. Qualification of the vendor is the responsibility of the transplanting facility. Characteristics of a quality product include its safety and efficacy, as well as availability. Qualification evaluates documentation provided by the vendor to support these claims, including standard operating procedures, results of inspections by outside agencies [e.g., FDA, American Association of Tissue Banks (AATB), Eye Bank Association of America (EBAA)], and certifications by these agencies.

Answer: *A*—Ensuring that the transplanting facility has a reliable, safe, and efficacious tissue supply is a key reason for supplier qualification. The least expensive (Answer B) tissue may not be the safest or most efficacious. While the qualification may help to clarify the relationship of the vendor to other tissue distributors (Answer C), this is not its purpose. Tracking and storing tissues (Answer D) are not intrinsic parts of vendor qualification, but are vital responsibilities of the hospital tissue service. Autologous tissue storage (Answer E) does not usually require a vendor.

5. JC standards for transplant safety require compliance by the hospital tissue service with which of the following elements?
   A. Existence of an order from the surgeon before bringing tissue to the operating room
   B. Informed consent from the recipient prior to the tissue implantation
   C. Tissue vendor notification of the identity of the patient receiving the tissue implant
   D. Investigation of adverse events suspected of being linked to tissue implants
   E. Billing of the recipient's insurance for the exact cost of the tissue

Concept: JC accredits healthcare organizations and their practices to ensure patient safety. Safe human tissue implantation is evaluated as part of this process.

Answer: *D*—Investigation of adverse events has a direct impact on the safety of future tissue implants from that manufacturer. Communication between the surgeon (Answer A) and the tissue service, or vendor and the tissue service, is not mandated by the Standards. To date, JC has not required informed consent (Answer B) prior to tissue implantation. Although they do require consent prior to a procedure, the specific detail of tissue implantation in the content of that consent is not required. JC requires that the tissue vendor be notified of the final disposition of the tissue, but they do not require that the tissue service share the identity of the patient receiving the implant (Answer C). They are also not concerned with reimbursement practices (Answer E).

6. Which of the following tissues can most easily be treated by the tissue vendor to reduce the infectious risk enough that the tissue may be considered sterile?
   A. Skin
   B. Cornea
   C. Achilles tendon
   D. Aortic valve
   E. Iliac vein

Concept: The standard for tissue sterility set by the AATB is less than 1 in 10^6 chance of a viable microorganism being present in the tissue. Methods to ensure sterility of tissues generally involve mechanical, chemical, or radiation treatments. These methods have the potential to damage live cells or delicate tissue structures. However, relatively acellular tissues are more likely to withstand such treatments. There are many different protocols for disinfecting bone and soft tissue. The FDA does not have a mandated sterilization process. Tissue banks may use a combination of techniques and often use patented protocols (Table 18.2).

Answer: *C*—Tissues offering structural support, such as bone, tendons, or sclera are used for reconstruction or repair purposes and sterilization methods can generally be used on these tissues without compromising biomechanical properties. If the replacement tissue relies on living cells to maintain its function (such as corneas and vascular tissues), milder aseptic processing methods are generally used. See Question 24 for further discussion of methods used for tissue disinfection and for an explanation of the incorrect answers.

TABLE 18.2 Comparison of Processing Techniques

	Aseptic	Ethylene oxide	Gamma irradiation	Chemical soaking
Kills bacteria	No	Yes	Yes	Yes
Kills fungi	No	Yes	Yes	Yes
Kills spores	No	Yes	Yes	No
Kills enveloped viruses (e.g., HIV, hepatitis A)	No	Yes	Yes (dose dependent)	Yes
Removes blood and lipids	Surface only	No	No	Surface only
Preserves strength	Yes	Yes	Decreases (dose dependent)	Yes
Preserves biocompatibility	Yes	Yes	Yes	Yes
Penetrates into tissue	Surface only	Thickness dependent	Full penetration	Surface only

7. Based on allograft-associated infectious disease reports to the FDA, infection transmission is the most common from which of the following types of allografts?
A. Cartilage
B. Bone
C. Tendons
D. Sclera
E. Heart valve

Concept: Tissues that can be sterilized or subjected to rigorous disinfection processes are less likely to transmit infectious disease.

Answer: *E*—Heart valves cannot be sterilized and are not always stored in disinfectant solutions, which may be the reason that they are the most common tissue type implicated in allograft-associated infections. Infections that have been associated with heart valve implants include bacteria (including *Mycobacterium tuberculosis*), hepatitis B, and fungi (including *Aspergillus fumigatus*). Studies have shown that disinfection with low dose antibiotics may still result in survival of infectious organisms.

The other choices (Answers A, B, C, and D) are incorrect because they can withstand rigorous disinfection and are therefore, less prone to infectious disease transmission.

8. Which of the following organisms has been reportedly transmitted by blood transfusion, tissue transplantation, and organ transplantation?
A. Rabies virus
B. Hepatitis C virus
C. Prions
D. *Mycobacterium tuberculosis*
E. *Babesia microti*

Concept: It is generally easier to eliminate infectious risk from tissues for transplantation than it is to eliminate these risk factors from blood products and/or solid organs. Tissues can generally tolerate more disinfection steps than organs or blood. In addition, the shelf-life of the tissue is longer, so the results of additional donor testing can be completed prior to dispensing the product.

Answer: *B*—Hepatitis C has been transmitted by blood transfusion, tissue transplantation, and organ transplantation. There has never been a documented transmission of rabies or tuberculosis from a blood transfusion (Answers A and D). *Babesia microti* causing babesiosis and prions causing Variant

Creutzfeldt-Jakob (vCJD) disease have never been reported to be transmitted from a transplanted organ (Answers C and E).

9. Vessels harvested at the time of organ donation may be stored for future use if they are not used during the transplantation of that organ. According to Organ Procurement and Transplantation Network (OPTN) policies, which of the following practices is permissible for these stored vessels?
   A. Implantation of a vessel from a donor with positive serology for hepatitis C into a recipient with hepatitis C
   B. Use of the vessel to repair a torn portal vein in a nontransplant patient
   C. Cryopreservation of the vessel
   D. Storage of the vessel for 21 days from harvest
   E. Implantation of a vessel with a positive bacterial culture

Concept: Though FDA does not regulate vessels harvested from organ donors, OPTN has specific rules governing their use. These rules are intended to ensure patient safety.

Answer: *E*—OPTN does not mandate bacterial culture of organ donor vessels at any point between recovery and implantation; thus, there are no rules prohibiting use of a vessel with a positive culture result if a culture has been obtained. OPTN does not allow storage of vessels with positive hepatitis C serology (Answer A). The vessels are stored at 2–8°C (Answer C) for up to 14 days (Answer D), and must be used only in patients who have received an organ transplant (Answer B).

10. Nonhematologic malignancies have been transmitted via transplantation of which of the following tissue types?
   A. Bone
   B. Skin
   C. Cornea
   D. Valves
   E. Tendons

Concept: Unlike blood, tissue transplant has been reported as the source of subsequent nonhematologic malignancy in the recipient. Thus, it is especially important to screen tissue donors for malignancy. The incidence of malignancy transmission is still an uncommon occurrence following tissue transplantation relative to organ transplantation.

Answer: *C*—Corneal transplants were the documented source of two cases of adenocarcinoma in recipients. None of the other tissues listed (Answers A, B, D, and E) have been reported to be the source of a malignancy.

11. Which of the following tissue types requires storage conditions of $\leq -135°C$ to maintain function for the maximal amount of time?
   A. Cornea
   B. Pericardium
   C. Cancellous bone
   D. Demineralized bone matrix (DBM)
   E. Pulmonary valve

Concept: Maintenance of manufacturer's recommended storage conditions is critical for maintaining graft function after implantation, and may require specialized storage equipment for the hospital tissue service (Table 18.3).

Answer: *E*—Studies have shown that cryopreservation of cardiac valves at $\leq -135°C$ allows similar clinical results to be achieved, compared to valves stored at 4°C. Additionally, cryopreservation permits storage times of the valves to be increased to years, rather than weeks if stored at 4°C.

TABLE 18.3 Common Storage Times and Conditions

Graft type	Storage condition	Storage time
Acellular dermal matrix	Room temperature	2 years
Cardiac tissue (e.g., valves, patches, veins, arteries)	≤ −100°C	2–5 years
Corneas for keratoplasty	2–8°C	14 days
Cornea for procedures other than keratoplasty	Room temperature	2 years
Demineralized bone matrix (e.g., putties, pastes, gels)	Ambient temperature (15–30°C)	2 years
Freeze dried bone	Ambient temperature (15–30°C)	2–5 years
Frozen soft tissues (e.g., tendons, skin)	<−40°C	5 years
Sclera	Room temperature	2 years

Cancellous bone, pericardium, and DBM (Answers B, C, and D) can be stored for long periods at room temperature without losing function. Corneas (Answer A) are either refrigerated at 2–8°C to retain function for cornea replacement, or stored at room temperature, if used for specialized procedures. No cryopreservation methods for longer term storage of corneas for cornea replacement have been validated.

12. What type of bone graft would be most efficacious for the repair of a large bone defect?
A. Autologous cancellous bone
B. Allogeneic cortical bone
C. Allogeneic cortical cancellous bone
D. Demineralized bone matrix (DBM)
E. Alloplastic graft (nonbiologic material)

Concept: Ideal surgical bone grafts should supply four elements for bone reformation: osteoconductive matrix (nonviable scaffolding conducive to bone growth), osteoinductive factors (growth factors provided by bone matrix, including bone morphogenetic proteins), osteogenic cells, and structural integrity. Autograft tissue is the ideal source based on these four elements, but the disadvantages to autografts are the finite quantity available, and the donor site morbidity. Table 18.4 provides differences in these four elements among different type of bone grafts.

TABLE 18.4 Different Bone Grafts and Their Properties

Graft	Osteoconduction	Osteoinduction	Osteogenic	Immediate strength
Autologous cancellous	4+	2+	3+	–
Autologous cortical	1+	±	±	2+
Freeze-dried cortical cancellous allograft	1+	±	–	+
Demineralized bone matrix	1+	–	–	–

Answer: C—Cortical cancellous bones offers the best of the properties of both cancellous and cortical bone. Since there is a large defect in this case, an allogeneic graft is required. Cancellous bone is more rapidly revascularized than cortical bone and it is osteoconductive, osteogenic, and offers a limited amount of osteoinductive growth factors (Answer A). While cortical bone (Answer B) is less biologically active, it provides more initial structural support than cancellous bone, which may be needed when dealing with a large defect. Demineralized bone matrix (Answer D) does not provide osteoinductive or osteogenic properties, nor immediate strength to the defect. Alloplastic grafts possess osteoconductive and structural properties, but do not provide osteoinductive or osteogenic properties (Answer E). Certain alloplastics may become brittle or have low tensile strength.

13. DBM differs from a cancellous bone allograft in that DBM contains more of which of the following?
A. Live mesenchymal stem cells
B. Hydroxyapatite
C. Calcium
D. Osteoinductive growth factors
E. Bone marrow

Concept: Demineralized bone matrix is allograft bone with inorganic material removed, leaving primarily collagen and osteoinductive growth factors [e.g., insulin growth factor (IGF)-1, transforming growth factor (TGF)-β1, and vascular endothelial growth factor (VEGF)]. To produce DBM, bone is pulverized and subjected to acidic extraction. Removal of the mineral decreases the mechanical strength of the allograft but increases its ability to promote new bone formation. Please see Table 18.4 for a comparison of the properties of DBM, cortical, and cancellous bone. DBM is often used in association with another structural supporting device, or as a cancellous or cortical bone graft extender.

Answer: *D*—DBM contains a higher concentration of osteoinductive growth factors compared to cancellous bone. There are no live cells (Answer A) in either DBM or cancellous bone following processing. Hydroxyapatite (Answer B) and calcium (Answer C) are part of the mineral content of bone, present in cancellous bone allograft but not DBM. DBM does not contain any bone marrow (Answer E).

14. Which of the following is an indication for using a fresh frozen bone allograft over a lyophilized (freeze-dried) allograft for a massive cortical defect?
A. Improved mechanical strength
B. Decreased immunogenicity
C. Improved osteoconductivity
D. Improved osteoinductivity
E. Sterility

Concept: Lyophilization involves the initial freezing of a bone graft followed by chemical and mechanical procedures to reduce the water content of the graft to as low as 5%–8%. The tissues can then be stored at room temperature, but require rehydration prior to implantation. Lyophilization destroys most of the osteoprogenitor cells, but cancellous and cortical structure is retained. Fresh frozen tissues are stored at −20 to −80°C, and simply require thawing and washing prior to implantation.

Answer: *A*—Mechanical strength is one of the four elements necessary for bone reformation, and is reduced when bone is lyophilized. Freezing and procedures to lyophilize bone allografts both reduce the immunogenicity (Answer B) of the graft. Lyophilization does not significantly reduce the healing properties of bone over freezing.

The risk of infectious disease transmission (Answer E) from fresh frozen allografts is greater than in tissue that has been processed for lyophilization; however, current donor eligibility criteria have significantly reduced the risk of infectious disease transmission from either type of bone. The lower risk of infectious disease transmission from lyophilized bone needs to be weighed against the loss of strength when replacing a large defect, especially if the defect involves a load bearing structure. Osteoconductivity and osteoinductivity (Answers C and D) are not different between fresh frozen and lyophilized tissue.

15. The manufacturer's instructions for use of a freeze-dried (lyophilized) tendon have been lost. The operating room nursing staff contacts the director of the hospital tissue service for guidance. Which of the following responses would be appropriate?

A. The tissue cannot be used without the instructions and should be discarded
B. Any portion of the tendon not implanted can be returned for future implantation
C. Isotonic solutions are recommended for rehydration
D. No antibiotics may be added to the rehydration solution
E. Manipulate the product continuously during rehydration

Concept: All purchased tissues are supplied with manufacturer's instructions for the preparation of the tissue prior to implantation. These instructions follow certain general principles based on tissue type, and if the instructions are missing, the tissue may still be used successfully if these principles are followed. Surgeons can override the tissue preparation instructions provided by the manufacturer, but the tissue service Medical Director is not in a position to do so.

Answer: *C*—Tissues can still be used if the instructions are missing as long as someone knowledgeable about their processing is able to provide guidance (Answer A). In this case, the tendon should be hydrated using an isotonic solution. Once a tissue is rehydrated, it cannot be stored for use in the future (Answer B). Rehydration should occur in an isotonic solution and antibiotics (Answer D) may be added to this solution, if the patient is not allergic to the antibiotic. The graft should not be manipulated during the initial rehydration period (Answer E), and it should be rinsed with a sterile irrigant following rehydration, prior to implantation.

16. An advantage of using a banked autograft rather than allograft or xenograft for a tissue implant is decreased risk of which of the following?

A. Bacterial contamination
B. Immune-mediated rejection
C. Clerical error
D. Storage malfunction
E. Inferior biomechanical properties

Concept: Autologous tissue storage programs have been very successful in allowing patients to avoid certain types of allografts, such as bone and skin. However, they depend on meticulous processing, labeling, and tracking to avoid introducing risks that are not present or minimized with the use of allograft.

Answer: *B*—The risk of immune-mediated rejection is eliminated with the use of autologous tissue. Bacterial contamination (Answer A) risk is higher with autografts due to the less controlled environment under which they are obtained. Loss due to storage malfunction (Answer D) is a risk avoided with allograft tissue, since the allograft is not uniquely matched to the patient, and therefore, can be replaced with tissue from a different donor. Allograft donors are carefully selected to provide tissues with high quality biomechanical properties (Answer E). Clerical errors (Answer C) would be expected to be comparable between allograft and autograft programs; however, the infectious risk from a clerical error in an autograft program could be higher than from the issue of an "incorrect" allograft.

17. For which allograft is the risk of immune-mediated rejection the highest?

A. Skin
B. Tendon
C. Bone
D. Cornea
E. Artery

Concept: More cellular grafts with higher human leukocyte antigen (HLA) antigen expression have a higher risk of rejection.

Answer: *A*—Skin is the most cellular of the tissues listed, and it contains many cells with high levels of HLA antigen expression, including T cells, Langerhans cells, and other dendritic cells. Thus, it has the highest risk of immune-mediated rejection. The cellular content of tendon and bone (Answers B and C) is much lower than that of skin, thus total Class I HLA antigen expression is lower. Furthermore, the density of lymphocytes and dendritic cells is much lower in tendon, bone, cornea, and artery (Answers D and E) than in skin, therefore Class II HLA antigen expression is substantially lower in these tissues. Finally, corneal transplants are placed into an immunosuppressive microenvironment, which further reduces corneal immune rejection risk.

18. The operating room sends an autologous bone flap to the hospital tissue service for storage. Which of the following deficiencies would most certainly lead to discard of the tissue?
A. *Staphylococcus epidermidis* is reported from a culture on the flap
B. The flap is delivered in a surgical towel with no container
C. Unknown HIV status of the patient
D. No statement "For Autologous Use only" on the label is present
E. No expiration date is recorded on the label

Concept: Autologous tissue is not subject to as many regulations as allografts. However, the tissue service must have confidence that the identity of the patient is clear, and that the tissue can be stored and reimplanted safely.

Answer: *B*—This tissue has been exposed to a nonsterile environment, and is not in an appropriate container validated to maintain tissue integrity. Rare skin contaminants in cultures from bone flaps have not been shown to correlate with subsequent infection with after implantation (Answer A). Viral serologies (Answer C) are not required prior to storage of autologous tissue. The labeling deficiencies listed could safely be added to the label by the tissue service (Answers D and E).

19. Which infectious disease test is required by FDA prior to storage for both allogeneic blood donors and anonymous sperm donors?
A. Anti-hepatitis C (HCV) antibodies
B. Nucleic acid test for hepatitis B (HBV)
C. Nucleic acid test for malaria
D. *Chlamydia trachomatis*
E. Anti-cytomegalovirus (CMV) antibodies

Concept: Regulatory and accrediting agencies require that blood and tissue be tested for multiple infectious diseases that are potentially transmissible by transfusion or implantation, and for which validated tests have been approved (Table 18.5). No infectious disease testing is required for donors of reproductive tissue designated for use in a sexually intimate partner, or donors of autologous tissue.

Answer: *A*—Anti-HCV testing is required for both blood and HCT/P donors. Nucleic acid testing for hepatitis B is not required for HCT/P donors (Answer B). The FDA has not approved any tests for malaria that may be used for blood or HCT/P donor eligibility determination. Malaria testing (Answer C) is not required for either of the products. Testing for *Chlamydia* (Answer D) must be performed for donors of reproductive HCT/Ps (unless they are sexually intimate partners of the recipient), and testing for CMV antibodies is required for donors of viable, leukocyte-rich HCT/Ps, which includes semen donors. The results of CMV testing (Answer E) do not necessarily make the donor ineligible, but the test must be performed.

20. A directed sperm donor who is not the sexually intimate partner of the intended recipient had incomplete retesting for Hepatitis B (sample tube broke at the testing laboratory). His testing at the time of donation was negative. According to AATB standards, labeling requirements for this product include which of the following?

TABLE 18.5 Infectious Disease Testing Required by the FDA for HCT/P Donors (As of 10/29/16)

Test	Blood	HCT/P
HBsAg	X	X
Anti-HBc	X	X
HBV NAT	X	
Anti-HCV	X	X
HCV NAT	X	X
Anti-HIV1	X	X
HIV-1 NAT	X	X
Anti-HIV2	X	X
Anti-HTLV I/II	X	X
Anti-*T. pallidum*	X	X
West Nile virus (WNV)	X	
Anti-CMV		X[a]
Chlamydia trachomatis		X[b]
Neisseria gonorrhoeae		X[b]

[a] *Donors of viable, leukocyte-rich HCT/P's only (includes hematopoietic progenitor cells and semen).*
[b] *Donors of reproductive HCT/P's from nonsexually intimate partners.*

A. "Warning: Advise Recipient of Communicable Disease Risks"
B. "Not evaluated for infectious substance"
C. Biohazard symbol
D. "For nonclinical use only"
E. "For use by sexually intimate partner only"

Concept: The failure to complete the second testing for infectious substances, for directed donors of reproductive tissues, does not necessarily mean that the product must be discarded. However, specific warning labels to point out the increased risk must be applied. The sample for testing must be collected from the donor at the time of recovery of the cells or tissue from the donor, or up to 7 days before or after the recovery. For donors of peripheral blood stem/progenitor cells, oocytes and bone marrow (if not exempted under CFR1271.3), the sample from the donor may be collected up to 30 days prior to recovery. Required infectious disease tests are listed in Table 18.5. If the donor tests positive for a relevant communicable disease agent or disease (RDCAD), the tissue is discarded, unless urgent medical need can be demonstrated. The FDA defines urgent medical need as "no comparable HCT/P is available and the recipient is likely to suffer death or serious morbidity without the HCT/P." This most commonly is used in the setting of allogeneic hematopoietic stem cell donors.

Answer: *B*—Evaluation for infectious substances was not completed, therefore, this statement applies. The tissue does not contain a known infectious disease risk, so "Warning: Advise Recipient of Communicable Disease Risks" (Answer A) and the biohazard label (Answer C) are not appropriate. The product may still be used as first intended, as long as the recipient is aware of the potential increased risk, therefore "For nonclinical use only" (Answer D) and "For use by sexually intimate partner only" (Answer E) statements do not apply.

21. As per AABB and FDA standards, maternal infectious disease testing must be performed as a surrogate for donor testing, if the tissue donor is an infant less than what age?

A. 1 day
B. 7 days
C. 28 days
D. 3 months
E. 6 months

Concept: Testing of an infant may not reflect their true infectious disease status. Maternal testing is more reliable in these circumstances.

Answer: *C*—Infants less than 1 month of age have immature immune systems that cannot rapidly develop antibodies to in utero infections. Thus, maternal testing will be a more reliable indicator of the newborns' infectious disease status. Though some maternal antibodies (generally IgG) can pass through the placenta to the infant, IgM antibodies, which are produced in the initial stages of infection, are unable to cross the placenta. All the other ages (Answers A, B, D, and E) are incorrect.

22. AATB standards require that a genetic history is elicited from all tissue donors in which of the following categories?
A. Neonatal allograft donors
B. Pediatric liver donors
C. Autologous donors
D. Oocyte donors
E. Living donors

Concept: The purpose of obtaining a genetic history on a tissue donor is to prevent a disease that may be undetected in the donor from being transmitted with the allograft.

Answer: *D*—Oocytes may harbor recessive mutations that may not be clinically evident in the oocyte donor but could be transmitted to the recipient (child that is derived from the oocyte). Genetic history is required only from donors of reproductive cells/tissue, therefore, all other choices (Answers A, B, C, and E) are incorrect.

23. AABB standards require that tissue services maintain records of the tissues they handle. These records must meet which of the following criteria?
A. Be maintained electronically
B. Be retained on site
C. Identify personnel who dispense the tissue
D. Be retained indefinitely
E. Contain infectious disease test results

Concept: Tissue records play a critical role in traceability of allograft tissues from donor to recipient. The nature of these records is largely left up to the tissue service; however, the content and length of storage of the records is specified to ensure that the tissues can be traced until the risk of recall is minimized.

Answer: *C*—Personnel who perform any critical step in tissue handling (receipt, preparation for dispensing, dispensing, preparation for implantation, return) must be documented in the records of the tissue service. There is no requirement for how the records are stored (electronically or on paper, or the location of storage) as long as they are readily retrievable (Answers A and B). They must be stored for a minimum of 10 years beyond the date of final disposition of the tissue (Answer D). There is no requirement for the tissue service to maintain infectious disease test results of allograft tissue (Answer E).

24. A chemical method acceptable for the disinfection of ocular tissue during processing is which of the following?

A. Glutaraldehyde
B. Quaternary ammonium compounds
C. Ultraviolet radiation
D. Povidone-iodine
E. Autoclaving

Concept: Chemical agents that damage the structural integrity of tissue are not acceptable as disinfection agents. Many chemical disinfection methods used by tissue banks are proprietary, but examples of solutions that have been used include povidone-iodine, ethanol, hexachlorophene, chlorhexidine gluconate, sodium thiosulfate, isopropyl alcohol, and antibiotic solutions. Harsher disinfection agents include peracetic acid, glutaraldehyde, formaldehyde, sodium hypochlorite (bleach), hydrogen peroxide, and quaternary ammonium compounds (e.g., benzalkonium chloride). These harsher agents remove more of the microbial burden but may lead to unacceptable levels of structural tissue damage, and death of viable cells in the tissue. Viable allografts are disinfected but not sterilized, as the sterilization process would also kill the viable cells in the tissue.

Answer: *D*—Povidone-iodine is a safe and effective chemical method of disinfection for ocular tissue. Glutaraldehyde (Answer A) and quaternary ammonium compounds (Answer B) risk tissue damage during disinfection and are not acceptable. Autoclaving (Answer E) and ultraviolet radiation (Answer C) are incorrect because they are not chemical disinfection methods, and would damage the ocular tissue.

25. Properties of an ideal cryoprotectant include which of the following?
A. Highly lipid-soluble
B. Able to form primarily intracellular ice crystals
C. Minimal depression of the freezing temperature
D. Able to form large quantities of ice crystals
E. Nontoxic at high concentrations

Concept: Cryoprotectants function by significantly raising the total osmolality of the suspension, which reduces the temperature at which ice forms and the quantity of ice that forms.

Answer: *E*—High concentrations of the cryoprotectant are needed to preserve viable tissue. The cryoprotectant should be water soluble (Answer A) and form primarily extracellular ice crystals (Answer B). It needs to be able to significantly depress the freezing temperature (Answer C) and minimize the amount of ice crystals formed (Answer D). In particular, intracellular ice crystal formation should be minimal as their presence would irreparably damage the cells.

26. As tissue service Medical Director, you are asked by the surgeon who harvested autologous skin to release a contaminated graft for implantation. Low-level contamination with which of the following organisms would pose the greatest risk to the patient?
A. *Streptococcus pyogenes*
B. *Propionibacterium acnes*
C. *Staphylococcus capitis*
D. *Corynebacterium striatum*
E. *Staphylococcus epidermidis*

Concept: Bacteria that are skin commensals are common contaminants in skin grafts. They are unlikely to cause a clinically significant infection. In general, the graft is safe to use if the bacterial burden is low (e.g., reported as rare or few colonies in the culture).

Answer: *A*—*Streptococcus pyogenes* is an infrequent skin commensal and is usually pathogenic. It is a cause of necrotizing fasciitis and sepsis as well as sequelae, such as rheumatic heart disease and glomerulonephritis. The remaining organisms are common skin commensals and are generally not pathogens (Answers B, C, D, and E).

27. Advantages gained by meshing a skin graft prior to implantation include which of the following?
   A. Increased area of wound coverage
   B. Smaller risk of residual scarring
   C. Less likelihood of wound desiccation
   D. Better long-term cosmetic appearance
   E. Smaller chance of allograft rejection

   Concept: Meshing is a process in which the skin graft is passed through two metal rollers to insert multiple fenestrations (holes) into the graft. Meshing for skin grafts can be performed for both autologous and allogeneic grafts. Meshing ratios vary from 1:1 to 9:1 expansion ratios. There are pros and cons to this process. Meshing allows the surface area of the harvested graft to be drastically increased. This avoids the need to harvest larger areas of skin and means there is a smaller donor site wound for autologous grafts.

   Answer: *A*—The main advantage for meshing a skin graft is the ability to use a small graft to cover a wound area larger than the site from which the graft was harvested. Meshed grafts lead to increased risk of wound desiccation (in the areas where holes were cut), with ultimately a poorer cosmetic appearance.

   The risk of allograft rejection is not altered by meshing (Answer E). Meshed skin is also useful for wounds that are expected to bleed or produce large amounts of serous fluid. The meshed graft allows the fluid to pass through and into the dressing and helps to prevent hematoma or seroma. However, this feature can also lead to increased risk of wound desiccation (Answer C) and potential scarring (Answer B). Meshed grafts maintain the meshed appearance after healing and thus, may be unsuitable for grafts on the hands, feet, and face (Answer D).

28. Which of the following graft types would be best suited for immediate treatment of the majority of the burns from a 3-year-old with full thickness burns of 45% BSA (body surface area), with burn sites including bilateral lower extremities, abdomen, buttocks, and back?
   A. Autologous full-thickness
   B. Autologous split-thickness
   C. Allogeneic full-thickness
   D. Allogeneic split-thickness
   E. Cultured epithelial-autograft

   Concept: Skin grafts are classified as either split-thickness or full-thickness. Split-thickness grafts consist of the epidermis and part of the dermis, while full-thickness grafts consist of the epidermis and the entire dermis. The best replacement for a full-thickness burn is an autologous full-thickness graft, but it is sometimes difficult to harvest enough full-thickness graft from the burn patient to cover the burns. For this reason, autologous split-thickness grafts are often used. Split-thickness grafts are best harvested from unburned areas that are not usually visible, such as the buttocks or upper thighs, since even though the epidermis will regenerate, the donor sites will not have a normal appearance once healed. Another source of autologous skin is cultured autologous epidermal cells. These cultured cells are costly and provide only a thin layer of cells that are very fragile, and only restore the epidermis. Cultured cells may not be available for 10 days or more, depending on the culture method and the culture facility schedule.

   Answer: *C*—The best skin replacement for a large full-thickness burn is one that restores both the dermal and epidermal layers and is available in a large enough quantity to cover the burn area. While full or split-thickness autologous grafts (Answers A and B) are preferable, the best harvest sites for these grafts in this patient are compromised due to the burn. Thus an allogeneic full-thickness graft is the best, and most cost effective, option to cover the majority of the burned area. While these grafts will ultimately be rejected, they play a crucial role in preventing tissue desiccation and infection, reducing energy requirements, and reducing wound pain. Split-thickness grafts (Answer D) and cultured autografts (Answer E) would not suffice in this case.

29. Which of the following allograft-associated adverse events should be reported to the tissue vendor by the hospital tissue service?

A. Failure of allograft effectiveness
B. Excessive increase in price of the tissue
C. Failure to maintain sufficient consigned tissue
D. Use of a non-FDA approved tissue by a surgeon
E. Surgical site infection from contaminated scrub solution

Concept: Adverse events fall into two categories: failure of graft efficacy or graft safety. Both of these should be reported to the tissue vendor so that they can improve their processes and potentially recall other tissue that might be similarly affected.

Answer: *A*—If an allograft fails to perform its intended function in the recipient, the tissue vendor should be notified. Price increases and failure to maintain sufficient consignment may warrant a conversation with the tissue vendor, but would not be considered adverse events. The use of a non-FDA approved tissue may be the result of participation in a study and therefore, acceptable. Surgical site infections that can clearly be attributed to a source other than the allograft are not reportable to the tissue vendor.

30. Which of the following is an advantage of using consigned tissue rather than directly purchased tissue?

A. More tissue service control over inventory levels
B. Less time spent in returning unused tissue
C. Less expense due to unused tissue expiration
D. Less tissue wastage in the operating room
E. Easier to track tissue disposition

Concept: Consigned tissue is provided by the tissue vendor to the hospital tissue service without charge, until the tissue is removed from inventory (implanted or wasted). Consignment arrangements allow the return of tissue to the vendor prior to its expiration, so that cost is not incurred by the hospital tissue service for unused or expired tissue.

Answer: *C*—Consignment tissue agreements lead to less expense because the tissue can be returned with no charge to the facility. The amount of allograft in the tissue service's inventory is controlled by an agreement between the vendor and the tissue service in consignment arrangements (Answer A). The vendor typically will not agree to consignment levels that are substantially above historical usage levels demonstrated by the hospital. Consigned tissue is also returned more often to the vendor (Answer B). The amount of tissue wasted in the operating room (Answer D) and tissue tracking (Answer E) is not affected by whether the tissue is initially consigned or purchased.

31. A frozen allograft is dispensed in a transport container containing dry ice that is validated to maintain the proper storage temperature for 24 h. The tissue is returned within the same validated transport container 24 h after the tissue was placed in the container. Dry ice is still present. The tissue appears frozen except at the extreme edges, which are slightly soft. The most appropriate handling of the graft at this point is to do which of the following?

A. Return the tissue to inventory if the internal temperature of the transport container is within the acceptable range
B. Return the tissue to inventory if the tissue packaging is intact and the tissue otherwise appears to be undamaged

C. Return the tissue to the distributor as being unacceptable
D. Complete the thawing of the graft and keep it refrigerated for a future surgical procedure
E. Discard the tissue

Concept: Tissues must be maintained at the processor's established temperature range during storage and transport and must meet acceptability criteria to be returned to inventory and dispensed to another patient.

Answer: *E*—This tissue is clearly damaged and should be discarded. Even if the internal temperature of the transport container is within the acceptable range, the tissue appears to have slightly thawed (Answers A, B, and D). The tissue may have been out of the transport container for an extended time, or the transport container may have been left open for a period of time, during which the entire tissue slightly thawed and refroze after the container was closed again. Tissue vendors will not accept returned (Answer C) tissue if acceptable storage conditions have not been maintained.

32. Eligibility of donors with which of the following conditions shall be evaluated by the tissue bank's Medical Director?
A. Systemic lupus erythematosus
B. Polycystic kidney disease
C. Endometriosis
D. Alzheimer's disease
E. Emphysema

Concept: Besides the required infectious disease testing for cell/tissue donation, donors should be screened for other conditions that could potentially cause adverse events in the recipient.

Answer: *A*—According to AATB standards, the Medical Director shall evaluate donors with a history of autoimmune disease for their suitability to donate cells and/or tissue. They should also be evaluated for ingestion of, or exposure to toxic substances. A history of rheumatoid arthritis, systemic lupus erythematosus, polyarteritis nodosa, sarcoidosis, or clinically significant metabolic bone disease precludes musculoskeletal tissue donation. The other listed conditions (Answers B, C, D, and E) will not affect tissue donation and do not require review by the Medical Director.

33. Which piece of information would provide the most accurate tracking of a tissue in the event of a recall?
A. Descriptive name of tissue
B. Tissue identification number
C. Lot number
D. Expiration date
E. Name of tissue processor

Concept: A "lot" is defined by AATB as all of the tissue produced from 1 donor at 1 time using 1 set of instruments and supplies. Individual grafts from that donor are then assigned a tissue identification number (defined as a unique combination of letters, numbers, and or/symbols assigned to the tissue and linked to a donor from which the complete history of the tissue from collection to implantation can be traced).

Tissue products may contain material from multiple lots of tissue, provided traceability to each donor is maintained. Certain information must be maintained at the time that a tissue is implanted to help with an investigation in the event of a tissue recall or suspected adverse event.
AATB requires that when a tissue is implanted, the following information must be maintained in the patient's medical record:

- Type of tissue
- Numeric or alphanumeric identifier
- Quantity

- Expiration date
- Date of use
- Personnel using the tissue
- Related adverse events if applicable

Answer: *B*—The tissue identification number is the most accurate way to track a specific piece of tissue during a recall investigation. While the lot number of a tissue (Answer C) identifies the tissue to a specific donor/donation, the tissue identification number or unique identifier identifies the specific tissue piece or package. The additional information of the tissue piece could help further an investigation by allowing tracking of the tissue through the steps following recovery. For example, different parts of the same lot may be processed separately (e.g., disinfected vs. sterilized), shipped and/or stored under different conditions, and dispensed/returned from/to storage varying number of times. All of these factors could variably impact the safety and efficacy of various parts of the lot of tissue. The other choices (Answers A, D, and E) are not very specific and are therefore incorrect.

34. AATB standards require that a physician must perform which of the following steps in tissue procurement?
A. Obtain informed consent from the donor or their next of kin
B. Ensure donor suitability
C. Conduct a medical, social, and sexual history
D. Harvest bone marrow from a living donor
E. Recover pericardium from a deceased donor

Concept: Individuals involved in the tissue procurement process must be trained and follow organizational standard operating procedures. However, there is no additional educational requirement mandated for most steps in the process.

Answer: *B*—A medical degree is required to establish donor suitability criteria (a role of the Medical Director), and to make a determination about the suitability of each donor from whom tissue is recovered. The Medical Director also must approve any exceptional releases of tissue, and the release of tissue for research. All the other choices (Answers A, C, D, and E) represent tasks that do require a medical degree.

Please answer Questions 35–38 based on the following scenario:

A 14-year-old boy is taken to the hospital after a car accident in which his parents were killed. He remains in critical condition for 3 days and is pronounced dead 4 days after the accident. The boy's legal guardian agrees to tissue donation. As the Medical Director of an AATB-accredited Tissue Bank that will be processing the tissues, the donor record comes to you for review and determination of donor eligibility.

35. You must defer the donation of aortic valves if which of the following is true?
A. Unable to get a complete genetic history from the legal guardian
B. Infectious disease testing did not include *Neisseria* species
C. Infectious disease testing did not include *Trypanosoma cruzi*
D. Infectious disease testing did not include HTLV-II
E. Unable to get a complete immunization record from the legal guardian

Concept: Donor suitability is ultimately determined by the Medical Director of the processing facility. Criteria used for establishing eligibility must follow AATB standards.

Answer: *C*—AATB standards require that heart valve donors be evaluated for the risk of Chagas' disease (*Trypanosoma cruzi*), due to the known risk of cardiac complications from this infection. Genetic testing/history (Answer A) and testing for *Neisseria gonorrhoeae* (Answer B) is required only for reproductive tissue donation, and HTLV-II (Answer D) testing is only required for the donation of leukocyte-rich tissues. An immunization history is not required (Answer E).

36. The boy received multiple blood transfusions and fluid infusions during his hospitalization. He is suitable for cell and/or tissue donation if he received which of the following?
A. 1500 mL of whole blood in the last 48 h; sample for infectious disease testing collected 60 h ago
B. 1100 mL of red blood cells in the last 48 h; sample for infectious disease testing collected 2 h ago
C. 3000 mL of whole blood in the last 48 h; sample for infectious disease testing collected 2 h ago
D. 2400 mL of red blood cells in the last 48 h; sample for infectious disease testing collected 2 h ago
E. 2500 mL of normal saline in the last 1 h; sample for infectious disease testing collected 30 min ago

Concept: Transfusions of blood products or infusions of colloidal or crystalloid solutions may cause a plasma dilution effect significant enough to alter the results of communicable disease testing (i.e., false negative).

Answer: *A*—The infusion of >2000 mL of whole blood, red blood cells, and/or colloids within 48 h prior to obtaining a blood sample for infectious disease testing (Answers C and D) or the infusion of >2000 mL of crystalloid within 1 h (Answer E), or any combination thereof is unacceptable as per AATB standards. In addition, for potential donors of 12 years of age or less, no transfusion or infusion prior to obtaining the sample for infectious disease testing is acceptable, unless an algorithm is utilized that evaluates the volumes administered in the 48 h before collecting the blood specimen to ensure that there has not been plasma dilution sufficient to affect the test results (Answer B). A pretransfusion sample collected within 14 days before (or after) tissue donation is acceptable for communicable disease testing.

37. The medical examiner orders an autopsy for the boy. Under which condition could this child still serve as a tissue donor?
A. None, an autopsy precludes tissue donation
B. The body was refrigerated within 2 h of cardiac death, autopsy is complete and the body is available for tissue retrieval 22 h after death
C. The body was not refrigerated after cardiac death, autopsy is complete and the body is available for tissue retrieval 17 h after death
D. Anal condylomata are discovered, raising a concern for sexual abuse of the donor
E. Mediastinal lymphadenopathy is identified, ultimately diagnosed as Hodgkin lymphoma

Concept: Limitations on the timing of refrigeration and initiation of tissue harvest are meant to ensure quality tissue donations.

Answer: *B*—If the body was refrigerated within 12 h of death, tissue retrieval must begin by 24 h after death. An autopsy (Answer A) performed in a timely manner does not preclude tissue donation, provided the findings at autopsy do not raise concern for the transmission of infectious disease or malignancy by the donated tissues. If the body was not refrigerated (Answer C), tissue retrieval must begin by 15 h after death. Discovery of anal condylomata (Answer D) or active hematologic malignancy (Answer E) would exclude tissue donation by the deceased individual.

38. The boy was deemed a suitable donor and multiple tissues were harvested and distributed to several hospitals. You receive notification from one hospital that a patient receiving a spinal bone spacer from this donor had a positive surgical wound culture. The most appropriate first step to take in response to this notification is to contact which of the following?
A. The hospital submitting the adverse event for further information regarding the circumstances of the adverse event
B. The other tissue centers that received grafts from this donor to see if there were other adverse events

C. The FDA with notification of the report
D. The tissue procurement organization that harvested the tissue for any indication of sterility problems during harvest
E. The physician performing the postmortem for additional information on potential risk factors

Concept: The FDA requires that adverse events be reported as soon as possible to ensure that such events are investigated properly.

Answer: C—Ideally, the FDA should be notified as soon as possible following adverse events, but absolutely must be notified within 15 calendar days. Establishments must also, as soon as practical, investigate all adverse reactions that are the subject of these 15-day reports and must submit follow-up reports within 15 calendar days of the receipt of new information or as requested by FDA. If additional information is not obtainable, a follow-up report may be required that describes briefly the steps taken to seek additional information and the reasons why it could not be obtained. The other choices (Answers A, B, D, and E) are part of the investigation, but would not be the first step.

End of Case

Please answer Questions 39–43 based on the following scenario:

A neurosurgeon in your facility has requested spinal spacers from the hospital tissue service's inventory for an upcoming surgery. The order is for multiple sizes and multiple quantities of each size as he is not certain which size will be needed until the time of implantation. Your centralized tissue service routinely stocks one "set" of the allografts on a consignment agreement with your supplier and grafts are replaced as used. The tissue service is also responsible for tracking the final disposition of tissues. The tissue supplier comes into your facility monthly and audits the spinal spacer inventory to reconcile their records.

39. Your tissue service issues the set of spinal spacers to the OR for the surgical procedure. After the case has started, the same surgeon places an emergent order for the set of spinal spacers for a procedure on a different patient in the same operating room immediately at the completion of the current procedure. Your service does not have another set of tissues in inventory to send for the second case. Which of the following is the most appropriate course of action?
A. Ask the OR staff to use the same tray for both cases without returning it to the tissue service between cases
B. Cancel the second surgery until replacement tissues can be received from the distributor
C. Have the tissue service document the identity and acceptability of tissues remaining from the first case before the second case begins
D. Ask the distributor's in-house representative, if they have additional tissue available to provide for the second case
E. Send a similar but nonidentical tissue tray to the OR for the second case

Concept: JC requires that tissue is tracked from receipt to final disposition, and that proper storage conditions for the tissue have been maintained. In this scenario, the centralized tissue service is responsible for ensuring that these functions are performed accurately (Table 18.6).

Answer: C—The centralized hospital tissue service must verify the final disposition of all tissues that have been dispensed to a patient. The best way to ensure this for the tissue service personnel is to inspect the remaining tissue after the completion of a surgical procedure in order to verify which tissue was used and to determine the acceptability of the remaining tissue, before dispensing it to another patient. If the tissue was provided directly to the surgical suite by a vendor's representative, the tissue service will be unable to assure accurate temperature and storage conditions. The other choices do not allow for proper inspection of the tissue (Answers A and D), or unnecessarily delay the patient's care (Answers B and E).

TABLE 18.6 AATB Labeling Requirements at the Time of Tissue/Cell Distribution

Tissue identifier number
Descriptive name of cells and/or tissue
Name(s) and address(es) of tissue bank(s) responsible for determining donor suitability[a]
Name(s) and address(es) of tissue bank(s) responsible for processing[a]
Name(s) and address(es) of tissue bank(s) responsible for distribution[a]
Expiration date (if applicable), including the month and year
Acceptable storage conditions, including recommended storage temperature and/or acceptable storage temperature range
Disinfection or sterilization procedure utilized (if applicable)
Preservative (if utilized) and/or method of preservation (is applicable)
Quantity of cells and/or tissue expressed as volume, weight, dimensions, or a combination of these units of measure, if applicable
Potential residues of processing, agents/solutions (e.g., antibiotics, ethanol, ethylene oxide, dimethyl sulfoxide)
Reference to the package insert
Biohazard and warning labels as applicable
Statements for autologous donations: "For Autologous Use Only" Patient's name and, if available, the name of the facility where the patient is being transplanted. Patient hospital registration number. If unavailable, social security number, birth date, or similar definitive identifier.

[a] *Should more than two banks be involved, the name of all banks is required but the address is only required for the bank determining the donor suitability.*

40. At the completion of the surgical case, the OR staff returns the tissue transport container to your tissue service. When your staff reviews the contents of the transport container against the list of grafts dispensed, all but two grafts have been returned unopened. Your staff reviews the implant log for the case in the patient's medical record and notes that only one of the grafts not returned is documented as being implanted. A conversation with the OR staff involved in the case is unable to establish the final disposition of the other graft. The best action to take to ensure patient safety in the event of a tissue recall is to take which of the following actions?

A. Record the final disposition in your log and with the supplier as wasted/discarded
B. Record the final disposition in your log and with the supplier as implanted in the patient
C. Record the final disposition in your log as wasted/discarded but inform the supplier at the monthly audit that the graft could have been implanted
D. Record the final disposition in your log and with the supplier as unaccounted for
E. Do not record a final disposition in your log or notify the supplier

Concept: Biovigilance is a process used to ensure the safety of tissue recipients. The FDA requires that adverse events occurring postimplantation that could be related to the implanted tissue must be reported to the tissue processor so that other recipients of grafts from that donor can be notified of a potential risk. In order for the notification of the recipient to occur, all tissues must be traceable from the donor to the recipient and vice versa.

Answer: *B*—Even if it is not possible to make a definite connection between the donor and the tissue, it is safer to err on the side of caution and track the graft as being possibly implanted in the patient so the patient can be contacted in the event of a potential recall. JC requires notification of the tissue processor of the final disposition of all tissues. The other choices (Answers A, C, D, and E) do not represent good practice for documentation of the disposition of the tissue.

41. Your facility receives a replacement graft for the tissues that the surgeon used from the spacer set, and the expiration date indicated on the packing slip does not match the expiration date on the tissue label. The most appropriate single action to take is which of the following?

A. Accept the allograft and log the expiration date from the package label in your records
B. Contact the distributor to determine which expiration date is correct
C. Modify the package label to match the expiration date recorded on the packing list
D. Discard the allograft due to labeling discrepancy
E. Accept the allograft and log the shorter of the two expiration dates in your records

Concept: AABB and JC standards require hospitals to log in and inspect all incoming tissues (Table 18.7). JC also requires that the transport temperature range is verified, for tissues that require a controlled environment. Finally, the tissue service must verify that package labels are complete, affixed, and legible. Accurate expiration dates must be available to assure that tissues are not used past the expiration date established by the tissue processor.

TABLE 18.7 Requirements for Receipt of Products by Hospital Tissue Services

AABB	JC
Each container used for collection, preservation, and storage shall be inspected to ensure that it is intact	Document the receipt of all tissues
Each container used for collection, preservation, and storage shall be inspected to ensure that the label is complete, affixed, and legible	
	Ensure package integrity is met
	Ensure transport temperature range was controlled and acceptable for tissues requiring a controlled environment

Answer: *B*—The distributor should be called to determine which expiration date is correct. It is not necessary to discard or reject the tissue if the distributor can provide written evidence of the correct expiration date (Answers A and D). If the discrepancy cannot be resolved, the tissue should be returned to the distributor (Answers C and E).

42. The storage temperature range on the allograft you received is 20–25°C. The room temperature for allograft storage in your facility often fluctuates outside of this range. In order to meet the standards for tissue storage and provide optimal patient care you should do which of the following?

A. Locate a device within your facility for tissue storage that can be maintained at 20–25°C
B. Leave the tissues in the current storage area, but monitor and note the temperature deviations and determine if there was a significant risk to the tissue
C. Have the tissue vendor's representative store the tissue and bring the tissue directly to the OR for the surgical procedure
D. Refuse to allow the tissue to be used in your facility
E. Ask the surgeon to sign a variance agreeing to use the tissue for a nonemergent surgery with knowledge of the out-of-range temperature storage

Concept: Hospital tissue services are required to follow the tissue processor's instructions for storage requirements that are listed in the package insert accompanying the tissue. Tissue storage equipment should be validated to be able to maintain desired temperatures and should have functional alarms.

Answer: *A*—The facility should make every effort to find an appropriate storage device before simply refusing to all the tissue to be used (Answer D). While monitoring temperatures and documenting excursions to defined temperature ranges is required, continuing to store tissues in areas that do not routinely meet established storage temperatures does not meet the standard requirements (Answer B). Allowing the distributor's representative to carry tissue into the OR bypasses the JC requirements for the hospital to verify that tissues have been maintained at acceptable temperature ranges during shipment and transport (Answer C). The surgeon should not be put in a position to sign for tissue stored outside of recommended temperature ranges (Answer E).

43. A month after the procedure, the supplier of the grafts informs you that three grafts that were consigned to your service have been recalled. Which of the following is the first step that should be taken by your tissue service?

A. Notify the surgeons who implanted the tissues
B. Contact the recipients of the tissues
C. Determine what testing/counseling is recommended for the recipients
D. Quarantine any recalled tissue in inventory
E. Generate a list of recipients of recalled tissue

Concept: Tissue recalls can occur for many reasons, of varying risk to the tissue or the recipient of the tissue. It is important to have a stepwise process in place for responding to recalls, and a robust tissue tracking system to identify the location and/or disposition of recalled tissue.

Answer: *D*—Until a determination can be made about the implications of a recall, tissue that is in inventory should be quarantined. Further investigation may (rarely) allow the tissue to be subsequently released for the implantation. The tissue tracking system should allow identification of all individuals who received the tissue(s) (Answers B and E), and in consultation with the surgeon who implanted the tissue (Answer A), individual risk assessment should be conducted. The patients can then be contacted if additional testing or counseling is deemed appropriate (Answer C).

End of Case

Suggested Reading

[1] 21 CFR Part 1271 Human cells, tissues, and cellular based products. US Department of Health and Human Services Food and Drug Administration, 2015.
[2] N. Dock, J.C. Osborne, S.A. Brubaker (Eds.), AATB Standards for Tissue Banking, thirteenth ed., AATB, McLean, VA, 2012 1.
[3] T. Eastlund, Bacterial infection transmitted by human allograft transplantation, Cell Tissue Bank 7 (2006) 147–166.
[4] A.B. Eisenbrey, (Ed.), Cell and tissue banking. Clin. Lab. Med. 25 (2005) 473–630.
[5] A.B. Eisenbrey, T. Eastlund (Eds.), Hospital Tissue Management: A Practitioner's Handbook, AABB, Bethesda, MD, 2008.
[6] D. Fehily, S.A. Brubaker, J.M. Kearney, L. Wolfinbarger (Eds.), Tissue and Cell Processing: An Essential Guide, Wiley-Blackwell, Hoboken, NJ, 2012.
[7] A.R. Gazdag, J.M. Lane, D. Glaser, R.A. Forester, Alternatives to autogenous bone graft: efficacy and indications, J. Am. Acad. Orthop. Surg. 3 (1995) 1–8.
[8] P.W. Ooley (Ed.), Standards for Blood Banks and Transfusion Services, thirtieth ed., AABB, Bethesda, MD, 2016.
[9] Organ Procurement and Transplantation Network Policy 16: Organ and Vessel Packaging, Labeling, Shipping and Storage. US Department of Health and Human Services Health Resources and Services Administration, 2015.
[10] C.N. Paul, Skin substitutes in burn care, Wounds 20 (2008) 203–205.
[11] The Joint Commission. Transplant Safety. Hospital Accreditation Standards. Oakbrook Terrace, IL: The Joint Commission, 2015.
[12] S. Vaishnav, C.T. Vangsness, New techniques in allograft tissue processing, Clin. Sports Med. 28 (2009) 127–141.
[13] R. Warwick, S. Brubaker (Eds.), Tissue and Cell Clinical Use: An Essential Guide, Wiley-Blackwell, Hoboken, NJ, 2012.
[14] R. Warwick, D. Fehily, S. Brubaker, T. Eastlund (Eds.), Tissue and Cell Donation: An Essential Guide, Wiley-Blackwell, Hoboken, NJ, 2009.

CHAPTER

19

Pathology Informatics

Rance C. Siniard, Seung Park***

*The University of North Carolina at Chapel Hill, Chapel Hill, NC, United States;
**Indiana University Health, Indianapolis, IN, United States

According to the Association for Pathology Informatics, the specialty involves "collecting, examining, reporting, and storing large, complex sets of data derived from tests performed in clinical laboratories, anatomic pathology laboratories, or research laboratories in order to improve patient care and enhance our understanding of disease-related processes." Data sets used in pathology informatics can be derived from anatomic and clinical pathology reports, clinical laboratory results, and even molecular data obtained through sequencing analysis. The need to process and store large amounts of data has been a driving force in the development of pathology informatics as a recognized pathology subspecialty. This chapter will provide a review of pathology informatics that is relevant to the practice of transfusion medicine, apheresis, and hemostasis.

1. A binary digit is the basic unit of digital information. How many binary digits (bits) does 1 byte refer to?
A. 2
B. 4
C. 6
D. 8
E. 10

Concept: A byte consists of 8 bits, and is the basis of the most common unit used to measure the capacity of storage devices (e.g., megabyte, gigabyte, terabyte, etc.). It is used for storing information, and can have a value of true or false, or off and on. It is typically represented by the values of 0 (for false or off) or 1 (for true or on).

Answer: *D*—A byte refers to 8 binary digits (bits). All other choices (Answers A, B, C, and E) are incorrect.

2. How many bytes are in 1 kilobyte according to the International Electrotechnical Commission (IEC)?
A. 500
B. 1000
C. 1012
D. 1024
E. 2048

Concept: A byte consists of 8 bits, and is the basis of the most common unit used to measure the capacity of storage devices. The International Electrotechnical Commission (IEC) is a nonprofit standards organization that publishes international standards for electrical, electronic, and related technologies.

Transfusion Medicine, Apheresis, and Hemostasis. http://dx.doi.org/10.1016/B978-0-12-803999-1.00019-5

Answer: *B*—This is the definition recommended by the IEC. This definition is used for data transfer rates, internal bus, hard drive and flash media transfer speeds, and for the capacities of most storage media (hard drives, flash media, and DVD-ROMs).

Answer D represents an unofficial definition of a kilobyte, and is used to represent Random Access Memory (RAM) capacity. This definition of a kilobyte is derived from the equation 1 kilobyte = 2^{10} bytes. This usage originated for multiples that needed to be expressed in powers of 2, but lacked a convenient unit prefix. The IEC has specified that the kilobyte should only be used to refer to 1000 bytes. However, 1 kilobyte being equal to 1024 bytes is still used unofficially in practice, but is not the recommended definition by the IEC. The other choices (Answers A, C, and E) are incorrect.

3. In terms of information systems and computer networks, which of the following is the best description of a "client"?
A. A device and/or software that inspects the network traffic that passes through it, and denies or permits passage of that traffic based on a set of predetermined rules
B. A device that routes or forwards information
C. An application or system that accesses a server
D. A device or software that accepts connections to service requests by sending back responses
E. A device that modulates an analog carrier signal to encode digital information

Concept: Networks are made up of various components, each of which has a defined function or purpose. These components include firewalls, routers, clients, servers, and modems.

Answer: *C*—A client establishes a connection to the server over a network, such as a local area network (LAN) or wide area network (WAN). Once the client request has been fulfilled, the connection to the server is terminated. A device or software package that inspects network traffic passing through it and denies or permits passage based on a set of predetermined rules is a firewall (Answer A). A router routes or forwards information to the appropriate location on the network (Answer B). A server is a device or software that accepts connections to service requests by sending back responses (Answer D). Finally, a device that modulates an analog carrier signal to encode digital information is a modem (Answer E).

4. Which of the following is true regarding the laboratory information system (LIS) in blood banks?
A. It is part of the hospital information system (HIS)
B. It is regulated by the CMS and inspected every 2 years by the CAP
C. Blood bank records stored in the LIS must be kept for a minimum of 5 years
D. It qualifies as a healthcare device and is regulated by the FDA
E. It directly supports billing for tests related to the blood bank

Concept: The LIS consists of a variety of modules, such as general chemistry, hematology, blood bank, microbiology, and others frequently connected via interfaces. It can be defined as a collection of data organized into a structural database architecture.

Answer: *D*—The LIS in blood banks qualifies as a healthcare device and is regulated by the FDA. The FDA inspects blood banks and the blood bank LIS (Answer B). The hospital information system (HIS) is physically separate from the LIS (Answer A). The HIS and LIS communicate through an interface engine. Electronic blood bank records must be stored for a minimum of 10 years per the *AABB Standards* (Answer C). The LIS doesn't directly support billing, but a billing system is connected to the LIS and other hospital information systems through an interface engine (Answer E).

5. What type of interface is shown in the following image (Fig. 19.1)?
A. Bidirectional
B. Unidirectional
C. Query bidirectional

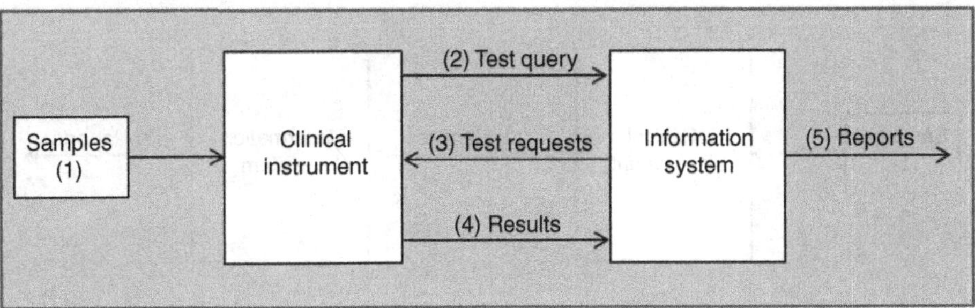

FIGURE 19.1 **Interface example.**

D. Distributed processing/query conversion
E. Custom

Concept: The LIS is a complex collection of systems that need to communicate in order to provide timely and accurate patient care. Interfaces are important components of the LIS that receive order information and assist in transmitting laboratory results to the appropriate location, such as to an in-house electronic medical record (EMR) or from a reference laboratory to a physician's clinic or hospital.

Answer: C—The interface type shown in the image is a query bidirectional interface. A query bidirectional interface is typically used with bar-coded labeled specimens. During specimen processing, the bar code label is scanned and the specimen ID is read. The analyzer then queries the host, requesting all of the information associated with the test. The host then looks up the specimen ID in the database and downloads the appropriate information. The key here is to note that the clinical instrument sends a test query to the information system, while the information system also queries the analyzer for additional test requests. Results are ultimately sent to the information system and then reports are generated.

A simple bidirectional interface is shown in the following image (Fig. 19.2).

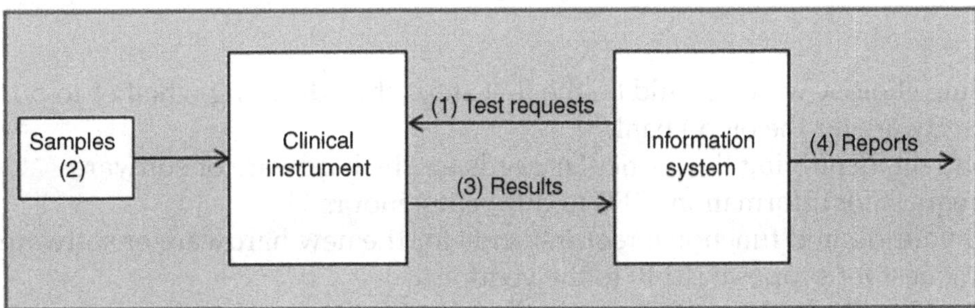

FIGURE 19.2 **Bidirectional interface.**

In a bidirectional interface (Answer A), true two-way communication is involved between the analyzer and the LIS with the absence of the additional test query from the clinical instrument to the information system. Bidirectional interfaces are usually found on instruments that can perform a wide variety of tests on each specimen.

The unidirectional interface (Answer B) is the traditional laboratory interface, in which the clinical instrument performs the test and then transmits the results directly to the LIS. From there, the LIS generates the reports. In contrast to a bidirectional interface, test requests are not sent back to the clinical instrument from the LIS (Fig. 19.3).

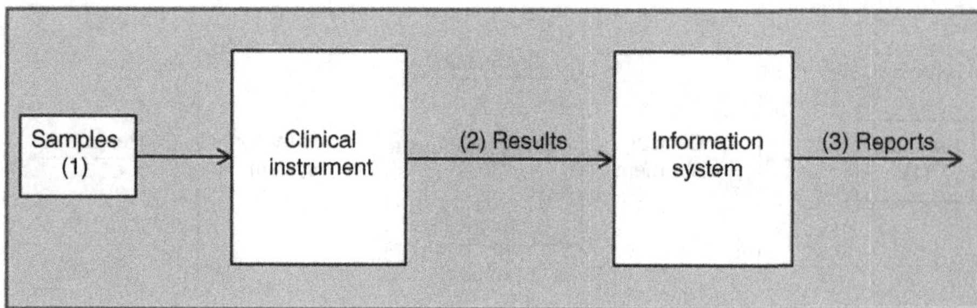

FIGURE 19.3 **Unidirectional interface.**

Distributed processing/query conversion (Answer D) results when an interface system, such as a PC-based interface is added to the system. In this type of interface, the host downloads information to the interface subsystem in advance and stores the downloaded information into access memory. This action speeds up the query process because the interface is able to quickly respond due to having already downloaded critical information for the laboratory test (Fig. 19.4).

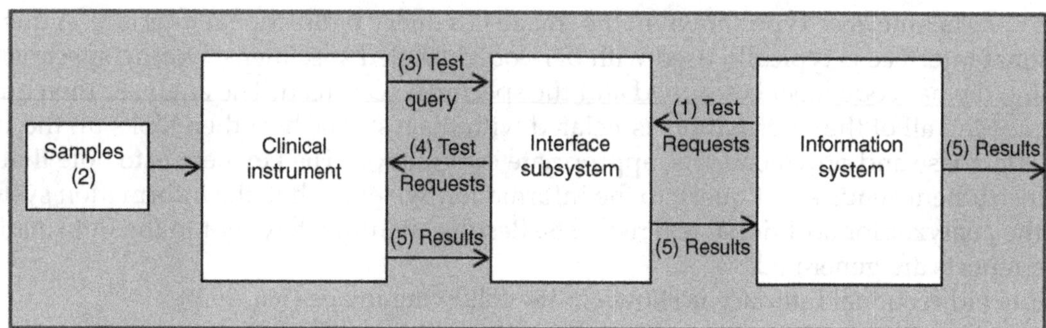

FIGURE 19.4 **Interface with distributed processing and query conversion.**

6. Of the following choices, which would be the first step when deciding whether to purchase new hardware or software for the blood bank?
   A. Write a contract identifying the particular needs for the hardware or software
   B. Submit a request for information (RFI) to different vendors
   C. Determine your desired functional requirements for the new hardware or software
   D. Submit a request for proposal (RFP) to the vendor
   E. Generate a unique identifier so that the new hardware or software can communicate with outside systems

Concept: One of the first steps in acquiring new hardware or software for any laboratory information system is determining the functional requirements—what tasks or goals the software is supposed to accomplish.

Answer: C—Of the choices offered, this choice would be the first step in the implementation of new hardware or software.

Once the functional requirements are agreed upon, then a RFI should be submitted to multiple vendors (Answer B). RFIs are used to receive more information on a particular product or products. After the RFI and functional requirements have been approved, the next step is to submit a RFP to one or more desired vendors (Answer D). RFP is equivalent to a "bid" offered by the vendor. Finally, once the desired vendor is chosen, the contract can then be negotiated (Answer A). A unique identifier generated by the laboratory information system in order to interface with outside systems would be one of the final steps in the implementation of new hardware or software (Answer E).

7. Blood product containers/bags use both 1D linear barcodes and 2D barcodes, while smaller blood collection tubes benefit from the use of 2D barcodes alone. Which of the following choices represent an advantage of 2D barcodes over linear 1D barcodes?
A. 2D barcodes can be read by all types of barcode scanners
B. 2D barcodes are scanned faster than 1D barcodes
C. 2D barcode labels are cheaper than 1D barcode labels
D. 2D barcodes have a single standard, while 1D barcodes have multiple standards
E. 2D barcodes have a higher information density than 1D barcodes

Concept: Bar codes play an important role in the labeling, tracking, and safety of blood and blood components. The international standard for the labeling of blood and blood components is ISBT-128.

Answer: *E*—2D barcodes have a higher information density than 1D barcodes. They are able to store more information and fit on smaller labels due to their smaller footprint. In addition, 2D barcodes have built-in error checking while 1D barcodes do not.

1D barcodes can be read by any type of barcode scanner, but 2D barcodes require a scanner specifically designed for 2D barcodes (Answer A). However, some 2D scanners can also scan 1D barcodes, which is termed backwards compatibility. In general, 2D barcodes scan at speeds slightly slower than 1D barcodes and though they are not yet equal, technological advances have closed the scanning speed gap in recent years (Answer B). 2D barcodes are more expensive than 1D barcodes (Answer C). Neither 2D nor 1D barcodes have a universal standard (Answer D).

8. The ISBT-128 labeling standard specifies that a certain type of 1D and 2D barcode be used on the labels of blood and blood components. Which of the following choices represent the correct barcode types for 1D and 2D barcodes according to ISBT-128 standards?
A. 1D barcode = Codabar; 2D barcode = Data Matrix
B. 1D barcode = GS1-128; 2D barcode = MaxiCode
C. 1D barcode = Code 128; 2D barcode = Data Matrix
D. 1D barcode = Data Matrix; 2D barcode = Code 128
E. 1D barcode = GS1-128; 2D barcode = Data Matrix

Concept: ISBT-128 recommends which types of 1D and 2D barcodes are acceptable for the use of labeling blood and blood components. The current standard recognizes Code 128 for 1D barcodes and Data Matrix for 2D barcodes.

Answer: *C*—ISBT-128 recommends the use of Code 128 barcodes for linear 1D barcodes, and 2D Data Matrix for 2D barcodes (Table 19.1).

Code 128 is a high-density barcode symbology, and can represent all 128 characters of the American Standard Code for Information Interchange (ASCII). A Code 128 barcode has six sections:

- Quiet zone
- Start character
- Encoded data
- Check character
- Stop character
- Quiet zone

TABLE 19.1 Various Types of Barcodes

Code 128	 Code 128
2D Data Matrix	
Codabar	 A2030405060B
MaxiCode	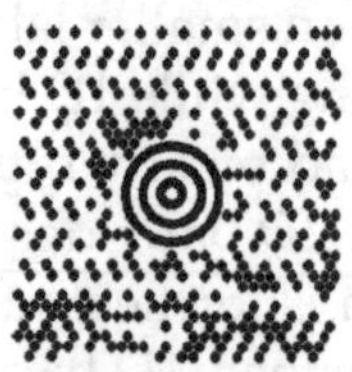
GS1-128 Barcode	 (12)34567

A Data Matrix code is a 2D matrix barcode with black and white modules arranged in a rectangular or square-shaped pattern. Error correction codes are built into Data Matrix barcodes so that if one or more individual modules are damaged, the code can still usually be read.

Codabar is a linear 1D barcode technology developed by Pitney-Bowes and was previously used before Code 128 on blood bank labels. It was designed to be read even when printed on low-resolution dot matrix printers. Codabar technology fell out of favor to Code 128 because more data can be stored in a smaller footprint with Code 128 barcodes.

GS1-128 is an application standard within Code 128 symbology. It uses a series of application identifiers, such as (12) (see earlier mentioned example) to include additional data.

MaxiCode is a bar code technology originally created and used by the United Parcel Service. Its most frequent application is in the tracking of packages, and it is available for use in the public domain.

9. Which of the following is a benefit of the International Society of Blood Transfusion 128 (ISBT-128) standard labeling?
   A. Sets a limit to only one bar code being read at a time, reducing the number of reading errors
   B. Uniformity of labeling by different collection centers, which allows better traceability of blood components
   C. Single-density coding of numeric characters, which permits encoding of more information in a given space
   D. Special characters of the ASCII character set are not included as these characters frequently cause reading errors
   E. Misreads never occur

Concept: The FDA requirements for labeling of blood and blood components are detailed in the *Guidelines for Uniform Labeling of Blood and Blood Components*, published in 1985. The FDA approved version 2.0.0 of the International Society for Blood Transfusion (ISBT) 128 symbology in 2006.

Answer: *B*—One of the key benefits to ISBT-128 labeling technology is that it creates an internationally-standardized system of labeling. This allows for better traceability of blood and blood components, even across international lines. Additional benefits of ISBT-128 include the following:

- Uniform labels applied on blood components
- Better traceability of components
- Built-in self-checking for errors
- Encodes the entire ASCII character set, including special characters
- Reduces the number of misreads as compared to older labeling technologies
- Double-density coding of numeric characters, which permits encoding of more information in a given space
- More detailed descriptions of blood components through the ability to store a larger number of product codes
- Has the ability to concatenate, or read more than one bar code at a time. This also serves as another built-in error checking device on each unit
- The ability to add or delete information for autologous donations
- Donor Identification Numbers (DINs) are uniform and static, and can track donation history back 100 years per donor

Currently, besides blood centers, the Foundation for the Accreditation of Cellular Therapy (FACT) also requires facilities that collect hematopoietic progenitor cells (HPCs) to have an active plan to move toward labeling HPC products using the ISBT-128 labeling system.

10. The ISBT-128-standardized label is shown later. Which of the following depicts the *product code* on the label shown in Fig. 19.5?

A. A
B. B
C. C
D. D
E. E

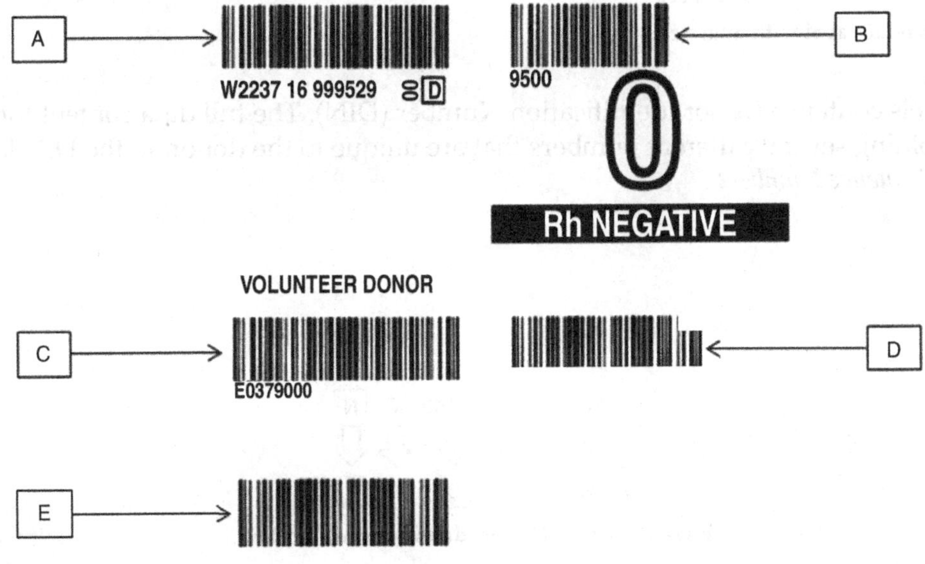

FIGURE 19.5 **ISBT-128 label.**

Concept: AABB Standards require that accredited facilities use ISBT-128 labels. The label must contain at minimum: the collection facility identification number, a lot number related to the donor, a product code, and the ABO group, and Rh type of the donor. This information must be presented in both eye-readable and machine-readable format. Additional special message labels may also be attached to blood component containers. These include "hold for further manufacturing," "for emergency use only," "for autologous use only," "not for transfusion," "irradiated," "biohazard," "from a therapeutic phlebotomy," and "screened for special factors." ISBT-128 allows for this incorporation of any special attributes associated with the blood component.

Answer: *C*—ISBT-128 labels are composed of a base label (not pictured) and a final container label. The base label includes the Container Manufacturer Identity, Catalog Number, and the Container Lot Number. The Final Container Label (shown later) will include at minimum the Donation Identification Number (Answer A), the ABO/Rh Blood Groups (Answer B), the Product Code (Answer C), and the expiration date and time (Answer D).

This information must be presented in both eye-readable and machine-readable format. Additional special message labels may also be attached to blood component containers. These include "hold for further manufacturing," "for emergency use only," "for autologous use only," "not for transfusion," "irradiated," "biohazard," "from a therapeutic phlebotomy," and "screened for special factors." ISBT-128 allows for this incorporation of any special attributes associated with the blood component (Fig. 19.6).

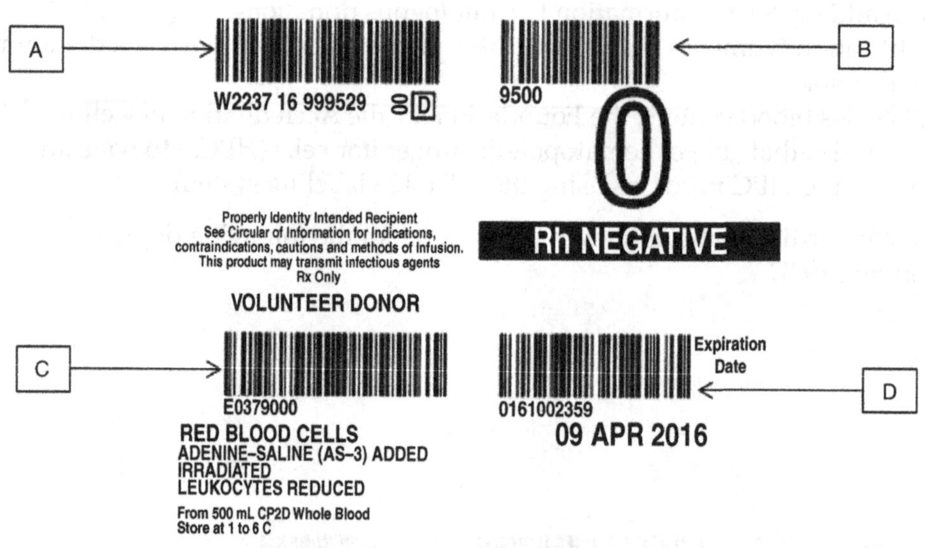

FIGURE 19.6 **ISBT-128 label with answers.**

11. ISBT-128 labels contain a Donor Identification Number (DIN). The full data content for the DIN is created by joining several different numbers that are unique to the donor. In the DIN later (Fig. 19.7), what is the *Sequence Number*?
 A. A
 B. B
 C. C
 D. D
 E. E

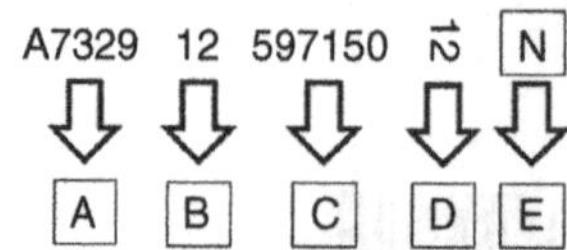

FIGURE 19.7 **Donor identification number.**

Concept: The Donation Identification Number (DIN) is a globally unique identifier that is assigned to each donation and each pooled product. The full data content for the DIN is composed of four parts: a facility identification number (FIN), a nominal year of collection, a serial number, and flag characters.

The FIN is a five-character code assigned by the International Council for Commonality in Blood Banking Automation (ICCBBA). Once assigned, the FIN is permanently associated with the facility in the Registered Facilities Database. The nominal year of collection has two digits. The purpose of this number is to ensure uniqueness of the DIN over a 100-year period; it is not intended to be used as a collection date. The next six numbers are a sequence number assigned by the facility. The next two characters, shown at a 90-degree angle rotated clockwise, are flag characters. The use of flag characters (for process control and/or check characters) is optional. The boxed character at the end of the DIN is the check character intended to confirm the accurate entry of the DIN when a manual keyboard entry is performed.

Answer: *C*—The sequence number is indicated by the letter (C) mentioned earlier. The other choices are incorrect as follows: Answer A represents the Facility Identification Number, Answer B is the nominal year of collection, Answer D is the flag character, and Answer E is the check character (Fig. 19.8).

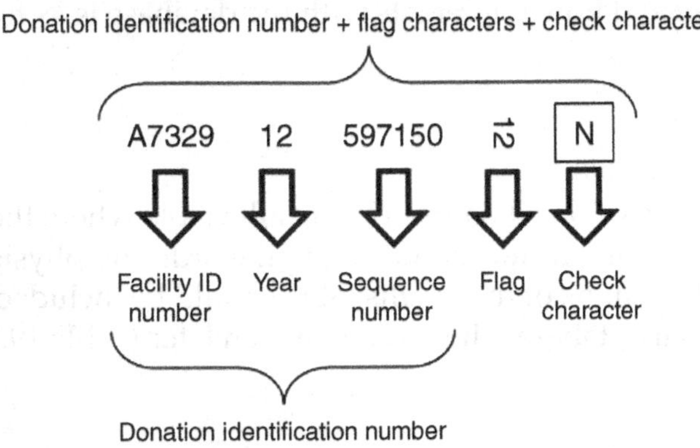

FIGURE 19.8 **Donor identification number explanation.**

12. Which of the following is true regarding radio frequency identification (RFID) systems for labeling in the blood bank?
    A. RFID labels can be updated as the product is modified, without requiring relabeling
    B. RFID labels require line-of-sight scanning
    C. RFID scanning is slower than traditional bar code scanning
    D. RFID labeling systems are designed to replace traditional bar codes on blood products
    E. RFID is more secure than ISBT-128 bar code labels

Concept: Radio frequency identification (RFID) systems provide the blood bank with real-time tracking and inventory management. RFID technology utilizes small memory-storage chips placed on the item, and readers that send and receive radio waves to detect the chips and their data. RFID labels do not replace standard bar codes and labels on blood products, but rather work in conjunction with them to augment the safety and speed of blood product usage/delivery.

Answer: *A*—One of the benefits of RFID labels is that they can be updated as the product is modified, without requiring relabeling. In addition, RFID labels can be read without requiring line-of-sight due to the technology using over-the-air radio-frequencies, in contrast to standard ISBT-128 bar codes (Answer B). In general, the scanning of RFID labels can be done much faster than standardized labels (Answer C). RFID labels do not replace standard bar codes and labels on blood products, but rather work in conjunction with them to augment the safety and speed of blood product usage/delivery (Answer D). RFID is neither more or less secure than ISBT-128 standardized labels (Answer E).

iTrace for Blood Centers, the first application to use Radio Frequency Identification (RFID) technology in blood establishments to assist in enhancing blood safety by preventing the release of unsuitable blood components, was cleared in May, 2013 by the US Food and Drug Administration.

13. Which of the following database architectures uses SQL as the main querying language and displays data in table format?

A. Parallel database
B. Virtual private database
C. Relational database
D. Object database
E. Correlation database

Concept: A relational database is organized based on the relational model of data that is displayed in tables. Relational database management systems (RDBMS) are software components used to maintain a relational database.

Answer: C—Virtually all relational database systems use SQL as the language for querying and maintaining the database. While the other database systems listed can use SQL as the programming interface, the main correlation is that "SQL = relational database = table architecture."

An example of SQL code and how it interacts with the database is as follows:

```
SELECT*
FROM Results
WHERE hemoglobin > 7
ORDER BY ordering_physician;
```

The earlier mentioned code selects data from the results table where the hemoglobin value is greater than 7 and then orders the results according to the ordering physician. The asterisk (*) in the select list indicates that all columns of the results table should be included in the result set.

An example of a relational database schema can be seen later (Table 19.2), displaying the table architecture:

TABLE 19.2 Example of a Relational Database

Patient information				
Paitent ID	**First name**	**Last name**	**Age**	**Sex**
P34645	Mary	Turner	46	F
P87345	Bill	Jones	68	M
P67897	Donald	Smith	22	M
Test order				
Order number	**Patient ID**	**Date**	**Physician**	**Test**
35856	P34645	11/15/2015	May	Hemoglobin
35857	P87345	11/15/2015	Orson	Hematocrit
Test results				
Order number	**Test**	**Result**		
35856	Hemoglobin	9		
35857	Hematocrit	28		
Reference range				
Test	**Adult male**	**Adult female**	**Child**	**Units**
Hemoglobin	14–18	12–16	11–16	g/dL
Hematocrit	42–52	37–47	31–43	%

A parallel database system (Answer A) improves performance through running simultaneous actions (parallelization), such as queries or the loading of data sets via the use of multiple database server processors. A virtual private database (Answer B) is part of a larger database such that only a small subset of data from the larger database is visible or accessible by a subset of users. These are often hosted on shared servers with numerous user accounts per server. In an object database (Answer D), information is presented in the form of objects. These objects can be manipulated through object-oriented programming. In contrast, in a relational database, the data is presented through tables. In a correlation database (Answer E), a value-based storage architecture is used such that each unique database entry is recorded 1 time and the indexing system is automatically generated.

14. Which of the following universal standards uses a database record with the following six fields: property, time, system, type of scale, type of method, and component?
A. PACS
B. LOINC
C. HL7
D. SNOMED
E. DICOM

Concept: According to loinc.org, Logical Observation Identifiers Names and Codes (LOINC) is a "common language (set of identifiers, names, and codes) for clinical and laboratory observations." It includes a database record with the following fields: property, time, system, type of scale, type of method, and component.

Answer: *B*—LOINC was created due to the need for a database system to be used for clinical care and management. It is publicly available at no cost. The purpose of LOINC is to facilitate the electronic exchange and acquisition of clinical test results with a universal and easily-adaptable format.

Each data record includes six fields for each test or measurement:
- Component—what is being measured or observed (e.g., sodium)
- Property—the specific characteristics of what is being measured or observed (e.g., length, color change, concentration)
- Time—the time over which the measurement or observation was made
- System—the type of specimen in which the measurement or observation was made (e.g., plasma, urine)
- Scale—quantitative, ordinal, nominal, or narrative
- Method—procedure or technology used to make the measurement or observation

PACS (Answer A) stands for Picture, Archiving, and Communication System and is used for medical imaging (such as in radiology). Health Level Seven International (HL7, Answer C) is a standard for the exchange and/or retrieval of health records at the application level. The Systematized Nomenclature of Medicine (SNOMED) is a systematic, electronic collection of medical terms (Answer D). Digital Imaging and Communications in Medicine (DICOM) is a standard for handling, storing, printing, and transmitting information in medical imaging. It includes a file format definition and a network communications protocol (Answer E).

15. A piece of equipment in the blood bank has malfunctioned, and you are called to provide more information on the nature of the problem. Instead of inserting the data into a database, the equipment is printing the following information to the output monitor:

```
<Patients>
  <Patient>
  <PatientID>398820</PatientID>
  <LastName>Smith</LastName>
  <FirstName>John</FirstName>
```

```
<BloodType>A</BloodType>
<RhType>Pos</RhType>
<AntibodyScreen>Neg</AntibodyScreen>
</Patient>
</Patients>
```

The earlier mentioned output is an example of what common data element (CDE)?

A. Hypertext markup language (HTML)
B. Cascading style sheets (CSS)
C. Extensible markup language (XML)
D. Hypertext preprocessor (PHP)
E. Structured query language (SQL)

Concept: A CDE is a standardized component of a data set. CDEs structure data in such a way to generate reproducible and consistent data entry while providing standardized formatting at the same time.

Answer: C—The output code given in the question is an example of Extensible Markup Language (XML). Extensible Markup Language (XML) is a way to describe data with the goal of integrating across incompatible formats and data standards. XML markup tags describe the data that is within each tag, and are a specific example of a type of CDE with attached metadata.

Hypertext Markup Language (HTML) is the standard markup language used to create web pages (Answer A). An example is shown as follows:

```
<!DOCTYPE html>
  <html>
    <body>
      <h1>This is a heading.</h1>
      <p>This is a paragraph.</p>
    </body>
  </html>
```

Cascading Style Sheets (CSS) is a style sheet language used for "styling" or the presentation of a document (typically a web page) that was written in a markup language, such as HTML (Answer B). An example of CSS is shown as follows:

```
<style>
  body {background-color:lightgrey;}
  h1 {color:blue;}
  p {color:green;}
</style>
```

This CSS code sets the background body color to light grey, the h1 (header 1) text to blue, and the p (paragraph) text color to green.

In short, hypertext preprocessor (PHP—derived from *P*ersonal *H*ome *P*age) is a server-side scripting language designed to produce web applications that dynamically generate web pages (Answer D). As a user requests a page, the PHP code on the page runs and executes operations before sending the final result to the user's browser. An example is shown as follows:

```
<?php
echo "This will print to the screen";
?>
```

Structured query language (SQL) is a programming language designed for managing data held in a relational database (Answer E). An example is shown as follows:

```
SELECT *
FROM Results
WHERE hemoglobin > 7
ORDER BY ordering_physician;
```

The earlier mentioned code selects data from the results table where the hemoglobin value is greater than 7 and then orders the results according to the ordering physician. The asterisk (*) in the select list indicates that all columns of the Results table should be included in the result set.

16. According to the Open Systems Interconnection (OSI) Model, which of the following refers to level 7?
A. Session
B. Application
C. Presentation
D. Network
E. Data Link

Concept: The Open System Interconnection (OSI) model is a networking framework that implements networking protocols in seven levels.

Answer: *B*—The Application Layer is considered level 7. This layer is particularly important in the health care setting due to Health Level Seven (see Question 17). It supports end-user applications. Examples include remote file access, resource sharing, and virtual terminals.

The Physical Layer (Layer 1) defines the physical and electrical blueprint of the data connection.

The Data Link Layer (Layer 2) transmits data between two different nodes that are connected by the physical layer (Answer E). The Network Layer (Layer 3) is involved in routing network traffic (Answer D). The Transport Layer (Layer 4) transfers data from a source to a destination through a network. The Session Layer (Layer 5) manages communication sessions, or continuous two-way data transmissions between two nodes (Answer A). The Presentation Layer (Layer 6) translates data between a networking service and an application. This translation can include data compression, encryption, decryption, or character encoding (Answer C).

17. Which of the following best describes Health Level Seven?
A. "An organization dedicated to providing a comprehensive framework and related standards for the exchange, integration, sharing, and retrieval of electronic health information"
B. "A common language (set of identifiers, names, and codes) for clinical and laboratory observations"
C. An act signed into law to "promote the adoption and meaningful use of health information technology"
D. A model that "standardizes communication of a computing system" and is organized into seven layers
E. United States legislation that "defines a comprehensive framework to protect government information, operations and assets against natural or man-made threats"

Concept: Health Level Seven International (HL7) is a not-for-profit, American National Standards Institute (ANSI)-accredited standards developing organization "dedicated to providing a comprehensive framework and related standards for the exchange, integration, sharing, and retrieval of electronic health information that supports clinical practice and the management, delivery, and evaluation of health services."

Answer: *A*—HL7 is an organization dedicated to provide a comprehensive framework and related standards for the exchange, integration, sharing, and retrieval of electronic health information. "Level Seven" refers to the seventh level of the International Organization for Standardization (ISO) communications model for Open Systems Interconnection (OSI)—the application level. The application level interfaces with application services. Answer (B) refers to Logical Observation Identifiers Names and Codes (LOINC). Answer (C) refers to The Health Information Technology for Economic and Clinical Health (HITECH) Act. Answer (D) refers to the Open Systems Interconnection model (OSI model). Answer (E) refers to the Federal Information Security Modernization Act (FISMA).

18. While troubleshooting your laboratory information system (LIS), you think the problem is related to the database structure. Based on Table 19.3, the database structure can best be defined as which of the following?

A. Flat file database
B. Cloud database
C. Correlation database
D. Key-value database
E. Relational database

TABLE 19.3 Database structure

Patient information				
Patient ID	**First name**	**Last name**	**Age**	**Sex**
P34645	Mary	Turner	46	F
P87345	Bill	Jones	68	M
P67897	Donald	Smith	22	M
Test order				
Order number	**Patient ID**	**Date**	**Physician**	**Test**
35856	P34645	11/15/2015	May	Hemoglobin
35857	P87345	11/15/2015	Orson	Hematocrit
Test result				
Order number	**Test**	**Result**		
35856	Hemoglobin	9		
35857	Hematocrit	28		
Reference range				
Test	**Adult male**	**Adult female**	**Child**	**Units**
Hemoglobin	14–18	12–16	11–16	g/dL
Hematocrit	42–52	37–47	31–43	%

Concept: Databases are collections of records that are organized into data fields. Databases that are organized into tables and use SQL as the main querying language are relational databases.

Answer: *E*—The database structure depicted in the question is a relational database. In a relational database, the data is organized in the columns and rows of a table. A unique key will identify each row. Virtually all relational database systems use SQL (Structured Query Language) as the language for querying and maintaining the database.

A flat file database (Answer A) is a database that is stored on a host computer. To access the data, the file must be first read into the computer's memory. A good example of a flat file database is a spreadsheet or data organized into a simple text file. A cloud database (Answer B) is a database that runs on a cloud computing platform. On such a platform, resources, such as server random access memory (RAM) or hard drive storage space can be changed on demand. A correlation database (Answer C), unlike relational database management systems, uses a value-based storage (VBS) architecture. In this architecture, each unique data value is stored only once. A key-value database (Answer D) is designed for storing, retrieving, and managing associative arrays—a data structure more commonly known as a *dictionary* or *hash*.

19. Regarding ICD and CPT codes, which of the following is correct?

A. ICD codes are used to enable exchange of health information across different systems while CPT codes are used to record a diagnosis

B. ICD codes represent a comprehensive electronic collection of healthcare terminology while CPT codes enable exchange of health information across different systems

C. ICD codes are used to enable exchange of health information across different systems while CPT codes represent a comprehensive electronic collection of healthcare terminology

D. ICD codes are used to code medical procedures while CPT codes are used to record a diagnosis

E. ICD codes are used to record a diagnosis while CPT codes are used to code medical procedures

Concept: The International Classification of Disease (ICD) is a widely recognized international system for recording diagnoses while Current Procedural Terminology (CPT) coding is a US standard for coding medical procedures, maintained and copyrighted by the American Medical Association (AMA).

Answer: *E*—ICD is a system for recording diagnoses while CPT coding is a US standard for coding medical procedures. Based on these descriptions, all the other choices (Answers A, B, C, and D) are incorrect.

ICD-10 is a more comprehensive and complex set of codes designed to address some of the issues of ICD-9, and the official transition to ICD-10 was made in October 2015. ICD-10 codes are longer than ICD-9 codes which reduce the risk of running out of available codes in the future. In addition, they are more detailed. This was absent in ICD-9.

CPT codes identify the services provided and are used by insurance companies to determine how much a physician will be paid for their services. CPT codes are managed by a CPT Editorial Panel, which meets every 4 months to discuss current issues related to new and emerging technologies. Logical Observation Identifiers Names and Codes (LOINC) was created in 1994 as a free, universal standard for laboratory and clinical observations. It also enables the exchange of health information across different systems. The Systematized Nomenclature of Medicine (SNOMED) is a comprehensive computer-accessible collection of medical terminology.

20. Which of the following choices represent an advantage of ICD-10 coding over ICD-9 coding?

A. As of the latest version, there are 268,000 existing ICD-10 codes, as opposed to the 13,000 in ICD-9

B. ICD-10 enables reporting of laterality, reflecting the importance of which side of the body or limb is the subject of the evaluation

C. There is no clear mapping between ICD-9 and ICD-10 code sets

D. The code set has been expanded from 3–5 positions in ICD-9 to 8 positions in ICD-10, allowing for an increase in specificity of the reporting

E. There are no placeholder characters in ICD-10, cutting down on the confusion of placeholder characters in ICD-9

Concept: The World Health Organization's (WHO) International Classification of Diseases (ICD) has been in place for over a century. The United States implemented ICD-9 in 1979, and as of 1 October, 2015, has transitioned over to ICD-10. ICD is a widely recognized international system for recording diagnoses.

Answer: *B*—One of the major improvements of ICD-10 over ICD-9 is the introduction of laterality. This will allow the code to reflect which side of the body is the purpose of the evaluation or intervention. This laterality (right or left) accounts for more than 40% of all billable codes. As of the latest ICD-10 version, there are approximately 68,000 codes (not 268,000) that are better able to capture the specificity of the billing code (Answer A). One of the most important concerns (and not advantages) in the transition from ICD-9 to ICD-10 coding is that there is no simple mapping or translation from the former to the latter (Answer C). While this statement is true, it is not correct in relation to the question.

The code set has been expanded from 3 to 5 positions in ICD-9 to 7 digits in ICD-10 (Answer D). For ICD-10:

- Digit 1 is alpha; digit 2 is numeric
- Digits 3–7 are alpha or numeric
- A decimal is placed after the third character

There are no placeholder characters in ICD-9, but ICD-10 incorporates these characters. The letter "X" serves as a placeholder when a code contains fewer than six characters and a seventh character applies. The placeholder character also allows for future expansion of the codes (Answer E).

21. A 59-year-old male with poorly-controlled type 2 diabetes mellitus and a past medical history significant for acute promyelocytic leukemia status post hematopoietic progenitor cell transplant presents to his primary care physician due to feeling fatigued. A complete blood count reveals hemoglobin of 6.7 g/dL and a hematocrit of 20%. The patient's BUN and creatinine are elevated and microalbuminuria is detected. No blasts are seen on peripheral smear. Upon further workup, his stool is found to be guaiac positive. The physician wishes to transfuse a unit of red blood cells and a type and screen is ordered. The antibody screen is positive in one out of three cells and anti-C is identified. Of the options listed later, choose the most appropriate and specific ICD-10 code for the antibody screen.

A. E11.21—Type 2 diabetes mellitus with diabetic nephropathy
B. C92.41—Acute promyelocytic leukemia in remission
C. Z01.84—Encounter for antibody response examination
D. R19.5—Occult blood in feces
E. K92.1—Blood in feces

Concept: Understanding the proper ICD-10 coding of why a particular test is being requested is becoming more important due to reimbursement. ICD-10 codes are more detailed than ICD-9 codes, and as such the most specific and accurate code should be selected.

Answer: *D*—Since the question mentions that the stool was guaiac positive, this indicates occult blood in feces and is the best ICD-10 code to use in this scenario. K92.1 is the ICD-10 code used for melena (Answer E). In addition, R19.5 is a Type 1 exclude from K92.1. A Type 1 exclude means that the two ICD-10 codes should never be used together. Thus, K19.5 and K92.1 should never be coded at the same time on the same encounter. Had the patient not had a stool guaiac test done but visible blood was noted in the stool, then K92.1 would be the most appropriate code to use.

Although the patient currently has type 2 diabetes mellitus with evidence of diabetic nephropathy as well as a past history of acute promyelocytic leukemia, these are not the appropriate codes to use for the antibody screen (Answers A and B). When deciding on which ICD-10 code to use, it is best to use the most specific code for why we are doing an antibody screen. Code Z01.84 is more appropriate coding for immunity testing, such as in an allergy workup (Answer C).

22. Your blood bank supervisor has been downloading patient records onto his smartphone for "backup purposes." The patient information includes first name, last name, social security number, address, medical record number, and primary diagnosis ICD-10 code. One evening while at a restaurant, his smartphone is stolen from his vehicle and an estimated 850 patient records from patients local to the area are compromised. Which of the following statements concerning this data breach is correct?

A. The hospital must post a notification on its website, and this notice must remain active for at least 60 days
B. The hospital must provide a toll-free number that remains active for at least 60 days
C. The local media must be notified no later than 60 days after discovery of the breach
D. The Secretary of the Department of Health and Human Services must be notified no later than 7 days after discovery of the breach
E. Individual notification by written form is not required for data breaches affecting fewer than 1,000 patients

Concept: The Health Information Technology for Economic and Clinical Health (HITECH) Act, passed in 2009 as part of the American Recovery and Reinvestment Act, updated notification requirements for breaches of protected health information that must be undertaken by entities covered by HIPAA (Health Insurance Portability and Accountability Act).

Answer: C—Covered entities that experience a breach affecting more than 500 residents of a State or jurisdiction are, in addition to notifying the affected individuals, required to provide notice to prominent media outlets serving the State or jurisdiction no later than 60 days following discovery of the breach. Individual notification is required in breaches of all sizes (Answer E).

Notification requirements include individual notice in written form by first-class mail, or alternatively, by e-mail if the affected individual has agreed to receive such notices electronically. If the covered entity has insufficient or out-of-date contact information for 10 or more individuals, the covered entity must provide substitute individual notice by either posting the notice on the home page of its web site for at least 90 days (Answer A) or by providing the notice in major print or broadcast media where the affected individuals likely reside. The covered entity must include a toll-free phone number that remains active for at least 90 days (Answer B) where individuals can learn if their information was involved in the breach.

Covered entities that experience a breach affecting more than 500 residents of a State or jurisdiction are, in addition to notify the affected individuals, required to provide notice to prominent media outlets serving the State or jurisdiction no later than 60 days following the discovery of the breach (Answer C). Additionally, covered entities must notify the Secretary of the Department of Health and Human Services without unreasonable delay and in no case later than 60 days following a breach (Answer D).

23. Which of the following factors contribute to the advancement of laboratory automation?
A. Higher calibration demands
B. Running a decreased number of analytes on a single system
C. Running multiple test methods on a single system
D. Longer turnaround times
E. Laboratory staff oversupply

Concept: Laboratory automation is a strategy with the goal of streamlining laboratory operations through the use of rapidly developing automation technologies. This field has seen rapid growth since the 1960s. Additional factors that help to advance laboratory automation include an increased number of different analytes and an increased volume of testing methods that are performed on a single system.

Total laboratory automation (TLA) can be described as a combination of multiple instruments coupled to specimen transport and handling systems, with the entire system being automated through a software interface.

Advantages of TLA include faster turnaround times, a reduction in the number of laboratory staff, and increased patient safety through the reduction of labeling and other errors in the preanalytical, analytical, and postanalytical phases.

Disadvantages of TLA include extensive planning for implementation, and the need for trained and highly technical personnel to operate and troubleshoot the system. Furthermore, additional floor space is needed for these systems, and there is a substantial initial financial investment. Finally, software interfacing between the automated system and existing laboratory information systems will often be a concern.

Answer: C—The ability to run multiple test methods on a single automated system is an important driver for laboratory automation. This ultimately increases efficiency while simultaneously reduces laboratory staffing requirements.

There are a number of other factors that help to drive automation in the laboratory, including the need for faster turnaround times (Answer D), shortages of laboratory staff (Answer E), decreased calibration demands (Answer A), faster start-up times, an increased number of different analytes ran on a single system (Answer B), 24-h per day uptime, a reduction in laboratory errors, and increased safety.

24. Which of the following tests would likely have the lowest rate of autoverification?
A. Gentamicin
B. Potassium
C. Albumin
D. BUN
E. Sodium

Concept: Autoverification is a method of using computer-based rules to verify laboratory results without manual intervention.

Answer: *A*—A study published in the Journal of Pathology Informatics looked at autoverification data over approximately 13 years and determined that the highest rates of autoverification occurred with the most frequently ordered tests, such as the basic metabolic panel (sodium, potassium, chloride, carbon dioxide, creatinine, blood urea nitrogen, calcium, and glucose; 99.6%), albumin (99.8%), and alanine aminotransferase (99.7%). The lowest rates of autoverification occurred with some therapeutic drug levels (gentamicin, lithium, and methotrexate) and with serum free light chains (kappa/lambda). The low autoverification rate in this subset of tests was mostly due to the need for dilutions and manual handling of the samples.

25. Which of the following is the best definition of a delta check?
A. The difference between the current laboratory value and the previous laboratory value for the same test on a different patient
B. The difference between the current laboratory value and the previous laboratory value for the same test on the same patient
C. The difference between the current laboratory value and the previous laboratory value for a different test on a different patient
D. The difference between the current laboratory value and the previous laboratory value for a different test on the same patient
E. The difference between the current laboratory value and the previous laboratory value for the same test on the same patient at least 30 days ago

Concept: A delta check compares a laboratory test result with results obtained on previous samples from the same patient. It is a method of quality control and is effective in detecting preanalytical error. Delta check is usually performed on parameters that show the least short-term biologic variability, such that deviations are highly unlikely to be caused by changes in patient status.

Answer: *B*—A delta check is an important quality control tool that can be programmed into the laboratory information system to detect error in results or to point out underlying pathology. For example, a delta check for mean corpuscular volume (MCV) would compare a patient's current MCV of 86 fL to their last value of 68 fL yesterday. Without any obvious evidence of transfusion and/or acute intravascular hemolysis, you would need to consider preanalytical error (such as wrong blood in tube) and/or an error with the analyzer.

26. Which of the following scenarios represent a clinical decision support (CDS) system for improving clinical decision making regarding transfusion?
A. Multidisciplinary teams and committees responsible for determining appropriate parameters for transfusion
B. Targeted lectures to the clinical staff on why a restrictive transfusion policy is more beneficial to patient care

C. Consultation with the Transfusion Medicine Medical Director on whether to order a unit of blood on a hemodynamically stable patient whose hemoglobin is 8.1 g/dL

D. A message on the order screen that notifies clinicians in real time if they are ordering a unit of RBCs for transfusion on patients whose hemoglobin is greater than 7 g/dL

E. Finding relevant peer-reviewed papers showing the advantages of a restrictive transfusion policy

Concept: Electronic medical records have allowed clinical decision support (CDS) to occur via an alert at the decision point simultaneously with physician order entry. In some studies, this has been shown to decrease the percentage of transfusions in patients with hemoglobin levels above 8 g/dL, as well as to decrease overall red blood cell utilization.

Answer: *D*—A common way that clinical decision support systems are implemented is by placing prompts in the ordering system at critical decision points. For example, when trying to place an order for a single unit of red blood cells for transfusion in a patient with a hemoglobin of 8.4 g/dL, the CDS would show a dialogue box reminding the clinician of the current laboratory values and ask them if they would like to cancel the transfusion or continue the order.

An example of a CDS system warning window is seen in Fig. 19.9.

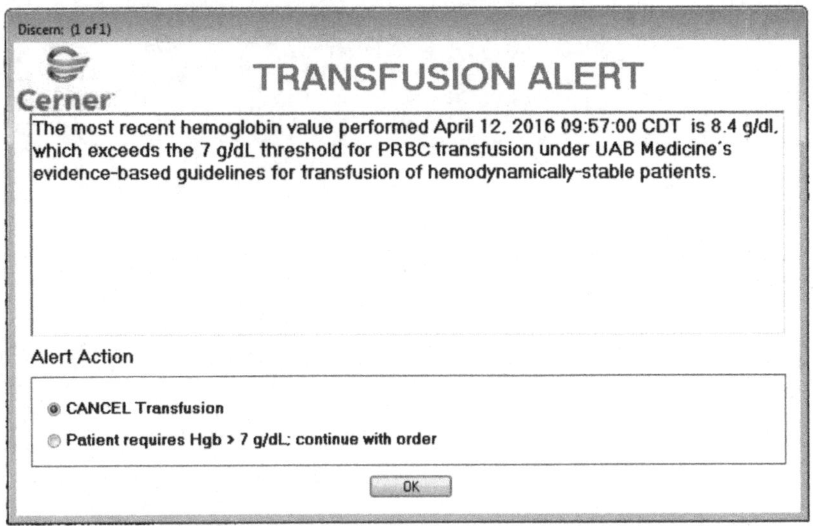

FIGURE 19.9 **CDS system warning example.**

While having teams and committees, targeted lectures, direct consultations, and researching the peer-reviewed literature can all have an impact on the improvement of proper blood utilization, a CDS is typically a software product and/or interfaced with an electronic ordering system.

Suggested Reading

[1] R.A. McPherson, M.R. Pincus, Henry's Clinical Diagnosis and Management by Laboratory Methods, Elsevier Health Sciences, (2011).

[2] M.K. Fung, B.J. Grossman, C.D. Hillyer, C.M. Westoff (Eds.), Technical Manual, eighteenth ed., AABB Press, Bethesda, MD, 2014.

[3] E.H. Shortliffe, J.J. Cimino, Biomedical Informatics: Computer Applications in Health Care and Biomedicine, Springer Science & Business Media, (2013).

[4] L. Williams, M. Fritsma, M.B. Marques, Quick Guide to Transfusion Medicine, second ed., AACC Press, (2014).

[5] L.T. Goodnough, N. Shah, The next chapter in patient blood management: real-time clinical decision support, Am. J. Clin. Pathol. 142 (2014) 741–747.

[6] M.D. Krasowski, S.R. Davis, D. Drees, C. Morris, J. Kulhavy, C. Crone, et al. Autoverification in a core clinical chemistry laboratory at an academic medical center, J. Pathol. Inform. 5 (2014) 13.

[7] United States Food and Drug Administration. Guidelines for the Uniform Labeling of Blood and Blood Components. 1985.
[8] International Council for Commonality in Blood Banking Automation, Inc. (ICCBBA). 2015.
[9] US Department of Health and Human Services. Breach Notification Rule. 2015.
[10] US Department of Health and Human Services. HITECH Act Enforcement Interim Final Rule. 2015.
[11] J. Selmyer, B. Cloutier, Interfacing the clinical laboratory: a primer for LIS managers, Med TechNet Online Services (1996).
[12] International Council for Commonality in Blood Banking Automation (ICCBBA). What is an ISBT 128 Donation Identification Number? TB-006 v1.1.0.

CHAPTER

20

Practical and Advanced Calculations in Transfusion Medicine, Apheresis, and Hemostasis

Huy P. Pham, Richard O. Francis***

*University of Alabama at Birmingham, Birmingham, AL, United States; **New York-Presbyterian Hospital—Columbia University, New York, NY, United States

Transfusion medicine specialists have to perform multiple calculations daily to ensure safe and effective treatments for their patients. Examples of these calculations include plasma volume estimation, coagulation factor dosages, and adjustment of anticoagulant in blood collection bags for smaller donors. This chapter will provide a review of practical calculations and their underlying principles. Most of these calculations may be used on a daily basis on a busy transfusion medicine service. However, some of the calculations will focus on more advanced concepts that, while not used on a daily basis, may help to integrate the knowledge across the fields and thus, enhance the skills.

There are many ways to estimate the blood volume of a patient. For the purpose of this chapter, unless otherwise stated, please use 70 mL/kg (for adults) and 80 mL/kg (for neonates) when calculating blood volume.

1. A 69-year-old woman with a history of acute myelogenous leukemia (AML) is brought to the emergency room by her husband. Because of severe shortness of breath, she is unable to give her own history. Her husband states that over the past 24 h she has been experiencing increasing shortness of breath, lethargy, and headache. A CBC with differential is checked and she is noted to have an elevated WBC count of 180×10^9/L and 80% blasts. The decision is made to perform leukocytapheresis and the apheresis service is consulted. The patient weights 50 kg and is 5 ft. 0 in. (1.5 m) tall.

The Nadler's formula, based on gender, is as follows:

TBV for males (L) = $(0.3669 \times \text{height}^3) + (0.03219 \times \text{weight}) + 0.6041$

TBV for females (L) = $(0.3561 \times \text{height}^3) + (0.03308 \times \text{weight}) + 0.1833$

Where:

Height is in meters

Weight is in kilograms

Using Nadler's formula to estimate the total blood volume, which of the following choices is the appropriate volume of whole blood that should be processed in order to perform a 1.5 to 2.0 volume procedure?

A. 0.5–3 L

B. 4.5–6 L

C. 8–10 L

Transfusion Medicine, Apheresis, and Hemostasis. http://dx.doi.org/10.1016/B978-0-12-803999-1.00020-1

D. 10–12 L
E. 15–20 L

Concept: Various methods can be used to calculate an estimated total blood volume (TBV). Several of these methods take into account various patient factors, such as body mass index, age, gender, and height. Nadler's equation, which uses the gender, height, and weight of the patient to calculate the estimated TBV, is the most commonly used in transfusion medicine and apheresis. It provides a more accurate estimate of the TBV comparing to the rough estimate of 70 mL/kg whole blood.

This patient is demonstrating symptoms of leukostasis due to the very high WBC and blast count. Therefore, leukocytapheresis will be performed and the recommended volume of whole blood that should be processed is 1.5–2 TBVs.

Answer: *B*—Using the above information:
TBV = (0.3561 × 1.5 m³) + (0.03308 × 50 kg) + 0.1833 = 3 L
For 1.5 TBV processed, the machine will process 1.5 × 3 L = 4.5 L
For 2 TBV processed, the machine will process 2 × 3 L = 6 L

Thus, the correct amount of whole blood to be processed is 4.5–6 L (1.5–2 TBVs) as recommended for this procedure. The other choices (Answers A, C, D, and E) are incorrect based on this formula and recommendation for leukocytapheresis.

2. A 23-year-old female is currently on the Labor & Delivery (L&D) floor with DIC and actively bleeding. Her current fibrinogen level is 75 mg/dL and her Hct is 36%. Her weight is 125 kg. Approximately, how many bags of cryoprecipitated AHF does she need if her obstetrician would like to increase her fibrinogen to 150 mg/dL?
A. 10
B. 17
C. 21
D. 28
E. 32

Concept: Although the FDA only requires a minimum of 150 mg/dL in each cryoprecipitated AHF bag, in reality, each bag of cryoprecipitated AHF contains ~250 mg/dL fibrinogen. Thus, the number of bags needed to raise the fibrinogen from an initial to final level can be calculated as:

$$\textit{Number of cryoprecipitated AHF bags needed} = \frac{(\textit{PV in mL}) * (\textit{fibrinogen}_{\textit{final}} - \textit{fibrinogen}_{\textit{initial}})}{(100 * 250)}$$

Answer: *B*—This patient's TBV is (70 mL/kg) × 125 kg = 8,750 mL. His PV is (1–0.36) × 8,750 mL = 5,600 mL. Applying the above formula, we have:

$$\textit{Number of cryoprecipitate bags needed} = \frac{(5{,}600\ \text{mL}) * (150-75)}{(100 * 250)} = 17\ \textit{bags}$$

In many institutions, instead of releasing each bag of cryoprecipitated AHF (~10–15 mL) separately, many choose to purchase 5-units prepooled cryoprecipitated AHF (~50–60 mL per prepooled bag) and release as such. If the pooling is done in a sterile fashion (i.e., closed system), then these will expire 6 h after thawing. If not, then they expire 4 h after thawing. In real world situation, the above formula is rarely used as the "typical" cryoprecipitated AHF dose for adults is 10 units (or two 5-units prepooled bags) and for children is 1 unit for every 10 kg. However, in patient's were minimal donor exposures is beneficial (e.g., transplant patients), this calculation could lead to less exposure to both infections and sensitizing antigens. The other choices (Answers A, C, D, and E) are incorrect based on this formula.

Please answer Questions 3–5 based on the following scenario:

A 37-year-old male with history of severe hemophilia A is admitted to the hospital for a scheduled cardiac bypass surgery next day. He takes prophylactic recombinant factor VIII (rFVIII) at home 3 times a week.

3. This patient's current weight is 60 kg and his factor VIII level is 20%. His hematoctit (Hct) is 45%. What dose of rFVIII do you recommend for this patient prior to his operation and for maintenance?

	Initial dose of rFVIII	Maintenance dose of rFVIII and frequency
A	1,848 IU	924 IU every 8–12 h
B	1,848 IU	1,848 IU every 8–12 h
C	1,848 IU	948 IU every 18–24 h
D	3,696 IU	1,848 IU every 8–12 h
E	3,696 IU	3,696 IU every 18–24 h

Concept: The most accurate way of calculating factor VIII dosage is based on plasma volume (PV) where PV = (1−Hct) × TBV. The amount of factor VIII concentrate to infuse can be calculated as follows:

$$Amount\ of\ factor\ VIII = (PV) * (Factor\ VIII_{goal} - Factor\ VIII_{initial})$$

In some instances, a quick common rule-of-thumb for factor VIII dosing is 1 IU/kg factor VIII concentrate will raise factor VIII level by 2%. However, this method may lead to over- or under-dosing of factor VIII, depending on the patient's hematocrit. It can be shown mathematically that if the patient's Hct is ~28% or more, the 'rule-of-thumb' calculation for factor VIII dosing overestimates the amount of factor necessary for the patient.

After administering the initial dose, *Factor VIII* is typically given for maintenance at half of the initial dose every 8–12 h. A peak and a trough level are generally recommended to help adjusting the doses.

If the patient has hemophilia B (Factor IX deficiency), then the initial dose will usually be 2 × PV × (Factor IX_{goal}–Factor $IX_{initial}$) because Factor IX has both intravascular and extravascular distribution. The frequency for Factor IX maintenance should be every 18–24 h. Again, a peak and a trough level are generally recommended to help adjust the maintenance dosages. Additionally, a quick common rule-of-thumb for Factor IX dosing as 1 IU/kg Factor IX will raise Factor IX level by 1%. However, similar to Factor VIII dosing, depending on the patient's initial Hct this way may also lead to over- or underdosing of Factor IX.

Of note, the rule-of-thumb dosage calculations are routinely used by many physicians and are the recommendations from the World Federation of Hemophilia. They are also present in the package insert of recombinant Factor VIII concentrate (Recombinate, Baxter, Westlake Village, CA). For recombinant Factor IX concentrate (BeneFIX, Pfizer, Philadelphia, PA), the dose is calculated as follows according to the package insert:

For adult patients (≥15 years): Dose of Factor IX required = 1.3 × (body weight in kg) × (desired increased in Factor IX in IU or %)

For pediatric patients (<15 years): Dose of Factor IX required = 1.4 × (body weight in kg) × (desired increased in Factor IX in IU or %)

Answer: *A*—This patient's estimated total blood volume (TBV) is (70 mL/kg) × 60 kg = 4,200 mL. His PV is (1−0.45) × 4,200 mL = 2,310 mL. Given that he will have cardiac bypass surgery, his Factor VIII level should be raised to 100% before the surgery. Thus, initial dose of factor VIII is (2,310) × (100%–20%) = 1,848 IU and the maintenance dose should be (1/2) × 1,848 IU = 924 IU every 8–12 h.

If the rule-of-thumb calculation is used, the initial dose will be (60 kg) × [(100%–20%)/2] = 2,400 IU of factor VIII, and thus, leading to a maintenance dose of 1,200 IU Factor VIII every 8–12 h (an overestimation of the PV-based calculation).

If this patient has hemophilia B, then the calculation is as follows:

Initial dose = 2 × (2,310) × (100%–20%) = 3,696 IU of factor IX

Maintenance dose = (1/2) × 3,696 = 1,848 IU of Factor IX every 18–24 h

Rule-of-thumb calculation for initial dose = (60 kg) × (100%–20%) = 4,800 IU of Factor IX, and thus, leading to a 2,400 IU factor IX every 18–24 h (an overestimation of the PV-based calculation). The other choices (Answers B, C, D, and E) are incorrect based on this formula and based on the type of hemophilia being treated (i.e., Hemophilia A in this case).

4. Approximately 5 years later, this patient was found to develop a Factor VIII inhibitor (at 120 Bethesda units [BU], last tested 1 week ago). He is now admitted to the hospital with cardiac tamponade after a fist fight. In the emergency department (ED), his laboratory results are as follows: prothrombin time (PT) 15 s, activated partial thromboplastin time (aPTT) 69 s, and fibrinogen 250 mg/dL. His current height is now 6 ft. 3 in. and his weight is 100 kg. The ED physician asks for advice regarding the type and dosage of blood product(s) and/or blood derivative(s) for this patient. The recommendation should be:

A. 9 mg recombinant activated factor VII (rFVIIa) every 2–3 h until bleeding stops
B. 3 mg rFVIIa every 2–3 h until bleeding stops
C. 9 mg rFVIIa every 8–12 h until bleeding stops
D. 3 mg rFVIIa every 2–3 h until bleeding stops
E. 9 mg rFVIIa now, and 5,000 IU factor VIII concentrates every 8–12 h until bleeding stops

Concept: This patient's inhibitor level is very high (at 120 BU). Thus, he is unlikely to respond to factor VIII concentrate. Therefore, a bypass agent (either rFVIIa or Factor Eight Inhibitor Bypass Agent [FEIBA]) should be used to treat his bleeding in addition to surgical intervention.

rFVIIa (NovoSeven, Novo Nordisk, Plainsboro, NJ) is approved by the US Food and Drug Administration (FDA) to treat the following:

1. congenital factor VII deficiency (15–30 μg/kg every 4–6 h until bleeding stops)
2. congenital factor VIII or IX deficiency with inhibitors (90 μg/kg every 2–3 h until bleeding stops)
3. acquired hemophilia (70–90 μg/kg every 2–3 h until bleeding stops)
4. Glanzmann's thrombasthenia with refractoriness to platelet transfusion (90 μg/kg every 2–6 h until bleeding stops)

Answer: *A*—Based on the rFVIIa package insert, this patient should receive 90 × 100 = 9,000 μg (or 9 mg rFVIIa) every 2–3 h until bleeding stops. The other choices (Answers B, C, D, and E) are incorrect based on this formula.

5. The Blood Bank realizes that only 5 mg rFVIIa are in the current inventory. Currently, there is a snow storm in the area, and thus, the next rFVIIa shipment may not arrive until the next morning. Thus, an alternative product, such as FEIBA, must be given. Please provide a dosage recommendation for FEIBA.

A. 5,000 IU every 2 h until bleeding stops
B. 5,000 IU every 4 h until bleeding stops
C. 10,000 IU every 4 h until bleeding stops
D. 10,000 IU every 8 h until bleeding stops
E. 10,000 IU every 12 h until bleeding stops

Concept: As explained in Question 2, FEIBA (FEIBA, Baxter Healthcare Corporation, Westlake Village, CA) may be used instead of rFVIIa in the treatment of congenital Factor VIII or IX with inhibitor. The maximum recommended dose for FEIBA is 200 IU/kg/day. If this dose is exceeded, the patient is at risk of disseminated intravascular coagulation (DIC) and/or other forms of thrombosis.

The initial dose of FEIBA is often based on the severity of bleeding. With mild bleeding a dose of 50 IU/kg may be sufficient. With severe bleeding, doses of 75–100 IU/kg are often chosen. Since there is a risk of thrombosis, 100 IU/kg is rarely used. The frequency for FEIBA can be between 6 and 12 h.

Answer: *E*—Only Answer E does not exceed the maximum dose of 200 IU FEIBA/kg/day; thus, the other choices (Answers A, B, C, and D) are incorrect.

6. A 65-year-old male is admitted to the ED with intracranial hemorrhage (ICH). He is on warfarin for stroke prophylaxis because of atrial fibrillation (AF). His other medical problems include diabetes mellitus, hyperlipidemia, chronic renal failure, congestive heart failure, and hepatitis C induced liver cirrhosis. He weighs 120 kg, and his international normalized ratio (INR) is currently 8.5. The neurosurgeon requests the blood bank resident's suggestion to reverse the effect of warfarin in this patient prior to emergent surgery. What is your recommendation?
A. 6 units of plasma and 10 mg intravenous (IV) vitamin K
B. 12 units of plasma and 10 mg IV vitamin K
C. 3,000 IU 4-factor prothrombin complex concentrate (PCC) and 10 mg IV vitamin K
D. 5,000 IU 4-factor PCC and 10 mg IV vitamin K
E. 7,000 IU 4-factor PCC and 10 mg IV vitamin K

Concept: 4-factor PCC (Kcentra, CSL Behring LLC, Kankakee, IL) has been approved by the FDA for the emergent reversal of warfarin. In addition to the vitamin K dependent coagulation factors (Factors II, VII, IX, and X), 4-factor PCC also contain protein C and S (natural anticoagulants). Since the half-life of Factor VII is short, vitamin K administration in addition to 4-factor PCC is also recommended to restart the liver's production of factors II, VII, IX, X, protein C and S. For 4-factor PCC, the dose is calculated as follows:

Pretreatment INR between 2 and 4: Give 25 IU/kg 4-factor PCC (maximum dose 2,500 IU)
Pretreatment INR between 4 and 6: Give 35 IU/kg 4-factor PCC (maximum dose 3,500 IU)
Pretreatment INR >6: Give 50 IU/kg 4-factor PCC (maximum dose 5,000 IU)

Elevation of INR alone is not an indication for administration of 4-factor PCC. In nonbleeding patients with a pretreatment INR of <10, treatment may simply consist of stopping the medication and/or administration of vitamin K.

Answer: *D*—This patient is bleeding and his pretreatment INR is 8.5; thus, he will receive 50 IU/kg 4-factor PCC to reverse the warfarin effect. However, per the package insert, the maximum dose is 5,000 IU, and he weighs more than 100 kg. Thus, the blood bank resident should recommend giving 5,000 IU 4-factor PCC. Additionally, since the half-life of factor VII in 4-factor PCC is short, and factor VII level has a significant effect on INR, IV vitamin K should be given along with 4-factor PCC to restart liver synthesis of the coagulation factors. The other choices (Answers A, B, C, and E) are incorrect based on this formula.

7. A 3-year-old male requires TPE for antibody-mediated rejection of the cardiac allograft. He weighs 15 kg, and his Hct is 22%. Given that your current apheresis machine has an extracorporeal volume (ECV) of 170 mL and an extracorporeal red cell volume (ERCV) of 68 mL, how should you instruct the apheresis staff in regards to priming of the apheresis machine?
A. No priming is necessary
B. RBC priming only
C. Plasma priming only
D. Lactated ringer priming only
E. 5% albumin priming only

Concept: In order to perform a safe apheresis procedure, at any given time, the ECV and the ERCV should not exceed 15% of the patient's TBV and RCV, respectively. In critically ill pediatric patients,

10% may be considered as the cutoff. If the ECV and/or the ERCV exceed 15% of the patient's TBV and RCV, respectively, then, priming of the apheresis device should be considered as following:

If only ECV > 15% of TBV, then priming with 5% albumin may be used alone (i.e., the patient only needs intravascular volume but not oxygen carrying capacity).

If ERCV > 15% of RCV with or without ECV > 15% of TBV, then RBCs alone or in combination with 5% albumin, should be used as prime fluid (i.e., the patient needs oxygen carrying capacity)

Answer: *B*—This patient's estimated TBV is (70 mL/kg) × 15 kg = 1,050 mL, and his RCV is (22%) × 1,050 = 231 mL. Therefore, using 15% as the cutoff, his maximum tolerable ECV and RCV are (15%) × 1,050 = 157.5 mL and (15%) × 231 = 34.65 mL, respectively. Both the device ECV and ERCV exceeds the maximum tolerable ECV and ERCV of the patient (i.e., 170 mL > 157.5 mL and 68 mL > 34.65 mL); thus, RBCs alone or in combination with 5% albumin, should be used as prime fluid. Some institutions use undiluted RBCs as prime fluid while others dilute RBCs with 5% albumin to a target Hct (either preset Hct, such as 30%, or to the patient's current Hct). If you prime the device with RBCs, then you should consider skipping the rinse back to avoid acutely increasing the fluid volume of a small patient with heart problems. Answer A is incorrect because priming is clearly indicated in this case. Answers C, D, and E are incorrect because priming with these fluids would be insufficient for this patients case (i.e., these fluid choices do not provide the oxygen carrying capacity).

8. A 4-year-old girl requires extracorporeal photopheresis (ECP) for the treatment of graft versus host disease (GVHD). She weighs 28 kg, has a single lumen port for ECP, and her peripheral veins are not suitable for access. Your ECP device can perform ECP using both single- and double-needle access with the following limitations:

Hematocrit (%)	Maximum ECV (mL)—single needle access	Maximum ECV (mL)—double needle access
28	500	450
30	450	400
32	400	350
34	350	300
36	300	250
38	250	200

Assuming that the Apheresis unit cannot do RBC priming for ECP, what is the minimum Hct that this patient must have in order for her to safely undergo ECP?

A. 30% with single needle access
B. 32% with double needle access
C. 34% with single needle access
D. 36% with double-needle access
E. 38% with single-needle access

Concept: As discussed in Question 7, in order to have a safe apheresis procedure, ECV cannot exceed 15% of the patient's TBV. ECP devices have different ECV requirements depending on the single vs. double access modes and the patient's pretreatment Hct. Double needle access allows for continuous flow, which results in smaller ECV requirements. Furthermore, a higher Hct allows the device to use less ECV to collect the buffy coat for photopheresis.

Answer: *E*—The patient's estimated TBV is (70 mL/kg) × 28 kg = 1,960 mL; thus, she can tolerate a maximum ECV of (15%) × 1,960 mL = 294 mL. Additionally, since the patient only has a single lumen port and no peripheral access options, she can only undergo ECP using the single needle access mode. From the device's characteristics, she needs a Hct of at least 38% for safe ECP procedure without RBC priming (250 mL ECV at Hct 38% is less than the maximum allowable ECV of 294 mL). The other choices (Answers A, B, C, and D) are strictly incorrect based on the chart and explanation above.

Please answer Questions 9–11 based on the following scenario:

A 56-year-old female has been admitted to the hospital with acute myeloid leukemia (AML). She was started on chemotherapy a few days ago. She currently has a fever (up to 102°F within the past 24 h) and epitaxis. Her medication list includes ceftriaxone and vancomycin. Her labs show a platelet count of $15 \times 10^9/L$ and her coagulation parameters within the reference range. She has received approximately 4–5 apheresis platelets [or single donor platelet (SDP)] units over the past 24 h without a durable increment in her platelet counts. She weighs 80 kg with a body surface area (BSA) of 1.5 m^2. Transfusion Medicine is consulted for recommendations.

9. What is your initial recommendation for transfusion or further laboratory testing based on the case scenario?
A. Recommend tranexamic acid (TXA) for treatment of DIC because it is the most likely cause of platelet refractoriness
B. Recommend 1-h post platelet count after administering 1 SDP unit to distinguish immune vs. non-immune causes of platelet refractoriness
C. Recommend 24-h post platelet count after administering 1 SDP unit to distinguish immune versus nonimmune causes of platelet refractoriness
D. Recommend antibiotic-dependent platelet antibody testing to distinguish vancomycin versus ceftriaxone as the cause of the platelet refractoriness
E. Recommend acetaminophen for treatment of fever because it is the cause of the platelet refractoriness

Concept: Platelet refractoriness can be immune, nonimmune-mediated, and/or both. Immune causes of platelet refractoriness include platelet antibody, human leukocyte antigen (HLA) antibody and/or both. Nonimmune causes include fevers, sepsis, medications (some can cause immune and some can cause nonimmune causes of platelet destruction), hepatosplenomegaly, DIC, and/or a combination of causes. To distinguish between immune versus nonimmune cause, a pretransfusion and a 1-h posttransfusion platelet count should be performed. A typical response to transfusion of SDP platelets is an increased of approximately $30–60,000 \times 10^9/L$.

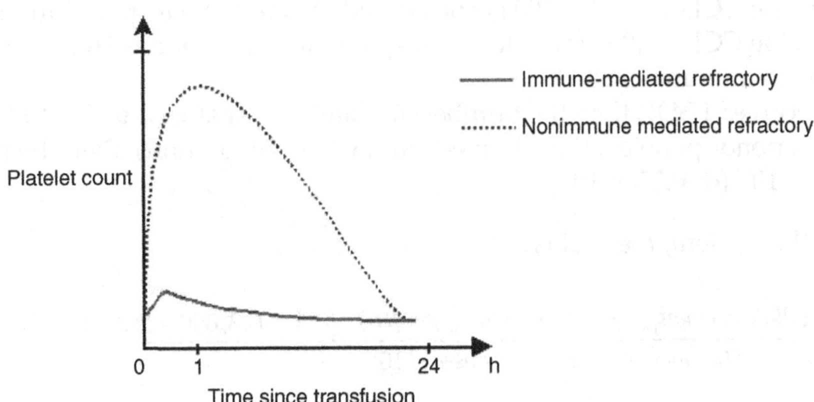

If the cause is immune-mediated, then the 1-h posttransfusion platelet count will essentially return to the pretransfusion platelet count because platelet antibodies destroy the platelets immediately. If the cause is nonimmune-mediated, then the 1-h platelet count will show an appropriate response but the platelet count at 24-h will be lower than the 1-h posttransfusion count. Since the treatment for immune- or nonimmune-mediated platelet refractoriness is different, performing 1-h post transfusion platelet count is highly recommended when encountering a patient requiring multiple SDP units daily without a durable response to platelet transfusion.

Answer: *B*—This patient has many risks for platelet refractoriness (fevers, antibiotics, and history of previous transfusion). Therefore, it is essential to distinguish between immune (antibody mediated) and nonimmune causes (fevers and antibiotics) by performing a 1-h platelet count after administering 1 SDP unit. There is no laboratory evidence that this patient is in DIC (Answer A). 24-h post platelet count (Answer C) would not be able to distinguish immune versus nonimmune causes of platelet refractoriness. Antibiotic-dependent platelet antibody testing to distinguish vancomycin versus ceftriaxone as the cause of the platelet refractoriness (Answer D) is not the first step in the management—it may be performed after ruling out immune mediated platelet destruction by doing the 1-h post platelet count. Treating the fevers with acetaminophen is not the first treatment for platelet refractoriness before ruling out immune mediated platelet destruction (Answer E).

10. A 1-h posttransfusion platelet count was performed. Prior to the transfusion of apheresis-derived platelets, the platelet count is 15×10^9/L, and posttransfusion, it is 23×10^9/L. A similar increase in her platelet count was noted after a second unit was transfused. What is the interpretation of this result based on the response to platelets and the information provided to you in the original case scenario?

A. Appropriate response at 1-h posttransfusion suggesting the cause is immune mediated
B. Appropriate response at 1-h posttransfusion suggesting the cause is nonimmune mediated
C. Appropriate response at 1-h posttransfusion suggesting the cause is due to HLA antibodies
D. Inappropriate response at 1-h posttransfusion suggesting the cause is nonimmune mediated
E. Inappropriate response at 1-h posttransfusion suggesting the cause is immune mediated

Concept: As discussed above, 1-h platelet count after administering 1 SDP unit can distinguish between immune versus nonimmune cause. A transfusion with 1 SDP unit typically increases the platelet count between 30 and 60 $\times 10^9$/L. A CCI is the formal calculation used to document platelet refractoriness. CCI is calculated as follows:

$$CCI = \frac{(Platelet\ count_{post} - Platelet\ count_{pre} per\ \mu L) * BSA}{(Number\ of\ Platelet\ transfused) * 10^{11}}$$

A 1-h posttransfusion CCI of 5,000–7,500 is considered an adequate response to platelet transfusion. A 1-h posttransfusion CCI < 5,000 for at least 2 sequential transfusion is considered platelet refractoriness due to immune-caused.

If the patient receives 1 SDP, then the number of platelets transfused is 3×10^{11}. If the patient receives 1 random donor platelets (i.e., derived from whole blood unit), then the number of platelets transfused is 5.5×10^{10} (or 0.55×10^{11}).

Answer: *E*—For this patient, the CCI is:

$$CCI = \frac{(Platelet\ count_{post} - Platelet\ count_{pre} per\ \mu L) * BSA}{(Number\ of\ Platelet\ transfused) * 10^{11}} = \frac{(23{,}000 - 15{,}000) * 1.5}{3} = 4{,}000$$

Thus, because the CCI <5,000 on 2 separate occasions, the cause of this patient's platelet refractoriness is likely immune in nature. Other choices (Answers A, B, C, and D) are wrong based on the above calculation and explanation.

11. It is determined that the cause of this patient's platelet refractoriness is due to HLA antibodies. Her HLA phenotype is A*02, A*04; B*15, B*35; Cw*04, Cw*07; DRB1*12, DRB1*13; DQB1*03, DRB1*06. The Donor Center currently has a platelet donor with the following HLA phenotype A*02, A*04;

B*15, B-; Cw*12, Cw-; DRB1*04, DRB1*13; DQB1*03, DRB1-. If the platelet unit from this donor is used to the patient, what is the grade for the level of HLA matching:

A. A
B. B1X
C. B1U
D. B2UX
E. C

Concept: HLA matched platelets are usually provided to patients with platelet refractoriness due to HLA antibody. Since platelets only have Class I HLA antigen on their surface, matching is typically done for HLA A and B antigen. The following is HLA match grade

Grade	Interpretation
A	4 antigens match
B1U	3 antigens match, 1 antigen unknown (possibly homozygosity)
B1X	3 antigens match, 1 antigen cross reactive
B2UX	1 antigen unknown, 1 antigen cross reactive
B2X	2 antigens match, 1 antigen cross reactive
C	3 antigens match, 1 antigen mismatch with no cross reactivity
D	2 antigens match, 2 antigens mismatch with no cross reactivity

In practice, grade C is usually no better than random platelet from the inventory. Unless the circumstances are dire, most Transfusion Medicine specialists will not order HLA match platelets unless they can get Grade A or B match.

Answer: C—Comparing the patient's and donor's HLA A and B antigen, the donor is a B1U match (i.e., 3 antigens match with 1 unknown antigen). The other choices (Answers A, B, D, and E) are incorrect based on the table above.

End of Case

12. A 32-year-old pregnant woman is being followed because she has an anti-c and the biological father of the child is known to be positive for the c antigen. An antibody titration study is performed today in the blood bank and the results are as follows:

Dilution	1:2	1:4	1:8	1:16	1:32	1:64	1:128	1:256	Score	Titer
Agglutination	4+	3+	2+	2+	1+	1+	w+	0	?	?

Given this information, what are the score and titer for this antibody titration study?

A. Score 48 and Titer 64
B. Score 64 and Titer 48
C. Score 40 and Titer 128
D. Score 12 and Titer 256
E. Score 4 and Titer 2

Concept: Pregnancy can be complicated by maternal alloantibodies that may attack fetal RBCs, if the fetal RBCs contain the corresponding antigen, leading to a risk for hemolytic disease of the fetus and newborn. Therefore, when maternal alloantibodies are detected during pregnancy, antibody titration studies are performed to determine the strength of the antibody. This information is used for determining the level of monitoring that is needed for adverse effects on the fetus. These titration studies are repeated in at least monthly intervals.

Serial dilutions of the patient serum that contains the antibody are tested against red cells that are positive for the antigen in question. The titer is the reciprocal of the highest dilution in which agglutination (not including weak reaction or reaction <1+) was observed.

A score can also be calculated in which each level of agglutination is assigned a given value and the score is the sum of the values. The agglutination and score values are as follows: 4+ = 12, 3+ = 10, 2+ = 8, 1+ = 5). In the case of alloimmunization against an Rh antigen, generally pregnancies in which antibody titers are 8 or lower can be managed by serial monitoring of the maternal antibody titers. If the titer is 16 or higher, however, fetal wellness assessment is necessary by ultrasonography to evaluate middle cerebral artery peak systolic velocity or serial amniocentesis for delta OD450 (measuring bilirubin). In addition, the results of the current specimen should be compared with prior specimens and a change in titer of 2 or more tubes or a change in score of 10 or more are considered significant.

Answer: *A*—Using the above information:

Correlating the agglutination to the scoring values: 4+ = 12, 3+ = 10, 2+ = 8, 2+ = 8, 1+ =5, 1+ = 5)
Score = 12 + 10 + 8 + 8 +5 +5 = 48
Titer = reciprocal of the highest dilution that showed agglutination: dilution of 1:64—Titer 64.
The other choices (Answers B, C, D, and E) are incorrect based on the above information and calculations.

13. A 17-year-old boy with a recent diagnosis of idiopathic thrombocytopenic purpura (ITP) presents to the ED. He states that over the past 24 h he has noticed bruising on his arms, bleeding from his gums when brushing his teeth this morning, and bleeding from his nose. Blood typing demonstrates that the patient is O, Rh positive, and CBC reveals hemoglobin of 13.5 g/dL and a decreased platelet count of 5×10^9/L. The patient has not had a splenectomy. The decision is made to treat with intravenous Rh immune globulin (IV RhIG) and the blood bank is called to help with calculating the appropriate initial dose. A body weight of 65 kg is provided for this patient. What is the correct initial dose of IV RhIG that should be administered to this patient?

A. 1,250 μg
B. 1,625 μg
C. 2,500 μg
D. 2,825 μg
E. 3,250 μg

Concept: The goal of treatment for patients with ITP is to achieve a platelet count that will support adequate hemostasis. Most patients with thrombocytopenia with no bleeding or only mild bleeding can be managed with observation alone. If it is deemed necessary to treat the patient, first line therapies include corticosteroids, IVIG, or IV RhIG). WinRho SDF (Cangene Corporation doing business as Emergent Biosolutions, Winnipeg, Manitoba, Canada) is an IV RhIg that is FDA approved for treatment of acute and chronic ITP in Rh positive patients who have a spleen. Although the exact mechanism is unknown, it has been proposed that the RhIG binds to the Rh(D)-antigen on the patient's RBCs, leading to immune-mediated clearance by the reticuloendothelial system, and presumably subsequent competition for clearance sights in the spleen is created, resulting in decreased clearance of antibody-coated platelets.

According to the package insert for WinRho SDF, in a patient with hemoglobin greater than 10 g/dL, the appropriate initial dose is 50 μg/kg (or 250 IU/kg). For patients with hemoglobin ≤ 10 g/dL, a lower dose [25–40 μg/kg (or 125–200 IU/kg)] should be used. A platelet count > 50 × 10^9/L is generally accepted as a threshold for satisfactory response. If the patient responds to therapy, and if subsequent therapy is required, then a dose of 25–60 μg/kg (or 125–300 IU/kg) may be used, depending on the patient's platelet count and hemoglobin level. The dosage and frequency of maintenance therapy depends on the patient's clinical status and laboratory parameters (platelet count, hemoglobin, and reticulocyte counts). If the patient does not respond, redosing is based on the hemoglobin level

If hemoglobin < 8 g/dL, then alternative therapies should be considered (i.e. no WinRho SDF should be given).

If 8 g/dL < hemoglobin < 10 g/dL, then redose at 25–40 μg/kg (or 125–200 IU/kg)

If hemoglobin is between 8 and 10 g/dL, then redose at 50–60 μg/kg (or 250–300 IU/kg)

Importantly, patients should be monitored for at least 8 h for the occurrence of adverse events that may include hemolysis, hypersensitivity reactions, febrile reactions, and transfusion-related acute lung injury. The FDA placed a black-box warning on this product indicating that intravascular hemolysis leading to death has been reported for using WinRho SDF in the treatment of ITP. Thus, the patient needs to be monitored closely for signs, symptoms, and laboratory evidence of intravascular hemolysis in the healthcare setting for at least 8 h. Dipstick urinalysis to monitor hemoglobinuria and hematuria should be performed at baseline, at 2 h, at 4 h, and before the end of the monitoring period.

Answer: *E*—Using the above information:

$$\textit{Initial dose}(\mu g) = 65\ kg \times 50\ \mu g/kg = 3{,}250\ \mu g$$

14. A 68-year-old man with multiple myeloma is preparing to undergo an autologous stem cell transplant. A request for hematopoietic progenitor cell (HPC) collection has been placed with a goal collection of 5 × 10^6 CD34+ cells per kg. The patient's weight is 70 kg, his hematocrit is 36%, his preharvest CD34+ cell count is 20/μL (20 × 10^6 /L), and the collection efficiency of the apheresis machine being used and the staff running it is 55%. What is the total volume of whole blood that needs to be processed in order to meet the collection goal of 5 × 10^6 CD34+ cells/kg?

A. 10 L
B. 16 L
C. 26 L
D. 32 L
E. 38 L

Concept: Estimation of the total amount of whole blood that must be processed for a HPC collection can be made when the weight of the donor in kilograms, the preharvest CD34+ cell count, CD34+ cell/kg collection goal, and collection efficiency (CE) of the apheresis machine being used are known. The CE is the percentage of cells that is collected from the total number of cells processed by the apheresis machine. A general rule that is often used in determining the amount of whole blood that needs to be processed is 3–4 TBVs. However, depending on the variables mentioned earlier, using a prediction formula to calculate the amount to be processed can improve the CD34+ cell yield. The formula for estimating the amount of whole blood to be processed for a HPC collection is:

$$\textit{Whole Blood to process} = \frac{(\textit{desired CD34+cell/kg}) * \textit{body weight in kg}}{(\textit{collection efficiency}) * (\textit{donor preharvest CD34+cell count})}$$

Answer: *D*—Using the above information:

$$Whole\,Blood\,to\,process = \frac{(5{,}000{,}000\,CD34+cell/kg)*70\,kg}{(0.55)*20{,}000{,}000\,CD34+cell/L} = 32\,L$$

This formula only provides an estimate. A postcollection CD34+ count of the products is always recommended. The other choices (Answers A, B, C, and E) are incorrect based on this formula.

15. A 45-year-old male undergoes therapeutic plasma exchange (TPE) for hyperviscosity due to Waldenstrom macroglobinemia. Assume the IgM monoclonal protein behaves as an ideal solute, after 1.2 plasma volume exchange, what is the percentage of IgM monoclonal protein left in the patient's body?
A. 25%
B. 30%
C. 37%
D. 63%
E. 70%

Concept: Assuming one compartmental model and the substance to be removed acting as an ideal solute, the fraction remained intravascular at any time during TPE is $x(t) = x_o e^{-v(t)}$ where $x(t)$ is the concentration of the substance of interest at time t, x_o is the initial concentration of the substance of interested, and $v(t)$ is the number of plasma volume exchange at time t.

Answer: *B*—Applying the above formula, at the end of 1.2 plasma volume exchange, we have: $x(t) = (100\%)\,e^{-1.2} = 30.1\%$. Thus, ~30% of the IgM monoclonal protein will be left at the end of 1.2-plasma volume exchange. Other choices (Answers A, C, D, and E) are wrong.

Using the same method, it can be shown that after 1, 1.5, and 2-plasma volume exchange, approximately 37%, 22%, and 14% of the initial amount of the ideal solute will remain, respectively.

16. A 2-kg-term newborn with a hematocrit of 35% requires a red blood cell (RBC) transfusion. You plan on transfusing 10 mL per kg of body weight of RBCs stored in citrate-phosphate-dextrose-adenine-1 (CPDA-1) solution. Once he receives RBC transfusion, what is his approximate Hct after a 20 mL transfusion?
A. 36%
B. 37%
C. 39%
D. 43%
E. 45%

Concept: The following formula is used to calculate the final Hct after a given volume of transfusion:

$$Hct_{final} = \frac{RCV_{initial} + (V_{transfused})(Hct_{transfused})}{TBV + V_{transfused}}$$

where TBV = total blood volume, $RCV_{initial}$ = initial red cell volume = $TBV \times Hct_{initial}$, and $V_{transfused}$ = volume transfusion.

For a normal adult, TBV can be estimated as 70 mL/kg. For term neonate, it is approximately 80 mL/kg. $Hct_{transfused}$ depends on the storage solution of the RBCs—for CPDA-1, it is ~70%; for additive solution (AS), it is ~60%.

Answer: *C*—Since this is a newborn, the TBV can be estimated as (80 mL/kg) × 2 kg = 160 mL. Thus, RCV is approximately 160 mL × 0.35 = 56 mL. Using the above formula, we have:

$$Hct_{final} = \frac{RCV_{initial} + (V_{transfused})(Hct_{transfused})}{TBV + V_{transfused}} = \frac{56 + 20*0.7}{160 + 20} = 0.39\,(or\,39\%)$$

17. If a technologist were to randomly screen the type O RBC inventory for units that are negative for the K, Fy^a, and Jk^b antigens, how many units would need to be screened to find two antigen negative units, if the antigen-negative frequencies are 90%, 35% and 30%, respectively?

A. 5
B. 16
C. 21
D. 30
E. 42

Concept: It is standard of practice to transfuse patients who have made alloantibodies to RBC antigens with RBC units that are negative for those antigens. Screening the un-phenotyped RBC inventory to find compatible RBCs for patients that have multiple alloantibodies can be extremely challenging.

$$Number\ of\ Units\ to\ screen = \frac{number\ of\ units\ to\ desired}{antigen - negative\ frequency}$$

When a patient has multiple antibodies, the antigen negative frequency is the product of the antigen-negative frequencies of each antigen (i.e., $\Pr(antigen - negative\,unit) = \Pi_1^n \Pr(antigen_i negative)$).

Answer: *C*—Using the above information:

$$Number\ of\ Units\ to\ screen = \frac{2}{0.9 * 0.35 * 0.30} = 21\ units$$

Please answer Questions 18–20 based on the following scenario:

A 23-year-old female at 30 weeks of gestation (second pregnancy) is admitted to the ED after a car accident. Per her obstetrics record, the patient is O Rh negative with anti-E (titer of 16) identified at 28 weeks of gestation. She received a vial of Rh immunoglobulins (RhIG) during her pre-natal visit 2 weeks ago. She is currently stable in the ED without any evidence of hemorrhage. Her weight is 75 kg and her labs are Hct 36% and platelet count 180 × 10^9/L.

18. Her current type and screen shows that she is O Rh negative and has anti-D and anti-E (titer of 16). A fetal screen test (i.e., rosette test) is performed and the result is negative. A Kleihauer Betke (KB) test is also performed and demonstrates 0% fetal hemoglobin is present. How many vial(s) of RhIG should the ED physician administer to this patient?

A. 0
B. 1
C. 2
D. 3
E. 4

Concept: The half-life if RhIG is 24 days. According to the American College of Obstetricians and Gynecologists (ACOG), if exogenous anti-D is present (indicating that RhIG is still in the system), no additional RhIG administration is necessary. Similarly, if delivery occurs within 3 weeks of the standard antenatal RhIG administration, the postnatal dose may be withheld in the absence of excessive fetomaternal bleeding.

Answer: *A*—This patient received RhIG 2 weeks ago, and RhIG is still persistent in her body evidenced by the presence of anti-D in her system. A negative fetal screen test and KB test indicates that there is no excess fetomaternal bleeding. Therefore, RhIG may be withheld in this case. If the history of RhIG administration is unclear, then it is almost impossible to distinguish the anti-D from RhIg versus real anti-D alloimmunization using the current routine blood bank tests. The other choices (Answers B, C, D, and E) are incorrect based on the information above.

19. Six weeks after the accident, the patient presented to L&D for delivery. Her current type and screen shows again that she is O Rh negative and has anti-D and anti-E (titer of 32). Postdelivery, a fetal screen test (i.e., rosette test) is performed and the result is positive. A Kleihauer Betke (KB) test is also performed and the result shows 25 fetal cells per 2,000 cells counted. How many vial(s) of RhIG should the physician administer to this patient?

A. 0
B. 1
C. 2
D. 3
E. 4

Concept: Postdelivery, if the fetal screen test (rosette) is negative, then only 1 vial of RhIG is necessary. However, a KB test must be performed if the rosette test is positive to quantify the amount of fetomaternal bleeding. From the KB results, number of RhIG vials can be calculated as following:

Step 1: Calculate the fetal hemorrhage from maternal TBV and KB results (if no weight is given, then use 5,000 mL)

$$Fetal\ hemorrhage\ (\text{in mL}) = \frac{(Maternal\ TBV) * (Fetal\ cells\ counted)}{Total\ cells\ counted}$$

Step 2: One 300-μg RhIG vial will cover 30 mL fetal whole blood. Therefore:

$$Number\ of\ RhIG\ vials\ required = \frac{Fetal\ hemorrhage\ (\text{in mL})}{30}$$

Due to the inherent imprecision of this formula, recommendations for dosage adjustment are as follows:

- If the calculated dose to the right of the decimal point is ≥ 0.5, the number of vials should be rounded up to the nearest whole number plus one vial
- If <0.5, then round down plus one vial
- Example: If the calculated dose results as 1.5, the number of vials administered will be 3. If the result is 1.4, then 2 vials will be given

Answer: *D*—Maternal TBV is (70 mL/kg) × 75 kg = 5,250

$$Fetal\ hemorrhage\ (\text{in mL}) = \frac{(Maternal\ TBV) * (Fetal\ cells\ counted)}{Total\ cells\ counted} = \frac{(5{,}250\ \text{mL}) * 25}{2{,}000} = 65.625$$

$$Number\ of\ RhIG\ vials\ required = \frac{Fetal\ hemorrhage\ (\text{in mL})}{30} = \frac{65.625}{30} = 2.19$$

Thus, this patient requires three RhIG vials and they should be given within 72 h from the delivery.

If there is no information on maternal weight, then a quick estimation of the number of RhIG vials required is (5/3) × (percentage of fetal cells in total number of cells counted in the KB test). This number will then be adjusted based on the above rule. The other choices (Answers A, B, C, and E) are incorrect based on this formula.

20. The patient's newborn boy (3 kg) is diagnosed with hemolytic disease of the fetus and newborn (HDFN). An exchange transfusion is requested. The newborn's current Hct is 35%. The pediatrician would like to do a 2-TBV exchange. You wish to perform this exchange with units of CPDA-1 whole blood. What are the volumes of RBC and plasma you should use to reconstitute the blood to achieve a final Hct of 50%?

	Volume of RBCs	Volume of plasma
A	400 mL irradiated, leukoreduced, O Rh negative RBCs (<7 days old) that are HgbS negative and CMV negative	100 mL AB plasma
B	400 mL irradiated, leukoreduced, O Rh negative RBCs (<7 days old) that are HgbS negative, CMV negative, and negative for E antigen	100 mL AB plasma
C	343 mL irradiated, leukoreduced, O Rh negative RBCs (<7 days old) that are HgbS negative, CMV negative, and negative for E antigen	137 mL O plasma
D	343 mL irradiated, leukoreduced, O Rh negative RBCs (<7 days old) that are HgbS negative and CMV negative	137 mL AB plasma
E	343 mL irradiated, leukoreduced, O Rh negative RBCs (<7 days old) that are HgbS negative, CMV negative, and negative for E antigen	137 mL AB plasma

Concept: Manual RBC exchange transfusion (2-TBV exchange) is typically performed in newborns with HDFN to remove bilirubin, maternal antigens and antibodies in the fetal circulation. Typically, O Rh negative RBCs that are also negative for the maternal antibodies are used to reconstitute along with AB plasma to obtain the final Hct requested. The volumes of the components can be calculated as following:

$$Volume\ of\ RBCs = \frac{(Volume\ whole\ blood\ requested) * (Hct\ whole\ blood\ requested)}{Hct\ of\ RBC\ unit\ used\ for\ reconstitution}$$

$$Volume\ of\ Plasma = Volume\ whole\ blood\ requested - Volume\ of\ RBCs$$

Answer: *E*—Using the above information, the newborn's TBV is (80 mL/kg) × (3 kg) = 240 mL. Therefore, the reconstituted whole blood for the exchange should be 2 × 240 = 480 mL. Hence,

$$Volume\ of\ RBCs = \frac{(Volume\ whole\ blood\ requested) * (Hct\ whole\ blood\ requested)}{Hct\ of\ RBC\ unit\ used\ for\ reconstitution} = \frac{(480\,\text{mL}) * 0.5}{0.7} = 342.9\,\text{mL}$$

$$Volume\ of\ Plasma = Volume\ whole\ blood\ requested - Volume\ of\ RBCs = 480 - 342.8 = 137.1\ \text{mL}$$

Furthermore, the mother has anti-D and E. Therefore, the RBCs use to reconstitute whole blood for the exchange should be irradiated, leukoreduced, O Rh negative RBCs (<7 days old) that are also HgbS negative, CMV negative, Rh negative, and negative for E antigen. The other choices (Answers A, B, C, and D) are incorrect based on these formulas.

End of Case

21. A 27 year-old female (blood type O Rh negative with negative antibody screen) is hospitalized with meningococcemia. She is in active DIC due to sepsis and is having epitaxis. Her platelet count is 19×10^9/L, and thus, the ICU team requests 1 unit of SDP. There is a major snow storm in the area, and the Blood Bank only has Rh positive SDPs. If RhIG is given after transfusing Rh positive SDPs, minimally, how many SDP units will 1 vial of RhIG cover?

A. 1
B. 3
C. 5
D. 7
E. 9

Concept: Besides giving to Rh negative woman during pregnancy and when having fetomaternal hemorrhage, RhIG may also be given to patients receiving Rh positive platelets to prevent Rh(D) alloimmunization. Apheresis techniques have improved significantly, and thus, the RBC contamination in a SDP unit is very small (~0.0004 mL on average, whereas whole blood derived platelets concentrates can contain ~0.5 mL RBCs. Nonetheless, for the purpose of calculating the number of SDP units that are covered by RhIG, a conservative method should be used. By the AABB standards, SDP units do not have to be crossmatched, unless they contain more than 2 mL RBCs.

Answer: *D*—Each vial of RhIG can cover 15 mL RBCs or 30 mL whole blood. Each SDP unit contains <2 mL RBCs; thus, minimally, each vial of RhIG can cover 15/2 or 7 units SDP. In reality, each vial of RhIG may cover many more SDP units than the 7 units calculated here.

Please answer Questions 22 and 23 based on the following scenario:

An 80-lb woman who is known to have a rare red blood cell antigen phenotype presents for blood donation at her local blood center. The blood center only has standard 450 mL phlebotomy bags.

22. Based upon her weight, what is the maximum amount of whole blood she can donate today?
A. 378 mL
B. 400 mL
C. 431 mL
D. 477 mL
E. 500 mL

Concept: In order to ensure safety to the donors, the FDA regulations state that the minimum weight of a donor must be 110 lbs. If the donor weighs less than 110 lbs, then the maximum allowable whole blood volume (including the amount of blood used for testing and discarded via the diversion pouch) is 10.5 mL/kg.

Answer: *A*—This donor only weighs 80 lbs (~36 kg); thus, the maximum amount of whole blood (including samples for testing) she can donate today is 36 × (10.5 mL/kg) = 378 mL. If a donor weighs more than 110 lbs, she can donate a maximum of 500 ± 50 mL whole blood, and most donors are able donate this volume at each donation. A unit is labeled as a "low-volume collection" when the amount collected into a 450 mL collection bag is between 300 and 404 mL, or when the amount collected into a 500 mL collection bag is between 333 and 449 mL, and the amount of anticoagulant in the collection bag is not adjusted. Low-volume allogeneic collections must be labeled as "RBCs low volume." Platelet, plasma, and cryoprecipitated AHF should not be prepared from such units.

23. The blood center staff is able to remove the appropriate amount of anticoagulant from the phlebotomy bag for this donor. Based on the maximum amount of blood that this donor is able to donate, how much anticoagulant does the staff need to remove from the primary bag, if the blood center uses the standard 450mL phlebotomy bag?
A. 63 mL
B. 51.5 mL
C. 31.5 mL
D. 17.2 mL
E. 0 mL

Concept: As discussed in Question 22, this donor weighs less than 110 lbs, and thus, less blood should be removed. The following describes the steps involved in calculating the amount of anticoagulant required for low-volume whole blood units using the standard 450 mL phlebotomy bags.

Step 1: Calculate the amount of blood that can be removed:

$$\textit{Volume to be removed}\,(\text{in mL}) = (450\ \text{mL})\left(\frac{\textit{Donor weight in lbs}}{110\ \text{lbs}}\right)$$

Step 2: Calculate the correct amount of anticoagulant:
For both citrate-phosphate-dextrose (CPD) and CPDA-1, the ratio of anticoagulant to whole blood should be 14 mL per 100 mL whole blood.

$$\textit{Volume of anticoagulant required} = (14\text{ mL})\left(\frac{\textit{Volume to be removed}\text{ (in mL)}}{100\text{ mL}}\right)$$

Step 3: Calculate the amount of anticoagulant needed to be removed from the standard phlebotomy bags. For the 450 mL-bag, it contains 63 mL anticoagulant.

$$\textit{Volume of anticoagulant to be removed} = 63\text{ mL} - \textit{Volume of anticoagulant required}$$

Answer: *D*—Since the facility only has standard 450 mL phlebotomy bags, the maximum amount of whole blood should be removed is $(450\text{ mL})\left(\frac{80\text{ lbs}}{110\text{ lbs}}\right) = 327.3\text{ mL}$. The volume of anticoagulant required is $(14\text{ mL})\left(\frac{327.3\text{ mL}}{100\text{ mL}}\right) = 45.8\text{ mL}$. Thus, the staff needs to remove $63\text{ mL} - 45.8\text{ mL} = 17.2\text{ mL}$ of anticoagulant from the standard 450 mL phlebotomy bag. The other choices (Answers A, B, C, and E) are incorrect based on these formulas.

End of Case

Please answer Questions 24–27 based on the following scenario:

A 16-year-old male with sickle cell disease (SCD) is admitted to the ED with shortness of breath, concerning for acute chest syndrome (ACS). He has multiple transfusions in the past (but none within the past 3 months), and his antibody panel currently shows that he is O Rh positive, and has anti-C, E, K, Js^a, Fy^a, Jk^b, and s. His weight is 60 kg. His current Hct is 22%.

24. If a technologist searches randomly the Rh negative inventory for RBC units for this patient, what is the odds of him finding 1 units that are negative for all the antibodies this patient has given that the probability of finding units that are negative for D, C, E, K, Fy^a, Jk^b, and s in the region are 15%, 30%, 70%, 90%, 35%, 30%, and 10%, respectively?

A. 0.00198
B. 0.00893
C. 0.00954
D. 0.0945
E. 0.104

Concept: The probability of finding antigen negative units is calculated by multiplying the percentage of the population that lacks each individual antigen times one another. This percentage is obtained by 1—the percent prevalence in the population.

Answer: *C*—The technologist searches randomly the Rh negative inventory for RBC units. It is commonly known that the majority of Rh negative units are rr (i.e., they lack the C and E antigen). Assuming that 100% of the Rh units lack the C and E antigen:

$$\begin{aligned}&\Pr(\textit{One RBC unit from Rh negative inventory that are negative for } \mathrm{C, E, K, Js^a, Fy^a}, Jk^b, \textit{and } s)\\&= \Pi \Pr(\textit{RBC negative for each antigen}) = (1 * 0.9 * 1 * 0.35 * 0.3 * 0.1)\\&= 0.00945\end{aligned}$$

(Js^a is a low frequency antigen; thus, virtually all RBC units (100%) will be negative for Js^a).

Hence, the odds is:

$$Odds = \frac{\Pr(event)}{1-\Pr(event)} = \frac{0.00945}{1-0.00945} = 0.00954$$

25. There is a major snow storm in the area; thus, it is not possible to procure more RBC units for this patient. Currently, the Blood Bank only has 2 AS-RBC units (250 mL each) that are compatible for this patient. If he receives simple transfusion (2 units), then what will his Hct and hemoglobin S (HgbS) likely be post transfusion?

	Posttransfusion Hct	Posttransfusion HgbS%
A	26%	75%
B	26%	Cannot determine from given information
C	Cannot determine from given information	70%
D	28%	75%
E	29%	Cannot determine from given information

Concept: From Question 1, $Hct_{final} = \frac{RCV_{initial} + (V_{transfused})(Hct_{transfused})}{TBV + V_{transfused}}$ is the formula to calculate Hct posttransfusion. Similar principle is applied to calculate the HgbS% level posttransfusion, and the formula is:

$$HgbS_{final} = \frac{(RCV * HgbS_{initial})}{RCV + (Hct_{transfused} * Volume\ transfused)}$$

Answer: *A*—This patient has not received transfusion over the past 3 months. Therefore, his HgbS% before transfusion is assumed to be 100%. AS units are used; thus, $Hct_{transfused}$ is ~60%. His $RCV_{initial}$ is TBV × (*Hct*) = (70 mL/kg) × (60 kg) × (0.22) = 924 mL. Hence:

$$Hct_{final} = \frac{RCV_{initial} + (V_{transfused})(Hct_{transfused})}{TBV + V_{transfused}} = \frac{924 + (2*250)*0.6}{4{,}200 + 2*250} = 0.26 \text{ (or 26\%)}$$

$$HgbS_{final} = \frac{(RCV * HgbS_{initial})}{RCV + (Hct_{transfused} * Volume\ transfused)} = \frac{924}{924 + 0.6*(2*250)} = 0.75 \text{ (or 75\%)}$$

26. This patient improved significantly after simple transfusion. During the hospitalization, he received a few more units. Two weeks later, he admitted to the ED with signs and symptoms of acute chest. His current labs show an Hct of 20% and HgbS level of 50%. The Blood Bank currently has 15 RBC units that are compatible for this patient; thus, both simple transfusion and RBC exchange are possibilities. Assuming that the patient can tolerate up to 1 L of simple transfusion of RBCs, if the goal is to reduce the HgbS level to ≤30%, then what should on-call apheresis physician recommend?

A. Simple transfusion may be able to achieve HgbS goal without a dramatic increase in iron storage
B. Simple transfusion may be able to achieve HgbS goal without a dramatic increase in blood viscosity
C. Simple transfusion may not be able to achieve HgbS goal without a dramatic increase in iron storage
D. Simple transfusion may not be able to achieve HgbS goal without a dramatic increase in blood viscosity
E. Cannot determine from the given information

Concept: In acute setting, such as acute chest, the NIH recommendations for post-transfusion or post-RBC exchange are HgbS ≤30% and Hct ≤30%. A Hct >30% may put the patient at risk for hyperviscosity. In many clinical settings, RBC exchange will be performed. However, if RBC exchange cannot be performed immediately, simple transfusion is the next best option, as long as the final Hct is not over 30%.

Answer: *B*—For this patient, the $RCV_{initial}$ is (70 mL/kg) × (60 kg) × 0.2 = 840 mL. Using the formula from Question 26, we know that $HgbS_{final} = \frac{(RCV * HgbS_{initial})}{RCV + (Hct_{transfused} * Volume\ transfused)}$

This equation can be rearranged to solve for the volume transfused if $HgbS_{final}$ is set at 30%.

$$Volume\ required = \frac{RCV * (HgbS_{initial} - HgbS_{final})}{(Hct_{tranfused}) * (HgbS_{final})} = \frac{840 * (0.5 - 0.3)}{0.6 * 0.3} = 933\,\text{mL}$$

We also need to check to ensure that with this amount of transfusion, the Hct_{final} will not exceed 30%.

$$Hct_{final} = \frac{RCV_{initial} + (V_{transfused})(Hct_{transfused})}{TBV + V_{transfused}} = \frac{840 + 933 * 0.6}{4,200 + 933} = 0.27 < 0.3$$

Therefore, simple transfusion (~4 units if each RBC unit is ~250 mL) is sufficient in this case to achieve both HgbS% and Hct goal posttransfusion assuming that the patient can tolerate this volume of transfusion. RBC exchange may not be necessary. The other choices (Answers A, C, D, and E) are incorrect based on the formulas and information above.

27. The hematologist decides to recommend chronic RBC exchange for this patient. He is now in the outpatient apheresis clinic. His current labs show an Hct of 24% and HgbS level of 62%. Based on your institutional protocol, the patient's HgbS post-procedure should be less than 20% and Hct should be ~27%–30%. Please calculate the appropriate fraction of cells remaining (FCR) for this procedure.

A. 32%
B. 39%
C. 48%
D. 67%
E. 310%

Concept: FCR denotes the percentage of the patient's "defect" RBCs at the end of the procedure comparing to the beginning of the procedure. Together with the initial and final Hct, it is part of the equation to calculate how many RBC units required for the procedure. FCR is calculated as:

$$FCR = \frac{Target\ HgbS\%\ at\ the\ end\ of\ the\ procedure}{HgbS\%\ at\ the\ beginning\ of\ the\ procedure}$$

Answer: *A*—Based on the above formula, FCR = 20/62 or 32%. The other choices (Answers B, C, D, and E) are incorrect based on this formula.

End of Case

28. A 25-year-old woman presents to the ED in labor. Of note, she was diagnosed with factor XI deficiency and history of excessive bleeding with dental extraction. Her last factor XI measurement is 20% a month ago. You receive a call from the obstetrician with the question of how to correct her coagulopathy prior to cesarian section to a factor XI level of ~50%. She weighs 75 kg, and her current Hct is 40%. What is the best recommendation in this scenario?

	Initial dose	Maintenance dose
A	945 IU factor XI concentrate	473 IU factor XI concentrate every other day
B	945 mL plasma	473 mL plasma every other day
C	945 mL plasma	473 mL plasma every day
D	1,890 mL plasma	945 mL plasma every other day
E	1,890 mL plasma	945 mL plasma every day

Concept: Factor XI deficiency is an autosomal recessive disorder. Unlike factor VIII or IX deficiency, Factor XI deficiency patient may or may not bleed excessively. With a baseline factor of ~20%, this patient has mild Factor XI deficiency. She has history of bleeding with surgical procedures; thus, her coagulopathy should be corrected. Currently, in the United States, there is no Factor XI concentrates available. Therefore, plasma is used to treat coagulopathy and/or bleeding in these patients. The half-life of Factor XI in plasma is ~2 days; thus, if necessary, maintenance doses may be given every other day.

Unlike dosing for Factor VIII and IX (which has factor concentrates), plasma for factor XI deficiency treatment has significant volume and should be factored in the calculation. Assuming plasma contains 100% factor XI, the initial plasma dose may be calculated as follows:

$$Factor\ XI_{final} = \frac{(Factor\ XI_{initial})(Initial\ PV) + (Volume\ transfused)}{(Initial\ PV) + (Volume\ transfused)}$$

$$Volume\ transfused = \frac{(Factor\ XI_{final} - Factor\ XI_{initial}) * (Initial\ PV)}{1 - Factor\ XI_{final}}$$

Answer: *D*—Using the above information, her TBV is (70 mL/kg) × 75 kg = 5,250 mL. Thus, her PV is (1−0.4) × 5,250 = 3,150 mL

$$Plasma\ dose\ (\text{mL}) = \frac{(3{,}150\ \text{mL})(0.5-0.2)}{1-0.5} = 1{,}890\ \text{mL}$$

Thus if an average unit of plasma has a volume of 250 mL, approximately 8 units of plasma would be required for the initial dose. This is a large volume transfusion, and thus, the patient should be monitored closely for signs and symptoms of volume overload. Obstetrician should be consulted regarding the risks versus benefits of using diuresis, such as furosemide, during the transfusion in this scenario.

Maintenance dose, should be ~945 mL plasma (~4 units) every other day. The other choices (Answers A, B, C, and E) are incorrect based on these formulas.

The following Questions (29 to 31) focus on advanced topics that may not appear on routine basis. Nonetheless, these will help to integrate your knowledge from previous chapters

29. An antibody identification panel is being performed for a 44-year-old woman who is scheduled to undergo a cholecystectomy. The technologist suspects that there is an antibody against the E antigen. Antibody panel results show that five cells negative for the E antigen reacted negatively, no cells that were positive for the E antigen reacted negatively, no cells negative for the E antigen reacted positively, and three cells positive for the E antigen reacted positively. The total number of red cells tested is eight. While the results appear to be straightforward the technologist asks you "how confident are we that an anti-E antibody is present in the patient serum and that we are not simply obtaining these results by chance?" Please calculate the *P*-value in order to answer this question.

A. 0.02
B. 0.05
C. 0.3
D. 0.5
E. 1

Concept: In many instances, nonspecific reactivity is noted on the antibody panel that does not match with known alloantibodies. The Fisher Exact Test is used to evaluate the probability that an antibody is present in a patient serum or plasma versus obtaining the result by chance. This is important because clinically significant red cell alloantibodies once they are detected will be taken into account for the rest of the patient's life leading to the need for providing antigen-negative RBCs. The target p-value is ≤ 0.05. It is from this target *P*-value that the general requirement of having at least three test cells known to be antigen-negative, give negative reactions, and at least three antigen-positive cells give positive reactions, in order to make the antibody identification. The equation for the Fisher Exact Test is:

$$\Pr(\textit{observed event is due to random chance}) = \frac{(A+B)!(C+D)!(A+C)!(B+D)!}{N!\ A!\ B!\ C!\ D!}$$

Where:
! = factorial
A = positive reactions when antigen is present
B = positive reactions when antigen is absent
C = negative reactions when antigen is present
D = negative reactions when antigen is absent
N = number of cells tested

Answer: *A*—Using the above information:

$$\Pr(\textit{observed event is due to random chance}) = \frac{(3+0)!\ (0+5)!\ (3+0)!\ (0+5)!}{8!\ 3!\ 0!\ 0!\ 5!}$$

$$\Pr(\textit{observed event is due to random chance}) = \frac{518,400}{29,030,400} = 0.02$$

Thus, since the *P*-value is below 0.05 (i.e., if it is due to random chance, it only happens 2% of the time), it is unlikely that these results are due to chance.

Please answer Questions 30 and 31 based on the information provided below:

The current FDA standard for approval of stored red blood cell products requires that at the end of the allowable storage period of 42 days for RBCs stored in additive solutions, the mean 24-h posttransfusion recovery as evaluated by in vivo, autologous chromium recovery studies must be $\geq 75\%$ and in vitro hemolysis must be $< 1\%$ (free hemoglobin in the supernatant in relation to total hemoglobin in the stored RBC product).

30. In the case of a RBC product that meets the FDA criteria by having a 75% 24-h posttransfusion recovery (75% of the transfused RBCs are still circulating 24 h after infusion), 25% of the infused RBCs will have been removed from the circulation in that 24-h period and destroyed. Since clearance/destruction of RBCs leads to the release of iron, how much iron will be released in 24 h if

25% of transfused RBCs are destroyed, when a patient is transfused with a 265 mL RBC unit that has a hematocrit of 60%?

A. 0.5 mg of iron
B. 5 mg of iron
C. 10 mg of iron
D. 20 mg of iron
E. 40 mg of iron

Concept: The removal of damaged and senescent RBCs is carried out by macrophages in the spleen. A consequence of ingestion of RBCs by macrophages is the degradation of hemoglobin eventually leading to the release of iron. Circulating RBCs have a normal lifespan of approximately 120 days and about 1/120th of the circulating RBCs are cleared each day (24-h period). Therefore, in an individual who has a total blood volume of 5 L (5,000 mL) and a hematocrit of 50%:

Volume of RBCs cleared in 24 h = (1/120) × (5,000 mL) × 50% = 21 mL
~ 1 mL RBC = ~ 1 × 10^{10} RBC = ~ 1 mg Fe
Thus approximately 21 mg Fe is normally released each day due to normal RBC turnover

Answer: *E*—Using the above information:

Volume of RBCs destroyed in 24 h if 25% of RBCs from a 265 mL pRBC unit with a 60% hematocrit are cleared: (265 mL) × 25% × 60% = 40 mL
~ 1 mL RBC = ~ 1 × 10^{10} RBC = ~ 1 mg Fe
Approximately 40 mg of iron is released.

Thus the amount of iron that is released from damaged RBCs in a single unit of RBCs with a 75% 24-h posttransfusion recovery (meeting the FDA standard) is approximately double the amount of iron that is released each day from normal RBC turnover. Other choices (Answers A, B, C, and D) are incorrect.

31. You are working as a medical consultant for a company that is manufacturing a new type of RBC storage solution that will still allow for storage for up to 42 days; however, preliminary data indicates that this storage solution allows for superior maintenance of RBC ATP levels during storage, when compared to the current additive solutions. Aware of the FDA criteria that at the end of the allowable storage period there must be less than 1% hemolysis, the company asks for your guidance in helping to calculate the hemolysis in one of its units. Several parameters were measured from an aliquot taken from the unit just prior to expiration: supernatant hemoglobin 200 mg/dL, hematocrit 56%, and total hemoglobin (RBCs and supernatant) 18.2 g/dL. What is the hemolysis in this unit of RBCs?

A. 0.2%
B. 0.5%
C. 0.8%
D. 1%
E. 2%

Concept: It is known that hemolysis does occur during the preparation and refrigerated storage of RBCs. In addition, the concentration of the free hemoglobin in the supernatant depends on the number of damaged RBCs as well as the volume of the fluid. Therefore, the degree of hemolysis is evaluated as the percent of free hemoglobin in relation to the total hemoglobin, taking the hematocrit of the product into account. To calculate the percent hemolysis, one needs to know the supernatant hemoglobin in milligram/deciliter, the hematocrit of the RBC unit, and the total hemoglobin of the unit in milligram/deciliter.

$$\textit{Percent hemolysis}\ (\%) = \frac{(100 - \textit{hematocrit}) * \textit{supernatant hemoglobin}\ (\mathrm{mg/dL})}{\textit{Total hemoglobin}\ (\mathrm{mg/dL})}$$

Answer: *B*—Using the above information:

$$\textit{Percent hemolysis}\ (\%) = \frac{(100-56) * 200\ \text{mg/dL}}{18,200\ \text{mg/dL}} = 0.5\%$$

Thus, this RBC unit would meet the FDA criteria of < 1% hemolysis. The other choices (Answers A, C, D, and E) are incorrect based on these formulas.

End of Case

Suggested Reading

[1] W. Ageno, Oral anticoagulant therapy: Antithrombotic therapy and prevention of thrombosis, 9th ed: American College of Chest Physicians evidence-based clinical practice guidelines, Chest 141 (2012) e44S–e88S.

[2] AABB, Standards for Blood Banks and Transfusion Services, twenty ninth ed., American Association of Blood Banks (AABB); 2014.

[3] Novo Nordisk Inc., NovoSeven RT Coagulation Factor VIIa (Recombinant) Prescribing information. Available from: http://www.novo-pi.com/novosevenrt.pdf.

[4] CSL Behring LLC. Kcentra Prothrombin Complex Concentrate Prescribing information. Available from: http://labeling.csl-behring.com/PI/US/Kcentra/EN/Kcentra-Prescribing-Information.pdf.

[5] Baxter Healthcare Corporation. FEIBA NF (Anti-inhibitor coagulant complex), Nanofiltered and vapor heated Prescribing information. Available from: http://www.feiba.com/us/forms/feiba_nf_pi.pdf.

[6] H.P. Pham, J. Schwartz, How do we approach a patient with symptoms of leukostasis requiring emergent leukocytapheresis, Transfusion 55 (10) (2015) 2306–2311.

[7] V.C. Hughes, Calculations in Transfusion Medicine: Clinical and Theoretical Applications, Duncan Rivers Press, St. George, UT, (2012).

[8] M.K. Fung (Ed.). Technical Manual, eighteenth ed., AABB, Bethesda, MD, 2014.

[9] H.P. Pham, M.C. Müller, L.A. Williams, N.P. Juffermans, Mathematical model and calculation to predict the effect of prophylactic plasma transfusion on change in INR in critically ill patients with coagulopathy, Transfusion 56 (4) (2016) 926–932.

[10] A.W. Bryan Jr., E.M. Staley, T. Kennell Jr., A.Z. Feldman, L.A. Williams, H.P. Phan, Plasma transfusion demystified: a review of the key factors influencing the response to plasma transfusion, Lab Med 48 (2) (2017) 108–112.

CHAPTER

21

Data Interpretation in Laboratory Medicine

Christopher A. Tormey, Eric A. Gehrie**, Huy P. Pham*[†], *R. Pat Bucy*[†], *Robin G. Lorenz*[†], *X. Long Zheng*[†], *Jeanne E. Hendrickson**

*Yale University, New Haven, CT, United States; **The Johns Hopkins Hospital, Baltimore, MD, United States; [†]University of Alabama at Birmingham, Birmingham, AL, United States

The interpretation and application of clinical and laboratory data is an essential component of the job of a laboratory medicine practitioner or Transfusion Medicine physicians. Such data may be generated from assays within one's own laboratory, or may arise in the form of published studies, with results incorporated into routine practice. The aim of this chapter is to provide a review of common scenarios faced by laboratorians and physicians where data interpretation is at the forefront of the clinical challenge.

Attention: Some diseases and scoring systems are used as examples in this chapter; however, intimate knowledge of these diseases and systems are not required to answer the questions. They are used to provide a reference for the statistical or data interpretation concept. Further, the data used for the questions in this chapter are only for illustration purpose—they may not be actual data, and thus, the conclusions (i.e., right answers in the questions) are only for the purpose of concept/calculation demonstration, and should not be used to make clinical decisions.

1. The blood bank has received materials from a reference center for the performance of quarterly proficiency testing, including a specimen for antibody screen testing. This antibody screen proficiency testing is performed by the lead technologist, and is also used in assessing the competency of several new employees who are training in the blood bank. The results of the proficiency testing for the antibody screen specimen are described in Table 21.1:

TABLE 21.1 Competency Results

Performing individual	Overall result	Agglutination score
Reference method (expected answer)	Positive	4+
Lead technologist	Positive	4+
Trainee #1	Positive	2+
Trainee #2	Positive	3+
Trainee #3	Positive	1+

Transfusion Medicine, Apheresis, and Hemostasis. http://dx.doi.org/10.1016/B978-0-12-803999-1.00021-3

Regarding the data generated in the "Agglutination Score" column, which of the following best describes the analytical error demonstrated?

A. An error of precision
B. An error of reagent sensitivity
C. A preanalytical error
D. A postanalytical error
E. No analytical error

Concept: Assessment of proficiency testing data is an important function of laboratory directors to ensure that their laboratories are releasing correct, high quality results. Proficiency testing is handled in the same way that patient specimens are handled and compared with a standard result. Because of the high reliability of such standardized tests, specimens used for proficiency testing can be valuable for training new employees and assessing competency for more experienced staff members. When discrepancies occur in testing (e.g., in data generated by two employees or two instruments), it is important to classify the type of discrepancy such that appropriate measures can be taken to correct the error(s) occurring.

Answer: *A*—Precision refers to thc rcproducibility of the test; thus, the best description of the type of error occurring in this case is an error of precision occur most often when an assay is performed repeatedly on the same specimen and different results (as compared to a standard) are obtained. Accuracy, on the other hand, refers to how close the measurement to the "true" value obtained from gold standard method. In this case, the results each employee obtained is also different from the standard; thus, the result is not accurate either. Regarding other options, there is no evidence supporting the lack of reagent sensitivity (Answer B) in this case. Preanalytical errors (Answer C) are those occurring before testing of a specimen, while postanalytical errors (Answer D) are those occurring after testing has been performed; neither types of error are applicable in this case. Answer E is wrong because there is an error in this case.

2. Upon review of the serologic records of patients with sickle cell disease that had been multiply transfused, the following data (Table 21.2) was reported in a recent publication. What conclusion can be made about factors that increase the risk of having at least one evanesced alloantibody?

A. Evanesce does not occur in patients with SCD
B. Evanesce does occur in the majority of patients with SCD, but it is not linked to any factors evaluated in the table
C. Evanesce does occur in the majority of patients with SCD, and the likelihood of not detecting an antibody is increased if patients are transfused at >1 hospital
D. Evanesce does occur in the majority of patients with SCD, and the likelihood of not detecting an antibody is increased if patients are transfused at only 1 hospital
E. Patients transfused at >1 hospital have a greater number of alloantibodies detected

TABLE 21.2 Study Results

	SCD patients
# patients with alloantibodies/# total patients transfused	66/150 (44%)
# patients with at least one evanesced alloantibody/total # patients with alloantibodies	42/66 (64%)
# patients with at least one evanesced alloantibody transfused at >1 hospital/total # patients with at least one evanesced alloantibodies	38/42 (86%)

Data with permission from S.K. Harm, et al., Am. J. Clin. Pathol., 141 (2014) 256–261.

Concept: RBC alloimmunization antibody identification is usually only performed right before a new transfusion event. However, it has been reported that nearly half of alloantibodies "evanesce" or decrease in titer to below the level of detection during a patient's lifetime. In an optimized transfusion plan, patients would not only be tested immediately prior to RBC transfusion, but would also have posttransfusion testing at regular periods to detect newly formed alloantibodies that might later evanesce.

Answer: *C*—A patient who has an unrecognized alloantibody (due to evanesce) is at a very high risk of a hemolytic transfusion reaction (HTR). If an SCD patient is always treated at the same hospital, then that blood bank will have records indicating all previously-detected alloantibodies and this risk of HTR due to evanesce of alloantibodies can be minimized. However, if an SCD patient (or any chronically transfused patient) is treated and transfused at multiple hospitals, the lack of a national database means that any alloantibody that evanesces below the level of detection will not be recognized and the patient will be at a much higher risk of complications due to RBC transfusion. The other choices (Answers A, B, D, and E) are incorrect based on the data presented in the table.

3. You are assigned the task of developing a guideline for apheresis procedures and indications at your facility according to the quality of available evidence. Among various disease states, clinicians are requesting that therapeutic plasma exchange (TPE) be performed for individuals with complex regional pain syndrome (CRPS) at your facility. A detailed review of the literature reveals that no randomized controlled trials or controlled trials with TPE have been performed for this indication. However, there are two case series (of about 40 patients in total), as well as several case reports detailing some success in treating this entity with TPE.

Based on the available data, what grade recommendation would you assign to this indication for TPE according to the scale developed by Guyatt and colleagues that were used as part of the Grade of Recommendation by ASFA (American Society for Apheresis) in its Guidelines for Therapeutic Apheresis?

A. Grade 1A
B. Grade 1B
C. Grade 2A
D. Grade 2C
E. Grade 3A

Concept: Performance of therapeutic apheresis is based not only on a patient's clinical scenario and the hypothetical benefit, but also can be stratified according to available evidence (i.e., applying the McLeod's criteria to evaluate an apheresis request). In order to help the requesting and/or apheresis physicians evaluate the role of apheresis as a treatment modality, ASFA has regularly published evidence-based guidelines. Of note, the ASFA Guidelines only contain the indications that currently have enough evidence to support (or not support) the use of therapeutic apheresis (TA). The ASFA Guidelines do not contain all the diseases that have been treated by TA and reported in the medical literature or the diseases that may be treated by TA. The role of TA in the treatment of diseases and indications in the newest guidelines are categorized based on the ASFA categories. Additionally, a recommendation grade based on the Grading of Recommendation Assessment, Development, and Evaluation (GRADE) system was also provided along with the ASFA category for each TA indication.

Answer: *D*—ASFA generates recommendations for therapeutic plasma exchange indications based in part on the methodological quality of supporting evidence. A Grade 2C recommendation is typically based on observational studies or case series, whereas the highest Grade (1A) recommendation would be based on randomized controlled trials (RCTs) without limitations or overwhelming evidence from observational studies. Based on the case reports and case series published to date, the level of evidence for TPE for CRPS is Grade 2C. The other choices (Answers A, B, C, and E) are incorrect based on the grading system (Table 21.3).

TABLE 21.3 Grading Recommendations

Recommendation grade	Description	Methodological quality of supporting evidence	Implications
1A	Strong recommendation, high-quality evidence	RCTs without important limitations or overwhelming evidence from observational studies	Strong recommendation, can apply to most patients in most circumstances without reservation
1B	Strong recommendation, moderate-quality evidence	RCTs with important limitations or exceptional strong evidence from observational studies	Strong recommendation, can apply to most patients in most circumstances without reservation
1C	Strong recommendation, low-quality evidence	Observational studies or case series	Strong recommendation, but may change when higher quality evidence becomes available
2A	Weak recommendation, high-quality evidence	RCTs without important limitations or overwhelming evidence from observational studies	Weak recommendation, best action may differ depending on circumstances
2B	Weak recommendation, moderate-quality evidence	RCTs with important limitations or exceptional strong evidence from observational studies	Weak recommendation, best action may differ depending on circumstances
2C	Weak recommendation, low-quality evidence	Observational studies or case series	Very weak recommendation; other alternatives may be equally reasonable

Modified with permission from Table 3 in Schwartz et al. J. Clin. Apher. 31 (2016), 149–338.

Please answer Questions 4 and 5 based on the following clinical scenario:

A pharmaceutical company has developed a new drug (FCZ) intended to increase the factor γ level in congenitally-deficient factor γ individuals. Fig. 21.1 shows the initial results from a phase I study, demonstrating the relationship between increasing doses in units/kilogram of FCZ (*x*-axis) and mean factor γ levels of study enrollees (*y*-axis) whose starting levels were approximately 0%.

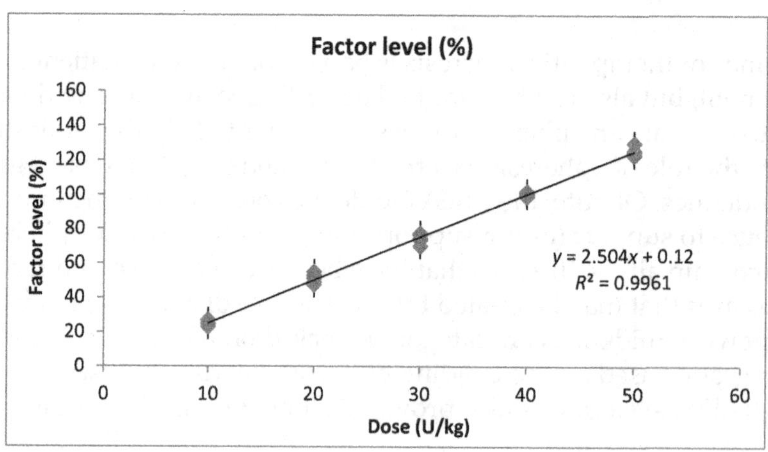

FIGURE 21.1 **Phase I study results.**

4. Which of the following best describes the relationship between FCZ and factor γ levels?
   A. Nonlinear relationship
   B. Inverse relationship
   C. Direct relationship
   D. Noncorrelating relationship
   E. No clear relationship between data points

Concept: Interpretation of data plots is an important part of the roles played by laboratorians and blood bankers. Trends in such plots can be useful for the prediction of patient outcomes, as well as applied to research undertakings.

Answer: *C*—In this case, the best interpretation of the data plot is a direct relationship, reflecting the fact that as the dose of FCZ increased, there was a similar increase in the amount of factor γ levels measured. Inverse relationship (Answer B) describes a relationship where one variable increases, the other variable decreases so that their product remains the same (i.e., the slope will be negative). Neither nonlinear relationship (Answer A) nor no relationship (Answer E) is the best description of Fig. 21.1. There is no such thing as noncorrelating relationship (Answer D).

5. Based on Fig. 21.1, if you wanted to achieve a factor γ level of 75% for your patient who has a level of 0%, what would be recommended as the initial dose for such an individual?
   A. 5 U/kg
   B. 10 U/kg
   C. 20 U/kg
   D. 30 U/kg
   E. 50 U/kg

   Concept: When the dependent variable (outcome) and independent variable (predictor) has a mathematical relationship, data can be extrapolated by examining a graphical plot itself as long as it is within range of the experiment.

   Answer: *D*—In this case, there is a direct, linear relationship between 10 U/kg and 50 U/kg of drug FCZ in relation to factor γ level. As such, by examining the plot we can establish the fact that in order to obtain a 75% factor level, our starting dose should be 30 U/kg. This dose is within the linearity range, and thus, the result is valid. The other choices (Answers A, B, C, and E) are wrong based on interpretation of Fig. 21.1.

End of Case

6. Serum samples from two different, previously transfused patients, are incubated with RBCs expressing the blood group antigen X (solid and dashed line histograms in Fig. 21.2). As a control for nonspecific binding, a serum specimen is also incubated with RBCs lacking blood group antigen X (shaded histogram in the Fig. 21.2). The secondary antibody used is a fluorescently-labeled, antihuman IgG.

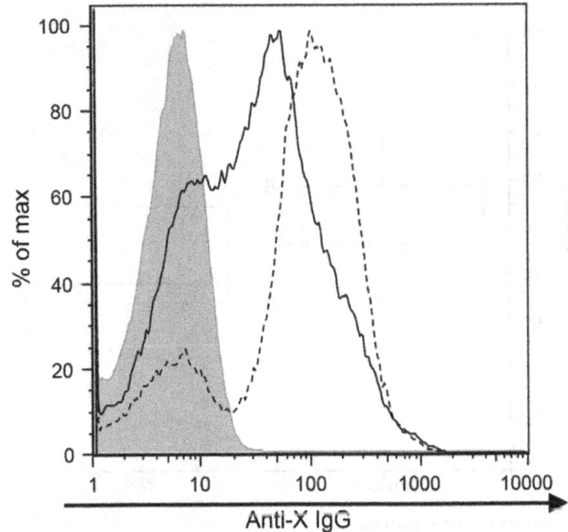

FIGURE 21.2 **Flow cytometric results for assessing antibody binding to antigen X.**

What is your interpretation of the data?

A. Neither patient has an IgG antibody against blood group antigen X
B. Both patients have an IgM antibody against blood group antigen X
C. The patient whose results are depicted by the solid histogram has a higher titer antiblood group antibody than the patient whose results are depicted by the dashed line histogram
D. The reactivity against RBCs expressing blood group antigen X cannot be judged, since there is minimal reactivity against RBCs not expressing this antigen
E. Both patients have an IgG antibody against blood group antigen X, and the results depicted by the dashed line histogram represent a higher titer/higher affinity antibody than the results depicted by the solid line histogram

Concept: Flow cytometric data can be used in the setting of hematological diseases, as well as the assessment of alloantibodies in immunohematology. It is important to be able to interpret flow cytometric data output, which can be represented in "dot plot" form or in histogram form, as in the Fig. 21.2 in this case.

Answer: *E*—In this case, while both patients are demonstrating a "right shift" relative to the negative control, the signal associated with dashed line histogram is demonstrating an antibody response reflecting either a higher titer or higher affinity for the target X antigen compared to the signal associated with the solid line. The secondary antibody is antihuman IgG; thus, it cannot detect IgM (Answer B). Both patients have antibodies against antigen X (Answer A) and that the patient whose results are depicted by the solid histogram has a lower titer antiblood group antibody than the patient whose results are depicted by the dashed line histogram (Answer C). Answer D is wrong—the negative control result is valid; thus, the reactivity against RBCs expressing blood group antigen X can be quantified.

Please answer Questions 7 and 8 based on the following clinical scenario:

7. You are reviewing a paper which presents heart rate data following plasma transfusion in two groups of patients (groups 1 and 2, as detailed in the *x*-axis of Fig. 21.3). The authors of the paper have presented this data in a "box-and-whiskers" plot. For the data presented in group 1, what do the lines corresponding to letters A and B in Fig. 21.3 represent?
A. The mean and median heart rate, respectively
B. The mean and mode heart rate, respectively
C. The median and mode heart rate, respectively
D. The highest and lowest heart rates observed, respectively
E. The 25th and 75th heart rate percentiles, respectively

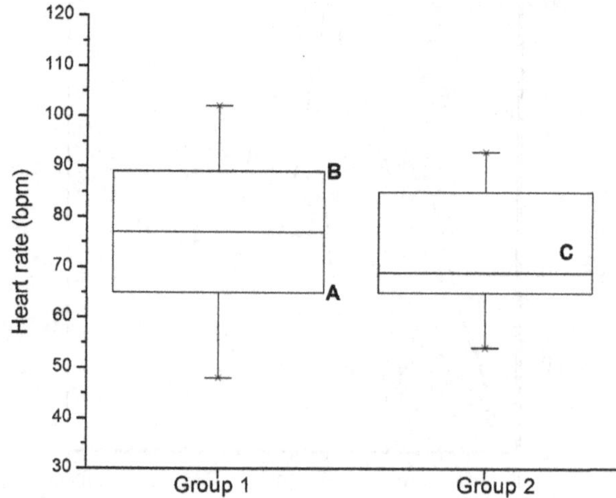

FIGURE 21.3 **Heart rate following plasma transfusion.**

Concept: Different plot types can be used to highlight different statistical concepts. The data presented here are in a "box-and-whiskers" plot, which helps to (in a single figure) highlight the data set's descriptive statistics. Further, it also helps to present the distribution of the data.

Answer: *E*—The "box" of the box and whisker plot represents the 25th (line A) and 75th (line B) percentiles. The line (line C) in the center of the "box" of a "box and whiskers" plot represents the median of the data set. The other choices (Answers A, B, C, and D) are incorrect.

8. In reference to the box-and-whiskers plots for groups 1 and 2 (Fig. 21.3), which of the following is most accurate?
A. The heart rate median for group 2 is higher than that of group 1
B. About 50% of the samples from group 1 have rates between 50 and 60
C. The observed heart rate range is larger for group 1 than group 2
D. About 50% of the samples from group 2 have rates between 65 and 90
E. The mean heart rates for groups 1 and 2 are identical

Concept: Different plot types can be used to highlight different statistical concepts. The data presented here are in a "box and whiskers" plot, which helps to (in a single figure) highlight the data set median, 25th and 75th percentiles, and the range of data points collected.

Answer: C—The height of the "box-and-whiskers" plot reflects the spread of the data. In this case, the "box-and-whiskers" plot is taller for group 1, indicating a wider data spread. The heart rate median for group 2 (~70) is less than that of group 1 (~75, Answer A). About 50% of the samples from group 1 have rates between 65 and 90 (between line A and line B) (Answer B). For group B, it is between 65 and 85 (Answer D). From the distribution, it appears that group A has higher heart rate than group B; thus, the mean heart rate is expected to be higher in group A than in group B (Answer E). Statistical test can be used to test to see if this difference is a true difference or is due to random chance; the plot alone cannot determine the statistical significant of the difference.

End of Case

Please answer Questions 9 and 10 based on the following clinical scenario:

A group of scientists is designing a study to examine a novel recombinant ADAMTS13 factor for the treatment of thrombotic thrombocytopenic purpura (TTP). Eligible patients will be enrolled in the study at the day of their diagnosis before they have received any other therapy and, via randomization, will be assigned to the experimental or standard/control apheresis treatment arm. The study is a multicenter, multinational investigation, with diverse patient populations. After randomization, patients in the experiment arm will be administered drug in addition to standard apheresis care, while control patients will get standard apheresis care + placebo.

One of the study outcomes is the percentage decrease in schistocytes, with the expectation that the drug will more rapidly lower schistocyte levels compared to standard treatment only. Schistocyte percentage will be assessed, in a blinded fashion, individually by three hematopathologists, who are unaware of the subject's treatment status. The investigators have not defined a method for assessing schistocyte percentage, but will allow each pathologist to determine this value according to their own criteria and expertise.

9. Which of the following research biases is most likely to occur in this study?
A. Measurement bias
B. Placebo bias
C. Survivor bias
D. Selection bias
E. No obvious bias based on design

Concept: When designing studies and trials, investigators must be aware of potential biases of their investigations. Such biases can sway results and lead to variations in study outcomes. It is also important to be aware of such biases when appraising trials and research papers as a reader of scientific and medical literature.

Answer: *A*—Given the study design, the most likely bias to occur is measurement bias. In this case, the lack of standardization in schistocyte assessment may yield highly variable results from case-to-case (Answer E). The remaining biases listed do not appear to be significantly impacting the study design. Placebo bias (Answer B) should not influence the assessment of schistocytes in this case. Based on enrollment design, there does not appear to be a survivor or selection bias (Answers C and D), as the study will be implemented before any other therapies are applied early in treatment (i.e., no survival bias), and subjects will be randomized to include patient from diverse geographical settings (i.e., no selection bias).

10. Given the earlier scenario, which of the following steps would *not* be helpful in minimizing the problematic bias encountered in this study?
A. Create clear definitions for what constitutes a "schistocyte" in this study
B. Eliminate the use of a placebo in this study group
C. Establish a set number of high power fields to be counted for schistocyte percentage
D. Employ only a single investigator to determine schistocyte percentages
E. Utilize an automated computer program for schistocyte identification

Concept: Measurement bias is most likely to occur when there is little or no standardization of the protocol used to make a scientific measurement or observation.

Answer: *B*—Eliminating a placebo in this study will not help to address to the measurement bias inherent in the study design. It may make the study design weaker since there will be no concurrent control group. All of the other choices (Answers A, C, D, and E) should help minimize the variability associated with assessment of a potentially subjective morphologic data point (i.e., schistocyte percentage analysis).

End of Case

Please answer Questions 11–13 using the following scenario:

Disease X is an inflammatory disease which was proposed to be caused by anti-υ antibody. At this time, disease X has a high morbidity rate. However, it has no definite treatment. Traditionally, medication S is used in these patients. Nonetheless, therapeutic plasma exchange (TPE) has also been attempted.

11. A retrospective study was conducted to evaluate the use of medication S versus TPE + medication S in disease X. The following is the data that has been collected (Tables 21.4a and b).

TABLE 21.4A Baseline Characteristics

Parameters	Group 1 (medication S only, $N = 20$)	Group 2 (medication S + TPE, $N = 40$)	*P*-value
Age (years) ± SD	32 ± 5	35 ± 8	0.13
Sex (% male)	50	55	0.85
Anti-υ antibody level ± SD (higher = worse)	950 ± 350	1400 ± 500	<0.01
Disease severity score ± SD (higher = worse)	4 ± 3	7 ± 5	0.02

TABLE 21.4B Outcomes at Day 14

Parameters	Group 1 (medication S only, $N = 20$)	Group 2 (medication S + TPE, $N = 40$)	P-value
Anti-υ antibody level ± SD (higher = worse)	400 ± 285	800 ± 500	<0.01
Disease severity score ± SD (higher = worse)	2.75 ± 1.8	5.15 ± 3.5	0.01

Which of the following statements is true regarding the earlier data?

A. Anti-υ antibody level should be used as a biomarker for the disease severity
B. Medication S + TPE is more efficacious than medication S alone in the treatment of disease X
C. Medication S + TPE is less efficacious than medication S alone in the treatment of disease X
D. Medication S + TPE is safer than medication S alone in the treatment
E. No comparison can be made regarding the efficacy between medication S + TPE versus medication S alone

Concept: Retrospective studies may be used at the beginning of patient care to investigate a potential treatment for a disease. Nonetheless, there are multiple weaknesses of such an approach, such as recall bias, selection bias, etc. Selection bias is the selection of patients for a study in such a way where proper randomization cannot be achieved. Statistical methods can reduce random errors, but they cannot reduce selection bias.

Answer: *E*—There is selection bias in this study. From Table 21.4a (baseline characteristics), it can be seen that the two groups are not the same at baseline. Specifically, the group receiving medication S + TPE tended to have more severe disease than the group receiving medication S only.

Although the results in Table 21.4b demonstrated that the patients in group 2 did worse than group 1 (by both the antibody level and severity score), it cannot be concluded that Medication S + TPE is more or less efficacious than medication S alone (Answers B and C), since group 2 already started worse than group 1. There is no data regarding safety (Answer D). Anti-υ antibody level appears to correlate with the disease severity score. However, the data for this correlation is not shown, and thus, at this time, the conclusion that anti-υ antibody level should be used as a biomarker for the disease cannot be made (Answer A).

12. The following data is for medication S.

- Half-life = 14 days
- Volume of distribution = 0.05 L/kg body weight
- Protein binding = 90%
- Excretion = renal (75%), liver (25%)

Based on the above information, which of the following is true regarding the effect of TPE on medication S?

A. TPE is predicted to remove medication S efficiently
B. TPE is predicted to remove medication S; however, the degree of removal is not significant
C. TPE modulates the effect of medication S by decreasing the fibrinogen level
D. TPE is predicted to not remove medication S
E. There is not enough data to conclude on the effect of TPE on medication S removal

Concept: Medication removal during TPE is a concern. However, there has been no extensive study on the effect of TPE on medication removal. Thus, generally, apheresis physicians recommend giving medications after TPE, if possible, to minimize any potential impact. If not, then checking the medication level post-TPE and supplementing the dosage may be necessary. In general, drugs with low volumes of distribution and/or high rates of protein binding are most likely to be removed during TPE.

Answer: *A*—Using the above pharmacokinetic data, medication S has a low volume of distribution and a high protein binding rate; thus, it is predicted to be removed significantly by TPE (Answers B, D, and E). There is no data regarding the interaction between medication S and fibrinogen; thus, it cannot be concluded that low fibrinogen level may modulate the effect of medication S (Answer C).

13. A different research group performed multivariable logistic regression on the likelihood of clinical improvement (which was defined as a decrease of 2 or more in the disease severity score post- vs. pretreatment) on a different group of patients. Table 21.5 contains the results of the regression.

TABLE 21.5 Multivariable Logistic Regression

Parameters	OR (for improvement at day 14)	*P*-value
Medication S + TPE versus medication S alone	1.15	0.15
Age (every 10 years)	0.89	0.06
Gender (female vs. male)	1.18	0.03
CRP (above reference range vs. in reference range)	0.85	0.21
Anti-υ antibody level (every 10 units)	0.75	0.04

Which of the following statement regarding the interpretation for the earlier results is correct?

A. The odds of having clinical improvement is significantly higher with medication S + TPE regimen than with medication S alone after adjusting for other variables
B. For every 10 years increase in age, the odds of having a clinical improvement is significantly higher after adjusting for other variables
C. Females are likely to have significant clinical improvement than male after adjusting for other variables
D. The higher the CRP level, the more likelihood of clinical improvement after adjusting for other variables
E. The higher the anti-υ antibody level, the more likelihood of clinical improvement after adjusting for other variables

Concept: Logistic regression is used to build a model when the outcome is categorical [such as improvement (yes or no) in this example]. It can be performed with one predictor (univariable analysis) or with multiple predictors (multivariable analysis). The predictor can be categorical (e.g., male or female) or continuous (e.g., anti-υ antibody level or age). Logistic regression will give the odds ratio (OR) for the outcome as well as the *P*-value with OR > 1 favors the outcome.

Answer: C—Using males as the reference, the OR for clinical improvement for females is 1.18 (*P*-value = 0.03); thus, the odds of having clinical improvement is 1.18 times higher for female comparing to male after adjusting for other variables, and this is statistically significant. Since the *P*-values for medication type, CRP and age are all >0.05, these are not significant predictors for clinical improvement (Answers A, B, and D). For anti-υ antibody, the OR for every 10 units increase is <1; thus, the higher the anti-υ antibody level, the less likelihood of clinical improvement after adjusting for other variables (Answer E).

End of Case

14. A common type of study is a large observational study to find correlations between the use of alternative management pathways and patient outcomes. However, the results of randomized controlled trials (RCT) are widely considered superior to observational studies. What is the primary rationale for such a conclusion?

A. RCTs typically isolate a single variable as the primary end-point of the study, while observational studies typically evaluate many different potential end-points, potentially leading to erroneous conclusions
B. In an observational study, participants and investigators are never blinded to the interventions under study, while RCTs always include such blinding
C. The role of various unacknowledged biases, particularly selection bias, in an observational study is overcome by randomizing application of an intervention in an otherwise homogeneous population of subjects
D. The design of an RCT with a single, predetermined primary end-point requires the statement of a specific hypothesis to be tested, whereas observational studies are rarely hypothesis-driven
E. RCTs are not considered superior to observational studies. Both are of equal value in clinical research

Concept: Randomized controlled trials (RCTs) are widely considered to be the gold standard of evidence based medicine. Observational studies suffer from the problem that correlation does not necessarily imply causation. The key element is that by randomizing a homogenous group of subjects to alternative interventions, the role of multiple different kinds of a bias (both unknown and known), particularly selection bias, are excluded.

Answer: C—The key difference between a RCT and an observational study is the randomization aspect, which eliminates both unknown and known bias at baseline. Although a sharp focus on a single end-point (Answer A) and formally stating a particular hypothesis (Answer D) are good practices, the elimination of selection bias is the primary advantage of an RCT design. Although many RCTs are conducted in a double-blind fashion, this characteristic is not the essential feature of the RCT design (Answer B). In many cases, blinding is not feasible, such as when one of the treatment options is surgical- or procedural-based. A well-designed RCT is usually considered to be superior to an observational study (Answer E).

15. Although RCTs are regarded as the gold standard in evidence-based medicine, the end-points measured in such trials are multiple. Assuming that all of the randomized trials outlined later are statistically significant, with excellent validity and generalizability, which one provides the best evidence that a new drug might be beneficial?

A. In a large RCT, using intention to treat analysis, subjects in the new drug group experienced decreased rates of the combined primary endpoint (nonfatal stroke, death, or elevated levels of a risk factor into the high-risk category for stroke)
B. In a large RCT, using per protocol analysis, subjects in the new drug group experienced decreased rates of the combined primary endpoint (nonfatal stroke, death, or elevated levels of a risk factor into the high-risk category for stroke)
C. In a large RCT, using intention to treat analysis, fewer subjects died from stroke in the new drug group than in the placebo group
D. In a large RCT, using intention to treat analysis, fewer subjects died for any reason in the new drug group than in the placebo group
E. In a large RCT, using per protocol analysis, a new drug lowers serum levels of a risk factor known to be associated with an increased risk of death from stroke

Concept: All of these outcomes suggest that the hypothetical new drug is effective, but the specific end-points used are different. It is generally considered that the more clinically important end-points used, the stronger the claim of effectiveness of a test intervention. Thus, different kinds of surrogate end-points are less certain than clinical outcome end-points. The validity of using a surrogate end-point is based on the practicality of completing the trial in an efficient fashion. Furthermore, if a surrogate biomarker is used, then it should be validated that it correlates with patients' important outcomes.

There are also two ways to analyze the results of a RCT: intention-to-treat and per protocol analysis. Intention-to-treat analysis means that all patients who were enrolled and randomized were included in the analysis in such a way that they were analyzed based on the groups they were allocated to regardless whether they were adhere to those group's treatment or not. Per-protocol analysis is a comparison of treatment groups that included the patients who completed the treatments originally allocated. Since randomization process should allocate both known and unknown biases equally among the groups, intention-to-treat analysis is the recommended method of analysis in order to avoid any bias. If done alone, per-protocol analysis may lead to biased results.

Answer: *D*—The correct order of strength of these end-points is: Answer D (all-cause mortality) > Answer C (disease-specific mortality) > Answers A and B (Composite) > Answer E (surrogate). Interestingly, there is a published study that examines the actual performance of a group of physicians in ascertaining these results and showed considerable errors in judgement on this comparison. Furthermore, intention-to-treat analysis has less biases comparing to per-protocol analysis. Other choices are wrong.

16. A randomized controlled trial is performed regarding the use of TPE to treat renal disease associated with multiple myeloma (MM). Investigators in the study enrolled 1150 dialysis-dependent patients with MM and creatinine levels >6.0 mg/dL in the TPE treatment arm and 1165 dialysis-dependent patients with MM and creatinine level >6.0 mg/dL in the observation arm. After 1 year of monthly TPE, investigators found no significant difference regarding dialysis dependence between experimental and control arms; however, the TPE group showed a mean creatinine of 7.7 mg/dL at the end of 12 months, versus a mean creatinine of 9.3 mg/dL in the control, nontreated group ($P < 0.001$; the study group deem *P*-values <0.05 as significant). What is the best interpretation of the study results from a clinical standpoint?

A. Given the highly significant *P*-values obtained, and the randomized-controlled nature of the trial, TPE seems clinically worthwhile for most patients with MM and advanced renal disease.
B. The study results suggest that chronic TPE would be most clinically worthwhile only for MM patients with creatinine values in excess of 7.7 mg/dL.
C. The study results suggest that chronic TPE would be most clinically worthwhile only for MM patients with creatinine values in excess of 9.3 mg/dL.
D. While a significant difference was attained in creatinine values, given that there was no difference in dialysis-dependence, chronic TPE is likely of no significant clinical benefit in MM.
E. The study was far too small to make any clear conclusions about the efficacy of TPE in the setting of renal disease and MM.

Concept: *P*-values are useful tools in helping to assess whether differences exist between populations/data sets and how likely the differences observed may be due to chance. However, *P*-values cannot be assessed in isolation and must be examined in the appropriate clinical and scientific context.

Answer: *D*—While the authors showed that creatinine values were significantly decreased in the treatment group as assessed by *P*-values, this change was not associated with a clinically-meaningful outcome regarding dialysis dependence. As such, and based on this large randomized controlled trial, there is no compelling evidence suggesting that monthly apheresis would impart any clinical benefit (Answers A, B, and C). The study is large (involved more than 2000 patients) and the results achieved statistically significant. However, it does not show clinical benefit. Thus, even if a larger study is conducted, the clinical significance is probably too small between the two groups and thus, the practical value of the interpretation is likely to remain the same (Answer E).

Please answer Questions 17 and 18 based on the following clinical scenario:

While covering the blood bank, you are consulted regarding premedication for prevention of a transfusion reaction. The patient is a 77-year-old man with myelodysplastic syndrome (MDS) who is dependent on chronic RBC transfusions and has had several definitive allergic reactions, but otherwise, no other adverse events, in the past 4 months. The clinical team would like to provide the patient with an antiallergic medication, as well as antifebrile medication, prior to transfusing this patient. You are aware of a recent randomized, controlled study that examined the benefits of premedication of 5000 adults undergoing RBC transfusion; 2500 received antiallergic and antifebrile medications prior to transfusion and 2500 received no premedication prior to transfusion; *P*-values <0.05 were deemed significant. The study reported the following:

Overall analysis—total patients (n = 5000):

- 0.52% (13/2500) allergic reaction rate (experimental) versus 0.6% (15/2500) allergic reaction rate (control); *P*-value = 0.7
- 0.32% febrile reaction rate (experimental) versus 0.36% febrile reaction rate (control); *P*-value = 0.81

Subset analysis—patients with history of allergic reactions (n = 600):

- 5% (15/300) reaction rate (experimental) versus 15% (45/300) reaction rate (control); *P*-value <0.0001

Subset analysis—patients with history of febrile reactions (n = 300):

- 6.7% (10/150) reaction rate (experimental) versus 7.3% (11/150) reaction rate (control); *P*-value = 0.82

17. Based *solely* on the clinical history and available data from this study, what is the most reasonable recommendation in for the current patient regarding transfusion reaction prevention?

A. Use of antifebrile medication is evidence-based, if the patient has a history of febrile reactions
B. Use of antiallergic medication is evidence-based, if the patient has a history of allergic reactions
C. Use of both antifebrile and antiallergic medications are evidence-based regardless of the past reaction history
D. Use of both antifebrile and antiallergic medications are not evidence based in any circumstance
E. Cannot use the study data since the investigation appears underpowered

Concept: It is possible to use clinical trial data in order to determine whether the available evidence supports a particular therapeutic approach in a given patient population.

Answer: *B*—Study data would indicate that in cases of patients having a history of allergic reactions, there was a significantly lower rate of allergic reactions. Febrile reactions appeared to be unaffected by premedication regimens regardless of the patient having a past-history of a febrile reaction. As such use of an antiallergic medication alone in this case is an evidence-based approach to prevent a subsequent reaction. The other choices (Answers A, C, and D) are incorrect based on the study findings. The study appears to be appropriately powered (Answer E).

18. Later that day you receive another call regarding premedication for the prevention of a transfusion reaction. In this case, the patient is a 23-year-old woman with severe postpartum anemia but no history of transfusions. The clinical team again asks your advice regarding the evidence to support the use of antiallergic as well as antifebrile medications prior to transfusion, in order to prevent adverse transfusion outcomes.

Based on the same clinical trial discussed earlier, what is the most reasonable recommendation in this case regarding transfusion reaction prevention?

A. With this patient's history, use of antifebrile medication is evidence-based, while use of antiallergic medication is not evidence-based
B. With this patient's history, use of antiallergic medication is evidence based, while use of antifebrile medication is not evidence-based
C. With this patient's history, use of both antifebrile and antiallergic medications are evidence-based
D. With this patient's history, use of both antifebrile and antiallergic medications are not evidence-based
E. Cannot use the study data since the investigation did not clearly indicate study of postpartum women

Concept: It is possible to use clinical trial data in order to determine whether the available evidence supports a particular therapeutic approach in a given patient population.

Answer: *D*—Study data would indicate that for total patients, allergic and febrile reactions appeared to be unaffected by premedication regimens and only those with a history of allergic reactions appeared to benefit from premedication. As such, the best recommendation would be to avoid any pretransfusion medication in this case, since the patient has no history of transfusion reactions. Although the data on the clinical trial did not mention whether or not postpartum women were included, it is unlikely that postpartum women would have a different outcome compared to other study participants regarding transfusion (Answer E). Based on this patient's clinical history and the findings of the study, all the other choices (Answers A, B, C) are incorrect.

End of Case

19. You are performing a study to determine the impact of a new additive solution (solution Z) on the development of blood group antibodies under the hypothesis that alloimmunization should not be increased with the new solution, as compared to RBC units prepared using, the current standard solution (AWS). Using a retrospective analysis of independent, randomly-selected patients, you have determined that of 756 individuals exposed to the additive solution Z, 27 have developed antibodies. By comparison, in a separate a control group of 800 randomly-selected individuals who were transfused with RBCs stored in AWS, 40 developed antibodies.

Which of the following is the best statistical test to employ to compare the proportion of individuals alloimmunized in your experimental group to the proportion alloimmunized in your control group?

A. Mann-Whitney U test
B. Chi-square test
C. Kruskal-Wallis test
D. Unpaired t test
E. Paired t test

Concept: Statistical tests can be utilized to help confirm or deny hypotheses about a data set. Major considerations before choosing a statistical test include whether the data set is randomly selected or not, whether the data represent a normal distribution or not, and what kind of comparison is sought (e.g., comparing two proportions, comparing means, comparing interval changes, etc.).

Answer: *B*—In the example listed, for comparing proportions between an experimental and control group of randomly-selected subjects, the Chi-square test is preferred for evaluating the null hypothesis (i.e., there is no difference between experimental and control groups). If the sample size is small, Fisher exact should be used (instead of Chi square). The remaining tests listed are typically utilized for nonproportional data comparisons from two or more data sets and are not appropriate for proportional comparisons. While discrete descriptions of each of these statistical methods is beyond the scope of this chapter, each of the wrong answer choices represent tests typically used to compare means (Answers A and C are for nonparametric distribution, while Answers D and E are for normal distribution). It is important to note that each of these methods must be strictly applied based on the data set and patient population being evaluated and cannot be used interchangeably.

20. An animal study is completed to investigate whether treatment with drug Y may be able to prevent anamnestic RBC alloantibody responses to the RBC antigen Z. Fifteen animals are given two transfusions, with the second transfusion occurring after drug Y is administered. The researchers also have data from control animals, transfused in the absence of drug Y. Assuming the outcomes are continuous in nature, what is the best statistical test to compare antibody responses in the same group of animals (i.e., those treated with drug or those not treated with drug), before and after the second transfusion? Note that the data are *not* parametric and are *not* normally distributed.

A. Students *t*-test
B. Paired *t*-test

C. Ordinary ANOVA
D. Wilcoxon matched pairs signed rank test
E. This type of a comparison cannot be done

Concept: The way in which data are analyzed impacts the conclusion of a study, with the correct statistical test being critical. There are statistical programs that can test the normality of your data, but you can also quickly check this yourself. If your data cleanly distribute into a smooth histogram format with a bell-shaped curve, then they are likely normally distributed. Thus, an informal way to test the normality of your data is to compare a histogram of your data to a normal probability curve.

Answer: *D*—The Wilcoxon matched pairs signed rank test allows for a comparison of two measurements in a given group, before and after a treatment. This test can be completed with nonparametric, nonnormally distributed data (Answer E). Because the data are not parametric and not normally distributed, a *t*-test (Answers A and B) cannot be performed. An ANOVA (Answer C) typically compares three or more groups.

21. When thinking about the design of a clinical study examining the safety of a new pathogen reduction technology, which of the following variables is most influential on statistical power?
A. The type of blood product being studied
B. The mean age of the study population
C. The sample size of the study population
D. The gender of the study population
E. The use of a placebo in parallel to the inactivation agent

Concept: The power of a statistical test is its ability to detect a postulated true different. Power is not the probability of finding a difference of size. For example, if the study design is to compare the new treatment versus standard of care, then power is the probability of correctly finding a statistically significant difference between two treatments and thus, declare a treatment benefit. Mathematically, power = 1−type II error [where type II error is the probability of failing to reject a null hypothesis (no treatment difference) when the alternative (there is a treatment difference) is true]. Type I error, on the other hand, is the probability of rejecting the null hypothesis when it is true. This is related to the *P*-value, which is the probability of obtaining an equal or more extreme test statistic in the direction of the alternative hypothesis than the observed one, assuming that the null hypothesis is true.

Answer: *C*—Of all the answers listed, the variable most influential on statistical power is sample size. In general, very small study sizes (e.g., studies involving only a handful of patients) are underpowered and measurement of a significant outcome can be difficult. Therefore, increasing sample size generally increases a study's power. There are four variables that related to each other—power, type I error, sample size, and effect size. If we know any of the three variables, then the fourth variable can be computed. The other variables mentioned (Answers A, B, D, and E), while important from a design standpoint, have less of an impact on the likelihood of generating a Type II error.

Please answer Questions 22 and 23 based on the following clinical scenario:

There have been a few small studies in animals and healthy volunteers demonstrating that substance X may be able to stimulate red blood cell (RBC) production. The Department of Laboratory Medicine in your hospital has collaborated with the Department of Orthopedics to create a biobank that stores all plasma from patients undergoing orthopedic surgery at your hospital. From this biobank, substance X was measured, and your research group decides to conduct a retrospective study to investigate the relationship between substance X and the need for RBC transfusion in all patients who underwent hip replacement procedures for the past 5 years.

22. Multivariable logistic regression is used to analyze the data. Your biostatistician modeled the odds of necessary transfusion and the following results were obtained.

Variable	OR (95% CI)	*P*-value
Gender (male vs. female)	7.4 (3–18.2)	0.001
Age (for every 10 year incremental)	3.3 (2.1–5.4)	0.03
Preoperative hemoglobin (for every 1 g/dL incremental)	0.4 (0.2–0.8)	0.03
Creatinine (for every 1 μg/dL incremental)	1.2 (0.5–2.7)	0.32
Substance X (for every 10 μg/dL incremental)	4.1 (1.2–13.5)	0.04

Which of the following statements regarding the above results is correct?

A. Logistic regression should not be used since the outcome is a continuous variable
B. Logistic regression should not be used since the predictor (substance X) is a continuous variable
C. Logistic regression is an appropriate statistical model, and the result regarding substance X and the odds of transfusion requirements is consistent with previous studies as described in the scenario
D. Logistic regression is an appropriate statistical model, and the study supports the use of exogenous substance X injection in patient to reduce transfusion requirement
E. Logistic regression is an appropriate statistical model, and the result regarding substance X and the odds of transfusion requirements is not consistent with previous studies as described in the scenario

Concept: It is important to understand which model to use to analyze the data in research. The choice of the analysis depends on the variable type of the outcome, which can be continuous variable or categorical variable. It does not depend on the nature of the predictor variables. A continuous variable is a variable that has infinite possible values. For example, patient's age, height, or weight are common continuous variables. A categorical variable (or nominal variable) is a variable that has two or more categories, but the categories are in no particular order. For example, sex (male vs. female) is a categorical variable. An ordinal variable is similar to categorical variable in a way that it also has two or more categories; however, in ordinal variable, there is a clear order for the categories. For example, clinical outcome (complete remission vs. partial remission vs. minimal remission vs. no remission) for a particular drug is an ordinal variable. In a majority of biomedical studies, if the distribution of the variables is approximately normal and/or the sample size is large, Fig. 21.4 can assist in choosing the appropriate statistical analysis.

			Outcome variable		
			Categorical		Continuous
			Dichotomous	>2 categories	
Predictors	Categorical	Dichotomous	(2×2 table) χ^2 test; 2-sample proportion *z*-test	χ^2 test	2-sample means *t*-test; ANOVA
		>2 categories	χ^2 test		ANOVA
	Continuous (or continuous & categorical)		Logistic regression	Polytomous regression	Correlation (for continuous predictors), regression

FIGURE 21.4 **Choice of common statistical tests for association.**

Answer: *E*—The outcome in the study is a categorical variable. It is a binary outcome—need red blood cell transfusion versus no need transfusion. Thus, logistic regression is appropriate (Answers A and B). From the results of the multivariate analysis, the OR for every 10 μg/dL incremental in substance X measurement is greater than 1. Since the biostatistician modeled the odds of necessary transfusion, it demonstrated the higher substance X level, the more likely the patient will need transfusion. Previous studies showed that substance X may be able to stimulate red blood cell production; thus, logically, higher level of substance X should reduce the likelihood of transfusion. Hence, the result of your retrospective study is not consistent with previous results (Answer C).

The choice of a statistical analysis depends on the type of outcome variable, and not depends on the predictor variable(s) (Answer B).

Finally, there must be multiple prospective studies that demonstrate the efficacy and safety of substance X before it can be widely used in patients (Answer D).

23. Since your group published the result in Question 22, other groups also reviewed the data in multiple patient populations and the results were conflicting. Furthermore, a pharmaceutical company has developed an active form of substance X and showed in a phase I study that this compound is safe and, at a dose of 30 mg daily, increased the hemoglobin level by 1 g/dL after a median of 3 days usage in healthy volunteers.

Given that transfusion of red blood cells (RBCs) is not without risks, a randomized control double blind clinical trial is designed to test the following hypothesis: At a daily dose of 30 mg, does the use of substance X noninferior to erythropoietin at a dose of 80,000 IU in reducing the need of RBC transfusion in patients undergoing hip replacement surgery? The result of this trial is shown as follows:

Variable	Relative risk (RR) for reducing transfusion need (95% CI)	*P*-value
Substance X versus erythropoietin	2.2 (1.5–2.9)	0.03

Assuming the trial is well-designed, which of the following statements is true?

A. Since the RR > 1, at the dosage and frequency stated, substance X is superior to erythropoietin in reducing red blood cell transfusion in patients undergoing hip replacement surgery

B. Since the RR > 1, at the dosage and frequency stated, substance X is equivalent to erythropoietin in reducing red blood cell transfusion in patients undergoing hip replacement surgery

C. Since the RR > 1, at the dosage and frequency stated, substance X is noninferior to erythropoietin in reducing red blood cell transfusion in patients undergoing hip replacement surgery

D. At the dosage and frequency stated, no conclusion can be reached regarding the efficacy of substance X versus erythropoietin. A larger sample is needed

E. At the dosage and frequency stated, no conclusion can be reached regarding the efficacy of substance X versus erythropoietin. Multivariable analysis should be performed before any conclusion can be reached

Concept: As described previously, a well-designed randomized controlled double blind clinical trial should be the gold standard to make treatment recommendations and changes in clinical practice. There are many different types of randomized controlled trials (Table 21.6 and Fig. 21.5)—the goal of each lies on the alternative hypothesis that the study aims to show. For example, the goal of a classical (or traditional) study is to demonstrate that there is a difference between the therapies. If the data fails to reject the null hypothesis, equality cannot be ruled out. For both equivalent and noninferior study, a margin δ is selected, the equivalent of the new treatment when there is evidence to demonstrate its efficacy is within δ from the standard treatment. Similarly, noninferiority is established if the data demonstrated that the new treatment is not more than δ less than that of the current treatment (if higher is better result) or that the new treatment is not more than δ more than that of the current treatment (if lower is better result). Hence, the

TABLE 21.6 Hypotheses Associated With Different Types of Study

Type of study	Null hypothesis	Alternative (or research) hypothesis
Traditional comparative (superior)	There is no difference between the therapies	There is a difference between the therapies
Equivalence	The therapies are not equivalent	The new therapy is equivalent to the current therapy (or standard of care)
Noninferiority	The new therapy is inferior to the current therapy (or standard of care)	The new therapy is not inferior to the current therapy (or standard of care)

Adapted with permission from E. Walker, A.S. Nowacki, J. Gen. Intern. Med. 26 (2011) 192–196.

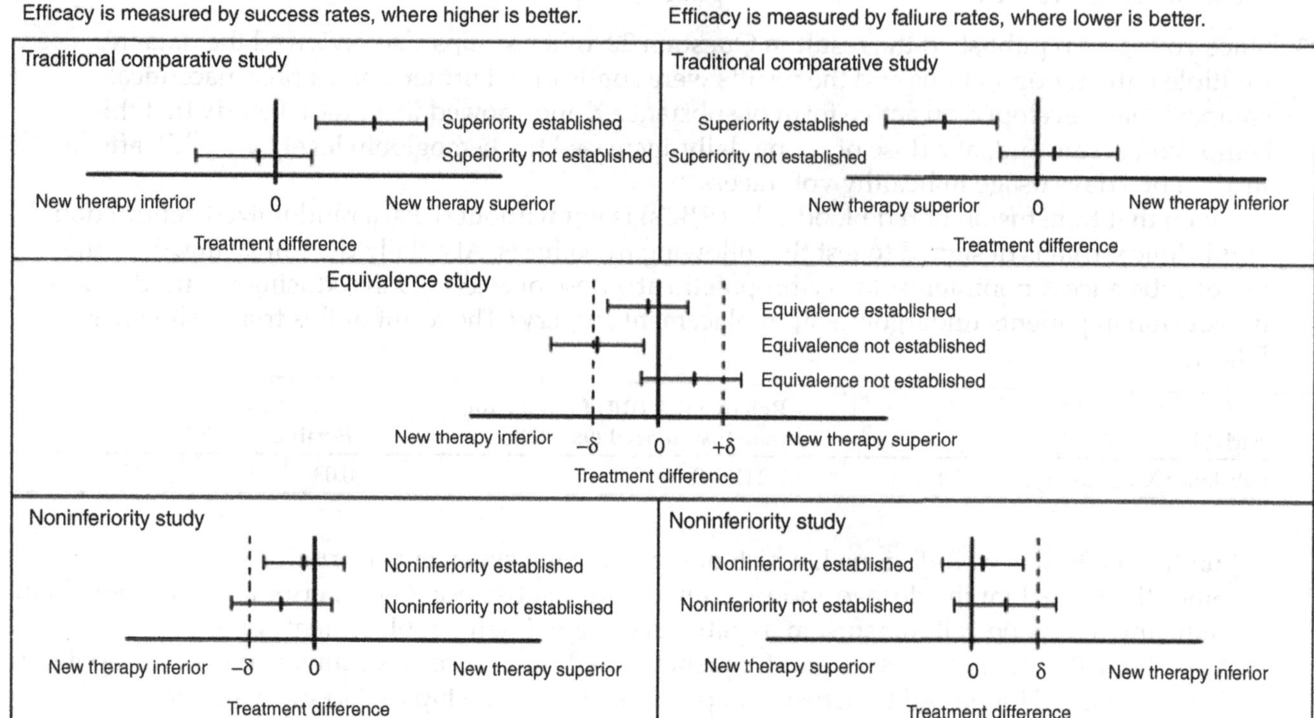

FIGURE 21.5 Statistical tests interpretation with regarding to different designs. *Source: Adapted from E. Walker, A.S. Nowacki, J. Gen. Intern. Med. 26 (2011) 192–196.*

determination of δ is the most essential step in designing an equivalent or noninferior study and it must be established prior to the data collection and analysis. It does not only affect the result of the test but it also establishes the scientific credibility of the study result. The steps in determining δ for such RCTs are beyond the scope of this chapter.

Answer: C—This is a noninferiority study; thus, the hypothesis tests whether substance X supplementation is noninferior to erythropoietin in reducing the need of red blood cell transfusion in patients undergoing hip replacement surgery. The *P*-value is <0.05 and the 95% confident interval does not contain 1; thus, there is enough evidence to reject the null hypothesis and accept the research hypothesis. In other words, at the dosage and frequency stated, substance X is noninferior to erythropoietin in reducing red blood cell transfusion in patients undergoing hip replacement surgery.

The design of the study is not superior or equivalent; thus, there is no data to support the conclusion of superiority or equivalent in efficacy between substance X and erythropoietin (Answers A and B). The sample size is large enough to demonstrate noninferiority for substance

X (Answer D). The conclusion can be reached from the hypothesis testing of the main research question (Answer E).

End of Case

Please answer Questions 24 and 25 based on the following clinical scenario:

A company is developing a multiplex set of serum biomarkers meant to identify nonsmall cell lung cancer. They have identified a panel of markers and an algorithm that shows 95% sensitivity and 95% specificity in a test cohort of 20 subjects with advanced stage cancer and 20 normal healthy subjects. The company has patented the panel of markers and algorithm. They then test this panel and algorithm on another 20 subjects with advanced stage cancer and 20 healthy controls and find a sensitivity of 90% and specificity of 90% for discriminating the cancer patients from normal. They publish these studies with an investigator from an academic institution that helped collecting the specimens used for the study. The manuscript concludes that the assay would be useful as a screening tool to detect nonsmall cell lung cancer in otherwise healthy subjects.

24. Which of the following additional features of this study would be most concerning about the potential for a significant conflict of interest being involved in the interpretation of these results?
A. The investigator discloses that he received a research grant from the company
B. The decrease in test accuracy in the validation cohort is not accounted for by the investigators
C. The study focused on subjects with advanced stage cancer rather than early small tumors that would be similar to those expected in a screening population
D. The study emphasizes that the test has a 90% positive predictive value which would be useful for screening
E. The study emphasizes that the test has a 90% negative predictive value which would be useful for screening

Concept: There are many potential sources of conflicts of interest in studies of novel diagnostic tests and therapeutic agents. The level of skepticism that readers maintain should not be reflective of the mere presence of a disclosed conflict of interest, but also evidence of a biased study design that works in the favor of the interests of the conflicted party.

Answer: *D*—The positive predictive value (PPV) is dependent on the prevalence of disease in the population tested. With 50% prevalence in this test population, the PPV is seemingly high. However, at a more realistic estimate of the presence of nonsmall cell lung cancer of around 0.1%, the PPV would be only 0.9%, which means roughly 99% of the positive tests would be false positives. This is clearly an unjustified implication, which furthers the interests of the company that holds a clear financial interest in the widespread use of the test as a screening tool. This clearly misleading implication of the study represents the most pervasive feature indicating a substantial conflict of interest.

The additional disclosure of a research grant to the supplier of the specimens for the study is hardly surprising and does not add much to the concern about the conflict of interest (Answer A). The decrease in the test accuracy in a validation cohort (Answer B) is another fully expected feature of such a study and not a significant additional issue in discerning a conflict of interest. The fact that the study focuses on advanced-stage cancer with presumably high tumor load almost certainly means that the test would perform less well with very small early stage tumors (Answer C) is a major concern, but there are practical reasons that the study as currently constructed is much more technically feasible. This issue mitigates to some degree the concern that the study design was planned primarily with the desired outcome in mind, which supports the interests in conflict. The purpose of screening is to identify individuals with disease; thus, a test with high sensitivity is desirable. Furthermore, without the exact prevalence of nonsmall cell lung cancer, the negative predictive value cannot be calculated from the sensitivity and specificity (Answer E).

25. Suppose that the maker of this test came to you as a scientific adviser. They want your expert judgement about what additional study would most pervasively justify the case that this test would be financially successful at the most inexpensive cost for the study?
   A. A study to further refine the composition of the panel of analytes and the algorithm to define a positive test using additional independent specimens collected in the same fashion
   B. Expanding the current study to include more normal subjects and those with COPD that may have a higher rate of false positive results
   C. A study using retrospective analysis of stored serum samples from participants in a previous study examining spiral CT scan to detect nonsmall cell lung cancer in smokers with a 5 year follow up for the development of lung cancer
   D. A prospective cohort study of a group of smokers in a COPD clinic with a 5 year follow up of all screened subjects for the development of lung cancer
   E. A prospective randomized trial of a group of smokers in a COPD clinic with a 5 year follow up of all screened subjects for the development of lung cancer

   Concept: One aspect of incisive analysis of scientific studies in the ability to look at the problem from the point of view of an investigator, even if you are not an active contributor to the field. If studies choose to examine a suboptimal research design that results in a positive set of findings concerning the importance of the work, but avoid a feasible study that would be fairly definitive, one must be suspicious of the potential of conflicts of interest in the study design. The attempt to balance real world feasibility and cost with the strength of the potential test of a particular hypothesis is a key feature of the investigator mind set.

   Answer: C—A retrospective study of stored specimens in a previously conducted study that would directly compare this test with the leading contender (i.e., spiral CT scan) to detect early lung cancer. Since only a small sample of these stored sera would be required, the ability to obtain such specimens would be potentially feasible and conducting the additional runs of the test would be relatively inexpensive. The time required would be largely the time involved in obtaining the specimens and analyzing the resultant data. Even a relatively small benefit of a two-step screening strategy of a serum test followed by a more expensive spiral CT scan might be superior to a spiral CT scan alone and justify substantial clinical use to the benefit of the company. Such a study might also show that this test does not add significant value to lung cancer detection, which might initially be perceived as bad for the company, but would more appropriately be seen as a way to cut their losses in the technology is actually not of benefit. Evidence of a desire by study designers to conduct a definitive test argues that although conflicts of interest might exist, that such conflicts are not the primary concern in the design of the study.

   A study to further refine the test (Answer A) might be relatively inexpensive, but marginal improvement in sensitivity and specificity would not make a major impact on the probability that the test would ultimately prove financially successful for the firm. In addition, modifying the current test format would forgo the potential advantage of already having obtained a patent on the current formulation of the test. Expanding the current set of test specimens to more normal subjects and those with COPD (Answer B) might also be relatively inexpensive, but would still not be a definitive study. A prospective study, either a cohort study or a randomized trial, (Answers D and E) would represent the ideal type of study that would most closely mimic the potential intended use of the test in practice, but such a study would be quite expensive and time consuming to conduct.

End of Case

26. An investigator examines the circulating lymphocyte population before and after extracorporeal photophoresis (ECP) used in the treatment of rejection of heart allografts. She finds that the relative frequency of CD8+ T effector cells goes down and the frequency of CD4+ T regulatory cells goes

up via testing performed at the same reference laboratory using standardized protocols. The conclusion of this manuscript is that these changes in lymphocyte frequencies are responsible for the immunosuppressive effect of ECP reported in other studies. What conclusion would you draw if you were one of the peer reviewers of this manuscript?

A. The changes observed appear to be technically valid and make physiological sense, so the manuscript should be accepted

B. You point out that since the authors did not actually demonstrate that the ECP procedure actually showed any significant immunosuppressive activity, the authors must move the conclusion about the linkage of the observed changes and immunosuppressive effect out of the abstract and results section, but can include this as a speculation in the discussion of the article

C. You are not sure how the patients were selected for ECP and if they have similar clinical and laboratory status at the beginning of ECP and thus, you recommend rejecting the paper

D. Since other data has shown that T cells undergo apoptosis during ECP and reinfusion of these cells resulting in inhibition of dendritic cells has been associated with immunosuppression, you recommend rejecting the paper since this other work was not cited

E. You are not sure that the flow cytometry analysis of the T cell subsets has been done consistently and the numbers of subjects studied was small (although the results were statistically significant), so you recommend rejecting the paper on these technical grounds

Concept: Although most journals have extensive review criteria, the approach of different reviewers can vary widely. One issue that comes up fairly frequently is a conclusion of the article, which involves a general pathophysiological point that was not actually examined by the data presented in the manuscript.

Answer: *B*—The most potent criticism of such a study is that the general conclusion, although potentially reasonable as a hypothesis, is not actually addressed, much less proven, by the data presented in the manuscript (Answer A). In this situation, authors should not be allowed to state such conclusions in the abstract or results section of the manuscript, but can make such speculation in the discussion section. However, many published papers have overly general conclusions stated in a definitive manner in the paper, so readers should beware this common pitfall in the literature. Although patient selection (Answer C) and repeatability in the methods (Answer E) are important, there is not enough data to conclude that there are problems with the methods or patient selection from the scenario. There is no evidence from the question stem supporting that the authors did not cite the previous work (Answer D)

27. Wilson's disease is a rare genetic disorder that affects about 1 in 30000 births worldwide. Thus far, the evidence for TPE in Wilson's disease is only based on case reports and small case series. A researcher would like to investigate the efficacy of TPE using all the available literature. He has two options of doing such study, metaanalysis and pooled-analysis. Which of the following is a major difference between metaanalysis study and pooled-analysis study?

A. A metaanalysis study is more expensive and more time consuming than pooled-analysis study

B. Generally, the investigators for each study included in the metaanalysis agree to participate in the study, while there is no need to contact the investigator of each study included in the pooled-analysis study

C. Obtaining primary data for each study is not necessary for a metaanalysis study while it is critical for a pooled-analysis study

D. Error in each individual study can be checked in a metaanalysis study while it cannot be checked in a pooled-analysis study

E. A metaanalysis is better to study rare disease comparing to a pooled-analysis study

Concept: Metaanalysis and pooled-analysis are two systematic methods using to summarize the published data. These are very useful for rare disease since it is likely that observations from multiple studies must be aggregated together in order to obtain clinically meaningful results.

Answer: C—There are differences between metaanalysis and pooled-analysis studies. Metaanalysis refers to the statistical analysis of a large collection of analytic results from individual studies for integration of findings. It can use to identify the heterogeneity between studies and to increase statistical power to provide a more estimate of the effect size. Pooled-analysis is similar to the traditional analysis except that individual parameters and outcomes from each study are combined, standardized, and then analyzed across all studies. Similar to metaanalysis, pooled-analysis has the benefit of increased power and can be used to examine rare diseases (Answer E). Furthermore, it can overcome some of the limitations of metaanalysis, such as its ability to harmonize exposures, covariates, and outcomes, as well as examine dose response and subgroup analyses. However, it is more expensive and time consuming than metaanalysis (Answer A). Since it has to use individual data from individual study, the investigators for each individual study must agree to be included in the pooled-analysis because they have to provide raw data. This is not the case with metaanalysis study (Answer B). Since the data must be standardized in pooled-analysis, potential error from each individual study can be checked (Answer D).

Suggested Reading

[1] N.M. Heddle, Evidence-based decision making in transfusion medicine, Vox Sang. 91 (2006) 214–220.
[2] N.M. Heddle, Clinical Research: Understanding the Methodology Toolboox, AABB Press, Bethesda, MD USA, (2012).
[3] D.S. Jacobs, W.R. DeMott, D.K. Oxley, Jacobs & DeMott Laboratory Test Handbook, fifth ed., Lexi-Comp Inc, Hudson, OH USA, (2001).
[4] R.K. Kandane-Rathnayake, J.C. Enticott, L.E. Phillips, Data distribution: normal or abnormal?, Transfusion 53 (2013) 257–259.
[5] R.K. Kandane-Rathnayake, J.C. Enticott, L.E. Phillips, Data distribution: normal or abnormal? Why it matters, Transfusion 53 (2013) 480–481.
[6] H. Motulsky, Intuitive Biostatistics, Oxford University Press, New York, NY, (1995).
[7] S.F. O'Brien, L. Osmond, Q.L. Yi, How do I interpret a p value?, Transfusion 55 (2015) 2778–2782.
[8] J. Schwartz, A. Padmanabhan, N. Aqui, R.A. Balogun, L. Connelly-Smith, M. Delaney, N.M. Dunbar, V. Witt, Y. Wu, B.H. Shaz, Guidelines on the use of therapeutic apheresis in clinical practice-evidence-based approach from the Writing Committee of the American Society for Apheresis: The seventh special issue, J. Clin. Apher. 31 (2016) 149–162.
[9] R.L. Wasserstein, N.A. Lazar, The ASA's statement on p-values: context, process, and purpose, Am. Stat. 70 (2016) 129–133.
[10] E. Walker, A.S. Nowacki, Understand equivalence and noninferior testing, J. Gen. Intern. Med. 26 (2011) 192–196.
[11] S.J. Pocock, G.W. Stone, The primary outcome fails—What next?, N. Engl. J. Med. 375 (8) (2016) 861–870.
[12] S. Greenland, S.J. Senn, K.J. Rothman, J.B. Carlin, C. Poole, S.N. Goodman, D.G. Altman, Statistical tests, P values, confidence intervals, and power: a guide to misinterpretations, Eur. J. Epidemiol. 31 (2016) 337–350.

CHAPTER

22

Posttest

Lance A. Williams, III, Helene DePalma**, Huy P. Pham**

*University of Alabama at Birmingham, Birmingham, AL, United States; **City University of New York, Jamaica, NY, United States

This chapter provides a formal assessment of the knowledge you obtained from this book. You should be able to complete the posttest in 80 min. The topics covered in this posttest have been covered throughout the book. You can compare your performance on the posttest with the pretest to identify areas for improvement.

Note: For the purpose of this test, unless otherwise stated, please use 70 mL/kg (for adults) and 80 mL/kg (for neonates) when calculate the blood volume of a person.

Place your answers in the spaces provided. Check your answers at the end of the chapter and give yourself a final score.

1.____	11.____	21.____	31.____	41.____
2.____	12.____	22.____	32.____	42.____
3.____	13.____	23.____	33.____	43.____
4.____	14.____	24.____	34.____	44.____
5.____	15.____	25.____	35.____	45.____
6.____	16.____	26.____	36.____	46.____
7.____	17.____	27.____	37.____	47.____
8.____	18.____	28.____	38.____	48.____
9.____	19.____	29.____	39.____	49.____
10.____	20.____	30.____	40.____	50.____

Number Correct _____/Number Incorrect _____ × 100% = Final Score __________%

Transfusion Medicine, Apheresis, and Hemostasis. http://dx.doi.org/10.1016/B978-0-12-803999-1.00022-5

1. What are the required doses for irradiation of blood products?
   A. 15 Gy to the center of the bag, 25 Gy to the remainder
   B. 2500 Gy to the center of the bag, 1500 Gy to the remainder
   C. 1500 Gy to the center of the bag, 2500 Gy to the remainder
   D. 25 Gy to the center of the bag, 15 Gy to the remainder
   E. 250 cGy to the center of the bag, 150 cGy to the remainder
2. After reviewing the results of an antibody panel, your blood bank technologist is convinced that a 32-year-old female who is 28 weeks pregnant has anti-D and anti-C. Of the following options, what is another possibility in this case?
   A. The patient may have anti-U
   B. The patient may have anti-Fy^3
   C. The patient may have anti-G
   D. The patient may have anti-Rh^{el}
   E. The patient may have anti-Rh27
3. A 25-year-old female is involved in a car accident at 35 weeks gestation. Her historical blood type is B Rh negative. The Emergency Medicine physician is concerned about a significant hemorrhage, so a Kleinhauer-Betke (KB) stain was ordered. The technologist observes 30 fetal cells out of 2000 total cells counted. What dose of Rh Immunoglobulin (RhIG) should be recommended for this case?
   A. 2.5 vials
   B. 3 vials
   C. 4 vials
   D. 5 vials
   E. No RhIG is necessary since RhIG is only indicated in Rh positive patients
4. A 32-year-old woman with severe menorrhagia is being evaluated for a hysterectomy in 7 weeks. Her complete blood count (CBC) reveals the following results: hemoglobin of 7.3 g/dL, a mean corpuscular volume (MCV) of 72 fL, and a platelet count of 250,000/µL. Her surgeon wishes to optimize her hemoglobin and hematocrit prior to her surgery. Which of the following is the best initial option to accomplish this?
   A. Advise the patient to donate autologous blood
   B. Advise the patient to ask her relatives to donate blood for her
   C. Start the patient on erythropoietin therapy
   D. Tell the surgeon to delay the surgery until 14 weeks
   E. Start the patient on PO iron therapy
5. A patient sample is submitted for antibody screen testing. Review the test results and select the appropriate interpretation.

	IS	37°C	IAT	CC
I	0	0	3+	NT
II	0	0	2+	NT
III	0	0	0	0

CC, Check cells; NT, not tested.

   A. The antibody screen should be reported as positive, with further testing to identify the likely IgG antibody.
   B. The antibody screen is should be reported as positive, with further testing to identify the likely IgM antibody.
   C. The antibody screen should be reported as negative, with no further testing indicated.

D. The antibody screen is invalid because the strength of reactivity is inconsistent. Repeat the testing.
E. The antibody screen is invalid due to the check cell reactions. Repeat the testing.

6. Review the antibody screen results below and select the response that correctly characterizes the nature/function of the antigen and the class/clinical significance of the corresponding antibody.

	Rh						MNS				Lu		P1	Lewis		Kell		Duffy		Kidd			LISS	LISS
	D	C	E	c	e	f	M	N	S	s	Lu^a	Lu^b	P1	Le^a	Le^b	K	k	Fy^a	Fy^b	Jk^a	Jk^b	IS	37C	IAT
1	+	+	0	0	+	0	+	+	+	+	0	+	+	+	0	0	+	0	+	0	+	0	0	2+
2	+	0	0	+	+	+	0	+	0	+	0	+	+	0	0	0	+	0	0	+	0	0	0	0√
3	0	+	0	+	+	+	0	+	0	+	0	+	0	0	+	0	+	0	0	+	0	0	0	0√
4	0	0	+	+	+	+	+	0	+	0	0	+	+	+	0	+	+	0	+	0	+	0	0	2+
5	0	0	0	+	+	+	+	+	+	0	0	+	0	0	+	0	+	+	+	+	0	0	0	0√
6	0	0	0	+	+	+	+	0	+	0	0	+	+	0	+	0	+	+	+	0	+	0	0	2+

A. Carbohydrate antigen found on glycophorin A; clinically significant IgM antibody
B. Protein antigen responsible for urea transport; clinically significant IgG antibody
C. Antigen is the receptor for *H. pylori*; clinically insignificant IgM antibody
D. Protein antigen; clinically insignificant IgM antibody
E. Carbohydrate antigen; clinically significant IgG antibody

7. Numerous studies have demonstrated that red blood cells (RBCs) stored for extended periods of time undergo structural changes, collectively termed "storage lesions." Which of the following represents one of the effects of storage lesions?
A. Increased extracellular potassium
B. Decreased membrane rigidity
C. Decreased phospholipid vesiculation
D. Increased 2,3-diphosphoglycerate (2,3 DPG) levels
E. Decreased extracellular hemoglobin

8. A trauma surgeon approaches you in the hall. His service has recently added a thromboelastogram (TEG) to their laboratory. He has noticed that in a lot of his patients that the TEG parameter for fibrinolysis is increased and asks for a recommendation. Which of the following could you suggest?
A. Cryoprecipitated AHF
B. Tranexamic acid
C. Prothrombin complex concentrate
D. Recombinant factor VIIa
E. Platelets

9. Antibody identification results for a patient indicate anti-E. The clinician has ordered four units of RBCs for transfusion. Group O, E- donor units are selected and tested by LISS antiglobulin crossmatch. One of the four units is 1+ incompatible by indirect antiglobulin technique. What is the most likely explanation?
A. The donor unit is not ABO compatible
B. The patient has an unexpected IgM antibody
C. The donor has an antibody to a low frequency IgM antibody
D. The donor unit has a positive direct antiglobulin test (DAT)
E. Antihuman globulin reagent was neutralized

10. You are reviewing a positive DAT of a newborn infant. The DAT is positive for IgG, yet the maternal antibody screen is negative. More results reveal that the mother is blood type O and the newborn is blood type A. Reviewing the medical record reveals that the newborn does have evidence of

significant hemolysis. The mother is Rh negative and received a dose of RhIG at 28 weeks gestation. What is the most likely explanation for these results?
A. The hemolysis is due to RhIG
B. The hemolysis is due to a D-variant
C. The hemolysis is due to IgM anti-A,B
D. The hemolysis is due to IgG anti-A,B
E. The hemolysis is due to an undetectable antibody

11. A neonatologist calls you about a newborn baby with widespread purpura. The baby's platelet count reveals thrombocytopenia (27,000/μL). He would like to test for antibodies against platelet antigens that are likely responsible for fetal/neonatal alloimmune thrombocytopenia (FNAIT), but is unsure of what to order. What would you recommend?
A. Anti-HPA-1a and HPA-3a
B. Anti-HPA-1a and HPA-5b
C. Anti-HPA-3a and HPA-5b
D. Anti-HLA class I
E. Anti-HLA class II

12. Which of the following is a product modification that may be used to prevent complications in a large volume RBC transfusion to a premature newborn that is blood type A Rh negative, especially if RBCs less than 7 days old are unavailable?
A. Leukoreduction
B. Phenotypically matched
C. Give O Rh negative units
D. Provide washed RBCs
E. Provide maternal RBCs

13. Which of the following is the potential shelf-life of a platelet that has been tested for bacterial contamination at the time of release?
A. 5 days
B. 7 days
C. 9 days
D. 11 days
E. 13 days

14. Which of the following product modifications renders a product "CMV safe"?
A. Irradiation
B. Washing
C. Leukoreduction
D. CMV adsorption
E. Volume reduction

Please answer Questions 15 and 16 based on the following clinical scenario:

You have an ICU patient that is scheduled to receive two units of RBCs. While receiving the second units, the patient develops severe respiratory distress, hypertension, and tachycardia. His other vital signs were normal. A transfusion reaction work-up is initiated and the resulting workup is negative. However, the pathology resident requests that the team order a chest X-ray and a brain naturetic peptide (BNP) level. The chest X-ray demonstrates bilateral pulmonary edema compared to a relatively normal X-ray before transfusion. The BNP level is markedly elevated.

15. What is the most likely diagnosis in this case?
A. Transfusion associated circulatory overload (TACO)
B. Transfusion associated dyspnea (TAD)
C. Transfusion related acute lung injury (TRALI)

D. Hemolytic transfusion reaction
E. Not related to transfusion

16. Unfortunately, the patient does not respond to treatment and died 3 h after the transfusion was discontinued. Which of the following actions is required by law?
A. Notify the Food and Drug Administration (FDA) by fax within 3 days
B. Notify the FDA by phone, fax, or email as soon as possible
C. Notify Department of Health & Human Services (HHS) by phone, fax, or email as soon as possible
D. Notify HHS by phone within 7 days
E. Notify the FDA in person within 3 days

End of Case

17. To prepare for a construction project in the laboratory, you have been asked to assess the records maintained in the file room. Based on the 30th edition of the AABB Standards for Blood Banks and Transfusion Services, which of the following records can be destroyed because the minimum retention timeframe has elapsed?
A. Blood product orders from 3 years ago
B. Physician requests for therapeutic apheresis from 6 years ago
C. Look-back investigations from 7 years ago
D. Evaluation of transfusion reactions from 8 years ago
E. Peer review of blood utilization from 2 years ago

18. The laboratory has purchased a second plasma thawer since the hospital just received the designation as the region level 1 trauma center. The new plasma thawer is the same manufacturer and model number as the current plasma thawer. Which statement is true regarding validation of the new piece of equipment?
A. The new plasma thawer must be validated prior to placing into service.
B. Since it is the exact manufacturer/model number as the current plasma thawer, no validation is required.
C. As long as the plasma thawer is an FDA-cleared device, no validation is required.
D. Only installation qualification is required. The validation from the original plasma thawer can satisfy other validation requirements.
E. If the original plasma thawer was validated less than 5 years ago, no validation is required for the new thawer.

19. The laboratory recently revised the massive transfusion protocol. There have been complaints from the Emergency Department that the response time has not met the timeframe described in the protocol for provision of products; however, the complaints are based on anecdotal examples. The laboratory initiates data collection for turnaround times for various stages of the process involved in the massive transfusion protocol. The term that best describes what the laboratory has implemented is which of the following?
A. Quality management system
B. Quality assurance
C. Quality indicator
D. Quality control
E. Qualification

20. A laboratory has selected and qualified a new kit to perform fetal bleed screening, to replace the kit previously in use. Following the revision of the Standard Operating Procedure (SOP), four technologists were trained to the new procedure. Which statement is correct regarding the competency assessment required?
A. Subscribe to a proficiency test and rotate to all four technologists during the first year following implementation
B. Six months after implementation, prepare an SOP quiz and administer to the technologists

C. Review the test records and result entry the first time the technologists perform the test
D. Before the new test method is implemented, perform a direct observation of the technologists
E. As long as both the current and new test kits are FDA approved, another competency assessment is not required before implementation of the new kit.

21. Which of the following variables has the most effect on the statistical power of a study?
A. Sample distribution
B. Sample size
C. Gender of the population
D. Number of doses given
E. Average age of participants

Please answer Questions 22–24 based on the following clinical scenario:

A 24-year-old male, weighing 83 kg presents to your Emergency Department with complaints of fatigue and dark urine for 7 days. His vital signs are unremarkable. A STAT CBC reveals hemoglobin of 5.5 g/dL, hematocrit of 16%, a platelet count of 7,000/μL, and lactate dehydrogenase (LDH) of 1245 U/L. The prothrombin time (PT), activated partial thrombolastin time (aPTT), and fibrinogen are within normal limits, as is the white blood cell (WBC) count. The DAT is also negative. The technologist comments that the peripheral smear has abundant schistocytes.

22. Based on this description, what disease process is at the top of your differential diagnosis?
A. Warm autoimmune hemolytic anemia
B. Typical hemolytic uremic syndrome (HUS)
C. Immune thrompocytopenia purpura (ITP)
D. Thrombotic thrombocytopenia purpura (TTP)
E. Malignant hypertension

23. You decide to perform an emergent 1-volume plasma exchange on this patient. Which of the following is the correct estimated plasma volume to exchange?
A. 4250 mL
B. 4410 mL
C. 4580 mL
D. 4770 mL
E. 4880 mL

24. Three years after the patient is recovered and discharged, you receive a notification from your blood supplier stating that a one of the donors of the plasma that your patient received during the treatment is now positive for an infectious disease marker. For which of the following diseases are you required to notify the patient for the need to be tested by their primary care physician?
A. Human immunodeficiency virus
B. Hepatitis B Virus
C. Hepatitis D Virus
D. West Nile Virus
E. Cytomegalovirus

End of Case

25. Advancements in labeling/barcode technology are aimed at making blood transfusion safer. Which of the following is an advantage of 2D barcodes over 1D barcodes?
A. 2D barcodes have a higher information density
B. 2D barcodes do not require special glasses to read
C. 2D barcodes can be scanned without having line of sight
D. 2D barcodes can be read by all types of barcode scanners.
E. 2D barcodes are more expensive

26. You are asked to perform a series of five plasma exchange procedures for the treatment of myasthenia gravis in a newly diagnosed 58-year-old female. After exchanging approximately 0.5 L, the patient experiences significant hypotension, but does not have any other complaints. The blood pressure is responsive to a fluid bolus, but drops again after the procedure is reinitiated. Which of the following is a potential cause of this reaction?
A. The patient is having an allergic reaction to the albumin
B. The patient is experiencing citrate toxicity
C. The patient is on lisinopril
D. The patient is on pressors
E. The patient is having an anaphylactic reaction to the citrate

27. Six months after undergoing a hematopoietic progenitor cell (HPC) transplant for the treatment of acute myeloid leukemia, a patient starts to notice tightening of her skin, sores in her mouth, and extremely dry eyes. Her bone marrow transplant physician diagnoses her with chronic graft-versus-host disease and started her on steroid treatment, to which she is unresponsive. What is another therapeutic option for this patient?
A. Extracorporeal photopheresis
B. Plasmapheresis
C. Plasma exchange
D. Antibiotic therapy
E. Donor lymphocyte infusion

28. A 48-year-old male with multiple myeloma is sent to you for HPC mobilization and collection. After stimulation with granulocyte colony stimulating factor, his CD34 count remains only 5/μL. Thus, the hematologist would like to try adding plerixafor as an additional mobilizing agent. Which of the following is the mechanism of action of plerixafor on HPC mobilization?
A. Inhibits HPC destruction
B. Inhibits HPC production
C. Inhibits CXCR-4 receptor on HPCs
D. Inhibits SDF-1α receptor on HPCs
E. Stimulate HPC production

29. A patient present's to your emergency department with diffuse alveolar hemorrhage (DAH), thought to be secondary to Goodpasture syndrome. You are asked to perform an urgent plasma exchange due to the risk of continued pulmonary hemorrhage. The patient's blood pressure is 90/60 and his heart rate is 100 bpm. Of the following answer choices, what will be your replacement fluid(s) for this procedure?
A. 50% plasma; 50% saline
B. 50% albumin; 50% cryoprecipitated AHF
C. 100% albumin
D. 100% plasma
E. 100% saline

Please answer Questions 30 and 31 based on the following clinical scenario:

You are consulted about a 32-year-old female who is delivering her third child. After delivery, the woman developed significant hemorrhage, but is refusing any blood due to her religious beliefs. She tells you that she is a Jehovah's Witness and does not wish to receive any blood product.

30. Of the following, according to her church guidance, which of the following products can she accept if she chooses?
A. Whole blood
B. Red blood cells
C. Platelets
D. Plasma
E. Factor concentrates

31. This pregnancy was unfortunately complicated by hemolytic disease of the newborn secondary to anti-D. After birth, the newborn is severely anemic, with hemoglobin of 6.4g/dL. The newborn has a strongly positive DAT (4+) and a high bilirubin of 29 mg/dL, which is not responsive to UV light therapy. The neonatologist states that they newborn will require an exchange transfusion, but the parents refuse due to their religious beliefs. Which of the following represents a viable option for treating this neonate?
A. Treat emergently without permission from the parents
B. Perform a plasma exchange to remove the antibody from the neonate
C. Seek an emergent injunction from the court in order to treat the neonate
D. Seek permission from the closest living aunt or uncle
E. Do not treat this neonate under any circumstances

End of Case

32. A 45-year-old male presents at the donor center. He is scheduled for knee replacement surgery in 3 weeks. He previously served in the military and was stationed for 6 months in Iraq 3 years ago. He has a history of surgery 6 months ago but received no blood transfusion. He is currently taking dutasteride. Today, his hemoglobin is 13.5 g/dL. Which of the following statements reflects the donor's eligibility?
A. He is eligible for both autologous and allogeneic donation
B. He is eligible for autologous donation but would be deferred as an allogeneic donor due to his hemoglobin level
C. He is eligible for autologous donation but would be deferred as an allogeneic donor due to travel history
D. He is eligible for autologous donation but would be deferred as an allogeneic donor due to medication
E. He is eligible for neither autologous nor allogeneic donation

33. A first time apheresis platelet donor is undergoing collection. Following some initial anxiety, the donor appears relaxed and the collection proceeds uneventfully. Thirty minutes later, the donor complains of shivering, muscle twitching, nausea, and foot cramping. What is the most appropriate intervention?
A. Increase the collection rate so the donor can complete the procedure more quickly
B. Remove the needle and apply pressure to the venipuncture site, followed by a cold compress
C. Lower the citrate infusion rate and provide oral calcium to the donor
D. Tilt donor bed back and elevate feet
E. Discontinue the procedure and ask the donor to breathe into a paper bag

34. Which of the following blood product storage temperatures and length of time is correct?
A. Octaplas plasma/1 year at ≤ −18°C
B. Red Blood Cells, Frozen/3 years at ≤65°C
C. CPDA-1 red blood cells/35 days at 1–6°C
D. CPD red blood cells with AS-1/21 days at 1–6°C
E. Cryoprecipitated AHF/2 years at ≤18°C

35. Select the combination that correctly matches the pathogen inactivation technology with the basis for the technology:

	Pathogen inactivation technology	Method
A	Octaplas plasma (Octapharma USA, Hoboken, NJ)	Amotosalen and 1% octoxynol for 1–1.5 h at +30°C (86°F)
B	INTERCEPT (Cerus Corporation, Concord, CA)	Amotosalen and UVA light
C	Mirasol, (Terumo BCT, Lakewood, CO, USA)	Vitamin B2 and UVA light
D	INTERCEPT (Cerus Corporation, Concord, CA)	1% tri(*n*-butyl) phosphate and UVA light
E	Mirasol, (Terumo BCT, Lakewood, CO, USA)	Vitamin B1 and UVB light

36. After receiving 30 mL of a unit of RBCs, a patient experiences anaphylaxis, eventually requiring intubation, and epinephrine for treatment. After recovery, you suspect that the patient may be IgA deficient, which is confirmed by testing. You also discover that the patient has anti-IgA. Which of the following is a blood product modification that can allow the patient to safely receive blood transfusions in the future?
A. Irradiation
B. Leukoreduction
C. Plasma extraction
D. Prewarming
E. Washing

37. Clinical Decision Support (CDS) systems are becoming more commonly used tools in blood management programs. Which of the following examples represents a possible CDS system?
A. The blood bank director prospectively audits transfusion decisions
B. The blood bank resident triages platelet orders
C. A computer pop-up alerts the physician to the most recent hemoglobin
D. A computer pop-up asks the physician to not transfuse the ordered unit
E. A computer pop-up alerts the physician to the current blood bank inventory

Please answer Questions 38 and 39 based on the following clinical scenario:

An 85-year-old male presents to the emergency department after a fall. There is a large hematoma on his left temple and he appears to be confused and in pain. His family states that he takes warfarin (5 mg/day) at home, but has been less reliable about his medications in recent years. A CT scan reveals intracranial hemorrhage and initial laboratory results are as follows:

Assay	**Result**
PT/INR	33 s/3.7
aPTT	37 s
Thrombin time (TT)	18 s
Fibrinogen	229 mg/dL
Platelet count	135,000/μL

38. Which of the following treatment options is recommended by the American College of Chest Physicians for this clinical scenario?
A. Give PO Vitamin K only
B. Give IV Vitamin K only
C. Give PCC and cryoprecipitated AHF
D. Give PCC and IV Vitamin K
E. Give PCC and PO Vitamin K

39. The attending physician asks for your help in dosing the PCC product in this case. Which of the following would be recommended for this 70 kg man?
A. 1750 units
B. 1875 units
C. 2000 units
D. 2500 units
E. 3000 units

End of Case

Please answer Questions 40 and 41 based on the following scenario:

As the new director of the tissue bank at your hospital, your chief financial officer (CFO) has given you the task of redesigning the structure and function of the service.

40. As a first order of business, he asks you to decide on whether to directly purchase tissue from the vendor or to get the tissue on consignment. Which of the following is a benefit of purchasing tissue on consignment?
A. Surgeons can waste more tissue without consequences
B. Less expense due to tissue expiration
C. Less expense due to higher inventory levels
D. More expense due to more tissue waste
E. Surgeons have more choice due to more tissue availability

41. Next, the CFO asks you about the advantages of a centralized versus a decentralized tissue bank. Which of the following is the best response?
A. Centralized tissue banks have less inventory to manage
B. Centralized tissue banks give surgeons quicker access to the tissue
C. Centralized tissue banks make it harder to track tissue disposition
D. Centralized tissue banks get less preferred pricing
E. Centralized tissue banks are managed by the surgeon

End of Case

42. An outside physician consults you regarding a patient that he suspects of having heparin induced thrombocytopenia (HIT). He tells you that the patient started taking heparin 10 days ago and the platelet count has dropped from 300,000/uL to 90,000/uL in that time. The patient has not yet developed a clot and is not on any other medications that can cause thrombocytopenia at this time, nor does he have any other reason for thrombocytopenia. Based on this information, what is the 4T score for the clinical likelihood of HIT?
A. 2
B. 4
C. 6
D. 8
E. 10

43. You are serving as the director of the HLA laboratory and receive a call from a renal transplant surgeon about the possibility of obtaining a crossmatch for a transplant tonight for a 45-year-old female hoping to receive her second kidney. A review of the patient's record reviews a calculated panel reactive antibody (cPRA) of 33% from a specimen collected 42 days ago. The surgeon is in a hurry to proceed. What is your recommendation?
A. Perform the crossmatch as requested
B. Perform the crossmatch as requested with two specimens from 42 days ago
C. Perform the crossmatch on a new specimen
D. Perform a virtual crossmatch
E. Recommend against transplant due to the patient's high level of sensitivity

44. Associations of HLA subclasses with various diseases are being discovered more often. Which of the following HLA types is associated with celiac disease?
A. HLA-DRB1
B. HLA-B27
C. HLA-DQ2
D. HLA-DQA1
E. HLA-DR3

45. 65-year-old male undergoes therapeutic plasma exchange (TPE) for hyperviscosity due to Waldenstrom macroglobinemia. Assuming that IgM monoclonal protein behaves as an ideal solute, after 1.5 plasma volume exchange, what is the percentage of IgM monoclonal protein left in the patient's body?
   A. 22%
   B. 27%
   C. 34%
   D. 63%
   E. 78%
46. A 16-year-old male received a bone marrow transplant from his brother. His brother is blood type AB, while the patient is blood type A. If the patient requires transfusion of RBCs in the peritransplant period (i.e., prior to converting to the donor's blood type), what is the blood type that can be provided that takes into account both compatibility and the inventory of the blood bank?
   A. Blood type B
   B. Blood type AB
   C. Blood type O
   D. Blood type A
   E. No blood can be given at this time
47. A 17-year-old burn victim decides to purse a skin autograft. Which of the following is an advantage of an autograft versus an allograft?
   A. Less risk of bacterial infections
   B. Less chance of clerical error
   C. More durable than allografts
   D. Superior function versus allografts
   E. Less chance of rejection
48. Two units of RBCs have been ordered for 68-year-old male patient due to postop bleeding from a hepatic resection 1 week ago. The patient's type and screen specimen was drawn 7 days ago, on the morning of surgery. The results revealed the patient is group O Rh positive, antibody screen negative. He was transfused 4 units of RBCs during the surgery, compatible by immediate spin crossmatch. Which statement is correct?
   A. A new specimen is not required, the previous specimen is valid through the end of day 7
   B. A new specimen is not required, the previous specimen is valid for 14 days
   C. Request a new specimen for type and screen testing and crossmatch. Assign a 3-day expiration date to the new specimen
   D. Request a new specimen for type and screen testing and crossmatch. Assign a 7-day expiration date to the new specimen
   E. Request a new specimen for type and screen testing and crossmatch. Assign a 14-day expiration date to the new specimen
49. A patient's ABO/Rh typing gives the following results. Choose the selection that provides the most probable explanation for these reactions.

Forward			Reverse			
Anti-A	**Anti-B**	**Anti-D**	**A_1 cells**	**B cells**	**Antibody screen**	**Auto control**
3+	4+	4+	1+	0	0	0

   A. Hypogammaglobulinemia has resulted in a missing reaction in the backtype
   B. The patient has a diagnosis of multiple myeloma that caused abnormal globulin production
   C. The patient is most likely a subgroup of B

D. The patient is most likely a subgroup of A with anti-A_1 detected in the plasma
E. The patient has an IgM antibody that has interfered with the forward typing

50. Select the combination that correctly matches the scenario with the appropriate product selection and crossmatch technique:

	Patient serological results	Product selection/crossmatch
A	O Rh positive, antibody screen negative, new patient (tested once)	Select O Rh positive unit and perform electronic crossmatch
B	A Rh positive, antibody screen negative, previous type and screen submitted 1 week ago with same results	Select AB Rh positive unit and perform an immediate spin crossmatch
C	B Rh positive, antibody screen negative, history of anti-Jk^a	Select B Rh positive unit and perform an electronic crossmatch
D	A Rh negative, antibody screen positive, anti-K identified	Select A Rh negative, K negative unit and perform an immediate spin crossmatch
E	O Rh negative, antibody screen positive, anti-D identified	Select O Rh negative unit and perform a full antiglobulin (extended) crossmatch

ANSWERS AND BRIEF EXPLANATIONS

1. **Answer**: *D*—25 Gy to the center of the bag, 15 Gy to the remainder. These requirements allow the irradiation to the entire unit. Refer to Chapter 8, Question 6 for more information.
2. **Answer**: *C*—The anti-D and anti-C combination could be mimicking the response of anti-G. This is often an important distinction in pregnant patients as they may benefit from Rh Immunoglobulin (RhIG). Refer to Chapter 15, Question 19 for more information.
3. **Answer**: *C*—4 vials should be administered. Refer to Chapter 8, Question 35 for more information and for an example of the calculation for this question/answer.
4. **Answer**: *E*—Start her on PO iron therapy. Often patients do not necessarily require blood transfusion for the treatment of anemia. Refer to Chapter 9, Question 5 for more information.
5. **Answer**: *E*—The check cells for cell III did not react, therefore, the antibody screen is invalid. Refer to Chapter 6, Question 1 for more information.
6. **Answer**: *B*—The protein containing the Kidd system antigens functions as a urea transporter. Kidd antibodies (anti-Jk^b in this example) is a clinically significant IgG antibody, expected to react by indirect antiglobulin technique. Refer to Chapter 6, Question 32 for more information.
7. **Answer**: *A*—Increased extracellular potassium is an example of a storage lesion. Refer to Chapter 9, Question 11 for more information.
8. **Answer**: *B*—Tranexamic acid is a recognized and often effective treatment for patients with increased fibrinolysis. Refer to Chapter 9, Question 16 for more information.
9. **Answer**: *D*—The positive DAT on the donor RBCs would cause an incompatible crossmatch by indirect antiglobulin technique. Refer to Chapter 6, Question 6 for more information.
10. **Answer**: *D*—The hemolysis is due to IgG anti-A,B, which is currently the most common cause of hemolytic disease of the fetus and newborn in the United States. Refer to Chapter 10, Question 2 for more information.
11. **Answer**: *B*—Anti-HPA-1a and HPA-5b are the most common antibodies in NAIT patients. Refer to Chapter 10, Question 30 for more information.
12. **Answer**: *D*—Provide washed RBCs, which will remove extracellular potassium. Refer to Chapter 10, Question 23 for more information.
13. **Answer**: *B*—7 days. Testing at the time of release is a new option that allows blood banks to extend the shelf-life of a platelet and avoid unnecessary wastage. Refer to Chapter 11, Questions 1 and 38 for more information.
14. **Answer**: *C*—Leukoreduction decreases the number of white blood cells in the product; thus, decreasing the risk of CMV transmission. Refer to Chapter 11, Question 3 for more information.
15. **Answer**: *A*—TACO (transfusion-associated circulatory overload). A diagnosis of TACO is consistent of the symptoms above and could necessitate treatment with furosemide to treat the patient. Transfusion associated dyspnea (TAD) should only be considered after TACO and TRALI have been ruled out. Refer to Chapter 12, Question 22 for more information.
16. **Answer**: *B*—When a death occurs as a possible consequence of transfusion, the FDA must be notified by any means necessary as soon as possible and a written report is required within 7 days of the event. Notification to the Department of Health & Human Services (HHS) is not necessary. Additionally, notification to the state department of health requirement varies by state. Refer to Chapter 12, Question 1 for more information.
17. **Answer**: *B*—The minimum retention time for physician requests for therapeutic apheresis is 5 years. Refer to Chapter 2, Questions 35–38; Chapter 3 Question 11 for more information.
18. **Answer**: *A*—The new plasma thawer must be validated before use. Since the manufacturer/model is the same, the validation plan that was originally developed can be followed. Refer to Chapter 2, Question 10; Chapter 3, Question 9 for more information.
19. **Answer**: *C*—Quality indicators measure data to monitor progress toward the quality goals of an organization. Refer to Chapter 3, Question 3 for more information.

20. **Answer**: *D*—Competency assessment is a separate process from training or proficiency testing and is required before independent performance of a moderate or high complexity test. Refer to Chapter 3, Question 7 for more information.
21. **Answer**: *B*—The sample size is an important contributor to the statistical power of a study. Refer to Chapter 21, Question 21 for more information.
22. **Answer**: *D*—The signs and symptoms in this case are suggestive of TTP, or another microangiopathic hemolytic anemia. Refer to Chapter 14, Question 18 for more information.
23. **Answer**: *E*—The plasma volume for this exchange is 4880 mL, based on the calculation as follows: (83 kg × 70 mL/kg) × (1-Hct) = 4880 mL. Of course, using Nadler's formula will result in a more accurate volume. Refer to Chapter 20, Question 1 for more information.
24. **Answer**: *A*—Notification that a donor tested positive for HIV at a later date requires the medical director of the blood bank to investigate the case and notify the recipient of the need for testing. If the recipient died, then a notification to the next-of-kin is required. Sometimes this occurs years after transfusion. Refer to Chapter 11, Question 10 for more information.
25. **Answer**: *A*—2D barcodes are more advanced than 1D barcodes and can hold/deliver more information. Refer to Chapter 19, Question 7 for more information.
26. **Answer**: *C*—A potential cause of this patient's symptoms is an ACE-inhibitor. The patient's medical record should be reviewed and unless the procedure is emergent, the medication should be discontinued for 24 h before restarting plasma exchange if possible. Refer to Chapter 14, Question 16 for more information.
27. **Answer**: *A*—Steroids are the initial treatment option for chronic GVHD; however, cases unresponsive to steroids may require more aggressive measures, such as extracorporeal photopheresis treatment. Refer to Chapter 17, Question 35 for more information.
28. **Answer**: *C*—By inhibiting the CXCR-4 receptor on stem cells; thus, preventing the binding to SDF-1 α. This inhibition allows the HPCs to leave the bone marrow and enter the peripheral blood stream. Refer to Chapter 17, Question 17 for more information.
29. **Answer**: *D*—100% plasma is often used as the replacement fluid for patients with significant DAH. If albumin alone were used, the resulting coagulopathy could worsen the patient's condition. Saline should not be used in this case since the patient's blood pressure is already low. Refer to Chapter 14, Question 3 for more information.
30. **Answer**: *E*—If she chooses, guidance from the church states that acceptance of fractions of blood products, such as factor concentrates or albumin, are left up to the discretion of the individual member, while accepting whole blood, red blood cells, platelets, or plasma is strongly discouraged. Refer to Chapter 15, Question 13 for more information.
31. **Answer**: *C*—In such cases, the parents do not have the right to withhold lifesaving care from their baby due to their own religious beliefs. A court injunction should be sought in order to save the neonates life. Refer to Chapter 15, Question 14 for more information.
32. **Answer**: *D*—The donor is not a candidate for allogeneic donation due to dutasteride (deferral for 6 months after the last dose). Refer to Chapter 4, Questions 2, 3, and 31 for more information.
33. **Answer**: *C*—The donor's symptoms are consistent with hypocalcemia due to the citrate anticoagulant. Refer to Chapter 14, Question 29 for more information.
34. **Answer**: *C*—Red cells collected in CPDA-1 can be stored up to 35 days at 1–6°C. Refer to Chapter 5, Question 15 and 31, Tables 5.2 and 5.3 for more information.
35. **Answer**: *B*—INTERCEPT (Cerus Corporation, Concord, CA) utilizes amotosalen, a synthetic psoralen, followed by illumination with UVA light. Refer to Chapter 5, Questions 31, 32, and 33 for more information.
36. **Answer**: *E*—Washing red blood cells 6 times often allows patients with anaphylaxis due to IgA deficiency (with anti-IgA) to safely receive blood transfusions. Platelets and plasma can obviously not be effectively/efficiently washed and thus, IgA deficient donors are often sought out if such products are required. Refer to Chapter 12, Question 5 for more information.

37. **Answer**: *C*—CDS systems are designed to prompt physicians with useful clinical or laboratory information at a critical decision point that may alter their decision to transfuse without adversely affecting the patient's care. Refer to Chapter 19, Question 26 for more information.
38. **Answer**: *D*—In patients on warfarin with serious bleeding the American College of Chest Physicians recommends treatment with either plasma or PCC and IV Vitamin K. Refer to Chapter 13, Questions 35 and 36 for more information.
39. **Answer**: *A*—Based on the clinical scenario and the INR, this patient should receive PCC at a dose of 25 units/kg, thus a dose of 1750 units is correct. Refer to Chapter 13, Table 13.9 for more information.
40. **Answer**: *B*—Consignment allows hospitals to return tissue that is unused before expiration. This saves money over directly purchased tissue, which must be discarded if not used. Refer to Chapter 18, Question 30 for more information.
41. **Answer**: *A*—Overall, centralized tissue banks provide a central location for tissue storage and distribution, which has advantages and disadvantages. Refer to Chapter 18, Question 3 for more information.
42. **Answer**: *C*—The 4T score for this patient is 6. He gets 2 points for the timing of the fall (with 5–10 days); 2 points for the degree of the fall (>50%); and 2 points for having no other reason for the thrombocytopenia. Refer to Chapter 13, Question 29 for more information.
43. **Answer**: *C*—In a patient that is already sensitized, a new specimen should be obtained the day of the possible surgery in order to provide the best information for potential compatibility or incompatibility. Refer to Chapter 16, Question 12 for more information.
44. **Answer**: *C*—HLA-DQ2 and HLA-DQ8 are known associations with celiac disease. Refer to Chapter 16, Question 24 for more information.
45. **Answer**: *A*—After 1.5 plasma volume exchange, approximately 22% of IgM monoclonal protein would be left in the patient's plasma, assuming that it behaves as an ideal solute. Refer to Chapter 20, Question 15 for more information.
46. **Answer**: *D*—Blood type A is the type that is compatible with both the donor and the recipient and takes into account the limited supply of type O blood in a typical blood bank. Refer to Chapter 17, Question 11 for more information.
47. **Answer**: *E*—Autografts do not have the risk of rejection due to incompatibility with the recipient as compared to allografts. Refer to Chapter 18, Question 16 for more information.
48. **Answer**: *C*—Following a transfusion of RBCs in the last 3 months (or history of an antibody or pregnancy), a type and screen specimen is valid for 3 days. Refer to Chapter 7, Question 2 for more information.
49. **Answer**: *D*—The reactions can be explained by an A_2B individual with anti-A1 detected in their plasma. Refer to Chapter 7, Question 5 for more information.
50. **Answer**: *E*—Clinically significant antibodies, such as anti-D, require an extended crossmatch. Refer to Chapter 7, Question 15 and 16 for more information.

Index

D

E

Q

R

U

V

Printed in the United States
By Bookmasters